Solar System Data

Body	Mass (kg)	Mean Radius (m)	Period (s)	Distance from Sun (m)
Mercury	3.18×10^{23}	2.43×10^6	7.60×10^6	5.79×10^{10}
Venus	4.88×10^{24}	6.06×10^6	1.94×10^7	1.08×10^{11}
Earth	5.98×10^{24}	6.37×10^6	3.156×10^7	1.496×10^{11}
Mars	6.42×10^{23}	3.37×10^6	5.94×10^7	2.28×10^{11}
Jupiter	1.90×10^{27}	6.99×10^7	3.74×10^8	7.78×10^{11}
Saturn	5.68×10^{26}	5.85×10^7	9.35×10^8	1.43×10^{12}
Uranus	8.68×10^{25}	2.33×10^7	2.64×10^9	2.87×10^{12}
Neptune	1.03×10^{26}	2.21×10^7	5.22×10^9	4.50×10^{12}
Pluto	$\approx 1.4 \times 10^{22}$	$\approx 1.5 \times 10^6$	7.82×10^9	5.91×10^{12}
Moon	7.36×10^{22}	1.74×10^6	—	—
Sun	1.991×10^{30}	6.96×10^8	—	—

Physical Data Often Used[a]

Average Earth-Moon distance	3.84×10^8 m
Average Earth-Sun distance	1.496×10^{11} m
Average radius of the Earth	6.37×10^6 m
Density of air (20°C and 1 atm)	1.29 kg/m³
Density of water (20°C and 1 atm)	1.00×10^3 kg/m³
Free-fall acceleration	9.80 m/s²
Mass of the Earth	5.98×10^{24} kg
Mass of the Moon	7.36×10^{22} kg
Mass of the Sun	1.99×10^{30} kg
Standard atmospheric pressure	1.013×10^5 Pa

[a] These are the values of the constants as used in the text.

Some Prefixes for Powers of Ten

Power	Prefix	Abbreviation	Power	Prefix	Abbreviation
10^{-18}	atto	a	10^1	deka	da
10^{-15}	femto	f	10^2	hecto	h
10^{-12}	pico	p	10^3	kilo	k
10^{-9}	nano	n	10^6	mega	M
10^{-6}	micro	μ	10^9	giga	G
10^{-3}	milli	m	10^{12}	tera	T
10^{-2}	centi	m	10^{15}	peta	P
10^{-1}	deci	d	10^{18}	exa	E

PHYSICS

For Scientists & Engineers

| Fourth Edition |

VOLUME 1

PHYSICS

For Scientists & Engineers

| Fourth Edition |

VOLUME 1

Raymond A. Serway

James Madison University

SAUNDERS GOLDEN SUNBURST SERIES

SAUNDERS COLLEGE PUBLISHING

Philadelphia   Fort Worth   Chicago   San Francisco

Montreal   Toronto   London   Sydney   Tokyo

Requests for permission to make copies of any part of the work should be mailed to: Permissions Department, Harcourt Brace & Company, 6277 Sea Harbor Drive, Orlando, Florida 32887-6777.

Text Typeface: New Baskerville
Composition and Layout: Progressive Information Technologies
Publisher: John Vondeling
Developmental Editor: Laura Maier
Senior Project Editor: Sally Kusch
Copy Editor: Charlotte Nelson
Managing Editor: Carol Field
Manager of Art and Design: Carol Bleistine
Associate Art Director: Sue Kinney
Art and Design Coordinator: Kathleen Flanagan
Text Designer: Rebecca Lemna
Cover Designer: Lawrence R. Didona
Text Artwork: Rolin Graphics
Photo Researcher: Sue Howard
Director of EDP: Tim Frelick
Production Manager: Charlene Squibb
Marketing Manager: Marjorie Waldron

Cover and title page: Rolling motion of a rigid body: disk by Richard Megna, © Fundamental Photographs, NYC

Printed in the United States of America

Physics for Scientists and Engineers, Volume I, Fourth edition

ISBN 0-03-015657-2

Library of Congress Catalog Card Number: 95-069566

6789012345   032   10   98765432

The publisher and author have gone to extreme measures in attempting to ensure the publication of an error-free text. The manuscript, galleys, and page proofs have been carefully checked by the author, the editors, and a battery of reviewers. While we realize that a 100% error-free text may not be humanly possible, Serway's *Physics for Scientists and Engineers* is very close. Confirmed in this belief, we are offering $5.00 for any first-time error you may find. (Note that we will only pay for each error the first time it is brought to our attention.) Please write to John Vondeling, Publisher.

Preface

Physics for Scientists and Engineers has been used successfully at over 700 colleges and universities over the course of three editions. This fourth edition has many new pedagogical features, and a major effort was made to improve clarity of presentation, precision of language, and accuracy throughout. Based on comments from users of the third edition and reviewers' suggestions, refinements have been added such as an increased emphasis on teaching concepts. The fourth edition has also integrated several new interactive software products that will be useful in courses using computer-assisted instruction.

This two-volume textbook is intended for a course in introductory physics for students majoring in science or engineering. The book is an extended version of *Physics for Scientists and Engineers* in that Volume II includes eight additional chapters covering selected topics in modern physics. This material on modern physics has been added to meet the needs of those universities that choose to cover the basic concepts of quantum physics and its application to atomic, molecular, solid state, and nuclear physics as part of their curriculum.

The entire contents of the text could be covered in a three-semester course, but it is possible to use the material in shorter sequences with the omission of selected chapters and sections. The mathematical background of the student taking this course should ideally include one semester of calculus. If that is not possible, the student should be enrolled in a concurrent course in introduction to calculus.

(Richard Megna/Fundamental Photographs)

OBJECTIVES

The main objectives of this introductory physics textbook are twofold: to provide the student with a clear and logical presentation of the basic concepts and principles of physics, and to strengthen an understanding of the concepts and principles through a broad range of interesting applications to the real world. In order to meet these objectives, emphasis is placed on sound physical arguments. At the same time, I have attempted to motivate the student through practical examples that demonstrate the role of physics in other disciplines including engineering, chemistry, and medicine.

CHANGES TO THE FOURTH EDITION

A number of changes and improvements have been made in preparing the fourth edition of this text. Many changes are in response to comments and suggestions offered by instructors and students using the third edition and by reviewers of the manuscript. The following represent the major changes in the fourth edition:

• **Line-by-Line Revision**   The entire text has been carefully edited to improve clarity of presentation and precision of language. We hope that the result is a book that is both accurate and enjoyable to read.

• **Organization**   The organization of the textbook is essentially the same as that of the third edition with one exception. Chapters 2 and 3 have been interchanged, so that the treatment of vectors precedes the discussion of motion in two dimensions, where vectors and their components are first used. Many sections have been streamlined or combined with other sections to allow for a more balanced presentation.

• Problems   A substantial revision of the end-of-chapter problems and questions was made in an effort to provide a greater variety and to reduce repetition. Approximately 25 percent of the problems (approximately 800), most of which are at the intermediate level, are new. The remaining problems have been carefully edited and reworded where necessary. All new problems are marked with an asterisk in the Instructors Manual. Solutions to approximately 25 percent of the problems are included in the Student Solutions Manual and Study Guide. These problems are identified by boxes around their numbers.

• Significant Figures   Significant figures in both worked examples and end-of-chapter problems have been handled with care. Most numerical examples and problems are worked out to either two or three significant figures, depending on the accuracy of the data provided.

• Visual Presentation   Most of the line art and many of the color photographs have been replaced or modified to improve the clarity of presentation, pedagogy, and visual appeal of the text. As in the third edition, color is used primarily for pedagogical purposes. A chart explaining the pedagogical use of color is included after the To the Student section following the preface.

NEW FEATURES IN THE FOURTH EDITION

• Integrated Software   The textbook is accompanied by two interactive software packages. *SD2000* is a self-contained software package of physics simulations and demonstrations that have been developed exclusively to accompany this textbook. Concepts and examples are presented and explained in an interactive format. Simulations developed for the Interactive Physics II™ program are keyed to appropriate worked-example problems and to selected end-of-chapter problems. Both packages are provided on disks and are described in more detail in the section dealing with ancillaries.

• Conceptual Examples   Approximately 150 conceptual examples have been included in this edition. These examples, which include reasoning statements, provide students with a means of reviewing the concepts presented in that section. The examples could also serve as models when students are asked to respond to end-of-chapter questions, which are largely conceptual in nature.

• Review Problems   Many chapters now include a multi-part review problem located prior to the list of end-of-chapter problems. The review problem requires the student to draw on numerous concepts covered in the chapter as well as those discussed in previous chapters. These problems can be used by students in preparing for tests, and by instructors in classroom discussions and review.

• Paired Problems   Several end-of-chapter problems have been paired with the same problem in symbolic form. For example, numerical Problem 9 may be followed by symbolic Problem 9A. If Problem 9 is assigned, Problem 9A can be used to test the student's understanding of the concepts used in solving the problem.

• Spreadsheet Problems   Most chapters will include several spreadsheet problems following the end-of-chapter problem sets. Spreadsheet modeling of physical phenomena enables the student to obtain graphical representations of physical quantities and perform numerical analyses of problems without the burden of having to learn a high-level computer language. Spreadsheets are particularly valuable in exploratory investigations; "what if" questions can be addressed easily and depicted graphically.

Level of difficulty in the spreadsheet problems, as with all end-of-chapter problems, is indicated by the color of the problem number. For the most straightforward problems (black) a disk with spreadsheet templates is provided. The student must enter the pertinent data, vary the parameters, and interpret the results. Intermediate level problems (blue) usually require students to modify an existing template to perform the required analysis. The more challenging problems (magenta) require students to develop their own spreadsheet templates. Brief instuctions on using the templates are provided in Appendix F.

COVERAGE

The material covered in this book is concerned with fundamental topics in classical physics and an introduction to modern physics. The book is divided into six parts. In the first volume, Part I (Chapters 1–15) deals with the fundamentals of Newtonian mechanics and the physics of fluids; Part II (Chapters 16–18) covers wave motion and sound; Part III (Chapters 19–22) is concerned with heat and thermodynamics. In the second volume, Part IV (Chapters 23–34) treats electricity and magnetism, Part V (Chapters 35–38) covers light and optics, and Part VI (Chapters 39–47) deals with relativity, quantum physics, and selected topics in modern physics. Each part opener includes an overview of the subject matter to be covered in that part and some historical perspectives.

TEXT FEATURES

Most instructors would agree that the textbook selected for a course should be the student's primary "guide" for understanding and learning the subject matter. Furthermore, a textbook should be easily accessible and should be styled and written for ease in instruction. With these points in mind, I have included many pedagogic features in the textbook which are intended to enhance its usefulness to both the student and instructor. These are as follows:

Style   As an aid for rapid comprehension, I have attempted to write the book in a style that is clear, logical, and engaging. The writing style is somewhat informal and relaxed, which I hope students will find appealing and enjoyable to read. New terms are carefully defined, and I have tried to avoid jargon.

Previews   Most chapters begin with a chapter preview, which includes a brief discussion of chapter objectives and content.

Important Statements and Equations   Most important statements and definitions are set in bold print for added emphasis and ease of review. Important equations are highlighted with a tan screen for review or reference.

Problem-Solving Strategies and Hints   I have included general strategies for solving the types of problems featured in both the examples and in the end-of-chapter problems. This feature will help students identify necessary steps in solving problems and eliminate any uncertainty they might have. Problem-solving strategies are highlighted by a light color screen for emphasis and ease of location.

Marginal Notes   Comments and marginal notes are used to locate important statements, equations, and concepts in the text.

Illustrations   The readability and effectiveness of the text material and worked examples are enhanced by the large number of figures, diagrams, photographs, and tables. Full color is used to add clarity to the artwork and to make it as realistic as possible. For example, vectors are color-coded, and curves in *xy*-plots are drawn in color. Three-dimensional effects are produced with the use of color airbrushed areas, where appropriate. The color photographs have been carefully selected, and their accompanying captions have been written to serve as an added instructional tool. Several chapter-opening photographs, particularly in the chapters on mechanics, include color-coded vector overlays that illustrate and present physical principles more clearly and apply them to real-world situations.

Mathematical Level   Calculus is introduced gradually, keeping in mind that a course in calculus is often taken concurrently. Most steps are shown when basic equations are developed, and reference is often made to mathematical appendices at the end of the text. Vector products are introduced later in the text where they are needed in physical applications. The dot product is introduced in Chapter 7, "Work and Energy." The cross product is introduced in Chapter 11, which deals with rotational dynamics.

Worked Examples   A large number of worked examples of varying difficulty are presented as an aid in understanding concepts. In many cases, these examples serve as models for solving the end-of-chapter problems. The examples are set off in a box, and the solution answers are highlighted with a light blue screen. Most examples are given titles to describe their content.

(Peter Aprahamian/Science Photo Library)

Worked-Example Exercises   Many of the worked examples are followed immediately by exercises with answers. These exercises are intended to make the textbook more interactive with the student and to immediately reinforce the student's understanding of concepts and problem-solving techniques. The exercises represent extensions of the worked examples.

Units   The international system of units (SI) is used throughout the text. The British engineering system of units (conventional system) is used only to a limited extent in the chapters on mechanics, heat, and thermodynamics.

Biographical Sketches   Throughout the text I have included short biographies of important scientists to add more historical emphasis and show the human side of the lives of scientists.

Optional Topics   Many chapters include special topic sections which are intended to expose the student to various practical and interesting applications of physical principles. These optional sections are labeled with an asterisk (*).

Summaries   Each chapter contains a summary which reviews the important concepts and equations discussed in that chapter.

Thought Questions   Questions requiring verbal answers are provided at the end of each chapter. Some questions provide the student with a means of self-testing the concepts presented in the chapter. Others could serve as a basis for initiating classroom discussions.

Problems   An extensive set of problems is included at the end of each chapter. Answers to odd-numbered problems are given at the end of the book; these pages have colored edges for ease of location. For the convenience of both the student and the instructor, about two thirds of the problems are keyed to specific sections of the chapter. The remaining problems, labeled "Additional Problems," are not keyed to specific sections. In my opinion, assignments should consist mainly of the keyed problems to help build self-confidence in students.

Usually, the problems within a given section are presented so that the straightforward problems (numbered in black print) are first, followed by problems of increasing difficulty.

For ease in identifying the intermediate-level problems, the problem number is printed in blue. I have also included a small number of challenging problems, which are indicated by a problem number printed in magenta.

Appendices and Endpapers Several appendices are provided at the end of the text, including the new appendix with instructions for problem-solving with spreadsheets. Most of the appendix material represents a review of mathematical techniques used in the text, including scientific notation, algebra, geometry, trigonometry, differential calculus, and integral calculus. Reference to these appendices is made throughout the text. Most mathematical review sections include worked examples and exercises with answers. In addition to the mathematical reviews, the appendices contain tables of physical data, conversion factors, atomic masses, and the SI units of physical quantities, as well as a periodic chart. Other useful information, including fundamental constants and physical data, planetary data, a list of standard prefixes, mathematical symbols, the Greek alphabet, and a table of standard abbreviations and symbols of units appears on the endpapers.

ANCILLARIES

The ancillary package has been updated and expanded in response to suggestions from users of the third edition. The most essential changes are an expansive set of interactive software, an updated test bank with greater emphasis on conceptual questions and open-ended problems, a new Student Solutions Manual and Study Guide with complete solutions to 25 percent of the text problems, a student's Pocket Guide, and a new spreadsheet supplement.

Interactive Software

Interactive Homework System

The World Wide Web (WWW) is the platform for an interactive homework system developed out of the University of Texas at Austin. This system, developed to coordinate with *Physics for Scientists and Engineers*, uses WWW, telnet, telephone, and Scantron submission of student work. The system has been class-tested at the University of Texas with over 1800 students participating each semester. Over 100,000 questions are answered electronically per month. Instructors at any university using Serway's *Physics for Scientists and Engineers* may establish access to this system by providing a class roster and making problem selections. Over 2000 algorithm-based problems are available; problem parameters vary from student to student, so that each student must do original work. All grading is done by computer, with results automatically posted on WWW. Students receive immediate right/wrong feedback, with multiple tries allowed for incorrect answers. When students answer incorrectly, they are automatically linked into text from the appropriate section of the fourth edition of Serway's *Physics for Scientists and Engineers with Modern Physics*.

A demo using the WWW interface is available at the URL **http://hw.ph.utexas.edu:80** by clicking on the demo link. Further information for instructors interested in importing the system to their institutions is available from **see@physics.utexas.edu**. The fourth edition of *Physics for Scientists and Engineers* will be linked into the system by January 1996.

SD2000 Interactive Software

This learning environment of physics simulations and demonstrations has been developed by Future Graph, Inc., exclusively to accompany this textbook. Its applications span all of the basic topics treated in the textbook. SD2000 is available on computer disk or CD-ROM in Macintosh and IBM Windows formats. The icon identifies examples and sections for which a simulation or demonstration exists.

Simulations   A collection of 10 powerful simulators allows students to model and bring to life an infinite number of physics problems. Students can model systems that include Kinetic motion, Collisions, Geometric optics, and Electric and magnetic fields, as well as laboratory tools such as Fourier synthesizers, Wave form generators, and Oscilloscopes. SD2000 boxes throughout the text identify how these simulators can be used to reinforce the concepts presented in the text. In modeling individualized simulations, students may investigate how varying the components of a situation will affect the outcome.

- Chapter 4: Motion, Section 4.4
- Chapter 9: Collisions, Section 9.5
- Chapter 16: Wave Motion, Section 16.4
- Chapter 18: Complex Waves—The Fourier Synthesizer, Section 18.8
- Chapter 21: Systems of Particles, Section 21.1
- Chapter 23: Motion in an Electric Field, Section 23.7
- Chapter 25: Mapping the Electric Field, Section 25.5
- Chapter 29: Motion of Charged Particles in Electric and Magnetic Fields, Section 29.5
- Chapter 33: The Oscilloscope, Section 33.5
- Chapter 36: Optical Instruments, Section 36.10

Demonstrations   Lessons derived from worked examples in the text of *Physics for Scientists and Engineers* allow students to investigate the results of changing parameters within the context of the example. Students can interactively explore physics through equations, calculations, graphs, tables, animations, and simulations. A complete list of demonstrations follows:

Chapter 2
Section 2.4
Example 2.3
Example 2.15

Chapter 3
Example 3.8

Chapter 4
Section 4.2
Section 4.3

Chapter 5
Example 5.6
Example 5.9
Example 5.15

Chapter 6
Example 6.4

Example 6.5
Example 6.9
Example 6.10

Chapter 7
Example 7.3
Example 7.5
Example 7.18

Chapter 8
Example 8.10

Chapter 9
Example 9.11
Example 9.16
Example 9.23

Chapter 10
Example 10.4

Chapter 11
Example 11.13

Chapter 13
Section 13.4
Section 13.6
Section 13.7

Chapter 14
Example 14.2
Example 14.4
Example 14.6
Example 14.7
Example 14.11

Chapter 16
Example 16.1
Example 16.3
Example 16.5

Chapter 18
Example 18.2
Example 18.3

Chapter 20
Example 20.6

Chapter 21
Example 21.1
Example 21.3
Section 21.6

Chapter 22
Example 22.4
Example 22.12

Interactive-Physics Simulations

Approximately 100 simulations by Ray Serway and Knowledge Revolution are available on computer disk in either Macintosh or IBM format to be used in conjunction with the highly acclaimed program *Interactive Physics II* from Knowledge Revolution. Most of these simulations are keyed to appropriate worked-example problems and to selected end-of-chapter problems. The remainder are demonstrations that complement concepts or applications discussed in the text. Simulations can be used in the classroom or laboratory to help students understand physics concepts by developing better visualization skills. The simulation is started by simply clicking the RUN button. The simulation engine calculates the motion of the defined system and displays it in smooth animation. The results can be displayed in graphical, digital, tabular, and bar-graph formats. The acquired data can also be exported to a spreadsheet of your choice for other types of analyses. The Interactive Physics Icon ▸ identifies the examples, problems, and figures for which a simulation exists. A complete list follows.

List of Interactive Physics Simulations

Chapter 2

Example 2.10
Example 2.12
Example 2.14
Example 2.15
Problem 2.46
Problem 2.49
Problem 2.72
Problem 2.76
Problem 2.80
Problem 2.81

Chapter 3

Example 3.8
Problem 3.50

Chapter 4

Example 4.2
Example 4.5
Example 4.6
Example 4.7
Example 4.11
Figure 4.5
Problem 4.17
Problem 4.55
Problem 4.58
Problem 4.66
Problem 4.82
Problem 4.84

Chapter 5

Example 5.8
Example 5.9

Example 5.12
Example 5.13
Example 5.14
Problem 5.18
Problem 5.37
Problem 5.38
Problem 5.42
Problem 5.47
Problem 5.55
Problem 5.70
Problem 5.73
Problem 5.74
Problem 5.76
Problem 5.83
Problem 5.84
Problem 5.87
Problem 5.88

Chapter 6

Example 6.1
Example 6.3
Problem 6.5
Problem 6.21
Problem 6.30

Chapter 7

Example 7.7
Example 7.8
Example 7.12
Figure 7.8
Problem 7.37
Problem 7.43

Problem 7.44
Problem 7.82
Problem 7.89

Chapter 8

Example 8.1
Example 8.3
Example 8.6
Example 8.8
Example 8.9
Problem 8.10
Problem 8.11
Problem 8.17
Problem 8.19
Problem 8.33
Problem 8.35
Problem 8.59
Problem 8.64
Problem 8.67

Chapter 9

Example 9.7
Example 9.11
Example 9.13
Example 9.14
Problem 9.66
Problem 9.72
Problem 9.83
Problem 9.87

Chapter 10

Example 10.11
Example 10.12

Example 10.15
Problem 10.51

Chapter 11

Problem 11.51
Problem 11.65

Chapter 12

Example 12.1
Example 12.3
Example 12.4
Problem 12.36
Problem 12.40
Problem 12.51

Chapter 13

Example 13.4
Example 13.5
Example 13.8
Figure 13.9
Problem 13.18
Problem 13.57
Problem 13.63

f(g) Scholar — Spreadsheet/Graphing, Calculator/Graphing Software

f(g) Scholar is a powerful, scientific/engineering spreadsheet software program with over 300 built-in math functions, developed by Future Graph, Inc. It uniquely integrates graphing calculator, spreadsheet, and graphing applications into one, and allows for quick and easy movement between the applications. Students will find many uses for f(g) Scholar across their science, math and engineering courses, including working through their laboratories from start to finished reports. Other features include a programming language for defining math functions, curve fitting, three-dimensional graphing and equation displaying. When bookstores order f(g) Scholar through Saunders College Publishing they can pass on our exclusive low price to the student.

Student Ancillaries

Student Solutions Manual and Study Guide by John R. Gordon, Ralph McGrew, Steve Van Wyk, and Ray Serway   The manual features detailed solutions to 25 percent of the end-of-chapter problems from the text. These are indicated with boxed problem numbers. The manual also features a skills section that reviews mathematical concepts and important notes from key sections of the text and provides a list of important equations and concepts.

Pocket Guide by V. Gordon Lind   This 5″ × 7″ notebook is a section-by-section capsule of the textbook that provides a handy guide for looking up important concepts, formulas, and problem-solving hints.

Discovery Exercises for Interactive Physics by Jon Staib   This workbook is designed to be used in conjunction with the Interactive Physics simulations previously described. The workbook consists of a set of exercises in which the student is required to fill in blanks, answer questions, construct graphs, predict results, and perform simple calculations. Each exercise is designed to teach at least one physical principle and/or to develop student's physical intuition. The workbook and templates can be used either as stand-alone tutorials, or in a laboratory setting.

Spreadsheet Templates   The Spreadsheet Template Disk contains spreadsheet files designed to be used with the end-of-chapter problems entitled Spreadsheet Problems. The files have been developed in Lotus 1-2-3 using the WK1 format. These can be used with most spreadsheet programs including all the recent versions of Lotus 1-2-3, Excel for Windows and Macintosh, Quattro Pro, and f(g) scholar. Over 30 templates are provided for the student.

(© Duomo/Steven E. Sutton, 1994)

Spreadsheet Investigations in Physics by Lawrence B. Golden and James R. Klein   This workbook with the accompanying disk illustrates how spreadsheets can be used for solving many physics problems and when spreadsheet analysis is useful. The workbook is divided into two parts. The first part consists of spreadsheet tutorials, while the second part is a short introduction to numerical methods. The tutorials include basic spreadsheet techniques emphasizing navigating the spreadsheet, entering data, constructing formulas, and graphing. The numerical methods include differentiation, integration, interpolation, and the solution of differential equations. Many examples and exercises are provided. Step-by-step instructions are given for constructing numerical models of selected physics problems. The exercises and examples used to illustrate the numerical methods are chosen from introductory physics and mathematics. The spreadsheet material is presented using Lotus 1-2-3 Release 2.x features, with specific sections devoted to features of other spreadsheet programs, including recent versions of Lotus 1-2-3 for Windows, Excel for Windows and the Macintosh, Quattro Pro, and f(g) Scholar.

Mathematical Methods for Introductory Physics with Calculus by Ronald C. Davidson, Princeton University   This brief book is designed for students who find themselves unable to keep pace in their physics class because of a lack of familiarity with the necessary mathematical tools. *Mathematical Methods* provides a brief overview of all the various types of mathematical topics that may be needed in an introductory-level physics course through the use of many worked examples and exercises.

So You Want to Take Physics: A Preparation for Scientists and Engineers by Rodney Cole   This text is useful to those students who need additional preparation before or during a course in physics. The book includes typical problems with worked-out solutions, and a review of techniques in mathematics and physics. The friendly, straightforward style makes it easier to understand how mathematics is used in the context of physics.

Practice Problems with Solutions   This collection of more than 500 level-1 problems taken from the third edition of *Physics for Scientists and Engineers* is available with full solutions. These problems can be used for homework assignments or student practice and drill exercises.

Challenging Problems in Physics by Boris Korsunsky   This set of 600 thought-provoking problems is meant to test the student's understanding of basic concepts and help them develop general approaches to solving physics problems.

Life Science Applications for Physics   This supplement, compiled by Jerry Faughn, provides examples, readings, and problems from the biological sciences as they relate to physics. Topics include "Friction in Human Joints," "Physics of the Human Circulatory System," "Physics of the Nervous System," and "Ultrasound and Its Applications." This supplement is useful in those courses having a significant number of pre-med students.

Physics Laboratory Manual by David Loyd   To supplement the learning of basic physical principles while introducing laboratory procedures and equipment, each chapter of the laboratory manual includes a pre-laboratory assignment, objectives, an equipment list, the theory behind the experiment, experimental procedure, calculations, graphs, and questions. In addition, a laboratory report is provided for each experiment so the student can record data, calculations, and experimental results.

(Ken Sakomoto, Black Star)

Instructor's Ancillaries

Instructor's Manual with Solutions by Ralph McGrew and Steve Van Wyk   This manual consists of complete, worked-out solutions to all the problems in the text and answers to even-numbered problems. The solutions to the new problems in the fourth edition are marked so the instructor can identify them. All solutions have been carefully reviewed for accuracy.

Computerized Test Bank by Jorge Cossio   Available for the IBM PC and Macintosh computers, this test bank contains over 2300 multiple-choice and open-ended problems and questions, representing every chapter of the text. The test bank enables the instructor to customize tests by rearranging, editing, and adding new questions. The software program prints each answer on a separate grading key. All questions have been reviewed for accuracy.

Printed Test Bank   This test bank is the printed version of the computerized test bank; it contains all of the multiple-choice questions and open-ended problems and questions from the software disk. Answers are also provided.

Interactive Physics Demonstrations by Ray Serway   A set of physics computer simulations that use the Interactive Physics II program is available for use in classroom presentations. These simulations are very useful to show animations of motion, and most are keyed to specific sections or examples in the textbook.

Saunders Physics Videodisc   Contains animations derived from SD2000 software and Interactive Physics II software, video clips demonstrating real-world applications of physics, and still images from the text of *Physics for Scientists and Engineer with Modern Physics*, fourth edition. The still images include most of the line art from the text with enlarged labels for better classroom viewing.

Physics Demonstration Videotape by J. C. Sprott of the University of Wisconsin, Madison   A unique two-hour video-cassette divided into 12 primary topics. Each topic contains between four and nine demonstrations for a total of 70 physics demonstrations.

Selected Solutions Transparency Masters   Selected worked-out solutions are identical to those included in the Student Solutions Manual and Study Guide. These can be used in the classroom when transferred to acetates.

Overhead Transparency Acetates   This collection of transparencies consists of more than 200 full-color figures from the text and features large print for easy viewing in the classroom.

Instructor's Manual to Accompany Challenging Problems for Physics by Boris Korsunsky   This book contains the answers and solutions to all 600 problems that appear in *Challenging Problems for Physics*. All problems are restated for convenience, along with the necessary diagrams.

Instructor's Manual for Physics Laboratory Manual by David Loyd   Each chapter contains a discussion of the experiment, teaching hints, answers to selected questions, and a post-laboratory-quiz with short answer and essay questions. A list of the suppliers of scientific equipment and a summary of the equipment needed for all the laboratory experiments in the manual are also included.

TEACHING OPTIONS

This book is structured in the following sequence of topics: Volume I includes classical mechanics, matter waves, and heat and thermodynamics, Volume II includes electricity and magnetism, light waves, optics, relativity, and modern physics. This presentation is a more traditional sequence, with the subject of matter waves presented before electricity and magnetism. Some instructors may prefer to cover this material after completing electricity and magnetism (after Chapter 34). The chapter on relativity was placed at the end of the text because this topic is often treated as an introduction to the era of "modern physics." If time permits, instructors may choose to cover Chapter 39 in Volume II after completing Chapter 14, which concludes the material on Newtonian mechanics.

For those instructors teaching a two-semester sequence, some sections and chapters could be deleted without any loss in continuity. I have labeled these with asterisks (*) in the Table of Contents and in the appropriate sections of the text. For student enrichment, some of these sections or chapters could be given as extra reading assignments.

ACKNOWLEDGMENTS

In preparing the fourth edition of this textbook, I have been guided by the expertise of the many people who reviewed part or all of the manuscript. Robert Bauman was instrumental in the reviewing process, checking the entire text for accuracy and offering numerous suggestions to improve clarity. I appreciate the assistance of Irene Nunes for skillfully editing and refining the language in the text. In addition, I would like to acknowledge the following scholars and express my sincere appreciation for their helpful suggestions, criticisms, and encouragement:

Edward Adelson, Ohio State University
Joel Berlinghieri, The Citadel
Ronald E. Brown, California Polytechnic State University–San Luis Obispo
Lt. Commander Charles Edmonson, U.S. Naval Academy
Phil Fraundorf, University of Missouri–St. Louis
Ken Ganeger, California State University–Dominiquez Hills
Alfonso Diaz-Jimínez, ADJOIN Research Center, Colombia
George Kottowar, Texas A & M
Raymond L. Kozub, Tennessee Technological University
Tom Moon, Montana Tech
Edward Mooney, Youngstown State University
Marvin Payne, Georgia Southern University
Sama'an Salem, California State University–Long Beach
John Sheldon, Florida International University
J. C. Sprott, University of Wisconsin–Madison
Larry Sudduth, Georgia Institute of Technology
David Taylor, Northwestern University
George Williams, University of Utah

I would also like to thank the following professors for their suggestions during the development of the prior editions of this textbook:

Reviewers

George Alexandrakis, University of Miami
Elmer E. Anderson, University of Alabama
Wallace Arthur, Fairleigh Dickinson
Duane Aston, California State University at Sacramento
Stephen Baker, Rice University
Richard Barnes, Iowa State University
Albert A. Bartlett
Stanley Bashkin, University of Arizona
Marvin Blecher, Virginia Polytechnic Institute and State University
Jeffrey J. Braun, University of Evansville
Kenneth Brownstein, University of Maine
William A. Butler, Eastern Illinois University
Louis H. Cadwell, Providence College
Bo Casserberg, University of Minnesota
Ron Canterna, University of Wyoming
Soumya Chakravarti, California State Polytechnic University
C. H. Chan, The University of Alabama in Huntsville
Edward Chang, University of Massachusetts, Amherst
Don Chodrow, James Madison University
Clifton Bob Clark, University of North Carolina at Greensboro
Walter C. Connolly, Appalachian State University
Hans Courant, University of Minnesota
David R. Currot
Lance E. De Long, University of Kentucky
James L. DuBard, Birmingham-Southern College
F. Paul Esposito, University of Cincinnati
Jerry S. Faughn, Eastern Kentucky University
Paul Feldker, Florissant Valley Community College
Joe L. Ferguson, Mississippi State University
R. H. Garstang, University of Colorado at Boulder
James B. Gerhart, University of Washington
John R. Gordon, James Madison University
Clark D. Hamilton, National Bureau of Standards
Mark Heald, Swarthmore College
Herb Helbig, Clarkson University
Howard Herzog, Broome Community College
Larry Hmurcik, University of Bridgeport
Paul Holoday, Henry Ford Community College
Jerome W. Hosken, City College of San Francisco
William Ingham, James Madison University
Mario Iona, University of Denver
Karen L. Johnston, North Carolina State University

Brij M. Khorana, Rose-Hulman Institute of Technology
Larry Kirkpatrick, Montana State University
Carl Kocher, Oregon State University
Robert E. Kribel, Jacksonville State University
Barry Kunz, Michigan Technological University
Douglas A. Kurtze, Clarkson University
Fred Lipschultz, University of Connecticut
Chelcie Liu
Francis A. Liuima, Boston College
Robert Long, Worcester Polytechnic Institute
Roger Ludin, California Polytechnic State University
Nolen G. Massey, University of Texas at Arlington
Howard McAllister
Charles E. McFarland, University of Missouri at Rolla
Ralph V. McGrew, Broome Community College
James Monroe, The Pennsylvania State University, Beaver Campus
Bruce Morgan, U.S. Naval Academy
Clem Moses, Utica College
Curt Moyer, Clarkson University
David Murdock, Tennessee Technological College
A. Wilson Nolle, The University of Texas at Austin
Thomas L. O'Kuma, San Jacinto College North
Fred A. Otter, University of Connecticut
George Parker, North Carolina State University
William F. Parks, University of Missouri, Rolla
Philip B. Peters, Virginia Military Institute
Eric Peterson, Highland Community College
Richard Reimann, Boise State University
Joseph W. Rudmin, James Madison University
Jill Rugare, DeVry Institute of Technology
C. W. Scherr, University of Texas at Austin
Eric Sheldon, University of Massachusetts–Lowell
John Shelton, College of Lake County
Stan Shepard, The Pennsylvania State University
A. J. Slavin
James H. Smith, University of Illinois at Urbana-Champaign
Richard R. Sommerfield, Foothill College
Kervork Spartalian, University of Vermont
J. C. Sprott
Robert W. Stewart, University of Victoria
James Stith, United States Military Academy
Charles D. Teague, Eastern Kentucky University
Edward W. Thomas, Georgia Institute of Technology

(Courtesy Jeanne Maier)

Carl T. Tomizuka, University of Arizona
Herman Trivilino, San Jacinto College North
Som Tyagi, Drexel University
Steve Van Wyk, Chapman College
Joseph Veit, Western Washington University
T. S. Venkataraman, Drexel University
Noboru Wada, Colorado School of Mines

James Walker, Washington State University
Gary Williams, University of California, Los Angeles
George Williams, University of Utah
William W. Wood
Edward Zimmerman, University of Nebraska, Lincoln
Earl Zwicker, Illinois Institute of Technology

I would like to thank Michael Carchidi for coordinating and contributing to the end-of-chapter problems in the fourth edition. I am very grateful to the following individuals for contributing many creative and interesting new problems to the text: Barry Gilbert, Rhode Island College; Boris Korsunsky, Northfield Mound Hermon School; Bo Lou, Ferris State University; and Roger Ludin, California Polytechnic State University–San Luis Obispo.

I appreciate the assistance of Steve Van Wyk and Ralph McGrew for their careful review of all new end-of-chapter problems and for the preparation of the answer section in the text and the Instructor's Manual. I am indebted to my colleague and friend John R. Gordon for his many contributions during the development of this text, for his continued encouragement and support, and for his expertise in preparing the Student Solutions Manual and Study Guide with the assistance of Ralph McGrew and Steve Van Wyk. Linda Miller is to be thanked for assisting in the preparation of the manuscript and for typesetting and proofreading the Student Solutions Manual and Study Guide. My thanks to Michael Rudmin for the contribution of illustrations to the Student Solutions Manual and Study Guide. Thanks to Sue Howard for locating many excellent photographs and to Jim Lehman and the late Henry Leap for providing numerous photographs of physics demonstrations.

My thanks to Larry Golden and James Klein for their development of end-of-chapter spreadsheet problems and the accompanying templates, as well as the supplement, Numerical Analysis: Spreadsheet Investigations in Physics.

The support package is becoming an ever more essential component of a textbook. I would like to thank the following individuals for authoring the ancillaries that accompany this text: Jorge Cossio for thoroughly reviewing and updating the test bank; Jon Staib for developing the Discovery Exercises workbook that accompanies the Interactive Physics simulations and for reviewing and fine tuning many of the simulations; Evelyn Patterson of the U.S. Air Force Academy for her insightful review and improvements to the Interactive Physics simulations; John Minnerly for converting the Interactive Physics files into an IBM Windows format; Bob Blitshtein and the staff at Future Graph, Inc., for their creation of the SD2000 software package to accompany this text; V. Gordon Lind for concepting and authoring the Pocket Guide; David Loyd for preparing the lab manual and accompanying instructor's manual; Boris Korsunsky for preparing a supplement of Challenging Problems for Physics; Jerry Faughn for compiling *Life Science Applications for Physics;* Ron Davidson for authoring *Mathematical Methods for Introductory Physics with Calculus;* and Rodney Cole for preparing the preparatory manual for physics, *So You Want to Take Physics.*

During the development of this textbook, I benefited from valuable discussions and communications with many people including Subash Antani, Gabe Anton, Randall Caton, Don Chodrow, Jerry Faughn, John R. Gordon, Herb Helbig, Lawrence Hmurcik, William Ingham, David Kaup, Len Ketelsen, Alfonso Diaz-Jiménez, Henry Leap, H. Kent Moore, Charles McFarland, Frank Moore, Clem Moses, Curt Moyer, William Parks, Dorn Peterson, Joe Rudmin, Joe Scaturro, Alex Serway, John Serway, Georgio Vianson, and Harold Zimmerman. Special recognition is due to my mentor and friend, Sam Marshall, a gifted teacher and scientist who helped me sharpen my writing skills while I was a graduate student.

Special thanks and recognition go to the professional staff at Saunders College Publishing for their fine work during the development and production of this text, especially Laura Maier, Developmental Editor; Sally Kusch, Senior Project Editor; Charlene Squibb, Production Manager; and Carol Bleistine, Manager of Art and Design. Thanks also to Tim Frelick, VP/Director of Editorial, Design, and Production, and to Margie Waldron, VP/Marketing, for their continued support of this project. I thank John Vondeling, Vice

President / Publisher, for his great enthusiasm for the project, his friendship, and his confidence in me as an author. I am most appreciative of the intelligent copyediting by Charlotte Nelson, the excellent artwork by Rolin Graphics, Inc., and the attractive design by Rebecca Lemna.

A special note of appreciation goes to the hundreds of students at Clarkson University who used the first edition of this text in manuscript form during its development. I also wish to thank the many users of the second and third editions who submitted suggestions and pointed out errors. With the help of such cooperative efforts, I hope to have achieved my main objective; that is, to provide an effective textbook for the student.

And last, I thank my wonderful family for continuing to support and understand my commitment to physics education.

Raymond A. Serway

James Madison University

June 1995

To The Student

I feel it is appropriate to offer some words of advice which should be of benefit to you, the student. Before doing so, I will assume that you have read the preface, which describes the various features of the text that will help you through the course.

HOW TO STUDY

Very often instructors are asked "How should I study physics and prepare for examinations?" There is no simple answer to this question, but I would like to offer some suggestions based on my own experiences in learning and teaching over the years.

First and foremost, maintain a positive attitude towards the subject matter, keeping in mind that physics is the most fundamental of all natural sciences. Other science courses that follow will use the same physical principles, so it is important that you understand and be able to apply the various concepts and theories discussed in the text.

CONCEPTS AND PRINCIPLES

It is essential that you understand the basic concepts and principles before attempting to solve assigned problems. This is best accomplished through a careful reading of the textbook before attending your lecture on that material. In the process, it is useful to jot down certain points which are not clear to you. Take careful notes in class, and then ask questions pertaining to those ideas that require clarification. Keep in mind that few people are able to absorb the full meaning of scientific material after one reading. Several readings of the text and notes may be necessary. Your lectures and laboratory work should supplement the text and clarify some of the more difficult material. You should reduce memorization of material to a minimum. Memorizing passages from a text, equations, and derivations does not necessarily mean you understand the material. Your understanding of the material will be enhanced through a combination of efficient study habits, discussions with other students and instructors, and your ability to solve the problems in the text. Ask questions whenever you feel it is necessary.

STUDY SCHEDULE

It is important to set up a regular study schedule, preferably on a daily basis. Make sure to read the syllabus for the course and adhere to the schedule set by your instructor. The lectures will be much more meaningful if you read the corresponding textual material before attending the lecture. As a general rule, you should devote about two hours of study time for every hour in class. If you are having trouble with the course, seek the advice of the instructor or students who have taken the course. You may find it necessary to seek further instruction from experienced students. Very often, instructors will offer review sessions in addition to regular class periods. It is important that you avoid the practice of delaying study until a day or two before an exam. More often than not, this will lead to disastrous results. Rather than staying up for an all-night session, it is better to review the basic concepts and equations briefly, followed by a good night's rest. If you feel in need of additional help in understanding the concepts, in preparing for exams, or in problem-solving, we suggest that you acquire a copy of the Student Solutions Manual and Study Guide which accompanies the text and should be available at your college bookstore.

USE THE FEATURES

You should make full use of the various features of the text discussed in the preface. For example, marginal notes are useful for locating and describing important equations and concepts, while important statements and definitions are highlighted in color. Many useful tables are contained in appendices, but most are incorporated in the text where they are used most often. Appendix B is a convenient review of mathematical techniques.

Answers to odd-numbered problems are given at the end of the text, and answers to end-of-chapter questions are provided in the study guide. Exercises (with answers), which follow some worked examples, represent extensions of those examples, and in most cases you are expected to perform a simple calculation. Their purpose is to test your problem-solving skills as you read through the text.

Problem-Solving Strategies and Hints are included in selected chapters throughout the text to give you additional information to help you solve problems. An overview of the entire text is given in the table of contents, while the index will enable you to locate specific material quickly. Footnotes are sometimes used to supplement the discussion or to cite other references on the subject. Most chapters include several problems that make use of spreadsheets. These are intended for those courses that place some emphasis on numerical methods. In some cases, spreadsheet templates are provided, while others require the modification of these templates or the creation of new templates.

After reading a chapter, you should be able to define any new quantities introduced in that chapter and to discuss the principles and assumptions that were used to arrive at certain key relations. The chapter summaries and the review sections of the study guide should help you in this regard. In some cases, it will be necessary to refer to the index of the text to locate certain topics. You should be able to correctly associate with each physical quantity a symbol used to represent that quantity and the unit in which the quantity is specified. Furthermore, you should be able to express each important relation in a concise and accurate prose statement.

PROBLEM SOLVING

R. P. Feynman, Nobel laureate in physics, once said, "You do not know anything until you have practiced." In keeping with this statement, I strongly advise that you develop the skills necessary to solve a wide range of problems. Your ability to solve problems will be one of the main tests of your knowledge of physics, and therefore you should try to solve as many problems as possible. It is essential that you understand basic concepts and principles before attempting to solve problems. It is good practice to try to find alternate solutions to the same problem. For example, problems in mechanics can be solved using Newton's laws, but very often an alternative method using energy considerations is more direct. You should not deceive yourself into thinking you understand the problem after seeing its solution in class. You must be able to solve the problem and similar problems on your own.

The method of solving problems should be carefully planned. A systematic plan is especially important when a problem involves several concepts. First, read the problem several times until you are confident you understand what is being asked. Look for any key words that will help you interpret the problem, and perhaps allow you to make certain assumptions. Your ability to interpret the question properly is an integral part of problem solving. You should acquire the habit of writing down the information given in a problem and deciding what quantities need to be found. You might want to construct a table listing quantities given and quantities to be found. This procedure is sometimes used in the worked examples of the text. After you have decided on the method you feel is appropriate for the situation, proceed with your solution. General problem-solving strategies of this type are included in the text and are highlighted by a light color screen.

I often find that students fail to recognize the limitations of certain formulas or physical laws in a particular situation. It is very important that you understand and remember the assumptions which underlie a particular theory or formalism. For example, certain equations in kinematics apply only to a particle moving with constant acceleration. These equations are not valid for situations in which the acceleration is not constant, such as the motion of an object connected to a spring or the motion of an object through a fluid.

General Problem-Solving Strategy

Most courses in general physics require the student to learn the skills of problem solving, and examinations are largely composed of problems that test such skills. This brief section describes some useful ideas which will enable you to increase your accuracy in solving problems, enhance your understanding of physical concepts, eliminate initial panic or lack of direction in approaching a problem, and organize your work. One way to help accomplish these goals is to adopt a problem-solving strategy. Many chapters in this text will include a section labeled "Problem-Solving Strategies and Hints" which should help you through the "rough spots."

In developing problem-solving strategies, five basic steps are commonly used.

- Draw a suitable diagram with appropriate labels and coordinate axes if needed.
- As you examine what is being asked in the problem, identify the basic physical principle (or principles) that are involved, listing the knowns and unknowns.
- Select a basic relationship or derive an equation that can be used to find the unknown, and solve the equation for the unknown symbolically.
- Substitute the given values along with the appropriate units into the equation.
- Obtain a numerical value for the unknown. The problem is verified and receives a check mark if the following questions can be answered properly: Do the units match? Is the answer reasonable? Is the plus or minus sign proper or meaningful?

One of the purposes of this strategy is to promote accuracy. Properly drawn diagrams can eliminate many sign errors. Diagrams also help to isolate the physical principles of the problem. Symbolic solutions and carefully labeled knowns and unknowns will help eliminate other careless errors. The use of symbolic solutions should help you think in terms of the physics of the problem. A check of units at the end of the problem can indicate a possible algebraic error. The physical layout and organization of your problem will make

EXAMPLE

A person driving in a car at a speed of 20 m/s applies the brakes and stops in a distance of 100 m. What was the acceleration of the car?

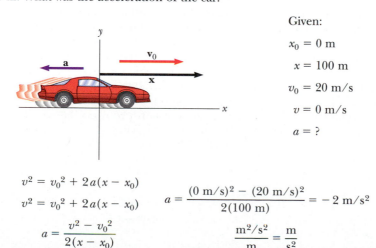

Given:

$$x_0 = 0 \text{ m}$$

$$x = 100 \text{ m}$$

$$v_0 = 20 \text{ m/s}$$

$$v = 0 \text{ m/s}$$

$$a = ?$$

$$v^2 = v_0{}^2 + 2a(x - x_0)$$

$$v^2 = v_0{}^2 + 2a(x - x_0)$$

$$a = \frac{v^2 - v_0{}^2}{2(x - x_0)}$$

$$a = \frac{(0 \text{ m/s})^2 - (20 \text{ m/s})^2}{2(100 \text{ m})} = -2 \text{ m/s}^2$$

$$\frac{\text{m}^2/\text{s}^2}{\text{m}} = \frac{\text{m}}{\text{s}^2}$$

the final product more understandable and easier to follow. Once you have developed an organized system for examining problems and extracting relevant information, you will become a more confident problem solver.

EXPERIMENTS

Physics is a science based upon experimental observations. In view of this fact, I recommend that you try to supplement the text through various types of "hands-on" experiments, either at home or in the laboratory. These can be used to test ideas and models discussed in class or in the text. For example, the common "Slinky" toy is excellent for studying traveling waves; a ball swinging on the end of a long string can be used to investigate pendulum motion; various masses attached to the end of a vertical spring or rubber band can be used to determine their elastic nature; an old pair of Polaroid sunglasses and some discarded lenses and a magnifying glass are the components of various experiments in optics; you can get an approximate measure of the free-fall acceleration by dropping a ball from a known height and measuring the time of its fall with a stopwatch. The list is endless. When physical models are not available, be imaginative and try to develop models of your own.

PEDAGOGICAL USE OF COLOR

The various colors that you will see in the illustrations of this text are used to improve clarity and understanding. Many figures with three-dimensional representations are air-brushed in various colors to make them as realistic as possible.

Most graphs are presented with curves plotted in either rust or blue, and coordinate axes are in black. Several colors are used in those graphs where many physical quantities are plotted simultaneously, or in those cases where different processes may be occurring and need to be distinguished.

Motional shading effects have been incorporated in many figures to remind the readers that they are dealing with a dynamic system rather than a static system. These figures will appear to be somewhat like a "multiflash" photograph of a moving system, with faint images of the "past history" of the system's path. In some figures, a broad, colored arrow is used to indicate the direction of motion of the system.

Color coding has been used in various parts of the book to identify specific physical quantities. The chart on p. xxiii should be a good reference when examining illustrations.

AN INVITATION TO PHYSICS

It is my sincere hope that you will find physics an exciting and enjoyable experience, and that you will profit from this experience, regardless of your chosen profession. Welcome to the exciting world of physics.

The scientist does not study nature because it is useful; he studies it because he delights in it, and he delights in it because it is beautiful. If nature were not beautiful, it would not be worth knowing, and if nature were not worth knowing, life would not be worth living.

Henri Poincaré

PEDAGOGICAL COLOR CHART

Part I (Chapters 1–15) : Mechanics

Displacement and
position vectors

Velocity vectors (**v**)
Velocity component vectors

Force vectors (**F**)
Force component vectors

Acceleration vectors (**a**)
Acceleration component vectors

Torque (**t**) and
angular momentum
(**L**) vectors

Linear or rotational
motion directions

Springs

Pulleys

Part IV (Chapters 23–34) : Electricity and Magnetism

Electric fields

Magnetic fields

Positive charges

Negative charges

Resistors

Batteries and other
dc power supplies

Switches

Capacitors

Inductors (coils)

Voltmeters

Ammeters

Galvanometers

ac generators

Ground symbol

Part V (Chapters 35–38) : Light and Optics

Light rays

Lenses and prisms

Mirrors

Objects

Images

Contents Overview

(Bruce Byers, FPG)

Contents

(David Madison, Tony Stone Images)

(James Stevenson/SPL/Photo Researchers)

(Daryl Torckler/Tony Stone Images)

(*Tom McHugh/Photo Researchers*)

(*Steve Niedorf/The Image Bank*)

(*Lois Moulton/Tony Stone Worldwide, Ltd.*)

Stonehenge is a circle of stone built by the ancient Britons on Salisbury Plain, England. Its orientation marks the seasonal rising and setting points of the Sun. For example, on the summer solstice, an observer at the center of the stone circle sees the Sun rising directly over a marker stone. (©John Serafin, Peter Arnold, Inc.)

Mechanics

Philosophy is written in that great book which ever lies before our gaze — I mean the Universe — but we cannot understand if we do not first learn the language and grasp the symbols in which it is written. The book is written in the mathematical language, and the symbols are triangles, circles, and other geometrical figures, without the help of which it is impossible to conceive a single word of it, and without which one wanders in vain through a dark labyrinth.

GALILEO GALILEI

Physics, the most fundamental physical science, is concerned with the basic principles of the Universe. It is the foundation upon which the other physical sciences — astronomy, chemistry, and geology — are based. The beauty of physics lies in the simplicity of the fundamental physical theories and in the manner in which just a small number of fundamental concepts, equations, and assumptions can alter and expand our view of the world around us.

The myriad physical phenomena in our world are a part of one or more of the following five areas of physics:

1. Classical mechanics, which is concerned with the motion of objects moving at speeds that are low compared to the speed of light
2. Relativity, which is a theory describing objects moving at any speed, even those whose speeds approach the speed of light
3. Thermodynamics, which deals with heat, work, temperature, and the statistical behavior of a large number of particles
4. Electromagnetism, which involves the theory of electricity, magnetism, and electromagnetic fields
5. Quantum mechanics, a theory dealing with the behavior of particles at the submicroscopic level as well as the macroscopic world

The first part of this textbook deals with mechanics, sometimes referred to as either classical mechanics or Newtonian mechanics. This is an appropriate place to begin an introductory text since many of the basic principles used to understand mechanical systems can later be used to describe such natural phenomena as waves and heat transfer. Furthermore, the laws of conservation of energy and momentum introduced in mechanics retain their importance in the fundamental theories that follow, including the theories of modern physics.

Today, mechanics is of vital importance to students from all disciplines. It is highly successful in describing the motions of material bodies, such as planets, rockets, and baseballs. In the first part of the text, we shall describe the laws of mechanics and examine a wide range of phenomena that can be understood with these fundamental ideas.

Physics and Measurement

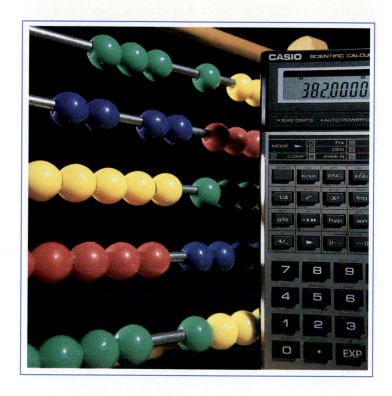

A modern pocket calculator together with a Chinese abacus. The abacus represents one of the first technological approaches to counting and calculating. There are evidences of its use as early as 3000 BC.   *(Sheila Terry/Photo Researchers)*

Physics is a fundamental science concerned with understanding the natural phenomena that occur in our Universe. Like all sciences, physics is based on experimental observations and quantitative measurements. The main objective of physics is to use the limited number of fundamental laws that govern natural phenomena to develop theories that can predict the results of future experiments. The fundamental laws used in developing theories are expressed in the language of mathematics, the tool that provides a bridge between theory and experiment.

When a discrepancy between theory and experiment arises, new theories and experiments must be formulated to remove the discrepancy. Many times a theory is satisfactory only under limited conditions; a more general theory might be satisfactory without such limitations. A classic example is Newton's laws of motion, which accurately describe the motion of bodies at normal speeds but do not apply to objects moving at speeds comparable to the speed of light. The special theory of relativity developed by Einstein successfully predicts the motion of objects at low speed and at speeds approaching the speed of light and hence is a more general theory of motion.

Classical physics, which means all of the physics developed prior to 1900, includes the theories, concepts, laws, and experiments in classical mechanics, thermodynamics, and electromagnetism. Galileo Galilei (1564–1642) made significant contributions to classical mechanics through his work on the motion of objects having constant acceleration. In the same era, Johannes Kepler (1571–1630) analyzed astronomical data to develop empirical laws for the motion of planetary bodies.

The most important contributions to classical mechanics were provided by Isaac Newton (1642–1727), who developed classical mechanics as a systematic theory and was one of the originators of the calculus as a mathematical tool. Although major developments in classical physics continued in the 18th century, thermodynamics and electricity and magnetism were not developed until the latter part of the 19th century, principally because the apparatus for controlled experiments was either too crude or unavailable until then. In this text we shall treat the various disciplines of classical physics in separate sections; however, we will see that the disciplines of mechanics and electromagnetism are basic to all the branches of classical and modern physics.

A new era in physics, usually referred to as *modern physics,* began near the end of the 19th century. Modern physics developed mainly because of the discovery that many physical phenomena could not be explained by classical physics. The two most important developments in this modern era were the theories of relativity and quantum mechanics. Einstein's theory of relativity completely revolutionized the traditional concepts of space, time, and energy. Among other things, Einstein's theory corrected Newton's laws of motion for describing the motion of objects moving at speeds comparable to the speed of light. The theory of relativity also assumes that the speed of light is the upper limit of the speed of an object or signal and shows the relationship between mass and energy. Quantum mechanics was formulated by a number of distinguished scientists to provide descriptions of physical phenomena at the atomic level.

Scientists are constantly working at improving our understanding of fundamental laws, and new discoveries are being made every day. In many research areas, there is a great deal of overlap between physics, chemistry, and biology, as well as engineering. Some of the most notable developments are (1) numerous space missions and the landing of astronauts on the Moon, (2) microcircuitry and high-speed computers, and (3) sophisticated imaging techniques used in scientific research and medicine. The impacts of such developments and discoveries on our society have indeed been great, and it is very likely that future discoveries and developments will be just as exciting and challenging and of great benefit to humanity.

1.1   STANDARDS OF LENGTH, MASS, AND TIME

The laws of physics are expressed in terms of basic quantities that require a clear definition. For example, such physical quantities as force, velocity, volume, and acceleration can be described in terms of more basic quantities that in themselves are defined in terms of measurements or comparison with established standards. In mechanics, the three basic quantities are length (L), mass (M), and time (T). All other physical quantities in mechanics can be expressed in terms of these three quantities.

Obviously, if we are to report the results of a measurement to someone who wishes to reproduce this measurement, a standard must be defined. It would be meaningless if a visitor from another planet were to talk to us about a length of 8 "glitches" if we do not know the meaning of the unit glitch. On the other hand, if someone familiar with our system of measurement reports that a wall is 2 meters high and our unit of length is defined as 1 meter, we then know that the height of the wall is twice our fundamental unit of length. Likewise, if we are told that a person has a mass of 75 kilograms and our unit of mass is defined as 1.0 kilogram, then that person is 75 times as massive as our basic unit of mass.[1]

In 1960, an international committee established a set of standards for these fundamental quantities. The system that was established is an adaptation of the metric system, and it is called the **International System (SI)** of units. The abbreviation SI comes from its French name "Système International." In this system, the units of length, mass, and time are the meter, kilogram, and second, respectively. (The SI system is closely related to the *mks* system which preceded it.) Other standard SI units established by the committee are those for temperature (the *kelvin*), electric current (the *ampere*), luminous intensity (the *candela*) and for the amount of substance (the *mole,* see Sec. 1.3). These seven units are the basic SI units. In the study of mechanics, however, we will be concerned only with the units of length, mass, and time. Definitions of units are under constant review and are changed from time to time.

Length

In A.D. 1120 the king of England decreed that the standard of length in his country would be the yard and that the yard would be precisely equal to the distance from the tip of his nose to the end of his outstretched arm. Similarly, the original standard for the foot adopted by the French was the length of the royal foot of King Louis XIV. This standard prevailed until 1799, when the legal standard of length in France became the meter, defined as one ten-millionth the distance from the equator to the North Pole along a longitudinal line that passes through Paris.

Many other systems have been developed in addition to those just discussed, but the advantages of the French system have caused it to prevail in most countries and in scientific circles everywhere. As recently as 1960, the length of the meter was defined as the distance between two lines on a specific platinum-iridium bar stored under controlled conditions. This standard was abandoned for several reasons, a principal one being that the limited accuracy with which the separation between the lines on the bar can be determined does not meet the current requirements of science and technology. Until recently, the meter was defined as 1 650 763.73 wavelengths of orange-red light emitted from a krypton-86 lamp. However, in October 1983, the **meter** was redefined as the distance traveled by light in vacuum during a time of 1/299 792 458 second. In effect, this latest definition establishes that the speed of light in vacuum is 299 792 458 meters per second.

Mass

The basic SI unit of mass, the **kilogram,** is defined as the mass of a specific platinum-iridium alloy cylinder kept at the International Bureau of Weights and Measures at Sèvres, France. This mass standard was established in 1887, and there has

---

[1] The need for assigning numerical values to various physical quantities through experimentation was expressed by Lord Kelvin (William Thomson) as follows: "I often say that when you can measure what you are speaking about, and express it in numbers, you should know something about it, but when you cannot express it in numbers, your knowledge is of a meagre and unsatisfactory kind. It may be the beginning of knowledge but you have scarcely in your thoughts advanced to the state of science."

(Left) The National Standard Kilogram No. 20, an accurate copy of the International Standard Kilogram kept at Sèvres, France, is housed under a double bell jar in a vault at the National Institute of Standards and Technology (NIST). *(Right)* The primary frequency standard (an atomic clock) at the NIST. This device keeps time with an accuracy of about 3 millionths of a second per year.   *(Photos courtesy of National Institute of Standards and Technology (NIST), U.S. Department of Commerce)*

been no change since that time because platinum-iridium is an unusually stable alloy. A duplicate is kept at the National Institute of Standards and Technology (NIST) in Gaithersburg, Md.

Time

Before 1960, the standard of time was defined in terms of the *mean solar day* for the year 1900.[2] The *mean solar second,* representing the basic unit of time, was originally defined as $(\frac{1}{60})(\frac{1}{60})(\frac{1}{24})$ of a mean solar day. The rotation of the Earth is now known to vary substantially with time, however, and therefore this motion is not a good one to use for defining a standard.

In 1967, consequently, the second was redefined to take advantage of the high precision obtainable in a device known as an *atomic clock.* In this device, the frequencies associated with certain atomic transitions (which are extremely stable and insensitive to the clock's environment) can be measured to a precision of one part in 10^{12}. This is equivalent to an uncertainty of less than one second every 30 000 years. Thus, in 1967 the SI unit of time, the *second,* was redefined using the characteristic frequency of a particular kind of cesium atom as the "reference clock": The basic SI unit of time, the **second,** is defined as 9 192 631 770 periods of the radiation from cesium-133 atoms.

---

[2] A solar day is the time interval between successive appearances of the Sun at the highest point it reaches in the sky each day.

TABLE 1.1 Approximate Values of Some Measured Lengths	
	Length (m)
Distance from Earth to most remote known quasar	1.4×10^{26}
Distance from Earth to most remote known normal galaxies	4×10^{25}
Distance from Earth to nearest large galaxy (M 31 in Andromeda)	2×10^{22}
Distance from Earth to nearest star (Proxima Centauri)	4×10^{16}
One lightyear	9.46×10^{15}
Mean orbit radius of the Earth about the Sun	1.5×10^{11}
Mean distance from Earth to Moon	3.8×10^{8}
Distance from the equator to the North Pole	1×10^{7}
Mean radius of the Earth	6.4×10^{6}
Typical altitude of satellite orbiting Earth	2×10^{5}
Length of a football field	9.1×10^{1}
Length of a housefly	5×10^{-3}
Size of smallest dust particles	1×10^{-4}
Size of cells of most living organisms	1×10^{-5}
Diameter of a hydrogen atom	1×10^{-10}
Diameter of an atomic nucleus	1×10^{-14}
Diameter of a proton	1×10^{-15}

Approximate Values for Length, Mass, and Time

Approximate values of various lengths, masses, and time intervals are presented in Tables 1.1 to 1.3. Note the wide range of values.[3] You should study these tables and get a feel for what is meant by a kilogram of mass (for example, the mass of an adult human is about 70 kilograms) or by a time interval of 10^8 seconds (one year is about 3×10^7 seconds). In addition to SI, there are two other systems of units you might run across in the literature. The *cgs system* was used in Europe before SI, and the *British engineering system* (sometimes called the conventional system) is still used in the United States despite acceptance of SI by the rest of the world. In the cgs system, the units of length, mass, and time are the centimeter (cm), gram (g), and second (s), respectively; in the British engineering system, the units of length, mass, and time are the foot (ft), slug, and second, respectively. Throughout most of this text we shall use SI units since they are almost universally accepted in science and industry. We will make some limited use of British engineering units in the study of classical mechanics.

TABLE 1.2 Masses of Various Bodies (Approximate Values)	
	Mass (kg)
Universe	1×10^{52}
Milky Way Galaxy	7×10^{41}
Sun	2×10^{30}
Earth	6×10^{24}
Moon	7×10^{22}
Horse	1×10^{3}
Human	7×10^{1}
Frog	1×10^{-1}
Mosquito	1×10^{-5}
Bacterium	1×10^{-15}
Hydrogen atom	1.67×10^{-27}
Electron	9.11×10^{-31}

CONCEPTUAL EXAMPLE 1.1

What types of natural phenomena could serve as alternative time standards?

Reasoning We could imagine any process with a beginning and an end to define the unit of time. For example, the duration of any biological process or the time it takes a fluid to empty from a container depends on many factors. To make the definition reproducible, each factor would have to be specified (such as the composition of the fluid, its temperature, and the dimensions and tilt of the container). Alternatively, a naturally isolated object like a flashing pulsar repeats the same flashing process in very precise time intervals. A "simpler" system is preferred as a time standard, such as a particular species of isolated atom or its nucleus.

---

[3] If you are unfamiliar with the use of powers scientific notation, you should review Section B.1 of the mathematical appendix at the back of this book.

TABLE 1.3 Approximate Values of Some Time Intervals

	Interval (s)
Age of the Universe	5×10^{17}
Age of the Earth	1.3×10^{17}
Average age of a college student	6.3×10^8
One year	3.2×10^7
One day (time for one revolution of Earth about its axis)	8.6×10^4
Time between normal heartbeats	8×10^{-1}
Period[a] of audible sound waves	1×10^{-3}
Period of typical radio waves	1×10^{-6}
Period of vibration of an atom in a solid	1×10^{-13}
Period of visible light waves	2×10^{-15}
Duration of a nuclear collision	1×10^{-22}
Time for light to cross a proton	3.3×10^{-24}

[a] Period is defined as the time interval of one complete vibration.

TABLE 1.4 Some Prefixes for SI Units

Power	Prefix	Abbreviation
10^{-24}	yocto	y
10^{-21}	zepto	z
10^{-18}	atto	a
10^{-15}	femto	f
10^{-12}	pico	p
10^{-9}	nano	n
10^{-6}	micro	μ
10^{-3}	milli	m
10^{-2}	centi	c
10^{-1}	deci	d
10^{1}	deka	da
10^{3}	kilo	k
10^{6}	mega	M
10^{9}	giga	G
10^{12}	tera	T
10^{15}	peta	P
10^{18}	exa	E
10^{21}	Zetta	Z
10^{24}	Yotta	Y

Some of the most frequently used prefixes for the various powers of ten and their abbreviations are listed in Table 1.4. For example, 10^{-3} m is equivalent to 1 millimeter (mm), and 10^3 m is 1 kilometer (km). Likewise, 1 kg is 10^3 g, and 1 megavolt (MV) is 10^6 volts.

1.2 THE BUILDING BLOCKS OF MATTER

A 1-kg cube of solid gold has a length of approximately 3.73 cm on a side. Is this cube nothing but wall-to-wall gold, with no empty space? If the cube is cut in half, the two resulting pieces still retain their chemical identity as solid gold. But what if the pieces are cut again and again, indefinitely? Will the smaller and smaller pieces always be the same substance, gold? Questions such as these can be traced back to early Greek philosophers. Two of them—Leucippus and Democritus—could not accept the idea that such cuttings could go on forever. They speculated that the process ultimately must end when it produces a particle that can no longer be cut. In Greek, *atomos* means ''not sliceable.'' From this comes our English word *atom* for the smallest, ultimate particle of matter. Elementary-particle physicists still engage in speculation and experimentation concerning the ultimate building blocks of matter.

Let us review briefly what is known about the structure of the world around us. It is useful to view the atom as a miniature Solar System with a dense positively charged nucleus occupying the position of the Sun and negatively charged electrons orbiting like the planets. This model of the atom enables us to understand some properties of the simpler atoms, such as hydrogen, but fails to explain many fine details of atomic structure.

Following the discovery of the nucleus in the early 1900s, the question arose: Does it have structure? That is, is the nucleus a single particle or a collection of particles? The exact composition of the nucleus is not known completely even today, but by the early 1930s a model evolved that helped us understand how the nucleus behaves. Specifically, scientists determined that occupying the nucleus are two basic entities, protons and neutrons. The *proton* carries a positive charge, and a specific element is identified by the number of protons in its nucleus. For instance, the nucleus of a hydrogen atom contains one proton, the nucleus of a helium atom contains two protons, and the nucleus of a uranium atom contains ninety-two protons.

The existence of *neutrons* was verified conclusively in 1932. A neutron has no charge and a mass about equal to that of a proton. One of its primary purposes is to act as a "glue" to hold the nucleus together. If neutrons were not present in the nucleus, the repulsive force between the positively charged particles would cause the nucleus to break apart.

But is this where the breaking down stops? Protons, neutrons, and a host of other exotic particles are now known to be composed of six different varieties of particles called **quarks,** which have been given the names of *up, down, strange, charmed, bottom,* and *top* (Fig. 1.1). The up, charmed, and top quarks have charges of $+2/3$ that of the proton, whereas the down, strange, and bottom quarks have charges of $-1/3$ that of the proton. The proton consists of two up quarks and one down quark, which you can easily show leads to the correct charge for the proton. Likewise, the neutron consists of two down quarks and one up quark, giving a net charge of zero.

1.3 DENSITY AND ATOMIC MASS

A property of any substance is its **density** ρ (Greek letter rho), defined as *mass per unit volume* (a table of the letters in the Greek alphabet is provided on the back endsheet of the textbook):

$$\rho \equiv \frac{m}{V} \tag{1.1}$$

For example, aluminum has a density of 2.70 g/cm^3, and lead has a density of 11.3 g/cm^3. Therefore, a piece of aluminum of volume 10.0 cm^3 has a mass of 27.0 g, while an equivalent volume of lead would have a mass of 113 g. A list of densities for various substances is given in Table 1.5.

The difference in density between aluminum and lead is due, in part, to their different *atomic masses;* the atomic mass of lead is 207 atomic mass units and that of aluminum is 27.0 atomic mass units. However, the ratio of atomic masses, $207/27.0 = 7.67$, does not correspond to the ratio of densities, $11.3/2.70 = 4.19$. The discrepancy is due to the difference in atomic spacings and atomic arrangements in the crystal structure of these two substances.

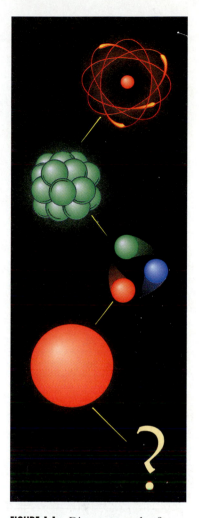

FIGURE 1.1 Distances at the frontier of nuclear physics are astonishingly short. An atom is so small that a single-file line of 250 000 of them would fit within the thickness of aluminum foil. The nucleus at the atom's center is a cluster of nucleons, each 100 000 times smaller than the atom itself. The three quarks inside each nucleon are even smaller. *(Courtesy of SURA, Inc.)*

TABLE 1.5 Densities of Various Substances	
Substance	Density ρ (kg/m^3)
Gold	19.3×10^3
Uranium	18.7×10^3
Lead	11.3×10^3
Copper	8.93×10^3
Iron	7.86×10^3
Aluminum	2.70×10^3
Magnesium	1.75×10^3
Water	1.00×10^3
Air	0.0012×10^3

All ordinary matter consists of atoms, and each atom is made up of electrons and a nucleus. Practically all of the mass of an atom is contained in the nucleus, which consists of protons and neutrons. Because different elements contain different numbers of protons and neutrons, the atomic masses of the various elements differ. The mass of a nucleus is measured relative to the mass of an atom of the carbon-12 isotope (this isotope of carbon has six protons and six neutrons).

The mass of ^{12}C is defined to be exactly 12 atomic mass units (u), where $1 \text{ u} = 1.660\ 540\ 2 \times 10^{-27}$ kg. In these units, the proton and neutron have masses of about 1 u. More precisely,

$$\text{mass of proton} = 1.0073 \text{ u}$$

$$\text{mass of neutron} = 1.0087 \text{ u}$$

One **mole** (mol) of a substance is that amount of it that consists of Avogadro's number, N_A, of molecules. Avogadro's number is defined so that one mole of carbon-12 atoms has a mass of exactly 12 g. Its value has been found to be $N_A = 6.02 \times 10^{23}$ molecules/mol. For example, one mole of aluminum has a mass of 27 g, and one mole of lead has a mass of 207 g. But one mole of aluminum contains the same number of atoms as one mole of lead, since there are 6.02×10^{23} atoms in one mole of *any* element. The mass per atom for a given element is then given by

---
The mass of an atom

$$m_{\text{atom}} = \frac{\text{atomic mass of the element}}{N_A} \tag{1.2}$$

For example, the mass of an aluminum atom is

$$m_{\text{Al}} = \frac{27 \text{ g/mol}}{6.02 \times 10^{23} \text{ atoms/mol}} = 4.5 \times 10^{-23} \text{ g/atom}$$

Note that 1 u is equal to N_A^{-1} g.

EXAMPLE 1.2   How Many Atoms in the Cube?

A solid cube of aluminum (density 2.7 g/cm³) has a volume of 0.20 cm³. How many aluminum atoms are contained in the cube?

Solution   Since density equals mass per unit volume, the mass of the cube is

$$m = \rho V = (2.7 \text{ g/cm}^3)(0.20 \text{ cm}^3) = 0.54 \text{ g}$$

To find the number of atoms, N, we can set up a proportion using the fact that one mole of aluminum (27 g) contains 6.02×10^{23} atoms:

$$\frac{6.02 \times 10^{23} \text{ atoms}}{27 \text{ g}} = \frac{N}{0.54 \text{ g}}$$

$$N = \frac{(0.54 \text{ g})(6.02 \times 10^{23} \text{ atoms})}{27 \text{ g}} = \boxed{1.2 \times 10^{22} \text{ atoms}}$$

1.4   DIMENSIONAL ANALYSIS

The word *dimension* has a special meaning in physics. It usually denotes the physical nature of a quantity. Whether a distance is measured in units of feet or meters or furlongs, it is a distance. We say its dimension is *length*.

The symbols we use in this book to specify length, mass, and time are L, M, and T, respectively. We shall often use brackets [ ] to denote the dimensions of a physical quantity. For example, the symbol we use for speed in this book is v, and in our notation the dimensions of speed are written $[v] = \text{L/T}$. As another exam-

TABLE 1.6   Dimensions of Area, Volume, Speed, and Acceleration

System	Area (L^2)	Volume (L^3)	Speed (L/T)	Acceleration (L/T^2)
SI	m^2	m^3	m/s	m/s^2
cgs	cm^2	cm^3	cm/s	cm/s^2
British engineering	ft^2	ft^3	ft/s	ft/s^2

ple, the dimensions of area, A, are $[A] = L^2$. The dimensions of area, volume, speed, and acceleration are listed in Table 1.6, along with their units in the three common systems. The dimensions of other quantities, such as force and energy, will be described as they are introduced in the text.

In many situations, you may have to derive or check a specific formula. Although you may have forgotten the details of the derivation, there is a useful and powerful procedure called *dimensional analysis* that can be used to assist in the derivation or to check your final expression. This procedure should always be used and should help minimize the rote memorization of equations. Dimensional analysis makes use of the fact that *dimensions can be treated as algebraic quantities.* That is, quantities can be added or subtracted only if they have the same dimensions. Furthermore, the terms on both sides of an equation must have the same dimensions. By following these simple rules, you can use dimensional analysis to help determine whether or not an expression has the correct form because the relationship can be correct only if the dimensions on each side of the equation are the same.

To illustrate this procedure, suppose you wish to derive a formula for the distance x traveled by a car in a time t if the car starts from rest and moves with constant acceleration a. In Chapter 2, we shall find that the correct expression is $x = \frac{1}{2}at^2$. Let us use dimensional analysis to check the validity of this expression.

The quantity x on the left side has the dimension of length. In order for the equation to be dimensionally correct, the quantity on the right side must also have the dimension of length. We can perform a dimensional check by substituting the dimensions for acceleration, L/T^2, and time, T, into the equation. That is, the dimensional form of the equation $x = \frac{1}{2}at^2$ is

$$L = \frac{L}{T^2} \cdot T^2 = L$$

The units of time cancel as shown, leaving the unit of length.

A more general procedure using dimensional analysis is to set up an expression of the form

$$x \propto a^n t^m$$

when n and m are exponents that must be determined and the symbol $\propto$ indicates a proportionality. This relationship is correct only if the dimensions of both sides are the same. Since the dimension of the left side is length, the dimension of the right side must also be length. That is,

$$[a^n t^m] = L = LT^0$$

Since the dimensions of acceleration are L/T^2 and the dimension of time is T, we have

$$(L/T^2)^n T^m = L$$

or

$$L^n T^{m-2n} = L$$

Since the exponents of L and T must be the same on both sides, we see that $n = 1$ and $m = 2$. Therefore, we conclude that

$$x \propto at^2$$

This result differs by a factor of 2 from the correct expression, which is $x = \frac{1}{2}at^2$. Because the factor $1/2$ is dimensionless, there is no way of determining this via dimensional analysis.

CONCEPTUAL EXAMPLE 1.3

Does dimensional analysis give any information on constants of proportionality that may appear in an algebraic expression? Explain.

Reasoning Dimensional analysis gives the units of the proportionality constant but gives no information about its numerical value. For example, experiments show that doubling or tripling the radius of a spherical water balloon makes its mass get eight or twenty-seven times larger. Its mass is proportional to the cube of its radius. Because $m \propto r^3$, we can write $m = kr^3$. Dimensional analysis shows that the proportionality constant k must have units kg/m^3, but to determine its value requires experimental data or geometrical reasoning.

EXAMPLE 1.4   Analysis of an Equation

Show that the expression $v = v_0 + at$ is dimensionally correct, where v and v_0 represent speeds, a is acceleration, and t is a time interval.

Solution   For the speed terms, we have from Table 1.6

$$[v] = [v_0] = L/T$$

The same table gives us L/T^2 for the dimensions of acceleration, and so the dimensions of at are

$$[at] = (L/T^2)(T) = L/T$$

Therefore the expression is dimensionally correct. (If the expression were given as $v = v_0 + at^2$, it would be dimensionally *incorrect*. Try it and see!)

EXAMPLE 1.5   Analysis of a Power Law

Suppose we are told that the acceleration of a particle moving with uniform speed v in a circle of radius r is proportional to some power of r, say r^n, and some power of v, say v^m. How can we determine the powers of r and v?

Solution   Let us take a to be

$$a = kr^n v^m$$

where k is a dimensionless constant. Knowing the dimensions of a, r, and v, we see that the dimensional equation must be

$$L/T^2 = L^n(L/T)^m = L^{n+m}/T^m$$

This dimensional equation is balanced under the conditions

$$n + m = 1 \quad \text{and} \quad m = 2$$

Therefore, $n = -1$, and we can write the acceleration

$$a = kr^{-1}v^2 = k\frac{v^2}{r}$$

When we discuss uniform circular motion later, we shall see that $k = 1$ if a consistent set of units is used. The constant k would not equal 1 if, for example, v were in km/h and you wanted a in m/s².

(Left) Conversion of miles to kilometers. *(Right)* This vehicle's speedometer gives speed readings in miles per hour and in kilometers per hour. Try confirming the conversion between the two sets of units for a few readings of the dial.   *(Paul Silverman, Fundamental Photographs)*

1.5 CONVERSION OF UNITS

Sometimes it is necessary to convert units from one system to another. Conversion factors between the SI and conventional units of length are as follows:

$$1 \text{ mile} = 1609 \text{ m} = 1.609 \text{ km} \qquad 1 \text{ ft} = 0.3048 \text{ m} = 30.48 \text{ cm}$$

$$1 \text{ m} = 39.37 \text{ in.} = 3.281 \text{ ft} \qquad 1 \text{ in.} = 0.0254 \text{ m} = 2.54 \text{ cm}$$

A more complete list of conversion factors can be found in Appendix A.

Units can be treated as algebraic quantities that can cancel each other. For example, suppose we wish to convert 15.0 in. to centimeters. Since 1 in. = 2.54 cm (exactly), we find that

$$15.0 \text{ in.} = (15.0 \text{ in.}) \left(2.54 \, \frac{\text{cm}}{\text{in.}} \right) = 38.1 \text{ cm}$$

EXAMPLE 1.6 The Density of a Cube

The mass of a solid cube is 856 g, and each edge has a length of 5.35 cm. Determine the density ρ of the cube in basic SI units.

Solution Since $1 \text{ g} = 10^{-3} \text{ kg}$ and $1 \text{ cm} = 10^{-2} \text{ m}$, the mass, m, and volume, V, in basic SI units are given by

$$m = 856 \text{ g} \times 10^{-3} \text{ kg/g} = 0.856 \text{ kg}$$

$$V = L^3 = (5.35 \text{ cm} \times 10^{-2} \text{ m/cm})^3$$
$$= (5.35)^3 \times 10^{-6} \text{ m}^3 = 1.53 \times 10^{-4} \text{ m}^3$$

Therefore

$$\rho = \frac{m}{V} = \frac{0.856 \text{ kg}}{1.53 \times 10^{-4} \text{ m}^3} = \boxed{5.59 \times 10^3 \text{ kg/m}^3}$$

1.6 ORDER-OF-MAGNITUDE CALCULATIONS

It is often useful to compute an approximate answer to a physical problem even where little information is available. Such results can then be used to determine whether or not a more precise calculation is necessary. These approximations are usually based on certain assumptions, which must be modified if more precision is needed. Thus, we shall sometimes refer to the order of magnitude of a certain quantity as the power of ten of the number that describes that quantity. If, for example, we say that a quantity increases in value by three orders of magnitude, this means that its value is increased by a factor of $10^3 = 1000$.

The spirit of order-of-magnitude calculations, sometimes referred to as "guesstimates" or "ball-park figures," is given in the following quotation: "Make an estimate before every calculation, try a simple physical argument . . . before every derivation, guess the answer to every puzzle. Courage: no one else needs to know what the guess is."[4]

EXAMPLE 1.7 Breaths in a Lifetime

Estimate the number of breaths taken during an average life span of 70 years.

Solution The only estimate we must make in this example is the average number of breaths that a person takes in 1 min. This number varies, depending on whether the person is exercising, sleeping, angry, serene, and so forth. We shall choose 8 breaths per minute as our estimate of the average.

The number of minutes in a year is

$$1 \text{ year} \times 365 \frac{\text{days}}{\text{year}} \times 24 \frac{\text{h}}{\text{day}} \times 60 \frac{\text{min}}{\text{h}} = 5.26 \times 10^5 \text{ min}$$

Thus, in 70 years there will be $(70)(5.26 \times 10^5 \text{ min}) = 3.68 \times 10^7$ min. At a rate of 8 breaths/min the individual would take about 3×10^8 breaths. You may want to check your own rate and repeat the above calculation.

EXAMPLE 1.8 How Many Atoms?

Estimate the number of atoms in 1 cm^3 of a solid.

Solution From Table 1.2 we note that the diameter of an atom is about 10^{-10} m. Thus, if in our model we assume that the atoms in the solid are solid spheres of this diameter, then the volume of each sphere is about 10^{-30} m^3 (more precisely, volume $= 4\pi r^3/3 = \pi d^3/6$, where $r = d/2$). Therefore,

since 1 cm^3 = 10^{-6} m^3, the number of atoms in the solid is of the order of $10^{-6}/10^{-30} = 10^{24}$ atoms.

A more precise calculation would require knowledge of the density of the solid and the mass of each atom. However, our estimate agrees with the more precise calculation to within a factor of 10.

EXAMPLE 1.9 How Much Gas Do We Use?

Estimate the number of gallons of gasoline used by all U.S. cars each year.

Solution Since there are about 240 million people in the United States, an estimate of the number of cars in the country is 120 million (assuming two cars and four people per family). We also estimate that the average distance traveled

per year is 10 000 miles. If we assume a gasoline consumption of 0.05 gal/mi, each car uses about 500 gal/year. Multiplying this by the total number of cars in the United States gives an estimated total consumption of 6×10^{10} gal. This number of gallons corresponds to a yearly consumer expenditure of over 60 billion dollars and is probably a low estimate because we haven't accounted for commercial consumption.

---

[4] E. Taylor and J. A. Wheeler, *Spacetime Physics,* San Francisco, W. H. Freeman, 1966, p. 60.

1.7 SIGNIFICANT FIGURES

When certain quantities are measured, the measured values are known only to within the limits of the experimental uncertainty. The value of the uncertainty can depend on various factors, such as the quality of the apparatus, the skill of the experimenter, and the number of measurements performed.

Suppose that in a laboratory experiment we are asked to measure the area of a rectangular plate using a meter stick as a measuring instrument. Let us assume that the accuracy to which we can measure a particular dimension of the plate is ±0.1 cm. If the length of the plate is measured to be 16.3 cm, we can claim only that its length lies somewhere between 16.2 cm and 16.4 cm. In this case, we say that the measured value has three significant figures. Likewise, if its width is measured to be 4.5 cm, the actual value lies between 4.4 cm and 4.6 cm. This measured value has only two significant figures. Note that the significant figures include the first estimated digit. Thus, we could write the measured values as 16.3 ± 0.1 cm and 4.5 ± 0.1 cm.

Suppose now that we would like to find the area of the plate by multiplying the two measured values. If we were to claim that the area is (16.3 cm)(4.5 cm) = 73.35 cm^2, our answer would be unjustifiable because it contains four significant figures, which is greater than the number of significant figures in either of the measured lengths. A good rule of thumb to use as a guide in determining the number of significant figures that can be claimed is as follows:

> When **multiplying** several quantities, the number of significant figures in the final answer is the same as the number of significant figures in the *least* accurate of the quantities being multiplied, where "least accurate" means "having the lowest number of significant figures." The same rule applies to division.

Applying this rule to the multiplication example above, we see that the answer for the area can have only two significant figures because the length of 4.5 cm has only two significant figures. Thus, all we can claim is that the area is 73 cm^2, realizing that the value can range between (16.2 cm)(4.4 cm) = 71 cm^2 and (16.4 cm)(4.6 cm) = 75 cm^2.

Zeros may or may not be significant figures. Those used to position the decimal point in such numbers as 0.03 and 0.0075 are not significant. Thus there are one and two significant figures, respectively, in these two values. When the positioning of zeros comes after other digits, however, there is the possibility of misinterpretation. For example, suppose the mass of an object is given as 1500 g. This value is ambiguous because we do not know whether the last two zeros are being used to locate the decimal point or whether they represent significant figures in the measurement. In order to remove this ambiguity, it is common to use scientific notation to indicate the number of significant figures. In this case, we would express the mass as 1.5×10^3 g if there are two significant figures in the measured value, 1.50×10^3 g if there are three significant figures, and 1.500×10^3 g if there are four. Likewise, 0.000 15 should be expressed in scientific notation as 1.5×10^{-4} if it has two significant figures or as 1.50×10^{-4} if it has three significant figures. The three zeros between the decimal point and the digit 1 in the number 0.000 15 are not counted as significant figures because they are present only to locate the decimal point. In general, a **significant figure** is a reliably known digit (other than a zero used to locate the decimal point).

For addition and subtraction, the number of decimal places must be considered when you are determining how many significant figures to report.

> When numbers are **added** or **subtracted,** the number of decimal places in the result should equal the smallest number of decimal places of any term in the sum.

For example, if we wish to compute $123 + 5.35$, the answer would be 128 and not 128.35. If we compute the sum $1.0001 + 0.0003 = 1.0004$, the result has five significant figures, even though one of the terms in the sum, 0.0003, has only one significant figure. Likewise, if we perform the subtraction $1.002 - 0.998 = 0.004$, the result has only one significant figure even though one term has four significant figures and the other has three. In this book, *most of the numerical examples and end-of-chapter problems will yield answers having either two or three significant figures.*

EXAMPLE 1.10   The Area of a Rectangle

A rectangular plate has a length of (21.3 ± 0.2) cm and a width of (9.80 ± 0.10) cm. Find the area of the plate and the uncertainty in the calculated area.

Solution

$$\text{Area} = \ell w = (21.3 \pm 0.2) \text{ cm} \times (9.80 \pm 0.10) \text{ cm}$$
$$\approx (21.3 \times 9.80 \pm 21.3 \times 0.10 \pm 0.2 \times 9.80) \text{ cm}^2$$
$$\approx \boxed{(209 \pm 4) \text{ cm}^2}$$

Note that the input data were given only to three significant figures, and so we cannot claim any more in our result. Furthermore, you should realize that the uncertainty in the product (2%) is *approximately* equal to the sum of the uncertainties in the length and width (each uncertainty is about 1%).

EXAMPLE 1.11   Installing a Carpet

A carpet is to be installed in a room whose length is measured to be 12.71 m (four significant figures) and whose width is measured to be 3.46 m (three significant figures). Find the area of the room.

Solution   If you multiply 12.71 m by 3.46 m on your calculator, you will get an answer of 43.9766 m². How many of these numbers should you claim? Our rule of thumb for multiplication tells us that you can claim only the number of significant figures in the least accurate of the quantities being measured. In this example, we have only three significant figures in our least accurate measurement, so we should express our final answer as 44.0 m². Note that in reducing 43.9766 to three significant figures for our answer, we used a general rule for rounding off numbers that states that the last digit retained is increased by 1 if the first digit dropped equals 5 or greater. Furthermore, if the last digit is 5, the result should be rounded to the nearest even number. (This helps avoid accumulation of errors.)

1.8   MATHEMATICAL NOTATION

Many mathematical symbols are used throughout this book, some of which you are surely aware of, such as the symbol $=$ to denote the equality of two quantities.

The symbol $\propto$ is used to denote a proportionality. For example, $y \propto x^2$ means that y is proportional to the square of x.

The symbol $<$ means *less than,* and $>$ means *greater than.* For example, $x > y$ means x is greater than y.

The symbol $\ll$ means *much less than*, and $\gg$ means *much greater than*.

The symbol $\approx$ is used to indicate that two quantities are *approximately equal* to each other.

The symbol $\equiv$ means *is defined as*. This is a stronger statement than a simple $=$.

It is convenient to use a symbol to indicate the change in a quantity. For example, Δx (read delta x) means the *change in the quantity x*. (It does not mean the product of Δ and x.) For example, if x_i is the initial position of a particle and x_f is its final position, then the *change in position* is written

$$\Delta x = x_f - x_i$$

We shall often have occasion to sum several quantities. A useful abbreviation for representing such a sum is the Greek letter Σ (capital sigma). Suppose we wish to sum a set of five numbers represented by x_1, x_2, x_3, x_4, and x_5. In the abbreviated notation, we would write the sum

$$x_1 + x_2 + x_3 + x_4 + x_5 \equiv \sum_{i=1}^{5} x_i$$

where the subscript i on a particular x represents any one of the numbers in the set. For example, if there are five masses in a system, m_1, m_2, m_3, m_4, and m_5, the total mass of the system $M = m_1 + m_2 + m_3 + m_4 + m_5$ could be expressed

$$M = \sum_{i=1}^{5} m_i$$

Finally, the *magnitude* of a quantity x, written $|x|$, is simply the absolute value of that quantity. The magnitude of x is *always positive*, regardless of the sign of x. For example, if $x = -5$, $|x| = 5$; if $x = 8$, $|x| = 8$.

A list of these symbols and their meanings is given on the back endsheet.

SUMMARY

The physical quantities we shall encounter in our study of mechanics can be expressed in terms of three fundamental quantities, length, mass, and time, which in the SI system have the units meters (m), kilograms (kg), and seconds (s), respectively.

The **density** of a substance is defined as its *mass per unit volume*. Different substances have different densities mainly because of differences in their atomic masses and atomic arrangements.

The number of atoms in one mole of any element or compound, called **Avogadro's number**, N_A, is 6.02×10^{23}.

It is often useful to use the method of *dimensional analysis* to check equations and to assist in deriving expressions.

QUESTIONS

1. In this chapter we described how the Earth's daily rotation on its axis was once used to define the standard unit of time. What other types of natural phenomena could serve as alternative time standards?

2. A hand is defined as 4 inches; a foot is defined as 12 inches. Why should the former be any less acceptable than the latter, which we use all the time?

3. Express the following quantities using the prefixes given in Table 1.4: (a) 3×10^{-4} m, (b) 5×10^{-5} s, (c) 72×10^2 g.

4. Suppose that two quantities *A* and *B* have different dimensions. Determine which of the following arithmetic operations *could* be physically meaningful: (a) *A* + *B*, (b) *A/B*, (c) *B* − *A*, (d) *AB*.

5. What level of accuracy is implied in an order-of-magnitude calculation?

6. Do an order-of-magnitude calculation for an everyday situation you might encounter. For example, how far do you walk or drive each day?

7. Estimate your age in seconds.

8. Estimate the masses of various objects around you in grams or in kilograms. If a scale is available, check your estimates.

9. Is it possible to use length, density, and time as three fundamental units rather than length, mass, and time? If so, what could be used as a standard of density?

10. An automobile tire is rated to last for 50 000 miles. Estimate the number of revolutions the tire will make in its lifetime.

11. Estimate the total length of all McDonald's french fries (laid end to end) sold in the United States in one year. How many round trips to the Moon would this equal?

PROBLEMS

Section 1.3 Density and Atomic Mass

1. Calculate the density of a solid cube that measures 5.00 cm on each side and has a mass of 350 g.

2. The mass of the planet Saturn is 5.64×10^{26} kg and its radius is 6.00×10^7 m. (a) Calculate its density. (b) If this planet were placed in a large enough ocean of water, would it float? Explain.

3. How many grams of copper are required to make a hollow spherical shell with an inner radius of 5.70 cm and an outer radius of 5.75 cm? The density of copper is 8.93 g/cm^3.

3A. How many grams of copper are required to make a hollow spherical shell having an inner radius r_1 (in cm) and an outer radius r_2 (in cm)? The density of copper is ρ (in g/cm^3).

4. The planet Jupiter has an average radius 10.95 times that of the average radius of the Earth and a mass 317.4 times that of the Earth. Calculate the ratio of Jupiter's mass density to the mass density of the Earth.

5. Calculate the mass of an atom of (a) helium, (b) iron, and (c) lead. Give your answers in atomic mass units and in grams. The atomic masses are 4, 56, and 207, respectively, for the atoms given.

6. A small cube of iron is observed under a microscope. The edge of the cube is 5.00×10^{-6} cm. Find (a) the mass of the cube and (b) the number of iron atoms in the cube. The atomic mass of iron is 56 u, and its density is 7.86 g/cm^3.

7. A structural I beam is made of steel. A view of its cross-section and its dimensions is shown in Figure P1.7. (a) What is the mass of a section 1.5 m long? (b) How many atoms are there in this section? The density of steel is 7.56×10^3 kg/m^3.

8. A flat circular plate of copper has a radius of 0.243 m and a mass of 62.0 kg. What is the thickness of the plate?

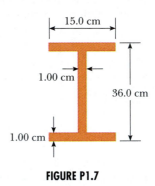

FIGURE P1.7

Section 1.4 Dimensional Analysis

9. Show that the expression $x = vt + \frac{1}{2}at^2$ is dimensionally correct, where *x* is a coordinate and has units of length, *v* is speed, *a* is acceleration, and *t* is time.

10. The displacement of a particle when moving under uniform acceleration is some function of the elapsed time and the acceleration. Suppose we write this displacement $s = ka^m t^n$, where *k* is a dimensionless constant. Show by dimensional analysis that this expression is satisfied if $m = 1$ and $n = 2$. Can this analysis give the value of *k*?

11. The square of the speed of an object undergoing a uniform acceleration *a* is some function of *a* and the displacement *s*, according to the expression $v^2 = ka^m s^n$, where *k* is a dimensionless constant. Show by dimensional analysis that this expression is satisfied only if $m = n = 1$.

12. The radius *r* of a circle inscribed in any triangle whose sides are *a*, *b*, and *c* is given by $r = [(s - a)(s - b)(s - c)/s]^{1/2}$, where *s* is an abbreviation for $(a + b + c)/2$. Check this formula for dimensional consistency.

13. Which of the equations below is dimensionally correct?

☐ indicates problems that have full solutions available in the Student Solutions Manual and Study Guide.

(a) $v = v_0 + ax$

(b) $y = (2 \text{ m})\cos(kx)$, where $k = 2 \text{ m}^{-1}$

14. The period T of a simple pendulum is measured in time units and is

$$T = 2\pi \sqrt{\frac{\ell}{g}}$$

where ℓ is the length of the pendulum and g is the free-fall acceleration in units of length divided by the square of time. Show that this equation is dimensionally correct.

15. The volume of an object as a function of time is calculated by $V = At^3 + B/t$, where t is time measured in seconds and V is in cubic meters. Determine the dimension of the constants A and B.

16. The consumption of natural gas by a company satisfies the empirical equation $V = 1.5t + 0.0080t^2$, where V is the volume in millions of cubic feet and t the time in months. Express this equation in units of cubic feet and seconds. Put the proper units on the coefficients. Assume a month is 30 days.

17. Newton's law of universal gravitation is

$$F = G\frac{Mm}{r^2}$$

Here F is the force of gravity, M and m are masses, and r is a length. Force has the SI units kg·m/s². What are the SI units of the constant G?

Section 1.5 Conversion of Units

18. Convert the volume 8.50 in.³ to m³, recalling that 1 in. = 2.54 cm and 1 cm = 10^{-2} m.

19. A rectangular building lot is 100.0 ft by 150.0 ft. Determine the area of this lot in m².

20. A classroom measures 40.0 m × 20.0 m × 12.0 m. The density of air is 1.29 kg/m³. What are (a) the volume of the room in cubic feet, and (b) the weight of air in the room in pounds?

21. A creature moves at a speed of 5.0 furlongs per fortnight (not a very common unit of speed). Given that 1.0 furlong = 220 yards and 1 fortnight = 14 days, determine the speed of the creature in m/s. (The creature is probably a snail.)

22. A section of land has an area of 1 square mile and contains 640 acres. Determine the number of square meters there are in 1 acre.

23. A solid piece of lead has a mass of 23.94 g and a volume of 2.10 cm³. From these data, calculate the density of lead in SI units (kg/m³).

24. A quart container of ice cream is to be made in the form of a cube. What should be the length of a side in cm? (Use the conversion 1 gallon = 3.786 liters.)

25. An astronomical unit (AU) is defined as the average distance between the Earth and Sun. (a) How many astronomical units are there in one lightyear?

(b) Determine the distance from Earth to the Andromeda galaxy in astronomical units.

26. The mass of the Sun is about 1.99×10^{30} kg, and the mass of a hydrogen atom, of which the Sun is mostly composed, is 1.67×10^{-27} kg. How many atoms are there in the Sun?

27. At the time of this book's printing, the U.S. national debt is about $4 trillion. (a) If payments were made at the rate of $1000/sec, how many years would it take to pay off the debt, assuming no interest were charged? (b) A dollar bill is about 15.5 cm long. If 4 trillion dollar bills were laid end to end around the Earth's equator, how many times would they encircle the Earth? Take the radius of the Earth at the equator to be 6378 km. (*Note:* Before doing any of these calculations, try to guess at the answers. You may be very surprised.)

28. A room measures 4.0 m × 4.0 m, and its ceiling is 2.5 m high. Is it possible to completely wallpaper the walls of this room with the pages of this book? Explain.

29. (a) Find a conversion factor to convert from mi/h to km/h. (b) Until recently, federal law mandated that highway speeds would be 55 mi/h. Use the conversion factor of part (a) to find this speed in km/h. (c) The maximum highway speed has been raised to 65 mi/h in some places. In km/h, how much increase is this over the 55 mi/h limit?

30. (a) How many seconds are there in a year? (b) If one micrometeorite (a sphere with a diameter of 1.00×10^{-6} m) strikes each square meter of the Moon each second, how many years would it take to cover the Moon to a depth of 1.00 m? (*Hint:* Consider a cubic box on the Moon 1.00 m on a side, and find how long it will take to fill the box.)

31. One gallon of paint (volume = 3.78×10^{-3} m³) covers an area of 25.0 m². What is the thickness of the paint on the wall?

32. A pyramid has a height of 481 ft and its base covers an area of 13.0 acres (1 acre = 43 560 ft²). If the volume of a pyramid is given by the expression $V = (1/3)Bh$, where B is the area of the base and h is the height, find the volume of this pyramid in cubic meters.

33. The pyramid described in Problem 32 contains approximately two million stone blocks that average 2.50 tons each. Find the weight of this pyramid in pounds.

34. Assuming that 70 percent of the Earth's surface is covered with water at an average depth of 1 mile, estimate the mass of the water on Earth in kilograms.

35. The diameter of our disk-shaped galaxy, the Milky Way, is about 1.0×10^5 lightyears. The distance to Andromeda, the galaxy nearest our own Milky Way, is about 2.0 million lightycars. If we represent the Milky Way by a dinner plate 25 cm in diameter, determine the distance to the next dinner plate.

36. The mean radius of the Earth is 6.37×10^6 m, and that of the Moon is 1.74×10^8 cm. From these data calculate (a) the ratio of the Earth's surface area to that of the Moon and (b) the ratio of the Earth's volume to that of the Moon. Recall that the surface area of a sphere is $4\pi r^2$ and the volume of a sphere is $\frac{4}{3}\pi r^3$.

37. From the fact that the average density of the Earth is 5.5 g/cm^3 and its mean radius is 6.37×10^6 m, compute the mass of the Earth.

38. Assume that an oil slick consists of a single layer of molecules and that each molecule occupies a cube 1.0 nm on a side. Determine the area of an oil slick formed by 1.0 m^3 of oil.

39. One cubic meter (1.00 m^3) of aluminum has a mass of 2.70×10^3 kg, and 1.00 m^3 of iron has a mass of 7.86×10^3 kg. Find the radius of a solid aluminum sphere that will balance a solid iron sphere of radius 2.00 cm on an equal-arm balance.

39A. Let ρ_{Al} represent the density of aluminum and ρ_{Fe} that of iron. Find the radius of a solid aluminum sphere that balances a solid iron sphere of radius r_{Fe} on an equal-arm balance.

Section 1.6   Order-of-Magnitude Calculations

40. Assuming 60 heartbeats a minute, estimate the total number of times the heart of a human beats in an average lifetime of 70 years.

41. Estimate the number of Ping-Pong balls that would fit into an average-size room (without being crushed).

42. Estimate the amount of motor oil used by all cars in the United States each year and its cost to the consumers.

43. When a droplet of oil spreads out on a smooth water surface, the resulting "oil slick" is approximately one molecule thick. An oil droplet of mass 9.00×10^{-7} kg and density 918 kg/m^3 spreads out into a circle of radius 41.8 cm on the water surface. What is the diameter of an oil molecule?

44. Approximately how many raindrops fall on a 1.0-acre lot during a 1.0-in. rainfall?

45. Army engineers in 1946 determined the distance from the Earth to the Moon by using radar. If the round trip from Earth to Moon and back again took the radar beam 2.56 s, what is the distance from the Earth to the Moon? (The speed of radar waves is 3.00×10^8 m/s.)

46. A billionaire offers to give you $1 billion if you can count it out using only one-dollar bills. Should you accept her offer? Assume you can count one bill every second, and be sure to allow for the fact that you need about 8 hours a day for sleeping and eating, and that right now you are probably at least 18 years old.

47. A high fountain of water is located at the center of a circular pool as in Figure P1.47. A student walks around the pool and estimates its circumference to be 150 m. Next, the student stands at the edge of the pool and uses a protractor to gauge the angle of elevation of the top of the fountain to be 55°. How high is the fountain?

FIGURE P1.47

48. Assume that you watch every pitch of every game of a 162-game major league baseball season. Approximately how many pitches would you see thrown?

49. Estimate the number of piano tuners living in New York City. This problem was posed by the physicist Enrico Fermi, who was well known for making order-of-magnitude calculations.

Section 1.7   Significant Figures

50. Determine the number of significant figures in the following numbers: (a) 23 cm, (b) 3.589 s, (c) 4.67×10^3 m/s, (d) 0.0032 m.

51. The radius of a circle is measured to be 10.5 ± 0.2 m. Calculate the (a) area and (b) circumference of the circle, and give the uncertainty in each value.

52. Carry out the following arithmetic operations: (a) the sum of the numbers 756, 37.2, 0.83, and 2.5; (b) the product 3.2×3.563; (c) the product $5.6 \times \pi$.

53. If the length and width of a rectangular plate are measured to be (15.30 ± 0.05) cm and (12.80 ± 0.05) cm, respectively, find the area of the plate and the approximate uncertainty in the calculated area.

54. The radius of a solid sphere is measured to be (6.50 ± 0.20) cm, and its mass is measured to be (1.85 ± 0.02) kg. Determine the density of the sphere in kg/m^3 and the uncertainty in the density.

55. How many significant figures are there in (a) 78.9 ± 0.2, (b) 3.788×10^9, (c) 2.46×10^{-6}, (d) 0.0053?

56. A farmer measures the distance around a rectangular field. The length of the long sides of the rectangle is found to be 38.44 m, and the length of the short sides is found to be 19.5 m. What is the total distance around the field?

57. A sidewalk is to be constructed around a swimming pool that measures (10.0 ± 0.1) m wide by (17.0 ± 0.1) m long. If the sidewalk is to measure (1.00 ± 0.01) m wide by (9.0 ± 0.1) cm thick, what volume of concrete is needed, and what is the approximate uncertainty of this volume?

Section 1.8 Mathematical Notation

58. Compute the value of $\Sigma_{i=1}^{4} x_i$ if $x_i = (2i + 1)$.

59. Determine whether the expression $\Sigma_{i=1}^{4} i^2$ is equal to $(\Sigma_{i=1}^{4} i)^2$.

ADDITIONAL PROBLEMS

60. In physics it is important to use mathematical approximations. Demonstrate for yourself that for small angles ($< 20°$)

$$\tan \alpha \approx \sin \alpha \approx \alpha \approx \frac{\pi \alpha'}{180°}$$

where α is in radians and α' is in degrees. Use a calculator to find the largest angle for which $\tan \alpha$ may be approximated by $\sin \alpha$ if the error is to be less than 10%.

61. A useful fact is that there are about $\pi \times 10^7$ s in one year. Use a calculator to find the percentage error in this approximation. *Note:* "percentage error" is defined as (difference/true value) $\times$ 100%.

62. Assume that there are 50 million passenger cars in the United States and that the average fuel consumption is 20 mi/gal of gasoline. If the average distance traveled by each car is 10 000 mi/yr, how much gasoline would be saved per year if average fuel consumption could be increased to 25 mi/gal?

63. One cubic centimeter (1.0 cm^3) of water has a mass of 1.0×10^{-3} kg. (a) Determine the mass of 1.0 m^3 of water. (b) Assuming biological substances are 98% water, estimate the mass of a cell that has a diameter of 1.0 μm, a human kidney, and a fly. Assume a kidney is roughly a sphere with a radius of 4.0 cm and a fly is roughly a cylinder 4.0 mm long and 2.0 mm in diameter.

64. The data in the following table represent measurements of the masses and dimensions of solid cylinders of aluminum, copper, brass, tin, and iron. Use these data to calculate the densities of these substances. Compare your results for aluminum, copper, and iron with those given in Table 1.5.

Substance	Mass (g)	Diameter (cm)	Length (cm)
Aluminum	51.5	2.52	3.75
Copper	56.3	1.23	5.06
Brass	94.4	1.54	5.69
Tin	69.1	1.75	3.74
Iron	216.1	1.89	9.77

65. Determine an order-of-magnitude estimation of (a) the number of words in this textbook, (b) the number of symbols (letters, numbers, Greek letters, mathematical operators, and so on) in this textbook, (c) the number of problems in this textbook.

SPREADSHEET PROBLEMS

S1. "On a Clear Day You Can See Forever" is the title song from a Broadway musical. How far can you really see? Assuming that light travels in straight lines, you could see the whole Earth from any point above the Earth's surface if the Earth were perfectly

THE WIZARD OF ID   **By Parker and Hart**

flat. Because the Earth is round, however, there is a maximum distance you can see from any given height. The location of the most distant visible point is called the *horizon*. Suppose your eyes are located at point P in Figure S1.1, where P is a height h above the Earth's surface. Assuming a perfectly smooth Earth of radius R, the horizon is located at point H. Point C is the center of the Earth, d is the straight-line distance from P to H, and s is the distance measured along the Earth's surface from a point directly beneath P to H. The Pythagorean theorem tells us that

$$(R + h)^2 = R^2 + d^2$$

from which we find

$$d = \sqrt{h^2 + 2hR}$$

From Figure S1.1, we see that

$$\cos \phi = \frac{R}{R + h}$$

$$s = R\phi$$

In the limit $h \ll R$, both d and s approach $\sqrt{2hR}$, which we can call L.

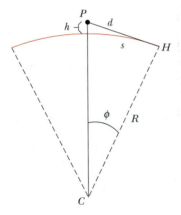

FIGURE S1.1

Spreadsheet 1.1 uses these equations to calculate h/R, d, ϕ, s, and L, for any given h and R. The angle ϕ is given in both radians and degrees. Use the spreadsheet to answer the following questions: (a) Suppose Jud and Spud are standing on the Earth and are gazing off into the distance. Jud's eyes are 2 m above the ground, and Spud's are 1.5 m above the ground. How far away are Jud's and Spud's horizons? Does it make any difference whether you use d, s, or L to answer this question? (b) If Jud and Spud were on the Moon ($R = 1740$ km), where would their horizons be?

S2. Suppose you are standing on top of New York's World Trade Center, 417 m high, looking out toward the Atlantic Ocean. (a) How far is your horizon? Use Spreadsheet 1.1. How high would you have to be above the Earth's surface to see from New York City to Paris? (Estimate the New York–Paris distance using a world map.) (b) How high would you have to be for d and s to differ by 1 percent? 10 percent? (c) Choose h such that $h = NR$. Set N successively equal to 10, 100, 1000, . . . , and observe what happens to s and ϕ. Explain your observations.

S3. Spreadsheet programs are useful in examining data graphically. Most spreadsheet programs can fit the best straight line (a regression line) to a set of data. (See Example 4 in Chapters 1 or 2 of *Computer Investigations* for instructions on applying the least-squares method to a set of data.) The following table gives experimental results for the measurement of the period T of a pendulum of length L. These data are consistent with an equation of the form $T = CL^n$, where C and n are constants and n is not necessarily an integer.

Length L(m)	Period T(s)
0.25	1.00
0.50	1.40
0.75	1.75
1.00	2.00
1.50	2.50
2.00	2.80

(a) Use the least-squares or the regression-line procedure of your spreadsheet program to find the best fit to the data. Since $T = CL^n$, we see that

$$\log T = \log C + n \log L$$

First calculate a column of $\log T$ values and another of $\log L$ values. Use $\log L$ as the independent variable and $\log T$ as the dependent variable. The least-squares fit finds the slope n and the intercept $\log C$. (b) Which data points deviate most from a straight-line plot of T versus L^n? (c) Is the experimental value of n found in part (a) consistent with a dimensional analysis of $T = CL^n$.

S4. The period T and orbit radius R for the motions of four moons of Jupiter are:

	Io	Europa	Ganymede	Calisto
Period, T(days)	1.77	3.55	7.16	16.69
R(km)	422 000	671 000	1 070 000	1 880 000

(a) These data can be fitted by the formula $T = CR^n$. Follow the procedures in Problem S3a to find C and n. (b) A fifth satellite, Amalthea, has a period of 0.50 days. Use $I = CR^n$ to find the radius for its orbit.

CHAPTER 2

Motion in One Dimension

An apple and a feather released from rest in a vacuum chamber fall at the same rate regardless of their mass difference. If we neglect air resistance, we can say that all objects fall to the Earth with the same constant acceleration of magnitude 9.80 m/s², as indicated by the violet arrows in this multiflash photograph. The velocity of the two objects increases linearly with time, as indicated by the red arrows. (© 1993 James Sugar/Black Star)

A s a first step in studying mechanics, it is convenient to describe motion in terms of space and time, ignoring for the present the agents that caused that motion. This portion of mechanics is called *kinematics.* In this chapter we consider motion along a straight line, that is, one-dimensional motion. Starting with the concept of displacement discussed in the previous chapter, we first define velocity and acceleration. Then, using these concepts, we study the motion of objects traveling in one dimension under a constant acceleration.

From everyday experience we recognize that motion represents the continuous change in the position of an object. In physics we are concerned with three types of motion: translational, rotational, and vibrational. A car moving down a highway is undergoing translational motion, the Earth's daily spin on its axis is an example of rotational motion, and the back-and-forth motion of a pendulum is an example of vibrational motion.

In this and the next few chapters, we are concerned only with translational motion. In many situations, we can treat the moving object as a particle, which in mathematics is defined as a point having no size. For example, if we wish to describe the motion of the Earth around the Sun, we can treat the Earth as a particle

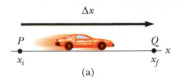

(a)

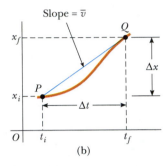

(b)

FIGURE 2.1 (a) A car moving to the right along a straight line taken to be the *x* axis. Because we are interested only in the car's translational motion, we can treat it as a particle. (b) Position-time graph for the motion of the "particle."

Displacement

Average velocity

---

and obtain reasonable accuracy in a prediction of the Earth's orbit. This approximation is justified because the radius of the Earth's orbit is large compared with the dimensions of the Earth and Sun. As an example, on a much smaller scale, it is possible to explain the pressure exerted by a gas on the walls of a container by treating the gas molecules as particles. Of course, the particle approach doesn't work in *all* situations. For instance, we cannot treat the Earth as a particle when we are examining its internal structure or when we are studying such phenomena as tides, earthquakes, and volcanic activity. On the other hand, we cannot consider gas molecules to be particles when we are studying properties that depend on molecular rotation and vibration. In general, we shall often find it both valid and convenient to treat a moving object as a particle.

2.1 DISPLACEMENT, VELOCITY, AND SPEED

The motion of a particle is completely known if the particle's position in space is known at all times. Consider a car (which we treat as a particle) moving along the *x* axis from a point *P* to a point *Q* as in Figure 2.1a. Let its position at point *P* be x_i at some time t_i, and let its position at point *Q* be x_f at time t_f. (The indices *i* and *f* refer to the initial and final values.) At times other than t_i and t_f, the position of the particle between these two points may vary as in Figure 2.1b. Such a plot is called a *position-time graph*. As the particle moves from position x_i to position x_f, its displacement is given by $x_f - x_i$. As mentioned in Chapter 1, we use the Greek letter delta (Δ) to denote the *change* in a quantity. Therefore, we write the change in the position of the particle (the displacement)

$$\Delta x \equiv x_f - x_i \tag{2.1}$$

From this definition, we see that Δx is positive if x_f is greater than x_i and negative if x_f is less than x_i.

> The **average velocity** $\bar{v}$ of the particle is defined as the ratio of its displacement Δx and the time interval Δt:
>
> $$\bar{v} \equiv \frac{\Delta x}{\Delta t} = \frac{x_f - x_i}{t_f - t_i} \tag{2.2}$$

From this definition,[1] we see that average velocity has dimensions of length divided by time (L/T)—m/s in SI units and ft/s in British engineering units. The average velocity is *independent* of the path taken between the points *P* and *Q*. This is true because the average velocity is proportional to the displacement, Δx, which depends only on the initial and final coordinates of the particle. It therefore follows that if a particle starts at some point and returns to the same point via any path, its average velocity for this trip is zero, because its displacement is zero.

Displacement should not be confused with the distance traveled, since the distance traveled for any motion is clearly nonzero. For instance, when a baseball player hits a home run as in Figure 2.2, he travels a distance of 360 ft in his trip

---

[1] In one-dimensional motion, all quantities describing the motion are scalars (having size only; see Chapter 3). Nevertheless, we introduce the term *velocity* here, rather than only *speed*, where the sign of the rate of motion is important.

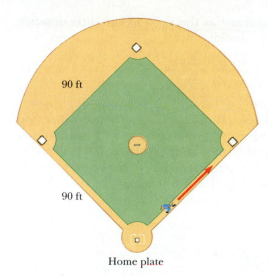

90 ft

90 ft

Home plate

FIGURE 2.2 Bird's-eye view of a baseball diamond. A batter who hits a home run travels 360 ft as he rounds the bases, but his displacement for the round trip is zero.

around the bases; however, his displacement is 0 ft because the player's final and initial positions are identical.

Average velocity gives us no details of the motion between points P and Q in Figure 2.1b. (How we evaluate the velocity at some instant in time is discussed in the next section.) The average velocity of a particle in one dimension can be positive or negative, depending on the sign of the displacement. (The time interval, Δt, is always positive.) If the coordinate of the particle increases in time (that is, if $x_f > x_i$), then Δx is positive and $\bar{v}$ is positive. This case corresponds to motion in the positive x direction. If the coordinate decreases in time ($x_f < x_i$), Δx is negative and hence $\bar{v}$ is negative. This case corresponds to motion in the negative x direction.

The average velocity can also be interpreted geometrically by drawing a straight line between the points P and Q in Figure 2.1b. This line forms the hypotenuse of a triangle of height Δx and base Δt. The slope of this line is the ratio $\Delta x / \Delta t$. Therefore, we see that the *average* velocity of the particle during the time interval t_i to t_f is equal to the slope of the straight line joining the initial and final points on the space-time graph.[2]

EXAMPLE 2.1   Calculate the Average Velocity

A particle moving along the x axis is located at $x_i = 12$ m at $t_i = 1$ s and at $x_f = 4$ m at $t_f = 3$ s. Find its displacement, average velocity, and average speed during this time interval.

Solution   The displacement is given by Equation 2.1:

$$\Delta x = x_f - x_i = 4 \text{ m} - 12 \text{ m} = \boxed{-8 \text{ m}}$$

The average velocity is, from Equation 2.2,

$$\bar{v} = \frac{\Delta x}{\Delta t} = \frac{x_f - x_i}{t_f - t_i} = \frac{4 \text{ m} - 12 \text{ m}}{3 \text{ s} - 1 \text{ s}} = -\frac{8 \text{ m}}{2 \text{ s}} = \boxed{-4 \text{ m/s}}$$

The displacement and average velocity are negative for this time interval because the particle has moved to the left, toward decreasing values of x. Its average speed for this trip is 4 m/s.

---

[2] Slope represents the ratio of the change in the quantity represented on the vertical axis to the change in the quantity represented on the horizontal axis.

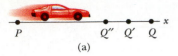

(a)

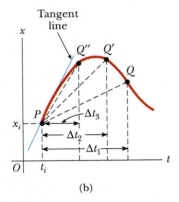

(b)

FIGURE 2.3 (a) As the car moves along the *x* axis, and *Q* is brought closer to *P*, the time it takes to travel the distance decreases. (b) Position-time graph for the "particle." As the time intervals get smaller and smaller, the average velocity for that interval, equal to the slope of the dashed line connecting *P* and the appropriate *Q*, approaches the slope of the line tangent at *P*. The instantaneous velocity at *P* is the slope of the blue tangent line at the time t_1.

Definition of instantaneous velocity

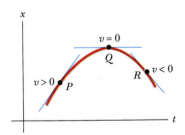

FIGURE 2.4 In this position-time graph, the velocity is positive at *P*, where the slope of the tangent line is positive, zero at *Q*, where the slope of the tangent line is zero, and negative at *R*, where the slope of the tangent line is negative.

The **average speed** of a particle is defined as the ratio of the total distance traveled to the total time it takes to travel that distance:

$$\text{Average speed} = \frac{\text{total distance}}{\text{total time}}$$

The SI unit of average speed, like velocity, is also meters per second. However, unlike average velocity, average speed has no direction, and hence carries no algebraic sign. Knowledge of the average speed of a particle tells you nothing about the details of the trip. For example, suppose it takes you 8.0 h to travel 280 km in your car. The average speed for your trip is 35 km/h. However, you most likely traveled at various speeds during the trip, and the average speed of 35 km/h could result from an infinite number of possible speed variations.

2.2   INSTANTANEOUS VELOCITY AND SPEED

We would like to be able to define the velocity of a particle at a particular instant of time, rather than just over a finite interval of time. The velocity of a particle at any instant of time — in other words, at some point on a space-time graph — is called the **instantaneous velocity**. This concept is especially important when the average velocity in different time intervals is *not* constant.

Consider the straight-line motion of a particle between two points *P* and *Q* on the *x* axis as in Figure 2.3a. As *Q* is brought closer and closer to *P*, the time needed to travel the distance gets progressively smaller. The average velocity for each time interval is the slope of the appropriate dotted line in the space-time graph shown in Figure 2.3b. As *Q* approaches *P*, the time interval approaches zero and the slope of the dotted line approaches that of the blue line tangent to the curve at *P*. The slope of this line is defined to be the *instantaneous velocity* at the time t_i. In other words,

the instantaneous velocity, *v*, equals the limiting value of the ratio $\Delta x/\Delta t$ as Δt approaches zero[3]:

$$v \equiv \lim_{\Delta t \to 0} \frac{\Delta x}{\Delta t} \tag{2.3}$$

In the calculus notation, this limit is called the *derivative* of *x* with respect to *t*, written *dx/dt*:

$$v \equiv \lim_{\Delta t \to 0} \frac{\Delta x}{\Delta t} = \frac{dx}{dt} \tag{2.4}$$

The instantaneous velocity can be positive, negative, or zero. When the slope of the position-time graph is positive, such as at *P* in Figure 2.4, *v* is positive. At point *R*, *v* is negative since the slope is negative. Finally, the instantaneous velocity is zero at the peak *Q* (the turning point), where the slope is zero.

---

[3] Note that the displacement, Δx, also approaches zero as Δt approaches zero. As Δx and Δt become smaller and smaller, the ratio $\Delta x/\Delta t$ approaches a value equal to the slope of the line tangent to the *x* versus *t* curve.

From here on, we use the word *velocity* to designate instantaneous velocity. When it is *average velocity* we are interested in, we always use the adjective *average.*

The **speed** of a particle is defined to be equal to the magnitude of its velocity. Speed has no direction associated with it and, hence, carries no algebraic sign. For example, if one particle has a velocity of + 25 m/s, and another particle has a velocity of − 25 m/s, both have a speed of 25 m/s. A car's speedometer indicates the speed of the car, not its velocity.

It is also possible to use a mathematical technique called integration to find the displacement of a particle if its velocity is known as a function of time. Because integration procedures may not be familiar to many students, the topic is treated in (optional) Section 2.6 for general interest and for those courses that cover this material.

Speed

EXAMPLE 2.2 Average and Instantaneous Velocity

A particle moves along the *x* axis. Its *x* coordinate varies with time according to the expression $x = -4t + 2t^2$, where *x* is in meters and *t* is in seconds. The position-time graph for this motion is shown in Figure 2.5. Note that the particle moves in the negative *x* direction for the first second of motion, is at rest at the moment $t = 1$ s, and then heads back in the positive *x* direction for $t > 1$ s. (a) Determine the displacement of the particle in the time intervals $t = 0$ to $t = 1$ s and $t = 1$ s to $t = 3$ s.

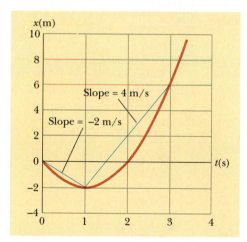

FIGURE 2.5 (Example 2.2) Position-time graph for a particle having an *x* coordinate that varies in time according to the expression $x = -4t + 2t^2$.

Solution   In the first time interval, we set $t_i = 0$ and $t_f = 1$ s. Since $x = -4t + 2t^2$, and using Equation 2.1, we get for the first displacement

$$\Delta x_{01} = x_f - x_i$$
$$= [-4(1) + 2(1)^2] - [-4(0) + 2(0)^2]$$
$$= \boxed{-2 \text{ m}}$$

In the second time interval we can set $t_i = 1$ s and $t_f = 3$ s. Therefore, the displacement in this interval is

$$\Delta x_{13} = x_f - x_i$$
$$= [-4(3) + 2(3)^2] - [-4(1) + 2(1)^2]$$
$$= \boxed{+8 \text{ m}}$$

These displacements can also be read directly from the position-time graph (Fig. 2.5).

(b) Calculate the average velocity in the time intervals $t = 0$ to $t = 1$ s and $t = 1$ s to $t = 3$ s.

Solution   In the first time interval, $\Delta t = t_f - t_i = 1$ s. Therefore, using Equation 2.2 and the results from (a) gives

$$\bar{v}_{01} = \frac{\Delta x_{01}}{\Delta t} = \frac{-2 \text{ m}}{1 \text{ s}} = \boxed{-2 \text{ m/s}}$$

In the second time interval, $\Delta t = 2$ s; therefore,

$$\bar{v}_{13} = \frac{\Delta x_{13}}{\Delta t} = \frac{8 \text{ m}}{2 \text{ s}} = \boxed{+4 \text{ m/s}}$$

These values agree with the slopes of the lines joining these points in Figure 2.5.

(c) Find the instantaneous velocity of the particle at $t = 2.5$ s.

average velocity using calculus

Solution   By measuring the slope of the position-time graph at $t = 2.5$ s, we find that $v = +6$ m/s. We could also use Equation 2.4 and the rules of differential calculus to find the velocity from the displacement:

$$v = \frac{dx}{dt} = \frac{d}{dt}(-4t + 2t^2) = 4(-1 + t) \text{ m/s}$$

Therefore, at $t = 2.5$ s,

$$v = 4(-1 + 2.5) = \boxed{+6 \text{ m/s}}$$

A review of basic operations in the calculus is provided in Appendix B.6.

Exercise   Do you see any symmetry in the motion? For example, does the speed ever repeat itself?

EXAMPLE 2.3   The Limiting Process

The position of a particle moving along the x axis varies in time according to the expression $x = (3 \text{ m/s}^2)t^2$, where x is in meters, and t is in seconds. Find the velocity at any time.

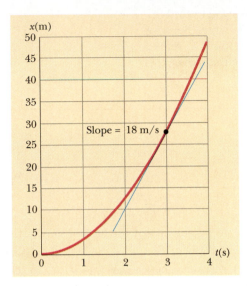

FIGURE 2.6   (Example 2.3) Position-time graph for a particle having an x coordinate that varies in time according to $x = 3t^2$. Note that the instantaneous velocity at $t = 3$ s equals the slope of the blue line tangent to the curve at this instant.

Reasoning and Solution   The position-time graph for this motion is shown in Figure 2.6. We can compute the velocity at any time t by using the definition of the instantaneous velocity (Eq. 2.3). If the initial coordinate of the particle at time t is $x_i = 3t^2$, then the coordinate at a later time $t + \Delta t$ is

$$x_f = 3(t + \Delta t)^2 = 3[t^2 + 2t\,\Delta t + (\Delta t)^2]$$
$$= 3t^2 + 6t\,\Delta t + 3(\Delta t)^2$$

Therefore, the displacement in the time interval Δt is

$$\Delta x = x_f - x_i = 3t^2 + 6t\,\Delta t + 3(\Delta t)^2 - 3t^2$$
$$= 6t\,\Delta t + 3(\Delta t)^2$$

The average velocity in this time interval is

$$\bar{v} = \frac{\Delta x}{\Delta t} = 6t + 3\,\Delta t$$

To find the instantaneous velocity, we take the limit of this expression as Δt approaches zero, as shown by Equation 2.3. In doing so, we see that the term $3\,\Delta t$ goes to zero, therefore

$$v = \lim_{\Delta t \to 0} \frac{\Delta x}{\Delta t} = (6 \text{ m/s}^2)t$$

Notice that this expression gives us the velocity at *any* general time t. It tells us that v is increasing linearly in time. It is then a straightforward matter to find the velocity at some specific time from the expression $v = (6 \text{ m/s}^2)t$. For example, at $t = 3.0$ s, the velocity is $v = (6 \text{ m/s}^2)(3.0) = +18$ m/s. Again, this can be checked from the slope of the graph (the blue line) at $t = 3.0$ s.

The limiting process can also be examined numerically. For example, we can compute the displacement and average velocity for various time intervals beginning at $t = 3.0$ s, using the expressions for Δx and $\bar{v}$. The results of such calculations are given in Table 2.1. As the time intervals get smaller and smaller, the average velocity approaches the value of the instantaneous velocity at $t = 3.0$ s, namely, $+18$ m/s.

TABLE 2.1   Displacement and Average Velocity for Various Time Intervals for the Function $x = (3 \text{ m/s}^2)t^2$ (the intervals begin at $t = 3.00$ s)

$\Delta t(\text{s})$	$\Delta x(\text{m})$	$\bar{v} = \Delta x/\Delta t(\text{m/s})$
1.00	21	21
0.50	9.75	19.5
0.25	4.69	18.8
0.10	1.83	18.3
0.05	0.9075	18.15
0.01	0.1803	18.03
0.001	0.018003	18.003

2.3   ACCELERATION

When the velocity of a particle changes with time, the particle is said to be *accelerating*. For example, the speed of a car will increase when you "step on the gas" and decreases when you apply the brakes. However, we need a more precise definition of acceleration than this.

Suppose a particle moving along the x axis has a velocity v_i at time t_i and a velocity v_f at time t_f, as in Figure 2.7a.

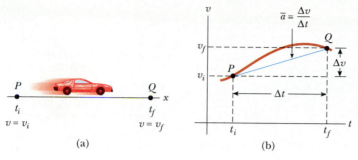

FIGURE 2.7 (a) A "particle" moving from P to Q has velocity v_i at $t = t_i$ and velocity v_f at $t = t_f$. (b) Velocity-time graph for the particle moving in a straight line. The slope of the blue straight line connecting P and Q is the average acceleration in the time interval $\Delta t = t_f - t_i$.

The **average acceleration** of the particle in the time interval $\Delta t = t_f - t_i$ is defined as the ratio $\Delta v / \Delta t$, where $\Delta v = v_f - v_i$ is the *change* in velocity in this time interval:

$$\bar{a} \equiv \frac{\Delta v}{\Delta t} = \frac{v_f - v_i}{t_f - t_i} \tag{2.5}$$

Average acceleration

Acceleration has dimensions of length divided by (time)^2, or L/T^2. Some of the common units of acceleration are meters per second per second (m/s^2) and feet per second per second (ft/s^2). As with velocity, we can use positive and negative signs to indicate the direction of the acceleration when the motion being analyzed is one dimensional.

In some situations, the value of the average acceleration may be different over different time intervals. It is therefore useful to define the **instantaneous acceleration** as the limit of the average acceleration as Δt approaches zero. This concept is analogous to the definition of instantaneous velocity discussed in the previous section. If we imagine that the point Q is brought closer and closer to the point P in Figure 2.7a, and take the limit of $\Delta v / \Delta t$ as Δt approaches zero, we get the *instantaneous acceleration:*

$$a \equiv \lim_{\Delta t \to 0} \frac{\Delta v}{\Delta t} = \frac{dv}{dt} \tag{2.6}$$

Instantaneous acceleration

That is, the instantaneous acceleration equals the derivative of the velocity with respect to time, which by definition is the slope of the velocity-time graph (Fig. 2.7b). One can interpret the derivative of the velocity with respect to time as the *time rate of change of velocity.* If a is positive, the acceleration is in the positive x direction, whereas negative a indicates acceleration in the negative x direction.

From now on we shall use the term *acceleration* to mean instantaneous acceleration.

Since $v = dx/dt$, the acceleration can also be written

$$a = \frac{dv}{dt} = \frac{d}{dt}\left(\frac{dx}{dt}\right) = \frac{d^2x}{dt^2} \tag{2.7}$$

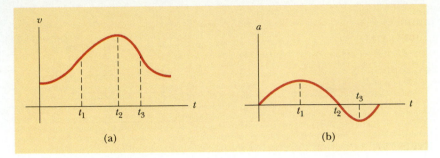

(a)   (b)

FIGURE 2.8   The instantaneous acceleration can be obtained from the velocity-time graph. (a) At each instant, the acceleration in the *a* versus *t* graph (b) equals the slope of the line tangent to the *v* versus *t* curve.

That is, in one-dimensional motion, the acceleration equals the *second derivative* of the *x* coordinate with respect to time.

Figure 2.8 shows how the acceleration-time curve can be derived from the velocity-time curve. The acceleration at any time is the slope of the velocity-time graph at that time. Positive values of the acceleration correspond to those points in Figure 2.8a where the velocity is increasing in the positive *x* direction. The acceler-

---

EXAMPLE 2.4   **Graphical Relations Between *x, v,* and *a***

The position of an object moving along the *x* axis varies with time as in Figure 2.9a. Let us use graphical procedures to obtain graphs of the velocity versus time and acceleration versus time for the object.

Reasoning and Solution   The velocity at any instant is the slope of the tangent to the *x-t* graph at that instant. Between $t = 0$ and $t = t_1$, the slope of the *x-t* graph increases uniformly, so the velocity increases linearly as in Figure 2.9b. Between t_1 and t_2, the slope of the *x-t* graph is constant, so the velocity remains constant. At t_4, the slope of the *x-t* graph is zero, so the velocity is zero at that instant. Between t_4 and t_5, the slope of the *x-t* graph is negative and decreases uniformly; therefore the velocity is negative and constant in this interval. In the interval t_5 to t_6, the slope of the *x-t* graph is still negative, and goes to zero at t_6. Finally, after t_6, the slope of the *x-t* graph is zero, so the object is at rest.

Similarly, the acceleration at any instant is the slope of the tangent to the *v-t* graph at that instant. The graph of acceleration versus time for this object is shown in Figure 2.9c. Note that the acceleration is constant and positive between 0 and t_1, where the slope of the *v-t* graph is positive; the acceleration is zero between t_1 and t_2 and for $t \geq t_6$, because the slope of the *v-t* graph is zero.

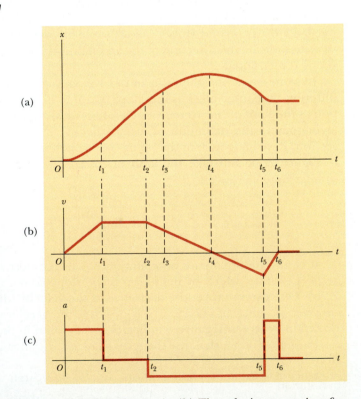

(a)

(b)

(c)

FIGURE 2.9   (Example 2.4) (a) Position-time graph for an object moving along the *x* axis. (b) The velocity versus time for the object is obtained by measuring the slope of the position-time graph at each instant. (c) The acceleration versus time for the object is obtained by measuring the slope of the velocity-time graph at each instant.

ation reaches a maximum at time t_1, when the slope of the velocity-time graph is a maximum. The acceleration then goes to zero at time t_2, when the velocity is a maximum (that is, when the velocity is momentarily not changing and the slope of the v versus t graph is zero). The acceleration is negative when the velocity in the positive x direction is decreasing in time, and it reaches its minimum value at time t_3.

CONCEPTUAL EXAMPLE 2.5

(a) If a car is traveling eastward, can its acceleration be westward? Explain. If a car is slowing down, can its acceleration be positive?

Reasoning   (a) Yes. This occurs when a car is slowing down, so that the direction of its acceleration is opposite to its direction of motion. (b) Yes. If the motion is in the direction chosen as negative, a positive acceleration causes a decrease in speed, often called a deceleration.

EXAMPLE 2.6   Average and Instantaneous Acceleration

The velocity of a particle moving along the x axis varies in time according to the expression $v = (40 - 5t^2)$ m/s, where t is in seconds. (a) Find the average acceleration in the time interval $t = 0$ to $t = 2.0$ s.

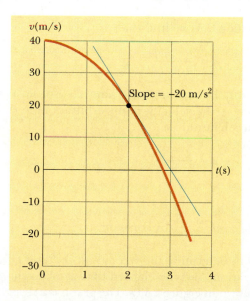

FIGURE 2.10   (Example 2.6) The velocity-time graph for a particle moving along the x axis according to the relation $v = (40 - 5t^2)$ m/s. Note that the acceleration at $t = 2$ s is equal to the slope of the blue tangent line at that time.

Solution   The velocity-time graph for this function is given in Figure 2.10. The velocities at $t_i = 0$ and $t_f = 2.0$ s are found by substituting these values of t into the expression given for the velocity:

$$v_i = (40 - 5t_i^2) \text{ m/s} = [40 - 5(0)^2] \text{ m/s} = +40 \text{ m/s}$$
$$v_f = (40 - 5t_f^2) \text{ m/s} = [40 - 5(2.0)^2] \text{ m/s} = +20 \text{ m/s}$$

Therefore, the average acceleration in the specified time interval $\Delta t = t_f - t_i = 2.0$ s is

$$\bar{a} = \frac{v_f - v_i}{t_f - t_i} = \frac{(20 - 40) \text{ m/s}}{(2.0 - 0) \text{ s}} = -10 \text{ m/s}^2$$

The negative sign is consistent with the fact that the slope of the line joining the initial and final points on the velocity-time graph is negative.

(b) Determine the acceleration at $t = 2.0$ s.

Solution   The velocity at time t is $v_i = (40 - 5t^2)$ m/s, and the velocity at time $t + \Delta t$ is

$$v_f = 40 - 5(t + \Delta t)^2 = 40 - 5t^2 - 10t\,\Delta t - 5(\Delta t)^2$$

Therefore, the change in velocity over the time interval Δt is

$$\Delta v = v_f - v_i = [-10t\,\Delta t - 5(\Delta t)^2] \text{ m/s}$$

Dividing this expression by Δt and taking the limit of the result as Δt approaches zero gives the acceleration at *any* time t:

$$a = \lim_{\Delta t \to 0} \frac{\Delta v}{\Delta t} = \lim_{\Delta t \to 0} (-10t - 5\,\Delta t) = -10t \text{ m/s}^2$$

Therefore, at $t = 2.0$ s, we find that

$$a = (-10)(2.0) \text{ m/s}^2 = -20 \text{ m/s}^2$$

This result can also be obtained by measuring the slope of the velocity-time graph at $t = 2.0$ s. Note that the acceleration is not constant in this example. Situations involving constant acceleration are treated in the next section.

So far we have evaluated the derivatives of a function by starting with the definition of the function and then taking the limit of a specific ratio. Those of you familiar with the calculus should recognize that there are specific rules for taking the derivatives of various functions. These rules, which are listed in Appendix B.6, enable us to evaluate derivatives quickly.

Suppose x is proportional to some power of t, such as

$$x = At^n$$

where A and n are constants. (This is a very common functional form.) The derivative of x with respect to t is given by

$$\frac{dx}{dt} = nAt^{n-1}$$

Applying this rule to Example 2.3, where $x = 3t^2$, we see that $v = dx/dt = 6t$, in agreement with our result when we took the limit explicitly. Likewise, in Example 2.4, where $v = 40 - 5t^2$, we find that $a = dv/dt = -10t$. (Note that the rate of change of any constant quantity is zero.)

 ## 2.4   ONE-DIMENSIONAL MOTION WITH CONSTANT ACCELERATION

If the acceleration of a particle varies in time, the motion can be complex and difficult to analyze. However, a very common and simple type of one-dimensional motion occurs when the acceleration is constant, or uniform. When the acceleration is constant, the average acceleration equals the instantaneous acceleration. Consequently, the velocity increases or decreases at the same rate throughout the motion.

If we replace $\bar{a}$ by a in Equation 2.5, we find that

$$a = \frac{v_f - v_i}{t_f - t_i}$$

For convenience, let $t_i = 0$ and t_f be any arbitrary time t. Also, let $v_i = v_0$ (the initial velocity at $t = 0$) and $v_f = v$ (the velocity at any arbitrary time t). With this notation, we can express the acceleration as

$$a = \frac{v - v_0}{t}$$

or

Velocity as a function of time

$$v = v_0 + at \qquad \text{(for constant } a\text{)} \qquad (2.8)$$

This expression enables us to determine the velocity at *any* time t if the initial velocity, the (constant) acceleration, and the elapsed time are known. A velocity-time graph for this motion is shown in Figure 2.11a. The graph is a straight line the slope of which is the acceleration, a, consistent with the fact that $a = dv/dt$ is a constant. Note that if the acceleration were negative, the slope of Figure 2.11a would be negative. If the acceleration is in the direction opposite to the velocity, then the particle is slowing down. From this graph and from Equation 2.8, we see that the velocity at any time t is the sum of the initial velocity, v_0, and the change in velocity, at.

The graph of acceleration versus time (Fig. 2.11b) is a straight line with a slope of zero, since the acceleration is constant.

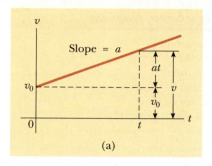
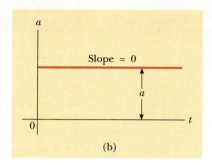
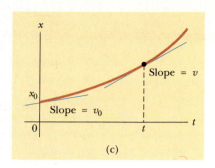

FIGURE 2.11   A particle moving along the x axis with constant acceleration a; (a) the velocity-time graph, (b) the acceleration-time graph, and (c) the position-time graph.

Because the velocity varies linearly in time according to Equation 2.8, we can express the average velocity in any time interval as the arithmetic mean of the initial velocity, v_0, and the final velocity, v:

$$\bar{v} = \frac{v_0 + v}{2} \qquad \text{(for constant } a\text{)} \tag{2.9}$$

Note that this expression is useful only when the acceleration is constant, that is, when the velocity varies linearly with time.

We can now use Equations 2.2 and 2.9 to obtain the displacement as a function of time. Again, we choose $t_i = 0$, at which time the initial position is $x_i = x_0$. This gives

$$\Delta x = \bar{v}\,\Delta t = \left(\frac{v_0 + v}{2}\right)t$$

$$x - x_0 = \tfrac{1}{2}(v + v_0)t \qquad \text{(for constant } a\text{)} \tag{2.10}$$

Displacement as a function of velocity and time

We can obtain another useful expression for the displacement by substituting Equation 2.8 into Equation 2.10:

$$x - x_0 = \tfrac{1}{2}(v_0 + v_0 + at)t$$

$$x - x_0 = v_0 t + \tfrac{1}{2}at^2 \qquad \text{(for constant } a\text{)} \tag{2.11}$$

Displacement as a function of time

The validity of this expression can be checked by differentiating it with respect to time:

$$v = \frac{dx}{dt} = \frac{d}{dt}\left(x_0 + v_0 t + \tfrac{1}{2}at^2\right) = v_0 + at$$

Finally, we can obtain an expression that does not contain the time by substituting the value of t from Equation 2.8 into Equation 2.10:

$$x - x_0 = \tfrac{1}{2}(v_0 + v)\left(\frac{v - v_0}{a}\right) = \frac{v^2 - v_0{}^2}{2a}$$

$$v^2 = v_0{}^2 + 2a(x - x_0) \qquad \text{(for constant } a\text{)} \tag{2.12}$$

Velocity as a function of displacement

TABLE 2.2 Kinematic Equations for Motion in a Straight Line Under Constant Acceleration	
Equation	**Information Given by Equation**
$v = v_0 + at$	Velocity as a function of time
$x - x_0 = \frac{1}{2}(v + v_0)t$	Displacement as a function of velocity and time
$x - x_0 = v_0 t + \frac{1}{2}at^2$	Displacement as a function of time
$v^2 = v_0^2 + 2a(x - x_0)$	Velocity as a function of displacement

Note: Motion is along the *x* axis. At $t = 0$, the position of the particle is x_0 and its velocity is v_0.

A position-time graph for motion under constant acceleration assuming positive *a* is shown in Figure 2.11c. Note that the curve representing Equation 2.11 is a parabola. The slope of the tangent to this curve at $t = 0$ equals the initial velocity, v_0, and the slope of the tangent line at any time *t* equals the velocity at that time.

If motion occurs in which the acceleration is *zero,* then we see that

$$\left.\begin{array}{l} v = v_0 \\ x - x_0 = vt \end{array}\right\} \quad \text{when } a = 0$$

That is, when the acceleration is zero, the velocity is a constant and the displacement changes linearly with time.

Equations 2.8 through 2.12 are *kinematic expressions that may be used to solve any problem in one-dimensional motion with constant acceleration*. Keep in mind that these relationships were derived from the definition of velocity and acceleration, together with some simple algebraic manipulations and the requirement that the acceleration be constant. It is often convenient to choose the initial position of the particle as the origin of the motion, so that $x_0 = 0$ at $t = 0$. In such a case, the displacement is simply *x*.

The four kinematic equations that are used most often are listed in Table 2.2 for convenience.

The choice of which kinematic equation or equations you should use in a given situation depends on what is known beforehand. Sometimes it is necessary to use two of these equations to solve for two unknowns, such as the displacement and velocity at some instant. For example, suppose the initial velocity, v_0, and acceleration, *a*, are given. You can then find (1) the velocity after a time *t* has elapsed, using $v = v_0 + at$, and (2) the displacement after a time *t* has elapsed, using $x - x_0 = v_0 t + \frac{1}{2}at^2$. You should recognize that the quantities that vary during the motion are velocity, displacement, and time.

You will get a great deal of practice in the use of these equations by solving a number of exercises and problems. Many times you will discover that there is more than one method for obtaining a solution.

CONCEPTUAL EXAMPLE 2.7

Can these equations of kinematics be used in a situation where the acceleration varies with time? Can they be used when the acceleration is zero?

Reasoning   The equations of kinematics cannot be used if the acceleration varies continuously. However, if the acceleration changes in steps, having one constant value for a while and then another constant value for some later time interval, the equations for constant-acceleration motion can be used to follow each section of the motion separately. If the acceleration is zero during some time interval, the velocity is constant, and the kinematic equations can be used; in this case, because $a = 0$, the equations become $v = v_0$, and $x - x_0 = vt$.

EXAMPLE 2.8   The Supercharged Sports Car

A certain automobile manufacturer claims that its super-deluxe sports car will accelerate from rest to a speed of 42.0 m/s in 8.00 s. Under the improbable assumption that the acceleration is constant: (a) determine the acceleration of the car in m/s².

Solution   First note that $v_0 = 0$ and the velocity after 8.00 s is $v = 42.0$ m/s. Because we are given v_0, v, and t, we can use $v = v_0 + at$ to find the acceleration:

$$a = \frac{v - v_0}{t} = \frac{42.0 \text{ m/s}}{8.00 \text{ s}} = \boxed{+ 5.25 \text{ m/s}^2}$$

In reality, this is an average acceleration, since it is unlikely that a car accelerates uniformly.

(b) Find the distance the car travels in the first 8.00 s.

Solution   Let the origin be at the original position of the car, so that $x_0 = 0$. Using Equation 2.10 we find that

$$x = \tfrac{1}{2}(v_0 + v)t = \tfrac{1}{2}(42.0 \text{ m/s})(8.00 \text{ s}) = \boxed{168 \text{ m}}$$

(c) What is the speed of the car 10.0 s after it begins its motion, assuming it continues to accelerate at the average rate of 5.25 m/s²?

Solution   Again we can use $v = v_0 + at$, this time with $v_0 = 0$, $t = 10.0$ s, and $a = 5.25$ m/s²:

$$v = v_0 + at = 0 + (5.25 \text{ m/s}^2)(10.0 \text{ s}) = \boxed{52.5 \text{ m/s}}$$

EXAMPLE 2.9   Accelerating an Electron

An electron in the cathode ray tube of a television set enters a region where it accelerates uniformly from a speed of 3.00×10^4 m/s to a speed of 5.00×10^6 m/s in a distance of 2.00 cm. (a) For what length of time is the electron in this region where it accelerates?

Solution   Taking the direction of motion to be along the x axis, we can use Equation 2.10 to find t, since the displacement and velocities are known:

$$x - x_0 = \tfrac{1}{2}(v_0 + v)t$$

$$t = \frac{2(x - x_0)}{v_0 + v} = \frac{2(2.00 \times 10^{-2} \text{ m})}{(3.00 \times 10^4 + 5.00 \times 10^6) \text{ m/s}}$$

$$= \boxed{7.95 \times 10^{-9} \text{ s}}$$

(b) What is the acceleration of the electron in this region?

Solution   To find the acceleration, we can use $v = v_0 + at$ and the results from (a):

$$a = \frac{v - v_0}{t} = \frac{(5.00 \times 10^6 - 3.00 \times 10^4) \text{ m/s}}{7.95 \times 10^{-9} \text{ s}}$$

$$= \boxed{6.25 \times 10^{14} \text{ m/s}^2}$$

We also could have used Equation 2.12 to obtain the acceleration, since the velocities and displacement are known. Try it! Although the acceleration is very large in this example, it occurs over a very short time interval and is a typical value for charged particles in accelerators.

EXAMPLE 2.10   Watch Out for the Speed Limit

A car traveling at a constant speed of 30.0 m/s (≈ 67 mi/h) passes a trooper hidden behind a billboard. One second after the speeding car passes the billboard, the trooper sets in chase after the car with a constant acceleration of 3.00 m/s². How long does it take the trooper to overtake the speeding car?

Reasoning   To solve this problem algebraically, we write expressions for the position of each vehicle as a function of time. It is convenient to choose the origin at the position of the billboard and take $t = 0$ as the time the trooper begins moving. At that instant, the speeding car has already traveled a distance of 30.0 m because it travels at a constant speed of

30.0 m/s. Thus, the initial position of the speeding car is $x_0 = 30.0$ m.

Solution   Because the car moves with constant speed, its acceleration is zero, and applying Equation 2.11 gives

$$x_C = 30.0 \text{ m} + (30.0 \text{ m/s})t$$

Note that at $t = 0$, this expression does give the car's correct initial position $x_C = x_0 = 30.0$ m.

For the trooper, who starts from the origin at $t = 0$, we have $x_0 = 0$, $v_0 = 0$, and $a = 3.00$ m/s². Hence, the position of the trooper as a function of time is

$$x_T = \tfrac{1}{2}at^2 = \tfrac{1}{2}(3.00 \text{ m/s}^2)t^2$$

The trooper overtakes the car at the instant that $x_T = x_C$, or

$$\tfrac{1}{2}(3.00 \text{ m/s}^2)t^2 = 30.0 \text{ m} + (30.0 \text{ m/s})t$$

This gives the quadratic equation

$$1.50t^2 - 30.0t - 30.0 = 0$$

whose positive solution is $t = 21.0$ s. (For help in solving quadratic equations, see Appendix B.2.) Note that in this time interval, the trooper travels a distance of about 660 m.

Exercise   This problem can also be solved graphically. On the same graph, plot the position versus time for each vehicle, and from the intersection of the two curves determine the time at which the trooper overtakes the speeding car.

2.5   FREELY FALLING OBJECTS

It is well known that all objects, when dropped, fall toward the Earth with nearly constant acceleration. There is a legend that Galileo first discovered this fact by observing that two different weights dropped simultaneously from the Leaning Tower of Pisa hit the ground at approximately the same time. Although there is some doubt that this particular experiment was carried out, it is well established that Galileo did perform many systematic experiments on objects moving on inclined planes. Through careful measurements of distances and time intervals, he was able to show that the displacement of an object starting from rest is proportional to the square of the time the object is in motion. This observation is consistent with one of the kinematic equations we derived for motion under constant acceleration (Eq. 2.11). Galileo's achievements in the science of mechanics paved the way for Newton in his development of the laws of motion.

You might want to try the following experiment. Drop a coin and a crumpled-up piece of paper simultaneously from the same height. In the absence of air resistance, both will experience the same motion and hit the floor at the same

Galileo performing demonstrations of balls rolling down a grooved inclined plane. As a ball rolled down the incline, Galileo carefully measured its position at the end of equal time intervals and showed that the displacement was proportional to the square of the elapsed time. This painting by Giuseppe Bezzouli is located in the Zoological Museum in Florence, Italy.   *(Art Resource)*

time. In a real (non-ideal) experiment, air resistance cannot be neglected. In the idealized case, where air resistance *is* neglected, such motion is referred to as *free-fall*. If this same experiment could be conducted in a good vacuum, where air friction is truly negligible, the paper and coin would fall with the same acceleration, regardless of the shape of the paper. This point is illustrated very convincingly on page 23, in the photograph of the apple and feather falling in a vacuum. On August 2, 1971, such an experiment was conducted by astronaut David Scott on the Moon (where air resistance is negligible). He simultaneously released a geologist's hammer and a falcon's feather, and in unison they fell to the lunar surface. This demonstration would have surely pleased Galileo!

We shall denote the *free-fall acceleration* by the symbol *g*. The value of *g* on Earth decreases with increasing altitude. Furthermore, there are slight variations in *g* with latitude. The free-fall acceleration is directed downward toward the center of the Earth. At the Earth's surface, the value of *g* is approximately 9.80 m/s², or 980 cm/s², or 32 ft/s². Unless stated otherwise, we shall use this value for *g* when doing calculations.

When we use the expression *freely falling object,* we do not necessarily refer to an object dropped from rest. A freely falling object is any object moving freely under the influence of gravity, *regardless* of its initial motion. Objects thrown upward or downward and those released from rest are all falling freely once they are released. Furthermore, it is important to recognize that any freely falling object experiences an acceleration directed *downward.* This is true regardless of the initial motion of the object.

> An object thrown upward and one thrown downward will both experience the same acceleration as an object released from rest. Once they are in free-fall, all objects have an acceleration downward, equal to the free-fall acceleration.

If we neglect air resistance and assume that the free-fall acceleration does not vary with altitude, then the vertical motion of a freely falling object is equivalent to motion in one dimension under constant acceleration. Therefore, our kinematic equations for constant acceleration can be applied. We shall take the vertical direction to be the *y* axis and call *y* positive upward. With this choice of coordinates, we can replace *x* by *y* in Equations 2.10, 2.11, and 2.12. Furthermore, since positive *y* is upward, the acceleration is negative (downward) and given by $a = -g$. The negative sign simply indicates that the acceleration is downward. With these substitutions, we get the following expressions[4]:

$$v = v_0 - gt \tag{2.13}$$

$$y - y_0 = \tfrac{1}{2}(v + v_0)t \qquad \text{(for constant } a = -g) \tag{2.14}$$

$$y - y_0 = v_0 t - \tfrac{1}{2}gt^2 \tag{2.15}$$

$$v^2 = v_0{}^2 - 2g(y - y_0) \tag{2.16}$$

You should note that the *negative sign for the acceleration is already included in these expressions.* Therefore, when using these equations in any free-fall problem, you should simply substitute $g = 9.80$ m/s².

[4] One can also take *y* positive downward, in which case $a = +g$. The results will be the same, regardless of the convention chosen.

Free-fall acceleration $g = 9.80$ m/s²

Definition of free-fall

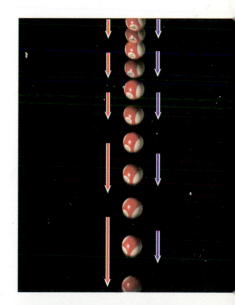

A multiflash photograph of a falling billiard ball. As the ball falls, the spacing between successive images increases, indicating that the ball accelerates downward. The motion diagram shows that the ball's velocity (red arrows) increases with time while its acceleration (violet arrows) remains constant. *(Richard Megna/Fundamental Photographs)*

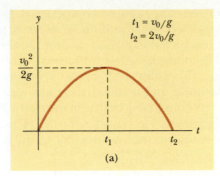
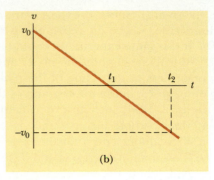

FIGURE 2.12   Graphs of (a) the displacement versus time and (b) the velocity versus time for a freely falling particle, where y and v are taken to be positive upward. Note the symmetry in (a) about $t = t_1$.

Consider the case of a particle thrown vertically upward from the origin with a velocity v_0. In this case, v_0 is positive and $y_0 = 0$. Graphs of the displacement and velocity as functions of time are shown in Figure 2.12. Note that the velocity is initially positive but decreases in time and goes to zero at the peak of the path. From Equation 2.13, we see that this peak occurs at the time $t_1 = v_0/g$. At this time, the displacement has its largest positive value, which can be calculated from Equation 2.15 with $t = t_1 = v_0/g$. This gives $y_{max} = v_0^2/2g$.

At the time $t_2 = 2t_1 = 2v_0/g$, we see from Equation 2.15 that the displacement is again zero, that is, the particle has returned to its starting point. Furthermore, at time t_2 the velocity is $v = -v_0$. (This follows directly from Eq. 2.13.) Hence, there is symmetry in the motion. In other words, both the displacement and the magnitude of the velocity repeat themselves in the time interval $t = 0$ to $t = 2v_0/g$.

In the examples that follow, we shall, for convenience, assume that $y_0 = 0$ at $t = 0$. This choice does not affect the solution to the problem. If y_0 is nonzero, then the graph of y versus t (Fig. 2.12a) is simply shifted upward or downward by an amount y_0, while the graph of v versus t (Fig. 2.12b) remains unchanged.

CONCEPTUAL EXAMPLE 2.11

A child throws a marble into the air with some initial velocity. Another child drops a ball at the same instant. Compare the accelerations of the two objects while they are in the air.

Reasoning   Once the objects leave the hand, both are freely falling objects, and hence both experience the same downward acceleration equal to the free-fall acceleration, $g = 9.80$ m/s^2.

CONCEPTUAL EXAMPLE 2.12

A ball is thrown upward. While the ball is in the air, (a) what happens to its velocity? (b) does its acceleration increase, decrease, or remain constant?

Reasoning   (a) The velocity of the ball changes continuously. As it travels upward, its speed decreases by 9.80 m/s during each second of its motion. When it reaches the peak of its motion, its speed becomes zero. As the ball moves downward, its speed increases by 9.80 m/s each second. (b) The acceleration of the ball remains constant while it is in the air, from the instant it leaves the hand until the instant before it strikes the ground. Its magnitude is the free-fall ac-

celeration, $g = 9.80$ m/s². (If the acceleration were zero at the peak when the velocity is zero, this would say that there would be no change in velocity thereafter, so the ball would stop at the peak, and remain there, which is not the case.)

CONCEPTUAL EXAMPLE 2.13 Try to Catch the Dollar Bill!

Emily challenges her best friend David to catch a dollar bill as follows. She holds the bill vertically, as in Figure 2.13, with the center of the bill between David's index finger and thumb. David must catch the bill after Emily releases it without moving his hand downward. Who would you bet on?

Reasoning   Place your bets on Emily. There is a time delay between the instant Emily releases the bill and the time David reacts and closes his fingers. The reaction time of most people is at best about 0.2 s. Since the bill is in free-fall, and undergoes a downward acceleration of 9.80 m/s², in 0.2 s it falls a distance of $\frac{1}{2}gt^2 \cong 0.2$ m $= 20$ cm. This distance is about twice the distance between the center of the bill and its top edge ($\cong 8$ cm). Thus, David will be unsuccessful. You might want to try this on one of your friends.

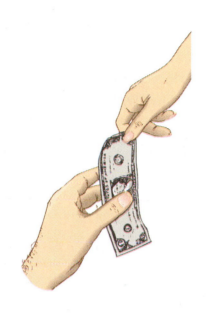

FIGURE 2.13   (Example 2.13) David is challenged to catch the dollar bill between his index finger and thumb after the bill is released by Emily. Is he able to meet the challenge?

EXAMPLE 2.14 Look Out Below!

A golf ball is released from rest from the top of a very tall building. Neglecting air resistance, calculate the position and velocity of the ball after 1.00, 2.00, and 3.00 s.

Solution   We choose our coordinates such that the starting point of the ball is at the origin ($y_0 = 0$ at $t = 0$) and remember that we have defined y to be positive upward. Since $v_0 = 0$, Equations 2.13 and 2.15 become

$$v = -gt = -(9.80 \text{ m/s}^2)t$$

$$y = -\tfrac{1}{2}gt^2 = -\tfrac{1}{2}(9.80 \text{ m/s}^2)t^2$$

where t is in seconds, v is in meters per second, and y is in meters. These expressions give the velocity and displacement at any time t after the ball is released. Therefore, at $t = 1.00$ s,

$$v = -(9.80 \text{ m/s}^2)(1.00 \text{ s}) = \boxed{-9.80 \text{ m/s}}$$

$$y = -\tfrac{1}{2}(9.80 \text{ m/s}^2)(1.00 \text{ s})^2 = \boxed{-4.90 \text{ m}}$$

At $t = 2.00$ s, we find that $v = -19.6$ m/s and $y = -19.6$ m. At $t = 3.00$ s, $v = -29.4$ m/s and $y = -44.1$ m. The minus signs for v indicate that the velocity is directed downward, and the minus signs for y indicate displacement in the negative y direction. (Because of air resistance, the actual speed of the ball is limited to approximately 30 m/s.)

Exercise   Calculate the position and velocity of the ball after 4.00 s.

Answer -78.4 m, -39.2 m/s.

EXAMPLE 2.15   Not a Bad Throw for a Rookie

A stone thrown from the top of a building is given an initial velocity of 20.0 m/s straight upward. The building is 50.0 m high, and the stone just misses the edge of the roof on its way down, as in Figure 2.14. Determine (a) the time needed for the stone to reach its maximum height, (b) the maximum height, (c) the time needed for the stone to return to the top of the building, (d) the velocity of the stone at this instant, and (e) the velocity and position of the stone at $t = 5.00$ s.

Solution   (a) To find the time necessary to reach the maximum height, use Equation 2.13, $v = v_0 - gt$, noting that $v = 0$ at maximum height:

$$20.0 \text{ m/s} - (9.80 \text{ m/s}^2) t_1 = 0$$

$$t_1 = \frac{20.0 \text{ m/s}}{9.80 \text{ m/s}^2} = \boxed{2.04 \text{ s}}$$

(b) This value of time can be substituted into Equation 2.15, to give the maximum height as measured from the position of the thrower:

$$y = v_0 t - \tfrac{1}{2} g t^2$$
$$y_{max} = (20.0 \text{ m/s})(2.04 \text{ s}) - \tfrac{1}{2}(9.80 \text{ m/s}^2)(2.04 \text{ s})^2$$

$$= \boxed{20.4 \text{ m}}$$

(c) When the stone is back at the height of the building top, the y coordinate is zero. Using Equation 2.15, with $y = 0$, we obtain

$$y = v_0 t - \tfrac{1}{2} g t^2$$
$$0 = (20.0 \text{ m/s}) t - (4.90 \text{ m/s}^2) t^2$$

This is a quadratic equation and has two solutions for t. The equation can be factored to give

$$t(20.0 - 4.90t) = 0$$

One solution is $t = 0$, corresponding to the time the stone starts its motion. The other solution is $t = 4.08$ s, which is the solution we are after.

(d) The value for t found in (c) can be inserted into Equation 2.13 to give

$$v = v_0 - gt = 20.0 \text{ m/s} - (9.80 \text{ m/s}^2)(4.08 \text{ s})$$

$$= \boxed{-20.0 \text{ m/s}}$$

Note that the velocity of the stone when it arrives back at its original height is equal in magnitude to its initial velocity but opposite in direction. This indicates that the motion is symmetric.

(e) From Equation 2.13, the velocity after 5.00 s is

$$v = v_0 - gt = 20.0 \text{ m/s} - (9.80 \text{ m/s}^2)(5.00 \text{ s})$$

$$= \boxed{-29.0 \text{ m/s}}$$

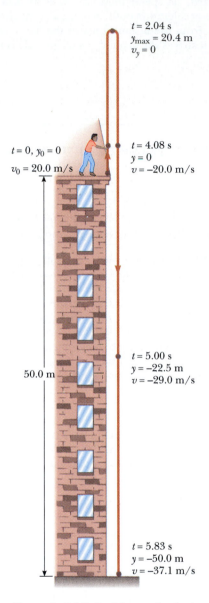

FIGURE 2.14   (Example 2.15) Position and velocity versus time for a freely falling particle thrown initially upward with a velocity of $v_0 = 20$ m/s.

We can use Equation 2.15 to find the position of the particle at $t = 5.00$ s:

$$y = v_0 t - \tfrac{1}{2} g t^2$$
$$= (20.0 \text{ m/s})(5.00 \text{ s}) - \tfrac{1}{2}(9.80 \text{ m/s}^2)(5.00 \text{ s})^2 = \boxed{-22.5 \text{ m}}$$

Exercise   Find (a) the velocity of the stone just before it hits the ground and (b) the total time the stone is in the air.

Answer   (a) -37.1 m/s (b) 5.83 s.

*2.6   KINEMATIC EQUATIONS DERIVED FROM CALCULUS

This is an optional section that assumes the reader is familiar with the techniques of integral calculus. If you have not studied integration in your calculus course as yet, this section should be skipped or covered at some later time after you become familiar with integration.

The velocity of a particle moving in a straight line can be obtained from a knowledge of its position as a function of time. Mathematically, the velocity equals the derivative of the coordinate with respect to time. It is also possible to find the displacement of a particle if its velocity is known as a function of time. In the calculus, this procedure is referred to as integration, or finding the antiderivative. Graphically, it is equivalent to finding the area under a curve.

Suppose the velocity versus time plot for a particle moving along the x axis is as shown in Figure 2.15. Let us divide the time interval $t_f - t_i$ into many small intervals of duration Δt_n. From the definition of average velocity, we see that the displacement during any small interval, such as the shaded one in Figure 2.15, is given by $\Delta x_n = \bar{v}_n \Delta t_n$, where $\bar{v}_n$ is the average velocity in that interval. Therefore, the displacement during this small interval is simply the area of the shaded rectangle. The total displacement for the interval $t_f - t_i$ is the sum of the areas of all the rectangles:

$$\Delta x = \sum_n \bar{v}_n \Delta t_n$$

where the sum is taken over all the rectangles from t_i to t_f. Now, as each interval is made smaller and smaller, the number of terms in the sum increases and the sum approaches a value equal to the area under the velocity-time graph. Therefore, in the limit $n \rightarrow \infty$, or $\Delta t_n \rightarrow 0$, we see that the displacement is given by

$$\Delta x = \lim_{\Delta t_n \rightarrow 0} \sum_n v_n \Delta t_n \qquad (2.17)$$

or

Displacement = area under the velocity-time graph

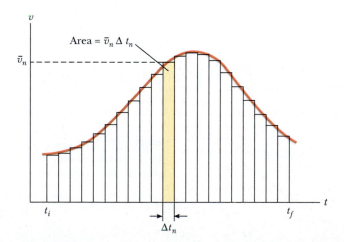

FIGURE 2.15   Velocity versus time for a particle moving along the x axis. The area of the shaded rectangle is equal to the displacement Δx in the time interval Δt_n, while the total area under the curve is the total displacement of the particle.

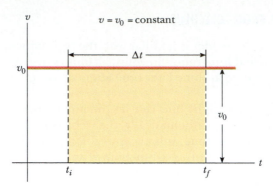

FIGURE 2.16 The velocity versus time curve for a particle moving with constant velocity v_0. The displacement of the particle during the time interval $t_f - t_i$ is equal to the area of the shaded rectangle.

Note that we have replaced the average velocity $\bar{v}_n$ by the instantaneous velocity v_n in the sum. As you can see from Figure 2.15, this approximation is clearly valid in the limit of very small intervals.

We conclude that if the velocity-time graph for motion along a straight line is known, the displacement during any time interval can be obtained by measuring the area under the curve corresponding to that time interval.

The limit of the sum in Equation 2.17 is called a **definite integral** and is written

Definite integral

$$\lim_{\Delta t_n \to 0} \sum_n v_n \, \Delta t_n = \int_{t_i}^{t_f} v(t) \, dt \qquad (2.18)$$

where $v(t)$ denotes the velocity at any time t. If the explicit functional form of $v(t)$ is known, and the limits are given, the integral can be evaluated.

If a particle moves with a constant velocity v_0 as in Figure 2.16, its displacement during the time interval Δt is simply the area of the shaded rectangle, that is,

$$\Delta x = v_0 \, \Delta t \qquad \text{(when } v = v_0 = \text{constant)}$$

As another example, consider a particle moving with a velocity that is proportional to t, as in Figure 2.17. Taking $v = at$, where a is the constant of proportionality (the acceleration), we find that the displacement of the particle during the time interval $t = 0$ to $t = t_1$ is the area of the shaded triangle in Figure 2.17:

$$\Delta x = \tfrac{1}{2}(t_1)(at_1) = \tfrac{1}{2}at_1^2$$

Kinematic Equations

We now make use of the defining equations for acceleration and velocity to derive two of our kinematic equations.

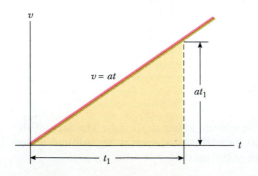

FIGURE 2.17 The velocity versus time curve for a particle moving with a velocity that is proportional to the time.

The defining equation for acceleration (Eq. 2.6),

$$a = \frac{dv}{dt}$$

may also be written in terms of an integral (or antiderivative) as

$$v = \int a \, dt + C_1$$

where C_1 is a constant of integration. For the special case where the acceleration is constant, this reduces to

$$v = at + C_1$$

The value of C_1 depends on the initial conditions of the motion. If we take $v = v_0$ when $t = 0$ and substitute these values into the last equation, we have

$$v_0 = a(0) + C_1$$

$$C_1 = v_0$$

Hence, we obtain the first kinematic equation (Eq. 2.8):

$$v = v_0 + at \qquad \text{(for constant } a\text{)}$$

Now let us consider the defining equation for velocity (Eq. 2.3):

$$v = \frac{dx}{dt}$$

We can write this in integral form as

$$x = \int v \, dt + C_2$$

where C_2 is another constant of integration. Since $v = v_0 + at$, this expression becomes

$$x = \int (v_0 + at) \, dt + C_2$$

$$x = \int v_0 \, dt + \int at \, dt + C_2$$

$$x = v_0 t + \tfrac{1}{2}at^2 + C_2$$

To find C_2, we make use of the initial condition that $x = x_0$ when $t = 0$. This gives $C_2 = x_0$. Therefore, we have

$$x - x_0 = v_0 t + \tfrac{1}{2} at^2 \qquad \text{(for constant } a\text{)}$$

This is a second equation of kinematics (Eq. 2.11). Recall that $x - x_0$ is equal to the displacement of the object, where x_0 is its initial position.

SUMMARY

As a particle moves along the x axis from some initial position x_i to some final position x_f, its **displacement** is

$$\Delta x \equiv x_f - x_i \tag{2.1}$$

The **average velocity** of a particle during some time interval is equal to the ratio of the displacement, Δx, and the time interval, Δt:

$$\bar{v} \equiv \frac{\Delta x}{\Delta t} \qquad (2.2)$$

The **average speed** of a particle is equal to the ratio of total distance it travels to the total time it takes to travel that distance.

The **instantaneous velocity** of a particle is defined as the limit of the ratio $\Delta x/\Delta t$ as Δt approaches zero. By definition, this limit equals the derivative of x with respect to t, or the time rate of change of the position:

$$v \equiv \lim_{\Delta t \to 0} \frac{\Delta x}{\Delta t} = \frac{dx}{dt} \qquad (2.3, 2.4)$$

The **speed** of a particle is defined to be equal to the magnitude of its velocity.

The **average acceleration** of a particle during some time interval is defined as the ratio of the change in its velocity, Δv, and the time interval, Δt:

$$\bar{a} \equiv \frac{\Delta v}{\Delta t} \qquad (2.5)$$

The **instantaneous acceleration** is equal to the limit of the ratio $\Delta v/\Delta t$ as Δt approaches 0. By definition, this limit equals the derivative of v with respect to t, or the time rate of change of the velocity:

$$a \equiv \lim_{\Delta t \to 0} \frac{\Delta v}{\Delta t} = \frac{dv}{dt} \qquad (2.6)$$

The slope of the tangent to the v versus t curve equals the instantaneous acceleration of the particle.

The **equations of kinematics** for a particle moving along the x axis with uniform acceleration a (constant in magnitude and direction) are

$$v = v_0 + at \qquad (2.8)$$

$$x - x_0 = \tfrac{1}{2}(v_0 + v)t \qquad (2.10)$$

$$x - x_0 = v_0 t + \tfrac{1}{2}at^2 \qquad (2.11)$$

$$v^2 = v_0{}^2 + 2a(x - x_0) \qquad (2.12)$$

An object falling freely in the presence of the Earth's gravity experiences a free-fall acceleration directed toward the center of the Earth. If air friction is neglected, and if the altitude of the motion is small compared with the Earth's radius, then one can assume that the free-fall acceleration, g, is constant over the range of motion, where g is equal to 9.80 m/s^2, or 32 ft/s^2. Assuming y is positive upward, the acceleration is given by $-g$ and the equations of kinematics for an object in free fall are the same as those given above, with the substitutions $x \to y$ and $a \to -g$.

QUESTIONS

1. Average velocity and instantaneous velocity are generally different quantities. Can they ever be equal for a specific type of motion? Explain.
2. If the average velocity is nonzero for some time interval, does this mean that the instantaneous velocity is never zero during this interval? Explain.
3. If the average velocity equals zero for some time interval Δt and if $v(t)$ is a continuous function, show that the instantaneous velocity must go to zero some time in this interval. (A sketch of x versus t might be useful in your proof.)
4. Is it possible to have a situation in which the velocity and

accclcration have opposite signs? If so, sketch a velocity-time graph to prove your point.

5. If the velocity of a particle is nonzero, can its acceleration ever be zero? Explain.

6. If the velocity of a particle is zero, can its acceleration ever be nonzero? Explain.

7. A stone is thrown vertically upward from the top of a building. Does the stone's displacement depend on the location of the origin of the coordinate system? Does the stone's velocity depend on the origin? (Assume that the coordinate system is stationary with respect to the building.) Explain.

8. A student at the top of a building of height h throws one ball upward with an initial speed v_0 and then throws a second ball downward with the same initial speed. How do the final velocities of the balls compare when they reach the ground?

9. Can the magnitude of the instantaneous velocity of an object ever be greater than the magnitude of its average velocity? Can it ever be less?

10. If the average velocity of an object is zero in some time interval, what can you say about the displacement of the object for that interval?

11. A rapidly growing plant doubles in height each week. At the end of the 25th day, the plant reaches the height of a building. At what time was the plant one-fourth the height of the building?

12. Two cars are moving in the same direction in parallel lanes along a highway. At some instant, the velocity of car A exceeds the velocity of car B. Does this mean that the acceleration of A is greater than that of B? Explain.

13. An apple is dropped from some height above the Earth's surface. Neglecting air resistance, how much does its speed increase each second during its fall?

PROBLEMS

Section 2.1   Displacement, Velocity, and Speed

1. The position of a car coasting down a hill was observed at various times and the results are summarized in the table below. Find the average velocity of the car during (a) the first second, (b) the last three seconds, and (c) the entire period of observation.

x(m)	0	2.3	9.2	20.7	36.8	57.5
t(s)	0	1.0	2.0	3.0	4.0	5.0

2. A motorist drives north for 35 min at 85 km/h and then stops for 15 min. He then continues north, traveling 130 km in 2.0 h. (a) What is his total displacement? (b) What is his average velocity?

3. The displacement versus time graph for a certain particle moving along the x axis is shown in Figure P2.3. Find the average velocity in the time intervals (a) 0 to 2 s, (b) 0 to 4 s, (c) 2 s to 4 s, (d) 4 s to 7 s, (e) 0 to 8 s.

4. A jogger runs in a straight line with an average velocity of +5.00 m/s for 4.00 min, and then with an average velocity of +4.00 m/s for 3.00 min. (a) What is her total displacement? (b) What is her average velocity during this time?

5. A person walks first at a constant speed of 5.0 m/s along a straight line from point A to point B and then back along the line from B to A at a constant speed of 3.0 m/s. (a) What is her average speed over the entire trip? (b) Her average velocity over the entire trip?

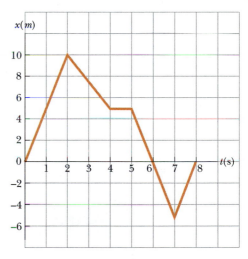

FIGURE P2.3

5A. A person walks first at a constant speed v_1 along a straight line from point A to point B, and then back along the line from B to A at a constant speed v_2. (a) What is her average speed over the entire trip? (b) Her average velocity over the entire trip?

6. A particle moves according to the equation $x = 10t^2$ where x is in meters and t is in seconds. (a) Find the average velocity for the time interval from 2.0 s to 3.0 s. (b) Find the average velocity for the time interval from 2.0 s to 2.1 s.

7. A car makes a 200-km trip at an average speed of 40 km/h. A second car starting 1.0 h later arrives at

---

□ indicates problems that have full solutions available in the Student Solutions Manual and Study Guide.

their mutual destination at the same time. What was the average speed of the second car for the period that it was in motion?

Section 2.2   Instantaneous Velocity and Speed

8. A speedy tortoise can run at 10.0 cm/s, and a hare can run 20 times as fast. In a race, they start at the same time, but the hare stops to rest for 2.0 min, and so the tortoise wins by a shell (20 cm). (a) How long does the race take? (b) What is the length of the race?

9. The position-time graph for a particle moving along the x axis is as shown in Figure P2.9. (a) Find the average velocity in the time interval $t = 1.5$ s to $t = 4.0$ s. (b) Determine the instantaneous velocity at $t = 2.0$ s by measuring the slope of the tangent line shown in the graph. (c) At what value of t is the velocity zero?

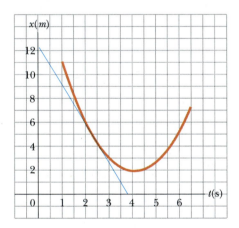

FIGURE P2.9

10. Two cars travel in the same direction along a straight highway, one at 55 mi/h and the other at 70 mi/h. (a) Assuming they start at the same point, how much sooner does the faster car arrive at a destination 10 miles away? (b) How far must the faster car travel before it has a 15-min lead on the slower car?

11. At $t = 1.0$ s, a particle moving with constant velocity is located at $x = -3.0$ m, and at $t = 6.0$ s, the particle is located at $x = 5.0$ m. (a) From this information, plot the position as a function of time. (b) Determine the velocity of the particle from the slope of this graph.

12. (a) Use the data in Problem 1 to construct a graph of position versus time. (b) By constructing tangents to the $x(t)$ curve, find the instantaneous velocity of the car at several instants. (c) Plot the instantaneous velocity versus time and, from this, determine the average acceleration of the car. (d) What is the initial velocity of the car?

13. Find the instantaneous velocity of the particle described in Figure P2.3 at the following times: (a) $t = 1.0$ s, (b) $t = 3.0$ s, (c) $t = 4.5$ s, and (d) $t = 7.5$ s.

14. The position-time graph for a particle moving along the z axis is as shown in Figure P2.14. Determine whether the velocity is positive, negative, or zero at times (a) t_1, (b) t_2, (c) t_3, (d) t_4.

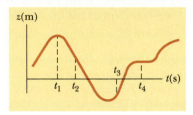

FIGURE P2.14

Section 2.3   Acceleration

15. A particle is moving with a velocity $v_0 = 60$ m/s at $t = 0$. Between $t = 0$ and $t = 15$ s, the velocity decreases uniformly to zero. What is the average acceleration during this 15-s interval? What is the significance of the sign of your answer?

16. An object moves along the x axis according to the equation $x(t) = (3.0t^2 - 2.0t + 3.0)$ m. Determine (a) the average speed between $t = 2.0$ s and $t = 3.0$ s, (b) the instantaneous speed at $t = 2.0$ s and at $t = 3.0$ s, (c) the average acceleration between $t = 2.0$ s and $t = 3.0$ s, and (d) the instantaneous acceleration at $t = 2.0$ s and $t = 3.0$ s.

17. A particle moves along the x axis according to the equation $x = 2.0t + 3.0t^2$, where x is in meters and t is in seconds. Calculate the instantaneous velocity and instantaneous acceleration at $t = 3.0$ s.

18. A particle moving in a straight line has a velocity of 8.0 m/s at $t = 0$. Its velocity at $t = 20$ s is 20.0 m/s. (a) What is its average acceleration in this time interval? (b) Can the average velocity be obtained from the information presented? Explain.

19. A particle starts from rest and accelerates as shown in Figure P2.19. Determine (a) the particle's speed at $t = 10$ s and at $t = 20$ s and (b) the distance traveled in the first 20 s.

20. The velocity of a particle as a function of time is shown in Figure P2.20. At $t = 0$, the particle is at $x = 0$. (a) Sketch the acceleration as a function of time. (b) Determine the average acceleration of the particle in the time interval $t = 2.0$ s to $t = 8.0$ s. (c) Determine the instantaneous acceleration of the particle at $t = 4.0$ s.

21. A particle moves along the x axis according to the equation $x = 2.0 + 3.0t - 1.0t$, where x is in meters and t is in seconds. At $t = 3.00$ s, find (a) the position of the particle, (b) its velocity, and (c) its acceleration.

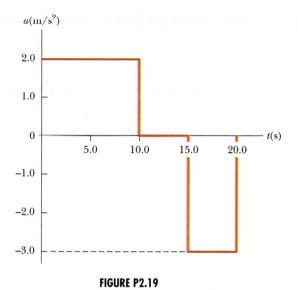

FIGURE P2.19

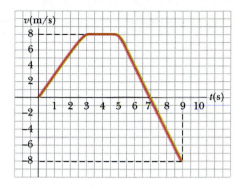

FIGURE P2.22

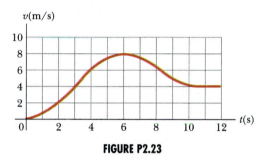

FIGURE P2.23

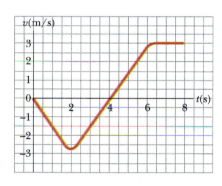

FIGURE P2.20

value of the acceleration and the time at which it occurs.

Section 2.4   One-Dimensional Motion with Constant Acceleration

22. A student drives a moped along a straight road as described by the speed-versus-time graph in Figure P2.22. Sketch this graph in the middle of a sheet of graph paper. (a) Directly above this graph, sketch a graph of the position versus time, aligning the time coordinates of the two graphs. (b) Sketch a graph of the acceleration versus time directly below the v-t graph, again aligning the time coordinates. On each graph, show the numerical values of x and a for all points of inflection. (c) What is the acceleration at $t = 6$ s? (d) Find the position (relative to the starting point) at $t = 6$ s. (e) What is the moped's final position at $t = 9$ s?

23. Figure P2.23 shows a graph of v versus t for the motion of a motorcyclist as she starts from rest and moves along the road in a straight line. (a) Find the average acceleration for the time interval $t_0 = 0$ to $t_1 = 6.0$ s. (b) Estimate the time at which the acceleration has its greatest positive value and the value of the acceleration at this instant. (c) When is the acceleration zero? (d) Estimate the maximum negative

24. A particle travels in the positive x direction for 10 s at a constant speed of 50 m/s. It then accelerates uniformly to a speed of 80 m/s in the next 5 s. Find (a) the average acceleration of the particle in the first 10 s, (b) its average acceleration in the interval $t = 10$ s to $t = 15$ s, (c) the total displacement of the particle between $t = 0$ and $t = 15$ s, and (d) its average speed in the interval $t = 10$ s to $t = 15$ s.

25. A body moving with uniform acceleration has a velocity of 12.0 cm/s when its x coordinate is 3.00 cm. If its x coordinate 2.00 s later is -5.00 cm, what is the magnitude of its acceleration?

26. The new BMW M3 can accelerate from zero to 60 mi/h in 5.6 s. (a) What is the resulting acceleration in m/s²? (b) How long would it take for the BMW to go from 60 mi/h to 130 mi/h at this rate?

27. The minimum distance required to stop a car moving at 35 mi/h is 40 ft. What is the minimum stopping distance for the same car moving at 70 mi/h, assuming the same rate of acceleration?

28. Figure P2.28 represents part of the performance data of a car owned by a proud physics student. (a) Calculate from the graph the total distance traveled. (b) What distance does the car travel between the

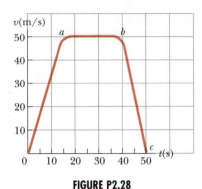

FIGURE P2.28

times $t = 10$ s and $t = 40$ s? (c) Draw a graph of its acceleration versus time between $t = 0$ and $t = 50$ s. (d) Write an equation for x as a function of time for each phase of the motion, represented by (i) *ab,* (ii) *bc,* (iii) *cd.* (e) What is the average velocity of the car between $t = 0$ and $t = 50$ s?

29. The initial speed of a body is 5.20 m/s. What is its speed after 2.50 s if it accelerates uniformly at (a) 3.00 m/s² and (b) at -3.00 m/s²?

30. A hockey puck sliding on a frozen lake comes to rest after traveling 200 m. If its initial velocity is 3.00 m/s, (a) what is its acceleration if that acceleration is assumed constant, (b) how long is it in motion, and (c) what is its speed after traveling 150 m?

31. A jet plane lands with a velocity of 100 m/s and can accelerate at a maximum rate of -5.0 m/s² as it comes to rest. (a) From the instant it touches the runway, what is the minimum time needed before it stops? (b) Can this plane land at a small airport where the runway is 0.80 km long?

32. A car and train move together along parallel paths at 25.0 m/s. The car then undergoes a uniform acceleration of -2.50 m/s² because of a red light and comes to rest. It remains at rest for 45.0 s, then accelerates back to a speed of 25.0 m/s at a rate of 2.50 m/s². How far behind the train is the car when it reaches the speed of 25.0 m/s, assuming that the train speed has remained at 25.0 m/s?

33. A drag racer starts her car from rest and accelerates at 10.0 m/s² for the entire distance of 400 m ($\frac{1}{4}$ mile). (a) How long did it take the car to travel this distance? (b) What is its speed at the end of the run?

34. An electron in a cathode ray tube (CRT) accelerates from 2.0×10^4 m/s to 6.0×10^6 m/s over 1.5 cm. (a) How long does the electron take to travel this distance? (b) What is its acceleration?

35. A particle starts from rest from the top of an inclined plane and slides down with constant acceleration. The inclined plane is 2.00 m long, and it takes 3.00 s for the particle to reach the bottom. Find (a) the acceleration of the particle, (b) its speed at the bottom

of the incline, (c) the time it takes the particle to reach the middle of the incline, and (d) its speed at the midpoint.

36. Two express trains started 5 min apart. Starting from rest, each is capable of a maximum speed of 160 km/h after uniformly accelerating over a distance of 2.0 km. (a) What is the acceleration of each train? (b) How far ahead is the first train when the second one starts? (c) How far apart are they when they are both traveling at maximum speed?

37. A teenager has a car that accelerates at 3.0 m/s² and decelerates at -4.5 m/s². On a trip to the store, he accelerates from rest to 12 m/s, drives at a constant speed for 5.0 s, and then comes to a momentary stop at the corner. He then accelerates to 18 m/s, drives at a constant speed for 20 s, decelerates for 8/3 s, continues for 4.0 s at this speed, and then comes to a stop. (a) How long does the trip take? (b) How far has he traveled? (c) What is his average speed for the trip? (d) How long would it take to walk to the store and back if he walked at 1.5 m/s?

38. A ball accelerates at 0.5 m/s² while moving down an inclined plane 9.0 m long. When it reaches the bottom, the ball rolls up another plane, where, after moving 15 m, it comes to rest. (a) What is the speed of the ball at the bottom of the first plane? (b) How long does it take to roll down the first plane? (c) What is the acceleration along the second plane? (d) What is the ball's speed 8.0 m along the second plane?

39. A car moving at a constant speed of 30.0 m/s suddenly stalls at the bottom of a hill. The car undergoes a constant acceleration of -2.00 m/s² (opposite its motion) while ascending the hill. (a) Write equations for the position and the velocity as functions of time, taking $x = 0$ at the bottom of the hill, where $v_0 = 30.0$ m/s. (b) Determine the maximum distance traveled by the car after it stalls.

40. An electron has an initial speed of 3.0×10^5 m/s. If it undergoes an acceleration of 8.0×10^{14} m/s², (a) how long will it take to reach a speed of 5.4×10^5 m/s and (b) how far has it traveled in this time?

41. Speedy Sue driving at 30 m/s enters a one-lane tunnel. She then observes a slow-moving van 155 m ahead traveling at 5.0 m/s. Sue applies her brakes but can decelerate only at 2.0 m/s² because the road is wet. Will there be a collision? If yes, determine how far into the tunnel and at what time the collision occurs. If no, determine the distance of closest approach between Sue's car and the van.

42. An indestructible bullet 2.00 cm long is fired straight through a board that is 10.0 cm thick. The bullet strikes the board with a speed of 420 m/s and emerges with a speed of 280 m/s. (a) What is the average acceleration of the bullet through the board? (b) What is the total time that the bullet is in contact with the board? (c) What thickness of boards

(calculated to 0.1 cm) would it take to stop the bullet?

43. Until recently, the world's land speed record was held by Colonel John P. Stapp, USAF. On March 19, 1954, he rode a rocket-propelled sled that moved down the track at 632 mi/h. He and the sled were safely brought to rest in 1.4 s. Determine (a) the negative acceleration he experienced and (b) the distance he traveled during this negative acceleration.

44. A hockey player is standing on his skates on a frozen pond when an opposing player skates by with the puck, moving with a uniform speed of 12.0 m/s. After 3.00 s, the first player makes up his mind to chase his opponent. If the first player accelerates uniformly at 4.00 m/s^2, (a) how long does it take him to catch the opponent? (b) How far has the first player traveled in this time? (Assume the opponent moves at constant speed.)

Section 2.5 Freely Falling Bodies

45. A woman is reported to have fallen 144 ft from the 17th floor of a building, landing on a metal ventilator box, which she crushed to a depth of 18.0 in. She suffered only minor injuries. Neglecting air resistance, calculate (a) the speed of the woman just before she collided with the ventilator, (b) her average acceleration while in contact with the box, and (c) the time it took to crush the box.

46. A ball is thrown directly downward with an initial speed of 8.00 m/s from a height of 30.0 m. When does the ball strike the ground?

47. A student throws a set of keys vertically upward to her sorority sister in a window 4.00 m above. The keys are caught 1.50 s later by the sister's outstretched hand. (a) With what initial velocity were the keys thrown? (b) What was the velocity of the keys just before they were caught?

48. A hot air balloon is traveling vertically upward at a constant speed of 5.00 m/s. When it is 21.0 m above the ground, a package is released from the balloon. (a) After it is released, for how long is the package in the air? (b) What is its velocity just before impact with the ground? (c) Repeat (a) and (b) for the case of the balloon descending at 5.00 m/s.

49. A ball is thrown vertically upward from the ground with an initial speed of 15.0 m/s. (a) How long does it take the ball to reach its maximum altitude? (b) What is its maximum altitude? (c) Determine the velocity and acceleration of the ball at $t = 2.00$ s.

50. A ball thrown vertically upward is caught by the thrower after 20.0 s. Find (a) the initial velocity of the ball and (b) the maximum height it reaches.

51. A baseball is hit such that it travels straight upward after being struck by the bat. A fan observes that it requires 3.00 s for the ball to reach its maximum height. Find (a) its initial velocity and (b) its maximum height. Ignore the effects of air resistance.

52. An astronaut standing on the Moon drops a hammer, letting it fall 1.00 m to the surface. The lunar gravity produces a constant acceleration of magnitude 1.62 m/s^2. Upon returning to Earth, the astronaut again drops the hammer, letting it fall to the ground from a height of 1.00 m with an acceleration of 9.80 m/s^2. Compare the times of fall in the two situations.

53. The height of a helicopter above the ground is given by $h = 3.00t^3$, where h is in meters and t is in seconds. After 2.00 s, the helicopter releases a small mailbag. How long after its release does the mailbag reach the ground?

54. A stone falls from rest from the top of a high cliff. A second stone is thrown downward from the same height 2.00 s later with an initial speed of 30.0 m/s. If both stones hit the ground simultaneously, how high is the cliff?

55. A daring stunt woman sitting on a tree limb wishes to drop vertically onto a horse galloping under the tree. The speed of the horse is 10.0 m/s, and the distance from the limb to the saddle is 3.00 m. (a) What must be the horizontal distance between the saddle and limb when the woman makes her move? (b) How long is she in the air?

*Section 2.6 Kinematic Equations Derived from Calculus

56. The speed of a bullet shot from a gun is given by $v = (-5.0 \times 10^7)t^2 + (3.0 \times 10^5)t$, where v is in meters/second and t is in seconds. The acceleration of the bullet just as it leaves the barrel is zero. (a) Determine the acceleration and position of the bullet as a function of time when the bullet is in the barrel. (b) Determine the length of time the bullet is accelerated while in the barrel. (c) Find the speed at which the bullet leaves the barrel. (d) What is the length of the barrel?

57. The position of a softball tossed vertically upward is described by the equation $y = 7.00t - 4.90t^2$, where y is in meters and t in seconds. Find (a) the ball's initial speed v_0 at $t_0 = 0$, (b) its velocity at $t = 1.26$ s, and (c) its acceleration.

58. A rocket sled for testing equipment under large accelerations starts at rest and accelerates according to the expression $a = (3 \text{ m/s}^3)t + 5.00 \text{ m/s}^2$. How far does the sled move in the time interval $t = 0$ to $t = 2.00$ s?

59. Automotive engineers refer to the time rate of change of acceleration as the "jerk." If an object moves in one dimension such that its jerk J is constant, (a) determine expressions for its acceleration $a(t)$, speed $v(t)$, and position $x(t)$, given that its initial acceleration, speed, and position are a_0, v_0,

and x_0, respectively. (b) Show that $a^2 = a_0{}^2 + 2J(v - v_0)$.

60. The acceleration of a marble in a certain fluid is proportional to the speed of the marble squared, and is given (in SI units) by $a = -3.00v^2$ for $v > 0$. If the marble enters this fluid with a speed of 1.50 m/s, how long will it take before the marble's speed is reduced to half of its initial value?

ADDITIONAL PROBLEMS

61. Another scheme to catch the roadrunner has failed, and a safe falls from rest from the top of a 25-m-high cliff toward Wiley Coyote, who is standing at the base. Wiley first notices the safe after it has fallen 15 m. How long does he have to get out of the way?

62. A motorist is traveling at 18.0 m/s when he sees a deer in the road 38.0 m ahead. (a) If the maximum negative acceleration of the vehicle is -4.50 m/s², what is the maximum reaction time Δt of the motorist that will allow him to avoid hitting the deer? (b) If his reaction time is 0.300 s, how fast will he be traveling when he reaches the deer?

63. An inquisitive physics student climbs a 50.0-m cliff that overhangs a calm pool of water. She throws two stones vertically downward 1.00 s apart and observes that they cause a single splash. The first stone has an initial velocity of 2.00 m/s. (a) At what time after release of the first stone do the two stones hit the water? (b) What initial velocity must the second stone have if they are to hit simultaneously? (c) What is the velocity of each stone at the instant they hit the water?

64. In a 100-m linear accelerator, an electron is accelerated to 1.0 percent of the speed of light in 40 m before it coasts 60 m to a target. (a) What is the electron's acceleration during the first 40 m? (b) How long does the total flight take?

65. A "superball" is dropped from a height of 2.00 m above the ground. On the first bounce, the ball reaches a height of 1.85 m, where it is caught. Find the velocity of the ball (a) just as it makes contact with the ground and (b) just as it leaves the ground on the bounce. (c) Neglecting the time the ball spends in contact with the ground, find the total time required for the ball to go from the dropping point to the point where it is caught.

66. A Cessna 150 aircraft has a lift-off speed of approximately 125 km/h. (a) What minimum constant acceleration does this require if the aircraft is to be airborne after a take-off run of 250 m? (b) What is the corresponding take-off time? (c) If the aircraft continues to accelerate at this rate, what speed will it reach 25.0 s after it begins to roll?

67. One runner covered the 100-m dash in 10.3 s. Another runner came in second at a time of 10.8 s. As-

suming that the runners traveled at their average speeds for the entire distance, determine the separation between them when the winner crossed the finish line.

68. A falling object requires 1.50 s to travel the last 30.0 m before hitting the ground. From what height above the ground did it fall?

69. A young woman named Kathy Kool buys a super-deluxe sports car that can accelerate at the rate of 4.90 m/s². She decides to test the car by dragging with another speedster, Stan Speedy. Both start from rest, but experienced Stan leaves 1.00 s before Kathy. If Stan moves with a constant acceleration of 3.50 m/s² and Kathy maintains an acceleration of 4.90 m/s², find (a) the time it takes Kathy to overtake Stan, (b) the distance she travels before she catches him, and (c) the velocities of both cars at the instant she overtakes him.

70. A hockey player takes a slap shot at a puck at rest on the ice. The puck glides over the ice for 10.0 ft without friction, at which point it runs over a concrete surface. The puck then accelerates opposite its motion at a uniform rate of -20.0 ft/s². If the velocity of the puck is 40.0 ft/s after traveling 100 ft from the point of impact, (a) what is the average acceleration imparted to the puck as it is struck by the hockey stick? (Assume that the time of contact is 0.0100 s.) (b) How far does the puck travel before stopping? (c) What is the total time the puck is in motion, neglecting contact time?

71. Two cars are traveling along a straight line in the same direction, the lead car at 25 m/s and the other at 30 m/s. At the moment the cars are 40 m apart, the lead driver applies the brakes so that her car accelerates at -2.0 m/s². (a) How long does it take for the lead car to stop? (b) Assuming that the chasing car brakes at the same time as the lead car, what must be the chasing car's minimum negative acceleration so as not to hit the lead car? (c) How long does it take for the chasing car to stop?

72. A motorist drives along a straight road at a constant speed of 15.0 m/s. Just as she passes a parked motorcycle police officer, the officer starts to accelerate at 2.00 m/s² to overtake her. Assuming the officer maintains this acceleration, (a) determine the time it takes the police officer to reach the motorist. Find (b) the speed and (c) the total displacement of the officer as he overtakes the motorist.

73. In 1987, Art Boileau won the Los Angeles Marathon, 26 mi and 385 yd, in 2 h, 13 min, and 9 s. (a) Find his average speed in meters per second and in miles per hour. (b) At the 21-mi marker, Boileau had a 2.50-min lead on the second-place winner, who later crossed the finish line 30.0 s after Boileau. Assume that Boileau maintained his constant average speed and that both runners were running at the same speed when Boileau passed the 21-mi marker. Find

the average acceleration (in meters per second squared) that the second-place runner had during the remainder of the race after Boileau passed the 21-mi marker.

74. A rock is dropped from rest into a well. (a) If the sound of the splash is heard 2.40 s later, how far below the top of the well is the water surface? The speed of sound in air (for the air temperature that day) was 336 m/s. (b) If the travel time for the sound is neglected, what percentage error is introduced when the depth of the well is calculated?

75. A train travels along a straight track between Stations 1 and 2 as shown in Figure P2.75. The engineer is instructed to start from rest at Station 1, accelerate uniformly between A and B, coast with a uniform speed between B and C, and then decelerate uniformly between C and D (at the same rate as between A and B) until the train stops at Station 2. If the distances AB, BC, and CD are all equal, and if it takes 5.00 min to travel between the two stations, determine how much of this 5.00-min period the train spends between points (i) A and B, (ii) B and C, and (iii) C and D.

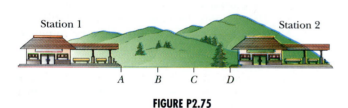

FIGURE P2.75

 76. A rocket is fired vertically upward with an initial velocity of 80.0 m/s. It accelerates upward at 4.00 m/s² until it reaches an altitude of 1000 m. At that point, its engines fail and the rocket goes into free-fall with acceleration −9.80 m/s². (a) How long is the rocket in motion? (b) What is its maximum altitude? (c) What is its velocity just before it collides with the Earth? (*Hint:* Consider the motion while the engine is operating separate from the free-fall motion.)

77. In a 100-m race, Maggie and Judy cross the finish line in a dead heat, both taking 10.2 s. Accelerating uniformly, Maggie takes 2.00 s and Judy 3.00 s to attain maximum speed, which they maintain for the rest of the race. (a) What is the acceleration of each sprinter? (b) What are their respective maximum speeds? (c) Which sprinter is ahead at the 6.00-s mark, and by how much?

78. A train travels in time in the following manner. In the first 60 min, it travels with a speed v; in the next 30 min it has a speed $3v$; in the next 90 min it travels with a speed $v/2$; in the final 120 min, it travels with a speed $v/3$. (a) Plot the speed-time graph for this trip.

(b) How far does the train travel? (c) What is the average speed of the train over the entire trip?

79. A commuter train can minimize the time t between two stations by accelerating at a rate $a_1 = 0.100$ m/s² for a time t_1 and then undergoing a negative acceleration $a_2 = -0.500$ m/s² as the engineer uses the brakes for a time t_2. Since the stations are only 1.00 km apart, the train never reaches its maximum speed. Find the minimum time of travel t and the time t_1.

 80. In order to protect his food from hungry bears, a boy scout raises his food pack, mass m, with a rope that is thrown over a tree limb of height h above his hands. He walks away from the vertical rope with constant speed v_s while holding the free end of the rope in his hands (Fig. P2.80). (a) Show that the speed v_p of the food pack is $x(x^2 + h^2)^{-1/2}v_s$, where x is the distance he has walked away from the vertical rope. (b) Show that the acceleration a_p of the food pack is $h^2(x^2 + h^2)^{-3/2}v_s^2$. (c) What values do the acceleration and speed have shortly after he leaves the vertical rope? (d) What values do the speed and acceleration approach as the distance x continues to increase?

80A. In Problem 80, let the height h equal 6.00 m and the speed v_s equal 2.00 m/s. Assume that the food pack starts from rest. (a) Tabulate and graph the speed-time graph. (b) Tabulate and graph the acceleration-time graph. (Let the range of time be from 0 s to 6.00 s and the time intervals be 0.50 s.)

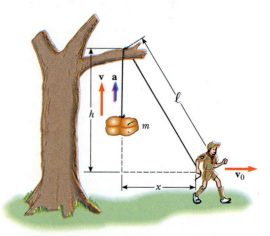

FIGURE P2.80

 81. Two objects A and B are connected by a rigid rod that has a length L. The objects slide along perpendicular guide rails, as shown in Figure P2.81. If A slides to the left with a constant speed v, find the velocity of B when $\alpha = 60°$.

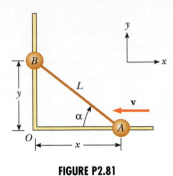

FIGURE P2.81

SPREADSHEET PROBLEMS

S1. Use Spreadsheet 2.1 to plot position and velocity as functions of time for one object traveling at constant velocity and another traveling with constant acceleration. Choose a variety of velocity and acceleration values, and view the graphs. Be sure to investigate zero and negative values. *Note:* There are two graphs associated with Spreadsheet 2.1, one for velocity versus time and one for position versus time.

S2. Spreadsheet 2.2 models the motion of a two-person 1.5-km race. For simplicity, we will assume that the runners can change speed instantaneously. Each runner can set his or her strategy. For instance, Racer 1 decides to lead initially, so his original speed, V11 = 7 m/s, is a little greater than Racer 2's, V21 = 6.5 m/s. Racer 2 realizes that she has fallen behind more than she wants, so she starts her "kick," at X2 = 250 m before the finish line. She is able to increase her speed, V22, to 10 m/s for 14 s. She then slows down to V23 = 8 m/s. During this time Racer 1 sees that his opponent is gaining, so at X1 = 150 m from the finish, he starts to "kick" at V12 = 11 m/s for 10 s. He then slows down to V13 = 6 m/s. (a) Who wins the race? (b) Devise a strategy so that Racer 1 beats the world's record of 3 min, 15 s, while Racer 2 finishes 0.2 s behind Racer 1. Your input data must be reasonable; neither runner is Superman.

S3. Spreadsheet 2.3 models the sport of drag racing. Two dragsters can have different accelerations A1 and A2, as well as different maximum speeds V1 and V2. Both cars start from rest at the same starting position. However, a time delay t' in starting times may be introduced if the cars are quite different. Enter the following data:

	Acceleration (ft/s^2)	Maximum Speed (ft/s)	Delay Time t' (s)
Car 1	5	300	
Car 2	6	250	0.5

(a) Which car wins a 1/4-mile race? (b) If the acceleration of Car 1 is increased to 5.2 ft/s^2, which car now wins?

S4. Modify Spreadsheet 2.1 to solve this problem: The State Police have set up a "speed trap" on the interstate highway. From a police car hidden behind a billboard, an officer with a radar gun measures a motorist's speed to be 35.0 m/s. Three seconds later, she alerts her partner, who is in another police car 100 m down the road. The second police car starts from rest accelerating at 2.00 m/s^2 in pursuit of the speeder 2.00 s after receiving the alert. (a) How much time elapses before the speeder is overtaken? (b) What is the police car's speed when it overtakes the speeder? (c) What is the distance from where the second police car was sitting to the point where the speeder is overtaken?

S5. As a demonstration, astronauts on a distant planet toss a rock into the air. With the aid of a high-speed camera, they record the height of the rock as a function of time as given in the table below: (a) Find the speed over each time interval by using a difference equation to approximate the differential. (b) Find the acceleration during each time interval by the same type of approximation.

Height of a Rock Versus Time for Problem S5										
Time (s)	0.00	0.25	0.50	0.75	1.00	1.25	1.50	1.75	2.00	2.25	2.50
Height (m)	5.00	5.75	6.40	6.94	7.38	7.72	7.96	8.10	8.13	8.07	7.90
Time (s)	2.75	3.00	3.25	3.50	3.75	4.00	4.25	4.50	4.75	5.00	
Height (m)	7.62	7.25	6.77	6.20	5.52	4.73	3.85	2.86	1.77	0.58	

Vectors

This interesting set of directional arrows was photographed in Estes Park, Colorado. A particular location relative to the position of the park is specified by the direction of the arrow and the distance to that location. This is one type of vector quantity. We examine it and other vector quantities in this chapter. *(Ray Serway)*

Physical quantities that have both numerical and directional properties are represented by vectors. Some examples of vector quantities are force, displacement, velocity, and acceleration. This chapter is primarily concerned with vector algebra and with some general properties of vector quantities. The addition and subtraction of vector quantities are discussed, together with some common applications to physical situations. Discussion of the products of vector quantities shall be delayed until these operations are needed.[1]

Vector quantities are used throughout this text, and it is therefore imperative that you master both their graphical and their algebraic properties.

---

[1] The dot, or scalar, product is discussed in Section 7.2, and the cross, or vector, product is introduced in Section 11.2.

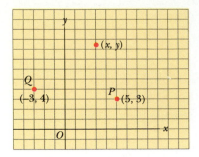

FIGURE 3.1 Designation of points in a cartesian coordinate system. Every point is labeled with coordinates (x, y).

3.1 COORDINATE SYSTEMS AND FRAMES OF REFERENCE

Many aspects of physics deal in some form or other with locations in space. For example, the mathematical description of the motion of an object requires a method for describing the position of the object at various times. Thus, it is fitting that we first discuss how to describe the position of a point in space. This description is accomplished by means of coordinates.

A point on a line can be described with one coordinate. A point in a plane is located with two coordinates, and three coordinates are required to locate a point in space. In general, a coordinate system used to specify locations in space consists of

- A fixed reference point O, called the origin
- A set of specified axes with appropriate scales and labels on the axes
- Instructions that tell us how to label a point in space relative to the origin and axes.

One convenient coordinate system that we shall use frequently is the *cartesian coordinate system*, sometimes called the *rectangular coordinate system*. Such a system in two dimensions is illustrated in Figure 3.1. An arbitrary point in this system is labeled with the coordinates (x, y). The positive x direction is arbitrarily defined to be to the right of the origin, and the positive y direction is arbitrarily defined to be upward from the origin. The negative x direction is to the left of the origin, and the negative y direction is downward from the origin. For example, the point P, which has coordinates $(5, 3)$, may be reached by first going 5 units to the right of the origin and then going 3 units above the origin. Similarly, the point Q has coordinates $(-3, 4)$, corresponding to going 3 units to the left of the origin and 4 units above the origin.

Sometimes it is more convenient to represent a point in a plane by its *plane polar coordinates*, (r, θ), as in Figure 3.2a. In this coordinate system, r is the distance from the origin to the point having cartesian coordinates (x, y) and θ is the angle between r and a fixed axis. This fixed axis is usually the positive x axis and θ is usually measured counterclockwise from it. From the right triangle in Figure 3.2b, we find $\sin \theta = y/r$ and $\cos \theta = x/r$. (A review of trigonometric functions is given in Appendix B.4.) Therefore, starting with plane polar coordinates, the cartesian coordinates can be obtained through the equations

$$x = r \cos \theta \tag{3.1}$$

$$y = r \sin \theta \tag{3.2}$$

Furthermore, the definitions of trigonometry tell us that

$$\tan \theta = \frac{y}{x} \tag{3.3}$$

and

$$r = \sqrt{x^2 + y^2} \tag{3.4}$$

These four expressions relating the coordinates (x, y) to the coordinates (r, θ) apply only when θ is defined as in Figure 3.2a — in other words, where positive θ is an angle measured *counterclockwise* from the positive x axis. If the reference axis for the polar angle θ is chosen to be other than the positive x axis or if the sense of increasing θ is chosen differently, then the corresponding expressions relating the two sets of coordinates will change. Scientific calculators provide conversions between cartesian and polar coordinates based on these standard conventions.

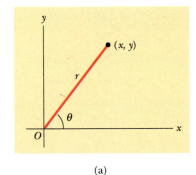

(a)

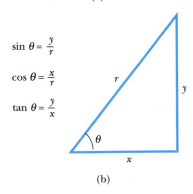

$$\sin \theta = \frac{y}{r}$$

$$\cos \theta = \frac{x}{r}$$

$$\tan \theta = \frac{y}{x}$$

(b)

FIGURE 3.2 (a) The plane polar coordinates of a point are represented by the distance r and the angle θ, where θ is measured counterclockwise from the positive x axis. (b) The right triangle used to relate (x, y) to (r, θ).

EXAMPLE 3.1   Polar Coordinates

The cartesian coordinates of a point in the xy plane are $(x, y) = (-3.50, -2.50)$ m, as in Figure 3.3. Find the polar coordinates of this point.

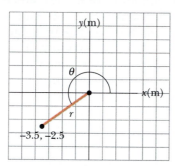

FIGURE 3.3   (Example 3.1).

Solution

$$r = \sqrt{x^2 + y^2} = \sqrt{(-3.50 \text{ m})^2 + (-2.50 \text{ m})^2} = \boxed{4.30 \text{ m}}$$

$$\tan \theta = \frac{y}{x} = \frac{-2.50 \text{ m}}{-3.50 \text{ m}} = 0.714$$

$$\theta = \boxed{216°}$$

Note that you must use the signs of x and y to find that the point lies in the third quadrant of the coordinate system. That is, $\theta = 216°$ and not $36°$.

3.2   VECTOR AND SCALAR QUANTITIES

The physical quantities we shall encounter in this text may be treated as either scalar quantities or vector quantities. A scalar quantity is one that is completely specified by a number with appropriate units. That is,

> A scalar quantity has only magnitude and no direction.

On the other hand, a vector quantity is a physical quantity that is completely specified by a number with appropriate units plus a direction. That is,

> A vector quantity has both magnitude and direction.

(a)

(b)

(a) The number of apples in the basket is one example of a scalar quantity. Can you think of other examples? *(Superstock)* (b) Jennifer pointing in the right direction. A vector is a physical quantity that must be specified by both magnitude and direction. *(Photo by Ray Serway)*

The number of apples in a basket is an example of a scalar quantity. If you are told there are 38 apples in the basket, this completes the required information; no specification of direction is required. Other examples of scalar quantities are temperature, volume, mass, and time intervals. The rules of ordinary arithmetic are used to manipulate scalar quantities.

Force is one example of a vector quantity. To describe completely the force on an object, we must specify both the direction of the applied force, a number to indicate the magnitude of the force, and often the line or point of application of the force. Velocity is another example of a vector quantity. If we wish to describe the velocity of a moving car, we must specify both its speed (say, 25 m/s) and the direction in which the car is moving (say, southwest). The rules of ordinary arithmetic cannot be used to manipulate vector quantities. Instead, we combine vectors according to special rules that are discussed in Sections 3.3 and 3.4.

A third example of a vector quantity is the displacement of a particle. Suppose the particle moves from some point O to the point P along a straight path, as in Figure 3.4. We represent this displacement by drawing an arrow from O to P, where the tip of the arrow represents the direction of the displacement and the length of the arrow represents the magnitude of the displacement. If the particle travels along some other path from O to P, such as the broken line in Figure 3.4, its displacement is still the arrow drawn from O to P. If the particle travels along any indirect path from O and P, its displacement is defined as being equivalent to the displacement for the direct path from O to P.

If a particle moves along the x axis from position x_i to position x_f, as in Figure 3.5, its displacement is given by $\Delta x = x_f - x_i$. (The indices i and f refer to the initial and final values.) We use the Greek letter delta (Δ) to denote the *change* in a quantity.

It is important to remember that the distance traveled by a particle is distinctly different from its displacement. The distance traveled (a scalar quantity) is the length of the path, which can be much greater than the magnitude of the displacement (see Fig. 3.4). The magnitude of any displacement is the shortest distance between the end points of the displacement vector.

In this text, we use boldface letters, such as **A**, to represent a vector quantity. Another common method for vector notation that you should be aware of is to use an arrow over the letter $\vec{A}$. The magnitude of the vector **A** is written A or, alternatively, $|\mathbf{A}|$. The magnitude of a vector has physical units, such as meters for displacement or meters per second for velocity.

3.3 SOME PROPERTIES OF VECTORS

Equality of Two Vectors

For many purposes, two vectors **A** and **B** may be defined to be equal if they have the same magnitude and point in the same direction. That is $\mathbf{A} = \mathbf{B}$, only if $A = B$ *and* they act along parallel directions. For example, all the vectors in Figure 3.6 are equal even though they have different starting points. This property allows us to translate a vector parallel to itself in a diagram without affecting the vector.

Addition

When two or more vectors are added together, *all* of them must have the same units. For example, it would be meaningless to add a velocity vector to a displacement vector since they are different physical quantities. Scalars also obey the same

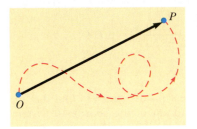

FIGURE 3.4 As a particle moves from O to P along an arbitrary path represented by the broken line, its displacement is a vector quantity shown by the arrow drawn from O to P.

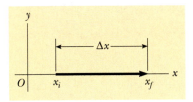

FIGURE 3.5 A particle moving along the x axis from x_i to x_f undergoes a displacement $\Delta x = x_f - x_i$.

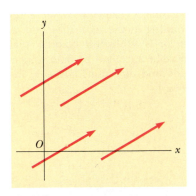

FIGURE 3.6 These four vectors are all equal since they have equal lengths and point in the same direction.

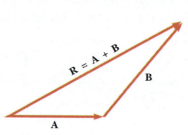

FIGURE 3.7 When vector **A** is added to vector **B**, the resultant **R** is the vector that runs from the tail of **A** to the tip of **B**.

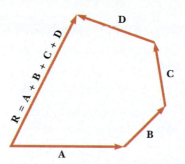

FIGURE 3.8 Geometric construction for summing four vectors. The resultant vector **R** is by definition the one that completes the polygon.

rule. For example, it would be meaningless to add time intervals and temperatures.

The rules for vector sums are conveniently described by geometric methods. To add vector **B** to vector **A**, first draw vector **A**, with its magnitude represented by a convenient scale, on graph paper and then draw vector **B** to the same scale with its tail starting from the tip of **A**, as in Figure 3.7. The *resultant* vector **R** = **A** + **B** is the vector drawn from the tail of **A** to the tip of **B**. This is known as the *triangle method of addition.*

Geometric constructions can also be used to add more than two vectors. This is shown in Figure 3.8 for the case of four vectors. The resultant vector sum **R** = **A** + **B** + **C** + **D** is *the vector that completes the polygon.* In other words, **R** is *the vector drawn from the tail of the first vector to the tip of the last vector.* Again, the order of the summation is unimportant.

An alternative graphical procedure for adding two vectors, known as the *parallelogram rule of addition,* is shown in Figure 3.9a. In this construction, the tails of the two vectors **A** and **B** are together and the resultant vector **R** is the diagonal of a parallelogram formed with **A** and **B** as its sides.

When two vectors are added, the sum is independent of the order of the addition. This can be seen from the geometric construction in Figure 3.9b and is known as the **commutative law of addition:**

$$\mathbf{A} + \mathbf{B} = \mathbf{B} + \mathbf{A} \qquad (3.5)$$

Commutative law

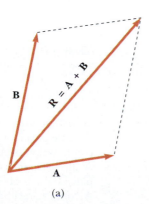

(a)

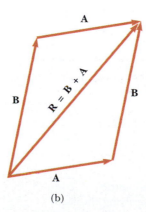

(b)

FIGURE 3.9 (a) In this construction, the resultant **R** is the diagonal of a parallelogram with sides **A** and **B**. (b) This construction shows that **A** + **B** = **B** + **A**. These geometric constructions verify the commutative law of addition for vectors.

If three or more vectors are added, their sum is independent of the way in which the individual vectors are grouped together. A geometric proof of this rule for three vectors is given in Figure 3.10. This is called the **associative law of addition:**

Associative law

$$A + (B + C) = (A + B) + C \tag{3.6}$$

Thus we conclude that *a vector quantity has both magnitude and direction and also obeys the laws of vector addition* as described in Figures 3.7 to 3.10.

Negative of a Vector

The negative of the vector **A** is defined as the vector that when added to **A** gives zero for the vector sum. That is, $A + (-A) = 0$. The vectors **A** and $-A$ have the same magnitude but point in opposite directions.

Subtraction of Vectors

The operation of vector subtraction makes use of the definition of the negative of a vector. We define the operation $A - B$ as vector $-B$ added to vector **A**:

$$A - B = A + (-B) \tag{3.7}$$

The geometric construction for subtracting two vectors is shown in Figure 3.11.

Multiplication of a Vector by a Scalar

If vector **A** is multiplied by a positive scalar quantity m, the product mA is a vector that has the same direction as **A** and magnitude mA. If m is a negative scalar quantity, the vector mA is directed opposite **A**. For example, the vector $5A$ is five times as long as **A** and points in the same direction as **A**; the vector $-\frac{1}{3}A$ is one third the length of **A** and points in the direction opposite **A**.

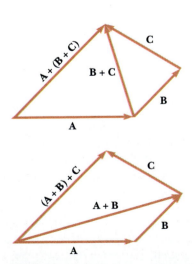

FIGURE 3.10 Geometric constructions for verifying the associative law of addition.

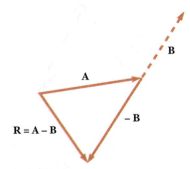

FIGURE 3.11 This construction shows how to subtract vector **B** from vector **A**. The vector $-B$ is equal in magnitude to vector **B** and points in the opposite direction. To subtract **B** and **A**, apply the rule of vector addition to the combination of **A** and $-B$: draw **A** along some convenient axis, place the tail of $-B$ at the tip of **A**, and the resultant is the difference $A - B$.

CONCEPTUAL EXAMPLE 3.2

If **B** is added to **A**, under what condition does the resultant vector **A** + **B** have a magnitude equal to $A + B$? Under what conditions is the resultant vector equal to zero?

Reasoning   The resultant has a magnitude $A + B$ when **A** is oriented in the same direction as **B**. The resultant vector **A** + **B** = 0 when **A** is oriented in the direction opposite to **B**, and when $A = B$.

EXAMPLE 3.3   A Vacation Trip

A car travels 20.0 km due north and then 35.0 km in a direction 60.0° west of north, as in Figure 3.12. Find the magnitude and direction of the car's resultant displacement.

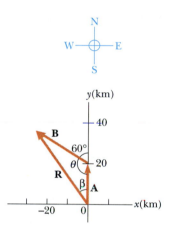

FIGURE 3.12   (Example 3.3) Graphical method for finding the resultant displacement vector **R** = **A** + **B**.

Reasoning   The problem can be solved geometrically using graph paper and a protractor, as shown in Figure 3.12.

The resultant displacement **R** is the sum of the two individual displacements **A** and **B**. The following is an algebraic solution.

Solution   The magnitude of **R** can be obtained using the law of cosines from trigonometry as applied to the obtuse triangle (Appendix B.4). Since $\theta = 180° - 60° = 120°$ and $R^2 = A^2 + B^2 - 2AB \cos \theta$, we find that

$$R = \sqrt{A^2 + B^2 - 2AB \cos \theta}$$
$$= \sqrt{(20.0)^2 + (35.0)^2 - 2(20.0)(35.0) \cos 120°} \ \text{km}$$

$$= \boxed{48.2 \ \text{km}}$$

The direction of **R** measured from the northerly direction can be obtained from the law of sines from trigonometry:

$$\frac{\sin \beta}{B} = \frac{\sin \theta}{R}$$

$$\sin \beta = \frac{B}{R} \sin \theta = \frac{35.0 \ \text{km}}{48.2 \ \text{km}} \sin 120° = 0.629$$

$$\beta = \boxed{38.9°}$$

Therefore, the resultant displacement of the car is 48.2 km in a direction 38.9° west of north.

3.4   COMPONENTS OF A VECTOR AND UNIT VECTORS

The geometric method of adding vectors is not the recommended procedure in situations where high precision is required or in three-dimensional problems. In this section, we describe a method of adding vectors that makes use of the *projections* of a vector along the axes of a rectangular coordinate system. These projections are called the **components** of the vector. Any vector can be completely described by its components.

Consider a vector **A** lying in the *xy* plane and making an arbitrary angle θ with the positive *x* axis, as in Figure 3.13. This vector can be expressed as the sum of two other vectors $\mathbf{A}_x$ and $\mathbf{A}_y$. From Figure 3.13, we see that the three vectors form a right triangle and $\mathbf{A} = \mathbf{A}_x + \mathbf{A}_y$. We shall often refer to the components of a vector **A**, written as A_x and A_y (without the boldface notation). The component A_x represents the projection of **A** along the *x* axis, and A_y represents the projection of **A**

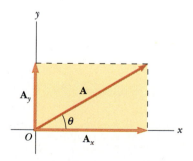

FIGURE 3.13   Any vector **A** lying in the *xy* plane can be represented by a vector lying along the *x* axis, $\mathbf{A}_x$, and a vector lying along the *y* axis, $\mathbf{A}_y$, where $\mathbf{A} = \mathbf{A}_x + \mathbf{A}_y$.

along the y axis. These components can be positive or negative. The component A_x is positive if $\mathbf{A}_x$ points along the positive x axis and is negative if $\mathbf{A}_x$ points along the negative x axis. The same is true for the component A_y.

From Figure 3.13 and the definition of sine and cosine, we see that cos $\theta = A_x/A$ and sin $\theta = A_y/A$. Hence, the components of **A** are

$$A_x = A \cos \theta \tag{3.8}$$

$$A_y = A \sin \theta \tag{3.9}$$

These components form two sides of a right triangle the hypotenuse of which has a magnitude A. Thus, it follows that the magnitude of **A** and its direction are related to its components through the expressions

$$A = \sqrt{A_x^2 + A_y^2} \tag{3.10}$$

and

$$\tan \theta = \frac{A_y}{A_x} \tag{3.11}$$

To solve for θ, we can write $\theta = \tan^{-1}(A_y/A_x)$, which is read "$\theta$ equals the angle whose tangent is the ratio A_y/A_x." *Note that the signs of the components A_x and A_y depend on the angle θ.* For example, if $\theta = 120°$, A_x is negative and A_y is positive. If $\theta = 225°$, both A_x and A_y are negative. Figure 3.14 summarizes the signs of the components when **A** lies in the various quadrants. When solving problems, you can specify a vector **A** with *either* the notation A_x, A_y or its magnitude and direction, A, θ.

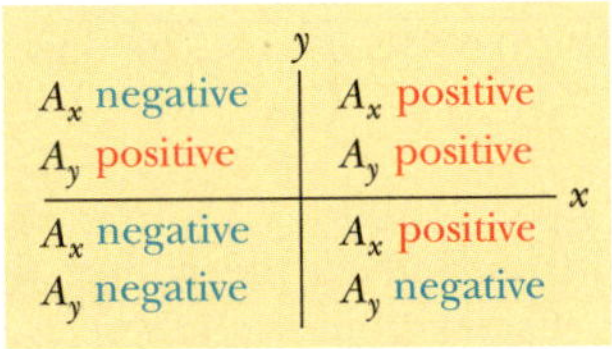

Suppose you are working a physics problem that requires resolving vectors into its components. If you choose reference axes or an angle other than the axes and angle shown in Figure 3.13, the components must be modified accordingly. In many applications, for example, it is convenient to express the components in a coordinate system having axes that are not horizontal and vertical but still perpendicular to each other. Suppose a vector **B** makes an angle θ' with the x' axis defined in Figure 3.15. The components of **B** along these axes are $B'_x = B \cos \theta'$ and $B'_y = B \sin \theta'$, as in Equations 3.8 and 3.9. The magnitude and direction of **B** are obtained from expressions equivalent to Equations 3.10 and 3.11. Thus, we can express the components of a vector in *any* coordinate system that is convenient for a particular situation.

The components of a vector are different when viewed from different coordinate systems. Furthermore, components can change with respect to a fixed coordinate system if the vector changes in magnitude, orientation, or both.

Unit Vectors

Vector quantities are often expressed in terms of unit vectors. A **unit vector** is a dimensionless vector having a magnitude of exactly one. Unit vectors are used to specify a given direction and have no other physical significance. They are used solely as a convenience in describing a direction in space. We shall use the symbols **i**, **j**, and **k** to represent unit vectors pointing in the positive x, y, and z directions, respectively. The unit vectors **i**, **j**, and **k** form a set of mutually perpendicular vectors in a right-handed coordinate system as shown in Figure 3.16a. The magnitude of each unit vector equals unity; that is, $|\mathbf{i}| = |\mathbf{j}| = |\mathbf{k}| = 1$.

Consider a vector **A** lying in the xy plane, as in Figure 3.16b. The product of the component A_x and the unit vector **i** is the vector $A_x\mathbf{i}$, which is parallel to the x axis

---

Components of the vector A

Magnitude of A

Direction of A

A_x negative A_y positive	A_x positive A_y positive
A_x negative A_y negative	A_x positive A_y negative

FIGURE 3.14   The signs of the components of a vector **A** depend on the quadrant in which the vector is located.

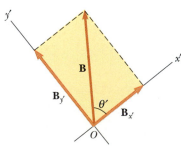

FIGURE 3.15   The component vectors of **B** in a coordinate system that is tilted.

and has magnitude A_x. (The vector $A_x\mathbf{i}$ is an alternative and more common way of representing A_x.) Likewise, $A_y\mathbf{j}$ is a vector of magnitude A_y parallel to the y axis. (Again, $A_y\mathbf{j}$ is an alternative way of representing A_y.) Thus, the unit-vector notation for the vector $\mathbf{A}$ is written

$$\mathbf{A} = A_x\mathbf{i} + A_y\mathbf{j} \tag{3.12}$$

For example, consider a point lying in the xy plane having cartesian coordinates (x, y) as in Figure 3.17. The point can be specified by the **position vector r**, which in unit-vector form is given by

$$\mathbf{r} = x\mathbf{i} + y\mathbf{j} \tag{3.13}$$

That is, the components of $\mathbf{r}$ are the coordinates x and y.

Now let us see how components are used to add vectors when the geometric method described in the preceding section is not appropriate. Suppose we wish to add vector $\mathbf{B}$ to vector $\mathbf{A}$, where $\mathbf{B}$ has components B_x and B_y. The procedure for performing this sum via the component method is to simply add the x and y components separately. The resultant vector $\mathbf{R} = \mathbf{A} + \mathbf{B}$ is therefore

$$\mathbf{R} = (A_x + B_x)\mathbf{i} + (A_y + B_y)\mathbf{j} \tag{3.14}$$

Since $\mathbf{R} = R_x\mathbf{i} + R_y\mathbf{j}$, we see that the components of the resultant vector are

$$R_x = A_x + B_x$$
$$R_y = A_y + B_y \tag{3.15}$$

The magnitude of $\mathbf{R}$ and the angle it makes with the x axis can then be obtained from its components using the relationships

$$R = \sqrt{R_x^2 + R_y^2} = \sqrt{(A_x + B_x)^2 + (A_y + B_y)^2} \tag{3.16}$$

and

$$\tan\theta = \frac{R_y}{R_x} = \frac{A_y + B_y}{A_x + B_x} \tag{3.17}$$

The procedure just described for adding two vectors $\mathbf{A}$ and $\mathbf{B}$ using the component method can be checked using a geometric construction, as in Figure 3.18.

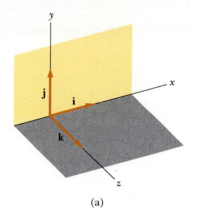

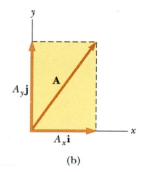

FIGURE 3.16 (a) The unit vectors $\mathbf{i}$, $\mathbf{j}$, and $\mathbf{k}$ are directed along the x, y, and z axes, respectively. (b) Vector $\mathbf{A} = A_x\mathbf{i} + A_y\mathbf{j}$ lying in the xy plane has components A_x and A_y.

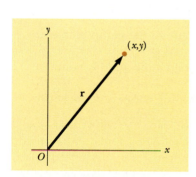

FIGURE 3.17 The point whose cartesian coordinates are (x, y) can be represented by the position vector $\mathbf{r} = x\mathbf{i} + y\mathbf{j}$.

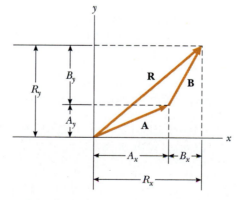

FIGURE 3.18 Geometric construction for the sum of two vectors showing the relationship between the components of the resultant $\mathbf{R}$ and the components of the individual vectors.

Again, you must take note of the *signs* of the components when using either the algebraic or the geometric method.

The extension of these methods to three-dimensional vectors is straight forward. If **A** and **B** both have x, y, and z components, we express them in the form

$$\mathbf{A} = A_x\mathbf{i} + A_y\mathbf{j} + A_z\mathbf{k} \tag{3.18}$$

$$\mathbf{B} = B_x\mathbf{i} + B_y\mathbf{j} + B_z\mathbf{k} \tag{3.19}$$

The sum of **A** and **B** is

$$\mathbf{R} = \mathbf{A} + \mathbf{B} = (A_x + B_x)\mathbf{i} + (A_y + B_y)\mathbf{j} + (A_z + B_z)\mathbf{k} \tag{3.20}$$

Thus, the resultant vector also has a z component $R_z = A_z + B_z$.

The component procedure just described can be used to sum up three or more vectors.

CONCEPTUAL EXAMPLE 3.4

If one component of a vector is not zero, can its magnitude be zero? Explain.

Reasoning   No. The magnitude of a vector **A** is equal to $\sqrt{A_x^2 + A_y^2 + A_z^2}$. Therefore, if any component is nonzero, **A** cannot be zero.

CONCEPTUAL EXAMPLE 3.5

If $\mathbf{A} + \mathbf{B} = 0$, what can you say about the components of the two vectors?

Reasoning $\mathbf{A} = -\mathbf{B}$, therefore the components of the two vectors must have opposite signs and equal magnitudes.

Problem-Solving Strategy
Adding Vectors

When two or more vectors are to be added, the following step-by-step procedure is recommended:

- Select a coordinate system.
- Draw a sketch of the vectors and label each one.
- Find the x and y components of all vectors.
- Find the resultant components (the algebraic sum of the components) in the x and y directions.
- Use the Pythagorean theorem to find the magnitude of the resultant vector.
- Use a suitable trigonometric function to find the angle the resultant vector makes with the x axis.

EXAMPLE 3.6   The Sum of Two Vectors

Find the sum of two vectors **A** and **B** lying in the xy plane and given by

$$\mathbf{A} = 2.0\mathbf{i} + 2.0\mathbf{j} \quad \text{and} \quad \mathbf{B} = 2.0\mathbf{i} - 4.0\mathbf{j}$$

Solution   Comparing the above expression for **A** with the general relation $\mathbf{A} = A_x\mathbf{i} + A_y\mathbf{j}$, we see that $A_x = 2.0$ and $A_y = 2.0$. Likewise, $B_x = 2.0$, and $B_y = -4.0$. Therefore, the resultant vector **R** is obtained by using Equation 3.20:

$$\mathbf{R} = \mathbf{A} + \mathbf{B} = (2.0 + 2.0)\mathbf{i} + (2.0 - 4.0)\mathbf{j} = 4.0\mathbf{i} - 2.0\mathbf{j}$$

or

$$R_x = 4.0 \qquad R_y = -2.0$$

The magnitude of **R** is given by Equation 3.16:

$$R = \sqrt{R_x^2 + R_y^2} = \sqrt{(4.0)^2 + (-2.0)^2} = \sqrt{20} = \boxed{4.5}$$

Exercise   Find the angle θ that **R** makes with the positive x axis.

Answer   330°.

EXAMPLE 3.7   The Resultant Displacement

A particle undergoes three consecutive displacements: $\mathbf{d_1} = (1.5\mathbf{i} + 3.0\mathbf{j} - 1.2\mathbf{k})$ cm, $\mathbf{d_2} = (2.3\mathbf{i} - 1.4\mathbf{j} - 3.6\mathbf{k})$ cm, and $\mathbf{d_3} = (-1.3\mathbf{i} + 1.5\mathbf{j})$ cm. Find the components of the resultant displacement and its magnitude.

Solution

$$\begin{aligned}\mathbf{R} &= \mathbf{d_1} + \mathbf{d_2} + \mathbf{d_3} \\ &= (1.5 + 2.3 - 1.3)\mathbf{i} + (3.0 - 1.4 + 1.5)\mathbf{j} \\ &\quad + (-1.2 - 3.6 + 0)\mathbf{k} \\ &= (2.5\mathbf{i} + 3.1\mathbf{j} - 4.8\mathbf{k}) \text{ cm}\end{aligned}$$

That is, the resultant displacement has components

$$R_x = \boxed{2.5 \text{ cm}} \quad R_y = \boxed{3.1 \text{ cm}} \quad \text{and} \quad R_z = \boxed{-4.8 \text{ cm}}$$

Its magnitude is

$$\begin{aligned}R &= \sqrt{R_x^2 + R_y^2 + R_z^2} \\ &= \sqrt{(2.5 \text{ cm})^2 + (3.1 \text{ cm})^2 + (-4.8 \text{ cm})^2} \\ &= \boxed{6.2 \text{ cm}}\end{aligned}$$

EXAMPLE 3.8   Taking a Hike

A hiker begins a trip by first walking 25.0 km southeast from her base camp. On the second day, she walks 40.0 km in a direction 60.0° north of east, at which point she discovers a forest ranger's tower. (a) Determine the components of the hiker's displacement for each day.

Solution   If we denote the displacement vectors on the first and second days by **A** and **B**, respectively, and use the camp as the origin of coordinates, we get the vectors shown in Figure 3.19. Displacement **A** has a magnitude of 25.0 km and is 45.0° southeast. From Equations 3.8 and 3.9, its components are

$$A_x = A\cos(-45.0°) = (25.0 \text{ km})(0.707) = \boxed{17.7 \text{ km}}$$

$$A_y = A\sin(-45.0°) = -(25.0 \text{ km})(0.707) = \boxed{-17.7 \text{ km}}$$

The negative value of A_y indicates that the y coordinate had decreased for this displacement—in other words, the hiker has walked in the negative y direction. The signs of A_x and A_y are also evident from Figure 3.19. The second displacement, **B**, has a magnitude of 40.0 km and is 60.0° north of east. Its rectangular components are

$$B_x = B\cos 60.0° = (40.0 \text{ km})(0.500) = \boxed{20.0 \text{ km}}$$

$$B_y = B\sin 60.0° = (40.0 \text{ km})(0.866) = \boxed{34.6 \text{ km}}$$

   (b) Determine the components of the hiker's resultant displacement **R** for the trip. Find an expression for **R** in terms of unit vectors.

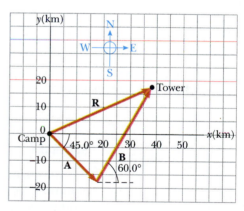

FIGURE 3.19   (Example 3.8) The total displacement of the hiker is the vector $\mathbf{R} = \mathbf{A} + \mathbf{B}$.

Solution   The resultant displacement for the trip, $\mathbf{R} = \mathbf{A} + \mathbf{B}$, has components given by Equation 3.15:

$$R_x = A_x + B_x = 17.7 \text{ km} + 20.0 \text{ km} = \boxed{37.7 \text{ km}}$$

$$R_y = A_y + B_y = -17.7 \text{ km} + 34.6 \text{ km} = \boxed{16.9 \text{ km}}$$

In unit-vector form, we can write the total displacement as $\mathbf{R} = (37.7\mathbf{i} + 16.9\mathbf{j})$ km.

Exercise   Determine the magnitude and direction of the total displacement.

Answer   41.3 km, 24.1° north of east from the base camp.

EXAMPLE 3.9   Let's Fly Away

A commuter airplane starts from an airport and takes the route shown in Figure 3.20. First, it flies to city A located 175 km in a direction 30.0° north of east. Next, it flies 150 km 20.0° west of north to city B. Finally, it flies 190 km west to city C. Find the location of city C relative to the location of the starting point.

Solution   As in the previous example, it is convenient to choose the coordinate system shown in Figure 3.20, where the x axis points to the east and the y axis points to the north. Let us denote the three consecutive displacements by the vectors **a**, **b**, and **c**. The first displacement **a** has a magnitude of 175 km and has components

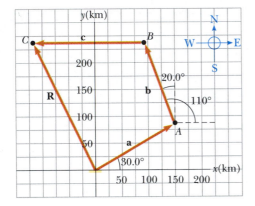

FIGURE 3.20   (Example 3.9) The airplane starts at the origin, flies first to *A*, then to *B*, and finally to *C*.

$$a_x = a\cos(30.0°) = (175\text{ km})(0.866) = 152\text{ km}$$
$$a_y = a\sin(30.0°) = (175\text{ km})(0.500) = 87.5\text{ km}$$

Displacement **b**, whose magnitude is 150 km, has components

$$b_x = b\cos(110°) = (150\text{ km})(-0.342) = -51.3\text{ km}$$
$$b_y = b\sin(110°) = (150\text{ km})(0.940) = 141\text{ km}$$

Finally, displacement **c**, whose magnitude is 190 km, has components

$$c_x = c\cos(180°) = (190\text{ km})(-1) = -190\text{ km}$$
$$c_y = c\sin(180°) = 0$$

Therefore, the components of the position vector **R** from the starting point to city C are

$$R_x = a_x + b_x + c_x = 152\text{ km} - 51.3\text{ km} - 190\text{ km}$$
$$= \boxed{-89.7\text{ km}}$$
$$R_y = a_y + b_y + c_y = 87.5\text{ km} + 141\text{ km} + 0$$
$$= \boxed{228\text{ km}}$$

In unit-vector notation, $\mathbf{R} = (-89.7\mathbf{i} + 228\mathbf{j})$ km. That is, city C can be reached from the starting point by first traveling 89.7 km due west and then traveling 228 km due north.

Exercise   Find the magnitude and direction of **R**.

Answer   245 km, 21.4° west of north.

SUMMARY

Vector quantities have both magnitude and direction and obey the vector law of addition. **Scalar quantities** have only magnitude and obey the laws of ordinary arithmetic.

Two vectors **A** and **B** can be added graphically using either the triangle method or the parallelogram rule. In the triangle method (Fig. 3.21a), the resultant vector

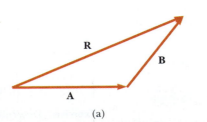

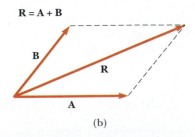

(a)   (b)

FIGURE 3.21   (a) Vector addition using the triangle method. (b) Vector addition using the parallelogram rule.

$\mathbf{R} = \mathbf{A} + \mathbf{B}$ runs from the tail of $\mathbf{A}$ to the tip of $\mathbf{B}$. In the parallelogram method (Fig. 3.21b), $\mathbf{R}$ is the diagonal of a parallelogram having $\mathbf{A}$ and $\mathbf{B}$ as its sides.

The x component, A_x, of the vector $\mathbf{A}$ is equal to its projection along the x axis of a coordinate system as in Figure 3.22, where $A_x = A \cos \theta$. Likewise, the y component, A_y, of $\mathbf{A}$ is its projection along the y axis, where $A_y = A \sin \theta$.

If a vector $\mathbf{A}$ has an x component equal to A_x and a y component equal to A_y, the vector can be expressed in unit-vector form as $\mathbf{A} = A_x\mathbf{i} + A_y\mathbf{j}$. In this notation, $\mathbf{i}$ is a unit vector pointing in the positive x direction and $\mathbf{j}$ is a unit vector in the positive y direction. Since $\mathbf{i}$ and $\mathbf{j}$ are unit vectors, $|\mathbf{i}| = |\mathbf{j}| = 1$.

The resultant of two or more vectors can be found by resolving all vectors into their x and y components, adding their resultant x and y components, and then using the Pythagorean theorem to find the magnitude of the resultant vector. The angle that the resultant vector makes with respect to the x axis can be found by use of a suitable trigonometric function.

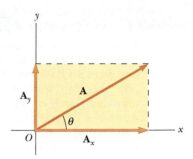

FIGURE 3.22   The x and y vector components of $\mathbf{A}$ are $\mathbf{A}_x$ and $\mathbf{A}_y$. The x and y rectangular components of $\mathbf{A}$ are $A_x = A \cos \theta$ and $A_y = A \sin \theta$.

QUESTIONS

1. A book is moved once around the perimeter of a rectangular tabletop of dimensions 1.0 m × 2.0 m. If the book ends up at its initial position, what is its displacement? What is the distance traveled?
2. Can the magnitude of a particle's displacement be greater than the distance traveled? Explain.
3. The magnitudes of two vectors $\mathbf{A}$ and $\mathbf{B}$ are $A = 5$ units and $B = 2$ units. Find the largest and smallest values possible for the resultant vector $\mathbf{R} = \mathbf{A} + \mathbf{B}$.
4. Vector $\mathbf{A}$ lies in the xy plane. For what orientations of $\mathbf{A}$ will both of its components be negative? For what orientations will its components have opposite signs?
5. If the component of vector $\mathbf{A}$ along the direction of vector $\mathbf{B}$ is zero, what can you conclude about the two vectors?

6. Can the magnitude of a vector have a negative value? Explain.
7. Which of the following are vectors and which are not: force, temperature, volume, ratings of a television show, height, velocity, age?
8. Under what circumstances would a nonzero vector lying in the xy plane have components that are equal in magnitude?
9. Is it possible to add a vector quantity to a scalar quantity? Explain.
10. Two vectors have unequal magnitudes. Can their sum be zero? Explain.

PROBLEMS

Section 3.1   Coordinate Systems and Frames of Reference

1. Two points in the xy plane have cartesian coordinates (2.00, −4.00) m and (−3.00, 3.00) m. Determine (a) the distance between these points and (b) their polar coordinates.
2. If the rectangular and polar coordinates of a point are $(2, y)$ and $(r, 30°)$, respectively, determine y and r.
3. The polar coordinates of a point are $r = 5.50$ m and $\theta = 240.0°$. What are the cartesian coordinates of this point?
4. Two points in a plane have polar coordinates (2.50 m, 30.0°) and (3.80 m, 120.0°). Determine (a) the cartesian coordinates of these points and (b) the distance between them.

5. A certain corner of a room is selected as the origin of a rectangular coordinate system. A fly is crawling on a wall adjacent to one of the axes. If the fly is located at a point having coordinates (2.00, 1.00) m, (a) how far is it from the corner of the room? (b) what is its location in polar coordinates?
6. If the polar coordinates of the point (x, y) are (r, θ), determine the polar coordinates for the points: (a) $(−x, y)$, (b) $(−2x, −2y)$, and (c) $(3x, −3y)$.
7. A point is located in a polar coordinate system by the coordinates $r = 2.50$ m and $\theta = 35.0°$. Find the cartesian coordinates of this point, assuming the two coordinate systems have the same origin.

☐ indicates problems that have full solutions available in the Student Solutions Manual and Study Guide.

Section 3.2   Vector and Scalar Quantities
Section 3.3   Some Properties of Vectors

8. An airplane flies 200 km due west from city A to city B and then 300 km in the direction of 30° north of west from city B to city C. (a) In straight-line distance, how far is city C from city A? (b) Relative to city A, in what direction is city C?

9. A surveyor estimates the distance across a river by the following method: standing directly across from a tree on the opposite bank, she walks 100 m along the riverbank, then sights across to the tree. The angle from her baseline to the tree is 35.0°. How wide is the river?

10. A pedestrian moves 6.00 km east and then 13.0 km north. Find the magnitude and direction of the resultant displacement vector using the graphical method.

11. A plane flies from base camp to lake A, a distance of 280 km at a direction of 20.0° north of east. After dropping off supplies, it flies to lake B, which is 190 km and 30.0° west of north from lake A. Graphically determine the distance and direction from lake B to the base camp.

12. Vector **A** has a magnitude of 8.00 units and makes an angle of 45.0° with the positive x axis. Vector **B** also has a magnitude of 8.00 units and is directed along the negative x axis. Using graphical methods, find (a) the vector sum **A** + **B** and (b) the vector difference **A** − **B**.

13. A person walks along a circular path of radius 5.00 m, around one half of the circle. (a) Find the magnitude of the displacement vector. (b) How far did the person walk? (c) What is the magnitude of the displacement if the person walks all the way around the circle?

14. A force **F**₁ of magnitude 6.00 units acts at the origin in a direction 30.0° above the positive x axis. A second force **F**₂ of magnitude 5.00 units acts at the origin in the direction of the positive y axis. Find graphically the magnitude and direction of the resultant force **F**₁ + **F**₂.

15. Each of the displacement vectors **A** and **B** shown in Figure P3.15 has a magnitude of 3.00 m. Find graphically (a) **A** + **B**, (b) **A** − **B**, (c) **B** − **A**, (d) **A** − 2**B**.

16. A dog searching for a bone walks 3.5 m south, then 8.2 m at an angle 30° north of east, and finally 15 m west. Find the dog's resultant displacement vector using graphical techniques.

17. A roller coaster moves 200 ft horizontally, then travels 135 ft at an angle of 30.0° above the horizontal. It then travels 135 ft at an angle of 40.0° below the horizontal. What is its displacement from its starting point? Use graphical techniques.

18. The driver of a car drives 3.00 km north, 2.00 km northeast (45.0° east of north), 4.00 km west, and then 3.00 km southeast (45.0° east of south). Where

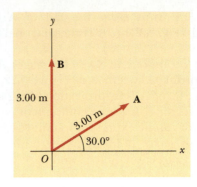

FIGURE P3.15

does he end up relative to his starting point? Work out your answer graphically. Check by using components. (The car is not near the North Pole or the South Pole.)

19. Find the horizontal and vertical components of the 100-m displacement of a superhero who flies from the top of a tall building following the path shown in Figure P3.19.

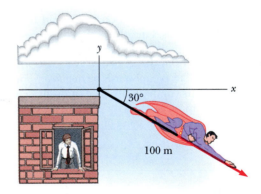

FIGURE P3.19

20. A person walks 25.0° north of east for 3.10 km. How far would she have to walk due north and due east to arrive at the same location?

21. Indiana Jones is trapped in a maze. To find his way out, he walks 10 m, makes a 90° right turn, walks 5.0 m, makes another 90° right turn, and walks 7.0 m. What is his displacement from his initial position?

22. While exploring a cave, a spelunker starts at the entrance and moves the following distances. She goes 75.0 m north, 250 m east, 125 m at an angle 30.0° north of east, and 150 m south. Find the resultant displacement from the cave entrance.

Section 3.4 Components of a Vector and Unit Vectors

23. A vector has an x component of -25.0 units and a y component of 40.0 units. Find the magnitude and direction of this vector.

24. Vector **B** has x, y, and z components of 4.00, 6.00, and 3.00 units, respectively. Calculate the magnitude of **B** and the angles that **B** makes with the coordinate axes.

25. Given the vectors $\mathbf{A} = 2.0\mathbf{i} + 6.0\mathbf{j}$ and $\mathbf{B} = 3.0\mathbf{i} - 2.0\mathbf{j}$, (a) sketch the vector sum $\mathbf{C} = \mathbf{A} + \mathbf{B}$ and the vector subtraction $\mathbf{D} = \mathbf{A} - \mathbf{B}$. (b) Find analytical solutions for **C** and **D** first in terms of unit vectors and then in terms of polar coordinates, with angles measured with respect to the $+x$ axis.

26. A displacement vector lying in the xy plane has a magnitude of 50.0 m and is directed at an angle of $120.0°$ to the positive x axis. What are the rectangular components of this vector?

27. Find the magnitude and direction of the resultant of three displacements having rectangular components $(3.00, 2.00)$ m, $(-5.00, 3.00)$ m, and $(6.00, 1.00)$ m.

28. Vector **A** has x and y components of -8.7 cm and 15 cm, respectively; vector **B** has x and y components of 13.2 cm and -6.6 cm, respectively. If $\mathbf{A} - \mathbf{B} + 3\mathbf{C} = 0$, what are the components of **C**?

29. Consider two vectors $\mathbf{A} = 3\mathbf{i} - 2\mathbf{j}$ and $\mathbf{B} = -\mathbf{i} - 4\mathbf{j}$. Calculate (a) $\mathbf{A} + \mathbf{B}$, (b) $\mathbf{A} - \mathbf{B}$, (c) $|\mathbf{A} + \mathbf{B}|$, (d) $|\mathbf{A} - \mathbf{B}|$, (e) the direction of $\mathbf{A} + \mathbf{B}$ and $\mathbf{A} - \mathbf{B}$.

30. A boy runs 3.0 blocks north, 4.0 blocks northeast, and 5.0 blocks west. Determine the length and direction of the displacement vector that goes from the starting point to his final position.

31. Obtain expressions for the position vectors having polar coordinates (a) 12.8 m, $150°$; (b) 3.30 cm, $60.0°$; (c) 22.0 in., $215°$.

32. Consider the displacement vectors $\mathbf{A} = (3\mathbf{i} + 3\mathbf{j})$ m, $\mathbf{B} = (\mathbf{i} - 4\mathbf{j})$ m, and $\mathbf{C} = (-2\mathbf{i} + 5\mathbf{j})$ m. Use the component method to determine (a) the magnitude and direction of the vector $\mathbf{D} = \mathbf{A} + \mathbf{B} + \mathbf{C}$, (b) the magnitude and direction of $\mathbf{E} = -\mathbf{A} - \mathbf{B} + \mathbf{C}$.

33. A particle undergoes the following consecutive displacements: 3.50 m south, 8.20 m northeast, and 15.0 m west. What is the resultant displacement?

34. A quarterback takes the ball from the line of scrimmage, runs backward for 10 yards, then sideways parallel to the line of scrimmage for 15 yards. At this point, he throws a forward pass 50 yards straight downfield perpendicular to the line of scrimmage. What is the magnitude of the football's resultant displacement?

35. A jet airliner moving initially at 300 mph to the east moves into a region where the wind is blowing at 100 mph in a direction $30.0°$ north of east. What are the new speed and direction of the aircraft?

36. A novice golfer on the green takes three strokes to sink the ball. The successive displacements are 4.00 m to the north, 2.00 m northeast, and 1.00 m $30.0°$ west of south. Starting at the same initial point, an expert golfer could make the hole in what single displacement?

37. Find the x and y components of the vectors **A** and **B** shown in Figure P3.15. Derive an expression for the resultant vector $\mathbf{A} + \mathbf{B}$ in unit-vector notation.

38. A particle undergoes two displacements. The first has a magnitude of 150 cm and makes an angle of $120.0°$ with the positive x axis. The *resultant* displacement has a magnitude of 140 cm and is directed at an angle of $35.0°$ to the positive x axis. Find the magnitude and direction of the second displacement.

39. The vector **A** has x, y, and z components of 8, 12, and -4 units, respectively. (a) Write a vector expression for **A** in unit-vector notation. (b) Obtain a unit-vector expression for a vector **B** one fourth the length of **A** pointing in the same direction as **A**. (c) Obtain a unit-vector expression for a vector **C** three times the length of **A** pointing in the direction opposite the direction of **A**.

40. Instructions for finding a buried treasure include the following: Go 75 paces at $240°$, turn to $135°$ and walk 125 paces, then travel 100 paces at $160°$. Determine the resultant displacement from the starting point.

41. Given the displacement vectors $\mathbf{A} = (3.0\mathbf{i} - 4.0\mathbf{j} + 4.0\mathbf{k})$ m and $\mathbf{B} = (2.0\mathbf{i} + 3.0\mathbf{j} - 7.0\mathbf{k})$ m, find the magnitudes of the vectors (a) $\mathbf{C} = \mathbf{A} + \mathbf{B}$ and (b) $\mathbf{D} = 2\mathbf{A} - \mathbf{B}$, also expressing each in terms of its rectangular components.

42. As it passes over Grand Bahama Island, the eye of a hurricane is moving in a direction $60.0°$ north of west with a speed of 41.0 km/h. Three hours later, it shifts due north, and its speed slows to 25.0 km/h. How far from Grand Bahama is the eye 4.50 h after it passes over the island?

43. Vector **A** has a negative x component 3.00 units in length and a positive y component 2.00 units in length. (a) Determine an expression for **A** in unit-vector notation. (b) Determine the magnitude and direction of **A**. (c) What vector **B** when added to **A** gives a resultant vector with no x component and a negative y component 4.00 units in length?

44. An airplane starting from airport A flies 300 km east, then 350 km $30.0°$ west of north, and then 150 km north to arrive finally at airport B. There is no wind on this day. (a) The next day, another plane flies directly from A to B in a straight line. In what direction should the pilot travel in this direct flight? (b) How far will the pilot travel in this direct flight?

45. Point A in Figure P3.45 is an arbitrary point along the line connecting the two points (x_2, y_2). Show that the coordinates of A are $(1 - f)x_1 + fx_2$, $(1 - f)y_1 + fy_2$.

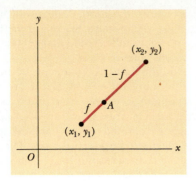

FIGURE P3.45

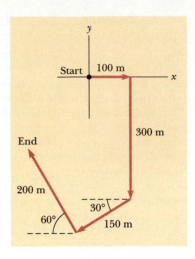

FIGURE P3.49

46. If $\mathbf{A} = (6.0\mathbf{i} - 8.0\mathbf{j})$ units, $\mathbf{B} = (-8.0\mathbf{i} + 3.0\mathbf{j})$ units, and $\mathbf{C} = (26.0\mathbf{i} + 19.0\mathbf{j})$ units, determine a and b so that $a\mathbf{A} + b\mathbf{B} + \mathbf{C} = 0$.

47. Three vectors are oriented as shown in Figure P3.47, where $|\mathbf{A}| = 20.0$ units, $|\mathbf{B}| = 40.0$ units, and $|\mathbf{C}| = 30.0$ units. Find (a) the x and y components of the resultant vector and (b) the magnitude and direction of the resultant vector.

50. The helicopter view in Figure P3.50 shows two people pulling on a stubborn mule. Find (a) the single force that is equivalent to the two forces shown, and (b) the force that a third person would have to exert on the mule to make the resultant force equal to zero.

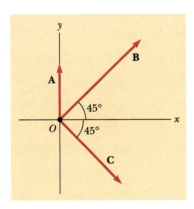

FIGURE P3.47

ADDITIONAL PROBLEMS

48. A vector is given by $\mathbf{R} = 2\mathbf{i} + \mathbf{j} + 3\mathbf{k}$. Find (a) the magnitudes of the x, y, and z components, (b) the magnitude of $\mathbf{R}$, and (c) the angles between $\mathbf{R}$ and the x, y, and z axes.

49. A person going for a walk follows the path shown in Figure P3.49. The total trip consists of four straight-line paths. At the end of the walk, what is the person's resultant displacement measured from the starting point?

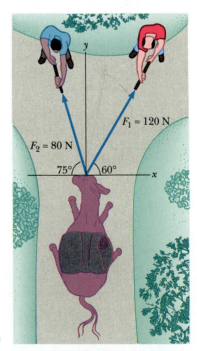

FIGURE P3.50

51. A pirate has buried his treasure on an island on which grow five trees located at the following points: $A(30 \text{ m}, -20 \text{ m})$, $B(60 \text{ m}, 80 \text{ m})$, $C(-10 \text{ m}, 10 \text{ m})$, $D(40 \text{ m}, -30 \text{ m})$, and $E(-70 \text{ m}, 60 \text{ m})$, all mea-

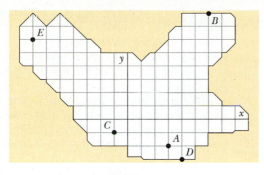

FIGURE P3.51

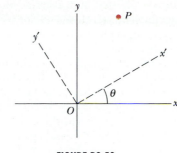

FIGURE P3.53

sured relative to some origin, as in Figure P3.51. His map instructs him to start at A and move toward B, but cover only one-half the distance between A and B. Then move toward C, covering one-third the distance between your current location and C. Then move toward D, covering one-fourth the distance between where you are and D. Finally move toward E, covering one-fifth the distance between you and E, stop and dig. (a) What are the coordinates of the point where his treasure is buried? (b) Rearrange the order of the trees (for instance, $B(30 \text{ m}, -20 \text{ m})$, $A(60 \text{ m}, 80 \text{ m})$, $E(-10 \text{ m}, 10 \text{ m})$, $C(40 \text{ m}, -30 \text{ m})$, and $D(-70 \text{ m}, 60 \text{ m})$), and repeat the calculation to show that the answer does not depend on the order of the trees. (*Hint:* See Problem 45.)

52. A rectangular parallelepiped has dimensions a, b, and c, as in Figure P3.52. (a) Obtain a vector expression for the face diagonal vector $\mathbf{R}_1$. What is the magnitude of this vector? (b) Obtain a vector expression for the body diagonal vector $\mathbf{R}_2$. What is the magnitude of this vector?

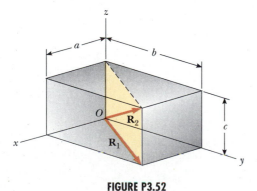

FIGURE P3.52

53. A point P is described by the coordinates (x, y) with respect to the normal cartesian coordinate system shown in Figure P3.53. Show that (x', y'), the coordinates of this point in the rotated coordinate system,

are related to (x, y) and the rotation angle θ by the expressions

$$x' = x \cos \theta + y \sin \theta$$

$$y' = -x \sin \theta + y \cos \theta$$

54. A point lying in the xy plane and having coordinates (x, y) can be described by the position vector $\mathbf{r} = x\mathbf{i} + y\mathbf{j}$. (a) Show that the displacement vector for a particle moving from (x_1, y_1) to (x_2, y_2) is given by $\mathbf{d} = (x_2 - x_1)\mathbf{i} + (y_2 - y_1)\mathbf{j}$. (b) Plot the position vectors $\mathbf{r}_1$ and $\mathbf{r}_2$ and the displacement vector $\mathbf{d}$, and verify by the graphical method that $\mathbf{d} = \mathbf{r}_2 - \mathbf{r}_1$.

SPREADSHEET PROBLEMS

S1. Use Spreadsheet 3.1 to find the total displacement corresponding to the sum of the three displacement vectors: $\mathbf{A} = 3.0\mathbf{i} + 4.0\mathbf{j}$, $\mathbf{B} = -2.3\mathbf{i} - 7.8\mathbf{j}$, and $\mathbf{C} = 5.0\mathbf{i} - 2.0\mathbf{j}$, where the units are in meters. (a) Give your answer in component form. (b) Give your answer in polar form. (c) Find a fourth displacement that returns you to the origin.

S2. A vendor must call on four customers once in every sales period. The four customers are in different cities, and the salesman wants to visit all customers in the least time. What path should he take?

To solve this problem, suppose you start at the origin and visit four different points (A, B, C, D) once before returning to the origin, always traveling in a straight line between points. The distance traveled depends upon the route taken. Spreadsheet 3.2 enables you to pick any four points A, B, C, D lying in a plane; the spreadsheet calculates the magnitude of the vector displacement for each straight-line portion of the trip and the total distance traveled. Find the path that gives the shortest total distance traveled.

Brute force can be used to find this path: try every path possible. To use Spreadsheet 3.2, enter the X and Y coordinates in the columns corresponding to the points A, B, C, D. The first site, the starting point, is labeled O, and the final site, which is also the starting position, is also labeled O. Note the total distance traveled and examine the graph that displays the path. To change the order in which the sites are visited, change the index order of the four non-origin sites and sort the block. For example, to visit *B* first, enter a "1" in the order column to the left of *B*; to visit *D* next, enter a "2" in the order column next to *D*, and so on; then sort the block containing the rows in which the points *A*, *B*, *C*, *D* appear in ascending order according to the order column. Note the total distance traveled and again examine the graph to see your new path. Repeat this process until you are convinced you have found the shortest path. (a) Find the shortest possible distance for visiting four points, when the locations are $(-10, 5)$, $(-8,$ $-7)$, $(1, 11)$, and $(12, 9)$; find the minimum distance necessary to visit these points. How many trials did you use to find this shortest distance? In what order should the vendor visit these four customers? (b) Choose any other four points and repeat part (a).

S3. Modify Spreadsheet 3.2 to include two more points. Choose any six points and use your spreadsheet to find the shortest path for visiting all six sites. How many trials did you need this time? This brute-force method is not practical when there are more than a handful of customers. For *N* sites, the method requires $(N-1)!/2$ trials. For example, if $N = 10$, you would need over 180 000 trials.

CHAPTER 4

Motion in Two Dimensions

The circus stuntmen being shot out of cannons are human projectiles. Neglecting air resistance, they move in parabolic paths until they land in a strategically placed net. What initial condition(s) will determine where the catching net should be placed?

(*Ringling Brothers Circus*)

I n this chapter we deal with the kinematics of a particle moving in a plane, which is two-dimensional motion. Some common examples of motion in a plane are the motion of projectiles and satellites and the motion of charged particles in uniform electric fields. We begin by showing that displacement, velocity, and acceleration are vector quantities. As in the case of one-dimensional motion, we derive the kinematic equations for two-dimensional motion from the fundamental definitions of displacement, velocity, and acceleration. As special cases of motion in two dimensions, we then treat constant-acceleration motion in a plane as well as uniform circular motion.

4.1 THE DISPLACEMENT, VELOCITY, AND ACCELERATION VECTORS

In Chapter 2 we found that the motion of a particle moving along a straight line is completely known if its coordinate is known as a function of time. Now let us extend this idea to motion in the *xy* plane. We begin by describing the position of a

71

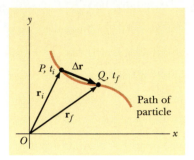

FIGURE 4.1 A particle moving in the *xy* plane is located with the position vector **r** drawn from the origin to the particle. The displacement of the particle as it moves from *P* to *Q* in the time interval $\Delta t = t_f - t_i$ is equal to the vector $\Delta \mathbf{r} = \mathbf{r}_f - \mathbf{r}_i$.

Average velocity

particle with a *position vector* **r**, drawn from the origin of some reference frame to the particle located in the *xy* plane, as in Figure 4.1. At time t_i, the particle is at point *P*, and at some later time t_f, the particle is at *Q*, where the indices *i* and *f* refer to initial and final values. As the particle moves from *P* to *Q* in the time interval $\Delta t = t_f - t_i$, the position vector changes from $\mathbf{r}_i$ to $\mathbf{r}_f$. As we learned in Chapter 2, the displacement of a particle is the difference between its final position and initial position. Therefore, the **displacement vector** for the particle of Figure 4.1 equals the difference between its final position vector and its initial position vector:

$$\Delta \mathbf{r} \equiv \mathbf{r}_f - \mathbf{r}_i \qquad (4.1)$$

The direction of $\Delta \mathbf{r}$ is indicated in Figure 4.1. As we see from Figure 4.1, the magnitude of the displacement vector is *less* than the distance traveled along the curved path.

> We define the **average velocity** of the particle during the time interval Δt as the ratio of the displacement to that time interval:
>
> $$\bar{\mathbf{v}} \equiv \frac{\Delta \mathbf{r}}{\Delta t} \qquad (4.2)$$

Since displacement is a vector quantity and the time interval is a scalar quantity, we conclude that the average velocity is a vector quantity directed along $\Delta \mathbf{r}$. Note that the average velocity between points *P* and *Q* is *independent of the path* between the two points. This is because the average velocity is proportional to the displacement, which in turn depends only on the initial and final position vectors and not on the path taken between those two points. As we did with one-dimensional motion, we conclude that if a particle starts its motion at some point and returns to this point via any path, its average velocity is zero for this trip since its displacement is zero.

Consider again the motion of a particle between two points in the *xy* plane, as in Figure 4.2. As the time intervals over which we observe the motion become smaller and smaller, the direction of the displacement approaches that of the line tangent to the path at the point *P*.

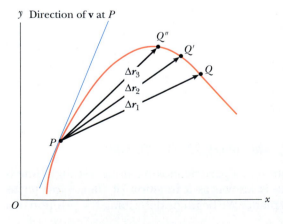

FIGURE 4.2 As a particle moves between two points, its average velocity is in the direction of the displacement vector $\Delta \mathbf{r}$. As the end point of the path is moved from *Q* to *Q'* to *Q''*, the respective displacements and corresponding time intervals become smaller and smaller. In the limit that the end point approaches *P*, Δt approaches zero and the direction of $\Delta \mathbf{r}$ approaches that of the line tangent to the curve at *P*. By definition, the instantaneous velocity at *P* is in the direction of this tangent line.

The **instantaneous velocity, v,** is defined as the limit of the average velocity, $\Delta \mathbf{r}/\Delta t$, as Δt approaches zero:

$$\mathbf{v} \equiv \lim_{\Delta t \to 0} \frac{\Delta \mathbf{r}}{\Delta t} = \frac{d\mathbf{r}}{dt} \qquad (4.3)$$

Instantaneous velocity

That is, the instantaneous velocity equals the derivative of the position vector with respect to time. The direction of the instantaneous velocity vector at any point in a particle's path is along a line that is tangent to the path at that point and in the direction of motion; this is illustrated in Figure 4.3. The magnitude of the instantaneous velocity vector is called the *speed*.

As a particle moves from one point to another along some path as in Figure 4.3, its instantaneous velocity vector changes from $\mathbf{v}_i$ at time t_i to $\mathbf{v}_f$ at time t_f.

The **average acceleration** of a particle as it moves from P to Q is defined as the ratio of the change in the instantaneous velocity vector, $\Delta \mathbf{v}$, to the elapsed time, Δt:

$$\bar{\mathbf{a}} \equiv \frac{\mathbf{v}_f - \mathbf{v}_i}{t_f - t_i} = \frac{\Delta \mathbf{v}}{\Delta t} \qquad (4.4)$$

Average acceleration

Since the average acceleration is the ratio of a vector quantity, $\Delta \mathbf{v}$, and a scalar quantity, Δt, we conclude that $\bar{\mathbf{a}}$ is a vector quantity directed along $\Delta \mathbf{v}$. As is indicated in Figure 4.3, the direction of $\Delta \mathbf{v}$ is found by adding the vector $-\mathbf{v}_i$ (the negative of $\mathbf{v}_i$) to the vector $\mathbf{v}_f$, since by definition $\Delta \mathbf{v} = \mathbf{v}_f - \mathbf{v}_i$.

The **instantaneous acceleration, a,** is defined as the limiting value of the ratio $\Delta \mathbf{v}/\Delta t$ as Δt approaches zero:

$$\mathbf{a} \equiv \lim_{\Delta t \to 0} \frac{\Delta \mathbf{v}}{\Delta t} = \frac{d\mathbf{v}}{dt} \qquad (4.5)$$

Instantaneous acceleration

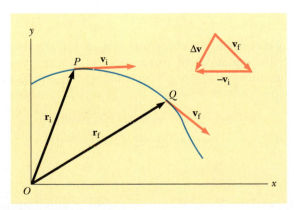

FIGURE 4.3   The average acceleration vector $\bar{\mathbf{a}}$ for a particle moving from P to Q is in the direction of the change in velocity, $\Delta \mathbf{v} = \mathbf{v}_f - \mathbf{v}_i$.

In other words, the instantaneous acceleration equals the derivative of the velocity vector with respect to time.

It is important to recognize that various changes may constitute acceleration of a particle. First, the magnitude of the velocity vector (the speed) may change with time as in straight-line (one-dimensional) motion. Second, only the direction of the velocity vector may change with time as its magnitude (speed) remains constant, as in curved-path (two-dimensional) motion. Finally, both the magnitude and the direction of the velocity vector may change.

CONCEPTUAL EXAMPLE 4.1

(a) Can an object accelerate if its speed is constant? (b) Can an object accelerate if its velocity is constant?

Reasoning   (a) Yes. Although its speed may be constant, the direction of its motion (that is, the direction of **v**) may change, causing an acceleration. For example, an object

moving in a circle with constant speed has an acceleration directed toward the center of the circle. (b) No. An object that moves with constant velocity has zero acceleration. Note that constant velocity means that both the direction and magnitude of **v** remain constant.

4.2   TWO-DIMENSIONAL MOTION WITH CONSTANT ACCELERATION

Let us consider two-dimensional motion during which the acceleration remains constant. That is, we assume that the magnitude and direction of the acceleration remain unchanged during the motion.

As we learned in Chapter 3, the motion of a particle can be determined by its position vector **r**. The position vector for a particle moving in the *xy* plane can be written

Position vector

$$\mathbf{r} = x\mathbf{i} + y\mathbf{j} \tag{4.6}$$

where *x*, *y*, and **r** change with time as the particle moves. If the position vector is known, the velocity of the particle can be obtained from Equations 4.3 and 4.6, which give

$$\mathbf{v} = \frac{d\mathbf{r}}{dt} = \frac{dx}{dt}\mathbf{i} + \frac{dy}{dt}\mathbf{j}$$

$$\mathbf{v} = v_x\mathbf{i} + v_y\mathbf{j} \tag{4.7}$$

Because **a** is assumed constant, its components a_x and a_y are also constants. Therefore, we can apply the equations of kinematics to the *x* and *y* components of the velocity vector. Substituting $v_x = v_{x0} + a_x t$ and $v_y = v_{y0} + a_y t$ into Equation 4.7 gives

$$\mathbf{v} = (v_{x0} + a_x t)\mathbf{i} + (v_{y0} + a_y t)\mathbf{j}$$
$$= (v_{x0}\mathbf{i} + v_{y0}\mathbf{j}) + (a_x\mathbf{i} + a_y\mathbf{j})t$$

Velocity vector as a function of time

$$\mathbf{v} = \mathbf{v}_0 + \mathbf{a}t \tag{4.8}$$

This result states that the velocity of a particle at some time *t* equals the vector sum of its initial velocity, $\mathbf{v}_0$, and the additional velocity **a***t* acquired in the time *t* as a result of its constant acceleration.

Similarly, from Equation 2.9 we know that the x and y coordinates of a particle moving with constant acceleration are

$$x = x_0 + v_{x0}t + \tfrac{1}{2}a_x t^2 \qquad \text{and} \qquad y = y_0 + v_{y0}t + \tfrac{1}{2}a_y t^2$$

Substituting these expressions into Equation 4.6 gives

$$\mathbf{r} = (x_0 + v_{x0}t + \tfrac{1}{2}a_x t^2)\mathbf{i} + (y_0 + v_{y0}t + \tfrac{1}{2}a_y t^2)\mathbf{j}$$
$$= (x_0\mathbf{i} + y_0\mathbf{j}) + (v_{x0}\mathbf{i} + v_{y0}\mathbf{j})t + \tfrac{1}{2}(a_x\mathbf{i} + a_y\mathbf{j})t^2$$

$$\boxed{\mathbf{r} = \mathbf{r}_0 + \mathbf{v}_0 t + \tfrac{1}{2}\mathbf{a}t^2} \qquad (4.9)$$

Position vector as a function of time

This equation implies that the displacement vector $\mathbf{r} - \mathbf{r}_0$ is the vector sum of a displacement $\mathbf{v}_0 t$, arising from the initial velocity of the particle, and a displacement $\tfrac{1}{2}\mathbf{a}t^2$, resulting from the uniform acceleration of the particle. Graphical representations of Equations 4.8 and 4.9 are shown in Figures 4.4a and 4.4b. For simplicity in drawing the figure, we have taken $\mathbf{r}_0 = 0$ in Figure 4.4b. That is, we assume that the particle is at the origin at $t = 0$. Note from Figure 4.4b that $\mathbf{r}$ is generally not along the direction of $\mathbf{v}_0$ or $\mathbf{a}$, because the relationship between these quantities is a vector expression. For the same reason, from Figure 4.4a we see that $\mathbf{v}$ is generally not along the direction of $\mathbf{v}_0$ or $\mathbf{a}$. Finally, if we compare the two figures, we see that $\mathbf{v}$ and $\mathbf{r}$ are not in the same direction.

Because Equations 4.8 and 4.9 are *vector* expressions, we may write their x and y component forms with $\mathbf{r}_0 = 0$:

$$\mathbf{v} = \mathbf{v}_0 + \mathbf{a}t \qquad \begin{cases} v_x = v_{x0} + a_x t \\ v_y = v_{y0} + a_y t \end{cases}$$

$$\mathbf{r} = \mathbf{v}_0 t + \tfrac{1}{2}\mathbf{a}t^2 \qquad \begin{cases} x = v_{x0}t + \tfrac{1}{2}a_x t^2 \\ y = v_{y0}t + \tfrac{1}{2}a_y t^2 \end{cases}$$

These components are illustrated in Figure 4.4. In other words, two-dimensional motion having constant acceleration is equivalent to two independent motions in the x and y directions having constant accelerations a_x and a_y.

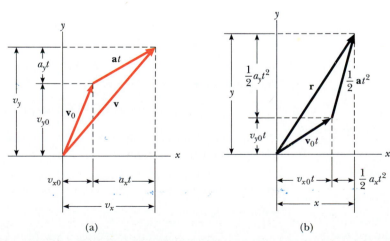

(a)　　　　　　　　　　　　(b)

FIGURE 4.4 Vector representations and rectangular components of (a) the velocity and (b) the displacement of a particle moving with a uniform acceleration **a**.

Multiflash exposure of a tennis player executing a backhand swing. Note that the ball follows a parabolic path characteristic of a projectile. Such photographs can be used to study the quality of sports equipment and the performance of an athlete. *(© Zimmerman, FPG International)*

EXAMPLE 4.2   Motion in a Plane

A particle starts from the origin at $t = 0$ with an initial velocity having an x component of 20 m/s and a y component of -15 m/s. The particle moves in the xy plane with an x component of acceleration only, given by $a_x = 4.0$ m/s². (a) Determine the components of velocity as a function of time and the total velocity vector at any time.

Solution   With $v_{x0} = 20$ m/s and $a_x = 4.0$ m/s², the equations of kinematics give

$$v_x = v_{x0} + a_x t = (20 + 4t) \text{ m/s}$$

Also, with $v_{y0} = -15$ m/s and $a_y = 0$,

$$v_y = v_{y0} = -15 \text{ m/s}$$

Therefore, using these results and noting that the velocity vector **v** has two components, we get

$$\mathbf{v} = v_x \mathbf{i} + v_y \mathbf{j} = \boxed{[(20 + 4.0t)\mathbf{i} - 15\mathbf{j}] \text{ m/s}}$$

We could also obtain this result using Equation 4.8 directly, noting that $\mathbf{a} = 4.0\mathbf{i}$ m/s² and $\mathbf{v}_0 = (20\mathbf{i} - 15\mathbf{j})$ m/s. Try it!

(b) Calculate the velocity and speed of the particle at $t = 5.0$ s.

Solution   With $t = 5.0$ s, the result from (a) gives

$$\mathbf{v} = \{[20 + 4(5.0)]\mathbf{i} - 15\mathbf{j}\} \text{ m/s} = \boxed{(40\mathbf{i} - 15\mathbf{j}) \text{ m/s}}$$

That is, at $t = 5.0$ s, $v_x = 40$ m/s and $v_y = -15$ m/s. Knowing these two components for this two-dimensional motion, we know the numerical value of the velocity vector. To determine the angle θ that **v** makes with the x axis, we use the fact that $\tan \theta = v_y / v_x$, or

$$\theta = \tan^{-1}\left(\frac{v_y}{v_x}\right) = \tan^{-1}\left(\frac{-15 \text{ m/s}}{40 \text{ m/s}}\right) = \boxed{-21°}$$

The speed is the magnitude of **v**:

$$v = |\mathbf{v}| = \sqrt{v_x^2 + v_y^2} = \sqrt{(40)^2 + (-15)^2} \text{ m/s} = \boxed{43 \text{ m/s}}$$

(*Note:* If you calculate v_0 from the x and y components of $\mathbf{v}_0$ you will find that $v > v_0$. Why?)

(c) Determine the x and y coordinates of the particle at any time t and the displacement vector at this time.

Solution   Since at $t = 0$, $x_0 = y_0 = 0$, Equation 2.11 gives

$$x = v_{x0}t + \tfrac{1}{2}a_x t^2 = \boxed{(20t + 2.0t^2) \text{ m}}$$

$$y = v_{y0}t = \boxed{(-15t) \text{ m}}$$

Therefore, the displacement vector at any time t is

$$\mathbf{r} = x\mathbf{i} + y\mathbf{j} = [(20t + 2.0t^2)\mathbf{i} - 15t\mathbf{j}] \text{ m}$$

Alternatively, we could obtain $\mathbf{r}$ by applying Equation 4.9 directly, with $\mathbf{v}_0 = (20\mathbf{i} - 15\mathbf{j})$ m/s and $\mathbf{a} = 4\mathbf{i}$ m/s². Try it!

Thus, for example, at $t = 5.0$ s, $x = 150$ m and $y = -75$ m, or $\mathbf{r} = (150\mathbf{i} - 75\mathbf{j})$ m. It follows that the distance of the particle from the origin to this point is the magnitude of the displacement:

$$|\mathbf{r}| = r = \sqrt{(150)^2 + (-75)^2} \text{ m} = 170 \text{ m}$$

Note that this is *not* the distance that the particle travels in this time! Can you determine this distance from the available data?

4.3 PROJECTILE MOTION

Anyone who has observed a baseball in motion (or, for that matter, any object thrown into the air) has observed projectile motion. The ball moves in a curved path when thrown at some angle with respect to the Earth's surface. This very common form of motion is surprisingly simple to analyze if the following two assumptions are made: (1) the free-fall acceleration, g, is constant over the range of motion and is directed downward,[1] and (2) the effect of air resistance is negligible.[2] With these assumptions, we find that the path of a projectile, which we call its *trajectory*, is *always* a parabola. *We use these assumptions throughout this chapter.*

If we choose our reference frame such that the y direction is vertical and positive upward, then $a_y = -g$ (as in one-dimensional free-fall) and $a_x = 0$ (because air friction is neglected). Furthermore, let us assume that at $t = 0$, the projectile leaves the origin ($x_0 = y_0 = 0$) with speed v_0, as in Figure 4.5. If the vector $\mathbf{v}_0$ makes

Assumptions of projectile motion

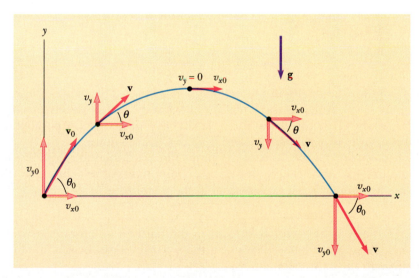

 FIGURE 4.5 The parabolic path of a projectile that leaves the origin with a velocity $\mathbf{v}_0$. The velocity vector $\mathbf{v}$ changes with time in both magnitude and direction. The change in the velocity vector is the result of acceleration in the negative y direction. The x component of velocity remains constant in time because there is no acceleration along the horizontal direction. Also, the y component of velocity is zero at the peak of the path.

[1] This approximation is reasonable as long as the range of motion is small compared with the radius of the Earth (6.4×10^6 m). In effect, this approximation is equivalent to assuming that the Earth is flat over the range of motion considered.

[2] This approximation is generally *not* justified, especially at high speeds. In addition, any spin imparted to a projectile, such as happens when a pitcher throws a curve ball, can give rise to some very interesting effects associated with aerodynamic forces.

an angle θ_0 with the horizontal, where θ_0 is the angle at which the projectile leaves the origin as in Figure 4.5, then from the definitions of the cosine and sine functions we have

$$\cos \theta_0 = v_{x0}/v_0 \quad \text{and} \quad \sin \theta_0 = v_{y0}/v_0$$

Therefore, the initial x and y components of velocity are

$$v_{x0} = v_0 \cos \theta_0 \quad \text{and} \quad v_{y0} = v_0 \sin \theta_0$$

Substituting these expressions into Equations 4.8 and 4.9 with $a_x = 0$ and $a_y = -g$ gives the velocity components and coordinates for the projectile at any time t:

Horizontal velocity component

$$v_x = v_{x0} = v_0 \cos \theta_0 = \text{constant} \tag{4.10}$$

Vertical velocity component

$$v_y = v_{y0} - gt = v_0 \sin \theta_0 - gt \tag{4.11}$$

Horizontal position component

$$x = v_{x0}t = (v_0 \cos \theta_0)t \tag{4.12}$$

Vertical position component

$$y = v_{y0}t - \tfrac{1}{2}gt^2 = (v_0 \sin \theta_0)t - \tfrac{1}{2}gt^2 \tag{4.13}$$

From Equation 4.10 we see that v_x remains constant in time and is equal to v_{x0}; there is no horizontal component of acceleration. For the y motion note that v_y and y are similar to Equations 2.12 and 2.14 for freely falling bodies. In fact, *all* of the equations of kinematics developed in Chapter 2 are applicable to projectile motion.

If we solve for t in Equation 4.12 and substitute this expression for t into Equation 4.13, we find that

$$y = (\tan \theta_0)x - \left(\frac{g}{2v_0{}^2 \cos^2 \theta_0} \right) x^2 \tag{4.14}$$

which is valid for the angles in the range $0 < \theta_0 < \pi/2$. This expression is of the form $y = ax - bx^2$, which is the equation of a parabola that passes through the origin. Thus, we have proved that the trajectory of a projectile is a parabola. Note that the trajectory is *completely* specified if v_0 and θ_0 are known.

One can obtain the speed, v, of the projectile as a function of time by noting that Equations 4.10 and 4.11 give the x and y components of velocity at any instant. Therefore, by definition, since v is equal to the magnitude of $\mathbf{v}$,

$$v = \sqrt{v_x{}^2 + v_y{}^2} \tag{4.15}$$

Also, since the velocity vector is tangent to the path at any instant, as shown in Figure 4.5, the angle θ that v makes with the horizontal can be obtained from v_x and v_y through the expression

$$\tan \theta = \frac{v_y}{v_x} \tag{4.16}$$

The vector expression for the position vector of the projectile as a function of time follows directly from Equation 4.9, with $\mathbf{a} = \mathbf{g}$:

$$\mathbf{r} = \mathbf{v}_0 t + \tfrac{1}{2}\mathbf{g}t^2$$

This expression gives the same information as the combination of Equations 4.12 and 4.13 and is plotted in Figure 4.6. Note that this expression for $\mathbf{r}$ is consistent with Equation 4.13, since the expression for $\mathbf{r}$ is a vector equation and $\mathbf{a} = \mathbf{g} = -g\mathbf{j}$ when the upward direction is taken to be positive.

Multiflash photograph of a tennis ball undergoing several bounces off a hard surface is an example of projectile motion. Note the parabolic path of the ball following each bounce. *(Richard Megna, Fundamental Photographs)*

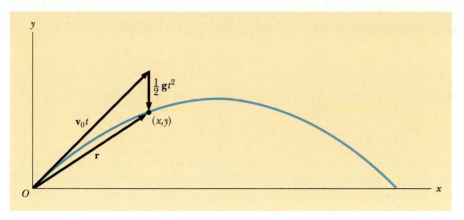

FIGURE 4.6   The displacement vector **r** of a projectile whose initial velocity at the origin is $\mathbf{v}_0$. The vector $\mathbf{v}_0 t$ would be the displacement of the projectile if gravity were absent, and the vector $\frac{1}{2}\mathbf{g}t^2$ is its vertical displacement due to its downward gravitational acceleration.

It is interesting to note that the motion of a particle can be considered the superposition of the term $\mathbf{v}_0 t$, which would be the displacement if no acceleration were present, and the term $\frac{1}{2}\mathbf{g}t^2$, which arises from the acceleration due to gravity. In other words, if there were no gravitational acceleration, the particle would continue to move along a straight path in the direction of $\mathbf{v}_0$. Therefore, the vertical distance $\frac{1}{2}\mathbf{g}t^2$ through which the particle "falls" off the straight path line is the same distance a freely falling body would fall during the same time interval. *We conclude that projectile motion is the superposition of two motions: (1) constant velocity motion in the initial direction and (2) the motion of a particle freely falling in the vertical direction under constant acceleration.*

Horizontal Range and Maximum Height of a Projectile

Let us assume that a projectile is fired from the origin at $t = 0$ with a positive v_y component, as in Figure 4.7. There are two special points that are interesting to analyze: the peak that has cartesian coordinates $(R/2, h)$, and the point having coordinates $(R, 0)$. The distance R is called the *horizontal range* of the projectile, and h is its *maximum height*. Let us find h and R in terms of v_0, θ_0, and g.

We can determine h by noting that at the peak, $v_y = 0$. Therefore, Equation 4.11 can be used to determine the time t_1 it takes to reach the peak:

$$t_1 = \frac{v_0 \sin \theta_0}{g}$$

Substituting this expression for t_1 into Equation 4.13 and replacing y with h gives h in terms of v_0 and θ_0:

$$h = (v_0 \sin \theta_0)\frac{v_0 \sin \theta_0}{g} - \frac{1}{2}g\left(\frac{v_0 \sin \theta_0}{g}\right)^2$$

$$h = \frac{v_0^2 \sin^2 \theta_0}{2g} \tag{4.17}$$

The range, R, is the horizontal distance traveled in twice the time it takes to reach the peak, that is, in a time $2t_1$. (This can be seen by setting $y = 0$ in Equation 4.13 and solving the quadratic for t. One solution of this quadratic is $t = 0$, and the

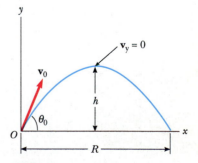

FIGURE 4.7   A projectile fired from the origin at $t = 0$ with an initial velocity $\mathbf{v}_0$. The maximum height of the projectile is h, and the horizontal range is R. At the peak of the trajectory, the particle has coordinates $(R/2, h)$.

second is $t = 2t_1$.) Using Equation 4.12 and noting that $x = R$ at $t = 2t_1$, we find that

$$R = (v_0 \cos \theta_0)2t_1 = (v_0 \cos \theta_0)\frac{2v_0 \sin \theta_0}{g} = \frac{2v_0^2 \sin \theta_0 \cos \theta_0}{g}$$

Since $\sin 2\theta = 2 \sin \theta \cos \theta$, R can be written in the more compact form

Range of projectile

$$R = \frac{v_0^2 \sin 2\theta_0}{g} \tag{4.18}$$

Keep in mind that Equations 4.17 and 4.18 are useful for calculating h and R only if v_0 and θ_0 are known and only for a symmetric path, as shown in Figure 4.7 (which means that only $\mathbf{v_0}$ has to be specified). The general expressions given by Equations 4.10 through 4.13 are the *more important* results, because they give the coordinates and velocity components of the projectile at *any* time t.

You should note that the maximum value of R from Equation 4.18 is given by $R_{\text{max}} = v_0^2/g$. This result follows from the fact that the maximum value of $\sin 2\theta_0$ is unity, which occurs when $2\theta_0 = 90°$. Therefore, we see that R is a maximum when $\theta_0 = 45°$, as you would expect if air friction is neglected.

Figure 4.8 illustrates various trajectories for a projectile of a given initial speed. As you can see, the range is a maximum for $\theta_0 = 45°$. In addition, for any θ_0 other than $45°$, a point with coordinates $(R, 0)$ can be reached by using either one of two complementary values of θ_0 such as $75°$ and $15°$. Of course, the maximum height and the time of flight will be different for these two values of θ_0.

CONCEPTUAL EXAMPLE 4.3

As a projectile moves in its parabolic path, is there any point along its path where the velocity and acceleration vectors are (a) perpendicular to each other? (b) parallel to each other?

Reasoning   (a) At the top of its flight, **v** is horizontal and **a** is vertical. This is the only point where the velocity and acceleration vectors are perpendicular. (b) If the object is thrown straight up or down, then **v** and **a** will be parallel throughout the downward motion. Otherwise, the velocity and acceleration vectors are never parallel.

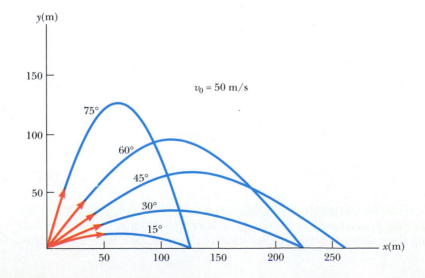

FIGURE 4.8   A projectile fired from the origin with an initial speed of 50 m/s at various angles of projection. Note that complementary values of θ_0 will result in the same value of x (range of the projectile).

EXAMPLE 4.4 The Long-Jump

A long-jumper leaves the ground at an angle of 20.0° to the horizontal and at a speed of 11.0 m/s. (a) How far does he jump? (Assume that his motion is equivalent to that of a particle.)

Solution The horizontal motion is described by Equation 4.12:

$$x = (v_0 \cos \theta_0)t = (11.0 \text{ m/s})(\cos 20.0°)t$$

The value of x can be found if t, the total time of the jump, is known. We are able to find t using Equation 4.11, $v_y = v_0 \sin \theta_0 - gt$, and by noting that at the top of the jump the vertical component of velocity goes to zero:

$$v_y = v_0 \sin \theta_0 - gt$$

$$0 = (11.0 \text{ m/s}) \sin 20.0° - (9.80 \text{ m/s}^2)t_1$$

$$t_1 = 0.384 \text{ s}$$

Note that t_1 is the time interval to reach the *top* of the jump. Because of the symmetry of the vertical motion, an identical time interval passes before the jumper returns to the ground. Therefore, the *total* time in the air is $t = 2t_1 = 0.768$ s. Substituting this value into the above expression for x gives

$$x = (11.0 \text{ m/s})(\cos 20.0°)(0.768 \text{ s}) = \boxed{7.94 \text{ m}}$$

(b) What is the maximum height reached?

Solution The maximum height reached is found using Equation 4.13 with $t = t_1 = 0.384$ s:

$$y_{max} = (v_0 \sin \theta_0)t_1 - \tfrac{1}{2}gt_1^2$$

(Example 4.4) In a long-jump event, Willie Banks can leap horizontal distances of at least 8 meters.   *(© R. Mackson/FPG)*

$$= (11.0 \text{ m/s})(\sin 20.0°)(0.384 \text{ s})$$
$$- \tfrac{1}{2}(9.80 \text{ m/s}^2)(0.384 \text{ s})^2$$

$$= \boxed{0.722 \text{ m}}$$

The assumption that the motion of the long-jumper is that of a particle is an oversimplification. Nevertheless, the values obtained are reasonable. Note that we also could have used Equations 4.17 and 4.18 to find the maximum height and horizontal range. However, the method used in our solution is more instructive.

EXAMPLE 4.5 It's a Bull's-Eye Every Time

In a very popular lecture demonstration, a projectile is fired at a target in such a way that the projectile leaves the gun at the same time the target is dropped from rest, as in Figure 4.9. Let us show that if the gun is initially aimed at the stationary target, the projectile hits the target.

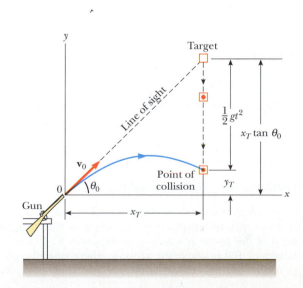

FIGURE 4.9 (Example 4.5) Schematic diagram of the projectile-and-target demonstration. If the gun is aimed directly at the stationary target and is fired at the same instant the target begins to fall, the projectile will hit the target. Both fall through the same vertical distance in a time t, because both experience the same acceleration, $a_y = -g$.

Reasoning and Solution We can argue that a collision will result under the conditions stated by noting that both the projectile and the target experience the same acceleration, $a_y = -g$, as soon as they are released. First, note from Figure 4.9 that the initial y coordinate of the target is $x_T \tan \theta_0$ and that it falls through a distance $\frac{1}{2}gt^2$ in a time t. Therefore, the y coordinate of the target as a function of time is, from Equation 4.14,

$$y_T = x_T \tan \theta_0 - \tfrac{1}{2}gt^2$$

Now if we write equations for x and y for the projectile path over time, using Equations 4.12 and 4.13 simultaneously, we get

$$y_p = x_p \tan \theta_0 - \tfrac{1}{2}gt^2$$

Thus, we see by comparing the two equations above that when $x_p = x_T$, $y_p = y_T$ and a collision results.

The result could also be arrived at with vector methods, using expressions for the position vectors for the projectile and target.

You should also note that a collision will *not* always take place. There is the further restriction that a collision results only when $v_0 \sin \theta_0 \geq \sqrt{gd/2}$, where d is the initial elevation of the target above the *floor*. If $v_0 \sin \theta_0$ is less than this value, the projectile will strike the floor before reaching the target.

EXAMPLE 4.6   That's Quite an Arm

A stone is thrown from the top of a building upward at an angle of 30.0° to the horizontal and with an initial speed of 20.0 m/s, as in Figure 4.10. If the height of the building is 45.0 m, (a) how long is the stone "in flight"?

Solution The initial x and y components of the velocity are

$$v_{x0} = v_0 \cos \theta_0 = (20.0 \text{ m/s})(\cos 30.0°) = 17.3 \text{ m/s}$$

$$v_{y0} = v_0 \sin \theta_0 = (20.0 \text{ m/s})(\sin 30.0°) = 10.0 \text{ m/s}$$

To find t, we can use $y = v_{y0}t - \frac{1}{2}gt^2$ (Eq. 4.13) with $y = -45.0$ m and $v_{y0} = 10.0$ m/s (we have chosen the top of the building as the origin, as in Figure 4.10):

$$-45.0 \text{ m} = (10.0 \text{ m/s})t - \tfrac{1}{2}(9.80 \text{ m/s}^2)t^2$$

Solving the quadratic equation for t gives, for the positive root, $t = 4.22$ s. Does the negative root have any physical meaning? (Can you think of another way of finding t from the information given?)

(b) What is the speed of the stone just before it strikes the ground?

Solution The y component of the velocity just before the stone strikes the ground can be obtained using the equation $v_y = v_{y0} - gt$ (Eq. 4.11) with $t = 4.22$ s:

$$v_y = 10.0 \text{ m/s} - (9.80 \text{ m/s}^2)(4.22 \text{ s}) = -31.4 \text{ m/s}$$

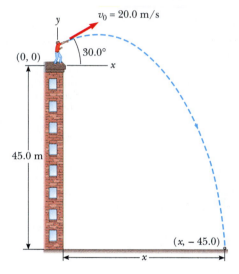

FIGURE 4.10   (Example 4.6)

Since $v_x = v_{x0} = 17.3$ m/s, the required speed is

$$v = \sqrt{v_x^2 + v_y^2} = \sqrt{(17.3)^2 + (-31.4)^2} \text{ m/s} = \boxed{35.9 \text{ m/s}}$$

Exercise   Where does the stone strike the ground?

Answer   73.0 m from the base of the building.

EXAMPLE 4.7   The Stranded Explorers

An Alaskan rescue plane drops a package of emergency rations to a stranded party of explorers, as shown in Figure 4.11. If the plane is traveling horizontally at 40.0 m/s at a height of 100 m above the ground, where does the package strike the ground relative to the point at which it is released?

Reasoning The coordinate system for this problem is selected as shown in Figure 4.11, with the positive x direction to the right and the positive y direction upward.

Consider first the horizontal motion of the package. The only equation available to us is $x = v_{x0}t$ (Eq. 4.12).

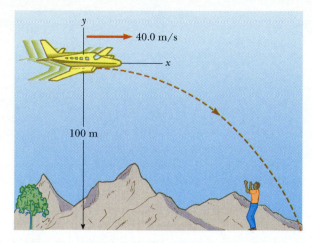

FIGURE 4.11 (Example 4.7) According to a ground observer, a package released from the rescue plane travels along the path shown. How does the path followed by the package appear to an observer on the plane (assumed to be moving at constant speed)?

The initial x component of the package velocity is the same as that of the plane when the package is released, 40.0 m/s. Thus, we have

$$x = (40.0 \text{ m/s}) t$$

If we know t, the length of time the package is in the air, we can determine x, the distance traveled by the package in the horizontal direction. To find t, we move to the equations for the vertical motion of the package. We know that at the instant the package hits the ground, its y coordinate is -100 m. We also know that the initial component of velocity of the package in the vertical direction, v_{y0}, is zero because the package was released with only a horizontal component of velocity.

Solution   From Equation 4.13, we have

$$y = -\tfrac{1}{2}gt^2$$

$$-100 \text{ m} = -\tfrac{1}{2}(9.80 \text{ m/s}^2)t^2$$

$$t^2 = 20.4 \text{ s}^2$$

$$t = 4.51 \text{ s}$$

The value for the time of flight substituted into the equation for the x coordinate gives

$$x = (40.0 \text{ m/s})(4.51 \text{ s}) = \boxed{180 \text{ m}}$$

The package hits the ground not directly under the drop point but 180 m to the right of that point.

Exercise   What are the horizontal and vertical components of the velocity of the package just before it hits the ground?

Answer $v_x = 40.0$ m/s; $v_y = -44.1$ m/s.

EXAMPLE 4.8 The End of the Ski Jump

A ski jumper travels down a slope and leaves the ski track moving in the horizontal direction with a speed of 25.0 m/s as in Figure 4.12. The landing incline below her falls off with a slope of 35.0°. (a) Where does she land on the incline?

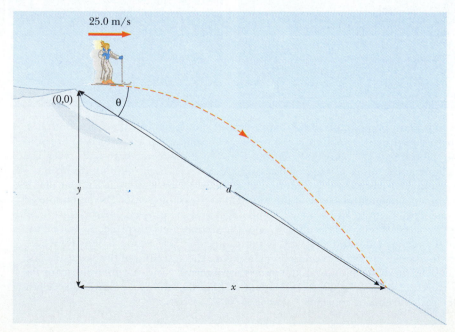

FIGURE 4.12   (Example 4.8)

Solution   It is convenient to select the origin ($x = y = 0$) at the beginning of the jump. Since $v_{x0} = 25.0$ m/s, and $v_{y0} = 0$ in this case, Equations 4.12 and 4.13 give

(1) $x = v_{x0}t = (25.0 \text{ m/s})t$

(2) $y = v_{y0}t - \frac{1}{2}gt^2 = -\frac{1}{2}(9.80 \text{ m/s}^2)t^2$

Taking d to be the distance she travels along the incline before landing, then from the right triangle in Figure 4.12, we see that her x and y coordinates at the point of landing are $x = d \cos 35.0°$ and $y = -d \sin 35.0°$. Substituting these relationships into (1) and (2) gives

(3) $d \cos 35.0° = (25.0 \text{ m/s})t$

(4) $-d \sin 35.0° = -\frac{1}{2}(9.80 \text{ m/s}^2)t^2$

Eliminating t from these equations gives $d = 109$ m. Hence, the x and y coordinates of the point at which she lands are

$x = d \cos 35.0° = (109 \text{ m}) \cos 35.0° = \boxed{89.3 \text{ m}}$

$y = -d \sin 35.0° = -(109 \text{ m}) \sin 35° = \boxed{-62.5 \text{ m}}$

Exercise   Determine how long the ski jumper is airborne and her vertical component of velocity just before she lands.

Answer   3.57 s; $v_y = -35.0$ m/s.

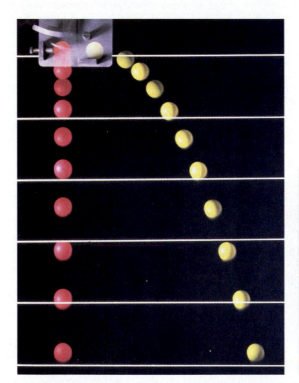

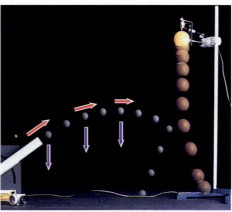

(Left) This multiflash photograph of two balls released simultaneously illustrates both free fall (red ball) and projectile motion (yellow ball). The yellow ball was projected horizontally, while the red ball was released from rest. Can you explain why both balls reach the floor simultaneously?   *(Richard Megna, Fundamental Photographs) (Right)* A multiflash photograph of a popular lecture demonstration in which a projectile is fired at a target that is being held by a magnet in the device at the top right of the photograph. The conditions of the experiment are that the gun is aimed at the target and the projectile leaves the gun at the instant the target is released from rest. Under these conditions the projectile will hit the target, independent of the initial speed of the projectile. The reason is that they both experience the same downward acceleration and hence the velocities of the projectile and target change by the same amount in the same time interval. Note that the velocity of the projectile (red arrows) changes in direction and magnitude, while its downward acceleration (violet arrows) remains constant.   *(Central Scientific Co.)*

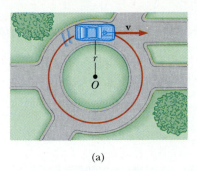

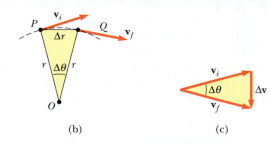

| (a) | (b) | (c) |

FIGURE 4.13   (a) An object moving along a circular path at constant speed experiences uniform circular motion. (b) As the particle moves from *P* to *Q*, the direction of its velocity vector changes from $\mathbf{v}_i$ to $\mathbf{v}_f$. (b) The construction for determining the direction of the change in velocity $\Delta\mathbf{v}$ that is toward the center of the circle.

4.4   UNIFORM CIRCULAR MOTION

Figure 4.13a shows an object moving in a circular path with constant linear speed *v*. Such motion is called **uniform circular motion**. It is often surprising to students to find that *even though an object moves ct a constant speed, it still has an acceleration*. To see why, consider the defining equation for average acceleration, $\bar{\mathbf{a}} = \Delta\mathbf{v}/\Delta t$ (Eq. 4.4).

Note that the acceleration depends on the change in the velocity vector. Because velocity is a vector quantity, there are two ways in which an acceleration can be produced, as mentioned in Section 4.1: by a change in the *magnitude* of the velocity and by a change in the *direction* of the velocity. It is the latter situation that is occurring for an object moving with constant speed in a circular path. The velocity vector is always tangent to the path of the object and perpendicular to the radius *r* of the circular path. We now show that the acceleration vector in uniform circular motion is always perpendicular to the path and always points toward the center of the circle. An acceleration of this nature is called a *centripetal acceleration* (center seeking), and its magnitude is

$$a_r = \frac{v^2}{r} \tag{4.19}$$

Magnitude of centripetal acceleration

where *r* is the radius of the circle.

To derive Equation 4.19, consider Figure 4.13b. Here an object is seen first at point *P* and then at point *Q*. The particle is at *P* at time t_i, and its velocity at that time is $\mathbf{v}_i$; it is at *Q* at some later time t_f, and its velocity at that time is $\mathbf{v}_f$. Let us also assume here that $\mathbf{v}_i$ and $\mathbf{v}_f$ differ only in direction; their magnitudes are the same (that is, $v_i = v_f = v$). In order to calculate the acceleration of the particle, let us begin with the defining equation for average acceleration (Eq. 4.4):

$$\bar{\mathbf{a}} = \frac{\mathbf{v}_f - \mathbf{v}_i}{t_f - t_i} = \frac{\Delta\mathbf{v}}{\Delta t}$$

This equation indicates that we must vectorially subtract $\mathbf{v}_i$ from $\mathbf{v}_f$, where $\Delta\mathbf{v} = \mathbf{v}_f - \mathbf{v}_i$ is the change in the velocity. Since $\mathbf{v}_i + \Delta\mathbf{v} = \mathbf{v}_f$, the vector $\Delta\mathbf{v}$ can be found using the vector triangle in Figure 4.13c. When Δt is very small, Δr and $\Delta\theta$ are also

Motion Simulator

Tthis simulator allows you to model a wide range of problems dealing with motion, including linear motion under constant acceleration and projectile motion. You will have the ability to vary an object's mass, as well as its initial position and initial velocity. In addition, you will be able to observe changes in the motion of an object in the presence or absence of a gravitational field and/or friction.

very small. In this case, $\mathbf{v}_f$ is almost parallel to $\mathbf{v}_i$ and the vector $\Delta\mathbf{v}$ is approximately perpendicular to them, pointing toward the center of the circle.

Now consider the triangle in Figure 4.13b, which has sides Δr and r. This triangle and the one in Figure 4.13c, which has sides Δv and v, are similar. (Two triangles are similar if the angle between any two sides is the same for both triangles and if the ratio of lengths of these sides is the same.) This enables us to write a relationship between the lengths of the sides:

$$\frac{\Delta v}{v} = \frac{\Delta r}{r}$$

This equation can be solved for Δv and the expression so obtained can be substituted into $\bar{a} = \Delta v/\Delta t$ (Eq. 4.4) to give

$$\bar{a} = \frac{v\,\Delta r}{r\,\Delta t}$$

Now imagine that points P and Q in Figure 4.13b become extremely close together. In this case $\Delta\mathbf{v}$ would point toward the center of the circular path, and because the acceleration is in the direction of $\Delta\mathbf{v}$, it too is toward the center. Furthermore, as P and Q approach each other, Δt approaches zero, and the ratio $\Delta r/\Delta t$ approaches the speed v. Hence, in the limit $\Delta t \to 0$, the magnitude of the acceleration is

$$a_r = \frac{v^2}{r}$$

Thus we conclude that in uniform circular motion, the acceleration is directed inward toward the center of the circle and has a magnitude given by v^2/r, where v is the speed of the particle and r is the radius of the circle. You should show that the dimensions of a_r are L/T^2. We shall return to the discussion of circular motion in Section 6.1.

4.5   TANGENTIAL AND RADIAL ACCELERATION

Let us consider the motion of a particle along a curved path where the velocity changes both in direction and in magnitude, as described in Figure 4.14. The velocity vector is always tangent to the path; however, in this situation, the acceleration vector **a** is at some angle to the path.

As the particle moves along the curved path in Figure 4.14, we see that the direction of the total acceleration vector, **a**, changes from point to point. This vector can be resolved into two component vectors: a radial component vector, $\mathbf{a}_r$, and a tangential component vector, $\mathbf{a}_t$. That is, the total acceleration vector can be written as the vector sum of these component vectors:

$$\mathbf{a} = \mathbf{a}_r + \mathbf{a}_t \qquad (4.20)$$

Total acceleration

The tangential acceleration arises from the change in the speed of the particle, and the projection of the acceleration along the direction of the velocity is

$$a_t = \frac{d|\mathbf{v}|}{dt} \qquad (4.21)$$

Tangential acceleration

The radial acceleration is due to the change in direction of the velocity vector and has an absolute magnitude given by

$$a_r = \frac{v^2}{r} \qquad (4.22)$$

Centripetal acceleration

where r is the radius of curvature of the path at the point in question. Since $\mathbf{a}_r$ and $\mathbf{a}_t$ are perpendicular component vectors of **a**, it follows that $a = \sqrt{a_r^2 + a_t^2}$. As in the case of uniform circular motion, $\mathbf{a}_r$ always points toward the center of curvature, as shown in Figure 4.14. Also, at a given speed, a_r is large when the radius of curvature is small (as at points P and Q in Fig. 4.14) and small when r is large (such

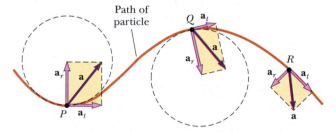

FIGURE 4.14   The motion of a particle along an arbitrary curved path lying in the *xy* plane. If the velocity vector **v** (always tangent to the path) changes in direction and magnitude, the component vectors of the acceleration **a** are a tangential vector $\mathbf{a}_t$ and a radial vector $\mathbf{a}_r$. The acceleration vectors are not necessarily to scale.

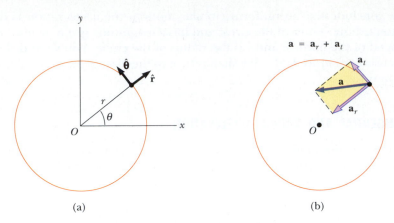

FIGURE 4.15   (a) Descriptions of the unit vectors $\hat{\mathbf{r}}$ and $\hat{\boldsymbol{\theta}}$. (b) The total acceleration $\mathbf{a}$ of a particle moving along a curved path (which at any instant is part of a circle of radius r) consists of a radial component $\mathbf{a}_r$ directed toward the center of rotation, and a tangential component $\mathbf{a}_t$. The tangential component vector $\mathbf{a}_t$ is zero if the speed is constant.

as at point R). The direction of $\mathbf{a}_t$ is either in the same direction as $\mathbf{v}$ (if v is increasing) or opposite $\mathbf{v}$ (if v is decreasing).

Note that in the case of uniform circular motion, where v is constant, $a_t = 0$ and the acceleration is always radial, as we described in Section 4.4. (Note that Eq. 4.22 is identical to Eq. 4.19.) In other words, uniform circular motion is a special case of motion along a curved path. Furthermore, if the direction of $\mathbf{v}$ doesn't change, then there is no radial acceleration and the motion is one dimensional ($a_r = 0$, $a_t \neq 0$).

It is convenient to write the acceleration of a particle moving in a circular path in terms of unit vectors. We do this by defining the unit vectors $\hat{\mathbf{r}}$ and $\hat{\boldsymbol{\theta}}$, where $\hat{\mathbf{r}}$ is a *unit vector along the radius vector directed radially outward* from the center of the circle, and $\hat{\boldsymbol{\theta}}$ is a *unit vector tangent to the circular path,* as in Figure 4.15a. The direction of $\hat{\boldsymbol{\theta}}$ is in the direction of increasing θ where θ is measured counterclockwise from the positive x axis. Note that both $\hat{\mathbf{r}}$ and $\hat{\boldsymbol{\theta}}$ "move along with the particle" and so vary in time relative to a stationary observer. Using this notation, we can express the total acceleration as

$$\mathbf{a} = \mathbf{a}_t + \mathbf{a}_r = \frac{d|\mathbf{v}|}{dt}\hat{\boldsymbol{\theta}} - \frac{v^2}{r}\hat{\mathbf{r}} \tag{4.23}$$

These vectors are described in Figure 4.15b. The negative sign on the v^2/r term for $\mathbf{a}_r$ indicates that the radial acceleration is always directed radially inward, *opposite* the unit vector $\hat{\mathbf{r}}$.

EXAMPLE 4.9   **The Swinging Ball**

A ball tied to the end of a string 0.50 m in length swings in a vertical circle under the influence of gravity, as in Figure 4.16. When the string makes an angle of $\theta = 20°$ with the vertical, the ball has a speed of 1.5 m/s. (a) Find the magnitude of the radial component of acceleration at this instant.

Solution   Since $v = 1.5$ m/s and $r = 0.50$ m, we find that

$$a_r = \frac{v^2}{r} = \frac{(1.5 \text{ m/s})^2}{0.50 \text{ m}} = \boxed{4.5 \text{ m/s}^2}$$

(b) When the ball is at an angle θ to the vertical, it has a tangential acceleration of magnitude $g \sin \theta$ (produced by the tangential component of the force mg.). Therefore, at $\theta = 20°$, $a_t = g \sin 20° = 3.4$ m/s². Find the magnitude and direction of the *total* acceleration at $\theta = 20°$.

Solution   Since $\mathbf{a} = \mathbf{a}_r + \mathbf{a}_t$, the magnitude of $\mathbf{a}$ at $\theta = 20°$ is

$$a = \sqrt{a_r^2 + a_t^2} = \sqrt{(4.5)^2 + (3.4)^2} \text{ m/s}^2 = \boxed{5.6 \text{ m/s}^2}$$

If ϕ is the angle between $\mathbf{a}$ and the string, then

$$\phi = \tan^{-1} \frac{a_t}{a_r} = \tan^{-1}\left(\frac{3.4 \text{ m/s}^2}{4.5 \text{ m/s}^2}\right) = \boxed{36°}$$

Note that all of the vectors—$\mathbf{a}$, $\mathbf{a}_t$, and $\mathbf{a}_r$—change in direction *and* magnitude as the ball swings through the circle. When the ball is at its lowest elevation ($\theta = 0$), $a_t = 0$, because there is no tangential component of g at this angle and a_r is a *maximum* because v is a maximum. If the ball has enough speed to reach its highest position ($\theta = 180°$), a_t is again zero but a_r is a minimum because v is now a minimum. Finally, in the two horizontal positions ($\theta = 90°$ and $270°$), $|a_t| = g$ and a_r has a value between its minimum and maximum values.

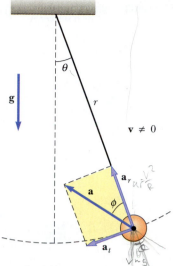

FIGURE 4.16   (Example 4.9) Motion of a ball suspended by a string of length r. The ball swings with nonuniform circular motion in a vertical plane, and its acceleration $\mathbf{a}$ has a radial component $\mathbf{a}_r$ and a tangential component $\mathbf{a}_t$.

4.6   RELATIVE VELOCITY AND RELATIVE ACCELERATION

In this section, we describe how observations made by different observers in different frames of reference are related to each other. We find that observers in different frames of reference may measure different displacements, velocities, and accelerations for a given particle. That is, two observers moving relative to each other generally do not agree on the outcome of a measurement.

For example, if two cars are moving in the same direction with speeds of 50 mi/h and 60 mi/h, a passenger in the slower car will measure the speed of the faster car relative to that of the slower car to be 10 mi/h. Of course, a stationary observer will measure the speed of the faster car to be 60 mi/h. This simple example demonstrates that velocity measurements differ in different frames of reference.

Next, suppose a person riding on a moving vehicle (observer A) throws a ball in such a way that it appears, in his frame of reference, to move first straight upward and then straight downward along the same vertical line, as in Figure 4.17a. However, a stationary observer (B) sees the path of the ball as a parabola, as illustrated in Figure 4.17b. Relative to the observer B, the ball has a vertical component of velocity (resulting from the initial upward velocity and the downward acceleration of gravity) *and* a horizontal component of velocity.

Another simple example is to imagine a package being dropped from an airplane flying parallel to the Earth with a constant velocity, which is the situation we studied in Example 4.5. An observer on the airplane would describe the motion of the package as a straight line toward the Earth. The stranded explorer on the ground, however, would view the trajectory of the package as a parabola. If the airplane continues to move horizontally with the same velocity, the package will hit the ground directly beneath the airplane (assuming that friction is neglected)!

In a more general situation, consider a particle located at the point P in Figure 4.18. Imagine that the motion of this particle is being described by two observers,

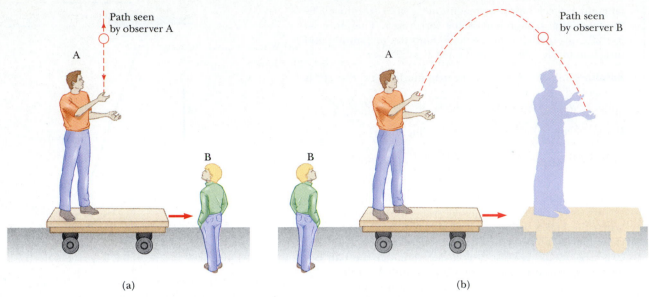

Path seen
by observer A

A

B

(a)

Path seen
by observer B

A

B

(b)

FIGURE 4.17   (a) Observer A in a moving vehicle throws a ball upward and sees it rise and fall in a straight-line path. (b) Stationary observer B sees a parabolic path for the same ball.

one in reference frame *S*, fixed relative to the Earth, and another in reference frame *S'*, moving to the right relative to *S* (and therefore relative to Earth) with a constant velocity **u**. (Relative to an observer in *S'*, *S* moves to the left with a velocity −**u**.) Where an observer stands in a reference frame is irrelevant in this discussion, but to be definite let us place both observers at the origin.

We label the position of the particle relative to the *S* frame with the position vector **r** and label its position relative to the *S'* frame with the vector **r'** both at some time *t*. If the origins of the two reference frames coincide at *t* = 0, then the vectors **r** and **r'** are related to each other through the expression **r** = **r'** + **u***t*, or

Galilean coordinate transformation

$$\mathbf{r}' = \mathbf{r} - \mathbf{u}t \qquad (4.24)$$

That is, after a time *t*, the *S'* frame is displaced to the right by an amount **u***t*.

If we differentiate Equation 4.24 with respect to time and note that **u** is constant, we get

$$\frac{d\mathbf{r}'}{dt} = \frac{d\mathbf{r}}{dt} - \mathbf{u}$$

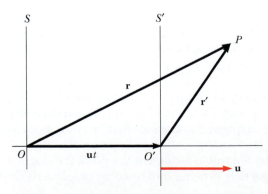

S   *S'*

P

r

r'

O   **u***t*   *O'*

u

FIGURE 4.18   A particle located at *P* is described by two observers, one in the fixed frame of reference, *S*, the other in the frame *S'*, which moves with a constant velocity **u** to the right. The vector **r** is the particle's position vector relative to *S*, and **r'** is its position vector relative to *S'*.

$$\mathbf{v}' = \mathbf{v} - \mathbf{u} \qquad (4.25)$$

Galilean velocity transformation

where $\mathbf{v}'$ is the velocity of the particle observed in the S' frame and $\mathbf{v}$ is its velocity observed in the S frame. Equations 4.24 and 4.25 are known as *Galilean transformation equations*. The equations relate the coordinates and velocity of a particle as measured, for example, in a frame fixed relative to the Earth to those measured in a frame moving with uniform motion relative to the Earth.

Although observers in the different reference frames measure different velocities for the particles, they measure the *same acceleration* when $\mathbf{u}$ is constant. This can be seen by taking the time derivative of Equation 4.25:

$$\frac{d\mathbf{v}'}{dt} = \frac{d\mathbf{v}}{dt} - \frac{d\mathbf{u}}{dt}$$

But $d\mathbf{u}/dt = 0$ because $\mathbf{u}$ is constant. Therefore, we conclude that $\mathbf{a}' = \mathbf{a}$ because $\mathbf{a}' = d\mathbf{v}'/dt$ and $\mathbf{a} = d\mathbf{v}/dt$. That is, *the acceleration of the particle measured by an observer in the Earth's frame of reference is the same as that measured by any other observer moving with constant velocity relative to the Earth frame.*

CONCEPTUAL EXAMPLE 4.10

A ball is thrown upward in the air by a passenger on a train that is moving with constant velocity. (a) Describe the path of the ball as seen by the passenger. Describe the path as seen by a stationary observer outside the train. (b) How would these observations change if the train were accelerating along the track?

Reasoning (a) The passenger sees the ball moving vertically up and then down with a velocity that changes with time. The outside observer sees the ball moving in a parabolic path, with constant horizontal velocity equal to that of the train, as well as a continuously changing vertical velocity. (b) If the train were accelerating horizontally forward, the passenger would see the ball accelerating horizontally backward as well as vertically downward. Relative to the passenger, the ball would move in a parabola with its axis inclined to the vertical. The outside observer would see pure parabolic motion for the ball, with a horizontal acceleration of zero, a vertical acceleration of $-g$, and a horizontal velocity equal to that of the train at the moment the ball was released.

EXAMPLE 4.11 A Boat Crossing a River

A boat heading due north crosses a wide river with a speed of 10.0 km/h relative to the water. The river has a uniform speed of 5.00 km/h due east relative to Earth. Determine the velocity of the boat relative to a stationary ground observer.

Solution We know

$\mathbf{v}_{br}$ = the velocity of the boat, b, relative to the river, r

$\mathbf{v}_{re}$ = the velocity of the river, r, relative to Earth, e

and we want $\mathbf{v}_{be}$, the velocity of the *boat* relative to *Earth*. The relationship between these three quantities is

$$\mathbf{v}_{be} = \mathbf{v}_{br} + \mathbf{v}_{re}$$

The terms in the equation must be manipulated as vector quantities; the vectors are shown in Figure 4.19. The quantity $\mathbf{v}_{br}$ is due north, $\mathbf{v}_{re}$ is due east, and the vector sum of the two, $\mathbf{v}_{be}$, is at an angle θ, as defined in Figure 4.19. Thus, the speed

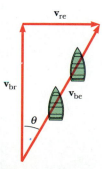

FIGURE 4.19 (Example 4.11)

of the boat relative to Earth can be found from the Pythago-rean theorem:

$$v_{be} = \sqrt{v_{br}^2 + v_{re}^2} = \sqrt{(10.0)^2 + (5.00)^2}\ \text{km/h} = \boxed{11.2\ \text{km/h}}$$

The direction of $\mathbf{v}_{be}$ is

$$\theta = \tan^{-1}\left(\frac{v_{re}}{v_{br}}\right) = \tan^{-1}\left(\frac{5.00}{10.00}\right) = 26.6°$$

Therefore, the boat will be traveling at a speed of 11.2 km/h in the direction 63.4° north of east relative to Earth.

Exercise   If the width of the river is 3.0 km, find the time it takes the boat to cross the river.

Answer   18 min.

EXAMPLE 4.12   Which Way Should We Head?

If the boat of the preceding example travels with the same speed of 10.0 km/h relative to the water and is to travel due north, as in Figure 4.20, what should be its heading?

Solution   As in the previous example, we know $\mathbf{v}_{br}$ and $\mathbf{v}_{re}$, and we want to find $\mathbf{v}_{be}$. The relationship between these three quantities, $\mathbf{v}_{be} = \mathbf{v}_{br} + \mathbf{v}_{re}$, is shown in Figure 4.20. That is, the boat must head upstream in order to be pushed directly northward across the river. The speed v_{be} can be found from the Pythagorean theorem:

$$v_{be} = \sqrt{v_{br}^2 - v_{re}^2} = \sqrt{(10.0)^2 - (5.00)^2}\ \text{km/h} = 8.66\ \text{km/h}$$

The direction of $\mathbf{v}_{be}$ is

$$\theta = \tan^{-1}\left(\frac{v_{re}}{v_{be}}\right) = \tan^{-1}\left(\frac{5.00}{8.66}\right) = \boxed{30.0°}$$

The boat must steer a course 30.0° west of north.

Exercise   If the width of the river is 3.0 km, how long does it take the boat to cross the river?

Answer   21 min.

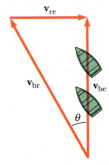

FIGURE 4.20   (Example 4.12)

---

*4.7   RELATIVE MOTION AT HIGH SPEEDS

The Galilean transformation equations, Equations 4.24 and 4.25, are valid *only* at particle speeds (relative to both observers) that are small compared with the speed of light. When the particle speed relative to either observer approaches the speed of light, these transformation equations must be replaced by the equations used by Einstein in his special theory of relativity.

Although we shall discuss the theory of relativity in Chapter 39, a forecast of some of its predictions is in order. As it turns out, the relativistic transformation equations reduce to the Galilean transformations when the particle speed is small compared with the speed of light. This is in keeping with the **correspondence principle** first proposed by Niels Bohr, which in effect states that, if an old theory accurately describes a number of physical phenomena, then any new theory *must* account for the same phenomena over the range of validity of the old theory.

You might wonder how one can test the validity of the transformation equations. The Galilean transformations are used in Newtonian mechanics, while Einstein's theory uses relativistic transformations. From experiments on high-speed particles, one finds that Newtonian mechanics *fails* at particle speeds approaching the speed of light. On the other hand, Einstein's theory of special relativity is in agreement with experiment at *all* speeds. Finally, Newtonian mechanics places no upper limit on speed of a particle. In contrast, the relativistic velocity transformation equation predicts that *no particle speed can exceed the speed of light.* Electrons and protons accelerated through very high voltages can acquire speeds close to the speed of light, but their speeds never reach this limiting value. Hence, experimental results are in complete agreement with the theory of relativity.

SUMMARY

If a particle moves with *constant* acceleration a and has velocity $\mathbf{v}_0$ and position $\mathbf{r}_0$ at $t = 0$, its velocity and position vectors at some later time t are

$$\mathbf{v} = \mathbf{v}_0 + \mathbf{a}t \tag{4.8}$$

$$\mathbf{r} = \mathbf{r}_0 + \mathbf{v}_0 t + \tfrac{1}{2}\mathbf{a}t^2 \tag{4.9}$$

For two-dimensional motion in the xy plane under constant acceleration, these vector expressions are equivalent to two-component expressions, one for the motion along x and one for the motion along y.

Projectile motion is two-dimensional motion under constant acceleration, where $a_x = 0$ and $a_y = -g$. In this case, if $x_0 = y_0 = 0$, the components of Equations 4.8 and 4.9 reduce to

$$v_x = v_{x0} = \text{constant} \tag{4.10}$$

$$v_y = v_{y0} - gt \tag{4.11}$$

$$x = v_{x0}\, t \tag{4.12}$$

$$y = v_{y0}\, t - \tfrac{1}{2}gt^2 \tag{4.13}$$

where $v_{x0} = v_0 \cos \theta_0$, $v_{y0} = v_0 \sin \theta_0$, v_0 is the initial speed of the projectile, and θ_0 is the angle $\mathbf{v}_0$ makes with the positive x axis. Note that these expressions give the velocity components (and hence the velocity vector) and the coordinates (and hence the position vector) at *any* time t that the projectile is in motion.

It is useful to think of projectile motion as the superposition of two motions: (1) uniform motion in the x direction, and (2) motion in the vertical direction subject to a constant downward acceleration of magnitude $g = 9.80 \text{ m/s}^2$. Hence, one can analyze the motion in terms of separate horizontal and vertical components of velocity, as in Figure 4.21.

A particle moving in a circle of radius r with constant speed v undergoes a

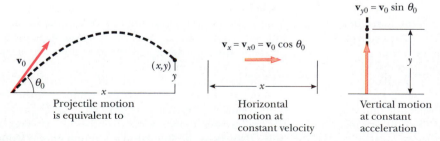

FIGURE 4.21 Analyzing projectile motion in terms of the horizontal and vertical components of velocity.

centripetal (or radial) acceleration, a_r, because the direction of **v** changes in time. The magnitude of a_r is

$$a_r = \frac{v^2}{r} \tag{4.19}$$

and its direction is always toward the center of the circle.

If a particle moves along a curved path in such a way that the magnitude and direction of **v** change in time, the particle has an acceleration vector that can be described by two component vectors: (1) a radial component vector, a_r, arising from the change in direction of **v**, and (2) a tangential component vector, a_t, arising from the change in magnitude of **v**. The magnitude of a_r is v^2/r, and the magnitude of a_t is $d|\mathbf{v}|/dt$.

The velocity of a particle, **v**, measured in a fixed frame of reference, S, is related to the velocity of the same particle, **v**′, measured in a moving frame of reference, $S′$, by

$$\mathbf{v}' = \mathbf{v} - \mathbf{u} \tag{4.25}$$

where **u** is the velocity of $S′$ relative to S.

QUESTIONS

1. If the average velocity of a particle is zero in some time interval, what can you say about the displacement of the particle for that interval?

2. If you know the position vectors of a particle at two points along its path and also know the time it took to get from one point to the other, can you determine the particle's instantaneous velocity? its average velocity? Explain.

3. Describe a situation in which the velocity of a particle is perpendicular to the position vector.

4. Explain whether or not the following particles have an acceleration: (a) a particle moving in a straight line with constant speed and (b) a particle moving around a curve with constant speed.

5. Correct the following statement: "The racing car rounds the turn at a constant velocity of 90 miles per hour."

6. Determine which of the following moving objects have an approximate parabolic trajectory: (a) a ball thrown in an arbitrary direction, (b) a jet airplane, (c) a rocket leaving the launching pad, (d) a rocket a few minutes after launch with failed engines, (e) a tossed stone moving to the bottom of a pond.

7. A student argues that as a satellite orbits the Earth in a circular path, the satellite moves with a constant velocity and therefore has no acceleration. The professor claims that the student is wrong because the satellite must have a centripetal acceleration as it moves in its circular orbit. What is wrong with the student's argument?

8. What is the fundamental difference between the unit vectors $\hat{r}$ and $\hat{\theta}$ defined in Figure 4.15 and the unit vectors **i** and **j**?

9. At the end of its arc, the velocity of a pendulum is zero. Is its acceleration also zero at this point?

10. If a rock is dropped from the top of a sailboat's mast, will it hit the deck at the same point whether the boat is at rest or in motion at constant velocity?

11. A stone is thrown upward from the top of a building. Does the stone's displacement depend on the location of the origin of the coordinate system? Does the stone's velocity depend on the location of the origin?

12. Is it possible for a vehicle to travel around a curve without accelerating? Explain.

13. A baseball is thrown with an initial velocity of (10**i** + 15**j**) m/s. When it reaches the top of its trajectory, what are (a) its velocity and (b) its acceleration? Neglect air resistance.

14. An object moves in a circular path with constant speed v. (a) Is the velocity of the object constant? (b) Is its acceleration constant? Explain.

15. A projectile is fired at some angle to the horizontal with some initial speed v_0, and air resistance is neglected. Is the projectile a freely falling body? What is its acceleration in the vertical direction? What is its acceleration in the horizontal direction?

16. A projectile is fired at an angle of 30° to the horizontal with some initial speed. What other projectile angle gives the same range if the initial speed is the same in both cases? Neglect air resistance.

17. A projectile is fired on the Earth with some initial velocity. Another projectile is fired on the Moon with the same initial velocity. Neglecting air resistance, which projectile has the greater range? Which reaches the greater altitude? (Note that the free-fall acceleration on the Moon is about 1.6 m/s².)

18. As a projectile moves through its parabolic trajectory,

which of these quantities, if any, remain constant: (a) speed, (b) acceleration, (c) horizontal component of velocity, (d) vertical component of velocity?

19. The maximum range of a projectile occurs when it is launched at an angle of 45° with the horizontal if air re-

sistance is neglected. If air resistance is not neglected, will this optimum angle be greater or less than 45°? Explain.

20. A passenger on a train that is moving with constant velocity drops a spoon. What is the acceleration of the spoon relative to (a) the train and (b) the Earth?

PROBLEMS

Review Problem

A ball is thrown with an initial speed v_0 at an angle θ_0 with the horizontal. The range of the ball is R, and the ball reaches a maximum height $R/6$. In terms of R and g, find (a) the time the ball is in motion, (b) the ball's speed at the peak of its path, (c) the initial vertical component of its velocity, (d) its initial speed, (e) the angle θ_0, (f) the maximum height that can be reached if the ball is thrown at the appropriate angle and at the speed found in (d), (g) the maximum range that can be reached if the ball is thrown at the appropriate angle and at the speed found in (d). (h) Suppose that two rocks are thrown from the same point at the same moment as in the figure below. Find the distance between them as a function of time. Assume that v_0 and θ_0 are given.

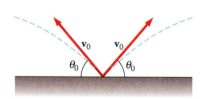

Section 4.1   The Displacement, Velocity, and Acceleration Vectors

1. Suppose that the trajectory of a particle is given by $r(t) = x(t)\mathbf{i} + y(t)\mathbf{j}$ with $x(t) = at^2 + bt$ and $y(t) = ct + d$, where a, b, c, and d are constants that have appropriate dimensions. What displacement does the particle undergo between $t = 1$ s and $t = 3$ s?

2. Suppose that the position vector for a particle is given as $r(t) = x(t)\mathbf{i} + y(t)\mathbf{j}$, with $x(t) = at + b$ and $y(t) = ct^2 + d$, where $a = 1.00$ m/s, $b = 1.00$ m, $c = 0.125$ m/s², and $d = 1.00$ m. (a) Calculate the average velocity during the time interval from $t = 2.00$ s to $t = 4.00$ s. (b) Determine the velocity and the speed at $t = 2.00$ s.

3. A motorist drives south at 20.0 m/s for 3.00 min, then turns west and travels at 25.0 m/s for 2.00 min, and finally travels northwest at 30.0 m/s for 1.00 min. For this 6.00-min trip, find (a) the result-

ant vector displacement, (b) the average speed, and (c) the average velocity.

4. A golf ball is hit off a tee at the edge of a cliff. The x and y coordinates of the golf ball versus time are given by the expressions $x = (18.0 \text{ m/s})t$ and $y = (4.00 \text{ m/s})t - (4.90 \text{ m/s}^2)t^2$. (a) Write a vector expression for the position $\mathbf{r}$ as a function of time t using the unit vectors $\mathbf{i}$ and $\mathbf{j}$. By taking derivatives, repeat for (b) the velocity vector $\mathbf{v}(t)$ and (c) the acceleration vector $\mathbf{a}(t)$. (d) Find the x and y coordinates of the ball at $t = 3.00$ s. Using the unit vectors $\mathbf{i}$ and $\mathbf{j}$, write expressions for (e) the velocity $\mathbf{v}$ and (f) the acceleration $\mathbf{a}$ at the instant $t = 3.00$ s.

Section 4.2   Two-Dimensional Motion with Constant Acceleration

5. At $t = 0$, a particle moving in the xy plane with constant acceleration has a velocity of $\mathbf{v}_0 = (3\mathbf{i} - 2\mathbf{j})$ m/s at the origin. At $t = 3$ s, its velocity is given by $\mathbf{v} = (9\mathbf{i} + 7\mathbf{j})$ m/s. Find (a) the acceleration of the particle and (b) its coordinates at any time t.

6. A particle starts from rest at $t = 0$ at the origin and moves in the xy plane with a constant acceleration of $\mathbf{a} = (2\mathbf{i} + 4\mathbf{j})$ m/s². After a time t has elapsed, determine (a) the x and y components of velocity, (b) the coordinates of the particle, and (c) the speed of the particle.

7. A fish swimming in a horizontal plane has velocity $\mathbf{v}_0 = (4.0\mathbf{i} + 1.0\mathbf{j})$ m/s at a point in the ocean whose position vector is $\mathbf{r}_0 = (10.0\mathbf{i} - 4.0\mathbf{j})$ m relative to a stationary rock at the shore. After the fish swims with constant acceleration for 20.0 s, its velocity is $\mathbf{v} = (20.0\mathbf{i} - 5.0\mathbf{j})$ m/s. (a) What are the components of the acceleration? (b) What is the direction of the acceleration with respect to the fixed x axis? (c) Where is the fish at $t = 25$ s and in what direction is it moving?

8. The position of a particle varies in time according to the expression $\mathbf{r} = (3.00\mathbf{i} - 6.00t^2\mathbf{j})$ m. (a) Find expressions for the velocity and acceleration as functions of time. (b) Determine the particle's position and velocity at $t = 1.00$ s.

9. The coordinates of an object moving in the xy plane vary with time according to the equations $x =$

− (5.0 m)sin(t)   and y = (4.0 m) − (5.0 m)cos(t), where t is in seconds. (a) Determine the components of the velocity and those of the acceleration at t = 0 s. (b) Write expressions for the position vector, the velocity vector, and the acceleration vector at any time t > 0. (c) Describe the path of the object in an xy plot.

Section 4.3   Projectile Motion

(Neglect air resistance in all problems.)

10. Jimmy is at the base of a hill, while Billy is 30 m up the hill. Jimmy is at the origin of an xy coordinate system, and the line that follows the slope of the hill is given by the equation, $y = 0.4x$, as shown in Figure P4.10. If Jimmy throws an apple to Billy at an angle of 50° with respect to the horizontal, with what speed must he throw the apple if it is to reach Billy?

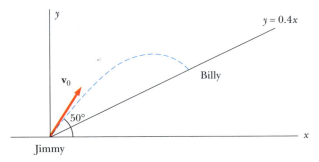

FIGURE P4.10

11. In a local bar, a customer slides an empty beer mug on the counter for a refill. The bartender is momentarily distracted and does not see the mug, which slides off the counter and strikes the floor 1.40 m from the base of the counter. If the height of the counter is 0.860 m, (a) with what speed did the mug leave the counter and (b) what was the direction of the mug's velocity just before it hit the floor?

12. A student decides to measure the muzzle speed of the pellets from her BB gun. She points the gun horizontally at a target placed on a vertical wall a distance x away from the gun. The shots hit the target a vertical distance y below the gun. (a) Show that the position of the pellet when traveling through the air is $y = Ax^2$, where A is a constant. (b) Express the constant A in terms of the initial speed and the free-fall acceleration. (c) If x = 3.0 m and y = 0.21 m, what is the speed of the BB?

13. A ball is thrown horizontally from the top of a building 35 m high. The ball strikes the ground at a point 80 m from the base of the building. Find (a) the time the ball is in flight, (b) its initial velocity, and (c) the x and y components of velocity just before the ball strikes the ground.

14. Superman is flying at treetop level near Paris when he sees the Eiffel Tower elevator start to fall (the cable snapped). His x-ray vision tells him Lois Lane is inside. If Superman is 1.00 km away from the tower, and the elevator falls from a height of 240 m, how long does he have to save Lois, and what must his average speed be?

14A. Superman is flying at treetop level near Paris when he sees the Eiffel Tower elevator start to fall (the cable snapped). His x-ray vision tells him Lois Lane is inside. If Superman is a distance d away from the tower and the elevator falls from a height h, how long does he have to save Lois, and what must his average speed be?

15. A soccer player kicks a rock horizontally off the edge of a 40.0-m-high cliff into a pool of water. If the player hears the sound of the splash 3.0 s after the kick, what was the initial speed? Assume the speed of sound in air is 343 m/s.

16. A baseball player throwing the ball in from the outfield usually allows it to take one bounce on the theory that the ball arrives sooner this way. Suppose the ball hits the ground at an angle θ and then rebounds at the same angle after the bounce but loses half its speed. (a) Assuming the ball is always thrown with the same initial speed, at what angle θ should it be thrown in order to go the same distance D with one bounce (solid path in Fig. P4.16) as one thrown upward at 45° and reaches its target without bouncing (dashed path in Fig. P4.16)? (b) Determine the ratio of the times for the one-bounce and the no-bounce throws.

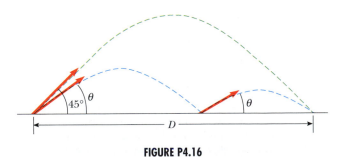

FIGURE P4.16

17. A place kicker must kick a football from a point 36.0 m (about 40 yards) from the goal and the ball must clear the crossbar, which is 3.05 m high. When kicked, the ball leaves the ground with a speed of 20.0 m/s at an angle of 53.0° to the horizontal. (a) By how much does the ball clear or fall short of clearing the crossbar? (b) Does the ball approach the crossbar while still rising or while falling?

18. A firefighter, 50.0 m away from a burning building, directs a stream of water from a firehose at an angle

of 30.0° above the horizontal as in Figure P4.18. If the initial speed of the stream is 40.0 m/s, at what height does the water strike the building?

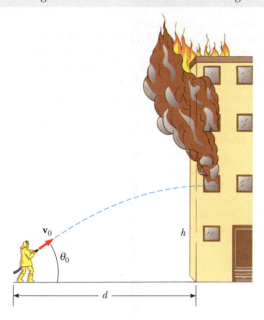

FIGURE P4.18

18A. A firefighter, a distance d from a burning building, directs a stream of water from a firehose at an angle of θ_0 above the horizontal as in Figure P4.18. If the initial speed of the stream is v_0, at what height h does the water strike the building?

19. An astronaut standing on the Moon fires a gun so that the bullet leaves the barrel initially moving in a horizontal direction. (a) What must the muzzle speed be if the bullet is to travel completely around the Moon and return to its original location? (b) How long is the bullet in flight? Assume that the free-fall acceleration on the Moon is one-sixth that on the Earth.

20. A rifle is aimed horizontally at the center of a large target 200 m away. The initial speed of the bullet is 500 m/s. (a) Where does the bullet strike the target? (b) To hit the center of the target, the barrel must be at an angle above the line of sight. Find the angle of elevation of the barrel.

21. During World War I, the Germans had a gun called Big Bertha that was used to shell Paris. The shell had an initial speed of 1.70 km/s at an initial inclination of 55.0° to the horizontal. In order to hit the target, adjustments were made for air resistance and other effects. If we ignore those effects, (a) how far away did the shell hit? (b) How long was it in the air?

22. One strategy in a snowball fight is to throw a snowball at a high angle over level ground. While your oppo-

nent is watching this first snowball, you throw a second one at a low angle timed to arrive at your opponent either before or at the same time as the first one. Assume both snowballs are thrown with a speed of 25 m/s. The first one is thrown at an angle of 70° with respect to the horizontal. (a) At what angle should the second snowball be thrown in order to arrive at the same point as the first? (b) How many seconds later should the second snowball be thrown after the first to arrive at your target at the same time?

23. A projectile is fired in such a way that its horizontal range is equal to three times its maximum height. What is the angle of projection?

24. A flea can jump a vertical height h. (a) What is the maximum horizontal distance it can jump? (b) What is the time in the air in both cases?

25. A cannon having a muzzle speed of 1000 m/s is used to destroy a target on a mountaintop. The target is 2000 m from the cannon horizontally and 800 m above the ground. At what angle, relative to the ground, should the cannon be fired? Ignore air friction.

26. A ball is tossed from an upper-story window of a building. The ball is given an initial velocity of 8.00 m/s at an angle of 20.0° below the horizontal. It strikes the ground 3.00 s later. (a) How far horizontally from the base of the building does the ball strike the ground? (b) Find the height from which the ball was thrown. (c) How long does it take the ball to reach a point 10.0 m below the level of launching?

Section 4.4   Uniform Circular Motion

27. If the rotation of the Earth increased to the point that the centripetal acceleration was equal to the gravitational acceleration at the equator, (a) what would be the tangential speed of a person standing at the equator, and (b) how long would a day be?

28. Young David, who slew Goliath, experimented with slings before tackling the giant. He found that with a sling of length 0.60 m, he could revolve the sling at the rate of 8.0 rev/s. If he increased the length to 0.90 m, he could revolve the sling only 6.0 times per second. (a) Which rate of rotation gives the larger linear speed? (b) What is the centripetal acceleration at 8.0 rev/s? (c) What is the centripetal acceleration at 6.0 rev/s?

29. An athlete rotates a 1.00-kg discus along a circular path of radius 1.06 m. The maximum speed of the discus is 20.0 m/s. Determine the magnitude of its maximum radial acceleration.

30. From information in the endsheets of this book, compute the radial acceleration of a point on the surface of the Earth at the equator.

31. The orbit of the Moon about the Earth is approximately circular, with a mean radius of 3.84×10^8 m.

It takes 27.3 days for the Moon to complete one revolution about the Earth. Find (a) the mean orbital speed of the Moon and (b) its centripetal acceleration.

32. In the spin cycle of a washing machine, the tub of radius 0.300 m rotates at a constant rate of 630 rev/min. What is the maximum linear speed with which water leaves the machine?

33. A ball on the end of a string is whirled around in a horizontal circle of radius 0.30 m. The plane of the circle is 1.2 m above the ground. The string breaks and the ball lands 2.0 m away from the point on the ground directly beneath the ball's location when the string breaks. Find the centripetal acceleration of the ball during its circular motion.

34. A tire 0.500 m in radius rotates at a constant rate of 200 rev/min. Find the speed and acceleration of a small stone lodged in the tread on the outer edge of the tire.

Section 4.5   Tangential and Radial Acceleration

35. Figure P4.35 represents, at a given instant, the total acceleration of a particle moving clockwise in a circle of radius 2.50 m. At this instant of time, find (a) the centripetal acceleration, (b) the speed of the particle, and (c) its tangential acceleration.

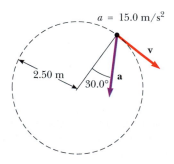

$a = 15.0 \text{ m/s}^2$

2.50 m

30.0°

FIGURE P4.35

36. A point on a rotating turntable 20.0 cm from the center accelerates from rest to 0.700 m/s in 1.75 s. At $t = 1.25$ s, find the magnitude and direction of (a) the centripetal acceleration, (b) the tangential acceleration, and (c) the total acceleration of the point.

37. A train slows down as it rounds a sharp, level turn, slowing from 90.0 km/h to 50.0 km/h in the 15.0 s that it takes to round the bend. The radius of the curve is 150 m. Compute the acceleration at the moment the train speed reaches 50.0 km/h.

38. A pendulum of length 1.00 m swings in a vertical plane (Fig. P4.16). When the pendulum is in the two horizontal positions $\theta = 90°$ and $\theta = 270°$, its speed

is 5.00 m/s. (a) Find the magnitude of the centripetal acceleration and tangential acceleration for these positions. (b) Draw vector diagrams to determine the direction of the total acceleration for these two positions. (c) Calculate the magnitude and direction of the total acceleration.

39. A student attaches a ball to the end of a string 0.600 m in length and then swings the ball in a vertical circle. The speed of the ball is 4.30 m/s at its highest point and 6.50 m/s at its lowest point. Find its acceleration at (a) its highest point and (b) its lowest point.

40. A ball swings in a vertical circle at the end of a rope 1.50 m long. When it is 36.9° past the lowest point on its way up, the ball's total acceleration is $(-22.5\mathbf{i} + 20.2\mathbf{j})$ m/s². For that instant, (a) sketch a vector diagram showing the components of its acceleration, (b) determine the magnitude of its centripetal acceleration, and (c) determine the magnitude and direction of its velocity.

Section 4.6   Relative Velocity and Relative Acceleration

41. Heather in her Corvette accelerates at the rate of $(3.0\mathbf{i} - 2.0\mathbf{j})$ m/s², while Jill in her Jaguar accelerates at $(1.0\mathbf{i} + 3.0\mathbf{j})$ m/s². They both start from rest at the origin of an xy coordinate system. After 5.0 s, (a) what is Heather's speed with respect to Jill, (b) how far apart are they, and (c) what is Heather's acceleration relative to Jill?

42. A motorist traveling west at 80.0 km/h is being chased by a police car traveling at 95.0 km/h. What is the velocity of (a) the motorist relative to the police car and (b) the police car relative to the motorist?

43. A river has a steady speed of 0.500 m/s. A student swims upstream a distance of 1.00 km and returns to the starting point. If the student can swim at a speed of 1.20 m/s in still water, how long does the trip take? Compare this with the time the trip would take if the water were still.

44. How long does it take an automobile traveling in the left lane at 60.0 km/h to overtake (pull alongside) a car traveling in the right lane at 40.0 km/h, if the cars' front bumpers are initially 100 m apart?

45. When the Sun is directly overhead, a hawk dives toward the ground at a speed of 5.00 m/s. If the direction of his motion is at an angle of 60.0° below the horizontal, calculate the speed of his shadow moving along the ground.

46. A boat crosses a river of width $w = 160$ m in which the current has a uniform speed of 1.50 m/s. The pilot maintains a bearing (i.e., the direction in which the boat points) perpendicular to the river and a throttle setting to give a constant speed of 2.00 m/s relative to the water. (a) What is the speed of the boat relative to a stationary shore observer? (b) How far

downstream from the initial position is the boat when it reaches the opposite shore?

47. The pilot of an airplane notes that the compass indicates a heading due west. The airplane's speed relative to the air is 150 km/h. If there is a wind of 30.0 km/h toward the north, find the velocity of the airplane relative to the ground.

48. Two swimmers, A and B, start at the same point in a stream that flows with a speed v. Both move at the same speed c relative to the stream, where $c > v$. A swims downstream a distance L and then upstream the same distance, while B swims directly perpendicular to the stream's flow a distance L and then back the same distance, so that both swimmers return back to the starting point. Which swimmer returns first? (*Note*: First guess at the answer.)

49. A car travels due east with a speed of 50.0 km/h. Rain is falling vertically relative to the Earth. The traces of the rain on the side windows of the car make an angle of 60.0° with the vertical. Find the velocity of the rain relative to (a) the car and (b) the Earth.

50. A child in danger of drowning in a river is being carried downstream by a current that has a speed of 2.50 km/h. The child is 0.600 km from shore and 0.800 km upstream of a boat landing when a rescue boat sets out. (a) If the boat proceeds at its maximum speed of 20.0 km/h relative to the water, what heading relative to the shore should the pilot take? (b) What angle does the boat velocity make with the shore? (c) How long does it take the boat to reach the child?

51. A bolt drops from the ceiling of a train car that is accelerating northward at a rate of 2.50 m/s². What is the acceleration of the bolt relative to (a) the train car? (b) the Earth?

52. A science student is riding on a flatcar of a train traveling along a straight horizontal track at a constant speed of 10.0 m/s. The student throws a ball into the air along a path that he judges to make an initial angle of 60.0° with the horizontal and to be in line with the track. The student's professor, who is standing on the ground nearby, observes the ball to rise vertically. How high does she see the ball rise?

ADDITIONAL PROBLEMS

53. At $t = 0$ a particle leaves the origin with a velocity of 6.00 m/s in the positive y direction. Its acceleration is given by $a = (2.00\mathbf{i} - 3.00\mathbf{j})$ m/s². When the particle reaches its maximum y coordinate, its y component of velocity is zero. At this instant, find (a) the velocity of the particle and (b) its x and y coordinates.

54. The speed of a projectile when it reaches its maximum height is one half the speed when the projectile is at half its maximum height. What is the initial projection angle?

55. A car is parked overlooking the ocean on an incline that makes an angle of 37.0° with the horizontal. The distance from where the car is parked to the bottom of the incline is 50.0 m, and the incline terminates at a cliff that is 30.0 m above the ocean surface. The negligent driver leaves the car in neutral, and the parking brakes are defective. If the car rolls from rest down the incline with a constant acceleration of 4.00 m/s², find (a) the speed of the car just as it reaches the cliff and the time it takes to get there, (b) the velocity of the car just as it lands in the ocean, (c) the total time the car is in motion, and (d) the position of the car relative to the base of the cliff just as it lands in the ocean.

56. A projectile is fired up an incline (incline angle ϕ) with an initial speed v_0 at an angle θ_0 with respect to the horizontal ($\theta_0 > \phi$), as shown in Figure P4.56. (a) Show that the projectile travels a distance d up the incline, where

$$d = \frac{2v_0^2 \cos \theta_0 \sin(\theta_0 - \phi)}{g \cos^2 \phi}$$

(b) For what value of θ_0 is d a maximum, and what is the maximum value?

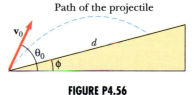

Path of the projectile

FIGURE P4.56

57. A batter hits a pitched baseball 1.00 m above the ground, imparting to the ball a speed of 40.0 m/s. The resulting line drive is caught on the fly by the left fielder 60.0 m from home plate with his glove 1.00 m above the ground. If the shortstop, 45.0 m from home plate and in line with the drive, were to jump straight up to make the catch instead of allowing the left fielder to make the play, how high above the ground would his glove have to be?

58. A basketball player 2.00 m tall throws to the basket from a horizontal distance of 10.0 m, as in Figure P4.58. If he shoots at a 40° angle with the horizontal, at what initial speed must he throw so that the ball goes through the hoop without striking the backboard?

59. A boy can throw a ball a maximum horizontal distance of 40.0 m on a level field. How far can he throw the same ball vertically upward? Assume that his muscles give the ball the same speed in each case.

59A. A boy can throw a ball a maximum horizontal dis-

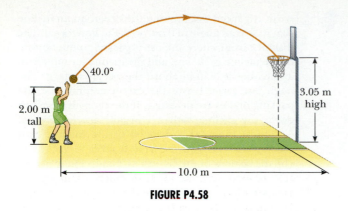

FIGURE P4.58

tance R on a level field. How far can he throw the same ball vertically upward? Assume that his muscles give the ball the same speed in each case.

60. The x and y coordinates of a particle are given by

$$x = 2.00 \text{ m} + (3.00 \text{ m/s})t \qquad y = x - (5.00 \text{ m/s}^2)t^2$$

How far from the origin is the particle at (a) $t = 0$; (b) $t = 2.00$ s?

61. A stone at the end of a sling is whirled in a vertical circle of radius 1.20 m at a constant speed $v_0 =$ 1.50 m/s as in Figure P4.61. The center of the string is 1.50 m above the ground. What is the range of the stone if it is released when the sling is inclined at $30.0°$ with the horizontal (a) at A? (b) at B? What is the acceleration of the stone (c) just before it is released at A? (d) just after it is released at A?

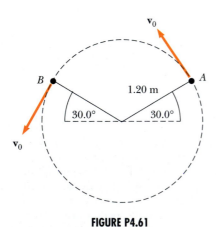

FIGURE P4.61

62. A truck is moving north with a constant speed of 10.0 m/s on a horizontal stretch of road. A boy riding on the back of the truck wishes to throw a ball while the truck is moving and to catch the ball after the truck has gone 20.0 m. (a) Neglecting air resistance, at what angle to the vertical should the ball be thrown? (b) What should be the initial speed of the

ball? (c) What is the shape of the path of the ball as seen by the boy? (d) An observer on the ground watches the boy throw the ball up and catch it. In this observer's fixed frame of reference, determine the general shape of the ball's path and the initial velocity of the ball.

63. A dart gun is fired while being held horizontally at a height of 1.00 m above ground level. With the gun at rest relative to the ground, the dart from the gun travels a horizontal distance of 5.00 m. A child holds the same gun in a horizontal position while sliding down a $45.0°$ incline at a constant speed of 2.00 m/s. How far will the dart travel if the gun is fired when it is 1.00 m above the ground?

64. A rocket is launched at an angle of $53.0°$ to the horizontal with an initial speed of 100 m/s. It moves along its initial line of motion with an acceleration of 30.0 m/s^2 for 3.00 s. At this time its engines fail and the rocket proceeds to move as a free body. Find (a) the maximum altitude reached by the rocket, (b) its total time of flight, and (c) its horizontal range.

65. A person standing at the top of a hemispherical rock of radius R kicks a ball (initially at rest on the top of the rock) so that its initial velocity is horizontal as in Figure P4.65. (a) What must its minimum initial speed be if the ball is never to hit the rock after it is kicked? (b) With this initial speed, how far from the base of the rock does the ball hit the ground?

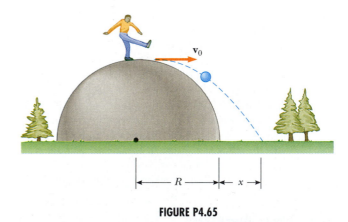

FIGURE P4.65

66. A home run in a baseball game is hit in such a way that the ball just clears a wall 21.0 m high, located 130 m from home plate. The ball is hit at an angle of $35.0°$ to the horizontal, and air resistance is negligible. Find (a) the initial speed of the ball, (b) the time it takes the ball to reach the wall, and (c) the velocity components and the speed of the ball when it reaches the wall. (Assume the ball is hit at a height of 1.00 m above the ground.)

67. A daredevil is shot out of a cannon at $45.0°$ to the horizontal with an initial speed of 25.0 m/s. A net is

located at a horizontal distance of 50.0 m from the cannon. At what height above the cannon should the net be placed in order to catch the daredevil?

68. The position of a particle as a function of time is described by

$$\mathbf{r} = (bt)\mathbf{i} + (c - dt^2)\mathbf{j} \qquad b = 2.00 \text{ m/s}$$

$$c = 5.00 \text{ m} \qquad d = 1.00 \text{ m/s}^2$$

(a) Express y in terms of x and sketch the trajectory of the particle. What is the shape of the trajectory? (b) Derive a vector relationship for the velocity. (c) At what time ($t > 0$) is the velocity vector perpendicular to the position vector?

69. A bomber flies horizontally with a speed of 275 m/s relative to the ground. The altitude of the bomber is 3000 m and the terrain is level. Neglect the effects of air resistance. (a) How far from the point vertically under the point of release does a bomb hit the ground? (b) If the plane maintains its original course and speed, where is it when the bomb hits the ground? (c) At what angle from the vertical at the point of release must the telescopic bomb sight be set so that the bomb hits the target seen in the sight at the time of release?

70. A football is thrown toward a receiver with an initial speed of 20.0 m/s at an angle of 30.0° above the horizontal. At that instant, the receiver is 20.0 m from the quarterback. In what direction and with what constant speed should the receiver run in order to catch the football at the level at which it was thrown?

71. A flea is at point A on a horizontal turntable 10.0 cm from the center. The turntable is rotating at $33\frac{1}{3}$ rev/min in the clockwise direction. The flea jumps vertically upward to a height of 5.00 cm and lands on the turntable at point B. Place the coordinate origin at the center of the turntable with the positive x axis fixed in space and passing through A. (a) Find the linear displacement of the flea. (b) Find the position of point A when the flea lands. (c) Find the position of point B when the flea lands.

72. A student who is able to swim at a speed of 1.50 m/s in still water wishes to cross a river that has a current of velocity 1.20 m/s toward the south. The width of the river is 50.0 m. (a) If the student starts from the west bank, in what direction should she head in order to swim directly across the river? How long will this trip take? (b) If she heads due east, how long will it take to cross the river? (*Note:* The student travels farther than 50.0 m in this case.)

73. A rifle has a maximum range of 500 m. (a) For what angles of elevation would the range be 350 m? What is the range when the bullet leaves the rifle (b) at 14.0°? (c) at 76.0°?

74. A river flows with a uniform velocity v. A person in a motorboat travels 1.00 km upstream, at which time a log is seen floating by. The person continues to travel upstream for 60.0 min at the same speed and then returns downstream to the starting point, where the same log is seen again. Find the velocity of the river. (*Hint:* The time of travel of the boat after it meets the log equals the time of travel of the log.)

75. An airplane has a velocity of 400 km/h due east relative to the moving air. At the same time, a wind blows northward with a speed of 75.0 km/h relative to Earth. (a) Find the airplane's velocity relative to Earth. (b) In what direction must the airplane head in order to move east relative to Earth?

76. A sailor aims a rowboat toward an island located 2.00 km east and 3.00 km north of her starting position. After an hour she sees the island due west. She then aims the boat in the opposite direction from which she was rowing, rows for another hour, and ends up 4.00 km east of her starting position. She correctly deduces that the current is from west to east. (a) What is the speed of the current? (b) Show that the boat's velocity relative to the shore for the first hour can be expressed as u = (4.00 km/h)i + (3.00 km/h)j, where i is directed east and j is directed north.

77. Two soccer players, Mary and Jane, begin running from approximately the same point at the same time. Mary runs in an easterly direction at 4.0 m/s, while Jane takes off in a direction 60° north of east at 5.4 m/s. (a) How long is it before they are 25 m apart? (b) What is the velocity of Jane relative to Mary? (c) How far apart are they after 4.0 s?

78. After delivering his toys in the usual manner, Santa decides to have some fun and slide down an icy roof, as in Figure P4.78. He starts from rest at the top of the roof, which is 8.00 m in length, and accelerates at the rate of 5.00 m/s². The edge of the roof is 6.00 m above a soft snow bank, which Santa lands on. Find (a) Santa's velocity components when he reaches the snow bank, (b) the total time he is in motion, and (c) the distance d between the house and the point where he lands in the snow.

FIGURE P4.78

79. A skier leaves the ramp of a ski jump with a velocity of 10 m/s, 15° above the horizontal, as in Figure P4.79.

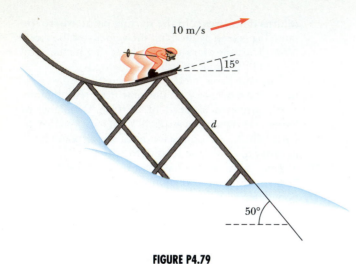

FIGURE P4.79

FIGURE P4.81

The slope is inclined at 50°, and air resistance is negligible. Find (a) the distance that the jumper lands down the slope and (b) the velocity components just before landing. (How do you think the results might be affected if air resistance were included? Note that jumpers lean forward in the shape of an airfoil with their hands at their sides to increase their distance. Why does this work?)

80. A golf ball leaves the ground at an angle θ and hits a tree while moving horizontally at height h above the ground. If the tree is a horizontal distance of b from the point of projection, show that (a) $\tan \theta = 2 \, h/b$. (b) What is the initial velocity of the ball in terms of b and h?

81. A truck loaded with cannonball watermelons stops suddenly to avoid running over the edge of a washed-out bridge (Fig. P4.81). The quick stop causes a number of melons to fly off the truck. One melon rolls over the edge with an initial speed $v_0 = 10$ m/s in the horizontal direction. What are the x and y coordinates of the melon when it splatters on the bank, if a cross-section of the bank has the shape of a parabola ($y^2 = 16x$ where x and y are measured in meters) with its vertex at the edge of the road?

82. An enemy ship is on the east side of a mountain island as shown in Figure P4.82. The enemy ship can maneuver to within 2500 m of the 1800-m-high mountain peak and can shoot projectiles with an initial speed of 250 m/s. If the western shoreline is horizontally 300 m from the peak, what are the distances from the western shore at which a ship can be safe from the bombardment of the enemy ship?

83. A hawk is flying horizontally at 10.0 m/s in a straight line 200 m above the ground. A mouse it was carrying is released from its grasp. The hawk continues on its path at the same speed for two seconds before attempting to retrieve its prey. To accomplish the retrieval, it dives in a straight line at constant speed and recaptures the mouse 3.0 m above the ground. Assuming no air resistance (a) find the diving speed of the hawk. (b) What angle did the hawk make with the horizontal during its descent? (c) For how long did the mouse "enjoy" free flight?

84. The determined coyote is out once more to try to capture the elusive road runner. The coyote wears a pair of Acme jet-powered roller skates, which provide a constant horizontal acceleration of 15 m/s² (Fig. P4.84). The coyote starts off at rest 70 m from the edge of a cliff at the instant the road runner zips by in the direction of the cliff. (a) If the road runner moves with constant speed, determine the minimum speed he must have in order to reach the cliff before the coyote. (b) If the cliff is 100 m above the base of a canyon, determine where the coyote lands in the

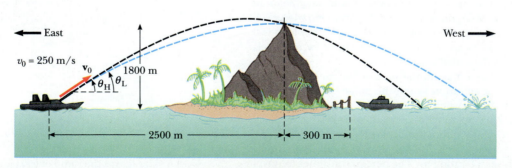

FIGURE P4.82

Coyoté Stupidus   Chicken Delightus

BEEP BEEP

FIGURE P4.84

canyon (assume his skates are still in operation when he is in "flight"). (c) Determine the coyote's velocity components just before he lands in the canyon. (As usual, the road runner is saved by making a sudden turn at the cliff.)

85. The Earth is 1.50×10^{11} m from the Sun and makes one revolution around the Sun in 3.16×10^7 s. The Moon is 3.84×10^8 m from the Earth and makes one revolution around the Earth in 2.36×10^6 s. Determine the velocity of the Moon relative to the Sun at the instant the Moon is heading directly toward the Sun, as in Figure P4.85.

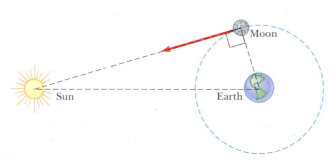

Moon

Sun   Earth

FIGURE P4.85

SPREADSHEET PROBLEMS

S1. The punter for the Los Angeles Rams needs to kick the football 40 to 45 yd (36.6 to 41.1 m) to keep the opposing team inside the 5-yd line. Use Spreadsheet 4.1 to estimate the angle at which he should kick the ball. *Note:* He must keep the ball in the air as long as possible to help prevent a runback (in other words, he wants a long "hang time"). The initial speed of the football should be in the range of 20 to 30 m/s.

S2. Modify Spreadsheet 4.1 by adding two columns to calculate the maximum height reached by the football in Problem S4.1 and the total time it is in the air.

S3. The x and y coordinates of a projectile are

$$x = x_0 + v_{x0}t$$
$$y = y_0 + v_{y0}t - \tfrac{1}{2}gt^2$$

where $v_{x0} = v_0 \cos \theta_0$ and $v_{y0} = v_0 \sin \theta_0$. Using these equations, set up a spreadsheet to calculate the x and y coordinates of a projectile and the components of its velocity v_x and v_y as functions of time. The initial speed and initial angle of the projectile should be input parameters. Use the graph function of the spreadsheet to plot the x and y coordinates as functions of time. Also graph v_x and v_y as functions of time. Use this spreadsheet to solve the following problem:

The New York Giants are tied with the Chicago Bears with only a few seconds left in the game. The Giants have the football, and call the place kicker into the game. He must kick the ball 52 yd (47.5 m) for a field goal. If the crossbar of the goal is 10 ft (3.05 m) high, at what angle and speed should he kick the ball to score the winning points? Is there only one solution to this problem? Solutions can be found by varying the parameters and studying the graph of y versus x.

The Wizard of Id   by Parker and Hart

The Laws of Motion

This calf-roping scene, taken in Steamboat, Colorado, is a standard rodeo event. The external forces acting on the horse are the force of friction between the horse and ground, the force of gravity, the tension force of the rope attached to the calf, the force of the cowboy on the horse, and the upward force of the ground. Can you identify the forces acting on the calf? *(FourbyFive, Inc.)*

In Chapters 2 and 4, we described the motion of particles based on the definitions of displacement, velocity, and acceleration. However, we would like to be able to answer specific questions related to the causes of motion, such as "What mechanism causes motion?" and "Why do some objects accelerate at a higher rate than others?" In this chapter, we use the concepts of force and mass to describe the change in motion of particles. We then discuss the three basic laws of motion, which are based on experimental observations and were formulated more than three centuries ago by Newton.

Classical mechanics describes the relationship between the motion of a body and the forces acting on that body. Classical mechanics deals only with objects that (a) are large compared with the dimensions of atoms ($\approx 10^{-10}$ m) and (b) move at speeds that are much less than the speed of light (3.00×10^8 m/s).

We learn in this chapter how it is possible to describe an object's acceleration in terms of the resultant force acting on the object and its mass. This force represents the interaction of the object with its environment. Mass is a measure of the object's inertia, that is, its tendency to resist an acceleration when a force acts on it.

We also discuss force laws, which describe the quantitative method of calculating the force on an object if its environment is known. We see that, although the force laws are simple in form, they successfully explain a wide variety of phenomena and experimental observations. These force laws, together with the laws of motion, are the foundations of classical mechanics.

5.1 THE CONCEPT OF FORCE

Everyone has a basic understanding of the concept of force from everyday experiences. When you push or pull an object, you exert a force on it. You exert a force when you throw or kick a ball. In these examples, the word *force* is associated with the result of muscular activity and some change in the state of motion of an object. Forces do not always cause an object to move, however. For example, as you sit reading this book, the force of gravity acts on your body, and yet you remain stationary. As a second example, you can push on a large block of stone and not be able to move it.

What force (if any) causes a distant star to drift freely through space? Newton answered such questions by stating that the change in velocity of an object is caused by forces. Therefore, if an object moves with uniform motion (constant velocity), no force is required to maintain the motion. Since only a force can cause a change in velocity, we can think of force as that which causes a body to accelerate.

Now consider a situation in which several forces act simultaneously on an object. In this case, the object accelerates only if the *net force* acting on it is not equal to zero. (We sometimes refer to the net force as either the *resultant force* or the *unbalanced force.*) *If the net force is zero, the acceleration is zero and the velocity of the object remains constant.* That is, if the net force acting on the object is zero, the object either remains at rest or continues to move with constant velocity. *When the velocity of a body is constant or when the body is at rest, it is said to be in equilibrium.*

Whenever a force is exerted on an object, the shape of the object can change. For example, when you squeeze a rubber ball or strike a punching bag with your fist, the objects are deformed to some extent. Even more rigid objects, such as an automobile, are deformed under the action of external forces. The deformations can be permanent if the forces are large enough, as in the case of a collision between vehicles.

In this chapter, we are concerned with the relationship between the force exerted on an object and the acceleration of that object. If you pull on a coiled spring, as in Figure 5.1a, the spring stretches. If the spring is calibrated, the distance it stretches can be used to measure the strength of the force. If you pull hard enough on a cart to overcome friction, as in Figure 5.1b, the cart moves. When a football is kicked, as in Figure 5.1c, the football is both deformed and set in motion. These are all examples of a class of forces called *contact forces.* That is, they represent the result of physical contact between two objects. Other examples of contact forces include the force exerted by gas molecules on the walls of a container and the force exerted by our feet on the floor.

Another class of forces, which do not involve physical contact between two objects but act through empty space, are known as *field forces.* The force of gravitational attraction between two objects is an example of this class of force, illustrated in Figure 5.1d. This gravitational force keeps objects bound to the Earth and gives rise to what we commonly call the *weight* of an object. The planets of our Solar System are bound under the action of gravitational forces. Another common example of a field force is the electric force that one electric charge exerts on another electric charge, as in Figure 5.1e. These charges might be an electron and proton forming a hydrogen atom. A third example of a field force is the force that a bar magnet exerts on a piece of iron, as shown in Figure 5.1f. The forces holding

A body accelerates due to an external force

Definition of equilibrium

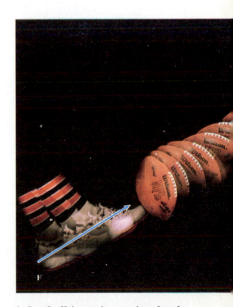

A football is set in motion by the contact force, F, on it due to the kicker's foot. The ball is deformed during the short time in contact with the foot. *(Ralph Cowan, Tony Stone Worldwide)*

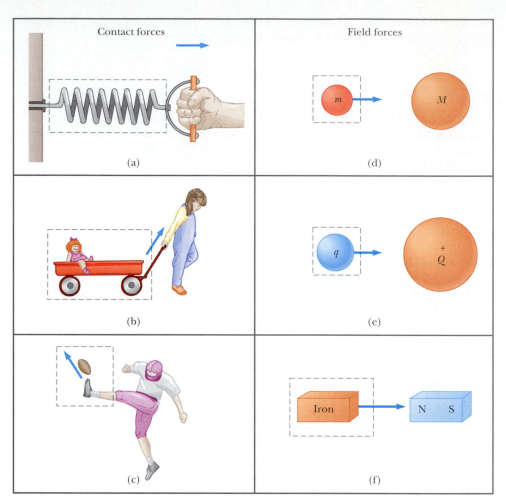

FIGURE 5.1   Some examples of forces applied to various objects. In each case a force is exerted on the object within the boxed area. Some agent in the environment external to the boxed area exerts the force on the object.

an atomic nucleus together are also field forces but are usually very short-range. They are the dominating interaction for particle separations of the order of 10^{-15} m.

Early scientists, including Newton, were uneasy with the concept of a force acting between two disconnected objects. To overcome this conceptual problem, Michael Faraday (1791–1867) introduced the concept of a *field*. According to this approach, when a mass m_1 is placed at some point P near a mass m_2, one can say that m_1 interacts with m_2 by virtue of the gravitational field that exists at P. The field at P is created by mass m_2. Likewise, a field exists at the position of m_2 created by m_1. In fact, all objects create a gravitational field in the space around them.

The distinction between contact forces and field forces is not as sharp as you may have been led to believe by the above discussion. When examined at the atomic level, all the forces we classify as contact forces turn out to be due to repulsive electrical (field) forces of the type illustrated in Figure 5.1e. Nevertheless, in developing models for macroscopic phenomena, it is convenient to use both classifications of forces. However, the only known *fundamental* forces in nature are all field forces: (1) gravitational attraction between objects, (2) electro-

Fundamental forces in nature

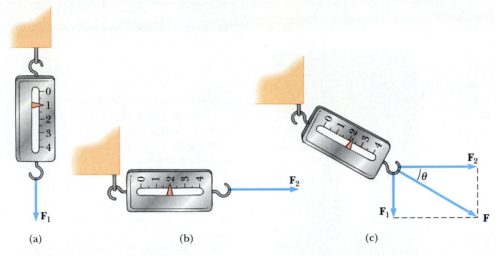

FIGURE 5.2 The vector nature of a force is tested with a spring scale. (a) The downward force F_1 elongates the spring 1 unit. (b) A downward force F_2 elongates the spring 2 units. (c) When F_2 is applied horizontally, and F_1 is downward, the combination of the two forces elongates the spring $\sqrt{1^2 + 2^2} = \sqrt{5}$ units.

magnetic forces between electric charges, (3) strong nuclear forces between subatomic particles, and (4) weak nuclear forces that arise in certain radioactive decay processes. In classical physics, we are concerned only with gravitational and electromagnetic forces.

It is convenient to use the deformation of a spring to measure force. Suppose a force is applied vertically to a spring that has a fixed upper end, as in Figure 5.2a. We can calibrate the spring by defining the unit force, F_1, as the force that produces an elongation of 1.00 cm. (Because force is a vector quantity, we use the bold-faced symbol **F**.) If a second downward force F_2 is applied whose magnitude is 2 units, as in Figure 5.2b, the scale elongation is 2.00 cm. This shows that the combined effect of the two forces that are collinear is the sum of the effect of the individual forces. Now suppose that the two forces are applied simultaneously with F_1 downward and F_2 horizontal, as in Figure 5.2c. In this case, the elongation of the spring is found to be $\sqrt{5} = 2.24$ cm. The single force, **F**, that would produce this same elongation is the vector sum of F_1 and F_2, as described in Figure 5.2c. That is, $|\mathbf{F}| = \sqrt{F_1^2 + F_2^2} = \sqrt{5}$ units, and its direction is $\theta = \arctan(-0.500) = -26.6°$. *Because forces are vectors, you must use the rules of vector addition to get the resultant force on a body.*

Springs that elongate in proportion to an applied force are said to obey *Hooke's law*. Such springs can be constructed and calibrated to measure the magnitude of unknown forces.

5.2 NEWTON'S FIRST LAW AND INERTIAL FRAMES

Before we state Newton's first law, consider the following simple experiment. Suppose a book is lying on a table. Obviously, the book remains at rest in the absence of any influences. Now imagine that you push the book with a horizontal force large enough to overcome the force of friction between book and table. The book can then be kept in motion with constant velocity if the force you apply is equal in magnitude to the force of friction and in the direction opposite the friction force.

Isaac Newton, a British physicist and mathematician, is regarded as one of the greatest scientists in history. Before the age of 30 he formulated the basic concepts and laws of motion, discovered the universal law of gravitation, and invented the calculus. Newton was able to explain the motions of the planets, the ebb and flow of the tides, and many special features of the motion of the Moon and the Earth. He also made many important discoveries in optics, showing, for example, that white light is composed of a spectrum of colors. His contributions to physical theories dominated scientific thought for two centuries and remain important today.

Newton was born prematurely on Christmas Day in 1642, shortly after his father's death. When he was three, his mother remarried and he was left in his grandmother's care. Because he was small in stature as a child, he was bullied by other children and took refuge in such solitary activities as the building of water clocks, kites carrying fiery lanterns, sundials, and model windmills powered by mice. His mother withdrew him from school at the age of 12 with the intention of turning him into a farmer. Fortu-

Isaac Newton

| 1 6 4 2 – 1 7 2 7 |

nately for later generations, his uncle recognized his scientific and mathematical abilities and helped send him to Trinity College in Cambridge.

In 1665, the year Newton completed his Bachelor of Arts degree, the university was closed because of the bubonic plague that was raging

through England. Newton returned to the family farm at Woolsthorpe to study. During this especially creative period, he laid the foundations of his work in mathematics, optics, motion, celestial mechanics, and gravity.

Newton was a very private person who studied alone and labored day and night in his laboratory, conducting experiments, performing calculations, and immersing himself in theological studies. His greatest single work, *Mathematical Principles of Natural Philosophy,* was published in 1687. In his later years he spent much of his time quarreling with other eminent minds, including the mathematician Gottfried Leibnitz, who worked independently on the development of calculus; Christian Huygens, who developed the wave theory of light; and Robert Hooke, who supported Huygens' theory. These disputes, the strain of his studies, and his work in alchemy, which involved mercury (a poison), caused him in 1692 to suffer severe depression. He was elected president of the Royal Society in 1703, and he retained that office until his death in 1727.

(Giraudon/Art Resource)

If the applied force exceeds the force of friction, the book accelerates. If you stop pushing, the book stops sliding after moving a short distance because the force of friction retards its motion. Now imagine pushing the book across a smooth, highly waxed floor. The book again comes to rest after you stop pushing, but not as quickly as before. Now imagine a floor so highly polished that friction is absent; in this case, the book, once set in motion, slides until it hits the wall. The forces that we have described are *external* forces, that is, forces exerted on the object by other objects.

Before about 1600, scientists felt that the natural state of matter was the state of rest. Galileo was the first to take a different approach to motion and the natural state of matter. He devised thought experiments, such as the one we just discussed for a book on a frictionless surface, and concluded that it is not the nature of an object to stop once set in motion: rather, it is its nature to resist changes in its motion. In his words, "Any velocity once imparted to a moving body will be rigidly maintained as long as the external causes of retardation are removed."

This new approach to motion was later formalized by Newton in a form that has come to be known as **Newton's first law of motion:**

An object at rest will remain at rest and an object in motion will continue in motion with constant velocity unless a net external force acts on the object. In this case, the wall of the building did not exert a large enough force on the moving train to stop it. *(Roger Viollet, Mill Valley, CA, University Science Books, 1982)*

> An object at rest remains at rest and an object in motion will continue in motion with a constant velocity (that is, constant speed in a straight line) unless it experiences a net external force.

Newton's first law

In simpler terms, we can say that *when the net force on a body is zero, its acceleration is zero.* That is, when $\Sigma \mathbf{F} = 0$, then $\mathbf{a} = 0$. From the first law, we conclude that an isolated body (a body that does not interact with its environment) is either at rest or moving with constant velocity.

Another example of uniform (constant velocity) motion on a nearly frictionless plane is the motion of a light disk on a column of air (the lubricant), as in Figure 5.3. If the disk is given an initial velocity, it coasts a great distance before stopping. This idea is used in the game of air hockey, where the disk makes many collisions with the walls before coming to rest.

Finally, consider a spaceship traveling in space and far removed from any planets or other matter. The spaceship requires some propulsion system to *change* its velocity. However, if the propulsion system is turned off when the spaceship reaches a velocity v, the spaceship coasts in space at a constant velocity and the astronauts get a free ride (that is, no propulsion system is required to keep them moving at the velocity v).

Inertial Frames

Newton's first law, sometimes called the *law of inertia*, defines a special set of reference frames called inertial frames. An **inertial frame** of reference is one in which Newton's first law is valid. Thus, an inertial frame of reference is an unaccel-

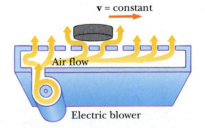

FIGURE 5.3 A disk moving on a column of air is an example of uniform motion, that is, motion in which the acceleration is zero and the velocity remains constant.

Inertial frame

erated frame. Any reference frame that moves with constant velocity relative to an inertial frame is itself an inertial frame. A reference frame that moves with constant velocity relative to the distant stars is the best approximation of an inertial frame. The Earth is not an inertial frame because of its orbital motion about the Sun and rotational motion about its own axis. As the Earth travels in its nearly circular orbit around the Sun, it experiences a centripetal acceleration of about 4.4×10^{-3} m/s^2 toward the Sun. In addition, since the Earth rotates about its own axis once every 24 hours, a point on the equator experiences an additional centripetal acceleration of 3.37×10^{-2} m/s^2 toward the center of the Earth. However, these accelerations are small compared with g and can often be neglected. In most situations *we shall assume that a set of nearby points on the Earth's surface constitutes an inertial frame.*

Thus, if an object is moving with constant velocity, an observer in one inertial frame (say, one at rest relative to the object) will claim that the acceleration and the resultant force on the object are zero. An observer in *any other* inertial frame will also find that $\mathbf{a} = 0$ and $\mathbf{F} = 0$ for the object. According to the first law, a body at rest and one moving with constant velocity are equivalent. In other words, the absolute motion of an object has no effect on its behavior. Unless stated otherwise, we shall write the laws of motion relative to an observer "at rest" in an inertial frame.

CONCEPTUAL EXAMPLE 5.1

Is it possible to have motion in the absence of a force?

Reasoning Motion requires no force. Newton's first law says that motion needs no cause but continues by itself. That is, an object in motion continues to move by itself in the absence of external forces. As an example, witness the motion of a meteoroid in outer space (imitated by a glider on an air track).

5.3 INERTIAL MASS

Inertia

If you attempt to change the velocity of an object, the object resists this change. **Inertia** is solely a property of an individual object; it is a measure of the response of an object to an external force. For instance, consider two large, solid cylinders of equal size, one balsa wood and the other steel. If you were to push the cylinders along a horizontal surface, the force required to give the steel cylinder some acceleration would be larger than the force needed to give the balsa wood cylinder the same acceleration. Therefore, we say that the steel cylinder has more inertia than the balsa wood cylinder.

Mass is used to measure inertia, and the SI unit of mass is the kilogram. The greater the mass of a body, the less that body accelerates (changes its state of motion) under the action of an applied force. For example, if a given force acting on a 3-kg mass produces an acceleration of 4 m/s^2, the same force when applied to a 6-kg mass will produce an acceleration of 2 m/s^2.

Let us now use this idea to obtain a quantitative description of the concept of mass. We begin by comparing the accelerations that a given force produces on different bodies. Suppose a force acting on a body of mass m_1 produces an acceleration $\mathbf{a}_1$, and the *same force* acting on a body of mass m_2 produces an acceleration

a_2. The ratio of the two masses is defined as the *inverse* ratio of the magnitudes of the accelerations produced by the same force:

$$\frac{m_1}{m_2} \equiv \frac{a_2}{a_1} \tag{5.1}$$

If one of these is a standard known mass—of say, 1 kg—the mass of an unknown can be obtained from acceleration measurements. For example, if the standard 1-kg mass undergoes an acceleration of 6 m/s^2 under the influence of some force, a 2-kg mass will undergo an acceleration of 3 m/s^2 under the action of the same force.

Mass is an inherent property of a body and is independent of the body's surroundings and of the method used to measure it. It is an experimental fact that *mass is a scalar quantity.* Finally, *mass is a quantity that obeys the rules of ordinary arithmetic.* That is, several masses can be combined in a simple numerical fashion. For example, if you combine a 3-kg mass with a 5-kg mass, their total mass is 8 kg. This result can be verified experimentally by comparing the acceleration of each object produced by a known force with the acceleration of the combined system using the same force.

Mass should not be confused with weight. *Mass and weight are two different quantities.* As we shall see below, the weight of a body is equal to the magnitude of the force exerted by the Earth on the body and varies with location. For example, a person who weighs 180 lb on Earth weighs only about 30 lb on the Moon. On the other hand, the mass of a body is the same everywhere, regardless of location. An object having a mass of 2 kg on Earth also has a mass of 2 kg on the Moon.

Mass and weight are different quantities

5.4 NEWTON'S SECOND LAW

Newton's first law explains what happens to an object when the resultant of all external forces on it is zero: it either remains at rest or moves in a straight line with constant speed. Newton's second law answers the question of what happens to an object that has a nonzero resultant force acting on it.

Imagine you are pushing a block of ice across a frictionless horizontal surface. When you exert some horizontal force **F**, the block moves with some acceleration **a**. If you apply a force twice as large, the acceleration doubles. Likewise, if the applied force is increased to 3**F**, the acceleration is tripled, and so on. From such observations, we conclude that *the acceleration of an object is directly proportional to the resultant force acting on it.*

As stated in the preceding section, the acceleration of an object also depends on its mass. This can be understood by considering the following set of experiments. If you apply a force **F** to a block of ice on a frictionless surface, the block undergoes some acceleration **a**. If the mass of the block is doubled, the same applied force produces an acceleration **a**/2. If the mass is tripled, the same applied force produces an acceleration **a**/3, and so on. According to this observation, we conclude that *the acceleration of an object is inversely proportional to its mass.*

These observations are summarized in **Newton's second law:**

> The acceleration of an object is directly proportional to the net force acting on it and inversely proportional to its mass.

Newton's second law

Thus we can relate mass and force through the following mathematical statement of Newton's second law[1]:

$$\sum \mathbf{F} = m\mathbf{a} \tag{5.2}$$

You should note that Equation 5.2 is a *vector* expression and hence is equivalent to the following three component equations:

Newton's second law—
component form

$$\sum F_x = ma_x \qquad \sum F_y = ma_y \qquad \sum F_z = ma_z \tag{5.3}$$

Units of Force and Mass

The SI unit of force is the **newton,** which is defined as the force that, when acting on a 1-kg mass, produces an acceleration of 1 m/s^2. From this definition and Newton's second law, we see that the newton can be expressed in terms of the following fundamental units of mass, length, and time:

Definition of newton

$$1 \text{ N} \equiv 1 \text{ kg} \cdot \text{m/s}^2 \tag{5.4}$$

The unit of force in the cgs system is called the **dyne** and is defined as that force that, when acting on a 1-g mass, produces an acceleration of 1 cm/s^2:

Definition of dyne

$$1 \text{ dyne} \equiv 1 \text{ g} \cdot \text{cm/s}^2 \tag{5.5}$$

In the British engineering system, the unit of force is the **pound,** defined as the force that, when acting on a 1-slug mass,[2] produces an acceleration of 1 ft/s^2:

Definition of pound

$$1 \text{ lb} \equiv 1 \text{ slug} \cdot \text{ft/s}^2 \tag{5.6}$$

Since $1 \text{ kg} = 10^3 \text{ g}$ and $1 \text{ m} = 10^2 \text{ cm}$, it follows that $1 \text{ N} = 10^5 \text{ dynes}$. It is left as a problem to show that $1 \text{ N} = 0.225 \text{ lb}$. The units of force, mass, and acceleration are summarized in Table 5.1.

TABLE 5.1   Units of Force, Mass, and Acceleration[a]

System of Units	Mass	Acceleration	Force
SI	kg	m/s^2	$\text{N} = \text{kg} \cdot \text{m/s}^2$
cgs	g	cm/s^2	$\text{dyne} = \text{g} \cdot \text{cm/s}^2$
British engineering	slug	ft/s^2	$\text{lb} = \text{slug} \cdot \text{ft/s}^2$

[a] $1 \text{ N} = 10^5 \text{ dyne} = 0.225 \text{ lb}$.

[1] Equation 5.2 is valid only when the speed of the particle is much less than the speed of light. We shall treat the relativistic situation in Chapter 39.

[2] The *slug* is the *unit of mass* in the British engineering system and is that system's counterpart of the SI *kilogram.* When we speak of going on a diet to lose a few pounds, we really mean that we want to lose a few slugs, that is, we want to reduce our mass. When we lose those few slugs, the force of gravity (pounds) on our reduced mass decreases and that is how we "lose a few pounds." Since most of the calculations in our study of classical mechanics are in SI units, the slug is seldom used in this text.

EXAMPLE 5.2   An Accelerating Hockey Puck

A hockey puck with a mass of 0.30 kg slides on the horizontal frictionless surface of an ice rink. Two forces act on the puck as shown in Figure 5.4. The force F_1 has a magnitude of 5.0 N, and F_2 has a magnitude of 8.0 N. Determine the magnitude and direction of the puck's acceleration.

Solution   The resultant force in the *x* direction is

$$\sum F_x = F_{1x} + F_{2x} = F_1 \cos 20° + F_2 \cos 60°$$
$$= (5.0\ \text{N})(0.940) + (8.0\ \text{N})(0.500) = 8.7\ \text{N}$$

The resultant force in the *y* direction is

$$\sum F_y = F_{1y} + F_{2y} = -F_1 \sin 20° + F_2 \sin 60°$$
$$= -(5.0\ \text{N})(0.342) + (8.0\ \text{N})(0.866) = 5.2\ \text{N}$$

Now we use Newton's second law in component form to find the *x* and *y* components of acceleration:

$$a_x = \frac{\sum F_x}{m} = \frac{8.7\ \text{N}}{0.30\ \text{kg}} = 29\ \text{m/s}^2$$

$$a_y = \frac{\sum F_y}{m} = \frac{5.2\ \text{N}}{0.30\ \text{kg}} = 17\ \text{m/s}^2$$

The acceleration has a magnitude of

$$a = \sqrt{(29)^2 + (17)^2}\ \text{m/s}^2 = \boxed{34\ \text{m/s}^2}$$

and its direction relative to the positive *x* axis is

$$\theta = \tan^{-1}(a_y/a_x) = \tan^{-1}(17/29) = \boxed{31°}$$

Exercise   Determine the components of a third force that, when applied to the puck, causes it to be in equilibrium.

Answer $F_x = -8.7\ \text{N}, F_y = -5.2\ \text{N}$.

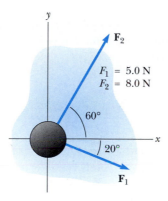

$F_1 = 5.0$ N
$F_2 = 8.0$ N

FIGURE 5.4   (Example 5.2) A hockey puck moving on a frictionless surface accelerates in the direction of the *resultant* force, $F_1 + F_2$.

5.5   WEIGHT

We are well aware of the fact that all objects are attracted to the Earth. The force exerted by the Earth on an object is called the **weight** of the object w. This force is directed toward the center of the Earth.[3]

We have seen that a freely falling body experiences an acceleration **g** acting toward the center of the Earth. Applying Newton's second law to the freely falling body of mass *m*, we have **F** = *m***a**. Because **F** = *m***g** and also **F** = *m***a**, it follows that **a** = **g** and **F** = **w**, or

$$\mathbf{w} = m\mathbf{g} \tag{5.7}$$

Since it depends on *g*, weight varies with geographic location. Bodies weigh less at higher altitudes than at sea level. (You should not confuse the italicized symbol *g* that we use for gravitational acceleration with the symbol g that is used as a symbol for grams.) This is because *g* decreases with increasing distance from the center of the Earth. Hence, weight, unlike mass, is not an inherent property of a body. For example, if a body has a mass of 70 kg, then its weight in a location where *g* = 9.80 m/s² is *mg* = 686 N (about 154 lb). At the top of a mountain where

---

[3] This statement ignores the fact that the mass distribution of the Earth is not perfectly spherical.

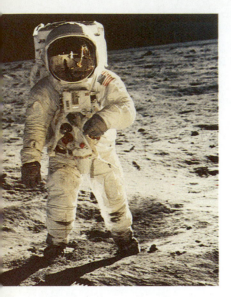

Astronaut Edwin E. Aldrin, Jr., walking on the Moon after the Apollo 11 lunar landing. The weight of the astronaut on the Moon is less than it is on Earth, but his mass remains the same. *(Courtesy of NASA)*

$g = 9.76$ m/s², this weight is 683 N. Therefore, if you want to lose weight without going on a diet, climb a mountain or weigh yourself at 30 000 ft during an airplane flight.

Because w = *m*g, we can compare the masses of two bodies by measuring their weights on a spring scale or balance. At a given location, the ratio of the weights of two bodies equals the ratio of their masses.

CONCEPTUAL EXAMPLE 5.3

A baseball of mass *m* is thrown upward with some initial speed. If air resistance is neglected, what is the force on the ball (a) when it reaches half its maximum height and (b) when it reaches its peak?

Reasoning   The only external force on the ball at *all* points in its trajectory is the downward force of gravity. The magnitude of this force is $w = mg$. For example, if the mass of the baseball is 0.15 kg, $w = 1.47$ N.

5.6   NEWTON'S THIRD LAW

> Newton's third law states that if two bodies interact, the force exerted on body 1 by body 2 is equal to and opposite the force exerted on body 2 by body 1:
>
> $$\mathbf{F}_{12} = -\mathbf{F}_{21} \tag{5.8}$$

This law, which is illustrated in Figure 5.5a, is equivalent to stating that *forces always occur in pairs* or that *a single isolated force cannot exist.* The force that body 1 exerts on body 2 is sometimes called the *action force,* while the force body 2 exerts on body 1 is called the *reaction force.* In reality, either force can be labeled the action or reaction force. *The action force is equal in magnitude to the reaction force and opposite in direction. In all cases, the action and reaction forces act on different objects.* For example, the force acting on a freely falling projectile is its weight, w = *m*g, which is the force exerted by the Earth on the projectile. The reaction to this force is the force

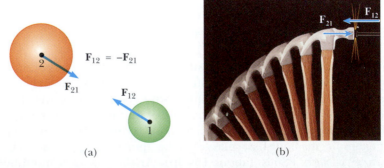

FIGURE 5.5   Newton's third law. (a) The force exerted by object 1 on object 2 is equal to and opposite the force exerted by object 2 on object 1. (b) The force exerted by the hammer on the nail is equal to and opposite the force exerted by the nail on the hammer. *(John Gillmoure, The Stock Market)*

of the projectile on the Earth, $w' = -w$. The reaction force, w', must accelerate the Earth toward the projectile just as the action force, w, accelerates the projectile toward the Earth. However, since the Earth has such a large mass, its acceleration due to this reaction force is negligibly small.

Another example of Newton's third law is shown in Figure 5.5b. The force exerted by the hammer on the nail (the action) is equal to and opposite the force exerted by the nail on the hammer (the reaction). You experience Newton's third law directly whenever you slam your fist against a wall or kick a football. You should be able to identify the action and reaction forces in these cases.

The weight of an object, w, is defined as the force the Earth exerts on the object. If the object is a TV at rest on a table, as in Figure 5.6a, the reaction force to w is the force the TV exerts on the Earth, w'. The TV does not accelerate because it is held up by the table. The table, therefore, exerts an upward action force, n, on the TV, called the **normal force**.[4] The normal force is the force that prevents the TV from falling through the table and can have any value needed, up to the point of breaking the table. The normal force balances the force of gravity acting on the TV and provides equilibrium. The reaction to n is the force exerted by the TV on the table, n'. Therefore, we conclude that

$$w = -w' \qquad \text{and} \qquad n = -n'$$

The forces n and n' have the same magnitude, which is the same as w unless the table has broken. Note that the forces acting on the TV are w and n, as shown in Figure 5.6b. The two reaction forces, w' and n', are exerted on objects other than

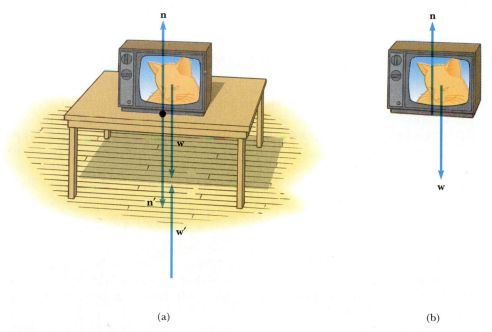

(a)   (b)

FIGURE 5.6 When a TV is at rest on a table, the forces acting on the TV are the normal force, n, and the force of gravity, w, as illustrated in (b). The reaction to n is the force exerted by the TV on the table, n'. The reaction to w is the force exerted by the TV on the Earth, w'.

---

[4] *Normal* in this context means *perpendicular*.

the TV. Remember, the two forces in an action-reaction pair always act on two different objects.

From the second law, we see that, because the TV is in equilibrium ($a = 0$), it follows that $w = n = mg$.

CONCEPTUAL EXAMPLE 5.4

If a small sports car collides head-on with a massive truck, which vehicle experiences the greater impact force? Which vehicle experiences the greater acceleration?

Reasoning   The car and truck experience forces that are equal in magnitude but in opposite directions. A calibrated spring scale placed between the colliding vehicles reads the same whichever way it faces. Because the car has the smaller mass, it stops with much greater acceleration.

CONCEPTUAL EXAMPLE 5.5

Is there any relation between the total force acting on an object and the direction in which it moves?

Reasoning   There is no relation between the total force on an object and the direction of its motion before or after the force acts. The force does determine the direction of the *change in motion*. Force describes what the rest of the Universe does to the object. The environment can push the object forward, backward, sideways, or not at all.

5.7   SOME APPLICATIONS OF NEWTON'S LAWS

In this section we apply Newton's laws to objects that are either in equilibrium ($a = 0$) or moving linearly under the action of constant external forces. We assume that the objects behave as particles so that we need not worry about rotational motions. We also neglect the effects of friction for those problems involving motion; this is equivalent to stating that the surfaces are *frictionless*. Finally, we usually neglect the mass of any ropes involved. In this approximation, the magnitude of the force exerted at any point along a rope is the same at all points along the rope. In problem statements, the terms *light* and *of negligible mass* are used to indicate that a mass is to be ignored when you work the problem. These two terms are synonymous in this context.

When we apply Newton's laws to a body, we are interested only in those external forces that *act on the body*. For example, in Figure 5.6 the only external forces acting on the TV are **n** and **w**. The reactions to these forces, **n'** and **w'**, act on the table and on the Earth, respectively, and therefore do not appear in Newton's second law as applied to the TV.

When an object is being pulled by a rope attached to it, the rope exerts a force on the object. In general, **tension** is a scalar and is defined as the magnitude of the force that the rope exerts on whatever is attached to it.

Consider a crate being pulled to the right on the frictionless, horizontal surface, as in Figure 5.7a. Suppose you are asked to find the acceleration of the crate and the force the floor exerts on it. First, note that the horizontal force being applied to the crate acts through the rope. The force that the rope exerts on the crate is denoted **T**. The magnitude of **T** is equal to the tension in the rope.

Tension

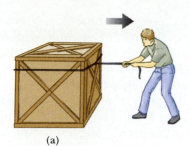

(a)

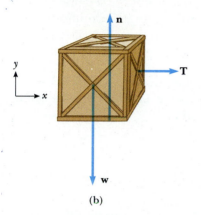

(b)

FIGURE 5.7   (a) A crate being pulled to the right on a frictionless surface. (b) The free-body diagram that represents the external forces acting on the crate.

The forces acting on the crate are illustrated in Figure 5.7b. In addition to the force **T**, this force diagram for the crate includes the force of gravity, **w**, and the normal force, **n**, exerted by the table on the crate. Such a force diagram is referred to as a **free-body diagram.** The construction of a correct free-body diagram is an important step in applying Newton's laws. The reactions to the forces we have listed, namely, the force exerted by the rope on the hand, the force exerted by the crate on the Earth, and the force exerted by the crate on the floor, are not included in the free-body diagram because they act on *other* bodies and not on the crate.

Free-body diagrams are important when applying Newton's laws

We can now apply Newton's second law in component form to the system. The only force acting in the x direction is **T**. Applying $\Sigma F_x = ma_x$ to the horizontal motion gives

$$\sum F_x = T = ma_x \quad \text{or} \quad a_x = \frac{T}{m}$$

There is no acceleration in the y direction. Applying $\Sigma F_y = ma_y$ with $a_y = 0$ gives

$$n - w = 0 \quad \text{or} \quad n = w$$

That is, the normal force is equal to and opposite the force of gravity.

If **T** is a constant force, then the acceleration, $a_x = T/m$, is also a constant. Hence, the equations of kinematics from Chapter 2 can be used to obtain the displacement, Δx, and velocity, v, as functions of time. Since $a_x = T/m = $ constant, these expressions can be written

$$\Delta x = v_0 t + \tfrac{1}{2}\left(\frac{T}{m}\right) t^2$$

$$v = v_0 + \left(\frac{T}{m}\right) t$$

where v_0 is the speed of the crate at $t = 0$.

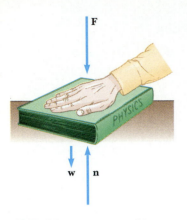

FIGURE 5.8 When one object pushes downward on another object with a force **F**, the normal force **n** is greater than the force of gravity. That is, $n = w + F$.

In the example just presented, the normal force **n** is equal in magnitude and opposite the force of gravity **w**. *This is not always the case.* For example, suppose you were to push down on a book with a force **F** as in Figure 5.8. In this case, the book has no motion in the *y* direction. Therefore $\Sigma F_y = 0$, which gives $n - w - F = 0$, or $n = w + F$. Other examples in which $n \neq w$ are presented later.

Consider a lamp of weight *w* suspended from a chain of negligible weight fastened to the ceiling, as in Figure 5.9a. The free-body diagram for the lamp (Fig. 5.9b) shows that the forces on it are the force of gravity **w** acting downward, and the force exerted by the chain on the lamp **T** acting upward.

If we apply the second law to the lamp, noting that **a** = 0, we see that because there are no forces in the *x* direction, $\Sigma F_x = 0$ provides no helpful information. The condition $\Sigma F_y = 0$ gives

$$\Sigma F_y = T - w = 0 \qquad \text{or} \qquad T = w$$

Note that **T** and **w** are *not* an action-reaction pair. The reaction force to **T** is **T′**, the downward force exerted by the lamp on the chain, as in Figure 5.9c. The ceiling exerts an equal and opposite force, **T′′**, on the chain.

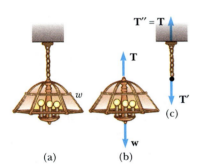

FIGURE 5.9 (a) A lamp of weight *w* suspended from a ceiling by a chain of negligible mass. (b) The forces acting on the lamp are the force of gravity, **w**, and the force exerted by the chain, **T**. (c) The forces acting on the chain are the force exerted by the lamp, **T′**, and the force exerted by the ceiling, **T″**.

Problem-Solving Strategy
Applying Newton's Laws

The following procedure is recommended when dealing with problems involving the application of Newton's laws:

- Draw a simple, neat diagram of the system.
- Isolate the object whose motion is being analyzed. Draw a free-body diagram for this object, that is, a diagram showing *all external forces acting on the object*. For systems containing more than one object, draw separate free-body diagrams for each object. Do *not* include in the free-body diagram forces that the object exerts on its surroundings.
- Establish convenient coordinate axes for each object and find the components of the forces along these axes. Apply Newton's second law, $\Sigma F = ma$, in *component* form. Check your dimensions to make sure that all terms have units of force.
- Solve the component equations for the unknowns. Remember that you must have as many independent equations as you have unknowns in order to obtain a complete solution.
- Check the predictions of your solutions for extreme values of the variables. You can often detect errors in your results by doing so.

EXAMPLE 5.6   A Traffic Light at Rest

A traffic light weighing 125 N hangs from a cable tied to two other cables fastened to a support, as in Figure 5.10a. The upper cables make angles of 37.0° and 53.0° with the horizontal. Find the tension in the three cables.

Reasoning We must construct two free-body diagrams in order to work this problem. The first of these is for the traffic light, shown in Figure 5.10b; the second is for the knot that holds the three cables together, as in Figure 5.10c. This knot is a convenient point to choose because all the forces we are interested in act through this point. Because the acceleration of the system is zero, we can use the condition that the net force on the light is zero, and the net force on the knot is zero.

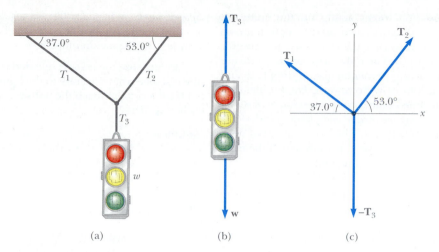

FIGURE 5.10 (Example 5.6) (a) A traffic light suspended by cables. (b) Free-body diagram for the traffic light. (c) Free-body diagram for the knot where the three cables are joined.

Solution First we construct a free-body diagram for the traffic light as in Figure 5.10b. The force exerted by the vertical cable, $\mathbf{T}_3$, supports the light, and so $T_3 = w = 125$ N. Next, we choose the coordinate axes as shown in Figure 5.10c and resolve the forces into their x and y components:

Force	x Component	y Component
T_1	$-T_1 \cos 37.0°$	$T_1 \sin 37.0°$
T_2	$T_2 \cos 53.0°$	$T_2 \sin 53.0°$
T_3	0	-125 N

The first condition for equilibrium gives us the equations

(1) $\sum F_x = T_2 \cos 53.0° - T_1 \cos 37.0° = 0$

(2) $\sum F_y = T_1 \sin 37.0° + T_2 \sin 53.0° - 125$ N $= 0$

From (1) we see that the horizontal components of $\mathbf{T}_1$ and $\mathbf{T}_2$

must be equal in magnitude, and from (2) we see that the sum of the vertical components of $\mathbf{T}_1$ and $\mathbf{T}_2$ must balance the weight of the light. We can solve (1) for T_2 in terms of T_1 to give

$$T_2 = T_1 \left(\frac{\cos 37.0°}{\cos 53.0°} \right) = 1.33 T_1$$

This value for T_2 can be substituted into (2) to give

$$T_1 \sin 37.0° + (1.33 T_1)(\sin 53.0°) - 125 \text{ N} = 0$$

$$T_1 = \boxed{75.1 \text{ N}}$$

$$T_2 = 1.33 T_1 = \boxed{99.9 \text{ N}}$$

Exercise In what situation will $T_1 = T_2$?

Answer When the supporting cables make equal angles with the horizontal support.

EXAMPLE 5.7   Crate on a Frictionless Incline

A crate of mass m is placed on a frictionless, inclined plane of angle θ, as in Figure 5.11a. (a) Determine the acceleration of the crate after it is released.

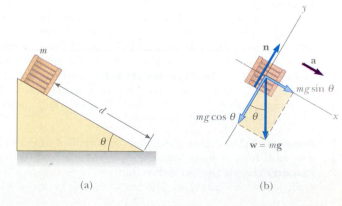

FIGURE 5.11 (Example 5.7) (a) A crate of mass m sliding down a frictionless incline. (b) The free-body diagram for the crate. Note that its acceleration along the incline is $g \sin \theta$.

Reasoning   Because the forces acting on the crate are known, Newton's second law can be used to determine its acceleration. First, we construct the free-body diagram for the crate as in Figure 5.11b. The only forces on the crate are the normal force, n, acting perpendicular to the plane and the weight, w, acting vertically downward. For problems of this type involving inclined planes, *it is convenient to choose the coordinate axes with x along the incline and y perpendicular to it.* Then, we replace the weight vector by a component of magnitude $mg \sin \theta$ along the positive x axis and one of the magnitude $mg \cos \theta$ in the negative y direction.

Solution   Applying Newton's second law to the crate in component form while noting that $a_y = 0$ gives

$$(1) \qquad \sum F_x = mg \sin \theta = ma_x$$

$$(2) \qquad \sum F_y = n - mg \cos \theta = 0$$

From (1) we see that the acceleration along the incline is provided by the component of weight directed down the incline:

$$(3) \qquad \boxed{a_x = g \sin \theta}$$

From (2) we conclude that the component of weight perpendicular to the incline is balanced by the normal force; that is, $n = mg \cos \theta$.

Note that the acceleration given by (3) is independent of the mass of the crate! It depends only on the angle of inclination and on g!

Special Cases   When $\theta = 90°$, $a = g$ and $n = 0$. This condition corresponds to the crate in free-fall. When $\theta = 0$, $a_x = 0$ and $n = mg$ (its maximum value).

(b) Suppose the crate is released from rest at the top, and the distance from the crate to the bottom is d. How long does it take the crate to reach the bottom, and what is its speed just as it gets there?

Solution   Since a_x = constant, we can apply Equation 2.10, $x - x_0 = v_{x0}t + \frac{1}{2}a_x t^2$, to the crate. Since the displacement $x - x_0 = d$ and $v_{x0} = 0$, we get

$$d = \tfrac{1}{2}a_x t^2$$

or

$$(4) \qquad t = \sqrt{\frac{2d}{a_x}} = \sqrt{\frac{2d}{g \sin \theta}}$$

Using Equation 2.12, $v_x^2 = v_{x0}^2 + 2a_x(x - x_0)$, with $v_{x0} = 0$, we find that

$$v_x^2 = 2a_x d$$

or

$$(5) \qquad v_x = \sqrt{2a_x d} = \sqrt{2gd \sin \theta}$$

Again, t and v_x are independent of the mass of the crate. This suggests a simple method of measuring g using an inclined air track or some other frictionless incline. Simply measure the angle of inclination, the distance traveled by the crate, and the time it takes to reach the bottom. The value of g can then be calculated from (4) and (5).

EXAMPLE 5.8   Atwood's Machine

When two unequal masses are hung vertically over a frictionless pulley of negligible mass as in Figure 5.12a, the arrangement is called *Atwood's machine.* The device is sometimes used in the laboratory to measure the gravitational field strength. Determine the magnitude of the acceleration of the two masses and the tension in the string.

Reasoning   The free-body diagrams for the two masses are shown in Figure 5.12b. Two forces act on each block: the upward force exerted by the string, T, and the downward force of gravity. Thus, the magnitude of the net force exerted on m_1 is $T - m_1 g$, while the magnitude of the net force exerted on m_2 is $T - m_2 g$. Because the blocks are connected by a string, their accelerations must be equal in magnitude. If we assume that $m_2 > m_1$, then m_1 must accelerate upward, while m_2 must accelerate downward.

Solution   When Newton's second law is applied to m_1, with a upward for this mass (because $m_2 > m_1$), we find (taking upward to be the positive y direction)

$$(1) \qquad \sum F_y = T - m_1 g = m_1 a$$

Similarly, for m_2 we find

$$(2) \qquad \sum F_y = T - m_2 g = -m_2 a$$

The negative sign on the right-hand side of (2) indicates that m_2 accelerates downward, in the negative y direction.

When (2) is subtracted from (1), T drops out and we get

$$-m_1 g + m_2 g = m_1 a + m_2 a$$

or

$$(3) \qquad \boxed{a = \left(\frac{m_2 - m_1}{m_1 + m_2} \right) g}$$

When (3) is substituted into (1), we get

$$(4) \qquad \boxed{T = \left(\frac{2m_1 m_2}{m_1 + m_2} \right) g}$$

The result for the acceleration, (3), can be interpreted as the ratio of the unbalanced force on the system to the total mass of the system.

Special Cases When $m_1 = m_2$, $a = 0$ and $T = m_1 g = m_2 g$, as we would expect for this balanced case. If $m_2 \gg m_1$, $a \approx g$ (a freely falling body) and $T \approx 2m_1 g$.

Exercise Find the magnitude of the acceleration and tension of an Atwood's machine in which $m_1 = 2.00$ kg and $m_2 = 4.00$ kg.

Answer $a = 3.27$ m/s^2, $T = 26.1$ N.

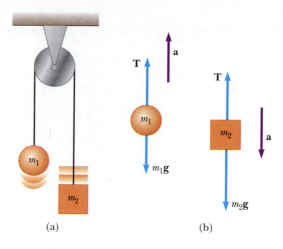
(a)   (b)

 FIGURE 5.12 (Example 5.8) Atwood's machine. (a) Two masses ($m_2 > m_1$) connected by a massless string over a frictionless pulley. (b) Free-body diagrams for m_1 and m_2.

EXAMPLE 5.9 Two Connected Objects

Two unequal masses are attached by a lightweight string that passes over a frictionless pulley of negligible mass as in Figure 5.13a. The block of mass m_2 lies on a frictionless incline of angle θ. Find the magnitude of the acceleration of the two masses and the tension in the string.

Reasoning and Solution Since the two masses are connected by a string (which we assume doesn't stretch), they both have accelerations of the same magnitude. The free-body diagrams for the two masses are shown in Figures 5.13b and 5.13c. Applying Newton's second law in component

form to m_1 while assuming that **a** is upward for this mass (and taking upward to be the positive y direction) gives

$$(1) \qquad \sum F_x = 0$$
$$(2) \qquad \sum F_y = T - m_1 g = m_1 a$$

Note that in order for this mass to accelerate upward, it is necessary that $T > m_1 g$.

For m_2 it is convenient to choose the positive x' axis along the incline as in Figure 5.13c. Here we choose positive acceleration to be down the incline, in the $+x'$ direction. Applying Newton's second law in component form to m_2 gives

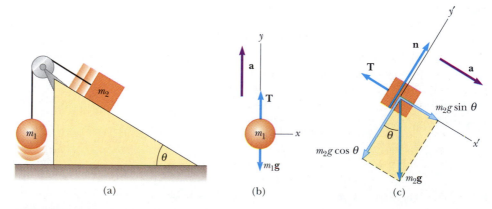
(a)   (b)   (c)

FIGURE 5.13 (Example 5.9) (a) Two masses connected by a massless string over a frictionless pulley. (b) Free-body diagram for m_1. (c) Free-body diagram for m_2 (the incline is frictionless).

(3) $\sum F_{x'} = m_2 g \sin \theta - T = m_2 a$

(4) $\sum F_{y'} = n - m_2 g \cos \theta = 0$

Expressions (1) and (4) provide no information regarding the acceleration. However, if we solve (2) and (3) simultaneously for a and T, we get

(5) $a = \dfrac{m_2 g \sin \theta - m_1 g}{m_1 + m_2}$

When this is substituted into (2), we find

(6) $T = \dfrac{m_1 m_2 g (1 + \sin \theta)}{m_1 + m_2}$

Note that m_2 accelerates down the incline only if $m_2 \sin \theta > m_1$ (that is, if a is in the direction we assumed). If $m_1 > m_2 \sin \theta$, the acceleration of m_2 is up the incline and downward for m_1. You should also note that the result for the acceleration, (5), can be interpreted as the resultant unbalanced force on the system divided by the total mass of the system.

Exercise   If $m_1 = 10.0$ kg, $m_2 = 5.00$ kg, and $\theta = 45.0°$, find the acceleration.

Answer $a = -4.22$ m/s^2, where the negative sign indicates that m_2 accelerates up the incline and m_1 accelerates downward.

EXAMPLE 5.10   One Block Pushes Another

Two blocks of masses m_1 and m_2 are placed in contact with each other on a frictionless, horizontal surface as in Figure 5.14a. A constant horizontal force F is applied to m_1 as shown. (a) Find the magnitude of the acceleration of the system.

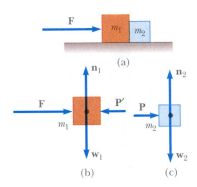

(a)

(b)   (c)

FIGURE 5.14   (Example 5.10).

Reasoning and Solution   Both blocks must experience the same acceleration because they are in contact with each other. Because F is the only horizontal force on the system (the two blocks), we have

$$\sum F_x \text{ (system)} = F = (m_1 + m_2) a$$

(1) $a = \dfrac{F}{m_1 + m_2}$

(b) Determine the magnitude of the contact force between the two blocks.

Reasoning and Solution   To solve this part of the problem, it is necessary to first construct a free-body diagram for

each block, as shown in Figures 5.14b and 5.14c, where the contact force is denoted by **P**. From Figure 5.14c, we see that the only horizontal force acting on m_2 is the contact force **P** (the force exerted by m_1 on m_2), which is directed to the right. Applying Newton's second law to m_2 gives

(2) $\sum F_x = P = m_2 a$

Substituting the value of a given by (1) into (2) gives

(3) $P = m_2 a = \left(\dfrac{m_2}{m_1 + m_2} \right) F$

From this result, we see that the contact force **P** is less than the applied force **F**. This is consistent with the fact that the force required to accelerate m_2 alone must be less than the force required to produce the same acceleration for the system of two blocks.

It is instructive to check this expression for P by considering the forces acting on m_1, as shown in Figure 15.14b. In this case, the horizontal forces acting on m_1 are the applied force **F** to the right, and the contact force **P′** to the left (the force exerted by m_2 on m_1). From Newton's third law, **P′** is the reaction to **P**, so that $|\mathbf{P'}| = |\mathbf{P}|$. Applying Newton's second law to m_1 gives

(4) $\sum F_x = F - P' = F - P = m_1 a$

Substituting the value of a from (1) into (4) gives

$$P = F - m_1 a = F - \dfrac{m_1 F}{m_1 + m_2} = \left(\dfrac{m_2}{m_1 + m_2} \right) F$$

This agrees with (3), as it must.

Exercise   If $m_1 = 4.00$ kg, $m_2 = 3.00$ kg, and $F = 9.00$ N, find the magnitude of the acceleration of the system and that of the contact force.

Answer $a = 1.29$ m/s^2, $P = 3.86$ N.

EXAMPLE 5.11 Weighing a Fish in an Elevator

A person weighs a fish of mass m on a spring scale attached to the ceiling of an elevator, as shown in Figure 5.15. Show that if the elevator accelerates in either direction, the spring scale gives a reading different from the weight of the fish.

Reasoning The external forces acting on the fish are the downward force of gravity, w, and the upward force, T, exerted on it by the scale. By Newton's third law, the tension T is also the reading of the spring scale. If the elevator is either at rest or moving at constant velocity, then the fish is not accelerating and $T = w = mg$. However, if the elevator accelerates in either direction, the tension is no longer equal to the weight of the fish.

Solution If the elevator accelerates upward with an acceleration a relative to an observer outside the elevator in an inertial frame (Fig. 5.15a), then the second law applied to the fish gives the total force on the fish:

$$(1) \qquad \sum F = T - w = ma \qquad \text{(if } a \text{ is upward)}$$

If the elevator accelerates downward as in Figure 5.15b, Newton's second law applied to the fish becomes

$$(2) \qquad \sum F = T - w = -ma \qquad \text{(if } a \text{ is downward)}$$

Thus, we conclude from (1) that the scale reading T is greater than the weight, w, if a is upward. From (2) we see that T is less than w if a is downward.

For example, if the weight of the fish is 40.0 N, and a is 2.00 m/s^2 *upward*, then the scale reading is

$$T = ma + mg = mg\left(\frac{a}{g} + 1\right)$$

$$= W\left(\frac{a}{g} + 1\right) = (40.0\text{ N})\left(\frac{2.00\text{ m/s}^2}{9.80\text{ m/s}^2} + 1\right)$$

$$= \boxed{48.2\text{ N}}$$

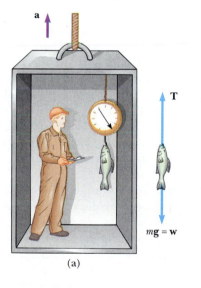

(a)

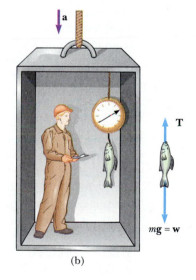

(b)

Observer in inertial frame

FIGURE 5.15 (Example 5.11) Apparent weight versus true weight. (a) When the elevator accelerates *upward* the spring scale reads a value *greater* than the weight of the fish. (b) When the elevator accelerates *downward* the spring scale reads a value *less* than the weight of the fish. When the elevator is accelerating, the spring scale reads the *apparent weight*.

If a is 2.00 m/s² *downward*, then

$$T = -ma + mg = mg\left(1 - \frac{a}{g}\right)$$

$$= W\left(1 - \frac{a}{g}\right) = (40.0 \text{ N})\left(1 - \frac{2.00 \text{ m/s}^2}{9.80 \text{ m/s}^2}\right)$$

$$= \boxed{31.8 \text{ N}}$$

Hence, if you buy a fish by weight in an elevator, make sure the fish is weighed while the elevator is at rest or acceler-

ating downward! Furthermore, note that from the information given here, one cannot determine the direction of motion of the elevator.

Special Cases   If the elevator cable breaks, then the elevator falls freely and $a = g$. Since $w = mg$, we see from (1) that the scale reading, T, is zero in this case; that is, the fish appears to be weightless. If the elevator accelerates downward with an acceleration greater than g, the fish (along with the person in the elevator) eventually hits the ceiling since the acceleration of the fish and person is still that of a freely falling body relative to an outside observer.

5.8   FORCES OF FRICTION

When a body is in motion either on a surface or through a viscous medium such as air or water, there is resistance to the motion because the body interacts with its surroundings. We call such resistance a **force of friction.** Forces of friction are very important in our everyday lives. They allow us to walk or run and are necessary for the motion of wheeled vehicles.

Consider a block on a horizontal table, as in Figure 5.16a. If we apply an external horizontal force F to the block, acting to the right, the block remains stationary if F is not too large. The force that counteracts F and keeps the block from moving acts to the left and is called the *frictional force,* f. As long as the block is not moving, $f = F$. Since the block is stationary, we call this frictional force the *force of static friction,* f_s. Experiments show that this force arises from contacting points that protrude beyond the general level of the surfaces, even for surfaces that are apparently very smooth, as in Figure 5.16a. (If the surfaces are clean and smooth at the atomic level, they are likely to weld together when contact is made.) The frictional force arises in part from one peak physically blocking the motion of a peak from the opposing surface and in part from chemical bonding of opposing points as they come into contact. If the surfaces are rough, bouncing is likely to occur, further complicating the analysis. Although the details of friction are quite complex at the atomic level, it ultimately involves the electrostatic force between atoms or molecules.

If we increase the magnitude of F, as in Figure 5.16b, the block eventually slips. When the block is on the verge of slipping, f_s is a maximum as shown by the graph in Figure 5.16c. When F exceeds $f_{s,\text{max}}$, the block moves and accelerates to the right. When the block is in motion, the retarding frictional force becomes less than $f_{s,\text{max}}$ (Fig. 5.16c). When the block is in motion, we call the retarding force the *force of kinetic friction,* f_k. The unbalanced force in the $+x$ direction, $F - f_k$, accelerates the block to the right. If $F = f_k$, the block moves to the right with constant speed. If the applied force F is removed, then the frictional force f acting to the left accelerates the block in the $-x$ direction and eventually brings it to rest.

Experimentally, one finds that, to a good approximation, both $f_{s,\text{max}}$ and f_k are *proportional to the normal force acting on the block.* The experimental observations can be summarized by the following empirical laws of friction:

Force of static friction

Force of kinetic friction

- The direction of the force of static friction between any two surfaces in contact is opposite the direction of any applied force and can have values

$$f_s \leq \mu_s n \tag{5.9}$$

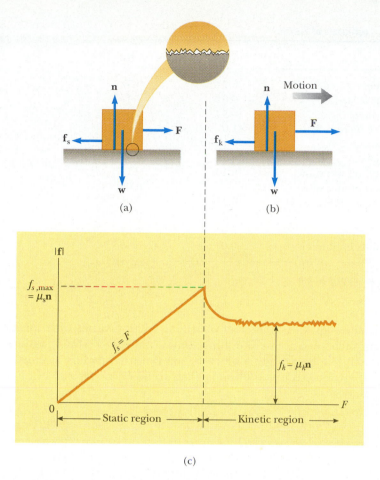

FIGURE 5.16 The direction of the force of friction, **f**, between a block and a rough surface is opposite the direction of the applied force, **F**. Due to the roughness of the two surfaces, contact is made only at a few points as illustrated in the "magnified" view. (a) The magnitude of the force of static friction equals the applied force. (b) When the magnitude of the applied force exceeds the force of kinetic friction, the block accelerates to the right. (c) A graph of the frictional force versus the applied force. Note that $f_{s,\text{max}} > f_k$.

where the dimensionless constant μ_s is called the **coefficient of static friction** and n is the magnitude of the normal force. The equality in Equation 5.9 holds when the block is on the verge of slipping, that is, when $f_s = f_{s,\text{max}} = \mu_s n$. The inequality holds when the applied force is less than this value.

- The direction of the force of kinetic friction acting on an object is opposite the direction of its motion and is given by

$$f_k = \mu_k n \qquad (5.10)$$

where μ_k is the **coefficient of kinetic friction**.

- The values of μ_k and μ_s depend on the nature of the surfaces, but μ_k is generally less than μ_s. Typical values of μ range from around 0.05 to 1.5. Table 5.2 lists some reported values.
- The coefficients of friction are nearly independent of the area of contact between the surfaces.

TABLE 5.2   Coefficients of Friction[a]

	μ_s	μ_k
Steel on steel	0.74	0.57
Aluminum on steel	0.61	0.47
Copper on steel	0.53	0.36
Rubber on concrete	1.0	0.8
Wood on wood	0.25–0.5	0.2
Glass on glass	0.94	0.4
Waxed wood on wet snow	0.14	0.1
Waxed wood on dry snow	—	0.04
Metal on metal (lubricated)	0.15	0.06
Ice on ice	0.1	0.03
Teflon on Teflon	0.04	0.04
Synovial joints in humans	0.01	0.003

[a] All values are approximate.

Finally, although the coefficient of kinetic friction varies with speed, we shall neglect any such variations. The approximate nature of the equations is easily demonstrated by trying to get a block to slip down an incline at constant speed. Especially at low speeds, the motion is likely to be characterized by alternate *stick* and *slip* episodes.

CONCEPTUAL EXAMPLE 5.12

A horse pulls a sled with a horizontal force, causing it to accelerate as in Figure 5.17a. Newton's third law says that the sled exerts an equal and opposite force on the horse. In view of this, how can the sled accelerate? Under what condition does the system (horse plus sled) move with constant velocity?

Reasoning   The motion of any object is determined by the external forces that act on it. In this situation, the horizontal forces exerted on the sled are the forward force exerted by the horse and the backward force of friction between sled and surface (Fig. 5.17b). When the forward force exerted on the sled exceeds the backward force, the resultant force on it is in the forward direction. This resultant force causes the sled to accelerate to the right. The horizontal forces that act on the horse are the forward force of friction between horse and surface and the backward force of the sled (Fig. 5.17c). The resultant of these two forces causes the horse to accelerate. When the forward force of friction acting on the horse balances the backward force of the sled, the system moves with constant velocity.

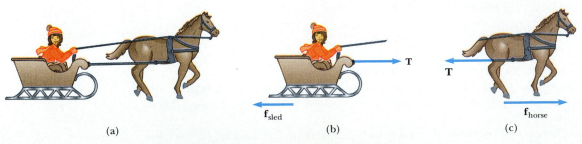

FIGURE 5.17   (Conceptual Example 5.12).

EXAMPLE 5.13 Experimental Determination of μ_s and μ_k

In this example we describe a simple method of measuring coefficients of friction. Suppose a block is placed on a rough surface inclined relative to the horizontal, as in Figure 5.18. The angle of the inclined plane is increased until the block slips. Let us show that by measuring the critical angle θ_c at which this slipping just occurs, we can obtain μ_s directly.

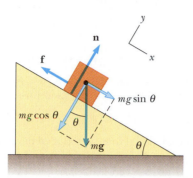

FIGURE 5.18 (Example 5.13) The external forces exerted on a block lying on a rough incline are the force of gravity, mg, the normal force, n, and the force of friction, f. For convenience, the force of gravity is resolved into a component along the incline, $mg \sin \theta$, and a component perpendicular to the incline, $mg \cos \theta$.

Solution The only forces acting on the block are the force of gravity, mg, the normal force, n, and the force of static friction, f_s. When we take x parallel to the plane and y per-

pendicular to it, Newton's second law applied to the block gives

Static case: (1) $\sum F_x = mg \sin \theta - f_s = 0$

(2) $\sum F_y = n - mg \cos \theta = 0$

We can eliminate mg by substituting $mg = n/\cos \theta$ from (2) into (1) to get

(3) $f_s = mg \sin \theta$

$= \left(\dfrac{n}{\cos \theta} \right) \sin \theta = n \tan \theta$

When the inclined plane is at the critical angle, θ_c, $f_s = f_{s,\,max} = \mu_s n$, and so at this angle, (3) becomes

$$\mu_s n = n \tan \theta_c$$

Static case: $\mu_s = \tan \theta_c$

For example, if the block just slips at $\theta_c = 20°$, then $\mu_s = \tan 20° = 0.364$. Once the crate starts to move at $\theta \geq \theta_c$, it accelerates down the incline and the force of friction is $f_k = \mu_k n$. However, if θ is reduced below θ_c, it may be possible to find an angle θ_c' such that the block moves down the incline with constant speed ($a_x = 0$). In this case, using (1) and (2) with f_s replaced by f_k gives

Kinetic case: $\mu_k = \tan \theta_c'$

where $\theta_c' < \theta_c$.

You should try this simple experiment using a coin as the block and a notebook as the inclined plane. Also, you can try taping two coins together to prove that you still get the same critical angles as with one coin.

EXAMPLE 5.14 The Sliding Hockey Puck

A hockey puck on a frozen pond is hit and given an initial speed of 20.0 m/s. If the puck always remains on the ice and slides 115 m before coming to rest, determine the coefficient of kinetic friction between the puck and the ice.

Reasoning The forces acting on the puck after it is in motion are shown in Figure 5.19. If we assume that the force of friction, f_k, remains constant, then this force produces a uniform negative acceleration of the puck. First, we find the acceleration using Newton's second law. Knowing the acceleration of the puck and the distance it travels, we can then use kinematics to find the coefficient of kinetic friction.

Solution Applying Newton's second law in component form to the puck gives

(1) $\sum F_x = -f_k = ma$

(2) $\sum F_y = n - mg = 0$ $(a_y = 0)$

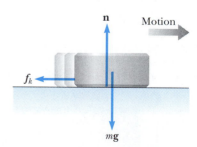

FIGURE 5.19 (Example 5.14) *After* the puck is given an initial velocity to the right, the only external forces acting on it are the force of gravity, mg, the normal force, n, and the force of kinetic friction, f_k.

But $f_k = \mu_k n$, and from (2) we see that $n = mg$. Therefore, (1) becomes

$$-\mu_k n = -\mu_k mg = ma$$

$$a = -\mu_k g$$

The negative sign means that the acceleration is to the left, corresponding to a negative acceleration of the puck. The acceleration is independent of the mass of the puck and is constant because we assume that μ_k remains constant.

Since the acceleration is constant, we can use Equation 2.12, $v^2 = v_0^2 + 2ax$, with the final speed $v = 0$.

$$v_0^2 + 2ax = v_0^2 - 2\mu_k gx = 0$$

$$\mu_k = \frac{v_0^2}{2gx}$$

$$\mu_k = \frac{(20.0 \text{ m/s})^2}{2(9.80 \text{ m/s}^2)(115 \text{ m})} = \boxed{0.177}$$

Note that μ_k has no dimensions.

EXAMPLE 5.15   Connected Objects with Friction

A mass m_1 on a rough, horizontal surface is connected to a second mass m_2 by a lightweight cord over a lightweight, frictionless pulley as in Figure 5.20a. A force of magnitude F at an angle θ with the horizontal is applied to m_1 as shown. The coefficient of kinetic friction between m_1 and the surface is μ. Determine the magnitude of the acceleration of the masses and the tension in the cord.

Reasoning   First we draw the free-body diagrams of m_1 and m_2 as in Figures 5.20b and 5.20c. Next, we apply Newton's second law in component form to each block, and make use of the fact that the magnitude of the force of kinetic friction is proportional to the normal force according to $f_k = \mu n$. Finally we solve for the acceleration in terms of the parameters given.

Solution   The applied force $\mathbf{F}$ has components $F_x = F \cos \theta$ and $F_y = F \sin \theta$. Applying Newton's second law to both masses and assuming the motion of m_1 is to the right, we get

Motion of m_1: $\quad \sum F_x = F \cos \theta - f_k - T = m_1 a$

(1) $\quad\quad\quad\quad \sum F_y = n + F \sin \theta - m_1 g = 0$

Motion of m_2: $\quad \sum F_x = 0$

(2) $\quad\quad\quad\quad \sum F_y = T - m_2 g = m_2 a$

But $f_k = \mu n$, and from (1), $n = m_1 g - F \sin \theta$ (note that in this case n is *not* equal to $m_1 g$); therefore

(3) $\quad\quad\quad f_k = \mu (m_1 g - F \sin \theta)$

That is, the frictional force is reduced because of the positive y component of $\mathbf{F}$. Substituting (3) and the value of T from (2) into (1) gives

$$F \cos \theta - \mu (m_1 g - F \sin \theta) - m_2 (a + g) = m_1 a$$

Solving for a, we get

(4) $\quad\boxed{a = \dfrac{F(\cos \theta + \mu \sin \theta) - g(m_2 + \mu m_1)}{m_1 + m_2}}$

We can find T by substituting this value of a into (2).

Note that the acceleration for m_1 can be either to the right or to the left,[5] depending on the sign of the numerator in (4). If the motion of m_1 is to the left, we must reverse the sign of f_k because the frictional force must oppose the motion. In this case, the value of a is the same as in (4) with μ replaced by $-\mu$.

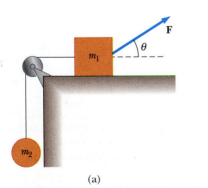

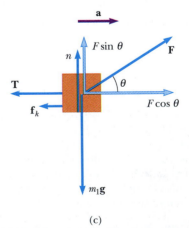

| (a) | (b) | (c) |

FIGURE 5.20   (Example 5.15) (a) The external force, $\mathbf{F}$, applied as shown can cause m_1 to accelerate to the right. (b) and (c) The free-body diagrams assuming that m_1 accelerates to the right while m_2 accelerates upward. The magnitude of the force of kinetic friction in this case is given by $f_k = \mu_k n = \mu_k (m_1 g - F \sin \theta)$.

[5] Equation (4) shows that, when $\mu m_1 > m_2$, there is a range of values of F for which no motion occurs at a given angle θ.

SUMMARY

Newton's first law states that a body at rest remains at rest and a body in uniform motion in a straight line maintains that motion unless an external resultant force acts on it. An **inertial frame** is one that is not accelerated.

Newton's second law states that the acceleration of an object is directly proportional to the resultant force acting on it and inversely proportional to its mass. If the mass of the body is constant, the net force equals the product of the mass and its acceleration, or $\Sigma\mathbf{F} = m\mathbf{a}$.

The **weight** of a body is equal to the product of its mass (a scalar quantity) and the free-fall acceleration, or $\mathbf{w} = m\mathbf{g}$.

Newton's third law states that if two bodies interact, the force exerted on body 1 by body 2 is equal to and opposite the force exerted on body 2 by body 1. Thus, an isolated force cannot exist in nature.

The **maximum force of static friction**, $f_{s,\max}$ between an object and a surface is proportional to the normal force acting on the object. In general, $f_s \leq \mu_s n$, where μ_s is the coefficient of static friction and $\mathbf{n}$ is the normal force. When an object slides over a surface, the force of kinetic friction, $\mathbf{f}_k$, is opposite the motion and is also proportional to the normal force. The magnitude of this force is given by $f_k = \mu_k n$, where μ_k is the coefficient of kinetic friction. Usually, $\mu_k < \mu_s$.

More on Free-Body Diagrams

In order to be successful in applying Newton's second law to a mechanical system you must first be able to recognize all the forces acting on the system. That is, you must be able to construct the correct free-body diagram. The importance of constructing the free-body diagram cannot be overemphasized. In Figure 5.21 a number of mechanical systems are presented together with their corresponding free-body diagrams. You should examine these carefully and then construct free-body diagrams for other systems described in the problems. When a system contains more than one element, it is important that you construct a free-body diagram for *each* element.

As usual, **F** denotes some applied force, $\mathbf{w} = m\mathbf{g}$ is the force of gravity, **n** denotes a normal force, **f** is the force of friction, and **T** is the force of the string on the object.

QUESTIONS

1. If gold were sold by weight, would you rather buy it in Denver or in Death Valley? If sold by mass, at which of the two locations would you prefer to buy it? Why?

2. A passenger sitting in the rear of a bus claims that he was injured when the driver slammed on the brakes, causing a suitcase to come flying toward the passenger from the front of the bus. If you were the judge in this case, what disposition would you make? Why?

3. A space explorer is in a spaceship moving through space far from any planet or star. She notices a large rock, taken as a specimen from an alien planet, floating around the cabin of the spaceship. Should she push it gently toward a storage compartment or kick it toward the compartment? Why?

4. How much does an astronaut weigh out in space, far from any planet?

5. A massive metal object on a rough metal surface may undergo contact welding to that surface. Discuss how this affects the frictional force between the object and the surface.

6. The observer in the elevator of Example 5.11 would claim that the "weight" of the fish is T, the scale reading. This

(continued on page 131)

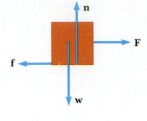

A block pulled to the right on a
rough, horizontal surface

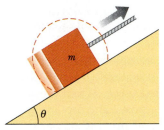

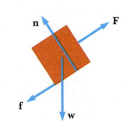

A block pulled up a rough incline

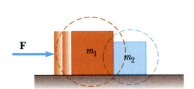

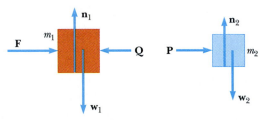

Two blocks in contact, pushed to the
right on a frictionless surface

Note: **P** = −**Q** because they are an action-reaction pair

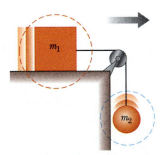

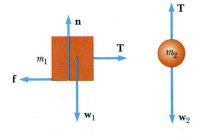

Two masses connected by a light cord. The
surface is rough and the pulley is frictionless

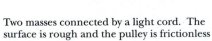

FIGURE 5.21   Various mechanical configurations *(left)* and the corresponding free-body diagrams *(right)*. The term *rough* here means only that the surface is not frictionless.

claim is obviously wrong. Why does this observation differ from that of a person outside the elevator at rest relative to the elevator?

7. Identify the action-reaction pairs in the following situations: a man takes a step; a snowball hits a woman in the back; a baseball player catches a ball; a gust of wind strikes a window.

8. A ball is held in a person's hand. (a) Identify all the external forces acting on the ball and the reaction to each. (b) If the ball is dropped, what force is exerted on it while it is falling? Identify the reaction force in this case. (Neglect air resistance.)

9. If a car is traveling westward with a constant speed of 20 m/s, what is the resultant force acting on it?

10. A large crate is placed on the bed of a truck without being tied to the truck. (a) As the truck accelerates forward, the crate remains at rest relative to the truck. What force causes the crate to accelerate? (b) If the truck driver slams on the brakes, what could happen to the crate?

11. A rubber ball is dropped onto the floor. What force causes the ball to bounce?

12. What is wrong with the statement, "Since the car is at rest, there are no forces acting on it"? How would you correct this sentence?

13. Suppose you are driving a car along a highway at a high speed. Why should you avoid slamming on your brakes if you want to stop in the shortest distance?

14. If you have ever taken a ride in an elevator of a high-rise building, you may have experienced the nauseating sensation of "heaviness" and "lightness" depending on the direction of acceleration. Explain these sensations. Are we truly weightless in free-fall?

15. The driver of a speeding empty truck slams on the brakes and skids to a stop through a distance d. (a) If the truck carried a heavy load such that its mass were doubled, what would be its skidding distance? (b) If the initial speed of the truck is halved, what would be its skidding distance?

16. Does it make sense to say that an object possesses force? Explain.

17. In an attempt to define Newton's third law, a student states that the action and reaction forces are equal to and opposite each other. If this is the case, how can there ever be a net force on an object?

18. In a tug-of-war between two athletes, each pulls on the rope with a force of 200 N. What is the tension in the rope?

19. If you push on a heavy box that is at rest, it requires some force **F** to start its motion. However, once it is sliding, it requires a smaller force to maintain that motion. Why?

20. What causes a rotary lawn sprinkler to turn?

21. The force of gravity is twice as great on a 20-N rock as on a 10-N rock. Why doesn't the 20-N rock have a greater free-fall acceleration?

PROBLEMS

Review Problem

Consider the three connected blocks shown in the diagram. If the inclined plane is frictionless, and the system is in equilibrium, find (in terms of m, g and θ) (a) the mass M and (b) the tensions T_1 and T_2. If the suspended mass is double that value found in part (a), find (c) the acceleration of each block, and (d) the tensions T_1 and T_2. If the coefficient of static friction between m and $2m$ and the inclined plane is μ_s, and the system is in equilibrium, find (e) the minimum value of M and (f) the maximum value of M. (g) Compare the values of T_2 when M has its minimum and maximum values.

Section 5.2 through Section 5.6

1. A force, F, applied to an object of mass m_1 produces an acceleration of 3.00 m/s². The same force applied to an object of mass m_2 produces an acceleration of 1.00 m/s². (a) What is the value of the ratio m_1/m_2? (b) If m_1 and m_2 are combined, find their acceleration under the action of F.

2. Three forces, given by $\mathbf{F}_1 = (-2.00\mathbf{i} + 2.00\mathbf{j})$ N, $\mathbf{F}_2 = (5.00\mathbf{i} - 3.00\mathbf{j})$ N, and $\mathbf{F}_3 = (-45.0\mathbf{i})$ N act on an object to give it an acceleration of magnitude 3.75 m/s². (a) What is the direction of the acceleration? (b) What is the mass of the object? (c) If the object is initially at rest, what is its speed after 10.0 s? (d) What are the velocity components of the object after 10.0 s?

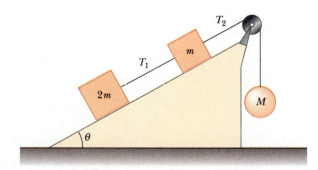

□ indicates problems that have full solutions available in the Student Solution Manual and Study Guide.

3. A time-dependent force, $F = (8.00i - 4.00tj)$ N (where t is in seconds), is applied to a 2.00-kg object initially at rest. (a) At what time will the object be moving with a speed of 15.0 m/s? (b) How far is the object from its initial position when its speed is 15.0 m/s? (c) Through what total displacement has the object traveled at this time?

4. A 3.00-kg particle starts from rest and moves a distance of 4.00 m in 2.00 s under the action of a single, constant force. Find the magnitude of the force.

5. A 5.0-g bullet leaves the muzzle of a rifle with a speed of 320 m/s. What average force is exerted on the bullet while it is traveling down the 0.82-m-long barrel of the rifle?

6. A pitcher releases a baseball of weight 1.4 N with a speed of 32 m/s by uniformly accelerating his arm for 0.090 s. If the ball starts from rest, (a) through what distance does it accelerate before release? (b) What average force is exerted on it to produce this acceleration?

6A. A pitcher releases a baseball of weight w with a speed of v by uniformly accelerating his arm for a time t. If the ball starts from rest, (a) through what distance does it accelerate before release? (b) What average force is exerted on it to produce this acceleration?

7. A 3.0-kg mass undergoes an acceleration given by $a = (2.0i + 5.0j)$ m/s^2. Find the resultant force, F, and its magnitude.

8. A freight train has mass of 1.5×10^7 kg. If the locomotive can exert a constant pull of 7.5×10^5 N, how long does it take to increase the speed of the train from rest to 80 km/h?

9. A person weighs 125 lb. Determine (a) her weight in newtons and (b) her mass in kilograms.

10. If the Earth's gravitational force causes a falling 60-kg student to accelerate downward at 9.8 m/s^2, determine the upward acceleration of the Earth during the student's fall. Take the mass of the Earth to be 5.98×10^{24} kg.

11. The average speed of a nitrogen molecule in air is about 6.7×10^2 m/s, and its mass is about 4.68×10^{-26} kg. (a) If it takes 3.0×10^{-13} s for a nitrogen molecule to hit a wall and rebound with the same speed but in an opposite direction, what is the average acceleration of the molecule during this time interval? (b) What average force does the molecule exert on the wall?

12. If a man weighs 875 N on Earth, what would he weigh on Jupiter, where the free-fall acceleration is 25.9 m/s^2?

13. On planet X, an object weighs 12 N. On planet B where the magnitude of the free-fall acceleration is $1.6g$, the object weighs 27 N. What is the mass of the object and what is the free-fall acceleration (in m/s^2) on planet X?

14. One or more external forces are exerted on each object enclosed in a dashed box shown in Figure 5.1. Identify the reaction to each of these forces.

15. A brick of weight w rests on top of a vertical spring of weight w_s. The spring rests on a table. (a) Draw a free-body diagram of the brick and label all forces acting on it. (b) Repeat (a) for the spring. (c) Identify all action-reaction pairs in the brick-spring-table-Earth system.

16. Forces of 10.0 N north, 20.0 N east, and 15.0 N south are simultaneously applied to a 4.00-kg mass. Obtain its acceleration.

17. A fire helicopter carries a 620-kg bucket of water at the end of a 20-m-long cable. Flying back from a fire at a constant speed of 40 m/s, the cable makes an angle of 40.0° with respect to the vertical. (a) Determine the force of air resistance on the bucket. (b) After filling the bucket with sea water, the helicopter returns to the fire at the same speed with the bucket now making an angle of 7.0° with the vertical. What is the mass of the water in the bucket?

18. Two forces F_1 and F_2 act on a 5.00-kg mass. If $F_1 = 20.0$ N and $F_2 = 15.0$ N, find the acceleration in (a) and (b) of Figure P5.18.

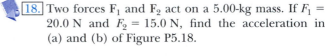
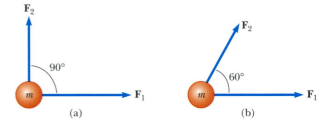

FIGURE P5.18

19. A constant force changes the speed of an 85-kg sprinter from 3.0 m/s to 4.0 m/s in 0.50 s. Calculate (a) the magnitude of the acceleration of the sprinter, (b) the magnitude of the force, and (c) the magnitude of the acceleration of a 58-kg sprinter experiencing the same force. (Assume linear motion.)

20. Besides its weight, a 2.80-kg object is subjected to one other constant force. The object starts from rest and in 1.20 s experiences a displacement of $(4.20$ m$)i - (3.30$ m$)j$, where the direction of j is the upward vertical direction. Determine the other force.

21. A 4.0-kg object has a velocity of $3.0i$ m/s at one instant. Eight seconds later, its velocity is $(8.0i + 10.0j)$ m/s. Assuming the object was subject to a constant net force, find (a) the components of the force and (b) its magnitude.

22. A barefoot field-goal kicker imparts a speed of 35 m/s to a football initially at rest. If the football has a mass of 0.50 kg and the time of contact with the ball

is 0.025 s, what is the force exerted by the ball on the foot?

23. A 2.0-ton truck provides an acceleration of 3.0 ft/s^2 to a 5.0-ton trailer. If the truck exerts the same force on the road while pulling a 15.0-ton trailer, what acceleration results?

24. An electron of mass 9.1×10^{-31} kg has an initial speed of 3.0×10^5 m/s. It travels in a straight line, and its speed increases to 7.0×10^5 m/s in a distance of 5.0 cm. Assuming its acceleration is constant, (a) determine the force on the electron and (b) compare this force with the weight of the electron.

25. Figure P5.25 shows the speed of a person's body, during a chin-up. Assuming the motion is vertical and the mass of the person (excluding the arms) is 64.0 kg, determine the magnitude of the force exerted on the body by the arms at various stages of the motion.

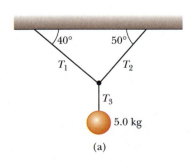

FIGURE P5.25

Section 5.7 Some Applications of Newton's Laws

26. Find the tension in each cord for the systems shown in Figure P5.26. (Neglect the mass of the cords.)

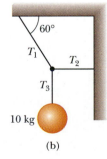

(a)

(b)

FIGURE P5.26

27. A 2.0-kg mass accelerates at 11 m/s^2 in a direction 30.0° north of east (Fig. P5.27). One of the two forces acting on the mass has a magnitude of 11 N

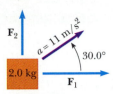

FIGURE P5.27

and is directed north. Determine the magnitude of the second force.

28. A 225-N weight is tied to the middle of a strong rope, and two people pull at opposite ends of the rope in an attempt to lift the weight. (a) What is the magnitude F of the force that each person must apply in order to suspend the weight as shown in Figure P5.28? (b) Can they pull in such a way as to make the rope horizontal? Explain.

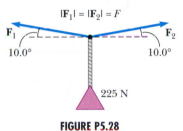

FIGURE P5.28

29. The distance between two telephone poles is 45 m. After a 1.0-kg bird lands on the telephone wire midway between the poles, the wire sags 0.18 m. What is the tension in the wire? Ignore the weight of the wire.

30. The systems shown in Figure P5.30 are in equilibrium. If the spring scales are calibrated in newtons, what do they read in each case? (Neglect the mass of the pulleys and strings, and assume the incline is frictionless.)

31. A bag of cement hangs from three wires as shown in Figure P5.31. Two of the wires make angles θ_1 and θ_2 with the horizontal. If the system is in equilibrium, (a) show that

$$T_1 = \frac{w \cos \theta_2}{\sin(\theta_1 + \theta_2)}$$

(b) Given that $w = 325$ N, $\theta_1 = 10°$, and $\theta_2 = 25°$, find the tensions T_1, T_2, and T_3 in the wires.

32. A woman is pulling her 25-kg suitcase at constant speed by pulling on a strap at an angle θ above the horizontal (Fig. P5.32). She pulls on the strap with a force of magnitude 35 N. A horizontal retarding force of 22 N also acts on the suitcase. (a) What is the

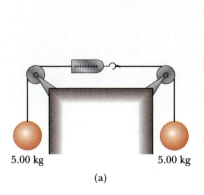

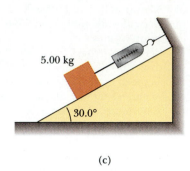

(a)   (b)   (c)

FIGURE P5.30

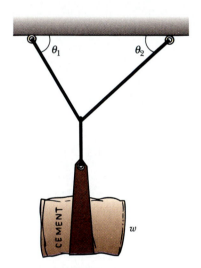

FIGURE P5.31

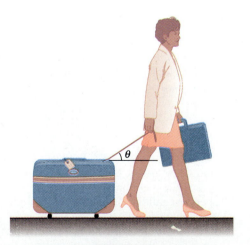

FIGURE P5.32

value of θ? (b) What normal force does the ground exert on the suitcase?

33. A block of mass $m = 2.0$ kg is held in equilibrium on an incline of angle $\theta = 60°$ by the horizontal force **F**, as shown in Figure P5.33. (a) Determine the value of F, the magnitude of **F**. (b) Determine the normal force exerted by the incline on the block (ignore friction).

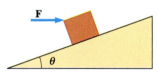

FIGURE P5.33

34. A rifle bullet with a mass of 12 g, traveling with a speed of 400 m/s, strikes a large wooden block, which it penetrates to a depth of 15 cm. Determine the magnitude of the retarding force (assumed constant) that acts on the bullet.

35. A simple accelerometer is constructed by suspending a mass m from a string of length L that is tied to the top of a cart. As the cart is accelerated the string system makes an angle of θ with the vertical. (a) Assuming that the mass of the string is negligible compared to m, derive an expression for the cart's acceleration in terms of θ and show that it is independent of the mass m and the length L. (b) Determine the acceleration of the cart when $\theta = 23°$.

36. The force of the wind on the sails of a sailboat is 390 N north. The water exerts a force of 180 N east. If the boat including crew has a mass of 270 kg, what are the magnitude and direction of its acceleration?

37. In the system shown in Figure P5.37, a horizontal force F_x acts on the 8.00-kg mass. (a) For what values of F_x does the 2.00-kg mass accelerate upward? (b) For what values of F_x is the tension in the cord

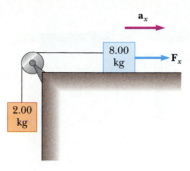

FIGURE P5.37

zero? (c) Plot the acceleration of the 8.00-kg mass versus F_x. Include values of F_x from -100 N to $+100$ N.

 38. Two masses, m_1 and m_2, situated on a frictionless, horizontal surface are connected by a massless string. A force, F, is exerted on one of the masses to the right (Fig. P5.38). Determine the acceleration of the system and the tension, T, in the string.

FIGURE P5.38

39. A small bug is placed between two blocks of masses m_1 and m_2 ($m_1 > m_2$) on a frictionless table. A horizontal force, F, can be applied to either m_1, as in Figure P5.39a, or m_2, as in Figure P5.39b. For which of these two cases does the bug have a greater chance of surviving? Explain. (*Hint:* Determine the contact force between the blocks in each case.)

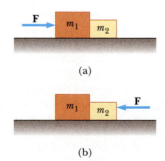

(a)

(b)

FIGURE P5.39

40. A block slides down a frictionless plane having an inclination of $\theta = 15°$. If the block starts from rest at the top and the length of the incline is 2.0 m, find

(a) the magnitude of the acceleration of the block and (b) its speed when it reaches the bottom of the incline.

41. A block of mass $m = 2.0$ kg is released from rest $h = 0.5$ m from the surface of a table, at the top of a $\theta = 30°$ incline as shown in Figure P5.41. The incline is fixed on a table of height $H = 2.0$ m, and the incline is frictionless. (a) Determine the acceleration of the block as it slides down the incline. (b) What is the speed of the block as it leaves the incline? (c) How far from the table will the block hit the floor? (d) How much time has elapsed between when the block is released and when it hits the floor? (e) Does the mass of the block affect any of the above calculations?

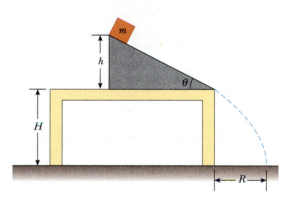

FIGURE P5.41

42. Two masses are connected by a massless string that passes over a massless pulley as in Figure 5.13. If the incline is also frictionless and if $m_1 = 2.00$ kg, $m_2 = 6.00$ kg, and $\theta = 55.0°$, find (a) the magnitude of the acceleration of the masses, (b) the tension in the string, and (c) the speed of each mass 2.00 s after they are released from rest.

43. A 72-kg man stands on a spring scale in an elevator. Starting from rest, the elevator ascends, attaining its maximum speed of 1.2 m/s in 0.80 s. It travels with this constant speed for the next 5.0 s. The elevator then undergoes a uniform acceleration in the negative y direction for 1.5 s and comes to rest. What does the spring scale register (a) before the elevator starts to move? (b) during the first 0.80 s? (c) while the elevator is traveling at constant speed? (d) during the time it is slowing down?

44. A ball of mass m is dropped (from rest) at the top of a building having height h. If a wind blowing along the side of the building exerts a constant horizontal force of magnitude F on the ball as it drops (Fig. P5.44), (a) show that the ball follows a straight-line path. (b) Does this mean that the ball falls with constant velocity? Explain. (c) If the ball is dropped with

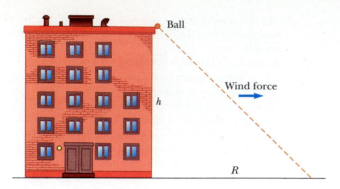

FIGURE P5.44

FIGURE P5.47

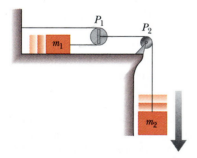

FIGURE P5.46

Section 5.8   Forces of Friction

 47. An 8.5-kg hanging block is connected by a string over a pulley to a 6.2-kg block sliding on a flat table (Fig. P5.47). If the coefficient of sliding friction is 0.20, find the tension in the string.

48. A 25-kg block is initially at rest on a horizontal surface. A horizontal force of 75 N is required to set the

an initial nonzero vertical velocity of v_0, will it still follow the path of a straight line? Explain. (d) Using $m = 10.0$ kg, $h = 10.0$ m, $F = 20.0$ N, and $v_0 = 4.00$ m/s downward, how far from the building will the ball hit the ground?

45. A net horizontal force $F = A + Bt^3$ acts on a 3.5-kg object, where $A = 8.6$ N and $B = 2.5$ N/s^3. What is the horizontal speed of this object 3.0 s after it starts from rest?

46. Mass m_1 on a frictionless horizontal table is connected to mass m_2 through a massless pulley P_1 and a massless fixed pulley P_2 as shown in Figure P5.46. (a) If a_1 and a_2 are the magnitudes of the accelerations of m_1 and m_2, respectively, what is the relationship between these accelerations? Find expressions for (b) the tensions in the strings and (c) the accelerations a_1 and a_2 in terms of m_1, m_2, and g.

block in motion. After it is in motion, a horizontal force of 60 N is required to keep the block moving with constant speed. Find the coefficients of static and kinetic friction from this information.

49. Assume the coefficient of friction between the wheels of a race car and the track is 1.00. If the car starts from rest and accelerates at a constant rate for 335 m, what is its speed at the end of the race?

50. What force must be exerted on block A in order for block B not to fall (Fig. P5.50). The coefficient of static friction between blocks A and B is 0.55, and the horizontal surface is frictionless.

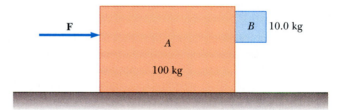

FIGURE P5.50

51. An ice skater moving at 12 m/s coasts to a halt in 95 m on an ice surface. What is the coefficient of friction between ice and skates?

52. A car is traveling at 50.0 mi/h on a horizontal highway. (a) If the coefficient of friction between road and tires on a rainy day is 0.10, what is the minimum distance in which the car will stop? (b) What is the stopping distance when the surface is dry and $\mu = 0.60$?

53. A boy drags his 60.0-N sled at constant speed up a 15° hill. He does so by pulling with a 25-N force on a rope attached to the sled. If the rope is inclined at 35° to the horizontal, (a) what is the coefficient of kinetic friction between sled and snow? (b) At the top of the hill, he jumps on the sled and slides down the hill. What is the magnitude of his acceleration down the slope?

54. A block moves up a 45° incline with constant speed under the action of a force of 15 N applied *parallel* to the incline. If the coefficient of kinetic friction is 0.30, determine (a) the weight of the block and (b) the minimum force required to allow it to move *down* the incline at constant speed.

55. Two blocks connected by a massless rope are being dragged by a horizontal force F (Fig. P5.38). Suppose that $F = 68$ N, $m_1 = 12$ kg, $m_2 = 18$ kg, and the coefficient of kinetic friction between each block and the surface is 0.10. (a) Draw a free-body diagram for each block. (b) Determine the tension, T, and the magnitude of the acceleration of the system.

56. A mass $M = 2.2$ kg is accelerated across a horizontal surface by a rope passing over a pulley, as shown in Figure P5.56. The tension in the rope is 10.0 N and the pulley is 10.0 cm above the top of the block. The coefficient of sliding friction is 0.40. (a) Determine the acceleration of the block when $x = 0.40$ m. (b) Find the value of x at which the acceleration becomes zero.

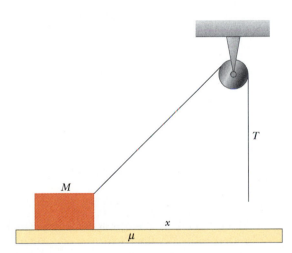

FIGURE P5.56

57. A 3.0-kg block starts from rest at the top of a 30.0° incline and slides 2.0 m down the incline in 1.5 s. Find (a) the magnitude of the acceleration of the block, (b) the coefficient of kinetic friction between block and plane, (c) the frictional force acting on the block, and (d) the speed of the block after it has slid 2.0 m.

58. A block slides on an incline having an inclination of θ with the horizontal. The coefficient of kinetic friction between block and plane is μ_k. (a) If the block accelerates down the incline, show that the magnitude of its acceleration is given by $a = g(\sin \theta - \mu_k \cos \theta)$. (b) If the block is projected up the incline, show that the magnitude of its acceleration is $a = -g(\sin \theta + \mu_k \cos \theta)$.

59. Three masses are connected on the table as shown in Figure P5.59. The table has a coefficient of sliding friction of 0.35. The three masses are 4.0 kg, 1.0 kg, and 2.0 kg, respectively, and the pulleys are frictionless. (a) Determine the acceleration of each block and their directions. (b) Determine the tensions in the two cords.

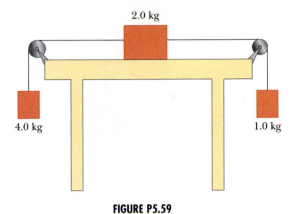

FIGURE P5.59

60. A crate of weight w is pushed by a force F on a horizontal floor. If the coefficient of static friction is μ_s and F is directed at angle ϕ below the horizontal, (a) show that the minimum value of F that will move the crate is

$$F = \frac{\mu_s w \sec \phi}{1 - \mu_s \tan \phi}$$

(b) Find the minimum value of F that can produce motion when $\mu_s = 0.40$, $w = 100$ N, and $\phi = 0°$, 15°, 30°, 45°, and 60°.

61. A block is placed on a plane inclined at 35° relative to the horizontal. If the block slides down the plane with an acceleration of magnitude $g/3$, determine the coefficient of kinetic friction between block and plane.

62. A crate is carried in a truck traveling horizontally at 15 m/s. If the coefficient of static friction between crate and truck is 0.40, determine the minimum stopping distance for the truck so that the crate will not slide.

63. An Olympic skier moving at 25 m/s down a 20° slope encounters a region of wet snow of coefficient of friction $\mu_k = 0.55$. How far down the slope does she travel before coming to a halt?

ADDITIONAL PROBLEMS

64. A car moving at 20.0 m/s brakes to a stop without skidding. The driver in the car behind the first, moving at 30.0 m/s, sees the brake lights, applies his

brakes after a 0.10-s delay, and brakes to a stop without skidding. Assume that $\mu_s = 0.75$ for both cars. Calculate the minimum distance between the cars at the instant the driver of the lead car applies the brakes if a rear-end collision is to be avoided. (Assume constant accelerations.)

65. A mass M is held in place by an applied force $\mathbf{F}$ and a pulley system as shown in Figure P5.65. The pulleys are massless and frictionless. Find (a) the tension in each section of rope, T_1, T_2, T_3, T_4, and T_5, and (b) the magnitude of $\mathbf{F}$.

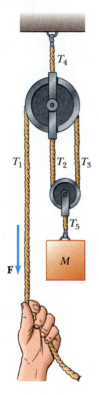

FIGURE P5.65

FIGURE P5.66

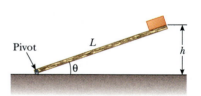

FIGURE P5.67

66. Alex remembered from high-school physics that pulleys can be used to aid in lifting heavy objects. Alex designed the frictionless pulley system shown in Figure P5.66 to lift a safe to a second-floor office. The safe weighs 400 lb, and Alex can pull with a force of 240 lb. (a) Will he be able to raise the safe? (b) What is the maximum weight he can lift using his pulley system? (*Note:* The large pulley is fastened by a yoke to the rope that Alex is pulling.)

67. As part of a laboratory investigation, a student wishes to measure the coefficients of friction between a block of metal and a wooden board. The board has length L, and the block is placed at one end of it. This end of the board is raised, and the block begins to slide when it is a distance of h above the lower end of the board, as in Figure P5.67. At this angle, the block slides down the length of the board in the time t. Determine (a) the coefficient of static friction between block and board, (b) the acceleration of the block, (c) the smallest angle that causes the block to move, and (d) the coefficient of kinetic friction between block and board.

68. About 200 years ago, Charles Coulomb invented the tribometer, a device used to investigate static friction. The instrument is represented schematically in Figure P5.68. To determine the coefficient of static friction, the hanging mass M is increased or decreased as necessary until m is on the verge of sliding. Prove that $\mu_s = M/m$.

69. A 2.00-kg aluminum block and a 6.00-kg copper block are connected by a light string over a frictionless pulley. They are allowed to move on a fixed steel block-wedge (of angle $\theta = 30.0°$) as shown in Figure P5.69. Determine (a) the acceleration of the two blocks and (b) the tension in the string.

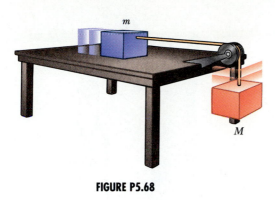

FIGURE P5.68

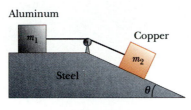

FIGURE P5.69

70. A block of mass $m = 2.00$ kg rests on the left edge of a block of length $L = 3.00$ m and mass $M = 8.00$ kg. The coefficient of kinetic friction between the two blocks is 0.300, and the surface on which the 8.00-kg block rests is frictionless. A constant horizontal force of magnitude $F = 10.0$ N is applied to the 2.00-kg block, setting it in motion as shown in Figure P5.70a. (a) How long will it take before this block makes it to the right side of the 8.00-kg block, as shown in Figure P5.70b? (*Note:* Both blocks are set in motion when **F** is applied.) (b) How far does the 8.00-kg block move in the process?

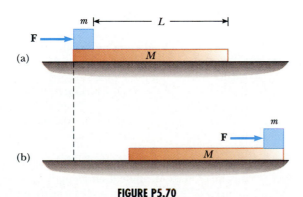

FIGURE P5.70

71. Three baggage carts of masses m_1, m_2, and m_3 are towed by a tractor of mass M along an airport apron. The wheels of the tractor exert a total frictional force **F** on the ground as shown (Fig. P5.71). In the following, express your answers in terms of F, M, m_1, m_2, m_3, and g. (a) What are the magnitude and direction of the horizontal force exerted on the tractor by the ground? (b) What is the smallest value of the coefficient of static friction that will prevent the wheels from slipping? Assume that each of the two drive wheels on the tractor bears $1/3$ of the tractor's weight. (c) What is the acceleration, **a**, of the system (tractor plus baggage carts)? (d) What are the tensions T_1, T_2, and T_3 in the connecting cables? (e) What is the net force on the cart of mass m_2?

FIGURE P5.71

72. In Figure P5.72, the man and the platform together weigh 750 N. Determine how hard the man would have to pull to hold himself off the ground. (Or is it impossible? If so, explain why.)

FIGURE P5.72

73. A 2.0-kg block is placed on top of a 5.0-kg block as in Figure P5.73. The coefficient of kinetic friction between the 5.0-kg block and the surface is 0.20. A horizontal force **F** is applied to the 5.0-kg block. (a) Draw a free-body diagram for each block. What force accel-

erates the 2.0-kg block? (b) Calculate the magnitude of the force necessary to pull both blocks to the right with an acceleration of 3.0 m/s². (c) Find the minimum coefficient of static friction between the blocks such that the 2.0-kg block does not slip under an acceleration of 3.0 m/s².

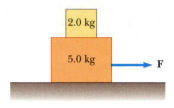

FIGURE P5.73

74. A 5.0-kg block is placed on top of a 10-kg block (Fig. P5.74). A horizontal force of 45 N is applied to the 10-kg block, and the 5.0-kg block is tied to the wall. The coefficient of kinetic friction between the moving surfaces is 0.20. (a) Draw a free-body diagram for each block and identify the action-reaction forces between the blocks. (b) Determine the tension in the string and the magnitude of the acceleration of the 10-kg block.

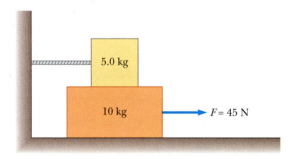

FIGURE P5.74

75. An inventive child named Brian wants to reach an apple in a tree without climbing the tree. Sitting in a chair connected to a rope that passes over a frictionless pulley (Fig. P5.75), he pulls on the loose end of the rope with such a force that the spring scale reads 250 N. His true weight is 320 N, and the chair weighs 160 N. (a) Draw free-body diagrams for Brian and the chair considered as separate systems, and another diagram for Brian and the chair considered as one system. (b) Show that the acceleration of the system is upward and find its magnitude. (c) Find the force that Brian exerts on the chair.

FIGURE P5.75

76. In Figure P5.76, a 500-kg horse pulls a sledge of mass 100 kg. The system (horse plus sledge) has a forward acceleration of 1.00 m/s² when the frictional force on the sledge is 500 N. Find (a) the tension in the connecting rope and (b) the magnitude and direction of the force of friction exerted on the horse. (c) Verify that the total forces of friction the Earth exerts on the system will give to the total system an acceleration of 1.00 m/s².

FIGURE P5.76

77. A block is released from rest at the top of a plane inclined at an angle of 45°. The coefficient of kinetic friction varies along the plane according to the relation $\mu_k = \sigma x$, where x is the distance along the plane measured in meters from the top and where $\sigma = 0.50$ m⁻¹. Determine (a) how far the block slides before coming to rest and (b) the maximum speed it attains.

78. A small block of mass m is initially at the bottom of an incline of mass M, angle θ, and length L, as shown in Figure P5.78a. Assume that all surfaces are friction-

less, and that a constant horizontal force of magnitude F is applied to the block so as to set it, *and the incline,* in motion. (a) Show that the mass m will reach the top of the incline (Fig. P5.78b) in time

$$t = \sqrt{\frac{2L[1 + (m/M)\sin^2\theta]}{(F/m)\cos\theta - g(1 + m/M)\sin\theta}}$$

(*Hint:* The block must always lie on the incline.) (b) How far does the incline travel in the process? (c) Does the expression in part (a) reduce to the expected result when $M \gg m$? Explain.

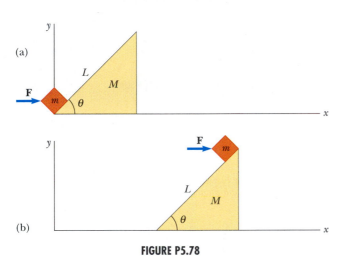

FIGURE P5.78

79. A wire *ABC* supports a body of weight w as shown in Figure P5.79. The wire passes over a fixed pulley at B and is firmly attached to a vertical wall at A. The line AB makes an angle ϕ with the vertical, and the pulley at B exerts on the wire a force of magnitude F inclined at angle θ with the horizontal. (a) Show that if the system is in equilibrium, $\theta = \phi/2$. (b) Show that $F = 2w\sin(\phi/2)$. (c) Sketch a graph of F as ϕ increases from $0°$ to $180°$.

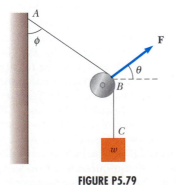

FIGURE P5.79

80. A mobile is formed by supporting four metal butterflies of equal mass m from a string of length L. The points of support are evenly spaced a distance ℓ apart as shown in Figure P5.80. The string forms an angle θ_1 with the ceiling at each end point. The center section of string is horizontal. (a) Find the tension in each section of string in terms of θ_1, m, and g. (b) Find the angle θ_2, in terms of θ_1, that the sections of string between the outside butterflies and the inside butterflies form with the horizontal. (c) Show that the distance D between the end points of the string is

$$D = \frac{L}{5}\left(2\cos\theta_1 + 2\cos\left[\tan^{-1}\left(\frac{1}{2}\tan\theta_1\right)\right] + 1\right)$$

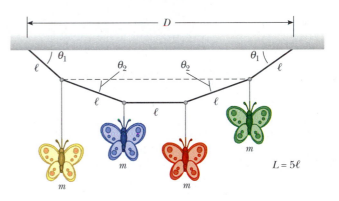

FIGURE P5.80

81. Forces $F_1 = (-6i - 4j)$ N and $F_2 = (-3i + 7j)$ N act on a 2-kg particle initially at rest at coordinates $(-2 \text{ m}, +4 \text{ m})$. (a) What are the components of the particle's velocity at $t = 10$ s? (b) In what direction is the particle moving at $t = 10$ s? (c) What displacement does the particle undergo during the first 10 s? (d) What are the coordinates of the particle at $t = 10$ s?

82. A student is asked to measure the acceleration of a crate on a frictionless inclined plane as in Figure 5.11 using an air track, a stopwatch, and a meter stick. The vertical height of the incline is 1.774 cm, and the total length of the incline is $d = 127.1$ cm. Hence, the angle of inclination θ is determined from the relationship $\sin\theta = 1.774/127.1$. The crate is released from rest at the top of the incline, and its displacement along the incline, x, is measured versus time, where $x = 0$ refers to the initial position of the crate. For x values of 10.0, 20.0, 35.0, 50.0, 75.0, and 100 cm, the measured times to undergo these displacements (averaged over six runs) are 1.02, 1.53, 2.01, 2.64, 3.30, and 3.75 s, respectively. Construct a graph of x versus t^2, and perform a linear least-

squares fit to the data. Determine the magnitude of the acceleration of the crate from the slope of this graph, and compare it with the value you would get using $a' = g \sin \theta$.

 83. What horizontal force must be applied to the cart shown in Figure P5.83 in order that the blocks remain stationary relative to the cart? Assume all surfaces, wheels, and pulley are frictionless. (*Hint:* Note that the force exerted by the string accelerates m_1.)

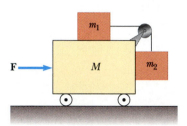

FIGURE P5.83

84. Initially the system of masses shown in Figure P5.83 is held motionless. All surfaces, pulley, and wheels are frictionless. Let the force F be zero and assume that m_2 can move only vertically. At the instant after the system of masses is released, find (a) the tension T in the string, (b) the acceleration of m_2, (c) the acceleration of M, and (d) the acceleration of m_1. (*Note:* The pulley accelerates along with the cart.)

85. The three blocks in Figure P5.85 are connected by massless strings that pass over frictionless pulleys. The acceleration of the system is 2.35 m/s² to the left and the surfaces are rough. Find (a) the tensions in the strings and (b) the coefficient of kinetic friction between blocks and surfaces. (Assume the same μ for both blocks.)

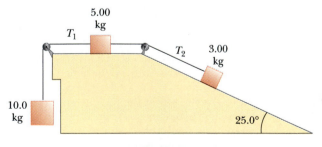

FIGURE P5.85

 86. In Figure P5.86, the coefficient of kinetic friction between the 2.00-kg and 3.00-kg blocks is 0.300. The

horizontal surface and the pulleys are frictionless, and the masses are released from rest. (a) Draw a free-body diagram for each block. (b) Determine the acceleration of each block. (c) Find the tension in the strings.

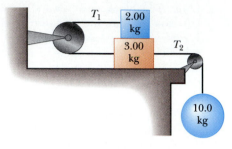

FIGURE P5.86

 87. Two blocks of mass 3.50 kg and 8.00 kg are connected by a massless string that passes over a frictionless pulley (Fig. P5.87). The inclines are frictionless. Find (a) the magnitude of the acceleration of each block and (b) the tension in the string.

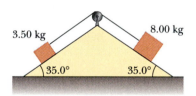

FIGURE P5.87

88. The system shown in Figure P5.87 has an acceleration of magnitude 1.5 m/s². Assume the coefficient of kinetic friction between block and incline is the same for both inclines. Find (a) the coefficient of kinetic friction and (b) the tension in the string.

89. A van accelerates down a hill (Fig. P5.89), going from rest to 30.0 m/s in 6.00 s. During the acceleration, a toy ($m = 100$ g) hangs by a string from the ceiling. The acceleration is such that the string remains perpendicular to the ceiling. Determine (a) the angle θ and (b) the tension in the string.

90. Before 1960 it was believed that the maximum attainable coefficient of static friction of an automobile tire was less than 1. Then about 1962, three companies independently developed racing tires with coefficients of 1.6. Since then, tires have improved, as illustrated in the following problem. According to the 1990 Guinness Book of Records, the fastest $\frac{1}{4}$ mile covered by a piston-engine car from a standing start is 4.96 s. This record elapsed time was set by

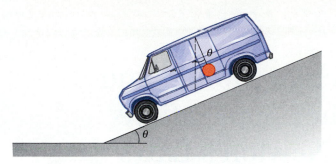

FIGURE P5.89

Shirley Muldowney in September 1989. (a) Assuming that the rear wheels nearly lifted the front wheels off the pavement, what minimum value of μ is necessary to achieve the record time? (b) Suppose Muldowney were able to double her engine power, keeping other things equal. How would this change affect the elapsed time?

91. A magician attempts to pull a tablecloth from under a 200-g mug located 30 cm from the edge of the cloth. If there is a frictional force of 0.10 N exerted on the mug by the cloth, and the cloth is pulled with a constant acceleration of magnitude 3.0 m/s², how far does the mug move on the tabletop before the cloth is completely out from under it? (*Hint:* The cloth moves more than 30 cm before it is out from under the mug!)

SPREADSHEET PROBLEMS

S1. An 8.4-kg mass slides down a fixed, frictionless inclined plane. Design and write a spreadsheet to determine the normal force exerted on the mass and its acceleration for a series of incline angles (measured from the horizontal) ranging from 0° to 90° in 5° increments. Use the graphing capability of your spreadsheet data to plot the normal force and the acceleration as functions of the incline angle. In the limiting cases of 0° and 90°, are your results consistent with the known behavior?

S2. A person must move a 65-kg crate resting on a level floor. The coefficient of static friction between crate and floor is 0.48. A force of magnitude F is applied at an angle θ with the horizontal. (a) Construct a spreadsheet that will enable you to calculate the force necessary to move the box for a sequence of angles. Take the angle θ as positive if the force has an upward component, and negative if the force has a downward component. Plot F versus θ, and from this graph determine the minimum force required to move the box. At what angle should the force be applied? (b) Investigate what happens when you change the coefficient of friction.

B.C.   By John Hart

CHAPTER 6

Circular Motion and Other Applications of Newton's Laws

The passengers on this corkscrew rollercoaster experience the thrill of various forces as they travel along the curved track. The forces on one of the passenger cars include the force exerted by the track, the force of gravity, and the force of air resistance. *(Robin Smith/Tony Stone Images)*

I n the previous chapter we introduced Newton's laws of motion and applied them to situations involving linear motion. In this chapter we apply Newton's laws of motion to circular motion. We also discuss the motion of an object when observed in an accelerated, or noninertial, frame of reference and the motion of an object through a viscous medium. Finally, we conclude this chapter with a brief discussion of the fundamental forces in nature.

6.1 NEWTON'S SECOND LAW APPLIED TO UNIFORM CIRCULAR MOTION

In Section 4.4 we found that a particle moving in a circular path of radius r with uniform speed v experiences an acceleration that has a magnitude

$$a_r = \frac{v^2}{r}$$

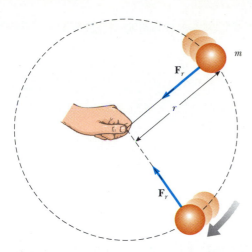

FIGURE 6.1 An overhead view of a ball moving in a circular path in a horizontal plane. A force **F**$_r$ directed toward the center of the circle keeps the ball moving in the circle with constant speed.

Because the velocity vector **v** changes its direction continuously during the motion, the acceleration vector **a**$_r$ is directed toward the center of the circle and hence is called **centripetal acceleration.** Furthermore, **a**$_r$ is *always* perpendicular to **v**.

Consider a ball of mass m tied to a string of length r and being whirled in a horizontal circular path on a table top as in Figure 6.1. Let us assume that the ball moves with constant speed. The ball tends to maintain the motion in a straight-line path; however, the string prevents this motion along a straight line by exerting a force on the ball to make it follow its circular path. This force is directed along the length of the string toward the center of the circle, as shown in Figure 6.1, and is an example of a class of forces called **central forces.** If we apply Newton's second law along the radial direction, we find that the required central force is

$$F_r = ma_r = m\frac{v^2}{r} \qquad (6.1)$$

Central force

Like the centripetal acceleration, the central force acts toward the center of the circular path followed by the particle. Because they act toward the center of rotation, central forces cause a change in the direction of the velocity. Central forces are no different from any other forces we have encountered. The term *central* is used simply to indicate that *the force is directed toward the center of a circle.* In the case of a ball rotating at the end of a string, the force exerted by the string on the ball is the central force. For a satellite in a circular orbit around the Earth, the force of gravity is the central force. The central force acting on a car rounding a curve on a flat road is the force of friction between the tires and the pavement, and so forth. In general, a body can move in a circular path under the influence of such forces as friction, the gravitational force, or a combination of forces.

Regardless of the example used, if the central force acting on an object should vanish, the object would no longer move in its circular path; instead, it would move along a straight-line path tangent to the circle. This idea is illustrated in Figure 6.2 for the case of the ball whirling in a circle at the end of a string. If the string breaks at some instant, the ball will move along the straight-line path tangent to the circle at the point where the string broke.

Let us consider some examples of uniform circular motion. In each case, be sure to recognize the external force (or forces) that causes the body to move in its circular path.

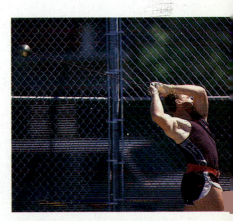

An athlete in the process of throwing the hammer. The force exerted by the chain is the central force. Only when the athlete releases the hammer will it move along a straight-line path tangent to the circle. *(Focus on Sports)*

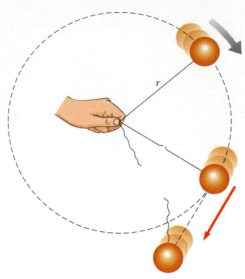

FIGURE 6.2 When the string breaks, the ball moves in the direction tangent to the circular path.

EXAMPLE 6.1   How Fast Can It Spin?

A ball of mass 0.500 kg is attached to the end of a cord whose length is 1.50 m. The ball is whirled in a horizontal circle as in Figure 6.2. If the cord can withstand a maximum tension of 50.0 N, what is the maximum speed the ball can have before the cord breaks?

Solution   Because the central force in this case is the force T exerted by the cord on the ball, Equation 6.1 gives

$$T = m\,\frac{v^2}{r}$$

Solving for v, we have

$$v = \sqrt{\frac{Tr}{m}}$$

The maximum speed that the ball can have corresponds to the maximum tension. Hence, we find

$$v_{max} = \sqrt{\frac{T_{max}r}{m}} = \sqrt{\frac{(50.0\ \text{N})(1.50\ \text{m})}{0.500\ \text{kg}}} = \boxed{12.2\ \text{m/s}}$$

Exercise   Calculate the tension in the cord if the speed of the ball is 5.00 m/s.

Answer   8.33 N.

EXAMPLE 6.2   The Conical Pendulum

A small body of mass m is suspended from a string of length L. The body revolves in a horizontal circle of radius r with constant speed v, as in Figure 6.3. (Since the string sweeps out the surface of a cone, the system is known as a *conical pendulum*.) Find the speed of the body and the period of revolution, T_p, defined as the time needed to complete one revolution.

Solution   The free-body diagram for the mass m is shown in Figure 6.3, where the force exerted by the string, T, has been resolved into a vertical component, $T\cos\theta$, and a component $T\sin\theta$ acting toward the center of rotation. Since the body does not accelerate in the vertical direction, the vertical component of T must balance the weight. Therefore,

(1) $T\cos\theta = mg$

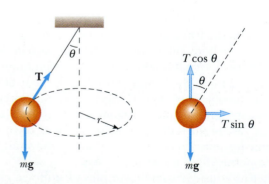

FIGURE 6.3 (Example 6.2) The conical pendulum and its free-body diagram.

Since the central force in this example is provided by the component $T \sin \theta$, from Newton's second law we get

$$(2) \qquad T \sin \theta = ma_r = \frac{mv^2}{r}$$

By dividing (2) by (1), we eliminate T and find that

$$\tan \theta = \frac{v^2}{rg}$$

But from the geometry, we note that $r = L \sin \theta$; therefore

$$v = \sqrt{rg \tan \theta} = \boxed{\sqrt{Lg \sin \theta \tan \theta}}$$

Since the ball travels a distance of $2\pi r$ (the circumference of the circular path) in a time equal to the period of revolution, T_p (not to be confused with the force T), we find

$$(3) \qquad T_p = \frac{2\pi r}{v} = \frac{2\pi r}{\sqrt{rg \tan \theta}} = \boxed{2\pi \sqrt{\frac{L \cos \theta}{g}}}$$

The intermediate algebraic steps used in obtaining (3) are left to the reader. Note that T_p is independent of m! If we take $L = 1.00$ m and $\theta = 20.0°$, we find using (3) that

$$T_p = 2\pi \sqrt{\frac{(1.00 \text{ m})(\cos 20.0°)}{9.80 \text{ m/s}^2}} = 1.95 \text{ s}$$

Is it physically possible to have a conical pendulum with $\theta = 90°$?

EXAMPLE 6.3   What Is the Maximum Speed of the Car?

A 1500-kg car moving on a flat, horizontal road negotiates a curve whose radius is 35.0 m as in Figure 6.4. If the coefficient of static friction between the tires and the dry pavement is 0.500, find the maximum speed the car can have in order to make the turn successfully.

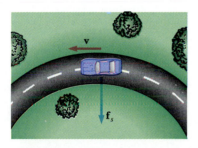

FIGURE 6.4   (Example 6.3) The force of static friction directed toward the center of the arc keeps the car moving in a circle.

Solution   In this case, the central force that enables the car to remain in its circular path is the force of static friction. Hence, from Equation 6.1 we have

$$(1) \qquad f_s = m \frac{v^2}{r}$$

The maximum speed that the car can have around the curve corresponds to the speed at which it is on the verge of skidding outwards. At this point, the friction force has its maximum value

$$f_{s\,max} = \mu n$$

Because the normal force equals the weight in this case, we find

$$f_{s\,max} = \mu mg = (0.500)(1500 \text{ kg})(9.80 \text{ m/s}^2) = 7350 \text{ N}$$

Substituting this value into (1), we find that the maximum speed is

$$v_{max} = \sqrt{\frac{f_{s\,max}\,r}{m}} = \sqrt{\frac{(7350 \text{ N})(35.0 \text{ m})}{1500 \text{ kg}}} = \boxed{13.1 \text{ m/s}}$$

Exercise   On a wet day, the car described in this example begins to skid on the curve when its speed reaches 8.00 m/s. What is the coefficient of static friction in this case?

Answer   0.187.

EXAMPLE 6.4   The Banked Exit Ramp

An engineer wishes to design a curved exit ramp for a toll road in such a way that a car will not have to rely on friction to round the curve without skidding. Suppose that a typical car rounds the curve with a speed of 30.0 mi/h (13.4 m/s) and that the radius of the curve is 50.0 m. At what angle should the curve be banked?

Reasoning   On a level road, the central force must be provided by a force of friction between car and road. However, if

the road is banked at an angle θ, as in Figure 6.5, the normal force, **n**, has a horizontal component $n \sin \theta$ pointing toward the center of the circular path followed by the car. We assume that only the component $n \sin \theta$ furnishes the central force. Therefore, the banking angle we calculate will be one for which *no frictional force is required*. In other words, a car moving at the correct speed (13.4 m/s) can negotiate the curve even on an icy surface.

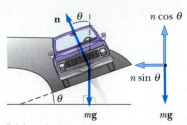

FIGURE 6.5   (Example 6.4) End view of a car rounding a curve on a road banked at an angle θ to the horizontal. The central force is provided by the horizontal component of the normal force when friction is neglected. Note that **n** is the *sum* of the forces that the road exerts on the wheels of the car.

Solution   Newton's second law written for the radial direction gives

$$(1) \qquad n \sin \theta = \frac{mv^2}{r}$$

The car is in equilibrium in the vertical direction. Thus, from $\Sigma F_y = 0$, we have

$$(2) \qquad n \cos \theta = mg$$

Dividing (1) by (2) gives

$$\tan \theta = \frac{v^2}{rg}$$

$$\theta = \tan^{-1}\left[\frac{(13.4 \text{ m/s})^2}{(50.0 \text{ m})(9.80 \text{ m/s}^2)}\right] = \boxed{20.1°}$$

If a car rounds the curve at a speed lower than 13.4 m/s, the driver will have to rely on friction to keep from sliding down the incline. A driver who attempts to negotiate the curve at a speed higher than 13.4 m/s will have to depend on friction to keep from sliding up the ramp.

Exercise   Write Newton's second law applied to the radial direction for the car in a situation in which a frictional force f is directed *down* the slope of the banked road.

Answer $n \sin \theta + f \cos \theta = \dfrac{mv^2}{r}$.

EXAMPLE 6.5   Satellite Motion

This example treats the problem of a satellite moving in a circular orbit around the Earth. In order to understand this problem, we must first note that the gravitational force between two particles having masses m_1 and m_2, separated by a distance r, is attractive and has a magnitude

$$F = G\frac{m_1 m_2}{r^2}$$

where $G = 6.672 \times 10^{-11} \text{ N·m}^2/\text{kg}^2$. This is Newton's law of gravity, which we shall discuss in detail in Chapter 14.

Now consider a satellite of mass m moving in a circular orbit around the Earth at a constant speed v and at an altitude h above the Earth's surface as in Figure 6.6. (a) Determine the speed of the satellite in terms of G, h, R_E (the radius of the Earth), and M_E (the mass of the Earth).

Solution   Because the only external force on the satellite is the force of gravity, which acts toward the center of the Earth, we have

$$F_r = G\frac{M_E m}{r^2}$$

From Newton's second law we get

$$G\frac{M_E m}{r^2} = m\frac{v^2}{r}$$

Solving for v and remembering that $r = R_E + h$, we get

$$(1) \qquad v = \sqrt{\frac{GM_E}{r}} = \boxed{\sqrt{\frac{GM_E}{R_e + h}}}$$

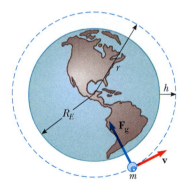

FIGURE 6.6   (Example 6.5) A satellite of mass m moving in a circular orbit of radius r and with constant speed v around the Earth. The central force **F** is provided by the gravitational force exerted by the Earth on the satellite.

(b) Determine the satellite's period of revolution, T_p (the time for one revolution about the Earth).

Solution   Since the satellite travels a distance of $2\pi r$ (the circumference of the circle) in a time T_p, we find using (1) that

$$(2) \qquad T_p = \frac{2\pi r}{v} = \frac{2\pi r}{\sqrt{GM_E/r}} = \boxed{\left(\frac{2\pi}{\sqrt{GM_E}}\right)r^{3/2}}$$

The planets move around the Sun in approximately circular orbits. The radii of these orbits can be calculated from (2) with M_E replaced by the mass of the Sun. If (2) is squared, we see that $T_p^2 \propto r^3$. The fact that the square of the period is proportional to the cube of the radius of the orbit was first recognized as an empirical relation based on planetary data. We shall return to this topic in Chapter 14.

Exercise A satellite is in a circular orbit at an altitude of 1000 km. The radius of the Earth is 6.37×10^6 m. Find the speed of the satellite and the period of its orbit.

Answer 7.35×10^3 m/s; 6.31×10^3 s = 105 min.

EXAMPLE 6.6 Let's Go Loop-the-Loop

A pilot of mass m in a jet aircraft executes a "loop-the-loop" maneuver as illustrated in Figure 6.7a. In this flying pattern, the aircraft moves in a vertical circle of radius 2.70 km at a *constant speed* of 225 m/s. Determine the force exerted by the seat on the pilot at (a) the bottom of the loop and (b) the top of the loop. Express the answers in terms of the weight of the pilot, mg.

Solution (a) The free-body diagram for the pilot at the bottom of the loop is shown in Figure 6.7b. The only forces acting on the pilot are the downward force of gravity, mg, and the upward force n_{bot} exerted by the seat. Since the net upward force that provides the centripetal acceleration has a magnitude $n_{bot} - mg$, Newton's second law for the radial direction gives

$$n_{bot} - mg = m\frac{v^2}{r}$$

$$n_{bot} = mg + m\frac{v^2}{r} = mg\left[1 + \frac{v^2}{rg}\right]$$

Substituting the values given for the speed and radius gives

$$n_{bot} = mg\left[1 + \frac{(225 \text{ m/s})^2}{(2.70 \times 10^3 \text{ m})(9.80 \text{ m/s}^2)}\right] = 2.91\,mg$$

Hence, the force exerted by the seat on the pilot is *greater* than his weight by a factor of 2.91. The pilot experiences an apparent weight that is greater than his true weight by the factor 2.91. This is discussed further in Section 6.4.

(b) The free-body diagram for the pilot at the top of the loop is shown in Figure 6.7c. At this point, both the weight and the force exerted by the seat on the pilot, n_{top}, act *downward*, so the net force downwards which provides the centripetal acceleration has a magnitude $n_{top} + mg$. Applying Newton's second law gives

$$n_{top} + mg = m\frac{v^2}{r}$$

$$n_{top} = m\frac{v^2}{r} - mg = mg\left[\frac{v^2}{rg} - 1\right]$$

$$n_{top} = mg\left[\frac{(225 \text{ m/s})^2}{(2.70 \times 10^3 \text{ m})(9.80 \text{ m/s}^2)} - 1\right]$$

$$= 0.911\,mg$$

In this case, the force exerted by the seat on the pilot is *less* than the true weight by a factor of 0.911. Hence, the pilot will feel lighter at the top of the loop.

Exercise Calculate the central force on the pilot if the aircraft is at point A in Figure 6.7a, midway up the loop.

Answer $n_A = 1.911\,mg$ directed to the right.

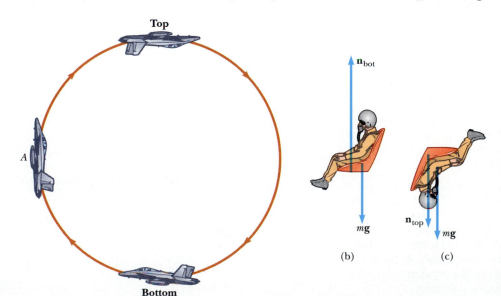

Top

A

Bottom

(a)

n_{bot}

mg

(b)

n_{top}

mg

(c)

FIGURE 6.7 (a) An aircraft executes a "loop-the-loop" maneuver as it moves in a vertical circle at constant speed. (b) Free-body diagram for the pilot at the bottom of the loop. In this position the pilot experiences an apparent weight that is greater than his true weight. (c) Free-body diagram for the pilot at the top of the loop.

Some examples of central forces acting during circular motion. (*Left*) Cyclists in the Tour de France negotiate a curve on a flat racing track. (*Right*) Passengers in a roller coaster at Knott's Berry Farm. What are the origins of the central forces in these two examples?   *(Left: Michel Gouverneur, Photo News, Gamma Sport; Right: Superstock)*

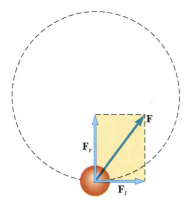

FIGURE 6.8   When the force acting on a particle moving in a circular path has a tangential component F_t, its speed changes. The total force on the particle in this case is the vector sum of the tangential force and the central force. That is, $\mathbf{F} = \mathbf{F}_t + \mathbf{F}_r$.

6.2   NONUNIFORM CIRCULAR MOTION

In Chapter 4 we found that if a particle moves with varying speed in a circular path, there is, in addition to the centripetal component of acceleration, a tangential component of magnitude dv/dt. Therefore, the force acting on the particle must also have a tangential and a radial component. That is, since the total acceleration is $\mathbf{a} = \mathbf{a}_r + \mathbf{a}_t$, the total force exerted on the particle is $\mathbf{F} = \mathbf{F}_r + \mathbf{F}_t$, as shown in Figure 6.8. The vector $\mathbf{F}_r$ is directed toward the center of the circle and is responsible for the centripetal acceleration. The vector $\mathbf{F}_t$ tangent to the circle is responsible for the tangential acceleration, which causes the speed of the particle to change with time. The following example demonstrates this type of motion.

EXAMPLE 6.7   Follow the Rotating Ball

A small sphere of mass m is attached to the end of a cord of length R, which rotates in a *vertical* circle about a fixed point O, as in Figure 6.9a. Let us determine the tension in the cord at any instant when the speed of the sphere is v and the cord makes an angle θ with the vertical.

Solution   First we note that the speed is *not* uniform because there is a tangential component of acceleration arising from the weight of the sphere. From the free-body diagram in Figure 6.9a, we see that the only forces acting on the sphere are the weight, $m\mathbf{g}$, and the force exerted by the cord, $\mathbf{T}$. Now we resolve $m\mathbf{g}$ into a tangential component, $mg \sin \theta$, and a radial component, $mg \cos \theta$. Applying Newton's second law to the forces in the tangential direction gives

$$\sum F_t = mg \sin \theta = ma_t$$

$$a_t = g \sin \theta$$

This component causes v to change in time, since $a_t = dv/dt$.

Applying Newton's second law to the forces in the radial direction and noting that both $\mathbf{T}$ and $\mathbf{a}_r$ are directed toward O, we get

$$\sum F_{r_i} = T - mg \cos \theta = \frac{mv^2}{R}$$

$$T = m\left(\frac{v^2}{R} + g \cos \theta\right)$$

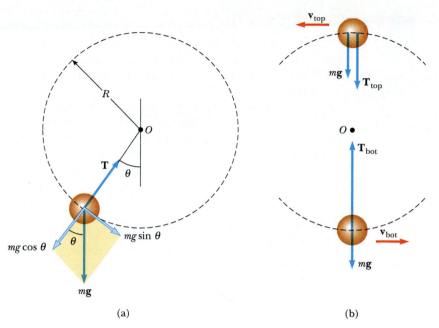

FIGURE 6.9 (Example 6.7) (a) Forces acting on a mass m connected to a string of length R and rotating in a vertical circle centered at O. (b) Forces acting on m when it is at the top and bottom of the circle. Note that the tension at the bottom is a maximum and the tension at the top is a minimum.

Limiting Cases At the top of the path, where $\theta = 180°$, we have $\cos 180° = -1$, and the tension equation becomes

$$T_{\text{top}} = m\left(\frac{v_{\text{top}}^2}{R} - g\right)$$

This is the *minimum* value of T. Note that at this point $a_t = 0$ and therefore the acceleration is radial and directed downward, as in Figure 6.9b.

At the bottom of the path, where $\theta = 0$, we see that since $\cos 0 = 1$,

$$T_{\text{bot}} = m\left(\frac{v_{\text{bot}}^2}{R} + g\right)$$

This is the *maximum* value of T. Again, at this point $a_t = 0$ and the acceleration is radial and directed upward.

Exercise At what orientation of the system would the cord most likely break if the average speed increased?

Answer At the bottom of the path, where T has its maximum value.

*6.3 MOTION IN ACCELERATED FRAMES

When Newton's laws of motion were introduced in Chapter 5, we emphasized that the laws are valid only when observations are made in an inertial frame of reference. In this section, we analyze how an observer in a noninertial frame of reference (one that is accelerating) would attempt to apply Newton's second law.

If a particle moves with an acceleration **a** relative to an observer in an inertial frame, then the inertial observer may use Newton's second law and correctly claim that $\Sigma\mathbf{F}_i = m\mathbf{a}$. If an observer in an accelerated frame (the noninertial observer) tries to apply Newton's second law to the motion of the particle, she or he must introduce **fictitious forces** to make Newton's second law work in that frame. These forces "invented" by the noninertial observer appear to be real forces in the accelerating frame. However, we emphasize that these fictitious forces do not exist when the motion is observed in an inertial frame. The fictitious forces are used only in an accelerating frame but do not represent "real" forces on the body. (By

Fictitious forces

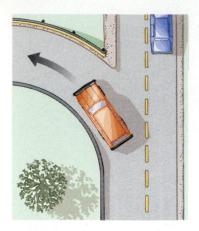

FIGURE 6.10   A car approaching a curved exit ramp. What causes the passenger on the right side to move toward the door?

real forces, we mean the interaction of the body with its environment.) If the fictitious forces are properly defined in the accelerating frame, then the description of motion in this frame will be equivalent to the description by an inertial observer who considers only real forces. Usually, motions are analyzed using inertial reference frames, but there are cases in which an accelerating frame is more convenient.

Suppose you are a passenger in a car accelerating along a straight line down a highway. What causes a building to appear to accelerate in the opposite direction? From this observation, you would have to imagine a force acting on the (stationary) building to cause its apparent acceleration.

In order to understand the motion of a system that is noninertial due to rotation, consider a car traveling along a highway at a high speed and approaching a curved exit ramp, as in Figure 6.10. As the car takes the sharp left turn onto the ramp, a person sitting in the passenger seat slides to the right across the seat and hits the door. At that point, the force exerted by the door keeps her from being ejected from the car. What causes the passenger to move toward the door? A popular, but improper, explanation is that some mysterious force pushes her outward. (This is often called the "centrifugal" force, but we shall not use this term since it often creates confusion.) The passenger invents this fictitious force in order to explain what is going on in her accelerated frame of reference. The driver of the car also experiences this effect, but holds on to the steering wheel to keep from sliding across the seat.

The phenomenon is correctly explained as follows. Before the car enters the ramp, the passenger is moving in a straight-line path. As the car enters the ramp and travels a curved path, the passenger tends to move along the original straight-line path. This is in accordance with Newton's first law: The natural tendency of a body is to continue moving in a straight line. However, if a sufficiently large central force (toward the center of curvature) acts on the passenger, she will move in a curved path along with the car. The origin of this central force is the force of friction between the passenger and the car seat. If this frictional force is not large enough, the passenger will slide across the seat as the car turns under her. Eventually, the passenger encounters the door, which provides a large enough central force to enable the passenger to follow the same curved path as the car. The passenger slides toward the door not because of some mysterious outward force but because *there is no central force large enough to allow her to travel along the circular path followed by the car.*

In summary, one must be very careful to distinguish real forces from fictitious ones in describing motion in an accelerating frame. An observer in a car rounding a curve is in an accelerating frame and invents a fictitious outward force to explain why he or she is thrown outward. A stationary observer outside the car, however, considers only real forces on the passenger. To this observer, the mysterious outward force *does not exist!* The only real external force on the passenger is the central (inward) force due either to friction or to the normal force exerted by the door.

EXAMPLE 6.8   Fictitious Forces in Linear Motion

A small sphere of mass m is hung by a cord from the ceiling of an accelerating boxcar, as in Figure 6.11. (When the boxcar is not accelerating, the cord is vertical, or $\theta = 0$.) According to the inertial observer at rest (Figure 6.11a), the forces on the sphere are force exerted by the cord **T** and the force of

gravity $m\mathbf{g}$. The inertial observer concludes that the acceleration of the sphere is the same as that of the boxcar and that this acceleration is provided by the horizontal component of **T**. Also, the vertical component of **T** balances the weight. Therefore, the inertial observer writes Newton's second law

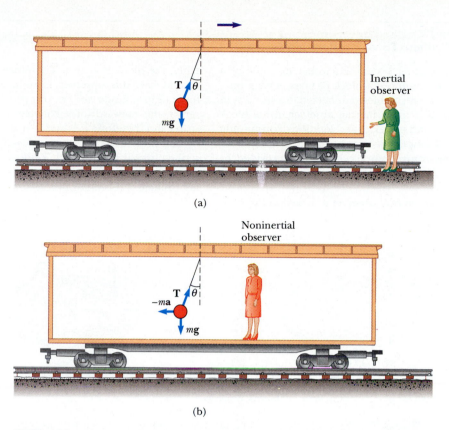

(a)

(b)

FIGURE 6.11 (Example 6.8) (a) A small sphere suspended from the ceiling of a boxcar accelerating to the right is deflected as shown. The inertial observer at rest outside the car claims that the acceleration of the sphere is provided by the horizontal component of T. (b) A noninertial observer riding in the car says that the net force on the sphere is zero and that the deflection of the string from the vertical is due to a fictitious force, $-m\mathbf{a}$, which balances the horizontal component of T.

as $\mathbf{T} + m\mathbf{g} = m\mathbf{a}$, which in component form becomes

Inertial observer $\begin{cases} (1) & \sum F_x = T\sin\theta = ma \\ (2) & \sum F_y = T\cos\theta - mg = 0 \end{cases}$

Thus, by solving (1) and (2) simultaneously, the inertial observer can determine the acceleration of the car through the relationship

$$a = g\tan\theta$$

Therefore, since the deflection of the string from the vertical serves as a measure of the acceleration of the car, *a simple pendulum can be used as an accelerometer.*

According to the noninertial observer riding in the car, described in Figure 6.11b, the cord still makes an angle θ

with the vertical, the sphere is at rest, and its acceleration is zero. Therefore, the noninertial observer introduces a fictitious force, $-m\mathbf{a}$, to balance the horizontal component of T and claims that the net force on the sphere is *zero!* In this noninertial frame of reference, Newton's second law in component form gives

Noninertial observer $\begin{cases} \sum F'_x = T\sin\theta - ma = 0 \\ \sum F'_y = T\cos\theta - mg = 0 \end{cases}$

These expressions are equivalent to (1) and (2); therefore, the noninertial observer gets the same mathematical results as the inertial observer. However, the physical interpretation of the deflection of the string differs in the two frames of reference. Note that even though a pendulum is used, it does not oscillate in this application.

EXAMPLE 6.9   Fictitious Force in a Rotating System

An observer in a rotating system is another example of a noninertial observer. Suppose a block of mass m lying on a horizontal, frictionless turntable is connected to a string as in Figure 6.12. According to an inertial observer, if the block rotates uniformly, it undergoes an acceleration of magnitude v^2/r, where v is its tangential speed. The inertial observer concludes that this centripetal acceleration is provided by the force exerted by the string, T, and writes Newton's second law $T = mv^2/r$.

According to a noninertial observer attached to the turn-table, the block is at rest. Therefore, in applying Newton's second law, this observer introduces a fictitious outward force of magnitude mv^2/r. According to the noninertial observer, this outward force balances the force exerted by the string and therefore $T - mv^2/r = 0$.

You should be careful when using fictitious forces to describe physical phenomena. Remember that fictitious forces are used *only* in noninertial frames of reference. When solving problems, it is often best to use an inertial frame.

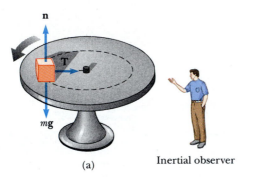
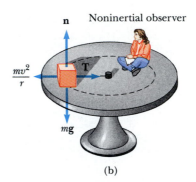

FIGURE 6.12 (Example 6.9) A block of mass m connected to a string tied to the center of a rotating turntable. (a) The inertial observer claims that the central force is provided by the force T exerted by the string on the block. (b) The noninertial observer claims that the block is not accelerating and therefore he introduces a fictitious force of magnitude mv^2/r, which acts outward and balances the force T.

*6.4   MOTION IN THE PRESENCE OF RESISTIVE FORCES

In the previous chapter we described the interaction between a moving object and the surface along which it moves. We completely ignored any interaction between the object and the medium through which it moves. Now let us consider the effect of a medium such as a liquid or gas. The medium exerts a **resistive force R** on the object moving through it. The magnitude of this force depends on such factors as the speed of the object, and the direction of **R** is always opposite the direction of motion of the object relative to the medium. Generally, the magnitude of the resistive force increases with increasing speed. Some examples are the air resistance associated with moving vehicles (sometimes called air drag) and the viscous forces that act on objects moving through a liquid.

The resistive force can have a complicated speed dependence. In the following discussions, we consider two situations. First, we assume that the resistive force is proportional to the speed; this is the case for objects that fall through a liquid with low speed and for very small objects, such as dust particles, that move through air. Second, we treat situations for which the resistive force is proportional to the square of the speed of the object; large objects, such as a skydiver moving through air in free-fall, experience such a force.

Resistive Force Proportional to Object Speed

If we assume that the resistive force acting on an object that is moving through a viscous medium is proportional to the object's velocity, then the resistive force can be expressed as

$$\mathbf{R} = -b\mathbf{v} \qquad (6.2)$$

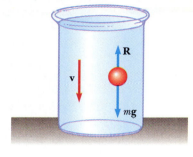

(a)

where **v** is the velocity of the object and b is a constant that depends on the properties of the medium and on the shape and dimensions of the object. If the object is a sphere of radius r, then b is proportional to r.

Consider a sphere of mass m released from rest in a liquid, as in Figure 6.13a. Assuming the only forces acting on the sphere are the resistive force, $-b\mathbf{v}$, and the weight, $m\mathbf{g}$, let us describe its motion.[1]

Applying Newton's second law to the vertical motion, choosing the downward direction to be positive, and noting that $\Sigma F_y = mg - bv$, we get

$$mg - bv = m\frac{dv}{dt}$$

where the acceleration is downward. Simplifying this expression gives

$$\frac{dv}{dt} = g - \frac{b}{m}v \qquad (6.3)$$

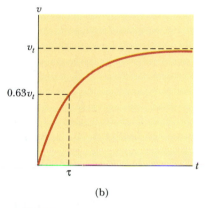

(b)

Equation 6.3 is called a *differential equation*, and the methods of solving such an equation may not be familiar to you as yet. However, note that initially, when $v = 0$, the resistive force is zero and the acceleration, dv/dt, is simply g. As t increases, the resistive force increases and the acceleration decreases. Eventually, the acceleration becomes zero when the resistive force equals the weight. At this point, the object reaches its **terminal speed,** v_t, and from then on it continues to move with zero acceleration. The terminal speed can be obtained from Equation 6.3 by setting $a = dv/dt = 0$. This gives

$$mg - bv_t = 0 \qquad \text{or} \qquad v_t = mg/b$$

The expression for v that satisfies Equation 6.3 with $v = 0$ at $t = 0$ is

$$v = \frac{mg}{b}(1 - e^{-bt/m}) = v_t(1 - e^{-t/\tau}) \qquad (6.4)$$

FIGURE 6.13 (a) A small sphere falling through a viscous fluid. (b) The speed-time graph for an object falling through a viscous medium. The object reaches a maximum, or terminal speed v_t, and τ is the time it takes to reach $0.63\,v_t$.

This function is plotted in Figure 6.13b. The time constant $\tau = m/b$ is the time it takes the object to reach 63.2% of its terminal speed. This can be seen by noting that when $t = \tau$, Equation 6.4 gives $v = 0.632v_t$. We can check that Equation 6.4 is a solution to Equation 6.3 by direct differentiation:

$$\frac{dv}{dt} = \frac{d}{dt}\left(\frac{mg}{b} - \frac{mg}{b}e^{-bt/m}\right) = -\frac{mg}{b}\frac{d}{dt}e^{-bt/m} = ge^{-bt/m}$$

Substituting this expression and Equation 6.4 into Equation 6.3 shows that our solution satisfies the differential equation.

Aerodynamic car. Streamlined bodies are used for sports cars and other vehicles to reduce air drag and increase fuel efficiency. (© *1992 Dick Kelley*)

---

[1] There is also a *buoyant* force acting on the submerged object, and this force is constant and equal to the weight of the displaced fluid. This force will only change the apparent weight of the sphere by a constant factor. We shall discuss such buoyant forces in Chapter 15.

By spreading their arms and legs and by keeping their bodies parallel to the ground, sky divers experience maximum air resistance resulting in a minimum terminal speed.   *(Guy Sauvage, Photo Researchers, Inc.)*

EXAMPLE 6.10   Sphere Falling in Oil

A small sphere of mass 2.00 g is released from rest in a large vessel filled with oil. The sphere reaches a terminal speed of 5.00 cm/s. Determine the time constant τ and the time it takes the sphere to reach 90% of its terminal velocity.

Solution   Since the terminal speed is given by $v_t = mg/b$, the coefficient b is

$$b = \frac{mg}{v_t} = \frac{(2.00\ \text{g})(980\ \text{cm/s}^2)}{5.00\ \text{cm/s}} = 392\ \text{g/s}$$

Therefore, the time τ is given by

$$\tau = \frac{m}{b} = \frac{2.00\ \text{g}}{392\ \text{g/s}} = \boxed{5.10 \times 10^{-3}\ \text{s}}$$

The speed of the sphere as a function of time is given by Equation 6.4. To find the time t it takes the sphere to reach a speed of $0.900 v_t$, we set $v = 0.900 v_t$ into this expression and solve for t:

$$0.900 v_t = v_t(1 - e^{-t/\tau})$$

$$1 - e^{-t/\tau} = 0.900$$

$$e^{-t/\tau} = 0.100$$

$$-\frac{t}{\tau} = -2.30$$

$$t = 2.30\tau = 2.30(5.10 \times 10^{-3}\ \text{s})$$

$$= 11.7 \times 10^{-3}\ \text{s} = \boxed{11.7\ \text{ms}}$$

Exercise   What is the sphere's speed through the oil at $t = 11.7$ ms? Compare this value with the speed the sphere would have if it were falling in a vacuum and so influenced only by gravity.

Answer   4.50 cm/s in oil versus 11.5 cm/s in free-fall.

Air Drag at High Speeds

For large objects moving at high speeds through air, such as airplanes, sky divers, and baseballs, the resistive force is approximately proportional to the square of the speed. In these situations, the magnitude of the resistive force can be expressed as

$$R = \tfrac{1}{2}D\rho A v^2 \tag{6.5}$$

where ρ is the density of air, A is the cross-sectional area of the falling object measured in a plane perpendicular to its motion, and D is a dimensionless empirical quantity called the *drag coefficient*. The drag coefficient has a value of about 0.5 for spherical objects but can be as high as 2 for irregularly shaped objects.

Consider an airplane in flight that experiences such a resistive force. Equation 6.5 shows that the force is proportional to the density of air and hence decreases with decreasing air density. Since air density decreases with increasing altitude, the resistive force on a jet airplane flying at a given speed must also decrease with increasing altitude. However, if at a given altitude the plane's speed is doubled, the resistive force increases by a factor of 4. In order to maintain this increased speed, the propulsive force must also increase by a factor of 4 and the power required (force times speed) must increase by a factor of 8.

Now let us analyze the motion of a mass in free-fall subject to an upward air resistive force whose magnitude is $R = \frac{1}{2}D\rho Av^2$. Suppose a mass m is released from rest from the position $y = 0$ as in Figure 6.14. The mass experiences two external forces: the downward force of gravity, mg, and the resistive force, $\mathbf{R}$, upward. (There is also an upward buoyant force that we neglect.) Hence, the magnitude of the net force is

$$F_{net} = mg - \tfrac{1}{2}D\rho Av^2 \tag{6.6}$$

Substituting $F_{net} = ma$ into Equation 6.6, we find that the mass has a downward acceleration of magnitude

$$a = g - \left(\frac{D\rho A}{2m}\right)v^2 \tag{6.7}$$

Again, we can calculate the terminal speed, v_t, using the fact that when the weight is balanced by the resistive force, the net force is zero and therefore the acceleration is zero. Setting $a = 0$ in Equation 6.7 gives

$$g - \left(\frac{D\rho A}{2m}\right)v_t^2 = 0$$

$$v_t = \sqrt{\frac{2mg}{D\rho A}} \tag{6.8}$$

Using this expression, we can determine how the terminal speed depends on the dimensions of the object. Suppose the object is a sphere of radius r. In this case, $A \propto r^2$ and $m \propto r^3$ (since the mass is proportional to the volume). Therefore, $v_t \propto \sqrt{r}$.

Table 6.1 lists the terminal speeds for several objects falling through air.

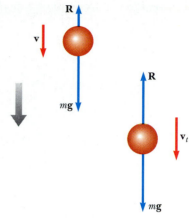

FIGURE 6.14 An object falling through air experiences a resistive drag force $\mathbf{R}$ and a gravitational force mg. The object reaches terminal speed (on the right) when the net force is zero, that is, when $\mathbf{R} = -mg$, or $R = mg$. Before this occurs, the acceleration varies with speed according to Equation 6.7.

TABLE 6.1   Terminal Speed for Various Objects Falling Through Air

Object	Mass (kg)	Surface Area (m²)	v_t(m/s)
Sky diver	75	0.70	60
Baseball (radius 3.7 cm)	0.145	4.2×10^{-3}	43
Golf ball (radius 2.1 cm)	0.046	1.4×10^{-3}	44
Hailstone (radius 0.50 cm)	4.8×10^{-4}	7.9×10^{-5}	14
Raindrop (radius 0.20 cm)	3.4×10^{-5}	1.3×10^{-5}	9.0

CONCEPTUAL EXAMPLE 6.11

Consider a sky diver falling through air before reaching her terminal speed. As the speed of the sky diver increases, what happens to her acceleration? What is her acceleration once she reaches terminal speed?

Reasoning The forces exerted on the sky diver are the downward force of gravity (her weight) and an upward force of air resistance, which is less than her weight before she reaches terminal speed. As her downward speed increases, the force of air resistance increases. The vector sum of the force of gravity and the force of air resistance gives a total force which decreases with time, so her acceleration decreases. Once she reaches terminal speed, the two forces balance each other, the total force is zero, and her acceleration is zero.

*6.5   NUMERICAL MODELING IN PARTICLE DYNAMICS [2]

As we have seen in this and the preceding chapter, the study of dynamics of a particle focuses on describing the position, velocity, and acceleration as functions of time. Cause-and-effect relationships exist between these quantities: A velocity causes position to change, an acceleration causes velocity to change, and an acceleration is the direct result of applied forces. Therefore, the study of motion usually begins with an evaluation of the net force on the particle.

In this section, we confine our discussion to one-dimensional motion, so bold-face notation will not be used for vector quantities. If a particle of mass m moves under the influence of a net force F, Newton's second law tells us that the acceleration of the particle is given by $a = F/m$. In general, we can then obtain the solution to a dynamics problem by using the following procedure:

1. Sum all the forces on the particle to get the net force F.
2. Use this force to determine the acceleration, using $a = F/m$.
3. Use this acceleration to determine the velocity from $dv/dt = a$.
4. Use this velocity to determine the position from $dx/dt = v$.

EXAMPLE 6.12   An Object Falling in a Vacuum

We can illustrate the procedure just described by considering a particle falling in a vacuum under the influence of the force of gravity as in Figure 6.15.

mg

FIGURE 6.15   An object falling in vacuum under the influence of gravity.

Reasoning and Solution   Applying Newton's second law, we set the sum of the external forces equal to the mass of the particle times its acceleration (taking upward to be the positive direction):

$$F = ma = -mg$$

Thus, $a = -g$, a constant. Because $dv/dt = a$, the resulting differential equation for velocity is $dv/dt = -g$, which may be integrated to give

$$v(t) = v_0 - gt$$

Then, since $dx/dt = v$, the position of the particle is obtained from another integration, which yields the well-known result

$$x(t) = x_0 + v_0 t - \tfrac{1}{2} at^2$$

In this expression, x_0 and v_0 represent the position and speed of the particle at $t = 0$.

The procedure just described is straightforward for some physical situations, such as the one described in the previous example. In the "real world," however, complications often arise that make analytical solutions for many practical situations difficult and perhaps beyond the mathematical abilities of most students

---

[2] The author is most grateful to Colonel James Head of the U.S. Air Force Academy for preparing this section. A more extensive treatment of this topic by Colonel Head is included in a supplement entitled *Numerical Methods in Physics*.

taking introductory physics. For example, the force may depend on the position of the particle, as in cases where variation in gravitational acceleration with height must be taken into account. The force may also vary with velocity, as we have seen in cases of resistive forces caused by motion through a liquid or gas. The force may depend on both position and velocity, as in the case of an object falling through air where the resistive force depends on velocity and on height (air density). In rocket motion, the mass changes with time, so even if the force is constant, the acceleration is not.

Another complication arises because the equations relating acceleration, velocity, position, and time are differential, not algebraic, equations. Differential equations are usually solved using integral calculus and other special techniques that introductory students may not have mastered.

So how does one proceed to solve real-world problems without advanced mathematics? One answer is to solve such problems on personal computers, using elementary numerical methods. The simplest of these is the Euler method, named after the Swiss mathematician Leonhard Euler (1707–1783).

The Euler Method

In the **Euler method** of solving differential equations, derivatives are approximated as finite differences. Considering a small increment of time, Δt, the relationship between speed and acceleration may be approximated as

$$a(t) = \frac{\Delta v}{\Delta t} = \frac{v(t + \Delta t) - v(t)}{\Delta t}$$

Then the speed of the particle at the end of the period Δt is approximately equal to the speed at the beginning of the period, plus the acceleration during the interval multiplied by Δt:

$$v(t + \Delta t) = v(t) + a(t)\,\Delta t \tag{6.9}$$

Since the acceleration is a function of time, this estimate of $v(t + \Delta t)$ will be accurate only if the time interval Δt is short enough that the change in acceleration during it is very small (as will be discussed later).

The position can be found in the same manner:

$$v(t) = \frac{x(t + \Delta t) - x(t)}{\Delta t}$$

so

$$x(t + \Delta t) = x(t) + v(t)\,\Delta t \tag{6.10}$$

It may be tempting to add the term $\frac{1}{2}a(\Delta t)^2$ to this result to make it look like the familiar kinematics equation, but this term is not included in the Euler method of integration because Δt is assumed to be so small that $(\Delta t)^2$ is nearly zero.

If the acceleration at any instant t is known, the particle's velocity and position at $(t + \Delta t)$ can be calculated from Equations 6.9 and 6.10. The calculation can then proceed in a series of finite steps to determine the velocity and position at any later time. The acceleration is determined by the net force acting on the object,

$$a(x, v, t) = \frac{F(x, v, t)}{m} \tag{6.11}$$

which may depend explicitly on the position, velocity, or time.

TABLE 6.2	The Euler Method for Solving Dynamics Problems			
Step	Time	Position	Velocity	Acceleration
0	t_0	x_0	v_0	$a_0 = F(x_0, v_0, t_0)/m$
1	$t_1 = t_0 + \Delta t$	$x_1 = x_0 + v_0 \Delta t$	$v_1 = v_0 + a_0 \Delta t$	$a_1 = F(x_1, v_1, t_1)/m$
2	$t_2 = t_1 + \Delta t$	$x_2 = x_1 + v_1 \Delta t$	$v_2 = v_1 + a_1 \Delta t$	$a_2 = F(x_2, v_2, t_2)/m$
3	$t_3 = t_2 + \Delta t$	$x_3 = x_2 + v_2 \Delta t$	$v_3 = v_2 + a_2 \Delta t$	$a_3 = F(x_3, v_3, t_3)/m$
.	.	.	.	.
.	.	.	.	.
.	.	.	.	.
n	t_n	x_n	v_n	a_n

It is convenient to set up the numerical solution to this kind of problem by numbering the steps and entering the calculations in a table. Table 6.2 illustrates how to do this in an orderly way.

The equations provided in the table can be entered into a spreadsheet and the calculations performed row by row to determine the velocity, position, and acceleration as functions of time. The calculations can also be carried out using a program written in BASIC, PASCAL, or FORTRAN or with commercially available mathematics packages for personal computers. Many small increments can be taken, and accurate results can usually be obtained with the help of a computer. Graphs of velocity versus time or position versus time can be displayed to help you visualize the motion.

The Euler method has the advantage that the dynamics is not obscured—the fundamental relationships of acceleration to force, velocity to acceleration, and position to velocity are clearly evident. Indeed, these relationships form the heart of the calculations. There is no need to use advanced mathematics, and the basic physics governs the dynamics.

The Euler method is completely reliable for infinitesimally small time increments, but for practical reasons a finite increment size must be chosen. In order for the finite difference approximation of Equation 6.9 to be valid, the time increment must be small enough that the velocity can be approximated as being constant during the increment. We can determine an appropriate size for the time increment by examining the particular problem that is being investigated. The criterion for the size of the time increment may need to be changed during the course of the motion. In practice, however, we usually choose a time increment appropriate to the initial conditions and use the same value throughout the calculations.

The size of the time increment influences the accuracy of the result, but unfortunately it is not easy to determine the accuracy of a solution by the Euler method without a knowledge of the correct analytical solution. One method of determining the accuracy of the numerical solution is to repeat the calculations with a smaller time increment and compare results. If the two calculations agree to a certain number of significant figures, you can assume that the results are correct to that precision.

6.6   THE FUNDAMENTAL FORCES OF NATURE

In the previous chapter, we described a variety of forces that are experienced in our everyday activities, such as the force of gravity that acts on all bodies at or near the Earth's surface and the force of friction as one surface slides over another.

Other forces we have encountered are the force exerted by a rope on an object, the normal force acting on an object in contact with some other object, and the resistive force as an object moves through a liquid or gas. Other forces we shall encounter include the restoring force in a deformed spring, the electrostatic force between two charged objects, and the magnetic force between a magnet and a piece of iron.

Forces also act in the atomic and subatomic world. For example, atomic forces within the atom are responsible for holding its constituents together, and nuclear forces act on different parts of the nucleus to keep its parts from separating.

Until recently, physicists believed that there were four fundamental forces in nature: the gravitational force, the electromagnetic force, the strong nuclear force, and the weak nuclear force.

The **gravitational force** mentioned in Example 6.5 is the mutual force of attraction between all masses. We have already encountered the gravitational force when describing the weight of an object. Although the magnitude of gravitational forces can be significant between macroscopic objects, they are the weakest of the four fundamental forces. For example, the gravitational force between the electron and proton in the hydrogen atom is only about 10^{-47} N, whereas the electrostatic force between the two particles is about 10^{-7} N. We shall learn more about the nature of the gravitational force in Chapter 14.

The **electromagnetic force** is an attraction or repulsion between two charged particles that may be in relative motion. Although much stronger than the gravitational force, the electromagnetic force between two charged elementary particles is of medium strength. The force that causes a rubbed comb to attract bits of paper and the force that a magnet exerts on an iron nail are examples of electromagnetic forces. It is interesting to note that essentially all forces in our macroscopic world (apart from the gravitational force) are manifestations of the electromagnetic force when examined closely. For example, friction forces, contact forces, tension forces, and forces in stretched springs or other deformed bodies are essentially the consequence of electromagnetic forces between charged particles in close proximity.

The **strong nuclear force** is responsible for the stability of nuclei. This force represents the "glue" that holds the nuclear constituents (called nucleons) together. It is the strongest of all the fundamental forces. For separations of about 10^{-15} m (a typical nuclear dimension), the strong nuclear force is one to two orders of magnitude stronger than the electromagnetic force. However, the strong nuclear force decreases rapidly with increasing separation and is negligible for separations greater than about 10^{-14} m.

Finally, the **weak nuclear force** is a short-range nuclear force that tends to produce instability in certain nuclei. Most radioactive decay reactions are caused by the weak nuclear force, which is about 12 orders of magnitude weaker than the electromagnetic force.

In 1979, physicists predicted that the electromagnetic force and the weak force are manifestations of one and the same force called the **electroweak** force. This prediction was confirmed experimentally in 1984.

Physicists and cosmologists believe that the fundamental forces of nature are closely related to the origin of the Universe. The Big Bang theory of the creation of the Universe states that the Universe erupted from a point-like singularity 15 to 20 billion years ago. According to this theory, the first moments ($\approx 10^{-10}$ s) after the Big Bang saw such extremes of energy that all four fundamental forces were unified. Scientists continue in their search for other possible connections among the four fundamental forces.

SUMMARY

Newton's second law applied to a particle moving in uniform circular motion states that the net force, which is a **central force**, is

$$F_r = ma_r = \frac{mv^2}{r} \tag{6.1}$$

The central force acting on an object that provides the centripetal acceleration could be the force of gravity (as in satellite motion), a force of friction, or a force exerted by a string.

A particle moving in nonuniform circular motion has both a centripetal acceleration and a nonzero tangential component of acceleration. In the case of a particle rotating in a vertical circle, the force of gravity provides the tangential acceleration and part or all of the centripetal acceleration.

An observer in a noninertial frame of reference must introduce **fictitious forces** when applying Newton's second law in that frame. If these fictitious forces are properly defined, the description of motion in the noninertial frame is equivalent to that made by an observer in an inertial frame. However, the observers in the two different frames will not agree on the causes of the motion.

A body moving through a liquid or gas experiences a **resistive force** that is speed dependent. This resistive force, which opposes the motion, generally increases with speed. The force depends on the shape of the body and the properties of the medium through which the body is moving. In the limiting case for a falling body, when the magnitude of the resistive force equals the weight ($a = 0$), the body reaches its **terminal speed.**

There are four fundamental forces in nature: the strong nuclear force, the electromagnetic force, the gravitational force, and the weak force.

QUESTIONS

1. Because the Earth rotates about its axis and revolves around the Sun, it is a noninertial frame of reference. Assuming the Earth is a uniform sphere, why would the apparent weight of an object be greater at the poles than at the equator?

2. Explain why the Earth bulges at the equator.

3. How would you explain the force that pushes a rider toward the side of a car as the car rounds a corner?

4. When an airplane does an inside "loop-the-loop" in a vertical plane, at what point does the pilot appear to be heaviest? What is the constraint force acting on the pilot?

5. A sky diver in free-fall reaches terminal speed. After the parachute is opened, what parameters change to decrease this terminal speed?

6. Why is it that an astronaut in a space capsule orbiting the Earth experiences a feeling of weightlessness?

7. Why does mud fly off a rapidly turning wheel?

8. A pail of water can be whirled in a vertical path such that none is spilled. Why does the water stay in, even when the pail is above your head?

9. Imagine that you attach a heavy object to one end of a spring and then whirl the spring and object in a horizon-tal circle (by holding the free end of the spring). Does the spring stretch? If so, why? Discuss this in terms of central force.

10. It has been suggested that rotating cylinders about 10 mi in length and 5 mi in diameter be placed in space and used as colonies. The purpose of the rotation is to simulate gravity for the inhabitants. Explain this concept for producing an effective gravity.

11. Why does a pilot tend to black out when pulling out of a steep dive?

12. Describe a situation in which an automobile driver can have a centripetal acceleration but no tangential acceleration.

13. Is it possible for a car to move in a circular path in such a way that it has a tangential acceleration but no centripetal acceleration?

14. Analyze the motion of a rock dropped into water in terms of its speed and acceleration as it falls. Assume that there is a resistive force acting on the rock that increases as the speed increases.

15. Centrifuges are often used in dairies to separate the cream from the milk. Which remains on the inside?

16. We often think of the brakes and the gas pedal on a car as

the devices that accelerate the car. Could a steering wheel also fall into this category? Explain.

17. Consider a small raindrop and a large raindrop falling through the atmosphere. Compare their terminal speeds. What are their accelerations when they reach terminal speed?

PROBLEMS

Section 6.1   Newton's Second Law Applied to Uniform Circular Motion

1. A toy car moving at constant speed completes one lap around a circular track (a distance of 200 m) in 25.0 s. (a) What is the average speed? (b) If the mass of the car is 1.50 kg, what is the magnitude of the central force that keeps it in a circle?

2. In a cyclotron (one type of particle accelerator), a deuteron (of atomic mass 2 u) reaches a final speed of 10% of the speed of light while moving in a circular path of radius 0.48 m. The deuteron is maintained in the circular path by a magnetic force. What magnitude of force is required?

3. The wheels of a certain roller coaster are both above and below the rails as shown in Figure P6.3 so that the car will not leave the rails. If the mass supported by this particular wheel is 320 kg and the radius of this section of track is 15 m, (a) what is the magnitude and direction of the force that the track exerts on the wheel when the speed of the car is 20 m/s? (b) What would be the net force exerted on a 60-kg person riding in the car? (c) What supplies this force?

FIGURE P6.3

4. In a hydrogen atom, the electron in orbit around the proton feels an attractive force of about 8.20×10^{-8} N. If the radius of the orbit is 5.30×10^{-11} m, what is the frequency in revolutions per second? See the front cover for additional data.

5. A 3.00-kg mass attached to a light string rotates on a horizontal, frictionless table. The radius of the circle is 0.800 m, and the string can support a mass of 25.0 kg before breaking. What range of speeds can the mass have before the string breaks?

6. A satellite of mass 300 kg is in a circular orbit around the Earth at an altitude equal to the Earth's mean radius (see Example 6.5). Find (a) the satellite's orbital speed, (b) the period of its revolution, and (c) the gravitational force acting on it.

7. While two astronauts were on the surface of the Moon, a third astronaut orbited the Moon. Assume

(Problem 6)   *(NASA/Peter Arnold, Inc.)*

the orbit to be circular and 100 km above the surface of the Moon. If the mass and radius of the Moon are 7.40×10^{22} kg and 1.70×10^6 m, respectively, determine (a) the orbiting astronaut's acceleration, (b) his orbital speed, and (c) the period of the orbit.

8. An automobile moves at constant speed over the crest of a hill. The driver moves in a vertical circle of radius 18.0 m. At the top of the hill, she notices that she barely remains in contact with the seat. Find the speed of the vehicle.

8A. An automobile moves at constant speed over the crest of a hill. The driver moves in a vertical circle of radius R. At the top of the hill, she notices that she barely remains in contact with the seat. Find the speed of the vehicle.

9. A crate of eggs is located in the middle of the flat bed of a pickup truck as the truck negotiates an unbanked curve in the road. The curve may be regarded as an arc of a circle of radius 35 m. If the coefficient of static friction between crate and truck is 0.60, what must be the maximum speed of the truck if the crate is not to slide?

10. A 55-kg ice skater is moving at 4.0 m/s when she grabs the loose end of a rope the opposite end of which is tied to a pole. She then moves in a circle of radius 0.80 m around the pole. (a) Determine the

☐ indicates problems that have full solutions available in the Student Solutions Manual and Study Guide.

force exerted by the rope on her arms. (b) Compare this force with her weight.

11. An air puck of mass 0.250 kg is tied to a string and allowed to revolve in a circle of radius 1.00 m on a horizontal, frictionless table. The other end of the string passes through a hole in the center of the table and a mass of 1.00 kg is tied to it. The suspended mass remains in equilibrium while the puck revolves. (a) What is the tension in the string? (b) What is the central force acting on the puck? (c) What is the speed of the puck?

12. The speed of the tip of the minute hand on a town clock is 1.75×10^{-3} m/s. (a) What is the speed of the tip of the second hand of the same length? (b) What is the centripetal acceleration of the tip of the second hand?

13. A coin placed 30.0 cm from the center of a rotating, horizontal turntable slips when its speed is 50.0 cm/s. (a) What provides the central force when the coin is stationary relative to the turntable? (b) What is the coefficient of static friction between coin and turntable?

Section 6.2   Nonuniform Circular Motion

14. A car traveling on a straight road at 9.00 m/s goes over a hump in the road. The hump may be regarded as an arc of a circle of radius 11.0 m. (a) What is the apparent weight of a 600-N woman in the car as she rides over the hump? (b) What must be the speed of the car over the hump if she is to experience weightlessness? (That is, her apparent weight is zero.)

15. A pail of water is rotated in a vertical circle of radius 1.00 m. What is the minimum speed of the pail at the top of the circle if no water is to spill out?

16. A hawk flies in a horizontal arc of radius 12.0 m at a constant speed of 4.00 m/s. (a) Find its centripetal acceleration. (b) It continues to fly along the same horizontal arc but increases its speed at the rate of 1.20 m/s². Find the acceleration (magnitude and direction) under these conditions.

17. A 40.0-kg child sits in a swing supported by two chains, each 3.00 m long. If the tension in each chain at the lowest point is 350 N, find (a) the child's speed at the lowest point and (b) the force of the seat on the child at the lowest point. (Neglect the mass of the seat.)

17A. A child of mass m sits in a swing supported by two chains, each of length R. If the tension in each chain at the lowest point is T, find (a) the child's speed at the lowest point and (b) the force of the seat on the child at the lowest point. (Neglect the mass of the seat.)

18. A 0.40-kg object is swung in a vertical circular path on a string 0.50 m long. If a constant speed of 4.0 m/s is maintained, what is the tension in the string when the object is at the top of the circle?

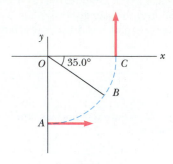

FIGURE P6.19

19. A car initially traveling eastward turns north by traveling in a circular path at uniform speed as in Figure P6.19. The length of the arc *ABC* is 235 m, and the car completes the turn in 36.0 s. (a) What is the acceleration when the car is at *B* located at an angle of 35.0°? Express your answer in terms of the unit vectors **i** and **j**. Determine (b) the car's average speed and (c) its average acceleration during the 36.0-s interval.

20. A roller-coaster vehicle has a mass of 500 kg when fully loaded with passengers (Fig. P6.20). (a) If the vehicle has a speed of 20.0 m/s at point *A*, what is the force exerted by the track on the vehicle at this point? (b) What is the maximum speed the vehicle can have at *B* and still remain on the track?

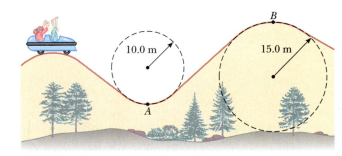

FIGURE P6.20

21. Tarzan ($m = 85.0$ kg) tries to cross a river by swinging from a vine. The vine is 10.0 m long, and his speed at the bottom of the swing (as he just clears the water) is 8.00 m/s. Tarzan doesn't know that the vine has a breaking strength of 1000 N. Does he make it safely across the river?

22. At the Six Flags Great America amusement park in Gurnee, Illinois, there is a roller coaster that incorporates some of the latest design technology and some basic physics. Each vertical loop, instead of being circular, is shaped like a teardrop (Fig. P6.22). The cars ride on the inside of the loop at the top, and the speeds are high enough to ensure that the cars remain on the track. The biggest loop is 40.0 m high,

FIGURE P6.22 *(Frank Cezus/FPG International)*

with a maximum speed of 31.0 m/s (nearly 70 mph) at the bottom. Suppose the speed at the top is 13.0 m/s and the corresponding centripetal acceleration is 2g. (a) What is the radius of the arc of the teardrop at the top? (b) If the total mass of the cars plus people is M, what force does the rail exert on it at the top? (c) Suppose the roller coaster had a loop of radius 20.0 m. If the cars have the same speed, 13.0 m/s at the top, what is the centripetal acceleration at the top? Comment on the normal force at the top in this situation.

*Section 6.3   Motion in Accelerated Frames

23. A merry-go-round makes one complete revolution in 12.0 s. If a 45.0-kg child sits on the horizontal floor of the merry-go-round 3.00 m from the center, find (a) the child's acceleration and (b) the horizontal force of friction that acts on the child. (c) What minimum coefficient of static friction is necessary to keep the child from slipping?

24. The Earth rotates about its axis with a period of 24 h. Imagine that the rotational speed can be increased. If an object at the equator is to have zero apparent weight, (a) what must the new period be? (b) By what factor would the speed of the object be increased when the planet is rotating at the higher speed? (*Hint:* See Problem 39 and note that the apparent weight of the object becomes zero when the normal force exerted on it is zero. Also, the distance traveled during one period of rotation is $2\pi R$, where R is the Earth's radius.)

25. A 0.500-kg object is suspended from the ceiling of

an accelerating boxcar as in Figure 6.11. If $a = 3.00$ m/s², find (a) the angle that the string makes with the vertical and (b) the tension in the string.

26. A 5.00-kg mass attached to a spring scale rests on a frictionless, horizontal surface as in Figure P6.26. The spring scale, attached to the front end of a boxcar, reads 18.0 N when the car is in motion. (a) If the spring scale reads zero when the car is at rest, determine the acceleration of the car while it is in motion. (b) What will the spring scale read if the car moves with constant velocity? (c) Describe the forces on the mass as observed by someone in the car and by someone at rest outside the car.

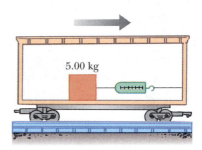

FIGURE P6.26

27. A person stands on a scale in an elevator. The maximum and minimum scale readings are 591 N and 391 N, respectively. Assume the magnitude of the acceleration is the same during starting and stopping, and determine (a) the weight of the person, (b) the person's mass, and (c) the acceleration of the elevator.

28. A plumb bob does not hang exactly along a line directed to the center of the Earth's rotation. How much does the plumb bob deviate from a radial line at 35° north latitude? Assume that the Earth is spherical.

*Section 6.4   Motion in the Presence of Resistive Forces

29. Assume that the resistive force exerted on a speed skater is $f = -kmv^2$, where k is a constant and m is the skater's mass. Show that, after finishing the race, the skater's speed as a function of time is $v(t) = v_f/(1 + ktv_f)$, where v_f is his speed at the finish line.

30. A small piece of Styrofoam packing material is dropped from a height of 2.00 m above the ground. Until the terminal speed is reached, the acceleration is given by $a = g - cv$. After falling 0.500 m, it reaches terminal speed, and the Styrofoam takes an extra 5.00 s to reach the ground. (a) What is the value of the constant c? (b) What is the acceleration at $t = 0$? (c) What is the acceleration when the speed is 0.150 m/s?

31. A motor boat cuts its engine when its speed is 10.0 m/s and coasts to rest. The equation governing the motion of the motorboat during this period is

$v = v_0 e^{-ct}$, where v is the speed at time t, v_0 is the initial speed, and c is a constant. At $t = 20.0$ s, the speed is 5.00 m/s. (a) Find the constant c. (b) What is the speed at $t = 40.0$ s? (c) Differentiate the expression for $v(t)$ and thus show that the acceleration of the boat is proportional to the speed at any time.

32. A fire helicopter carries a 620-kg bucket at the end of a cable 20.0 m long as in Figure P6.32. As the helicopter flies to a fire at a constant speed of 40.0 m/s, the cable makes an angle of 40.0° with respect to the vertical. The bucket presents a cross-sectional area of 3.80 m² in a plane perpendicular to the air moving past it. Determine the drag coefficient assuming that the resistive force is proportional to the square of the bucket's speed.

40.0 m/s

20.0 m

40.0°

620 kg

FIGURE P6.32

33. A small, spherical bead of mass 3.00 g is released from rest at $t = 0$ in a bottle of liquid shampoo. The terminal speed is observed to be 2.00 cm/s. Find (a) the value of the constant b in Equation 6.4, (b) the time, τ, it takes to reach $0.630 v_t$, and (c) the value of the resistive force when the bead reaches terminal speed.

ADDITIONAL PROBLEMS

34. Suppose the boxcar of Figure 6.11 is moving with constant acceleration a up a hill that makes an angle ϕ with the horizontal. If the plumb bob makes an angle of θ with the perpendicular to the ceiling, what is a?

35. In the Bohr model of the hydrogen atom, the speed of the electron is approximately 2.2×10^6 m/s. Find (a) the central force acting on the electron as it revolves in a circular orbit of radius 0.53×10^{-10} m and (b) the centripetal acceleration of the electron.

36. A 9.00-kg object starting from rest moves through a viscous medium and experiences a resistive force $\mathbf{R} = -b\mathbf{v}$, where $\mathbf{v}$ is the velocity of the object. If the object's speed reaches one-half its terminal speed in 5.54 s, (a) determine the terminal speed. (b) At what time is the speed of the object three-fourths the ter-

minal speed? (c) How far has the object traveled in the first 5.54 s of motion?

37. Consider a conical pendulum with an 80.0-kg bob on a 10.0-m wire making an angle of 5.00° with the vertical (Fig. 6.3). Determine (a) the horizontal and vertical components of the force exerted by the wire on the pendulum and (b) the radial acceleration of the bob.

38. An air puck of mass 0.25 kg is tied to a string and allowed to revolve in a circle of radius 1.0 m on a frictionless horizontal table. The other end of the string passes through a hole in the center of the table, and a mass of 1.0 kg is tied to it (Fig. P6.38). The suspended mass remains in equilibrium while the puck on the tabletop revolves. What are (a) the tension in the string, (b) the central force exerted on the puck, and (c) the speed of the puck?

38A. An air puck of mass m_1 is tied to a string and allowed to revolve in a circle of radius R on a frictionless horizontal table. The other end of the string passes through a hole in the center of the table, and a mass m_2 is tied to it (Fig. P6.38). The suspended mass remains in equilibrium while the puck on the tabletop revolves. (a) What is the tension in the string? (b) What is the central force acting on the puck? (c) What is the speed of the puck?

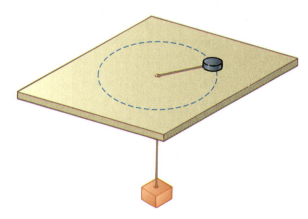

FIGURE P6.38

39. Because the Earth rotates about its axis, a point on the equator experiences a centripetal acceleration of 0.0340 m/s², while a point at the poles experiences no centripetal acceleration. (a) Show that at the equator the gravitational force on an object (the true weight) must exceed the object's apparent weight. (b) What is the apparent weight at the equator and at the poles of a person having a mass of 75.0 kg? (Assume the Earth is a uniform sphere and take $g = 9.800$ N/kg.)

40. A piece of putty is initially located at point A on the rim of a grinding wheel rotating about a horizontal

axis. The putty is dislodged from point A when the diameter through A is horizontal. It then rises vertically and returns to A the instant the wheel completes one revolution. (a) Find the speed of a point on the rim of the wheel in terms of the free-fall acceleration and the radius R of the wheel. (b) If the mass of the putty is m, what is the magnitude of the force that held it to the wheel?

41. A string under a tension of 50.0 N is used to whirl a rock in a horizontal circle of radius 2.50 m at a speed of 20.4 m/s. The string is pulled in and the speed of the rock increases. When the string is 1.00 m long and the speed of the rock is 51.0 m/s, the string breaks. What is the breaking strength (in newtons) of the string?

42. A car rounds a banked curve as in Figure 6.5. The radius of curvature of the road is R, the banking angle is θ, and the coefficient of static friction is μ. (a) Determine the range of speeds the car can have without slipping up or down the road. (b) Find the minimum value for μ such that the minimum speed is zero. (c) What is the range of speeds possible if $R = 100$ m, $\theta = 10°$, and $\mu = 0.10$ (slippery conditions)?

43. A model airplane of mass 0.75 kg flies in a horizontal circle at the end of a 60-m control wire, with a speed of 35 m/s. Compute the tension in the wire if it makes a constant angle of 20° with the horizontal. The forces exerted on the airplane are the pull of the control wire, its own weight, and aerodynamic lift, which acts at 20° inward from the vertical as shown in Figure P6.43.

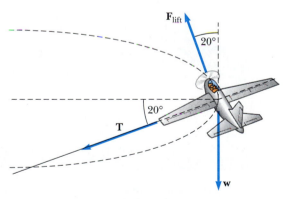

FIGURE P6.43

44. A child's toy consists of a small wedge that has an acute angle θ (Fig. P6.44). The sloping side of the wedge is frictionless, and the wedge is spun at a constant speed by rotating a rod that is firmly attached to it at one end. Show that, when the mass m rises up the wedge a distance L, the speed of the mass is

$$v = \sqrt{gL \sin \theta}$$

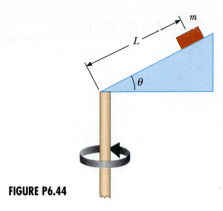

FIGURE P6.44

45. The pilot of an airplane executes a constant-speed loop-the-loop maneuver in a vertical plane. The speed of the airplane is 300 mi/h, and the radius of the circle is 1200 ft. (a) What is the pilot's apparent weight at the lowest point if his true weight is 160 lb? (b) What is his apparent weight at the highest point? (c) Describe how the pilot could experience weightlessness if both the radius and the speed can be varied. (*Note:* His apparent weight is equal to the force that the seat exerts on his body.)

46. In order for a satellite to move in a stable circular orbit at a constant speed, its centripetal acceleration must be inversely proportional to the square of the radius r of the orbit. (a) Show that the tangential speed of a satellite is proportional to $r^{-1/2}$. (b) Show that the time required to complete one orbit is proportional to $r^{3/2}$.

47. An 1800-kg car passes over a bump in a road that follows the arc of a circle of radius 42 m as in Figure P6.47. (a) What force does the road exert on the car as the car passes the highest point of the bump if the car travels at 16 m/s? (b) What is the maximum speed the car can have as it passes the highest point before losing contact with the road?

47A. A car of mass m passes over a bump in a road that follows the arc of a circle of radius R as in Figure P6.47. (a) What force does the road exert on the car as the car passes the highest point of the bump if the car travels at a speed v? (b) What is the maximum speed the car can have as it passes this point before losing contact with the road?

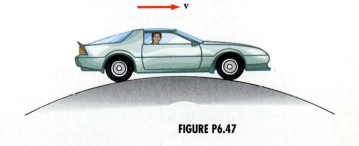

FIGURE P6.47

48. A student builds and calibrates an accelerometer, which she uses to determine the speed of her car around a certain highway curve. The accelerometer is a plumb bob with a protractor that she attaches to the roof of her car. A friend riding in the car with her observes that the plumb bob hangs at an angle of 15.0° from the vertical when the car has a speed of 23.0 m/s. (a) What is the centripetal acceleration of the car rounding the curve? (b) What is the radius of the curve? (c) What is the speed of the car if the plumb bob deflection is 9.0° while rounding the same curve?

49. An amusement park ride consists of a large vertical cylinder that spins about its axis fast enough that any person inside is held up against the wall when the floor drops away (Fig. P6.49). The coefficient of static friction between person and wall is μ_s, and the radius of the cylinder is R. (a) Show that the maximum period of revolution necessary to keep the person from falling is $T = (4\pi^2 R\mu_s/g)^{1/2}$. (b) Obtain a numerical value for T if $R = 4.00$ m and $\mu_s = 0.400$. How many revolutions per minute does the cylinder make?

and block are 0.45 (kinetic) and 0.52 (static), what is the maximum speed of the disk in revolutions per minute without the block or penny sliding on the disk?

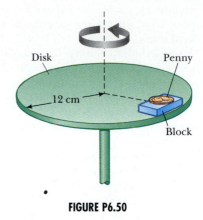

FIGURE P6.50

51. Figure P6.51 shows a Ferris wheel that rotates four times each minute and has a diameter of 18.0 m. (a) What is the centripetal acceleration of a rider? What force does the seat exert on a 40.0-kg rider (b) at the lowest point of the ride and (c) at the highest point of the ride? (d) What force (magnitude and direction) does the seat exert on a rider when the rider is halfway between top and bottom?

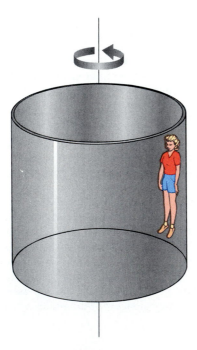

FIGURE P6.49

50. A penny of mass 3.1 g rests on a small 20.0-g block supported by a spinning disk (Fig. P6.50). If the coefficients of friction between block and disk are 0.75 (static) and 0.64 (kinetic) while those for the penny

FIGURE P6.51   *(Color Box/FPG)*

52. An amusement park ride consists of a rotating circular platform 8.00 m in diameter from which 10.0-kg seats are suspended at the end of 2.50-m massless chains (Fig. P6.52). When the system rotates, the chains make an angle $\theta = 28.0°$ with the vertical. (a) What is the speed of each seat? (b) If a child of mass 40.0 kg sits in a seat, what is the tension in the chain?

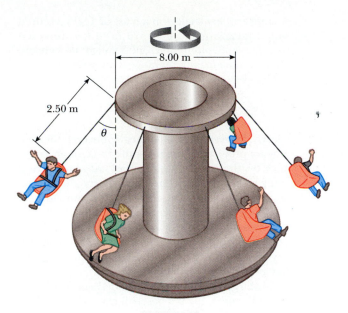

FIGURE P6.52

53. A stream of air moving at speed v exerts a resistive force on a sphere of radius r. The magnitude of the resistive force (in newtons) is $F = arv + br^2v^2$, where v is in meters per second, r is in meters, and a and b are constants with appropriate SI units. Their numerical values are $a = 3.10 \times 10^{-4}$ and $b = 0.870$. Using this formula, find the terminal speed for water droplets falling under their own weight in air, taking these values for the drop radii: (a) 10.0 μm, (b) 100 μm, and (c) 1.00 mm. Note that for (a) and (c) you can obtain accurate answers without solving a quadratic equation by considering which of the two contributions to the air resistance is dominant and ignoring the lesser contribution.

SPREADSHEET PROBLEMS

S1. In studying the drag forces that act on a falling object, it is often assumed that the magnitude of the drag force is $R = bv^n$. The net force acting on the object in this case is

$$F_{\text{net}} = mg - bv^n$$

where m is the mass of the object, and b and n are constants. When the net force goes to zero, the object reaches its terminal speed v_t:

$$v_t = \left(\frac{mg}{b}\right)^{\frac{1}{n}}$$

Spreadsheet 6.1 numerically integrates the force equation for $n = 1$ and calculates the position and speed of the falling object. In addition, it integrates the force equation without air resistance. Choose the

initial position and initial speed to be zero, and enter these values in the spreadsheet. (a) Set $b = 0.200$ and use $m = 0.100$, 1.00, and 10.0 kg to find v_T numerically in each case as follows. Plot v as a function of time, t. For t large enough, v approaches v_t. Compare your results for v_t obtained this way with the corresponding results calculated from the formula for v_t. *Note:* You may need to change the time increment to reach the terminal speed. (b) Change b to 0.500 and repeat part (a).

S2. Use Spreadsheet 6.1 with $m = 1$ kg, $v_0 = 0$, $x_0 = 100$ m, and $b = 0.2$. (a) Compare the position of the object with air resistance and without air resistance at times $t = 0$ s, 0.5 s, 1.0 s, 5.0 s, and 10 s. (b) Do the same for $m = 2$ kg and 10 kg. What is the effect of changing the mass?

S3. Members of a skydiving club were given the following data to use in planning their jumps. In the table, d is the distance fallen from rest by a sky diver in a "free-fall stable spread position" versus the time of fall t. (a) Enter the data into a spreadsheet and convert the distances in feet to distances in meters. (b) Graph d (in meters) versus t. (c) Determine the value of the terminal speed v_t by finding the slope of the linear portion of the curve. Use a least-squares fit to determine the slope.

t (s)	d (ft)
1	16
2	62
3	1138
4	242
5	366
6	504
7	652
8	808
9	971
10	1138
11	1309
12	1483
13	1657
14	1831
15	2005
16	2179
17	2353
18	2527
19	2701
20	2875

S4. A 50.0-kg box at rest on a level floor must be moved. The coefficient of static friction between box and floor is 0.600. A force of magnitude F is applied at an angle θ with the horizontal. (a) Using a spreadsheet, calculate the force F necessary to move the box for a sequence of angles θ. Take the angle θ as positive if the force has an upward component, negative if it has a downward component. Plot F versus θ. From the graph, find the minimum force necessary to move the box. At what angle should it be applied? (b) Investigate what happens when you change the coefficient of friction.

S5. It is "common knowledge" that when air resistance is *not* negligible, a heavier object always hits the ground before a lighter object when both are dropped simultaneously. (a) Use Spreadsheet 6.1 to test this prediction, with $m = 0.100$ kg, 1.00 kg, and 5.00 kg and $b = 0.200$. This corresponds to dropping objects of the same size and shape but with each having a different mass, such as a cork ball, a wooden ball, and a lead ball. (b) Investigate what happens if the objects have an initial velocity. Use different values (both positive and negative) for the initial velocity. Make a table of the times it takes the objects to hit the ground when they are thrown from a height of 200 m. In the spreadsheet, $x = 0$ is the ground. Does the heavier object always hit first?

S6. Modify Spreadsheet 6.1 so that $n = 2$ (see Problem S1). Then the net force in Newton's second law has the form

$$F_{net} = mg - b\mathbf{v} \cdot |\mathbf{v}|$$

since the drag force term must always be in the opposite direction from the velocity. Compare your results for both position and speed versus time for $n = 1$ and $n = 2$. Note that the value of b should also change as n is changed. Pick values for b such that the terminal speeds for $n = 1$ and $n = 2$ are the same for each mass. Plot the $n = 1$ and $n = 2$ solutions on the same graphs and compare them to the case where air resistance can be neglected. Vary x_0, v_0, and m, and recompare.

S7. If the drag force due to air resistance for the sky diver in Problem S3 can be represented as $R = bv^n$, the acceleration of the falling sky diver is

$$a = g\left[1 - \left(\frac{v}{v_t}\right)^n\right]$$

The most commonly used empirical values of n are $n = 1$ and $n = 2$. Which choice best fits the sky diver data in Problem S3? To answer this question, use your value of the terminal speed from Problem S3 and solve numerically for the speed and position of the sky diver as functions of time for both cases, $n = 1$ and $n = 2$. Plot the position-versus-time curves for both cases along with the sky diver data on the same graph.

S8. A large artillery shell with a mass of 44.0 kg is fired with a muzzle speed $v_0 = 570$ m/s. It is reported that the maximum horizontal range is 15.0 km. However, the maximum range if air resistance were negligible would be 33.0 km according to

$$R_{max} = \frac{v_0{}^2}{g}$$

Clearly, air resistance is not negligible even for such a heavy projectile. Assuming the force of air resistance on the shell is $f = -bv|v|$, do a numerical study to find an appropriate value for b such that $R_{max} = 15.0$ km. (You could modify Spreadsheet 6.1 to include two-dimensional motion.) *Note:* The maximum range when air resistance is included does not necessarily occur at $\theta_0 = 45°$. You will have to vary θ_0 for each value of b to find R_{max} as a function of b.

S9. A 0.142-kg baseball has a terminal speed of 42.5 m/s (95 mph). (a) If a baseball experiences a drag force of magnitude $R = Kv^2$, what is the value of K? (b) What is the magnitude of the drag force when the speed of the baseball is 36.0 m/s? (c) Set up a spreadsheet (or modify Spreadsheet 6.1) to determine the motion of a baseball thrown vertically upward at an initial speed of 36.0 m/s. What maximum height does the ball reach? How long is it in the air? What is its speed just before it hits the ground?

S10. A 50.0-kg parachutist jumps from an airplane and falls to Earth with a drag force of magnitude $R = Kv^2$. Take $K = 0.200$ kg/m with the parachute closed and $K = 20.0$ kg/m with the parachute open. (a) Determine the terminal speed of the parachutist before and after the parachute is opened. (b) Set up a spreadsheet to determine the position and speed of the parachutist as functions of time. Assume the jumper began the descent at an altitude of 1000 m, and fell in free fall for 10 s before opening the parachute. (*Hint:* When the parachute is opened there is a sudden large acceleration; a smaller time step may be necessary in this region.)

S11. A 10.0-kg projectile is launched with an initial speed of 150 m/s at an elevation angle of 35.0° with the horizontal. The resistive force acting on the projectile is $\mathbf{R} = b\mathbf{v}$, where $b = 15.0$ kg/s. (a) Set up a spreadsheet to determine the horizontal and vertical positions of the projectile as functions of time. (b) Find the range of this projectile. (c) Determine the elevation angle that gives the maximum range for this projectile. (*Hint:* Adjust the elevation angle by trial and error to find the greatest range.)

CHAPTER 7

Work and Energy

These cyclists are working hard and expending energy as they pedal uphill in Marin County, California. *(David Madison/Tony Stone Images)*

The concept of energy is one of the most important concepts in both contemporary science and engineering practice. In everyday usage, we think of energy in terms of fuel for transportation and heating, electricity for lights and appliances, and the foods that we consume. However, these ideas do not really define energy. They tell us only that fuels are needed to do a job and that those fuels provide us with something we call energy.

Energy is present in the Universe in various forms, including mechanical energy, electromagnetic energy, chemical energy, thermal energy, and nuclear energy. Furthermore, one form of energy can be converted to another. For example, when an electric motor is connected to a battery, chemical energy is converted to electrical energy, which in turn is converted to mechanical energy. The transformation of energy from one form to another is an essential part of the study of physics, engineering, chemistry, biology, geology, and astronomy. When energy is changed from one form to another, its total amount remains the same. Conservation of energy says that although the form of energy may be changed, if an object (or system) loses energy, that same amount of energy appears in another object (or in the surroundings).

In this chapter, we shall first introduce the concept of work. Work is done by a force acting on an object when the point of application of that force moves through some distance and the force has a component along the line of motion. We then define kinetic energy, which is energy associated with the motion of an object. We shall see that the concepts of work and energy can be applied to the dynamics of a mechanical system without resorting to Newton's laws. However, it is important to note that the work-energy concepts are based upon Newton's laws and therefore do not involve any new physical principles.

Although the approach we shall use provides the same results as Newton's laws in describing the motion of a mechanical system, the general ideas of the work-energy concept can be applied to a wide range of phenomena in the fields of electromagnetism and atomic and nuclear physics. In addition, in a complex situation the "energy approach" can often provide a much simpler analysis than the direct application of Newton's second law.

This alternative method of describing motion is especially useful when the force acting on a particle is not constant. In this case, the acceleration is not constant, and we cannot apply the simple kinematic equations we developed in Chapter 2. Often, a particle in nature is subject to a force that varies with the position of the particle. Such forces include gravitational forces and the force exerted on an object attached to a spring. We shall describe techniques for treating such systems with the help of an extremely important development called the *work-energy theorem,* which is the central topic of this chapter.

7.1   WORK DONE BY A CONSTANT FORCE

Almost all of the terms we have used thus far—velocity, acceleration, force, and so on—have conveyed the same meaning in physics as they do in everyday life. Now, however, we encounter a term whose meaning in physics is distinctly different from its everyday meaning. That new term is **work**. Consider a particle that undergoes a displacement **s** along a straight line while acted on by a constant force **F,** which makes an angle θ with **s**, as in Figure 7.1.

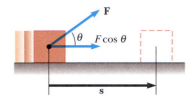

FIGURE 7.1   If a force acting on an object undergoes a displacement **s**, the work done by the force **F** is $(F \cos \theta)\, s$.

Work done by a constant force

> The work *W* done by an agent exerting a constant force is the product of the component of the force in the direction of the displacement and the magnitude of the displacement of the force.
>
> $$W = Fs \cos \theta \qquad (7.1)$$

From this definition, we see that a force does no work on a particle if the particle does not move. That is, if $s = 0$, Equation 7.1 gives $W = 0$. For example, if a person pushes against a brick wall, a force is exerted on the wall but the person does no work on the wall provided the point of application of the force does not move. Also note from Equation 7.1 that the work done by a force is zero when the force is perpendicular to the displacement. That is, if $\theta = 90°$, then $W = 0$ and $\cos 90° = 0$. For example, in Figure 7.2, the work done by the normal force and the work done by the force of gravity are zero because both forces are perpendicular to the displacement and have zero components in the direction of **s**.

The sign of the work also depends on the direction of **F** relative to **s**. The work done by the applied force is positive when the vector associated with the component $F \cos \theta$ is in the same direction as the displacement. For example, when an object is lifted, the work done by the applied force is positive because the lifting force is upward, that is, in the same direction as the displacement. In this situation, the work done by the gravitational force is negative.

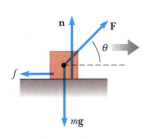

FIGURE 7.2   When an object is displaced horizontally, the normal force **n** and the force of gravity, *m***g**, do no work. The force of friction (box on floor) does some work on the floor.

Does the weight lifter do any work as he holds the weight on his shoulders? Does he do any work as he raises the weight? *(Gerard Vandystadt/Photo Researchers)*

When the vector associated with the component $F \cos \theta$ is in the direction opposite the displacement, W is negative. The factor $\cos \theta$ that appears in the definition of W (Eq. 7.1) automatically takes care of the sign. In the case of the object being lifted, for instance, the work done by the gravitational force is negative. It is important to note that work is an energy transfer; if energy is transferred *to* the system (object), W is positive; if energy is transferred *from* the system, W is negative.

If an applied force **F** acts along the direction of the displacement, then $\theta = 0$, and $\cos 0 = 1$. In this case, Equation 7.1 gives

$$W = Fs$$

Work is a scalar quantity, and its units are force multiplied by length. Therefore, the SI unit of work is the **newton·meter** (N·m). Another name for the newton·meter is the **joule** (J). The unit of work in the cgs system is the **dyne·cm,** also called the **erg;** the unit in the British engineering system is the **ft·lb.** These are summarized in Table 7.1. Note that $1 \text{ J} = 10^7$ ergs.

In general, a particle may be moving with a constant or varying velocity under the influence of several forces. In that case, since work is a scalar quantity, the total work done as the particle undergoes some displacement is the algebraic sum of the amounts of work done by each of the forces.

TABLE 7.1   Units of Work in the Three Common Systems of Measurement

System	Unit	Alternate Name
SI	newton·meter (N·m)	joule (J)
cgs	dyne·centimeter (dyne·cm)	erg
British engineering	foot·pound (ft·lb)	foot·pound (ft·lb)

EXAMPLE 7.1   Mr. Clean

A man cleaning his apartment pulls a vacuum cleaner with a force of magnitude $F = 50$ N. The force makes an angle of 30° with the horizontal as shown in Figure 7.3. The vacuum cleaner is displaced 3.0 m to the right. Calculate the work done by the 50-N force.

Solution   Using the definition of work (Equation 7.1), we have

$$W_F = (F \cos \theta)s = (50 \text{ N})(\cos 30°)(3.0 \text{ m})$$

$$= 130 \text{ N} \cdot \text{m} = \boxed{130 \text{ J}}$$

Note that the normal force **n**, the weight mg, and the upward component of the applied force, $(50 \text{ N})\sin 30°$, do no work because they are perpendicular to the displacement.

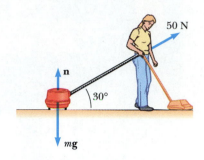

FIGURE 7.3   (Example 7.1) A vacuum cleaner being pulled at an angle of 30° with the horizontal.

Exercise   Find the work done by the man on the vacuum cleaner if he pulls it 3.0 m with a horizontal force of 32 N.

Answer   96 J.

CONCEPTUAL EXAMPLE 7.2

A person lifts a cement block of mass m a vertical height h, and then walks horizontally a distance d while holding the block, as in Figure 7.4. Determine the work done by the person and by the force of gravity in this process.

Reasoning   Assuming that the person lifts the block with a force of magnitude equal to the weight of the block, mg, the work done by the person during the vertical displacement is mgh, since the force in this case is in the direction of the displacement. The work done by the person during the horizontal displacement of the block is zero since the applied force in this process is perpendicular to the displacement. Thus, the net work done by the person is mgh. The work done by the force of gravity during the vertical displacement of the block is $-mgh$, because this force is opposite the displacement. The work done by the force of gravity is zero during the horizontal displacement because this force is also perpendicular to the displacement. Hence, the net work done by the force of gravity is $-mgh$. The net work done on the block is zero $(+mgh - mgh = 0)$. The kinetic energy of the block does not change.

FIGURE 7.4   (Conceptual Example 7.2) A person lifts a cement block of mass m a vertical height h and then walks horizontally a distance d.

7.2   THE SCALAR PRODUCT OF TWO VECTORS

It is convenient to express Equation 7.1 in terms of a **scalar product** of the two vectors **F** and **s**. We write this scalar product **F·s**. Because of the dot symbol, the scalar product is often called the *dot product*. Thus, we can express Equation 7.1 as a scalar product:

$$W = \mathbf{F} \cdot \mathbf{s} = Fs \cos \theta \qquad (7.2)$$

In other words, $\mathbf{F} \cdot \mathbf{s}$ (read "F dot s") is a shorthand notation for $Fs \cos \theta$.

> In general, the scalar product of any two vectors **A** and **B** is a scalar quantity equal to the product of the magnitudes of the two vectors and the cosine of the angle θ between them:
>
> $$\mathbf{A} \cdot \mathbf{B} \equiv AB \cos \theta \qquad (7.3)$$

where A is the magnitude of **A**, B is the magnitude of **B**, and θ is the smaller angle between **A** and **B**, as in Figure 7.5. Note that **A** and **B** need not have the same units.

In Figure 7.5, $B \cos \theta$ is the projection of **B** onto **A**. Therefore, Equation 7.3 says that $\mathbf{A} \cdot \mathbf{B}$ is the product of the magnitude of **A** and the projection of **B** onto **A**.[1]

From Equation 7.3 we also see that the scalar product is *commutative*. That is,

$$\mathbf{A} \cdot \mathbf{B} = \mathbf{B} \cdot \mathbf{A}$$

Finally, the scalar product obeys the *distributive law of multiplication,* so that

$$\mathbf{A} \cdot (\mathbf{B} + \mathbf{C}) = \mathbf{A} \cdot \mathbf{B} + \mathbf{A} \cdot \mathbf{C}$$

The dot product is simple to evaluate from Equation 7.3 when **A** is either perpendicular or parallel to **B**. If **A** is perpendicular to **B** ($\theta = 90°$), then $\mathbf{A} \cdot \mathbf{B} = 0$. (The equality $\mathbf{A} \cdot \mathbf{B} = 0$ also holds in the more trivial case when either **A** or **B** is zero.) If **A** and **B** point in the same direction ($\theta = 0°$), then $\mathbf{A} \cdot \mathbf{B} = AB$. If **A** and **B** point in opposite directions ($\theta = 180°$), then $\mathbf{A} \cdot \mathbf{B} = -AB$. The scalar product is negative when $90° < \theta < 180°$.

The unit vectors **i**, **j**, and **k**, which were defined in Chapter 3, lie in the positive $x, y,$ and z directions, respectively, of a right-handed coordinate system. Therefore, it follows from the definition of $\mathbf{A} \cdot \mathbf{B}$ that the scalar products of these unit vectors are

$$\mathbf{i} \cdot \mathbf{i} = \mathbf{j} \cdot \mathbf{j} = \mathbf{k} \cdot \mathbf{k} = 1 \qquad (7.4)$$

$$\mathbf{i} \cdot \mathbf{j} = \mathbf{i} \cdot \mathbf{k} = \mathbf{j} \cdot \mathbf{k} = 0 \qquad (7.5)$$

Two vectors **A** and **B** can be expressed in component vector form as

$$\mathbf{A} = A_x\mathbf{i} + A_y\mathbf{j} + A_z\mathbf{k}$$

$$\mathbf{B} = B_x\mathbf{i} + B_y\mathbf{j} + B_z\mathbf{k}$$

Therefore Equations 7.4 and 7.5 reduce the scalar product of **A** and **B** to

$$\mathbf{A} \cdot \mathbf{B} = A_xB_x + A_yB_y + A_zB_z \qquad (7.6)$$

In the special case where $\mathbf{A} = \mathbf{B}$, we see that

$$\mathbf{A} \cdot \mathbf{A} = A_x^2 + A_y^2 + A_z^2 = A^2$$

---

[1] This is equivalent to stating that $\mathbf{A} \cdot \mathbf{B}$ equals the product of the magnitude of **B** and the projection of **A** onto **B**.

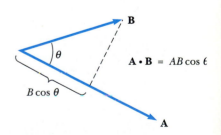

$$\mathbf{A} \cdot \mathbf{B} = AB \cos \theta$$

FIGURE 7.5 The scalar product $\mathbf{A} \cdot \mathbf{B}$ equals the magnitude of **A** multiplied by the projection of **B** onto **A**.

EXAMPLE 7.3   The Scalar Product

The vectors **A** and **B** are given by $\mathbf{A} = 2\mathbf{i} + 3\mathbf{j}$ and $\mathbf{B} = -\mathbf{i} + 2\mathbf{j}$. (a) Determine the scalar product $\mathbf{A} \cdot \mathbf{B}$.

Solution

$$\mathbf{A} \cdot \mathbf{B} = (2\mathbf{i} + 3\mathbf{j}) \cdot (-\mathbf{i} + 2\mathbf{j})$$
$$= -2\mathbf{i} \cdot \mathbf{i} + 2\mathbf{i} \cdot 2\mathbf{j} - 3\mathbf{j} \cdot \mathbf{i} + 3\mathbf{j} \cdot 2\mathbf{j}$$
$$= -2 + 6 = \boxed{4}$$

where we have used the facts that $\mathbf{i} \cdot \mathbf{i} = \mathbf{j} \cdot \mathbf{j} = 1$ and $\mathbf{i} \cdot \mathbf{j} = \mathbf{j} \cdot \mathbf{i} = 0$. The same result is obtained using Equation 7.6 directly, where $A_x = 2$, $A_y = 3$, $B_x = -1$, and $B_y = 2$.

(b) Find the angle θ between **A** and **B**.

Solution   The magnitudes of **A** and **B** are given by

$$A = \sqrt{A_x^2 + A_y^2} = \sqrt{(2)^2 + (3)^2} = \sqrt{13}$$
$$B = \sqrt{B_x^2 + B_y^2} = \sqrt{(-1)^2 + (2)^2} = \sqrt{5}$$

Using Equation 7.3 and the result from (a) gives

$$\cos \theta = \frac{\mathbf{A} \cdot \mathbf{B}}{AB} = \frac{4}{\sqrt{13}\sqrt{5}} = \frac{4}{\sqrt{65}}$$

$$\theta = \cos^{-1} \frac{4}{8.06} = \boxed{60°}$$

EXAMPLE 7.4   Work Done by a Constant Force

A particle moving in the xy plane undergoes a displacement $\mathbf{s} = (2.0\mathbf{i} + 3.0\mathbf{j})$ m while a constant force $\mathbf{F} = (5.0\mathbf{i} + 2.0\mathbf{j})$ N acts on the particle. (a) Calculate the magnitude of the displacement and that of the force.

Solution

$$s = \sqrt{x^2 + y^2} = \sqrt{(2.0)^2 + (3.0)^2} = \boxed{3.6 \text{ m}}$$

$$F = \sqrt{F_x^2 + F_y^2} = \sqrt{(5.0)^2 + (2.0)^2} = \boxed{5.4 \text{ N}}$$

(b) Calculate the work done by **F**.

Solution   Substituting the expressions for **F** and **s** into Equation 7.2 and using Equations 7.4 and 7.5, we get

$$W = \mathbf{F} \cdot \mathbf{s} = (5.0\mathbf{i} + 2.0\mathbf{j}) \cdot (2.0\mathbf{i} + 3.0\mathbf{j}) \text{ N} \cdot \text{m}$$
$$= 5.0\mathbf{i} \cdot 2.0\mathbf{i} + 5.0\mathbf{i} \cdot 3.0\mathbf{j} + 2.0\mathbf{j} \cdot 2.0\mathbf{i} + 2.0\mathbf{j} \cdot 3.0\mathbf{j}$$
$$= 10 + 0 + 0 + 6 = 16 \text{ N} \cdot \text{m} = \boxed{16 \text{ J}}$$

Exercise   Calculate the angle between **F** and **s**.

Answer   35°.

7.3   WORK DONE BY A VARYING FORCE

Consider a particle being displaced along the x axis under the action of a varying force, as in Figure 7.6. The particle is displaced in the direction of increasing x from $x = x_i$ to $x = x_f$. In such a situation, we cannot use $W = (F \cos \theta)s$ to calculate

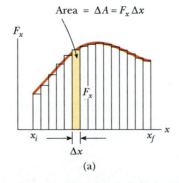
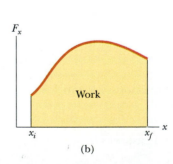

FIGURE 7.6   (a) The work done by the force F_x for the small displacement Δx is $F_x \Delta x$, which equals the area of the shaded rectangle. The total work done for the displacement from x_i to x_f is approximately equal to the sum of the areas of all the rectangles. (b) The work done by the variable force F_x as the particle moves from x_i to x_f is *exactly* equal to the area under this curve.

the work done by the force, because this relationship applies only when **F** is constant in magnitude and direction. However, if we imagine that the particle undergoes a very small displacement Δx, shown in Figure 7.6a, then the x component of the force, F_x, is approximately constant over this interval, and we can express the work done by the force for this small displacement as

$$W_1 = F_x \Delta x$$

This is just the area of the shaded rectangle in Figure 7.6a. If we imagine that the F_x versus x curve is divided into a large number of such intervals, then the total work done for the displacement from x_i to x_f is approximately equal to the sum of a large number of such terms:

$$W \cong \sum_{x_i}^{x_f} F_x \Delta x$$

If the displacements are allowed to approach zero, then the number of terms in the sum increases without limit but the value of the sum approaches a definite value equal to the area under the curve bounded by F_x and the x axis:

$$\lim_{\Delta x \to 0} \sum_{x_i}^{x_f} F_x \Delta x = \int_{x_i}^{x_f} F_x \, dx$$

This definite integral is numerically equal to the area under the F_x versus x curve between x_i and x_f. Therefore, we can express the work done by F_x for the displacement of the object from x_i to x_f as

$$W = \int_{x_i}^{x_f} F_x \, dx \qquad (7.7)$$

Work done by a varying force

This equation reduces to Equation 7.1 when $F_x = F \cos \theta$ is constant.

If more than one force acts on a particle, the total work done is just the work done by the resultant force. For systems that do not act as particles, work must be found for each force separately. If we express the resultant force in the x direction as ΣF_x, then the *net work* done as the particle moves from x_i to x_f is

$$W_{\text{net}} = \int_{x_i}^{x_f} \left(\sum F_x \right) dx \qquad (7.8)$$

EXAMPLE 7.5 Calculating Total Work Done from a Graph

A force acting on a particle varies with x as shown in Figure 7.7. Calculate the work done by the force as the particle moves from $x = 0$ to $x = 6.0$ m.

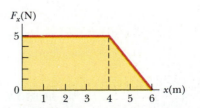

Solution The work done by the force is equal to the area under the curve from $x = 0$ to $x = 6.0$ m. This area is equal to the area of the rectangular section from $x = 0$ to $x = 4.0$ m plus the area of the triangular section from $x = 4.0$ m to $x = 6.0$ m. The area of the rectangle is $(4.0)(5.0)$ N·m $= 20$ J, and the area of the triangle is $\frac{1}{2}(2.0)(5.0)$ N·m $= 5.0$ J. Therefore, the total work done is 25 J.

FIGURE 7.7 (Example 7.5) The force acting on a particle is constant for the first 4.0 m of motion and then decreases linearly with x from $x = 4.0$ m to $x = 6.0$ m. The net work done by this force is the area under this curve.

Work Done by a Spring

A common physical system for which the force varies with position is shown in Figure 7.8. A block on a horizontal, frictionless surface is connected to a spring. If the spring is stretched or compressed a small distance from its unstretched,

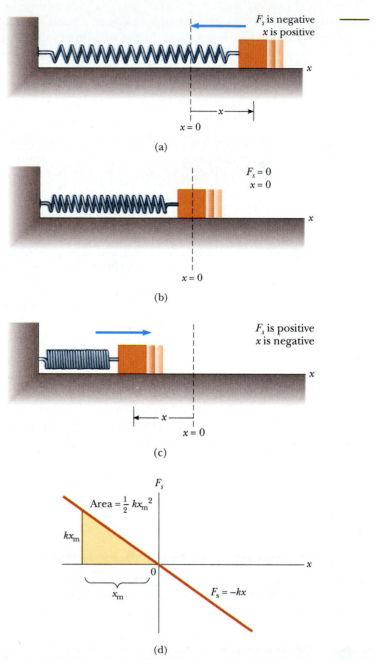

FIGURE 7.8   The force exerted by a spring on a block varies with the block's displacement from the equilibrium position $x = 0$. (a) When x is positive (stretched spring), the spring force is to the left. (b) When x is zero, the spring force is zero (natural length of the spring). (c) When x is negative (compressed spring), the spring force is to the right. (d) Graph of F_s versus x for the mass-spring system. The work done by the spring force as the block moves from $-x_m$ to 0 is the area of the shaded triangle, $\frac{1}{2}kx_m^2$.

or equilibrium, configuration, the spring will exert a force on the block given by

$$F_s = -kx \tag{7.9}$$

where x is the displacement of the block from its unstretched ($x = 0$) position and k is a positive constant called the *force constant* of the spring. As we learned in Chapter 5, this force law for springs is known as **Hooke's law**. Note that Hooke's law is valid only in the limiting case of small displacements. The value of k is a measure of the stiffness of the spring. Stiff springs have large k values, and soft springs have small k values.

The negative sign in Equation 7.9 signifies that the force exerted by the spring is always directed *opposite* the displacement. For example, when $x > 0$ as in Figure 7.8a, the spring force is to the left, or negative. When $x < 0$ as in Figure 7.8c, the spring force is to the right, or positive. Of course, when $x = 0$ as in Figure 7.8b, the spring is unstretched and $F_s = 0$. Since the spring force always acts toward the equilibrium position, it is sometimes called a *restoring force*. Suppose the spring is compressed so that the block is displaced a distance $-x_m$ from equilibrium. After it is released, the block moves from $-x_m$ through zero to $+x_m$. If the spring is stretched until the block is at x_m and then released, the block moves from $+x_m$ through zero to $-x_m$. The details of the ensuing oscillating motion will be given in Chapter 13.

Suppose that the block is pushed to the left a distance x_m from equilibrium and then released. Let us calculate the work done by the spring force as the block moves from $x_i = -x_m$ to $x_f = 0$. Applying Equation 7.7 assuming the block may be treated as a particle, we get

$$W_s = \int_{x_i}^{x_f} F_s \, dx = \int_{-x_m}^{0} (-kx) \, dx = \tfrac{1}{2} k x_m{}^2 \tag{7.10}$$

where we have used the indefinite integral $\int x \, dx = x^2/2$. That is, the work done by the spring force is positive because the spring force is in the same direction as the displacement (both are to the right). The positive value of W_s confirms that energy is transferred from the spring to the block. However, if we consider the work done by the spring force as the block moves from $x_i = 0$ to $x_f = x_m$, we find that $W_s = -\tfrac{1}{2} k x_m{}^2$, since for this part of the motion the displacement is to the right and the spring force is to the left. Therefore, the *net* work done by the spring force as the block moves from $x_i = -x_m$ to $x_f = x_m$ is *zero*.

Figure 7.8d is a plot of F_s versus x. The work calculated in Equation 7.10 is the area of the shaded triangle in Figure 7.8d, corresponding to the displacement from $-x_m$ to 0. Because the triangle has base x_m and height kx_m, its area is $\tfrac{1}{2} k x_m{}^2$, the work done by the spring as given by Equation 7.10.

If the block undergoes an arbitrary displacement from $x = x_i$ to $x = x_f$, the work done by the spring force is

$$W_s = \int_{x_i}^{x_f} (-kx) \, dx = \frac{1}{2} k x_i{}^2 - \frac{1}{2} k x_f{}^2 \tag{7.11}$$

From this equation we see that the work done by the spring force is zero for any motion that ends where it began ($x_i = x_f$). We shall make use of this important result in Chapter 8, where we describe the motion of this system in more detail.

Equations 7.10 and 7.11 describe the work done by the spring force exerted on the block. Now let us consider the work done on the spring by an *external agent* as the spring is stretched very slowly from $x_i = 0$ to $x_f = x_m$, as in Figure 7.9. This

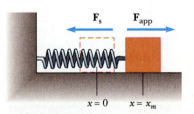

FIGURE 7.9 A block being pulled from $x = 0$ to $x = x_m$ on a frictionless surface by a force F_{app}. If the process is carried out very slowly, the applied force is equal to and opposite the spring force at all times.

work can be easily calculated by noting that the *applied force,* F_{app}, is equal to and opposite the spring force, F_s, at any value of the displacement, so that $F_{app} = -(-kx) = kx$. Therefore, the work done by this applied force (the external agent) is

$$W_{F_{app}} = \int_0^x F_{app}\, dx = \int_0^x kx\, dx = \tfrac{1}{2}kx^2$$

You should note that this work is equal to the negative of the work done by the spring force for this displacement. For example, if a spring of force constant 80 N/m is compressed 3.0 cm from equilibrium, the work done by the spring force as the block moves from $x_i = -3.0$ cm to its unstretched position, $x_f = 0$, is 3.6×10^{-2} J.

EXAMPLE 7.6 *Measuring k for a Spring*

A common technique used to measure the force constant of a spring is described in Figure 7.10. The spring is hung vertically and then a mass m is attached to the lower end of the spring. The spring stretches a distance d from its equilibrium position under the action of the "load" mg. Since the spring force is upward, it must balance the weight mg downward when the system is at rest. In this case, we can apply Hooke's law to give $|F_s| = kd = mg$, or

$$k = \frac{mg}{d}$$

For example, if a spring is stretched 2.0 cm by a suspended mass of 0.55 kg, the force constant of the spring is

$$k = \frac{mg}{d} = \frac{(0.55\ \text{kg})(9.80\ \text{m/s}^2)}{2.0 \times 10^{-2}\ \text{m}} = 2.7 \times 10^2\ \text{N/m}$$

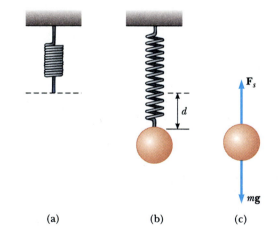

(a)   (b)   (c)

FIGURE 7.10 (Example 7.6) Determining the force constant k of a helical spring. The elongation d of the spring is due to the attached weight mg. Because the spring force balances the weight, it follows that $k = mg/d$.

7.4 KINETIC ENERGY AND THE WORK-ENERGY THEOREM

Solutions using Newton's second law can be difficult if the forces in the problem are complex. An alternative approach that enables us to understand and solve such motion problems is to relate the speed of the particle to its displacement under the influence of some net force. As we shall see in this section, if the work done by the net force on a particle can be calculated for a given displacement, the change in the particle's speed will be easy to evaluate.

Figure 7.11 shows a particle of mass m moving to the right under the action of a constant net force F. Because the force is constant, we know from Newton's second law that the particle will move with a constant acceleration a. If the particle is displaced a distance s, the net work done by the force F is

$$W_{net} = Fs = (ma)s \tag{7.12}$$

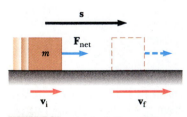

FIGURE 7.11 A particle undergoing a displacement and change in velocity under the action of a constant net force F.

In Chapter 2 we found that the following relationships are valid when a particle undergoes constant acceleration:

$$s = \tfrac{1}{2}(v_i + v_f)t \qquad a = \frac{v_f - v_i}{t}$$

where v_i is the speed at $t = 0$ and v_f is the speed at time t. Substituting these expressions into Equation 7.12 gives

$$W_{\text{net}} = m\left(\frac{v_f - v_i}{t}\right)\tfrac{1}{2}(v_i + v_f)t$$

$$W_{\text{net}} = \tfrac{1}{2}mv_f^2 - \tfrac{1}{2}mv_i^2 \tag{7.13}$$

The quantity $\tfrac{1}{2}mv^2$ represents the energy associated with the motion of a particle. It is so important that it has been given a special name — **kinetic energy.** The kinetic energy, K, of a particle of mass m moving with a speed v is defined as

$$K \equiv \tfrac{1}{2}mv^2 \tag{7.14}$$

Kinetic energy is a scalar quantity and has the same units as work. For example, a 2.0-kg mass moving with a speed of 4.0 m/s has a kinetic energy of 32 J. Table 7.2 lists the kinetic energies for various objects. We can think of kinetic energy as energy associated with the motion of a body. It is often convenient to write Equation 7.14 in the form

$$W_{\text{net}} = K_f - K_i = \Delta K \tag{7.15}$$

That is,

the work done by the constant net force $\mathbf{F}_{\text{net}}$ in displacing a particle equals the change in kinetic energy of the particle.

Equation 7.15 is an important result known as the **work-energy theorem.** For convenience, it was derived under the assumption that the net force acting on the particle was constant. You should note that when energy is transferred to a particle

Kinetic energy is energy associated with the motion of a body

Work-energy theorem

TABLE 7.2   Kinetic Energies for Various Objects

Object	Mass (kg)	Speed (m/s)	Kinetic Energy (J)
Earth orbiting the Sun	5.98×10^{24}	2.98×10^4	2.65×10^{33}
Moon orbiting the Earth	7.35×10^{22}	1.02×10^3	3.82×10^{28}
Rocket moving at escape speed[a]	500	1.12×10^4	3.14×10^{10}
Automobile at 55 mi/h	2000	25	6.3×10^5
Running athlete	70	10	3.5×10^3
Stone dropped from 10 m	1.0	14	9.8×10^1
Golf ball at terminal speed	0.046	44	4.5×10^1
Raindrop at terminal speed	3.5×10^{-5}	9.0	1.4×10^{-3}
Oxygen molecule in air	5.3×10^{-26}	500	6.6×10^{-21}

[a] Escape speed is the minimum speed an object must reach near the Earth's surface in order to escape the Earth's gravitational force.

(as work), it must appear as kinetic energy of the particle because that is the *only* form of energy that can change for a particle.

Now we shall show that the work-energy theorem is valid even when the force is varying. If the resultant force acting on a body in the *x* direction is ΣF_x, then Newton's second law states that $\Sigma F_x = m\mathbf{a}$. Thus, we can use Equation 7.8 and express the net work done as

$$W_{\text{net}} = \int_{x_i}^{x_f} \left(\sum F_x \right) dx = \int_{x_i}^{x_f} ma\, dx$$

Because the resultant force varies with *x*, the acceleration and speed also depend on *x*. We can now use the following chain rule to evaluate W_{net}:

$$a = \frac{dv}{dt} = \frac{dv}{dx}\frac{dx}{dt} = v\frac{dv}{dx}$$

Substituting this into the expression for W_{net} gives

$$W_{\text{net}} = \int_{x_i}^{x_f} mv\frac{dv}{dx}\, dx = \int_{v_i}^{v_f} mv\, dv$$

$$W_{\text{net}} = \tfrac{1}{2}mv_f^2 - \tfrac{1}{2}mv_i^2 \tag{7.16}$$

where the limits of the integration were changed because the variable was changed from *x* to *v*. Thus, we conclude that the work done on a particle by the net force acting on it is equal to the change in the kinetic energy of the particle.

The work-energy theorem also says that the speed of the particle will increase if the net work done on it is positive, because the final kinetic energy will be greater than the initial kinetic energy. The speed will decrease if the net work is negative because the final kinetic energy will be less than the initial kinetic energy. The speed and kinetic energy of a particle change only if work is done on the particle by some external force.

Consider the relationship between the work done on a particle and the change in its kinetic energy as expressed by Equation 7.15. Because of this connection, we can also think of kinetic energy as the work the particle can do in coming to rest. For example, suppose a hammer is on the verge of striking a nail, as in Figure 7.12. The moving hammer has kinetic energy and can do work on the nail. The work done on the nail appears as the product *Fs*, where *F* is the average force exerted on the nail by the hammer and *s* is the distance the nail is driven into the wall.

Situations Involving Kinetic Friction

Suppose that an object of mass *m* sliding on a horizontal surface is pulled with a constant horizontal external force **F** to the right and a kinetic frictional force **f** acts to the left, where **F** > **f**. In this case, the net force is to the right as in Figure 7.13,

Work done on a particle equals the change in its kinetic energy

FIGURE 7.12 The hammer has kinetic energy associated with its motion and can do work on the nail, driving it into the wall.

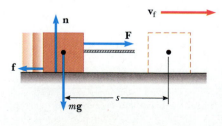

FIGURE 7.13 A mass *m* on a horizontal surface undergoes a displacement *s*. **F** is a constant horizontal external force and **f** is the force of kinetic friction directed toward the left.

and we might believe that we could find the net work done on the object as it undergoes a displacement **s** to the right by evaluating

$$W_{net} = (\mathbf{F} - \mathbf{f}) \cdot \mathbf{s} = Fs - fs$$

However, the object is not a particle, and it is incorrect to say that $-fs$ is the work done by the frictional force on the object. The frictional force acts on the object at the microscopic level through a series of microscopic displacements. The work done by kinetic friction depends on both the displacement of the object and on the details of the motion between the initial and final positions. In fact, the work done by kinetic friction on an extended object cannot be explicitly evaluated because friction forces and their individual displacements are very complex.

Now suppose that a block moving on a horizontal surface and given an initial horizontal velocity $\mathbf{v}_i$ slides a distance s before reaching a final velocity $\mathbf{v}_f$ as in Figure 7.14. The external force that causes the block to undergo an acceleration in the negative x direction is the force of kinetic friction **f** acting to the left, opposite the motion. The initial kinetic energy of the block is $\frac{1}{2}mv_i^2$ and its final kinetic energy is $\frac{1}{2}mv_f^2$. The change in kinetic energy of the block is equal to $-fs$. This can be shown by applying Newton's second law to the block. (Newton's second law gives the acceleration of the center of mass of any object regardless of how or where the forces act.) Since the net force on the block in the x direction is the friction force, Newton's second law gives $-f = ma$. Multiplying both sides of this expression by s, and using the equation $v_f^2 - v_i^2 = 2as$ for motion under constant acceleration gives $-fs = (ma)s = \frac{1}{2}mv_f^2 - \frac{1}{2}mv_i^2$ or

$$\Delta K = -fs \qquad (7.17)$$

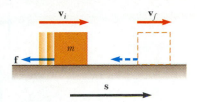

FIGURE 7.14   A block sliding to the right on a horizontal surface slows down in the presence of a force of kinetic friction acting to the left. The initial velocity of the block is $\mathbf{v}_i$, and its final velocity is $\mathbf{v}_f$. The normal force and force of gravity are not included in the diagram because they are perpendicular to the direction of motion and therefore do not influence the change in velocity of the block.

Loss in kinetic energy due to friction

This result says that the loss in kinetic energy of the block is equal to $-fs$, which corresponds to the energy dissipated by the force of kinetic friction. Part of this energy is transferred to internal energy of the block, and part is transferred from the block to the surface.[2] In effect, the loss in kinetic energy of the block results in an increase in internal energy of both the block and surface in the form of thermal (heat) energy. For example, if the loss in kinetic energy of the block is 300 J, and 100 J appears as an increase in internal energy of the block, then the remaining 200 J must have been transferred from the block to the surface.

EXAMPLE 7.7   A Block Pulled on a Frictionless Surface

A 6.0-kg block initially at rest is pulled to the right along a horizontal, frictionless surface by a constant, horizontal force of 12 N, as in Figure 7.15a. Find the speed of the block after it has moved 3.0 m.

Solution   The weight of the block is balanced by the normal force, and neither of these forces does work since the displacement is horizontal. Since there is no friction, the resultant external force is the 12-N force. The work done by this force is

---

[2] For more details on energy transfer situations involving forces of kinetic friction, see B. A. Sherwood and W. H. Bernard, *American Journal of Physics*, 52:1001, 1984, and R. P. Bauman, *The Physics Teacher*, 30:264, 1992.

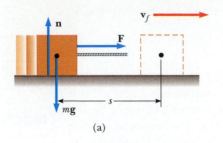
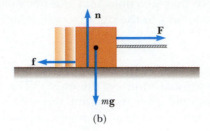

FIGURE 7.15   (a) Example 7.7. (b) Example 7.8.

$$W = Fs = (12 \text{ N})(3.0 \text{ m}) = 36 \text{ N} \cdot \text{m} = 36 \text{ J}$$

Using the work-energy theorem and noting that the initial kinetic energy is zero, we get

$$W = K_f - K_i = \tfrac{1}{2} m v_f^2 - 0$$

$$v_f^2 = \frac{2W}{m} = \frac{2(36 \text{ J})}{6.0 \text{ kg}} = 12 \text{ m}^2/\text{s}^2$$

$$v_f = \boxed{3.5 \text{ m/s}}$$

Exercise   Find the acceleration of the block, and determine its final speed using the kinematic equation $v_f^2 = v_i^2 + 2as$.

Answer $a = 2.0 \text{ m/s}^2$; $v_f = 3.5 \text{ m/s}$.

EXAMPLE 7.8   A Block Pulled on a Rough Surface

Find the final speed of the block described in Example 7.7 if the surface is rough and the coefficient of kinetic friction is 0.15.

Reasoning   In this case, we must use Equation 7.17 to calculate the change in kinetic energy, ΔK. The net force exerted on the block is the sum of the applied 12-N force and the frictional force, as in Figure 7.15b. Since the frictional force is in the direction opposite the displacement, it must be subtracted.

Solution   The magnitude of the frictional force is $f = \mu n = \mu mg$. Therefore the net force acting on the block is

$$F_{\text{net}} = F - \mu mg = 12 \text{ N} - (0.15)(6.0 \text{ kg})(9.80 \text{ m/s}^2)$$
$$= 12 \text{ N} - 8.82 \text{ N} = 3.18 \text{ N}$$

Multiplying this constant force by the displacement, and using Equation 7.17, gives

$$\Delta K = F_{\text{net}} s = (3.18 \text{ N})(3.0 \text{ m}) = 9.54 \text{ J} = \tfrac{1}{2} m v_f^2$$

using the information that $v_i = 0$. Therefore,

$$v_f^2 = \frac{2(9.54 \text{ J})}{6.0 \text{ kg}} = 3.18 \text{ m}^2/\text{s}^2$$

$$v_f = \boxed{1.8 \text{ m/s}}$$

Exercise   Find the acceleration of the block from Newton's second law, and determine its final speed using kinematics.

Answer $a = 0.53 \text{ m/s}^2$; $v_f = 1.8 \text{ m/s}$.

CONCEPTUAL EXAMPLE 7.9

A team of furniture movers wishes to load a truck using a ramp from the ground to the rear of the truck. One of the movers claims that less work would be required to load the truck if the length of the ramp were increased, reducing the angle of the ramp with respect to the horizontal. Is his claim valid? Explain.

Reasoning   His claim is not valid. Although less force is required with a longer ramp, the force must act over a larger distance and do the same amount of work. Suppose a refrigerator attached to a frictionless wheeled dolly is rolled up a ramp at constant speed as in Figure 7.16. The normal force acting 90° to the motion does no work. Because $\Delta K = 0$ in

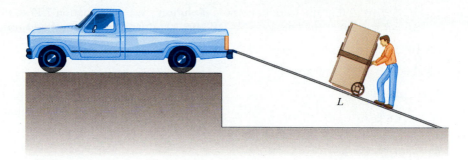

FIGURE 7.16 (Conceptual Example 7.9) A refrigerator attached to a frictionless wheeled dolly is moved up a ramp at constant speed. Does the amount of work required depend on the length of the ramp?

this case, the work-energy theorem gives

$$W_{net} = W_{by\ movers} + W_{by\ gravity} = 0$$

The work done by gravity equals the weight of the refrigera-tor times the vertical height through which it is displaced times cos 180°, or $W_{by\ gravity} = -mgh$. Thus, the movers, however long the ramp, must do work mgh on the refrigerator.

CONCEPTUAL EXAMPLE 7.10

A car traveling at a speed v skids a distance d after its brakes lock. Estimate how far it will skid if its brakes lock when its initial speed is $2v$. What happens to the car's kinetic energy as it stops?

Reasoning   Let us assume that the force of kinetic friction between car and road surface is constant and the same in both cases. The net force times the displacement of the center of mass is equal to the initial kinetic energy of the car. If the speed is doubled as in this example, the kinetic energy of the car is quadrupled. For a given applied force (in this case, the frictional force), the distance traveled is four times as great when the initial speed is doubled, so the estimated distance it skids is $4d$. The kinetic energy of the car is changed into internal energy associated with the tires, brake pads, and road as they heat up.

CONCEPTUAL EXAMPLE 7.11

In most situations we have encountered in this chapter, frictional forces tend to reduce the kinetic energy of an object. However, frictional forces can sometimes increase an object's kinetic energy. Describe a few situations in which friction causes an increase in kinetic energy.

Reasoning   If a crate is located on the bed of a truck, and the truck accelerates to the east, the static friction force ex-erted on the crate by the truck acts to the east to give the crate the same acceleration as the truck (assuming the crate doesn't slip). Another example is a car that accelerates because of the frictional forces exerted on the car's tires by the road. These forces act in the direction of the car's motion, and the sum of these forces causes an increase in the car's kinetic energy.

EXAMPLE 7.12   A Mass-Spring System

A block of mass 1.6 kg is attached to a spring that has a force constant of 1.0×10^3 N/m as in Figure 7.8. The spring is compressed a distance of 2.0 cm, and the block is released from rest. (a) Calculate the speed of the block as it passes through the equilibrium position $x = 0$ if the surface is fric-tionless.

Solution   We use Equation 7.10 to find the work done by the spring with $x_m = -2.0$ cm $= -2.0 \times 10^{-2}$ m:

$$W_s = \tfrac{1}{2}kx_m{}^2 = \tfrac{1}{2}(1.0 \times 10^3 \,\text{N/m})(-2.0 \times 10^{-2}\,\text{m})^2 = 0.20\,\text{J}$$

Using the work-energy theorem with $v_i = 0$ gives

$$W_s = \tfrac{1}{2}mv_f^2 - \tfrac{1}{2}mv_i^2$$

$$0.20\,\text{J} = \tfrac{1}{2}(1.6\,\text{kg})v_f^2 - 0$$

$$v_f^2 = \frac{0.40\,\text{J}}{1.6\,\text{kg}} = 0.25\,\text{m}^2/\text{s}^2$$

$$v_f = \boxed{0.50\,\text{m/s}}$$

(b) Calculate the speed of the block as it passes through the equilibrium position if a constant frictional force of 4.0 N retards its motion.

Solution   We use Equation 7.17 to calculate the kinetic energy lost due to friction and add this to the kinetic energy found in the absence of friction. Considering only the frictional force, the kinetic energy lost due to friction is

$$-fs = -(4.0\,\text{N})(2.0 \times 10^{-2}\,\text{m}) = -0.08\,\text{J}$$

The final kinetic energy, without this loss, was found in part (a) to be 0.20 J. Therefore, the final kinetic energy in the presence of friction is

$$K_f = 0.20\,\text{J} - 0.08\,\text{J} = 0.12\,\text{J} = \tfrac{1}{2}mv_f^2$$

$$\tfrac{1}{2}(1.6\,\text{kg})v_f^2 = 0.12\,\text{J}$$

$$v_f^2 = \frac{0.24\,\text{J}}{1.6\,\text{kg}} = 0.15\,\text{m}^2/\text{s}^2$$

$$v_f = \boxed{0.39\,\text{m/s}}$$

CONCEPTUAL EXAMPLE 7.13

An Earth satellite is in a circular orbit at an altitude of 500 km. Explain why the work done by the gravitational force acting on the satellite is zero. Using the work-energy theorem, what can you conclude about the speed of the satellite?

Reasoning   As the satellite moves in a circular orbit about the Earth as in Figure 7.17, its velocity is tangent to the circular path. Its incremental displacement $d\mathbf{s}$ in any small time interval is always at 90° to the inward gravitational force, which always acts toward the center of the Earth. Now cos 90° = 0, therefore $\mathbf{F} \cdot d\mathbf{s} = 0$. So as the satellite turns through any angle or through many revolutions, the total work done on it by the gravitational force is always zero. The work-energy theorem says that the net work done on a particle during any displacement is equal to the change in its kinetic energy. Because the net work done on the satellite is zero, the change in its kinetic energy is zero, and its speed remains constant.

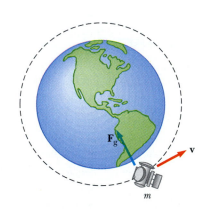

FIGURE 7.17   (Conceptual Example 7.13) An Earth satellite in a circular orbit. What is the work done by the gravitational force acting on the satellite?

7.5   POWER

From a practical viewpoint, it is interesting to know not only the work done on an object but also the rate at which the work is being done. The time rate of doing work is called **power**.

   If an external force is applied to an object (which we assume acts as a particle), and if the work done by this force is W in the time interval Δt, then the **average power** during this interval is defined as

Average power

$$\overline{P} \equiv \frac{W}{\Delta t}$$

The work done on the object contributes to increasing the energy of the object. A more general definition of power is the *time rate of energy transfer*. The **instantaneous** power is the limiting value of the average power as Δt approaches zero:

$$P \equiv \lim_{\Delta t \to 0} \frac{W}{\Delta t} = \frac{dW}{dt}$$

where we have represented the infinitesimal value of the work done by dW (even though it is not a change and therefore not a differential). We find from Equation 7.2 that $dW = \mathbf{F} \cdot d\mathbf{s}$. Therefore, the instantaneous power can be written

$$P = \frac{dW}{dt} = \mathbf{F} \cdot \frac{d\mathbf{s}}{dt} = \mathbf{F} \cdot \mathbf{v} \qquad (7.18)$$

Instantaneous power

where we have used the fact that $\mathbf{v} = d\mathbf{s}/dt$.

The SI unit of power is joules per second (J/s), also called a *watt* (W) (after James Watt):

$$1 \text{ W} = 1 \text{ J/s} = 1 \text{ kg} \cdot \text{m}^2/\text{s}^3$$

The watt

The symbol W for watt should not be confused with the symbol W for work.

A unit of power in the British engineering system is the horsepower (hp):

$$1 \text{ hp} = 550 \text{ ft} \cdot \text{lb/s} = 746 \text{ W}$$

A unit of energy (or work) can now be defined in terms of the unit of power. One kilowatt hour (kWh) is the energy converted or consumed in 1 h at the constant rate of 1 kW. The numerical value of 1 kWh is

$$1 \text{ kWh} = (10^3 \text{ W})(3600 \text{ s}) = 3.60 \times 10^6 \text{ J}$$

The kilowatt hour is a unit of energy

It is important to realize that a kilowatt hour is a unit of energy, not power. When you pay your electric bill, you are buying energy, and the amount of electricity used by an appliance is usually expressed in kilowatt hours. For example, an electric bulb rated at 100 W would "consume" 3.60×10^5 J of energy in 1 h.

EXAMPLE 7.14   Power Delivered by an Elevator Motor

An elevator has a mass of 1000 kg and carries a maximum load of 800 kg. A constant frictional force of 4000 N retards its motion upward, as in Figure 7.18. (a) What must be the minimum power delivered by the motor to lift the elevator at a constant speed of 3.00 m/s?

Solution   The motor must supply the force T that pulls the elevator upward. From Newton's second law and from the fact that $a = 0$ since v is constant, we get

$$T - f - Mg = 0$$

where M is the *total* mass (elevator plus load), equal to 1800 kg. Therefore,

$$T = f + Mg$$
$$= 4.00 \times 10^3 \text{ N} + (1.80 \times 10^3 \text{ kg})(9.80 \text{ m/s}^2)$$
$$= 2.16 \times 10^4 \text{ N}$$

Using Equation 7.18 and the fact that T is in the same direction as v gives

$$P = \mathbf{T} \cdot \mathbf{v} = Tv$$
$$= (2.16 \times 10^4 \text{ N})(3.00 \text{ m/s}) = 6.49 \times 10^4 \text{ W}$$
$$= 64.9 \text{ kW} = \boxed{87.0 \text{ hp}}$$

(b) What power must the motor deliver at any instant if it is designed to provide an upward acceleration of 1.00 m/s²?

Solution   Applying Newton's second law to the elevator gives

$$T - f - Mg = Ma$$
$$T = M(a + g) + f$$
$$= (1.80 \times 10^3 \text{ kg})(1.00 + 9.80)\text{m/s}^2 + 4.00 \times 10^3 \text{ N}$$
$$= 2.34 \times 10^4 \text{ N}$$

Therefore, using Equation 7.18 we get for the required power

$$P = Tv = \;(2.34 \times 10^4 v)\; \text{W}$$

where v is the instantaneous speed of the elevator in meters per second. Hence, the power required increases with increasing speed.

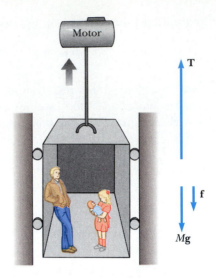

FIGURE 7.18   (Example 7.14) A motor exerts an upward force, equal in magnitude to the tension T, on the elevator. A frictional force **f** and the force of gravity M**g** act downward.

CONCEPTUAL EXAMPLE 7.15

In the first part of the previous example, the motor delivers power to lift the elevator, yet the elevator moves at constant speed. A student analyzing this situation claims that according to the work-energy theorem, if the speed of the elevator remains constant, the work done on it is zero. The student concludes that the power that must be delivered by the motor must also be zero. How would you explain this apparent paradox?

Reasoning   The student has attempted to apply the work-energy theorem to a system that is not acting as a particle (there is significant friction). Applying Newton's law, it is the *net* force on the system, times the displacement (of the center of mass), that is equal to the change in the kinetic energy of the system. In this case, there are three forces acting on the elevator: the upward force **T** exerted by the cable, the downward force of gravity, and the downward frictional force (see Fig. 7.18). The elevator moves at constant speed (zero acceleration) when the upward force is balanced by the sum of the two downward forces ($T = Mg + f$). The power that is supplied by the motor is equal to Tv, which is *not* zero. Part of the energy supplied by the motor in some time interval is used to increase its potential energy and part is lost due to the frictional force. So there is no paradox.

*7.6   ENERGY AND THE AUTOMOBILE

Automobiles powered by gasoline engines are very inefficient machines. Even under ideal conditions, less than 15% of the available energy in the fuel is used to power the vehicle. The situation is much worse under stop-and-go driving in the city. This section uses the concepts of energy, power, and friction to analyze automobile fuel consumption.

Many mechanisms contribute to energy loss in a typical automobile. About two thirds of the energy available from the fuel is lost in the engine. This energy ends up in the atmosphere, partly via the exhaust system and partly via the cooling system. As we shall see in Chapter 22, the large power loss in the exhaust and cooling system is not easy to overcome because of some fundamental laws of thermodynamics. Approximately 10% of the available energy is lost to friction in the transmission, drive shaft, wheel and axle bearings, and differential. Friction in other moving parts dissipates approximately 6% of the energy, and 4% is used to operate fuel and oil pumps and such accessories as power steering and air conditioning. Finally, approximately 13% of the available energy is used to propel the automobile. This energy is used mainly to overcome road friction and air resistance.

TABLE 7.3	**Frictional Forces and Power Requirements for a Typical Car**				

$v\,(\mathrm{m/s})$	$n\,(\mathrm{N})$	$f_r\,(\mathrm{N})$	$f_a\,(\mathrm{N})$	$f_t\,(\mathrm{N})$	$P = f_t v\,(\mathrm{kW})$
0	14 200	227	0	227	0
8.9	14 100	226	51	277	2.5
17.8	13 900	222	204	426	7.6
26.8	13 600	218	465	683	18.3
35.9	13 200	211	830	1041	37.3
44.8	12 600	202	1293	1495	66.8

In this table, n is the normal force, f_r is road friction, f_a is air friction, f_t is total friction, and P is the power delivered to the wheels.

Let us examine the power required to overcome road friction and air drag. The coefficient of rolling friction μ between tires and road is about 0.016. For a 1450-kg car, the weight is 14 200 N and the force of rolling friction is $\mu n = \mu w = 227\ \mathrm{N}$. As the speed of the car increases, there is a small reduction in the normal force as a result of a reduction in air pressure as air flows over the top of the car. This causes a slight reduction in the force of rolling friction f_r with increasing speed.

Now let us consider the effect of the resistive force that results from air moving past the various surfaces of the car. For large objects, the resistive force associated with air friction is proportional to the square of the speed (in meters per second) (Section 6.4) and is given by Equation 6.5:

$$f_a = \tfrac{1}{2} DA\rho v^2$$

where D is the drag coefficient, A is the cross-sectional area of the moving object, and ρ is the density of air. This expression can be used to calculate the f_a values in Table 7.3 using $D = 0.50$, $\rho = 1.293\ \mathrm{kg/m^3}$, and $A \approx 2\ \mathrm{m^2}$.

The magnitude of the total frictional force, f_t, is the sum of the rolling friction force and the air resistive force:

$$f_t = f_r + f_a \approx \text{constant} + \tfrac{1}{2} DA\rho v^2 \tag{7.19}$$

At low speeds, road resistance is the predominant resistive force, but at high speeds air drag predominates, as shown in Table 7.3. Road friction can be reduced by reducing tire flexing (increase the air pressure slightly above recommended values) and using radial tires. Air drag can be reduced by using a smaller cross-sectional area and streamlining the car. Although driving a car with the windows open does create more air drag, resulting in a 3% decrease in mileage, driving with the windows closed and the air conditioner running results in a 12% decrease in mileage.

The total power needed to maintain a constant speed v is $f_t v$, and this is the power that must be delivered to the wheels. For example, from Table 7.3 we see that at $v = 26.8\ \mathrm{m/s}$, the required power is

$$P = f_t v = (683\ \mathrm{N})\left(26.8\ \frac{\mathrm{m}}{\mathrm{s}}\right) = 18.3\ \mathrm{kW}$$

This can be broken into two parts: (1) the power needed to overcome road friction, $f_r v$, and (2) the power needed to overcome air drag, $f_a v$. At $v = 26.8\ \mathrm{m/s}$,

these have the values

$$P_r = f_r v = (218 \text{ N}) \left(26.8 \frac{\text{m}}{\text{s}}\right) = 5.84 \text{ kW}$$

$$P_a = f_a v = (465 \text{ N}) \left(26.8 \frac{\text{m}}{\text{s}}\right) = 12.5 \text{ kW}$$

Note that $P = P_r + P_a$.

On the other hand, at $v = 44.8$ m/s (100 mi/h), we find that $P_r = 9.05$ kW, $P_a = 57.9$ kW, and $P = 67.0$ kW. This shows the importance of air drag at high speeds.

EXAMPLE 7.16   Gas Consumed by a Compact Car

A compact car has a mass of 800 kg, and its efficiency is rated at 18%. (That is, 18% of the available fuel energy is delivered to the wheels.) Find the amount of gasoline used to accelerate the car from rest to 60 mi/h (27 m/s). Use the fact that the energy equivalent of one gallon of gasoline is 1.3×10^8 J.

Solution   The energy required to accelerate the car from rest to a speed v is its kinetic energy, $\frac{1}{2}mv^2$. For this case,

$$E = \tfrac{1}{2}mv^2 = \tfrac{1}{2}(800 \text{ kg})\left(27 \frac{\text{m}}{\text{s}}\right)^2 = 2.9 \times 10^5 \text{ J}$$

If the engine were 100% efficient, each gallon of gasoline would supply 1.3×10^8 J of energy. Since the engine is only 18% efficient, each gallon delivers only $(0.18)(1.3 \times 10^8 \text{ J}) = 2.3 \times 10^7$ J. Hence, the number of gallons used to accelerate the car is

$$\text{Number of gal} = \frac{2.9 \times 10^5 \ \dfrac{\text{J}}{}}{2.3 \times 10^7 \ \text{J/gal}} = \boxed{0.013 \text{ gal}}$$

At this rate, a gallon of gas would be used after 77 such accelerations. This demonstrates the severe energy requirements for extreme stop-and-start driving.

EXAMPLE 7.17   Power Delivered to Wheels

Suppose the car described in Example 7.16 has a mileage rating of 35 mi/gal when traveling at 60 mi/h. How much power is delivered to the wheels?

Solution   The car consumes 60/35 = 1.7 gal/h. Using the fact that each gallon is equivalent to 1.3×10^8 J, we find that the total power used is

$$P = \frac{(1.7 \text{ gal/h})(1.3 \times 10^8 \text{ J/gal})}{3.6 \times 10^3 \text{ s/h}}$$

$$= \frac{2.2 \times 10^8 \text{ J}}{3.6 \times 10^3 \text{ s}} = 62 \text{ kW}$$

Since 18% of the available power is used to propel the car, the power delivered to the wheels is $(0.18)(62 \text{ kW}) = 11$ kW. This is about one-half the value obtained for the large 1450-kg car discussed in the text. Size is clearly an important factor in power-loss mechanisms.

EXAMPLE 7.18   Car Accelerating Up a Hill

Consider a car of mass m accelerating up a hill, as in Figure 7.19. Assume that the magnitude of the resistive force is

$$|\mathbf{f}| = (218 + 0.70v^2) \text{ N}$$

where v is the speed in meters per second. Calculate the power the engine must deliver to the wheels.

Solution   The forces on the car are shown in Figure 7.19, where $\mathbf{F}$ is the force of static friction that propels the car and the remaining forces have their usual meaning. Newton's

second law applied to the motion along the road surface gives

$$\sum F_x = F - |\mathbf{f}| - mg \sin\theta = ma$$

$$F = ma + mg \sin\theta + |\mathbf{f}|$$

$$= ma + mg \sin\theta + (218 + 0.70v^2)$$

Therefore, the power required for propulsion is

$$P = Fv = mva + mvg \sin\theta + 218v + 0.70v^3$$

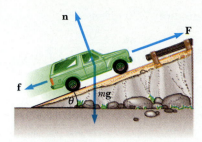

FIGURE 7.19 (Example 7.18)

nally, the term $0.70v^3$ is the power needed to overcome air drag.

If we take $m = 1450$ kg, $v = 27$ m/s ($= 60$ mi/h), $a = 1.0$ m/s^2, and $\theta = 10°$, the various terms in P are calculated to be

$$mva = (1450 \text{ kg})(27 \text{ m/s})(1.0 \text{ m/s}^2)$$
$$= 39 \text{ kW} = 52 \text{ hp}$$

$$mvg \sin \theta = (1450 \text{ kg})(27 \text{ m/s})(9.80 \text{ m/s}^2)(\sin 10°)$$
$$= 67 \text{ kW} = 89 \text{ hp}$$

$$218v = 218(27) = 5.9 \text{ kW} = 7.9 \text{ hp}$$

$$0.70v^3 = 0.70(27)^3 = 14 \text{ kW} = 18 \text{ hp}$$

where mva represents the power the engine must deliver to accelerate the car. If the car moves at constant speed, this term is zero and the power requirement is reduced. The term $mvg \sin \theta$ is the power required to overcome the force of gravity as the car moves up the incline. This term would be zero for motion on a horizontal surface. The term $218v$ is the power required to counterbalance rolling friction. Fi-

Hence, the total power required is 126 kW, or 167 hp. Note that the power requirements for traveling at constant speed on a horizontal surface are only 20 kW, or 26 hp (the sum of the last two terms). Furthermore, if the mass is halved (as in compact cars), the power required is also reduced by almost the same factor.

*7.7 KINETIC ENERGY AT HIGH SPEEDS

The laws of Newtonian mechanics are valid only for describing the motion of particles moving at speeds that are small compared with the speed of light in a vacuum, c ($\approx 3 \times 10^8$ m/s). When the particle speeds are comparable to c, the equations of Newtonian mechanics must be replaced by the more general equations predicted by the theory of relativity. One consequence of the theory of relativity is that the kinetic energy of a particle of mass m moving with a speed v is no longer given by $K = mv^2/2$. Instead, one must use the relativistic form of the kinetic energy:

$$K = mc^2 \left(\frac{1}{\sqrt{1 - (v/c)^2}} - 1 \right) \qquad (7.20)$$

Relativistic kinetic energy

According to this expression, speeds greater than c are not allowed because K would be imaginary for $v > c$. Furthermore, as v approaches c, K approaches ∞. This is consistent with experimental observations on subatomic particles, which have shown that no particles travel at speeds greater than c. (That is, c is the ultimate speed.) From the point of view of the work-energy theorem, v can only approach c, since it would take an infinite amount of work to attain the speed $v = c$.

All formulas in the theory of relativity must reduce to those in Newtonian mechanics at low particle speeds. It is instructive to show that this is the case for the kinetic energy relationship by analyzing Equation 7.20 when v is small compared to c. In this case, we expect that K should reduce to the Newtonian expression. We can check this by using the binomial expansion applied to the quantity $[1 - (v/c)^2]^{-1/2}$, with $v/c \ll 1$. If we let $x = (v/c)^2$, the expansion gives

$$\frac{1}{(1 - x)^{1/2}} = 1 + \frac{x}{2} + \frac{3}{8}x^2 + \cdots$$

Making use of this expansion in Equation 7.20 gives

$$K = mc^2 \left(1 + \frac{v^2}{2c^2} + \frac{3}{8} \frac{v^4}{c^4} + \cdots - 1 \right)$$

$$= \frac{1}{2} mv^2 + \frac{3}{8} m \frac{v^4}{c^2} + \cdots$$

$$\approx \frac{1}{2} mv^2 \quad \text{for} \quad \frac{v}{c} \ll 1$$

Thus, we see that the relativistic kinetic energy expression does indeed reduce to the Newtonian expression for speeds that are small compared with *c*. We shall return to the subject of relativity in more depth in Chapter 39.

SUMMARY

The **work** done by a *constant* force **F** acting on a particle is defined as the product of the component of the force in the direction of the particle's displacement and the magnitude of the displacement. If **F** makes an angle θ with the displacement **s**, the work done by **F** is

$$W \equiv Fs \cos \theta \tag{7.1}$$

The **scalar**, or dot, **product** of two vectors **A** and **B** is defined by the relationship

$$\mathbf{A} \cdot \mathbf{B} \equiv AB \cos \theta \tag{7.3}$$

where the result is a scalar quantity and θ is the angle between the directions of the two vectors. The scalar product obeys the commutative and distributive laws.

The **work** done by a *varying* force acting on a particle moving along the *x* axis from x_i to x_f is

$$W \equiv \int_{x_i}^{x_f} F_x \, dx \tag{7.7}$$

where F_x is the component of force in the *x* direction. If there are several forces acting on the particle, the net work done by all forces is the sum of the individual amounts of work done by each force.

The **kinetic energy** of a particle of mass *m* moving with a speed *v* (where *v* is small compared with the speed of light) is

$$K \equiv \tfrac{1}{2} mv^2 \tag{7.14}$$

The **work-energy theorem** states that the net work done on a particle by external forces equals the change in kinetic energy of the particle:

$$W_{\text{net}} = K_f - K_i = \tfrac{1}{2} mv_f^2 - \tfrac{1}{2} mv_i^2 \tag{7.16}$$

The **instantaneous power** is defined as the time rate of energy transfer. If an agent applies a force **F** to an object moving with a velocity **v**, the power delivered by that agent is

$$P \equiv \frac{dW}{dt} = \mathbf{F} \cdot \mathbf{v} \tag{7.18}$$

QUESTIONS

1. When a particle rotates in a circle, a central force acts on it directed toward the center of rotation. Why is it that this force does no work on the particle?
2. Is there any direction associated with the dot product of two vectors?
3. If the dot product of two vectors is positive, does this imply that the vectors must have positive rectangular components?
4. As the load on a spring hung vertically is increased, one would not expect the F_s versus x curve to always remain linear as in Figure 7.8d. Explain qualitatively what you would expect for this curve as m is increased.
5. Can kinetic energy be negative? Explain.
6. If the speed of a particle is doubled, what happens to its kinetic energy?
7. What can be said about the speed of a particle if the net work done on it is zero?
8. Can the average power ever equal the instantaneous power? Explain.
9. In Example 7.18, does the required power increase or decrease as the force of friction is reduced?
10. An automobile sales representative claims that a "souped-up" 300-hp engine is a necessary option in a compact car (instead of a conventional 130-hp engine). Suppose you intend to drive the car within speed limits ($\leqslant 55$ mi/h) and on flat terrain. How would you counter this sales pitch?
11. One bullet has twice the mass of a second bullet. If both are fired so that they have the same speed, which has more kinetic energy? What is the ratio of the kinetic energies of the two bullets?

12. When a punter kicks a football, is he doing any work on the ball while his toe is in contact with it? Is he doing any work on the ball after it loses contact with his toe? Are there any forces doing work on the ball while it is in flight?
13. Discuss the work done by a pitcher throwing a baseball. What is the approximate distance through which the force acts as the ball is thrown?
14. Estimate the time it takes you to climb a flight of stairs. Then approximate the power required to perform this task. Express your answer in horsepower.
15. Cite two examples in which a force is exerted on an object without doing any work on the object.
16. Two sharpshooters fire 0.30-caliber rifles using identical shells. The barrel of rifle A is 2.00 cm longer than that of rifle B. Which rifle will have the higher muzzle speed? (*Hint:* The force of the expanding gases in the barrel accelerates the bullets.)
17. As a simple pendulum swings back and forth, the forces acting on the suspended mass are the force of gravity, the tension in the supporting cord, and air resistance. (a) Which of these forces, if any, does no work on the pendulum? (b) Which of these forces does negative work at all times during its motion? (c) Describe the work done by the force of gravity while the pendulum is swinging.
18. The kinetic energy of an object depends on the frame of reference in which its motion is measured. Give an example to illustrate this point.

PROBLEMS

Review Problem

Two constant forces act on a 5.0-kg object moving in the *xy* plane as shown in the figure. The force F_1 has a magnitude of 25 N and makes an angle of 35° with the *x* axis, while the force F_2 has a magnitude of 42 N and makes an angle of 150° with the *x* axis. At $t = 0$, the object is at the origin and has a velocity of $(4.0i + 2.5j)$ m/s. Find (a) the components of the applied forces and expressions for the applied forces in unit vector notation; (b) a graphical solution for the resultant force on the object; (c) the components of the resultant force and an expression for the resultant force in unit vector notation; (d) the magnitude and direction of the object's acceleration, and an expression for the acceleration in unit vector notation; (e) the components of the velocity and an expression for the velocity in unit vector notation at $t = 3.0$ s; (f) the coordi-

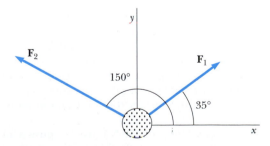

nates and position vector of the object at $t = 3.0$ s; (g) the kinetic energy of the object at $t = 3.0$ s; and (h) the net work done on the object treating it as a particle as it moves from the origin to its location at $t = 5.0$ s.

Section 7.1   Work Done by a Constant Force

1. If a person lifts a 20.0-kg bucket from a well and does 6.00 kJ of work, how deep is the well? Assume the speed of the bucket remains constant as it is lifted.

2. A raindrop ($m = 3.35 \times 10^{-5}$ kg) falls vertically at constant speed under the influence of gravity and air resistance. After the drop has fallen 100 m, what is (a) the work done by gravity and (b) the energy dissipated by air resistance?

3. A block of mass 2.5 kg is pushed 2.2 m along a frictionless horizontal table by a constant 16.0-N force directed 25° below the horizontal. Determine the work done by (a) the applied force, (b) the normal force exerted by the table, (c) the force of gravity, and (d) the net force on the block.

4. Two balls having masses $m_1 = 10.0$ kg and $m_2 = 8.0$ kg hang from a frictionless pulley, as in Figure P7.4. (a) Determine the work done by the force of gravity on each ball separately as the 10.0-kg mass is displaced downward by 0.50 m. (b) What is the total work done on each ball, including the work done by the force of the string? (c) Comment on any relationship you have discovered connecting these quantities.

m_1

m_2

FIGURE P7.4

5. A cheerleader lifts his 50.0-kg partner straight up off the ground a distance of 0.60 m before releasing her. If he does this 20 times, how much work has he done?

6. A team of dogs drags a 100-kg sled 2.0 km over a horizontal surface at a constant speed. If the coefficient of friction between sled and snow is 0.15, find (a) the work done by the dogs and (b) the energy lost due to friction.

7. A horizontal force of 150 N is used to push a 40.0-kg box 6.00 m on a rough, horizontal surface. If the box moves at constant speed, find (a) the work done by the 150-N force, (b) the energy lost due to friction, and (c) the coefficient of kinetic friction.

8. A 15-kg block is dragged over a rough, horizontal surface by a 70-N force acting 20° above the horizontal. The block is displaced 5.0 m, and the coefficient of kinetic friction is 0.30. Find the work done by (a) the 70-N force, (b) the normal force, and (c) the force of gravity. (d) What is the energy lost due to friction?

8A. A block of mass m is dragged over a rough, horizontal surface by a force F acting at an angle θ above the horizontal. The block is displaced a distance d, and the coefficient of kinetic friction is μ_k. Find the work done by (a) the force F, (b) the normal force, and (c) the force of gravity. (d) What is the energy lost due to friction?

9. If you push a 40.0-kg crate at a constant speed of 1.40 m/s across a horizontal floor ($\mu_k = 0.25$), at what rate (a) is work being done on the crate by you and (b) is energy dissipated by the frictional force?

10. Batman, whose mass is 80.0 kg, is holding on to the free end of a 12.0-m rope, the other end of which is fixed to a tree limb above. He is able to get the rope in motion as only Batman knows how, eventually getting it to swing enough that he can reach a ledge when the rope makes a 60° angle with the vertical. How much work was done against the force of gravity in this maneuver?

11. A cart loaded with bricks has a total mass of 18.0 kg and is pulled at constant speed by a rope. The rope is inclined at 20.0° above the horizontal, and the cart moves 20.0 m on a horizontal surface. The coefficient of kinetic friction between ground and cart is 0.500. (a) What is the tension in the rope? (b) How much work is done on the cart by the rope? (c) What is the energy lost due to friction?

11A. A cart loaded with bricks has a total mass m and is pulled at constant speed by a rope. The rope is inclined at an angle θ above the horizontal, and the cart moves a distance d on a horizontal surface. The coefficient of kinetic friction between ground and cart is μ_k. (a) What is the tension in the rope? (b) How much work is done on the cart by the rope? (c) What is the energy lost due to friction?

Section 7.2   The Scalar Product of Two Vectors

12. For $\mathbf{A} = 4\mathbf{i} + 3\mathbf{j}$ and $\mathbf{B} = -\mathbf{i} + 3\mathbf{j}$, find (a) $\mathbf{A} \cdot \mathbf{B}$ and (b) the angle between $\mathbf{A}$ and $\mathbf{B}$.

13. Vector $\mathbf{A}$ extends from the origin to a point having polar coordinates (7, 70°) and vector $\mathbf{B}$ extends from the origin to a point having polar coordinates (4, 130°). Find $\mathbf{A} \cdot \mathbf{B}$.

□ indicates problems that have full solutions available in the Student Solutions Manual and Study Guide.

13A. Vector **A** extends from the origin to a point having polar coordinates (r_1, θ_1), and vector **B** extends from the origin to a point having polar coordinates (r_2, θ_2). Find **A** · **B**.

14. Vector **A** has a magnitude of 5.00 units, and **B** has a magnitude of 9.00 units. The two vectors make an angle of 50.0° with each other. Find **A** · **B**.

15. Show that $\mathbf{A} \cdot \mathbf{B} = A_x B_x + A_y B_y + A_z B_z$. (*Hint:* Write **A** and **B** in unit vector form and use Eqs. 7.4 and 7.5.)

16. For **A** = 3**i** + **j** − **k**, **B** = −**i** + 2**j** + 5**k**, and **C** = 2**j** − 3**k**, find **C** · (**A** − **B**).

17. A force **F** = (6**i** − 2**j**) N acts on a particle that undergoes a displacement **s** = (3**i** + **j**) m. Find (a) the work done by the force on the particle and (b) the angle between **F** and **s**.

18. Vector **A** is 2.0 units long and points in the positive *y* direction. Vector **B** has a negative *x* component 5.0 units long, a positive *y* component 3 units long, and no *z* component. Find **A** · **B** and the angle between the vectors.

19. A force **F** = (3.00**i** + 4.00**j**) N acts on a particle. The angle between **F** and the displacement vector **s** is 32.0°, and 100.0 J of work is done by **F**. Find **s**.

20. Find the angle between **A** = −5**i** − 3**j** + 2**k** and **B** = −2**j** − 2**k**.

21. Using the definition of the scalar product, find the angles between (a) **A** = 3**i** − 2**j** and **B** = 4**i** − 4**j**; (b) **A** = −2**i** + 4**j** and **B** = 3**i** − 4**j** + 2**k**; (c) **A** = **i** − 2**j** + 2**k** and **B** = 3**j** + 4**k**.

Section 7.3 Work Done by a Varying Force

22. A force **F** = (4.0*x***i** + 3.0*y***j**) N acts on a particle as the object moves in the *x* direction from the origin to *x* = 5.0 m. Find the work done on the object by the force.

23. A particle is subject to a force F_x that varies with position as in Figure P7.23. Find the work done by the force on the body as it moves (a) from *x* = 0 to *x* = 5.0 m, (b) from *x* = 5.0 m to *x* = 10 m, and (c) from *x* = 10 m to *x* = 15 m. (d) What is the total work done by the force over the distance *x* = 0 to *x* = 15 m?

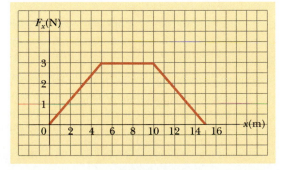

FIGURE P7.23

24. The force acting on a particle varies as in Figure P7.24. Find the work done by the force as the particle moves (a) from *x* = 0 to *x* = 8.0 m, (b) from *x* = 8.0 m to *x* = 10 m, and (c) from *x* = 0 to *x* = 10 m.

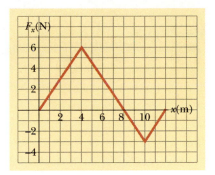

FIGURE P7.24

25. An archer pulls her bow string back 0.400 m by exerting a force that increases uniformly from zero to 230 N. (a) What is the equivalent spring constant of the bow? (b) How much work is done in pulling the bow?

25A. An archer pulls her bow string back a distance *d* by exerting a force that increases uniformly from zero to *F*. (a) What is the equivalent spring constant of the bow? (b) How much work is done in pulling the bow?

26. A 100-g bullet is fired from a rifle having a barrel 0.60 m long. Assuming the origin is placed where the bullet begins to move, the force (in newtons) exerted on the bullet by the expanding gas is $15\,000 + 10\,000x - 25\,000x^2$ where *x* is in meters. (a) Determine the work done by the gas on the bullet as the bullet travels the length of the barrel. (b) If the barrel is 1.00 m long, how much work is done, and how does this value compare to the work calculated in (a)?

27. A 6000-kg freight car rolls along rails with negligible friction. The car is brought to rest by a combination of two coiled springs, as illustrated in Figure P7.27. Both springs obey Hooke's law, with $k_1 = 1600$ N/m and $k_2 = 3400$ N/m. After the first spring compresses a distance of 30.0 cm, the second spring (acting with the first) increases the force so that there is additional compression, as shown in the graph. If the car is brought to rest 50.0 cm after first contacting the two-spring system, find the car's initial speed.

28. A marine in the jungle finds himself in the middle of a swamp. The force F_x he must exert in the *x* direction as he struggles to get out is $F_x = (1000 - 50.0x)$ N, where *x* is in meters. (a) Sketch the graph of F_x versus *x*. (b) What is the average force he exerts in traveling from 0 to *x*? (c) If he travels *x* = 20.0 m to

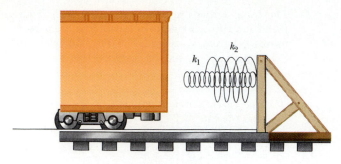

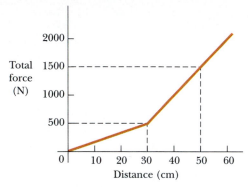

FIGURE P7.27

get completely free of the swamp, how much energy does he expend against the swamp?

29. A small mass *m* is pulled to the top of a frictionless half-cylinder (of radius *R*) by a cord passing over the top of the cylinder, as illustrated in Figure P7.29. (a) If the mass moves at a constant speed, show that $F = mg \cos \theta$. (*Hint:* If the mass moves at constant speed, the component of its acceleration tangent to the cylinder must be zero at all times.) (b) By directly integrating $W = \int \mathbf{F} \cdot d\mathbf{s}$, find the work done in moving the mass at constant speed from the bottom to the top of the half-cylinder.

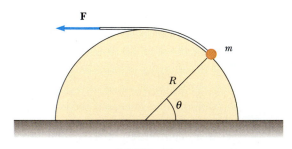

FIGURE P7.29

30. The force required to stretch a Hooke's-law spring varies from zero to 50.0 N as we stretch the spring by

moving one end 12.0 cm from its unstressed position. (a) Find the force constant of the spring. (b) Find the work done in stretching the spring.

31. If it takes 4.00 J of work to stretch a Hooke's-law spring 10.0 cm from its unstressed length, determine the extra work required to stretch it an additional 10.0 cm.

31A. If it takes work *W* to stretch a Hooke's-law spring a distance *d* from its unstressed length, determine the extra work required to stretch it an additional distance *d*.

Section 7.4   Kinetic Energy and the Work-Energy Theorem

32. A 0.600-kg particle has a speed of 2.00 m/s at point *A* and kinetic energy of 7.50 J at *B*. What is (a) its kinetic energy at *A*? (b) its speed at *B*? (c) the total work done on the particle as it moves from *A* to *B*?

33. A picture tube in a certain television set is 36 cm long. The electrical force accelerates an electron in the tube from rest to 1% the speed of light over this distance. Determine (a) the kinetic energy of the electron as it strikes the screen at the end of the tube, (b) the magnitude of the average electrical force acting on the electron over this distance, (c) the magnitude of the average acceleration of the electron over this distance, and (d) the time of flight.

34. A 7.00-kg bowling ball moves at 3.00 m/s. How fast must a 46-g golf ball move so that the two balls have the same kinetic energy?

35. A mechanic pushes a 2500-kg car from rest to a speed *v*, doing 5000 J of work in the process. During this time, the car moves 25.0 m. Neglecting friction between car and road, (a) what is the final speed, **v**, of the car? (b) What is the horizontal force exerted on the car?

35A. A mechanic pushes a car of mass *m* from rest to a speed *v*, doing work *W* in the process. During this time, the car moves a distance *d*. Neglecting friction between car and road, (a) what is the final speed, *v*, of the car? (b) What is the horizontal force exerted on the car?

36. A 3.0-kg mass has an initial velocity $\mathbf{v}_0 = (6.0\mathbf{i} - 2.0\mathbf{j})$ m/s. (a) What is its kinetic energy at this time? (b) Find the change in its kinetic energy if its velocity changes to $(8.0\mathbf{i} + 4.0\mathbf{j})$ m/s. (*Hint:* Remember that $v^2 = \mathbf{v} \cdot \mathbf{v}$.)

37. A 40-kg box initially at rest is pushed 5.0 m along a rough, horizontal floor with a constant applied horizontal force of 130 N. If the coefficient of friction between box and floor is 0.30, find (a) the work done by the applied force, (b) the energy lost due to friction, (c) the change in kinetic energy of the box, and (d) the final speed of the box.

38. A 4.0-kg particle is subject to a force that varies with

position as shown in Figure P7.23. The particle starts from rest at $x = 0$. What is its speed at (a) $x = 5.0$ m, (b) $x = 10$ m, (c) $x = 15$ m?

39. A 15.0-g bullet is accelerated in a rifle barrel 72.0 cm long to a speed of 780 m/s. Use the work-energy theorem to find the average force exerted on the bullet while it is being accelerated.

40. A bullet with a mass of 5.00 g and a speed of 600 m/s penetrates a tree to a depth of 4.00 cm. (a) Use energy considerations to find the average frictional force that stops the bullet. (b) Assuming that the frictional force is constant, determine how much time elapsed between the moment the bullet entered the tree and the moment it stopped.

40A. A bullet with a mass m and speed v penetrates a tree to a depth d. (a) Use energy considerations to find the average frictional force that stops the bullet. (b) Assuming that the frictional force is constant, determine how much time elapsed between the moment the bullet entered the tree and the moment it stopped.

41. A block of mass m hangs on the end of a cord and is connected to a block of mass M by the pulley arrangement shown in Figure P7.41. Using energy considerations, (a) find an expression for the speed of m as a function of the distance it has fallen. Assume that the blocks are initially at rest and there is no friction. (b) Repeat (a) assuming sliding friction (coefficient μ_k) between M and the table. (c) Show that the result obtained in (b) reduces to that obtained in (a) in the limit as μ_k goes to zero.

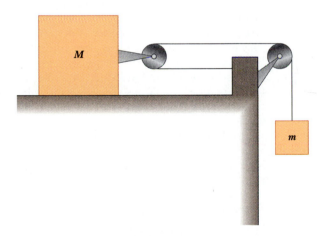

FIGURE P7.41

42. A 5.0-kg steel ball is dropped onto a copper plate from a height of 10.0 m. If the ball leaves a dent 0.32 cm deep, what is the average force exerted on the ball by the plate during the impact?

 43. An Atwood's machine (Fig. 5.12) supports masses of 0.20 kg and 0.30 kg. The masses are held at rest beside each other and then released. Neglecting friction, what is the speed of each mass the instant they have both moved 0.40 m?

44. A 2.0-kg block is attached to a spring of force constant 500 N/m as in Figure 7.8. The block is pulled 5.0 cm to the right of equilibrium and released from rest. Find the speed of the block as it passes through equilibrium if (a) the horizontal surface is frictionless and (b) the coefficient of friction between block and surface is 0.35.

45. A sled of mass m is given a kick on a frozen pond, imparting to it an initial speed $v_i = 2$ m/s. The coefficient of kinetic friction between sled and ice is $\mu_k = 0.10$. Use energy considerations to find the distance the sled moves before stopping.

46. A block of mass 12.0 kg slides from rest down a frictionless 35.0° incline and is stopped by a strong spring with $k = 3.00 \times 10^4$ N/m. The block slides 3.00 m from the point of release to the point where it comes to rest against the spring. When the block comes to rest, how far has the spring been compressed?

47. A crate of mass 10.0 kg is pulled up a rough incline with an initial speed of 1.50 m/s. The pulling force is 100 N parallel to the incline, which makes an angle of 20.0° with the horizontal. The coefficient of kinetic friction is 0.400, and the crate is pulled 5.00 m. (a) How much work is done by gravity? (b) How much energy is lost due to friction? (c) How much work is done by the 100-N force? (d) What is the change in kinetic energy of the crate? (e) What is the speed of the crate after being pulled 5.00 m?

48. A block of mass 0.60 kg slides 6.0 m down a frictionless ramp inclined at 20° to the horizontal. It then travels on a rough horizontal surface where $\mu_k = 0.50$. (a) What is the speed of the block at the end of the incline? (b) What is its speed after traveling 1.00 m on the rough surface? (c) What distance does it travel on this horizontal surface before stopping?

49. A 4.00-kg block is given an initial speed of 8.00 m/s at the bottom of a 20.0° incline. The frictional force that retards its motion is 15.0 N. (a) If the block is directed up the incline, how far does it move before stopping? (b) Will it slide back down the incline?

50. A time-varying net force acting on a 4.0-kg particle causes the particle to have a displacement given by $x = 2.0t - 3.0t^2 + 1.0t^3$, where x is in meters and t is in seconds. Find the work done on the particle in the first 3.0 s of motion.

51. A 3.00-kg block is moved up a 37.0° incline under the action of a constant horizontal force of 40.0 N. The coefficient of kinetic friction is 0.100, and the block is displaced 2.00 m up the incline. Calculate (a) the work done by the 40.0-N force, (b) the work done by gravity, (c) the energy lost due to friction,

and (d) the change in kinetic energy of the block. (*Note:* The applied force is *not* parallel to the incline.)

52. A 4.0-kg block attached to a string 2.0 m in length rotates in a circle on a horizontal surface. (a) If the surface is frictionless, identify all the forces on the block and show that the work done by each force is zero for any displacement of the block. (b) If the coefficient of friction between block and surface is 0.25, find the energy lost due to friction in each revolution.

Section 7.5   Power

53. A 700-N marine in basic training climbs a 10.0-m vertical rope at a constant speed in 8.00 s. What is his power output?

54. Water flows over a section of Niagara Falls at a rate of 1.2×10^6 kg/s and falls 50 m. How many 60-W bulbs can be lit with this power?

55. A 650-kg elevator starts from rest. It moves upward for 3.00 s with constant acceleration until it reaches its cruising speed, 1.75 m/s. (a) What is the average power of the elevator motor during this period? (b) How does this power compare with its power while it moves at its cruising speed?

56. A 200-kg crate is pulled along a level surface by an engine. The coefficient of friction between crate and surface is 0.40. (a) How much power must the engine deliver to move the crate at 5.0 m/s? (b) How much work is done by the engine in 3.0 min?

57. A 1500-kg car accelerates uniformly from rest to 10 m/s in 3.0 s. Find (a) the work done on the car in this time, (b) the average power delivered by the engine in the first 3.0 s, and (c) the instantaneous power delivered by the engine at $t = 2.0$ s.

57A. A car of mass m accelerates uniformly from rest to a speed v in a time t. Find (a) the work done on the car in this time, (b) the average power delivered by the engine during this time, and (c) the instantaneous power delivered by the engine at some time less than t, neglecting drag.

58. A certain automobile engine delivers 30.0 hp $(2.24 \times 10^4$ W) to its wheels when moving at 27.0 m/s $(\approx 60$ mi/h). What is the resistive force acting on the automobile at that speed?

59. An outboard motor propels a boat through the water at 10.0 mi/h. The water resists the forward motion of the boat with a force of 15.0 lb. How much power is delivered through the propeller?

60. A car of weight 2500 N operating at a rate of 130 kW develops a maximum speed of 31 m/s on a level, horizontal road. Assuming that the resistive force (due to friction and air resistance) remains constant, (a) what is the car's maximum speed on an incline of 1 in 20 (i.e., if θ is the angle of the incline with the horizontal, sin $\theta = 1/20$)? (b) What is its power output on a 1-in-10 incline if the car is traveling at 10 m/s?

61. If a certain horse can maintain 1.0 hp of output for 2.0 h, how many 70.0-kg bundles of shingles can the horse hoist (via some pulley arrangement) to the roof of a house 8.0 m tall, assuming 70% efficiency?

62. A force **F** acts on a particle of mass *m*. The particle starts at rest at $t = 0$. (a) Show that the instantaneous power delivered by the force at any time t is $(F^2/m)t$. (b) If $F = 20.0$ N and $m = 5.00$ kg, what is the power delivered at $t = 3.00$ s?

*Section 7.6   Energy and the Automobile

63. The horsepower required to keep an aerodynamic car of mass 1500 kg moving at 30.0 mi/h is about 10.0 hp. Assuming that the total retarding force due to friction, air resistance, and so forth is proportional to the square of the car's speed, what is the horsepower required to keep the car moving at 60.0 mi/h?

64. A passenger car carrying two people has a fuel economy of 25 mi/gal. It travels 3000 miles. A jet airplane making the same trip with 150 passengers has a fuel economy of 1.0 mi/gal. Compare the fuel consumed per passenger for the two modes of transportation.

65. A compact car of mass 900 kg has an overall motor efficiency of 15%. (That is, 15% of the energy supplied by the fuel is transformed into kinetic energy of the car.) (a) If burning one gallon of gasoline supplies 1.34×10^8 J of energy, find the amount of gasoline used in accelerating the car from rest to 55 mph. (b) How many such accelerations will one gallon provide? (c) The mileage claimed for the car is 38 mi/gal at 55 mph. What power is delivered to the wheels (to overcome frictional effects) when the car is driven at this speed?

66. Suppose the empty car described in Table 7.3 has a fuel economy of 6.40 km/liter (15 mi/gal) when traveling at 26.8 m/s (60 mi/h). Assuming constant efficiency, determine the fuel economy of the car if the total mass of passengers plus driver is 350 kg.

67. When an air conditioner is added to the car described in Problem 66, the additional output power required to operate the air conditioner is 1.54 kW. If the fuel economy is 6.40 km/liter without the air conditioner, what is it when the air conditioner is operating?

*Section 7.7   Kinetic Energy at High Speeds

68. An electron moves with a speed of $0.995c$. (a) What is its kinetic energy? (b) If you use the classical expression to calculate its kinetic energy, what percentage error would result?

69. A proton in a high-energy accelerator moves with a

speed of $c/2$. Use the work-energy theorem to find the work required to increase its speed to (a) $0.75c$, (b) $0.995c$.

ADDITIONAL PROBLEMS

70. Diatomic molecules exert attractive forces on each other at large distances and replusive forces at short distances. For many molecules the Lennard-Jones law is a good approximation to the magnitude of these intermolecular forces:

$$F = F_0 \left[2\left(\frac{\sigma}{r}\right)^{13} - \left(\frac{\sigma}{r}\right)^7 \right]$$

where r is the center-to-center distance between the atoms in the molecule, σ is a length parameter, and F_0 is the force when $r = \sigma$. For an oxygen molecule, $F_0 = 9.6 \times 10^{-11}$ N and $\sigma = 3.5 \times 10^{-10}$ m. Determine the work done by this force from $r = 4.0 \times 10^{-10}$ m to $r = 9.0 \times 10^{-10}$ m.

71. A bartender slides a bottle of rye on the horizontal counter to a customer 7.0 m away. With what speed did she release the bottle if the coefficient of sliding friction is 0.10 and the bottle comes to rest in front of the customer?

72. When a spring is stretched near its elastic limit, the spring force satisfies the equation $F = -kx + \beta x^3$. If $k = 10$ N/m and $\beta = 100$ N/m^3, calculate the work done by this force when the spring is stretched 0.10 m.

73. A particle of mass m moves with a constant acceleration $\mathbf{a}$. If the initial position vector and velocity of the particle are $\mathbf{r}_0$ and $\mathbf{v}_0$, respectively, show that its speed v at any time satisfies the equation

$$v^2 = v_0^2 + 2\mathbf{a} \cdot (\mathbf{r} - \mathbf{r}_0)$$

where $\mathbf{r}$ is the position vector of the particle at that same time.

74. The direction of an arbitrary vector $\mathbf{A}$ can be completely specified with the angles α, β, γ that the vector makes with the x, y, and z axes, respectively. If $\mathbf{A} = A_x\mathbf{i} + A_y\mathbf{j} + A_z\mathbf{k}$, (a) find expressions for $\cos\alpha$, $\cos\beta$, and $\cos\gamma$ (these are known as *direction cosines*) and (b) show that these angles satisfy the relationship $\cos^2\alpha + \cos^2\beta + \cos^2\gamma = 1$. (*Hint:* Take the scalar product of $\mathbf{A}$ with $\mathbf{i}$, $\mathbf{j}$, and $\mathbf{k}$ separately.)

75. A 4.0-kg particle moves along the x axis. Its position varies with time according to $x = t + 2.0t^3$, where x is in meters and t is in seconds. Find (a) the kinetic energy at any time t, (b) the acceleration of the particle and the force acting on it at time t, (c) the power being delivered to the particle at time t, and (d) the work done on the particle in the interval $t = 0$ to $t = 2.0$ s.

76. An Atwood's machine has a 3.00-kg mass and a 2.00-kg mass at the ends of the string (Fig. 5.12). The

2.00-kg mass is released from rest on the floor, 4.00 m below the 3.00-kg mass. (a) If the pulley is frictionless, what will be the speed of the masses when they pass each other? (b) Suppose that the pulley does not rotate and the string must slide over it. If the total frictional force between the pulley and string is 5.00 N, what are their speeds when the masses pass each other?

77. A 2100-kg pile driver is used to drive a steel I beam into the ground. The pile driver falls 5.0 m before contacting the beam, and it drives the beam 12 cm into the ground before coming to rest. Using energy considerations, calculate the average force the beam exerts on the pile driver while the pile driver is brought to rest.

78. A particle of mass m is attached to two identical springs on a horizontal frictionless table as in Figure P7.78. Both springs have spring constant k. (a) If the particle is pulled a distance x along a direction perpendicular to the initial configuration of the springs, show that the force exerted on the particle due to the springs is

$$\mathbf{F} = -2kx\left(1 - \frac{L}{\sqrt{x^2 + L^2}}\right)\mathbf{i}$$

(b) Determine the amount of work done by this force in moving the particle from $x = A$ to $x = 0$.

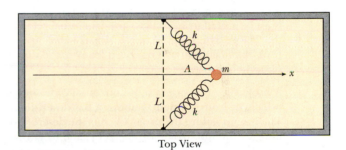

Top View

FIGURE P7.78

79. A 200-g block is pressed against a spring of force constant 1.40 kN/m until the block compresses the spring 10.0 cm. The spring rests at the bottom of a ramp inclined at $60.0°$ to the horizontal. Use energy considerations to determine how far up the incline the block moves before it stops (a) if there is no friction between block and ramp and (b) if the coefficient of kinetic friction is 0.400.

80. A block of mass m is attached to a massless spring of force constant k as in Figure 7.8. The spring is compressed a distance d from its position and released from rest. (a) If the block comes to rest when it first reaches the equilibrium position, what is the coeffi-

cient of friction between block and surface? (b) If the block first comes to rest when the spring is stretched a distance $d/2$ from equilibrium, what is μ?

81. A 0.400-kg particle slides on a horizontal, circular track 1.50 m in radius. It is given an initial speed of 8.00 m/s. After one revolution, its speed drops to 6.00 m/s because of friction. (a) Find the energy lost due to friction in one revolution. (b) Calculate the coefficient of kinetic friction. (c) What is the total number of revolutions the particle makes before stopping?

82. A horizontal string is attached to a 0.25-kg mass lying on a rough, horizontal table. The string passes over a light frictionless pulley and a 0.40-kg mass is then attached to its free end. The coefficient of sliding friction between the 0.25-kg mass and table is 0.20. Use energy considerations to determine (a) the speed of the masses after each has moved 20 m from rest and (b) the mass that must be added to the 0.25-kg mass so that, given an initial velocity, the masses continue to move at a constant speed. (c) What mass must be removed from the 0.40-kg mass to accomplish the same thing as in (b)?

82A. A string is attached to a mass m_1 lying on a rough, horizontal table. The string passes over a light frictionless pulley and a mass m_2 is then attached to its free end. The coefficient of sliding friction between m_1 and table is μ_k. Use energy considerations to determine (a) the speed of the masses after each has moved a distance d from rest and (b) the mass that must be added to m_1 so that, given an initial velocity, the masses continue to move at a constant speed. (c) What mass must be removed from m_2 to accomplish the same thing as in (b)?

83. A projectile of mass m is shot horizontally with initial velocity $\mathbf{v}_i$ from a height h above a flat desert floor. The instant before the projectile hits the desert floor find (a) the work done on the projectile by gravity, (b) the change in kinetic energy of the projectile since it was fired, and (c) the kinetic energy of the projectile.

84. A cyclist and her bicycle have a combined mass of 75 kg. She coasts down a road inclined at 2.0° with the horizontal at 4.0 m/s and coasts down another road inclined at 4.0° at 8.0 m/s. She then holds on to a moving vehicle and coasts on a level road. What power must the vehicle expend to maintain her speed at 3.0 m/s? Assume that the force of air resistance is proportional to her speed, and assume that other frictional forces remain constant.

85. A 60.0-kg load is raised by a two-pulley arrangement as shown in Figure P7.85. How much work is done by the force **F** to raise the load 3.00 m if there is a frictional force of 20.0 N in each pulley? (The pulleys do not rotate, but the rope slides across each surface.)

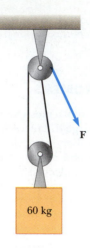

FIGURE P7.85

86. A small sphere of mass m hangs from a string of length L as in Figure P7.86. A variable horizontal force **F** is applied to the sphere in such a way that it moves slowly from the vertical position until the string makes an angle θ with the vertical. Assuming the sphere is always in equilibrium, (a) show that $F = mg \tan \theta$. (b) Show that the work done by **F** is $mgL(1 - \cos \theta)$. (*Hint:* Note that $s = L\theta$, and so $ds = L \, d\theta$.)

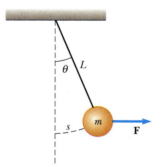

FIGURE P7.86

87. To operate one of the generators in the Grand Coulee Dam in Washington, 7.24×10^8 W of mechanical power is required. This power is provided by gravity as it performs work on the water while it falls 87 m to the generator. The kinetic energy acquired during the fall is given up turning the generator. (a) Prove that the available water power is mgh/t, where m is the mass of water falling through the height h during the time t. (b) A maximum of about 60% can be extracted and keep the water flowing. Calculate the flow rate in kilograms per second. (c) Determine the volume of water needed to power this generator for one day. (d) If the water for one day were stored in a circular lake 10 m deep, what would the radius of the lake be?

88. Power windmills turn in response to the force of high-speed drag. For a sphere moving through a fluid, the resistive force F_R is proportional to r^2v^2 where r is the radius of the sphere and v is the fluid speed. The power developed, $P = F_R v$, is proportional to r^2v^3. The power developed by a windmill can be expressed as $P = ar^2v^3$ where r is the windmill radius, v wind speed, and $a = 2.00$ W·s³/m⁵. For a home windmill with $r = 1.50$ m, calculate the power delivered to the generator if (a) $v = 8.00$ m/s and (b) $v = 24.0$ m/s. For comparison, a typical home needs about 3.0 kW of electric power. (*Note:* This representation ignores system efficiency—which is about 25%.)

89. The ball launcher in a pinball machine has a spring that has a force constant of 1.20 N/cm (Fig. P7.89). The surface on which the ball moves is inclined 10.0° with respect to the horizontal. If the spring is initially compressed 5.00 cm, find the launching speed of a 100-g ball when the plunger is released. Friction and the mass of the plunger are negligible.

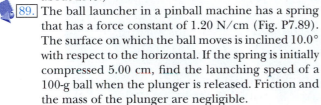

FIGURE P7.89

90. Suppose a car is modeled as a cylinder moving with a speed v, as in Figure P7.90. In a time Δt, a column of air of mass Δm must be moved a distance $v\Delta t$ and hence must be given a kinetic energy $\frac{1}{2}(\Delta m)v^2$. Using this model, show that the power loss due to air resistance is $\frac{1}{2}\rho Av^3$ and the resistive force is $\frac{1}{2}\rho Av^2$, where ρ is the density of air.

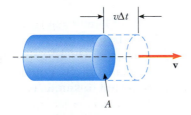

FIGURE P7.90

SPREADSHEET PROBLEMS

S1. When different weights are hung on a spring, the spring stretches to different lengths, as shown in the table below. (a) Use a spreadsheet to plot the length of the spring versus applied force. Use the spreadsheet's least-squares fitting features to determine the straight line that best fits the data. (You may not want to use all the data points.) (b) From the slope of the

least-squares fit line, find the spring constant k. (c) If the spring is extended to 105 mm, what force does it exert on the suspended weight?

F(N)	L(m)
2.0	15
4.0	32
6.0	49
8.0	64
10	79
12	98
14	112
16	126
18	149
20	175
22	190

S2. A 0.178-kg particle moves along the x axis from $x = 12.8$ m to $x = 23.7$ m under the influence of a force

$$F = \frac{375}{x^3 + 3.75x}$$

where F is in newtons and x is in meters. Use numerical integration and set up a spreadsheet to determine the total work done by this force during this displacement. Your calculations should have an accuracy of at least 2%.

S3. A box of mass m is pushed up a curved inclined plane as in Figure PS7.3. The equation of the curve is $y = 5 - x^2$. The applied force F remains constant in magnitude but changes direction so that it is always parallel to the curve. (a) Show that the work done by the applied force is

$$W = \int_a^b \mathbf{F} \cdot d\mathbf{s} = \int_a^b \frac{Fdx}{\cos\theta} = \int_a^b \frac{Fdx}{\cos[\tan^{-1}(-2x)]}$$

(b) Use a rectangular trapezoidal numerical integration scheme to calculate the work done by the applied force as the box is moved from $x = -2$ m to $x = 0$ m if $F = 100$ N.

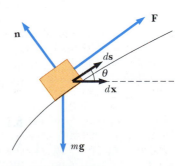

FIGURE PS7.3

CHAPTER 8

Potential Energy and Conservation of Energy

Twin Falls on the Island of Kauai, Hawaii. The gravitional potential energy of the water at the top of the falls is converted to kinetic energy at the bottom. In many locations, this mechanical energy is used to produce electrical energy. *(Bruce Byers, FPG)*

In Chapter 7 we introduced the concept of kinetic energy, which is the energy associated with the motion of an object. In this chapter we introduce another form of mechanical energy, *potential energy*, which is the energy associated with the position or configuration of an object. Potential energy can be thought of as stored energy that can be converted to kinetic energy or other forms of energy.

The potential energy concept can be used only when dealing with a special class of forces called *conservative forces*. When only internal conservative forces, such as gravitational or spring forces, act within a system, the kinetic energy gained (or lost) by the system as its members change their relative positions is compensated by an equal energy loss (or gain) in potential energy.

8.1 POTENTIAL ENERGY

An object with kinetic energy can do work on another object, as illustrated by a moving hammer driving a nail into a wall. Now we shall see that an object can also do work because of the energy resulting from its *position* in space.

As an object falls in a gravitational field, the field exerts a force on it in the direction of its motion, doing work on it, and thereby increasing its kinetic energy. Consider a brick dropped from rest directly above a nail in a board that is lying horizontally on the ground. When the brick is released, it falls toward the ground gaining speed and therefore gaining kinetic energy. Due to its position in space, the brick has potential energy (it has the *potential* to do work), which is converted into kinetic energy as it falls. When the brick reaches the ground, it does work on the nail, driving it into the board. The energy that an object has due to its position in space is called **gravitational potential energy**. It is energy held by the gravitational field and transferred to the object as it falls.

Let us now derive an expression for the gravitational potential energy of an object at a given location. To do this, consider a block of mass m at an initial height y_i above the ground, as in Figure 8.1. Neglecting air resistance, as the block falls, the only force that does work on it is the gravitational force, $m\mathbf{g}$. The work done by the gravitational force as the block undergoes a downward displacement $\mathbf{s}$ is the product of the downward force times the displacement, or

$$W_g = (m\mathbf{g}) \cdot \mathbf{s} = (-mg\mathbf{j}) \cdot (y_f - y_i)\mathbf{j} = mgy_i - mgy_f$$

We now define the quantity mgy to be the gravitational potential energy, U_g:

$$U_g \equiv mgy \qquad (8.1)$$

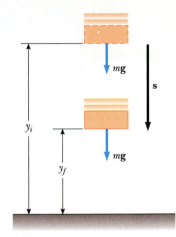

FIGURE 8.1 The work done by the gravitational force as the block falls from y_i to y_f is equal to $mgy_i - mgy_f$.

Gravitational potential energy

Thus, the gravitational potential energy associated with an object at any point in space is the product of the object's weight and its vertical coordinate. The origin of the coordinate system could be located at the surface of the Earth or at any other convenient point.

If we substitute U for the mgy terms in the expression for W_g, we have

$$W_g = U_i - U_f \qquad (8.2)$$

From this result, we see that the work done on any object by the gravitational force — that is, the energy transferred to the object from the gravitational field — is equal to the initial value of the potential energy minus the final value of potential energy. For convenience, the potential energy is often "assigned to" the object, which includes the gravitational field as part of the system. However, when this choice is made, one must be careful *not* to include the work done by the gravitational field on the object; because it is part of the system, the field no longer exerts an external force.

The units of gravitational potential energy are the same as those of work. That is, potential energy may be expressed in joules, ergs, or foot-pounds. Potential energy, like work and kinetic energy, is a scalar quantity.

Note that the gravitational potential energy associated with an object depends only on the vertical height of the object above the surface of the Earth. From this result, we see that the work done by the force of gravity on an object as it falls vertically to the Earth is the same as if it starts at the same point and slides down a frictionless incline to the Earth. Also note that Equation 8.1 is valid only for objects near the surface of the Earth, where $\mathbf{g}$ is approximately constant.[1]

In working problems involving gravitational potential energy, it is always necessary to set the gravitational potential energy equal to zero at some location. The choice of zero level is completely arbitrary, because the important quantity is the

[1] The assumption that the force of gravity is constant is a good one as long as the vertical displacement is small compared with the Earth's radius.

difference in potential energy, and this difference is independent of the choice of zero level.

It is often convenient to choose the surface of the Earth as the reference position for zero potential energy, but again this is not essential. Often, the statement of the problem suggests a convenient level to choose.

8.2 CONSERVATIVE AND NONCONSERVATIVE FORCES

Forces found in nature can be divided into two categories: conservative and nonconservative. We shall describe the properties of conservative and nonconservative forces separately in this section.

Conservative Forces

Properties of a conservative force

> A force is conservative if the work it does on a particle moving between any two points is independent of the path taken by the particle. Furthermore, the work done by a conservative force exerted on a particle moving through any closed path is zero.

The force of gravity is conservative. As we learned in the previous section, the work done by the gravitational force on an object moving between any two points near the Earth's surface is

$$W_g = mgy_i - mgy_f$$

From this we see that W_g depends only on the initial and final coordinates of the object and, hence, is independent of path. Furthermore, W_g is zero when the object moves over any closed path (where $y_i = y_f$).

We can associate a potential energy function with any conservative force. In the previous section, the potential energy function associated with the gravitational force was found to be

$$U_g \equiv mgy$$

Gravitational potential energy is the energy stored in the gravitational field when the object is lifted against the field.

Potential energy functions can be defined only for conservative forces. In general, the work, W_c, done on an object by a conservative force is equal to the initial value of the potential energy associated with the object minus the final value:

Work done by a conservative force

$$W_c = U_i - U_f \tag{8.3}$$

Another example of a conservative force is the force of a spring on an object attached to the spring, where the spring force is given by $F_s = -kx$. In the previous chapter, we learned that the work done by the spring is

$$W_s = \tfrac{1}{2}kx_i^2 - \tfrac{1}{2}kx_f^2$$

where the initial and final coordinates of the block are measured from its equilibrium position, $x = 0$. Again we see that W_s depends only on the initial and final x coordinates of the object and is zero for any closed path. The **elastic potential** energy function associated with the spring force is defined by

$$U_s \equiv \tfrac{1}{2}kx^2 \qquad (8.4)$$

Elastic potential energy

The elastic potential energy can be thought of as the energy stored in the deformed spring (one that is either compressed or stretched from its equilibrium position). To visualize this, consider Figure 8.2, which shows an undeformed spring on a frictionless, horizontal surface. When the block is pushed against the spring (Fig. 8.2b), compressing the spring a distance x, the elastic potential energy stored in the spring is $kx^2/2$. When the block is released from rest, the spring snaps back to its original length and the stored elastic potential energy is transformed into kinetic energy of the block (Fig. 8.2c). The elastic potential energy stored in the spring is zero whenever the spring is undeformed ($x = 0$). Energy is stored in the spring only when the spring is either stretched or compressed. Furthermore, the elastic potential energy is a maximum when the spring has reached its maximum compression or extension (that is, when $|x|$ is a maximum). Finally, since the elastic potential energy is proportional to x^2, we see that U_s is always positive in a deformed spring. If the spring and object connected to it are taken together as the system, then no work is done as the spring changes length because the forces are internal.

Nonconservative Forces

A force is **nonconservative** if it causes a change in mechanical energy. For example, if you move an object on a horizontal surface, returning to the same location and same state of motion, but you found that it was necessary to do a net

Properties of a nonconservative force

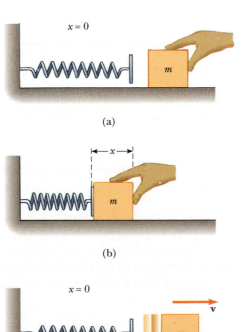

$x = 0$

m

(a)

x

m

(b)

$x = 0$

m

$\mathbf{v}$

(c)

FIGURE 8.2   (a) An undeformed spring on a frictionless horizontal surface. (b) A block of mass m is pushed against the spring, compressing the spring a distance x. (c) The block is released from rest and the elastic potential energy stored in the spring is transferred to the block in the form of kinetic energy.

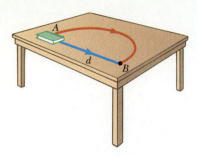

FIGURE 8.3   The loss in mechanical energy due to the force of friction depends on the path taken as the book is moved from *A* to *B*. The loss in mechanical energy is greater along the red path compared with the blue path.

amount of work on the object, then something must have dissipated that energy transferred to the object. That dissipative force is recognized as friction between the surface and object. Friction is a dissipative or "nonconservative" force. By contrast, if the object is lifted, work is required, but that energy is recovered when the object is lowered. The gravitational force is a nondissipative or "conservative" force.

Suppose you displace a book between two points on a table. If the book is displaced in a straight line between two points *A* and *B* in Figure 8.3, the loss in mechanical energy due to friction is simply $-fd$, where *d* is the distance between the points. However, if you move the book along *any other* path between the two points, the loss in mechanical energy due to friction is *greater* (in absolute value) than $-fd$. For example, the loss in mechanical energy due to friction along the semicircular path in Figure 8.3 is equal to $-f(\pi d/2)$, where *d* is the diameter of the circle.

8.3   CONSERVATIVE FORCES AND POTENTIAL ENERGY

Because work done is a function only of a particle's initial and final coordinates, we can define a **potential energy function** *U* such that the work done by a conservative force equals the decrease in the potential energy of the particle. The work done by a conservative force **F** as a particle moves along the *x* axis is[2]

$$W_c = \int_{x_i}^{x_f} F_x \, dx = -\Delta U = U_i - U_f \tag{8.5}$$

where F_x is the component of **F** in the direction of the displacement. That is, *the work done by a conservative force equals the negative of the change in the potential energy associated with that force,* where the change in the potential energy is defined as $\Delta U = U_f - U_i$. For example, the work done by the gravitational field on an object, as the object is lowered in the field, is $U_i - U_f$, which is positive, showing that energy has been transferred from the gravitational field to the object. This energy may appear as kinetic energy of the falling object or may be transferred to something else. We can also express Equation 8.5 as

$$\Delta U = U_f - U_i = -\int_{x_i}^{x_f} F_x \, dx \tag{8.6}$$

Therefore ΔU is negative when F_x and dx are in the same direction, as when an object is lowered in a gravitational field or a spring pushes an object toward equilibrium.

The term *potential energy* implies that the object has the potential, or capability, of either gaining kinetic energy or doing work when released from some point under the influence of gravity. It is often convenient to establish some particular location, x_i, as a reference point and measure all potential energy differences with respect to it. We can then define the potential energy function as

$$U_f(x) = -\int_{x_i}^{x_f} F_x \, dx + U_i \tag{8.7}$$

---

[2] For a general displacement, the work done in two or three dimensions also equals $U_i - U_f$, where $U = U(x, y, z)$. We write this formally as $W = \int_i^f \mathbf{F} \cdot d\mathbf{s} = U_i - U_f$.

Furthermore, as we discussed earlier, the value of U_i is often taken to be zero at some arbitrary reference point. It really doesn't matter what value we assign to U_i, since any nonzero value only shifts $U_f(x)$ by a constant, and only the *change* in potential energy is physically meaningful. If the conservative force is known as a function of position, we can use Equation 8.7 to calculate the change in potential energy of a body as it moves from x_i to x_f. It is interesting to note that in the one-dimensional case, a force is always conservative if it is a function of position only. This is generally not the case for motion involving two- or three-dimensional displacements.

The amount of mechanical energy dissipated by a nonconservative force depends on the path as an object moves from one position to another and can also depend on the object's speed or on other quantities. Because the work done by a nonconservative force is not simply a function of the initial and final coordinates, there is no potential energy function associated with a nonconservative force.

8.4 CONSERVATION OF ENERGY

An object held at some height h above the floor has no kinetic energy, but as we learned earlier, there is an associated gravitational potential energy equal to mgh relative to the floor if the gravitational field is included as part of the system. If the object is dropped, it falls to the floor; as it falls, its speed and thus its kinetic energy increase, while the potential energy decreases. If factors such as air resistance are ignored, whatever potential energy the object loses as it moves downward appears as kinetic energy. In other words, the sum of the kinetic and potential energies, called the *mechanical energy E,* remains constant in time. This is an example of the principle of **conservation of energy.** For the case of an object in free-fall, this principle tells us that any increase (or decrease) in potential energy is accompanied by an equal decrease (or increase) in kinetic energy.

Since the total mechanical energy E is defined as the sum of the kinetic and potential energies, we can write

$$E = K + U \tag{8.8}$$

Total mechanical energy

Therefore, we can apply conservation of energy in the form $E_i = E_f$, or

$$K_i + U_i = K_f + U_f \tag{8.9}$$

The mechanical energy of an isolated system remains constant

Conservation of energy requires that the total mechanical energy of a system remains constant in any isolated system of objects that interact only through conservative forces.

It is important to note that Equation 8.9 is valid provided no energy is added to or removed from the system. Furthermore, there must be no nonconservative forces within the system.

If more than one conservative force acts on the object, then a potential energy function is associated with each force. In such a case, we can apply the principle of conservation of mechanical energy for the system as

$$K_i + \sum U_i = K_f + \sum U_f \tag{8.10}$$

where the number of terms in the sums equals the number of conservative forces acting on the system. For example, if a mass connected to a spring oscillates vertically, two conservative forces act on it: the spring force and the force of gravity. (We discuss this situation later, in a worked example.)

If the force of gravity is the *only* force acting on an object, then the total mechanical energy of the object is constant. Therefore, the principle of conservation of energy for an object in free-fall can be written

---
The mechanical energy for a freely falling body remains constant
---

$$\tfrac{1}{2}mv_i^2 + mgy_i = \tfrac{1}{2}mv_f^2 + mgy_f \tag{8.11}$$

Likewise, if the only force acting on an object is the conservative spring force for which the potential energy is $U_s = \tfrac{1}{2}kx^2$, the principle of conservation of energy gives

---
The mechanical energy for a mass-spring system remains constant
---

$$\tfrac{1}{2}mv_i^2 + \tfrac{1}{2}kx_i^2 = \tfrac{1}{2}mv_f^2 + \tfrac{1}{2}kx_f^2 \tag{8.12}$$

CONCEPTUAL EXAMPLE 8.1

A mass is connected to a massless spring that is suspended vertically from the ceiling as in Figure 8.4. If the mass is displaced downward from its equilibrium position and released, it will oscillate up and down. If air resistance is neglected, will the total mechanical energy of the system (mass plus spring) be conserved? How many forms of potential energy are there for this situation?

Reasoning   Yes, the total mechanical energy of the system is conserved because the only forces acting are conservative: the force of gravity and the spring force. There are two forms of potential energy in this case, gravitational potential energy and elastic potential energy stored in the spring. The sum of the kinetic energy, the gravitational potential energy, and the elastic potential energy remains constant for the system.

FIGURE 8.4   (Conceptual Example 8.1) A mass *m* connected to a massless spring suspended vertically from a support. What forms of potential energy are associated with the mass-spring system when the mass is displaced downward?

CONCEPTUAL EXAMPLE 8.2

Discuss the energy transformations that occur during the pole vault event pictured in the multiflash photograph. Ignore rotational motion.

Reasoning   As an athlete runs, chemical energy in the body is converted mostly into thermal energy but also into kinetic energy. The kinetic energy of the athlete just before ascending becomes elastic potential energy stored in the bent pole at the bottom, then gravitational potential energy in the elevated body of the vaulter, and then kinetic energy as he falls. When the vaulter lands on the pad, this energy turns into additional thermal energy, mostly in the pad.

(Conceptual Example 8.2) Multiflash photograph of a pole vault event. How many forms of energy can you identify in this picture?   *(© Harold E. Edgerton, Courtesy of Palm Press Inc.)*

CONCEPTUAL EXAMPLE 8.3

Three identical balls are thrown from the top of a building, all with the same initial speed. The first ball is thrown horizontally, the second at some angle above the horizontal, and the third at some angle below the horizontal as in Figure 8.5. Neglecting air resistance, describe their motions and compare the speeds of the balls as they reach the ground.

Reasoning   The first and third balls speed up after they are thrown, while the second ball first slows down and then speeds up after reaching its peak. The paths of all three are parabolas. The three take different times to reach the ground. However, all have the same impact speed because all start with the same kinetic energy and undergo the same change in gravitational potential energy. In other words, $E_{\text{total}} = \frac{1}{2}mv^2 + mgh$ is the same for all three balls.

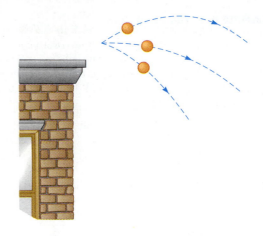

FIGURE 8.5   (Conceptual Example 8.3) Three identical balls are thrown with the same initial speed from the top of a building.

EXAMPLE 8.4   Ball in Free-Fall

A ball of mass m is dropped from a height h above the ground as in Figure 8.6. (a) Neglecting air resistance, determine the speed of the ball when it is at a height y above the ground.

Reasoning   Since the ball is in free-fall, the only force acting on it is the gravitational force. Therefore, we can use the principle of constancy of mechanical energy. Initially, the ball has potential energy and no kinetic energy. As it falls, its total energy (the sum of kinetic and potential energies) remains constant and equal to its initial potential energy.

Solution   When the ball is released from rest at a height h above the ground, its kinetic energy is $K_i = 0$ and its potential energy is $U_i = mgh$, where the y coordinate is measured from ground level. When the ball is at a distance y above the ground, its kinetic energy is $K_f = \frac{1}{2}mv_f^2$ and its potential energy relative to the ground is $U_f = mgy$. Applying Equation 8.9, we get

$$K_i + U_i = K_f + U_f$$

$$0 + mgh = \tfrac{1}{2}mv_f^2 + mgy$$

$$v_f^2 = 2g(h - y)$$

$$v_f = \boxed{\sqrt{2g(h - y)}}$$

(b) Determine the speed of the ball at y if it is given an initial speed v_i at the initial altitude h.

Solution   In this case, the initial energy includes kinetic energy equal to $\frac{1}{2}mv_i^2$ and Equation 8.11 gives

$$\tfrac{1}{2}mv_i^2 + mgh = \tfrac{1}{2}mv_f^2 + mgy$$

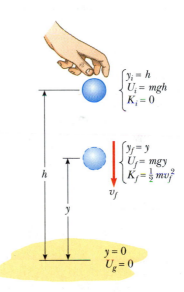

FIGURE 8.6   (Example 8.4) A ball is dropped from a height h above the ground. Initially, its total energy is potential energy, equal to mgh relative to the ground. At the elevation y, its total energy is the sum of kinetic and potential energies.

$$v_f^2 = v_i^2 + 2g(h - y)$$

$$v_f = \boxed{\sqrt{v_i^2 + 2g(h - y)}}$$

This result is consistent with the expression $v_y^2 = v_{y0}{}^2 - 2g(y - y_0)$, from kinematics, where $y_0 = h$. Furthermore, this result is valid even if the initial velocity is at an angle to the horizontal (the projectile situation).

EXAMPLE 8.5   The Pendulum

A pendulum consists of a sphere of mass m attached to a light cord of length L as in Figure 8.7. The sphere is released from rest when the cord makes an angle θ_0 with the vertical, and the pivot at P is frictionless. (a) Find the speed of the sphere when it is at the lowest point, b.

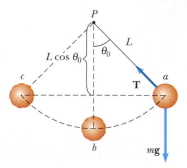

FIGURE 8.7   (Example 8.5) If the sphere is released from rest at the angle θ_0, it will never swing above this position during its motion. At the start of the motion, position a, its energy is entirely potential. This initial potential energy is all transformed into kinetic energy at the lowest elevation, position b. As the sphere continues to move along the arc, the energy again becomes entirely potential energy at position c.

Reasoning   The only force that does work on m is the force of gravity, since the force of tension is always perpendicular to each element of the displacement and hence does no work. Since the force of gravity is a conservative force, the total mechanical energy is constant. Therefore, as the pendulum swings, there is a continuous transfer between potential and kinetic energy. At the instant the pendulum is re-

leased, the energy is entirely potential energy. At point b, the pendulum has kinetic energy but has lost some potential energy. At point c, the pendulum has regained its initial potential energy and its kinetic energy is again zero.

Solution   If we measure the y coordinates from the center of rotation, then $y_a = -L \cos \theta_0$ and $y_b = -L$. Therefore, $U_a = -mgL \cos \theta_0$ and $U_b = -mgL$. Applying the principle of constancy of mechanical energy gives

$$K_a + U_a = K_b + U_b$$

$$0 - mgL \cos \theta_0 = \tfrac{1}{2}mv_b^2 - mgL$$

$$(1) \qquad v_b = \boxed{\sqrt{2gL(1 - \cos \theta_0)}}$$

(b) What is the tension T in the cord at b?

Solution   Since the force of tension does no work, it cannot be determined using the energy method. To find T_b, we can apply Newton's second law to the radial direction. First, recall that the centripetal acceleration of a particle moving in a circle is equal to v^2/r directed toward the center of rotation. Since $r = L$ in this example, we get

$$(2) \qquad \sum F_r = T_b - mg = ma_r = mv_b^2/L$$

Substituting (1) into (2) gives for the tension at point b

$$(3) \qquad T_b = mg + 2mg(1 - \cos \theta_0) = \boxed{mg(3 - 2\cos \theta_0)}$$

Exercise   A pendulum of length 2.00 m and mass 0.500 kg is released from rest when the cord makes an angle of 30.0° with the vertical. Find the speed of the sphere and the tension in the cord when the sphere is at its lowest point.

Answer   2.29 m/s; 6.21 N.

8.5   CHANGES IN MECHANICAL ENERGY WHEN NONCONSERVATIVE FORCES ARE PRESENT

In real physical systems, nonconservative forces, such as friction, are usually present. Such forces remove mechanical energy from the system. Therefore, the total mechanical energy is not constant. In general, we cannot calculate the work done by nonconservative forces, but we can find the change in kinetic energy of the system, acted upon by a net force, using the net-force equation:

$$\int \mathbf{F}_{net} \cdot d\mathbf{x} = \Delta K \qquad \qquad (8.13)$$

Since the change in kinetic energy can be the result of many types of force, it is convenient to separate ΔK into three parts:

1. The change in kinetic energy due to internal conservative forces, $\Delta K_{int\text{-}c}$
2. The change in kinetic energy due to internal nonconservative forces, $\Delta K_{int\text{-}nc}$

3. The change in kinetic energy due to external forces (conservative or nonconservative), ΔK_{ext}.

The first of these is $\Delta K_{int\text{-}c} = -\Delta U$. This is simply an exchange of potential energy into kinetic energy within the system. The second term, $\Delta K_{int\text{-}nc}$, may be positive or negative: if it represents internal friction, it will be negative; however, it can be positive (as in the case of a muscle operating on biochemical energy). The last term, ΔK_{ext}, is the change in kinetic energy of the system due to external forces. Thus, we have

$$\Delta K = \Delta K_{int\text{-}c} + \Delta K_{int\text{-}nc} + \Delta K_{ext}$$

or

$$\Delta K + \Delta U = \Delta K_{int\text{-}nc} + \Delta K_{ext} \tag{8.14}$$

From this we see that if there are no internal nonconservative forces and no external forces acting on the system, then the right side of Equation 8.14 is zero, and the sum $K + U$ is constant, which is consistent with Equation 8.9. If internal nonconservative forces are present, they may increase or decrease the kinetic energy. Similarly, external forces may either increase or decrease the kinetic energy of the system.

Problem-Solving Strategies
Conservation of Energy

Many problems in physics can be solved using the principle of conservation of energy. In this chapter we applied the principle to special cases in which the total energy of the system is constant and any loss of mechanical energy is determined as loss of kinetic energy from the net-force equation. The following procedure should be used when you apply this principle:

- Define your system, which may consist of more than one object and may or may not include fields, springs, or other sources of potential energy.
- Select a reference position for the zero point of potential energy (both gravitational and elastic), and use this position throughout your analysis. If there is more than one conservative force, write an expression for the potential energy associated with each force.
- Remember that if friction or air resistance is present, mechanical energy *is not constant*.
- If mechanical energy is constant, write the total initial energy, E_i, at some point as the sum of the kinetic and potential energies at that point. Then write an expression for the total final energy, $E_f = K_f + U_f$, at the final point. Since mechanical energy is constant, equate the two total energies and solve for the unknown.
- If external or frictional forces are present (and thus mechanical energy is not constant), first write expressions for the total initial and total final energies. In this case, however, the total final energy differs from the total initial energy, the difference being the amount of energy dissipated by nonconservative forces. That is, apply Equation 8.13.

EXAMPLE 8.6   Crate Sliding Down a Ramp

A 3.00-kg crate slides down a ramp at a loading dock. The ramp is 1.00 m in length and inclined at an angle of 30.0°, as shown in Figure 8.8. The crate starts from rest at the top, experiences a constant frictional force of magnitude 5.00 N, and continues to move a short distance on the flat floor. Use energy methods to determine the speed of the crate when it reaches the bottom of the ramp.

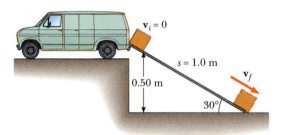

FIGURE 8.8   (Example 8.6) A crate slides down a ramp under the influence of gravity. Its potential energy decreases while its kinetic energy increases.

Solution   Since $v_i = 0$, the initial kinetic energy is zero. If the y coordinate is measured from the bottom of the ramp, then $y_i = 0.500$ m. Therefore, the total mechanical energy of the system at the top is all potential energy:

$$U_i = mgy_i = (3.00 \text{ kg}) \left(9.80 \, \frac{\text{m}}{\text{s}^2} \right) (0.500 \text{ m}) = 14.7 \text{ J}$$

When the crate reaches the bottom, the potential energy is *zero* because the elevation of the crate is $y_f = 0$. Therefore,

the total mechanical energy at the bottom is all kinetic energy,

$$K_f = \tfrac{1}{2} m v_f^2$$

However, we cannot say that $U_i = K_f$ in this case, because there is an external nonconservative force that removes mechanical energy from the system: the force of friction. In this case, $\Delta K_{\text{ext}} = -fs$, where s is the displacement along the ramp. (Remember that the forces normal to the ramp do no work on the crate because they are perpendicular to the displacement.) With $f = 5.00$ N and $s = 1.00$ m, we have

$$\Delta K_{\text{ext}} = -fs = (-5.00 \text{ N})(1.00 \text{ m}) = -5.00 \text{ J}$$

This says that some mechanical energy is lost because of the presence of the retarding frictional force. Applying Equation 8.13 gives

$$-fs = \tfrac{1}{2} m v_f^2 - mgy_i$$

$$\tfrac{1}{2} m v_f^2 = 14.7 \text{ J} - 5.00 \text{ J} = 9.70 \text{ J}$$

$$v_f^2 = \frac{19.4 \text{ J}}{3.00 \text{ kg}} = 6.47 \text{ m}^2/\text{s}^2$$

$$v_f = \boxed{2.54 \text{ m/s}}$$

Exercise   Use Newton's second law to find the acceleration of the crate along the ramp and the equations of kinematics to determine the final speed of the crate.

Answer   3.23 m/s²; 2.54 m/s.

Exercise   If the ramp is assumed to be frictionless, find the final speed of the crate and its acceleration along the ramp.

Answer   3.13 m/s; 4.90 m/s².

EXAMPLE 8.7   Motion on a Curved Track

A child of mass m takes a ride on an irregularly curved slide of height $h = 6.00$ m, as in Figure 8.9. The child starts from rest at the top. (a) Determine the speed of the child at the bottom, assuming no friction is present.

Reasoning   The normal force, **n**, does no work on the child since this force is always perpendicular to each element of the displacement. Furthermore, since there is no friction, mechanical energy is constant; that is, $K + U = $ constant.

Solution   If we measure the y coordinate from the bottom of the slide, then $y_i = h$, $y_f = 0$, and we get

$$K_i + U_i = K_f + U_f$$

$$0 + mgh = \tfrac{1}{2} m v_f^2 + 0$$

$$v_f = \sqrt{2gh}$$

FIGURE 8.9   (Example 8.7) If the slide is frictionless, the speed of the child at the bottom depends only on the height of the slide.

Note that the result is the same as it would be if the child fell vertically through a distance h! In this example, $h = 6.00$ m, giving

$$v_f = \sqrt{2gh} = \sqrt{2\left(9.80\,\frac{\text{m}}{\text{s}^2}\right)(6.00\text{ m})} = \boxed{10.8\text{ m/s}}$$

(b) If a frictional force acts on the child, how much mechanical energy is dissipated by this force? Assume that $v_f = 8.00$ m/s and $m = 20.0$ kg.

Solution In this case $\Delta K_{\text{ext}} \neq 0$ and mechanical energy is *not* constant. We can use Equation 8.13 to find the loss of kinetic energy due to friction, assuming the final speed at the bottom is known:

$$\Delta K_{\text{ext}} = E_f - E_i = \tfrac{1}{2}mv_f^2 - mgh$$
$$\Delta K_{\text{ext}} = \tfrac{1}{2}(20.0\text{ kg})(8.00\text{ m/s})^2$$
$$- (20.0\text{ kg})\left(9.80\,\frac{\text{m}}{\text{s}^2}\right)(6.00\text{ m})$$
$$= \boxed{-536\text{ J}}$$

Again, ΔK_{ext} is negative since friction is removing kinetic energy from the system. Note, however, that because the slide is curved, the normal force changes in magnitude and direction during the motion. Therefore, the frictional force, which is proportional to n, also changes during the motion. Do you think it would be possible to determine μ from these data?

EXAMPLE 8.8   Let's Go Skiing

A skier starts from rest at the top of a frictionless incline of height 20.0 m as in Figure 8.10. At the bottom of the incline, the skier encounters a horizontal surface where the coefficient of kinetic friction between the skis and snow is 0.210. How far does the skier travel on the horizontal surface before coming to rest?

Solution First, let us calculate the speed of the skier at the bottom of the incline. Since the incline is frictionless, the mechanical energy remains constant and we find

$$v = \sqrt{2gh} = \sqrt{2(9.80\text{ m/s}^2)(20.0\text{ m})} = 19.8\text{ m/s}$$

Now we apply Equation 8.13 as the skier moves along the rough horizontal surface. The change in kinetic energy along the horizontal is $\Delta K_{\text{ext}} = -fs$, where s is the horizontal displacement. Therefore,

$$\Delta K_{\text{ext}} = -fs = K_f - K_i$$

To find the distance the skier travels before coming to rest, we take $K_f = 0$. Since $v_i = 19.8$ m/s, and the friction force is $f = \mu n = \mu mg$, we get

$$-\mu mgs = -\tfrac{1}{2}mv_i^2$$

or

$$s = \frac{v_i^2}{2\mu g} = \frac{(19.8\text{ m/s})^2}{2(0.210)(9.80\text{ m/s}^2)} = \boxed{95.2\text{ m}}$$

Exercise Find the horizontal distance the skier travels before coming to rest if the incline also has a coefficient of kinetic friction equal to 0.210.

Answer 40.3 m.

20.0 m

20.0°

FIGURE 8.10   (Example 8.8).

EXAMPLE 8.9   The Spring-Loaded Popgun

The launching mechanism of a toy gun consists of a spring of unknown spring constant (Fig. 8.11a). When the spring is compressed 0.120 m, the gun is able to launch a 35.0-g pro-jectile to a maximum height of 20.0 m when fired vertically from rest. (a) Neglecting all resistive forces, determine the spring constant.

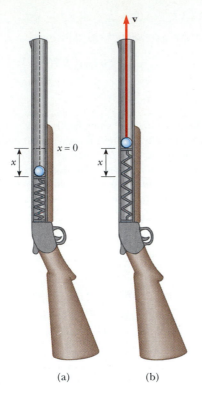

(a)   (b)

FIGURE 8.11   (Example 8.9).

Reasoning   Since the projectile starts from rest, the initial kinetic energy in the system is zero. If the zero point for the gravitational potential energy is taken at the lowest position of the projectile, then the initial gravitational potential energy is also zero. The mechanical energy of this system remains constant because there are no nonconservative forces present.

Solution   The total initial energy of the system is the elastic potential energy stored in the spring, which is $kx^2/2$. Since

the projectile rises to a maximum height $h = 20.0$ m, the final gravitational potential energy is mgh, its final kinetic energy is zero, and the final elastic potential energy is zero. Because the mechanical energy of the system is constant, we find

$$\tfrac{1}{2}kx^2 = mgh$$

$$\tfrac{1}{2}k(0.120 \text{ m})^2 = (0.0350 \text{ kg})(9.80 \text{ m/s}^2)(20.0 \text{ m})$$

$$k = \boxed{953 \text{ N/m}}$$

(b) Find the speed of the projectile as it moves through the equilibrium position of the spring (where $x = 0$) as shown in Figure 8.11b.

Solution   Using the same reference level for the gravitational potential energy as in part (a), we see that the initial energy of the system is still the elastic potential energy $kx^2/2$. The energy of the system when the projectile moves through the unstretched position of the spring consists of the kinetic energy of the projectile, $mv^2/2$, and the gravitational potential energy of the projectile, mgx. Hence, conservation of energy in this case gives

$$\tfrac{1}{2}kx^2 = \tfrac{1}{2}mv^2 + mgx$$

Solving for v gives

$$v = \sqrt{\frac{kx^2}{m} - 2gx}$$

$$v = \sqrt{\frac{(953 \text{ N/m})(0.120 \text{ m})^2}{(0.0350 \text{ kg})} - 2(9.80 \text{ m/s}^2)(0.120 \text{ m})}$$

$$= \boxed{19.7 \text{ m/s}}$$

Exercise   What is the speed of the projectile when it is at a height of 10.0 m?

Answer   14.0 m/s.

EXAMPLE 8.10   Mass-Spring Collision

A mass of 0.80 kg is given an initial velocity $v_i = 1.2$ m/s to the right and collides with a light spring of force constant $k = 50$ N/m, as in Figure 8.12. (a) If the surface is frictionless, calculate the initial maximum compression of the spring after the collision.

Reasoning   Before the collision, the mass has kinetic energy and the spring is uncompressed, so the energy stored in the spring is zero. Thus, the total energy of the system (mass plus spring) before the collision is $\tfrac{1}{2}mv_i^2$. After the collision, and when the spring is fully compressed, the mass is at rest and has zero kinetic energy, while the energy stored in the spring has its maximum value, $\tfrac{1}{2}kx_f^2$. The total mechanical

energy of the system is constant since no nonconservative forces act on the system.

Solution   Because mechanical energy is constant, the kinetic energy of the mass before the collision must equal the maximum energy stored in the spring when it is fully compressed, or

$$\tfrac{1}{2}mv_i^2 = \tfrac{1}{2}kx_f^2$$

$$x_f = \sqrt{\frac{m}{k}}\, v_i = \sqrt{\frac{0.80 \text{ kg}}{50 \text{ N/m}}}\,(1.2 \text{ m/s}) = \boxed{0.15 \text{ m}}$$

(b) If a constant force of friction acts between block and surface with $\mu = 0.50$ and if the speed of the block just as it

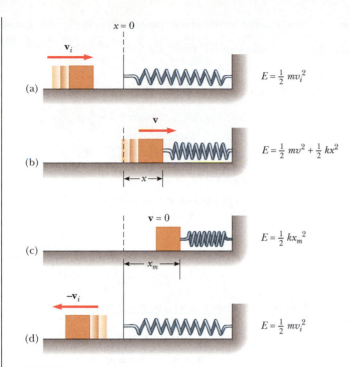

FIGURE 8.12   (Example 8.10) A block sliding on a smooth, horizontal surface collides with a light spring. (a) Initially the mechanical energy is all kinetic energy. (b) The mechanical energy is the sum of the kinetic energy of the block and the elastic potential energy in the spring. (c) The energy is entirely potential energy. (d) The energy is transformed back to the kinetic energy of the block. The total energy remains constant throughout the motion.

collides with the spring is $v_i = 1.2$ m/s, what is the maximum compression in the spring?

Solution   In this case, mechanical energy is *not* conserved because of friction. The magnitude of the frictional force is

$$f = \mu n = \mu mg = 0.50(0.80 \text{ kg})\left(9.80\, \frac{\text{m}}{\text{s}^2}\right) = 3.9 \text{ N}$$

Therefore, the loss in kinetic energy due to friction as the block is displaced from $x_i = 0$ to $x_f = x$ is

$$\Delta K_{ext} = -fx = (-3.92x) \text{ J}$$

Substituting this into Equation 8.14 gives

$$\Delta K_{ext} = (0 + \tfrac{1}{2}kx^2) - (\tfrac{1}{2}mv_i^2 + 0)$$

$$-3.92x = \tfrac{1}{2}(50)x^2 - \tfrac{1}{2}(0.80)(1.2)^2$$

$$25x^2 + 3.92x - 0.576 = 0$$

Solving the quadratic equation for x gives $x = 0.092$ m and $x = -0.25$ m. The physically meaningful root is $x = 0.092$ m = 9.2 cm. The negative root is meaningless because the block must be to the right of the origin when it comes to rest. Note that 9.2 cm is less than the distance obtained in the frictionless case (a). This result is what we expect because friction retards the motion of the system.

EXAMPLE 8.11   Connected Blocks in Motion

Two blocks are connected by a light string that passes over a frictionless pulley as in Figure 8.13. The block of mass m_1 lies on a horizontal surface and is connected to a spring of force constant k. The system is released from rest when the spring is unstretched. If m_2 falls a distance h before coming to rest, calculate the coefficient of kinetic friction between m_1 and the surface.

Reasoning and Solution   In this situation there are two forms of potential energy to consider: gravitational and elastic. We can write Equation 8.14 as

$$(1) \qquad \Delta K_{ext} = \Delta K + \Delta U_g + \Delta U_s$$

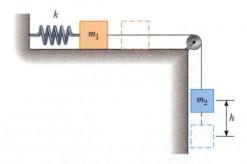

FIGURE 8.13   (Example 8.11) As m_2 moves from its highest elevation to its lowest, the system loses gravitational potential energy but gains elastic potential energy in the spring. Some mechanical energy is lost because of friction between m_1 and the surface.

where ΔU_g is the change in the gravitational potential energy and ΔU_s is the change in the elastic potential energy of the system. In this situation, $\Delta K = 0$ because the initial and final speeds of the system are zero. Also, the loss in kinetic energy due to friction is

(2) $\Delta K_{ext} = -fh = -\mu m_1 gh$

The change in the gravitational potential energy is associated only with m_2 since the vertical coordinate of m_1 does not change. Therefore, we get

(3) $\Delta U_g = U_f - U_i = -m_2 gh$

where the coordinates have been measured from the lowest position of m_2.

The change in the elastic potential energy in the spring is

(4) $\Delta U_s = U_f - U_i = \frac{1}{2}kh^2 - 0$

Substituting (2), (3), and (4) into (1) gives

$$-\mu m_1 gh = -m_2 gh + \frac{1}{2}kh^2$$

$$\mu = \frac{m_2 g - \frac{1}{2}kh}{m_1 g}$$

This set-up represents a way of measuring the coefficient of kinetic friction between an object and some surface.

EXAMPLE 8.12   One Way to Lift an Object

Two blocks are connected by a massless cord that passes over a frictionless pulley and a frictionless peg as in Figure 8.14. One end of the cord is attached to a mass $m_1 = 3.00$ kg that is a distance $R = 1.20$ m from the peg. The other end of the cord is connected to a block of mass $m_2 = 6.00$ kg resting on a table. From what angle θ (measured from the vertical) must the 3.00-kg mass be released in order to just begin to lift the 6.00-kg block off the table?

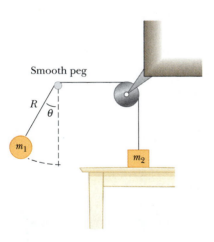

FIGURE 8.14   Example 8.12.

Reasoning   It is necessary to use several concepts to solve this problem. First, we use conservation of energy to find the speed of the 3.00-kg mass at the bottom of the circular path as a function of θ and the radius of the path. Next, we apply Newton's second law to the 3.00-kg mass at the bottom of its path to find the tension as a function of the given parameters. Finally, we note that the 6.00-kg block lifts off the table when the upward force exerted on it by the cord exceeds the force of gravity acting on the block. This procedure enables us to find the required angle.

Solution   Applying conservation of energy to the 3.00-kg mass gives

$$K_i + U_i = K_f + U_f$$

(1) $0 + m_1 gy_i = \frac{1}{2}m_1 v^2 + 0$

where v is the speed of the 3.00-kg mass at the bottom of its path. (Note that $K_i = 0$ since the 3.00-kg mass starts from rest and $U_f = 0$ because the bottom of the circle is the zero level of potential energy.) From the geometry in Figure 8.14, we see that $y_i = R - R\cos\theta = R(1 - \cos\theta)$. Using this relation in (1) gives

(2) $v^2 = 2gR(1 - \cos\theta)$

Now we apply Newton's second law to the 3.00-kg mass when it is at the bottom of the circular path:

$$T - m_1 g = m_1 \frac{v^2}{R}$$

(3) $T = m_1 g + m_1 \frac{v^2}{R}$

This same force is transmitted to the 6.00-kg block, and if it is to be just lifted off the table, the normal force on it becomes zero, and we require that $T = m_2 g$. Using this condition, together with (2) and (3), gives

$$m_2 g = m_1 g + m_1 \frac{2gR(1 - \cos\theta)}{R}$$

Solving for θ, and substituting in the given parameters, we get

$$\cos\theta = \frac{3m_1 - m_2}{2m_1} = \frac{3(3.00\ \text{kg}) - 6.00\ \text{kg}}{2(3.00\ \text{kg})} = \frac{1}{2}$$

$$\theta = 60.0°$$

Exercise   If the initial angle is $\theta = 40.0°$, find the speed of the 3.00-kg mass and the tension in the cord when the 3.00-kg mass is at the bottom of its circular path.

Answer   2.35 m/s; 43.2 N.

8.6 RELATIONSHIP BETWEEN CONSERVATIVE FORCES AND POTENTIAL ENERGY

In the previous sections we saw that one mode of energy storage is as potential energy, which is related to the configuration, or coordinates, of a system. Potential energy functions are associated only with conservative forces. If an object or field does work on some external object, energy is transferred from the object or field to the external object. The work done is

$$\int \mathbf{F} \cdot d\mathbf{x} = -\Delta U$$

That is, the energy transferred as work decreases the potential energy of the system from which the energy came. (In this expression, $\mathbf{F}$ is the force exerted by the object or field on the external object, and $d\mathbf{x}$ is the displacement of the point of application of the force.)

If there is an infinitesimal displacement, $d\mathbf{x}$, we can express the infinitesimal change in potential energy of the system, dU, as

$$dU = -F_x \, dx$$

Therefore, the conservative force is related to the potential energy function through the relationship[3]

$$F_x = -\frac{dU}{dx} \qquad (8.15)$$

Relationship between force and potential energy

That is, *the conservative force equals the negative derivative of the potential energy with respect to x.*

We can easily check this relationship for the two examples already discussed. In the case of the deformed spring, $U_s = \frac{1}{2}kx^2$, and therefore

$$F_s = -\frac{dU_s}{dx} = -\frac{d}{dx}\left(\tfrac{1}{2}kx^2\right) = -kx$$

which corresponds to the restoring force in the spring. Since the gravitational potential energy function is $U_g = mgy$, it follows from Equation 8.15 that $F_g = -mg$.

We now see that U is an important function because the conservative force can be derived from it. Furthermore, Equation 8.15 should clarify the fact that adding a constant to the potential energy is unimportant.

Often we define our system to include all the interacting parts (for example, a spring and the mass on which it acts, or a gravitational field and a mass). Then no energy is transferred to or from the system, so

$$\Delta E = \Delta K + \Delta U = 0$$

$$\Delta K = -\Delta U$$

---

[3] In three dimensions, the expression is $\mathbf{F} = -\mathbf{i}\dfrac{\partial U}{\partial x} - \mathbf{j}\dfrac{\partial U}{\partial y} - \mathbf{k}\dfrac{\partial U}{\partial z}$, where $\dfrac{\partial U}{\partial x}$, etc., are partial derivatives. In the language of vector calculus, $\mathbf{F}$ equals the negative of the gradient of the scalar quantity $U(x, y, z)$.

*8.7   ENERGY DIAGRAMS AND THE EQUILIBRIUM OF A SYSTEM

The motion of a system can often be understood qualitatively through its potential energy curve. Consider the potential energy function for the mass-spring system, given by $U_s = \frac{1}{2}kx^2$. This function is plotted versus x in Figure 8.15a. The force is related to U through Equation 8.15:

$$F_s = -\frac{dU_s}{dx} = -kx$$

That is, the force is equal to the negative of the slope of the U versus x curve. When the mass is placed at rest at the equilibrium position ($x = 0$), where $F = 0$, it will remain there unless some external force acts on it. If the spring is stretched from equilibrium, x is positive and the slope dU/dx is positive; therefore F_s is negative and the mass accelerates back toward $x = 0$. If the spring is compressed, x is negative and the slope is negative; therefore F_s is positive and again the mass accelerates toward $x = 0$.

From this analysis, we conclude that the $x = 0$ position is one of **stable equilibrium.** That is, any movement away from this position results in a force directed back toward $x = 0$. In general, *positions of stable equilibrium correspond to those points for which U(x) has a minimum value.*

From Figure 8.15 we see that if the mass is given an initial displacement x_m and released from rest, its total energy initially is the potential energy stored in the spring, $\frac{1}{2}kx_m^2$. As motion commences, the system acquires kinetic energy and loses an equal amount of potential energy. Because the total energy must remain constant, the mass oscillates between the two points $x = \pm x_m$, called the *turning points*. In fact, because there is no energy loss (no friction), the mass will oscillate between $-x_m$ and $+x_m$ forever. (We discuss these oscillations further in Chapter 13.) From an energy viewpoint, the energy of the system cannot exceed $\frac{1}{2}kx_m^2$; therefore the

Stable equilibrium

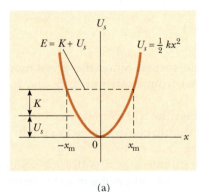

(a)

(b)

FIGURE 8.15   (a) Potential energy as a function of x for the mass-spring system shown in (b). The mass oscillates between the turning points, which have the coordinates $x = \pm x_m$. Note that the restoring force of the spring always acts toward $x = 0$, the position of stable equilibrium.

mass must stop at these points and, because of the spring force, accelerate toward $x = 0$.

Another simple mechanical system that has a position of stable equilibrium is a ball rolling about in the bottom of a bowl. If the ball is displaced from its lowest position, it will always tend to return to that position when it is released.

Now consider an example where the U versus x curve is as shown in Figure 8.16. Once again, $F_x = 0$ at $x = 0$, and so the particle is in equilibrium at this point. However, this is a position of **unstable equilibrium** for the following reason. Suppose the particle is displaced to the right ($x > 0$). Because the slope is negative for $x > 0$, $F_x = -dU/dx$ is positive and the particle accelerates away from $x = 0$. Now suppose that the particle is displaced to the left ($x < 0$). In this case, the force is negative because the slope is positive for $x < 0$, and the particle again accelerates away from the equilibrium position. The $x = 0$ position in this situation is one of unstable equilibrium because for any displacement from this point, the force pushes the particle farther away from equilibrium. In fact, the force pushes the particle toward a position of lower potential energy. A ball placed on the top of an inverted bowl in the shape of a hemisphere is in a position of unstable equilibrium. If the ball is displaced slightly from the top and released, it will surely roll off the bowl. In general, *positions of unstable equilibrium correspond to those points for which U(x) has a maximum value.*

Finally, a situation may arise where U is constant over some region and hence $F = 0$. This is called a position of **neutral equilibrium.** Small displacements from this position produce neither restoring nor disrupting forces. A ball lying on a flat horizontal surface is an example of an object in neutral equilibrium.

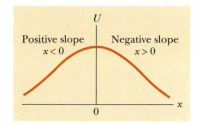

FIGURE 8.16 A plot of U versus x for a particle that has a position of unstable equilibrium, located at $x = 0$. For any finite displacement of the particle, the force on the particle is directed away from $x = 0$.

Neutral equilibrium

8.8 CONSERVATION OF ENERGY IN GENERAL

We have seen that the total mechanical energy of a system is constant when only conservative forces act within the system. Furthermore, we can associate a potential energy function with each conservative force. On the other hand, as we saw in Section 8.5, mechanical energy is lost when nonconservative forces such as friction are present.

In the study of thermodynamics we shall find that mechanical energy can be transformed to internal energy of the system. For example, when a block slides over a rough surface, the mechanical energy lost is transformed into internal energy temporarily stored in the block and the surface, as evidenced by a measurable increase in the block's temperature. We shall see that on a submicroscopic scale, this internal energy is associated with the vibration of atoms about their equilibrium positions. Such internal atomic motion has kinetic and potential energy. Therefore, if we include this increase in the internal energy of the system in our energy expression, the total energy is conserved.

This is just one example of how you can analyze an isolated system and always find that its total energy does not change, as long as you account for all forms of energy. That is, *energy can never be created or destroyed. Energy may be transformed from one form to another, but the total energy of an isolated system is always constant.* From a universal point of view, we can say that the *total energy of the Universe is constant:* If one part of the Universe gains energy in some form, another part must lose an equal amount of energy. No violation of this principle has been found.

Total energy is always conserved

*8.9   MASS-ENERGY EQUIVALENCE

This chapter has been concerned with the important principle of energy conservation and its application to various physical phenomena. Another important principle, **conservation of mass,** says that, in any kind of physical or chemical process, *mass is neither created nor destroyed.* That is, the mass before the process equals the mass after the process.

For centuries, it appeared to scientists that energy and mass were two quantities that were separately conserved. However, in 1905 Einstein made the incredible discovery that mass, or inertia, of any system is a measure of the total energy of the system. Hence, *energy and mass are related concepts.* The relationship between the two is given by Einstein's most famous formula

$$E = mc^2 \tag{8.16}$$

where c is the speed of light and E is the energy equivalent of a mass m. As we shall see in Chapter 39, mass increases with speed; however, this dependence is insignificant for $v \ll c$. Hence, the masses we use for describing situations in our everyday experiences are always taken to be rest masses.

The energy associated with even a small amount of matter is enormous. For example, the energy of 1 kg of any substance is

$$E = mc^2 = (1 \text{ kg})(3 \times 10^8 \text{ m/s})^2 = 9 \times 10^{16} \text{ J}$$

This is equivalent to the energy content of about 15 million barrels of crude oil (about one day's consumption in the United States)! If this energy could easily be released, as useful work, our energy resources would be unlimited.

In reality, only a small fraction of the energy contained in a material sample can be released through chemical or nuclear processes. The effects are largest in nuclear reactions, where fractional changes in energy, and hence mass, of approximately 10^{-3} are routinely observed. A good example is the enormous energy released when the uranium-235 nucleus splits into two smaller nuclei. This happens because the ^{235}U nucleus is more massive than the sum of the masses of the product nuclei. The awesome nature of the energy released in such reactions is vividly demonstrated in the explosion of a nuclear weapon.

Expressed briefly, Equation 8.16 tells us that *energy has mass.* Whenever the energy of an object changes in any way, its mass changes. If ΔE is the change in energy of an object, its change in mass is

$$\Delta m = \frac{\Delta E}{c^2} \tag{8.17}$$

Any time energy ΔE in any form is supplied to an object, the change in the mass of the object is $\Delta m = \Delta E / c^2$. Because c^2 is so large, however, the changes in mass in any ordinary mechanical experiment or chemical reaction are too small to be detected.

*8.10   QUANTIZATION OF ENERGY

As you may have learned in your chemistry course, all ordinary matter consists of atoms, and each atom consists of a collection of electrons and a nucleus, which in turn contains smaller particles, each with a distinctive electric charge. Thus, on the fine scale of the atomic world, mass comes in discrete quantities corresponding to the atomic masses. In the language of modern physics, we say that certain quantities such as electric charge are *quantized*.

As we shall learn in the extended version of this text, many more physical quantities are quantized. The quantized nature of energy in bound systems is especially important in the atomic and subatomic world. For example, let us consider the energy levels of the hydrogen atom (an electron orbiting around a proton). The atom can occupy only certain energy levels, called *quantum states,* as shown in Figure 8.17a. The atom cannot have any energy values lying between these quantum states. The lowest energy level, E_0, is called the *ground state* of the atom. The ground state corresponds to the state an isolated atom usually occupies. The atom can move to higher energy states by absorbing energy from some external source or by colliding with other atoms. The highest energy on the scale shown in Figure 8.17a, E_∞, corresponds to the energy of the atom when the electron is completely removed from the proton and is called the *ionization energy*. Note that the energy levels get closer together at the high end of the scale. Above this energy level, the energies are continuous.

Next, consider a satellite in orbit about the Earth. If you were asked to describe the possible energies the satellite could have, it would be reasonable (but incorrect) to say it could have any arbitrary energy value. Just like the hydrogen atom, however, *the energy of the satellite is also quantized*. If you were to construct an energy level diagram for the satellite showing its allowed energies, the levels would be so close to one another, as in Figure 8.17b, that it would be impossible to tell that they were not continuous. In other words, we have no way of experiencing quantization of energy in the macroscopic world; hence we can ignore it in describing everyday experiences.

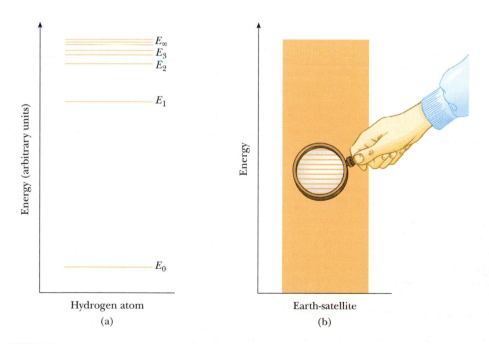

Hydrogen atom
(a)

Earth-satellite
(b)

FIGURE 8.17   Energy-level diagrams: (a) Quantum states of the hydrogen atom. The lowest state, E_0, is the ground state. (b) The energy levels of an Earth satellite are also quantized but are so close together that they cannot be distinguished from each other.

SUMMARY

The **gravitational potential energy** of a particle of mass m that is elevated a distance y near the Earth's surface is

$$U_g \equiv mgy \qquad (8.1)$$

The **elastic potential energy** stored in a spring of force constant k is

$$U_s \equiv \tfrac{1}{2}kx^2 \qquad (8.4)$$

A force is **conservative** if the work it does on a particle is independent of the path the particle takes between two points. Alternatively, a force is conservative if the work it does is zero when the particle moves through an arbitrary closed path and returns to its initial position. A force that does not meet these criteria is said to be **nonconservative.**

A **potential energy** function U can be associated only with a conservative force. If a conservative force **F** acts on a particle that moves along the x axis from x_i to x_f, the change in the potential energy of the particle equals the negative of the work done by that force:

$$U_f - U_i = -\int_{x_i}^{x_f} F_x \, dx \qquad (8.6)$$

The **total mechanical energy of a system** is defined as the sum of the kinetic energy and potential energy:

$$E \equiv K + U \qquad (8.8)$$

If no external forces do work on a system, and there are no nonconservative forces, the total mechanical energy of the system is constant:

$$K_i + U_i = K_f + U_f \qquad (8.9)$$

The change in total mechanical energy of a system equals the change in the kinetic energy due to internal nonconservative forces, $\Delta K_{\text{int-nc}}$, plus the change in kinetic energy due to all external forces, ΔK_{ext}:

$$\Delta K + \Delta U = \Delta K_{\text{int-nc}} + \Delta K_{\text{ext}} \qquad (8.14)$$

QUESTIONS

1. A bowling ball is suspended from the ceiling of a lecture hall by a strong cord. The bowling ball is drawn away from its equilibrium position and released from rest at the tip of the demonstrator's nose. If the demonstrator remains stationary, explain why she will not be struck by the ball on its return swing. Would the demonstrator be safe if she pushed the ball as she released it?

2. Can gravitational potential energy ever be negative? Explain.

3. One person drops a ball from the top of a building, while another person at the bottom observes its motion. Will these two people agree on the value of the ball's potential energy? on its change in potential energy? on its kinetic energy?

4. When a person runs in a track event at constant velocity, is any work done? (*Note:* Although the runner moves with constant velocity, the legs and arms accelerate.) How does air resistance enter into the picture? Does the center of mass of the runner move horizontally?

5. Our body muscles exert forces when we lift, push, run, jump, and so forth. Are these forces conservative?

6. If three conservative forces and one nonconservative force act on a system, how many potential energy terms appear in the problem?

7. Consider a ball fixed to one end of a rigid rod with the other end pivoted on a horizontal axis so that the rod can rotate in a vertical plane. What are the positions of stable and unstable equilibrium?

8. Is it physically possible to have a situation where $E - U < 0$?

9. What would the curve of U versus x look like if a particle were in a region of neutral equilibrium?

10. Explain the energy transformations that occur during (a) the pole vault, (b) the shot put, (c) the high jump. What is the source of energy in each case?

11. Discuss all the energy transformations that occur during the operation of an automobile.

12. A ball is thrown straight up into the air. At what position is its kinetic energy a maximum? At what position is its gravitational potential energy a maximum?

13. Consider the Earth to be a perfect sphere. By how much does your potential energy change when you (a) walk from the North Pole to the equator? (b) drop through a "tunnel" from the North Pole to the South Pole passing through the center of the Earth?

14. Does a single external force acting on a particle necessarily change (a) its kinetic energy? (b) its velocity?

15. In the high jump, is any portion of the athlete's kinetic energy converted to potential energy during the jump?

16. In the pole vault or high jump, why does the athlete attempt to keep his or her center of gravity as low as possible (consistent with passing over the bar) near the top of the jump?

17. A right circular cone can be balanced on a horizontal surface in three ways. Sketch these three equilibrium configurations and identify them as being stable, unstable, or neutral.

PROBLEMS

Review Problem

A block of mass m rests on top of a frictionless inclined plane of mass $3m$ and height h as shown below. The block slides down the plane and moves along a horizontal surface until it stops. The coefficient of kinetic friction between the block and horizontal surface is μ_k. At the same time, the plane slides to the left without friction and collides with a spring of force constant k. Find (a) the initial potential energy of the block, (b) the kinetic energy of the system when the block leaves the plane, (c) the speed of the block when it leaves the plane, (d) the speed of the plane when the block leaves the plane, (e) the distance the block travels before it stops, (f) the time it takes the block to stop, (g) the maximum compression of the spring caused by the inclined plane, and (h) the maximum value of the force exerted by the spring on the inclined plane. (i) Find the distance the block travels before it stops if the inclined plane is attached to the table.

Section 8.2 Conservative and Nonconservative Forces

1. A 4.00-kg particle moves from the origin to the position having coordinates $x = 5.00$ m and $y = 5.00$ m under the influence of gravity acting in the negative y direction (Fig. P8.1). Using Equation 7.2, calculate the work done by gravity in going from O to C along (a) OAC, (b) OBC, (c) OC. Your results should all be identical. Why?

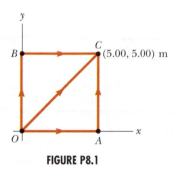

FIGURE P8.1

2. (a) Starting with Equation 7.2 for the definition of work done by a constant force, show that any constant force is conservative. (b) As a special case, suppose a particle of mass m is under the influence of force $\mathbf{F} = (3\mathbf{i} + 4\mathbf{j})$ N and moves from O to C in Figure P8.1. Calculate the work done by $\mathbf{F}$ along the three paths OAC, OBC, and OC. (Again, your three answers should be identical.)

3. Under the influence of gravity, a block of mass m

□ indicates problems that have full solutions available in the Student Solutions Manual and Study Guide.

slides down a quarter-circular track on which friction (coefficient $= \mu_k$) is present. The radius of the track is R. (a) Show that the change in mechanical energy of the block is $mgR(1 - \mu)$. (b) If the change in mechanical energy of the block is 42.0 J, determine the work done by the conservative forces and the energy dissipated by the nonconservative forces. Assume that $m = 2.0$ kg and $R = 3.2$ m. (c) What is μ_k?

4. A single conservative force acting on a particle varies as $\mathbf{F} = (- Ax + Bx^2)\mathbf{i}$ N, where A and B are constants and x is in meters. (a) Calculate the potential energy associated with this force, taking $U = 0$ at $x = 0$. (b) Find the change in potential energy and change in kinetic energy as the particle moves from $x = 2.0$ m to $x = 3.0$ m.

5. A force acting on a particle moving in the xy plane is $\mathbf{F} = (2y\mathbf{i} + x^2\mathbf{j})$ N, where x and y are in meters. The particle moves from the origin to a final position having coordinates $x = 5.0$ m and $y = 5.0$ m, as in Figure P8.1. Calculate the work done by $\mathbf{F}$ along (a) OAC, (b) OBC, (c) OC. (d) Is $\mathbf{F}$ conservative or nonconservative? Explain.

Section 8.3   Conservative Forces and Potential Energy
Section 8.4   Conservation of Energy

6. A 4.0-kg particle moves along the x axis under the influence of a single conservative force. If the work done on the particle is 80.0 J as it moves from the point $x = 2.0$ m to $x = 5.0$ m, find (a) the change in its kinetic energy, (b) the change in its potential energy, and (c) its speed at $x = 5.0$ m if it starts at rest at $x = 2.0$ m.

7. A single conservative force $F_x = (2.0x + 4.0)$ N acts on a 5.0-kg particle, where x is in meters. As the particle moves along the x axis from $x = 1.0$ m to $x = 5.0$ m, calculate (a) the work done by this force, (b) the change in the potential energy of the particle, and (c) its kinetic energy at $x = 5.0$ m if its speed at $x = 1.0$ m is 3.0 m/s.

8. At time t_i, the kinetic energy of a particle is 30 J and its potential energy is 10 J. At some later time t_f, its kinetic energy is 18 J. (a) If only conservative forces act on the particle, what are its potential energy and its total energy at time t_f? (b) If the potential energy at time t_f is 5 J, are there any nonconservative forces acting on the particle? Explain.

9. A single constant force $\mathbf{F} = (3.0\mathbf{i} + 5.0\mathbf{j})$ N acts on a 4.0-kg particle. (a) Calculate the work done by this force if the particle moves from the origin to the point having the vector position $\mathbf{r} = (2.0\mathbf{i} - 3.0\mathbf{j})$ m. Does this result depend on the path? Explain. (b) What is the speed of the particle at $\mathbf{r}$ if its speed at the origin is 4.0 m/s? (c) What is the change in its potential energy?

 10. A 5.0-kg mass is attached to a light cord that passes over a massless, frictionless pulley. The other end of the cord is attached to a 3.5-kg mass as in Figure

P8.10. Use conservation of energy to determine the final speed of the 5.0-kg mass after it has fallen (starting from rest) 2.5 m.

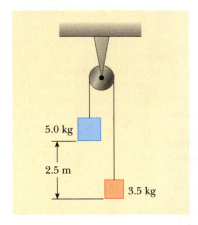

FIGURE P8.10

 11. A bead slides without friction around a loop-the-loop (Fig. P8.11). If the bead is released from a height $h = 3.50R$, what is its speed at point A? How large is the normal force on it if its mass is 5.00 g?

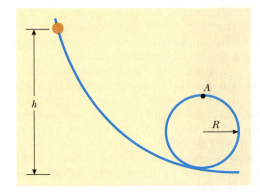

FIGURE P8.11

12. A particle of mass 0.500 kg is shot from P as shown in Figure P8.12 with an initial velocity $\mathbf{v}_0$ having a horizontal component of 30.0 m/s. The particle rises to a maximum height of 20.0 m above P. Using conservation of energy, determine (a) the vertical component of $\mathbf{v}_0$, (b) the work done by the gravitational force on the particle during its motion from P to B, and (c) the horizontal and the vertical components of the velocity vector when the particle reaches B.

13. A rocket is launched at an angle of 53° to the horizontal from an altitude h with a speed v_0. (a) Use energy methods to find its speed when its altitude is $h/2$. (b) Find the x and y components of velocity when the rocket's altitude is $h/2$, using the fact that $v_x = v_{x0} = $ constant (since $a_x = 0$) and the results to part (a).

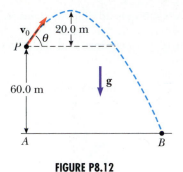

FIGURE P8.12

14. A 2.0-kg ball is attached to a 10-lb (44.5 N) fish line. The ball is released from rest at the horizontal position ($\theta = 90°$). At what angle θ (measured from the vertical), does the line break?

15. A 20.0-kg cannon ball is fired from a cannon at muzzle speed of 1000 m/s and at an angle of 37.0° with the horizontal. A second ball is fired at an angle of 90.0°. Use the conservation of mechanical energy to find, for each ball, (a) the maximum height reached and (b) the total mechanical energy at the maximum height.

16. A ball of mass m is spun in a vertical circle having radius R. The ball has a speed v_0 at its highest point. Take zero potential energy at the lowest point and use the angle θ measured with respect to the vertical as shown in Figure P8.16. (a) Derive an expression for the speed v at any time as a function of R, θ, v_0, and g. (b) What minimum speed v_0 is required to keep the ball moving in a circle? (c) Does the equation you derived in part (a) account for the result found in part (b)? Explain.

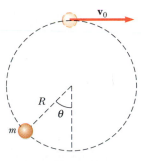

FIGURE P8.16

17. Two masses are connected by a light string passing over a light frictionless pulley as shown in Figure P8.17. The 5.0-kg mass is released from rest. Using the law of conservation of energy, (a) determine the speed of the 3.0-kg mass just as the 5.0-kg mass hits the ground. (b) Find the maximum height to which the 3.0-kg mass rises.

17A. Two masses are connected by a light string passing over a light frictionless pulley as in Figure P8.17. The mass m_1 is released from rest. Using the law of conservation of energy, (a) determine the speed of m_2 just as m_1 hits the ground. (b) Find the maximum height to which m_2 rises.

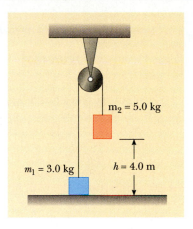

FIGURE P8.17

18. A child slides down the frictionless slide shown in Figure P8.18. In terms of R and H, at what height h will he lose contact with the section of radius R?

FIGURE P8.18

Section 8.5 Changes in Mechanical Energy When Nonconservative Forces Are Present

19. A 5.0-kg block is set into motion up an inclined plane with an initial speed of 8.0 m/s (Fig. P8.19). The block comes to rest after traveling 3.0 m along the plane, which is inclined at an angle of 30° to the horizontal. Determine (a) the change in the block's kinetic energy, (b) the change in its potential energy, (c) the frictional force exerted on it (assumed to be constant). (d) What is the coefficient of kinetic friction?

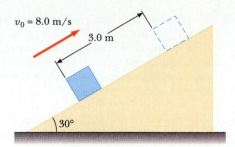

FIGURE P8.19

20. A 3.0-kg block starts at a height $h = 60$ cm on a plane that has an inclination angle of 30° as in Figure P8.20. Upon reaching the bottom, the block slides along a horizontal surface. If the coefficient of friction on both surfaces is $\mu_k = 0.20$, how far does the block slide on the horizontal surface before coming to rest? (*Hint:* Divide the path into two straight-line parts.)

20A. A block of mass m starts at a height h on a plane that has an inclination angle θ as in Figure P8.20. Upon reaching the bottom, the block slides along a horizontal surface. If the coefficient of friction on both surfaces is μ_k, how far does the block slide on the horizontal surface before coming to rest? (*Hint:* Divide the path into two straight-line parts.)

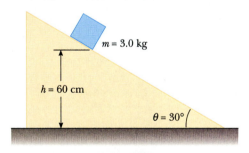

FIGURE P8.20

21. A parachutist of mass 50.0 kg jumps out of an airplane at a height of 1000 m and lands on the ground with a speed of 5.00 m/s. How much energy was lost to air friction during this jump?

22. A 0.500-kg bead slides on a curved wire, starting from rest at point A in Figure P8.22. The segment from A to B is frictionless, and the segment from B to C is rough. (a) Find the speed of the bead at B. (b) If the bead comes to rest at C, find the energy lost due to friction as it moves from B to C.

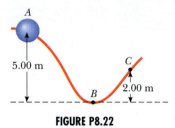

FIGURE P8.22

23. In the 1800's French engineer Hector Horeau proposed a design for a tunnel across the English Channel. Railroad cars would be able to roll freely through the tunnel until they ran out of kinetic energy, and then a steam engine would pull them the rest of the way (Fig. P8.23). Assume that the 32.0-km-long tunnel was designed with a 6.0° slope for the first 1.00 km at each end and ran level the rest of the way. Let the coefficient of rolling friction be 0.1 (a small value). (a) What is the change in potential energy of a 4000-kg railroad car as it rolls down the incline? (b) If the car starts from rest, what is its speed when it reaches the level track? (c) How far into the tunnel does the car get before it stops? (d) How feasible was this design?

24. An 80.0-kg skydiver jumps out of an airplane at an altitude of 1000 m and opens the parachute at an altitude of 200 m. (a) Assuming that the total retarding force on the diver is constant at 50.0 N with the parachute closed and constant at 3600 N with the parachute open, what is her speed when she lands? (b) Do you think the skydiver will get hurt? Explain. (c) At what height should the parachute be opened so that the speed of the diver just as she hits the ground is 5.00 m/s? (d) How realistic is the assumption that the total retarding force is constant? Explain.

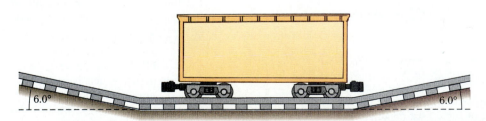

FIGURE P8.23

25. The coefficient of friction between the 3.0-kg mass and surface in Figure P8.25 is 0.40. The system starts from rest. What is the speed of the 5.0-kg mass when it has fallen 1.5 m?

25A. The coefficient of friction between m_1 and surface in Figure P8.25 is μ. The system starts from rest. What is the speed of m_2 when it has fallen a distance h?

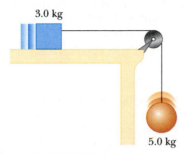

3.0 kg

5.0 kg

FIGURE P8.25

26. A toy gun uses a spring to project a 5.3-g soft rubber sphere. The spring constant is 8.0 N/m, the barrel of the gun is 15 cm long, and there is a constant frictional force of 0.032 N between barrel and projectile. With what speed is the projectile launched from the barrel if the spring is compressed 5.0 cm?

27. A block slides down a curved frictionless track and then up an inclined plane as in Figure P8.27. The coefficient of kinetic friction between block and incline is μ_k. Use energy methods to show that the maximum height reached by the block is

$$y_{\text{max}} = \frac{h}{1 + \mu_k \cot(\theta)}$$

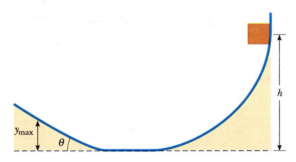

y_{max}

θ

h

FIGURE P8.27

28. A 1.5-kg mass is first held 1.2 m above a relaxed massless spring that has a spring constant of 320 N/m and then dropped onto the spring. (a) How far does the spring compress? (b) The same experiment is repeated on the Moon, where $g = 1.63$ m/s². (c) Repeat part (a), but this time assume that a constant

0.70-N air-resistance acts on the mass during the fall.

29. In the dangerous "sport" of bungee-jumping, a daring student jumps from a balloon with a specially designed elastic cord attached to his ankles. The unstretched length of the cord is 25.0 m, the student weighs 700 N, and the balloon is 36.0 m above the surface of a river. Calculate the required force constant of the cord if the student is to stop safely 4.00 m above the river.

30. A 3.0-kg mass starts from rest and slides a distance d down a frictionless 30° incline, where it contacts an unstressed spring of negligible mass as in Figure P8.30. The mass slides an additional 0.20 m as it is brought momentarily to rest by compressing the spring ($k = 400$ N/m). Find the initial separation d between mass and spring.

30A. A mass m starts from rest and slides a distance d down a frictionless incline of angle θ, where it contacts an unstressed spring of negligible mass as in Figure P8.30. The mass slides an additional distance x as it is brought momentarily to rest by compressing the spring (force constant k). Find the initial separation d between mass and spring.

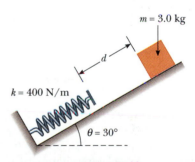

$m = 3.0$ kg

d

$k = 400$ N/m

$\theta = 30°$

FIGURE P8.30

31. An 8.00-kg block travels on a rough, horizontal surface and collides with a spring as in Figure 8.12. The speed of the block *just before* the collision is 4.00 m/s. As the block rebounds to the left with the spring uncompressed, its speed as it leaves the spring is 3.00 m/s. If the coefficient of kinetic friction between block and surface is 0.400, determine (a) the energy lost due to friction while the block is in contact with the spring and (b) the maximum distance the spring is compressed.

32. A child's pogo stick (Fig. P8.32) stores energy in a spring ($k = 2.5 \times 10^4$ N/m). At position A ($x_1 = -0.10$ m) the spring compression is a maximum and the child is momentarily at rest. At position B ($x = 0$) the spring is relaxed and the child is moving upward. At position C, the child is again momentarily at rest at the top of the jump. Assuming that the combined

mass of child and pogo stick is 25 kg, (a) calculate the total energy of the system if both potential energies are zero at $x = 0$, (b) determine x_2, (c) calculate the speed of the child at $x = 0$, (d) determine the value of x for which the kinetic energy of the system is a maximum, and (e) obtain the child's maximum upward speed.

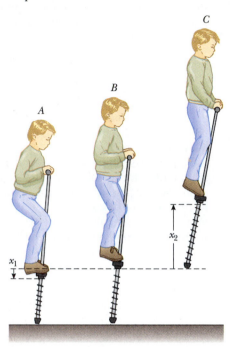

FIGURE P8.32

33. A block of mass 0.250 kg is placed on top of a vertical spring of constant $k = 5000$ N/m and pushed downward, compressing the spring 0.100 m. After the block is released it travels upward and then leaves the spring. To what maximum height above the point of release does it rise?

33A. A block of mass m is placed on top of a vertical spring of constant k and pushed downward, compressing the spring a distance d. After the block is released it travels upward and then leaves the spring. To what maximum height h above the point of release does it rise?

34. A block of mass 2.0 kg is kept at rest as it compresses a horizontal massless spring ($k = 100$ N/m) by 10 cm. As the block is released, it travels 0.25 m on a rough horizontal surface before stopping. Calculate the coefficient of kinetic friction between surface and block.

35. A 10.0-kg block is released from point A in Figure P8.35. The track is frictionless except for the portion BC, of length 6.00 m. The block travels down the track, hits a spring of force constant $k = 2250$ N/m, and compresses it 0.300 m from its equilibrium position before coming to rest momentarily. Determine the coefficient of kinetic friction between surface BC and block.

36. A 120-g mass is attached to the end of an unstressed vertical spring ($k = 40$ N/m) and then dropped. (a) What is its maximum speed? (b) How far does it drop before coming to rest momentarily?

36A. A mass m is attached to the end of an unstressed vertical spring with force constant k and then dropped. (a) What is its maximum speed? (b) How far does it drop before coming to rest momentarily?

Section 8.6   Relationship Between Conservative Forces and Potential Energy

37. The potential energy of a two-particle system separated by a distance r is $U(r) = A/r$, where A is a constant. Find the radial force $\mathbf{F}_r$ in terms of A and r.

38. A 3.00-kg block moving along the x axis is acted upon by a single force that varies with the block's position according to the equation $F_x = ax^2 + b$, where $a = 5.00$ N/m² and $b = -2.50$ N. At $x = 1.0$ m, the block is moving to the right at 4.0 m/s. Determine its speed at $x = 2.0$ m.

39. A potential energy function for a two-dimensional force is of the form $U = 3x^3y - 7x$. Find the force that acts at the point (x, y).

*Section 8.7   Energy Diagrams and the Equilibrium of a System

40. For the potential energy curve shown in Figure P8.40, (a) determine whether the force F_x is positive, negative, or zero at the five points indicated. (b) In-

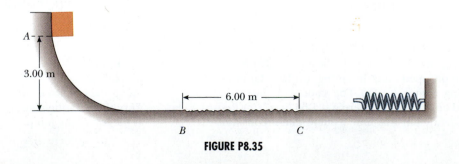

FIGURE P8.35

dicate points of stable, unstable, and neutral equilibrium. (c) Sketch the curve of F_x versus x from $x = 0$ to $x = 8.0$ m.

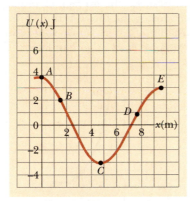

FIGURE P8.40

41. A particle of mass m is suspended between two identical springs on a horizontal frictionless tabletop as shown in Figure P8.41. Both springs have spring constant k. (a) If the particle is pulled a distance x along a direction perpendicular to the initial configuration of the springs, show that its potential energy due to the springs is

$$U(x) = kx^2 + 2kL(L - \sqrt{x^2 + L^2})$$

(*Hint:* See Problem 78 of Chapter 7.) (b) Plot $U(x)$ versus x and identify all equilibrium points. Assume that $L = 1.20$ m and $k = 40.0$ N/m. (c) If the particle is pulled 0.500 m to the right and then released, what is its speed when it reaches $x = 0$?

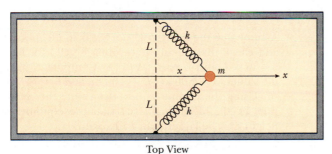

Top View

FIGURE P8.41

42. A hollow pipe has one or two weights attached to its inner surface as shown in Figure P8.42. Characterize each configuration as being stable, unstable, or neutral equilibrium and explain each of your choices.

43. A particle of mass $m = 5.00$ kg is released from point A on the frictionless track shown in Figure P8.43. Determine (a) the particle's speed at points B and C and (b) the net work done by the force of gravity in moving the particle from A to C.

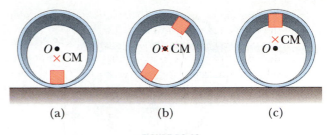

FIGURE P8.42

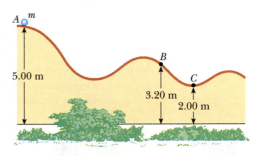

FIGURE P8.43

*Section 8.9   Mass-Energy Equivalence

44. Find the energy equivalence of (a) an electron of mass 9.11×10^{-31} kg, (b) a uranium atom of mass 4.0×10^{-25} kg, (c) a paper clip of mass 2.0 g, and (d) the Earth of mass 5.99×10^{24} kg.

45. The expression for the kinetic energy of a particle moving with speed v is given by Equation 7.20, which can be written as $K = \gamma mc^2 - mc^2$, where $\gamma = [1 - (v/c)^2]^{-1/2}$. The term γmc^2 is the total energy of the particle, and the term mc^2 is its rest energy. A proton moves with a speed of $0.990c$, where c is the speed of light. Find (a) its rest energy, (b) its total energy, and (c) its kinetic energy.

ADDITIONAL PROBLEMS

46. A 200-g particle is released from rest at point A along the horizontal diameter on the inside of a frictionless, hemispherical bowl of radius $R = 30.0$ cm (Fig. P8.46). Calculate (a) its gravitational potential energy at point A relative to point B, (b) its kinetic energy at point B, (c) its speed at point B, and (d) its kinetic energy and potential energy at point C.

47. The particle described in Problem 46 (Fig. P8.46) is released from rest at A, and the surface of the bowl is rough. The speed of the particle at B is 1.50 m/s. (a) What is its kinetic energy at B? (b) How much energy is lost due to friction as the particle moves from A to B? (c) Is it possible to determine μ from these results in any simple manner? Explain.

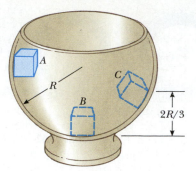

FIGURE P8.46

48. A child's toy consists of a piece of plastic attached to a spring (Fig. P8.48). The spring is compressed 2.0 cm, and the toy is released. If the mass of the toy is 100 g and it rises to a maximum height of 60 cm, estimate the force constant of the spring.

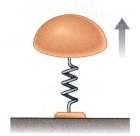

FIGURE P8.48

49. A child slides without friction from a height *h* along a curved water slide (Fig. P8.49). She is launched from a height *h*/5 into the pool. Determine her maximum airborne height *y* in terms of *h* and *θ*.

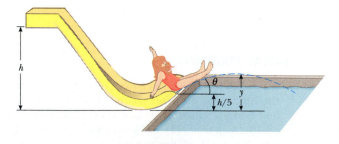

FIGURE P8.49

50. A particle of mass *m* starts from rest and slides down a frictionless track as in Figure P8.50. It leaves the track horizontally, striking the ground as indicated in the sketch. Determine *h*.
51. The masses of the javelin, the discus, and the shot are 0.80 kg, 2.0 kg, and 7.2 kg, respectively, and record throws in the track events using these objects are about 89 m, 69 m, and 21 m, respectively. Neglecting air resistance, (a) calculate the minimum initial kinetic energies that produce these throws and (b) estimate the average force exerted on each object during the throw, assuming the force acts over a distance of 2.0 m. (c) Do your results suggest that air resistance is an important factor?

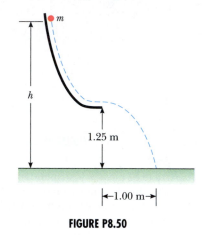

FIGURE P8.50

52. A ball having mass *m* is connected by a string of length *L* to a pivot point and held in place in a vertical position. A constant wind force of magnitude *F* is blowing from left to right as in Figure P8.52a. (a) If the ball is released from rest, show that the maximum height *H* it reaches, as measured from its initial height, is

$$H = \frac{2L}{1 + (mg/F)^2}$$

Assume that the string does not break in the process, and check that the above formula is valid for both $0 \le H \le L$ and $L \le H < 2L$. (b) Compute *H* using the values *m* = 2.00 kg, *L* = 2.00 m, and *F* = 14.7 N. (c) Using these same values, determine the *equilibrium* height of the ball. (d) Can the equilibrium height ever be larger than *L*? Explain.

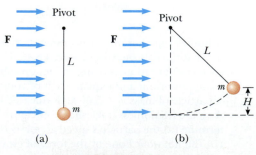

FIGURE P8.52

53. Prove that the following forces are conservative and find the change in potential energy corresponding to each, taking $x_i = 0$ and $x_f = x$: (a) $F_x = ax + bx^2$, (b) $F_x = Ae^{\alpha x}$. (a, b, A, and α are all constants.)

54. A bobsled makes a run down an ice track starting at a point on the track that is a vertical distance of 150 m above ground level. If friction is neglected, what is its speed at the bottom of the hill?

55. A 2.00-kg block situated on a rough incline is connected to a spring of negligible mass having a spring constant of 100 N/m (Fig. P8.55). The block is released from rest when the spring is unstretched, and the pulley is frictionless. The block moves 20.0 cm down the incline before coming to rest. Find the coefficient of kinetic friction between block and incline.

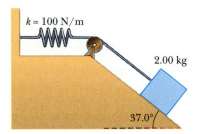

FIGURE P8.55

56. Suppose the incline is frictionless for the system described in Problem 55 (Fig. P8.55). The block is released from rest with the spring initially unstretched. (a) How far does it move down the incline before coming to rest? (b) What is its acceleration at its lowest point? Is the acceleration constant? (c) Describe the energy transformations that occur during the descent.

57. A ball whirls around in a vertical circle at the end of a string. If the ball's total energy remains constant, show that the tension in the string at the bottom is greater than the tension at the top by six times the weight of the ball.

58. A pendulum made of a string of length L and a sphere swing in the vertical plane. The string hits a peg located a distance d below the point of suspension (Fig. P8.58). (a) Show that if the pendulum is

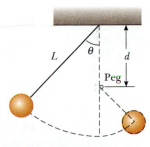

FIGURE P8.58

released from a height below that of the peg, it will return to this height after striking the peg. (b) Show that if the pendulum is released from the horizontal position ($\theta = 90°$) and is to swing in a complete circle centered on the peg, then the minimum value of d must be $3L/5$.

59. A 20.0-kg block is connected to a 30.0-kg block by a string that passes over a frictionless pulley. The 30.0-kg block is connected to a spring that has negligible mass and a force constant of 250 N/m, as in Figure P8.59. The spring is unstretched when the system is as shown in the figure, and the incline is frictionless. The 20.0-kg block is pulled 20.0 cm down the incline (so that the 30.0-kg block is 40.0 cm above the floor) and released from rest. Find the speed of each block when the 30.0-kg block is 20.0 cm above the floor (that is, when the spring is unstretched).

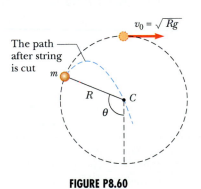

FIGURE P8.59

60. Consider a ball swinging in a vertical plane, with speed $v_0 = \sqrt{Rg}$ at the top of the circle, as in Figure P.860. At what angle θ should the string be cut so that the ball travels through the center of the circle.

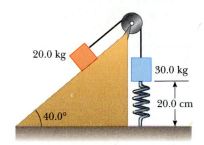

FIGURE P8.60

61. A uniform chain of length 8.0 m initially lies stretched out on a horizontal table. (a) If the coefficient of static friction between chain and table is 0.60, show that the chain begins to slide off the table when 3.0 m of it hangs over the edge. (b) Determine the speed of the chain as all of it leaves the table,

given that the coefficient of kinetic friction between chain and table is 0.40.

62. Jane, whose mass is 50.0 kg, needs to swing across a river (of width D) filled with crocodiles in order to save Tarzan from danger. However, she must swing into a constant horizontal wind force **F** on a vine having length L and initially making an angle θ with the vertical (Fig. P8.62). Taking $D = 50.0$ m, $F = 110$ N, $L = 40.0$ m, and $\theta = 50.0°$, (a) with what minimum speed must Jane begin her swing in order to just make it to the other side? (b) Once the rescue is complete, Tarzan and Jane must swing back across the river. With what minimum speed must they begin their swing? Assume that Tarzan has a mass of 80.0 kg.

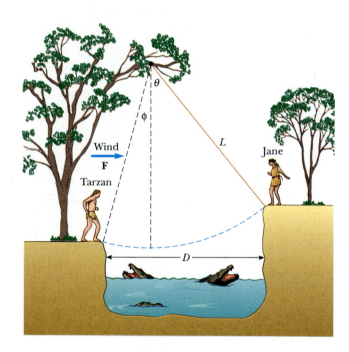

FIGURE P8.62

63. A 3.50-kg particle moves along the x direction under the influence of a force described by the potential energy function $U = (4.70 \text{ J/m}) |x|$, where x is the position of the particle in meters measured from the origin as in Figure P8.63. The total energy of the particle is 15.0 J. (a) Determine the distance it travels from the origin before reversing direction. (b) Find its maximum speed.

64. A 5.0-kg block free to move on a horizontal, frictionless surface is attached to a spring. The spring is compressed 0.10 m from equilibrium and released. The speed of the block is 1.2 m/s when it passes the equilibrium position of the spring. The same experiment is now repeated with the frictionless surface replaced

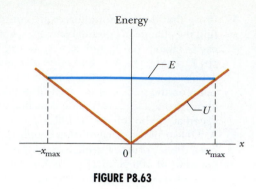

FIGURE P8.63

by a surface for which $\mu_k = 0.30$. Determine the speed of the block at the equilibrium position of the spring.

65. A block of mass 0.500 kg is pushed against a horizontal spring of negligible mass, compressing the spring a distance of Δx (Fig. P8.65). The spring constant is 450 N/m. When released, the block travels along a frictionless, horizontal surface to point B, the bottom of a vertical circular track of radius $R = 1.00$ m, and continues to move up the track. The speed of the block at the bottom of the track is $v_B = 12$ m/s, and the block experiences an average frictional force of 7.00 N while sliding up the track. (a) What is Δx? (b) What is the speed of the block at the top of the track? (c) Does the block reach the top of the track, or does it fall off before reaching the top?

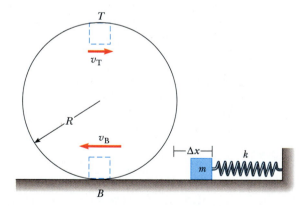

FIGURE P8.65

66. Two identical massless springs, both of constant $k = 200$ N/m, are fixed at opposite ends of a level track. A 5.00-kg block is pressed against the left spring, compressing it by 0.150 m. The block (initially at rest) is then released, as shown in Figure P8.66a. The entire track is frictionless *except* for the section between A and B. Given that the coefficient of kinetic friction between block and track along AB is $\mu_k = 0.080$, and given that the length of AB is 0.250 m,

(a) determine the maximum compression of the spring on the right (Fig. P8.66b). (b) Determine where the block eventually comes to rest, as measured from A (Fig. P8.66c).

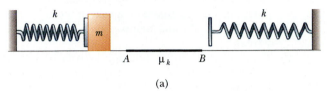

(a)

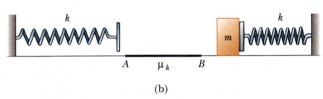

(b)

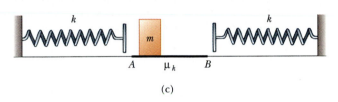

(c)

FIGURE P8.66

67. A 50-kg block and 100-kg block are connected by a string as in Figure P8.67. The pulley is frictionless and of negligible mass. The coefficient of kinetic friction between the 50-kg block and incline is $\mu_k = 0.25$. Determine the change in the kinetic energy of the 50-kg block as it moves from C to D, a distance of 20 m.

67A. A block of mass m_1 and a block of mass m_2 are connected by a string as in Figure P8.67. The pulley is frictionless and of negligible mass. The coefficient of kinetic friction between m_1 and incline is μ_k. Determine the change in the kinetic energy of m_1 as it moves from C to D, a distance d.

68. A pinball machine launches a 100-g ball with a spring-driven plunger (Fig. P8.68). The game board is inclined at 8° above the horizontal. Find the force constant k of the spring that will give the ball a speed of 80 cm/s when the plunger is released from rest with the spring compressed 5.0 cm from its relaxed position. Assume that the plunger's mass and frictional effects are negligible.

69. A 1.0-kg mass slides to the right on a surface having a coefficient of friction $\mu = 0.25$ (Fig. P8.69). The mass has a speed of $v_i = 3.0$ m/s when contact is made with a spring that has a spring constant $k = 50$ N/m. The mass comes to rest after the spring has

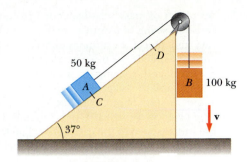

FIGURE P8.67

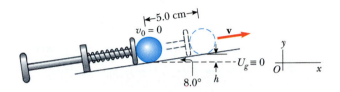

FIGURE P8.68

been compressed a distance d. The mass is then forced toward the left by the spring and continues to move in that direction beyond the unstretched position. Finally the mass comes to rest a distance D to the left of the unstretched spring. Find (a) the compressed distance d, (b) the speed v at the unstretched position when the system is moving to the left, and (c) the distance D where the mass comes to rest.

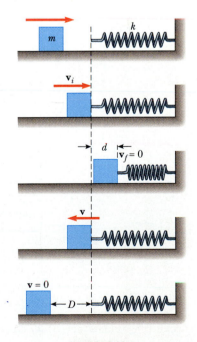

FIGURE P8.69

70. An object of mass m is suspended from the top of a cart by a string of length L, as in Figure P8.70a. The cart and object are initially moving to the right at constant speed v_0. The cart comes to rest after colliding with and sticking to a bumper, as in Figure P8.70b, and the suspended object swings through an angle θ. (a) Show that the cart speed is $v_0 = \sqrt{2gL(1 - \cos\theta)}$. (b) If $L = 1.2$ m and $\theta = 35°$, find the initial speed of the cart. (*Hint:* The force exerted by the string on the object does no work on it.)

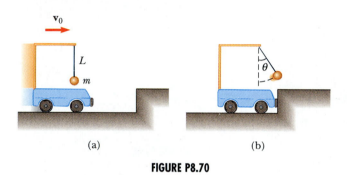

(a)   (b)

FIGURE P8.70

SPREADSHEET PROBLEMS

S1. The potential energy function of a particle is

$$U(x) = \tfrac{1}{2}kx^2 + bx^3 + c$$

where $k = 300$ N/m, $b = -12.0$ N/m^2, and $c = -1000$ J. Use Spreadsheet 8.1 to plot the function from $x = -5.00$ m to $x = +15.0$ m. Describe the general motion of a particle under the influence of this potential energy function. How does the motion change when the initial energy of the object is increased?

S2. Using Spreadsheet 8.1, numerically differentiate the potential energy function given in Problem S8.1. Find the equilibrium points. Are they stable or unstable?

S3. The potential energy function associated with the force between two atoms is modeled by the Lennard-Jones potential

$$U(x) = 4\mathcal{E}\left[\left(\frac{\sigma}{x}\right)^{12} - \left(\frac{\sigma}{x}\right)^6\right]$$

In this model, there are two adjustable parameters, σ and $\mathcal{E}$, that are determined from experiments (σ is a range parameter and $\mathcal{E}$ is the well depth). Modify Spreadsheet 8.1 to plot $U(x)$ versus x for $\sigma = 0.263$ nm and $\mathcal{E} = 1.51 \times 10^{-22}$ J. (These are typical values for helium-helium interactions.) Numerically integrate $U(x)$ to find the force F_x.

CHAPTER 9

Linear Momentum and Collisions

As a result of the collision between the bowling ball and pin, part of the ball's momentum is transferred to the pin. Consequently, the pin acquires momentum and kinetic energy, while the ball loses momentum and kinetic energy. However, the total momentum of the system (ball and pin) remains constant. *(Ben Rose/The Image Bank)*

Consider what happens when a golf ball is struck by a club. The ball is given a very large initial velocity as a result of the collision; consequently, it is able to travel more than a hundred meters through the air. The ball experiences a large change in velocity and a correspondingly large acceleration. Furthermore, because the ball experiences this acceleration over a very short time interval, the average force on it during the collision is very large. By Newton's third law, the club experiences a reaction force that is equal to and opposite the force on the ball. This reaction force produces a change in velocity of the club. Because the club is much more massive than the ball, however, the change in velocity of the club is much less than the change in velocity of the ball.

One of the main objectives of this chapter is to enable you to understand and analyze such events. As a first step, we shall introduce the concept of *momentum,* a term that is used in describing objects in motion. For example, a very massive football player is often said to have a great deal of momentum as he runs down the field. A much less massive player, such as a halfback, can have equal or greater momentum if his speed is greater than that of the more massive player. This follows from the fact that momentum is defined as the product of mass and velocity.

The concept of momentum leads us to a second conservation law, that of conservation of momentum. This law is especially useful for treating problems that involve collisions between objects and for analyzing rocket propulsion. The concept of the center of mass of a system of particles is also introduced, and we shall see that the motion of a system of particles can be represented by the motion of one representative particle located at the center of mass.

9.1   LINEAR MOMENTUM AND ITS CONSERVATION

The linear momentum of a particle of mass m moving with a velocity $\mathbf{v}$ is defined to be the product of the mass and velocity:

$$\mathbf{p} \equiv m\mathbf{v} \tag{9.1}$$

Definition of linear momentum of a particle

Linear momentum is a vector quantity since it equals the product of a scalar, m, and a vector, $\mathbf{v}$. Its direction is along $\mathbf{v}$, and it has dimensions of ML/T. The SI unit for linear momentum is the kg·m/s.

If a particle is moving in an arbitrary direction, $\mathbf{p}$ will have three components and Equation 9.1 is equivalent to the component equations

$$p_x = mv_x \qquad p_y = mv_y \qquad p_z = mv_z \tag{9.2}$$

As you can see from its definition, the concept of momentum provides a quantitative distinction between heavy and light particles moving at the same velocity. For example, the momentum of a bowling ball moving at 10 m/s is much greater than that of a tennis ball moving at the same speed. Newton called the product $m\mathbf{v}$ *quantity of motion*, perhaps a more graphic description than *momentum*, which comes from the Latin word for movement.

Using Newton's second law of motion, we can relate the linear momentum of a particle to the resultant force acting on the particle: *The time rate of change of the linear momentum of a particle is equal to the resultant force acting on the particle.*[1] That is,

$$\mathbf{F} = \frac{d\mathbf{p}}{dt} \tag{9.3}$$

From Equation 9.3 we see that if the resultant force is zero, the time derivative of the momentum is zero, and therefore the linear momentum[2] of the particle must be constant. In other words, the linear momentum of a particle is *constant* when $\mathbf{F} = 0$. Of course, if the particle is *isolated*, then by necessity, $\mathbf{F} = 0$ and $\mathbf{p}$ remains unchanged.

Conservation of Momentum for a Two-Particle System

Consider two particles that can interact with each other but are isolated from their surroundings (Fig. 9.1). That is, the particles may exert a force on each other, but no external forces are present. It is important to note the impact of Newton's third

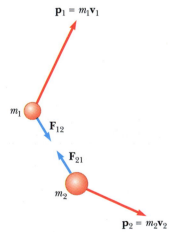

$\mathbf{p}_1 = m_1\mathbf{v}_1$

m_1

$\mathbf{F}_{12}$

$\mathbf{F}_{21}$

m_2

$\mathbf{p}_2 = m_2\mathbf{v}_2$

FIGURE 9.1   At some instant, the momentum of m_1 is $\mathbf{p}_1 = m_1\mathbf{v}_1$ and the momentum of m_2 is $\mathbf{p}_2 = m_2\mathbf{v}_2$. Note that $\mathbf{F}_{12} = -\mathbf{F}_{21}$.

[1] The formula $\mathbf{F} = d\mathbf{p}/dt$ is valid in relativity provided we use the relationship $\mathbf{p} = m\mathbf{v}/(1 - v^2/c^2)^{1/2}$ for the momentum. We shall return to the relativistic treatment of motion in Chapter 39.

[2] In this chapter, the terms "momentum" and "linear momentum" mean the same thing. Later, in Chapter 11, we will use the term "angular momentum" when dealing with rotational motion.

law on this analysis. Recall from Chapter 5 that Newton's third law states that the forces on these two particles are equal in magnitude and opposite in direction. Thus, if an internal force (say a gravitational force) acts on particle 1, then there must be a second internal force, equal in magnitude but opposite in direction, which acts on particle 2.

Suppose that at some instant, the momentum of particle 1 is $\mathbf{p}_1$ and that of particle 2 is $\mathbf{p}_2$. Applying Newton's second law to each particle, we can write

$$\mathbf{F}_{12} = \frac{d\mathbf{p}_1}{dt} \quad \text{and} \quad \mathbf{F}_{21} = \frac{d\mathbf{p}_2}{dt}$$

where $\mathbf{F}_{12}$ is the force exerted on particle 1 by particle 2 and $\mathbf{F}_{21}$ is the force exerted on particle 2 by particle 1. (These forces could be gravitational forces, or have some other origin. This really isn't important for the present discussion.) Newton's third law tells us that $\mathbf{F}_{12}$ and $\mathbf{F}_{21}$ are equal in magnitude and opposite in direction. That is, they form an action-reaction pair, $\mathbf{F}_{12} = -\mathbf{F}_{21}$. We can express this condition as

$$\mathbf{F}_{12} + \mathbf{F}_{21} = 0$$

or as

$$\frac{d\mathbf{p}_1}{dt} + \frac{d\mathbf{p}_2}{dt} = \frac{d}{dt}(\mathbf{p}_1 + \mathbf{p}_2) = 0$$

Since the time derivative of the total momentum ($\mathbf{p}_{tot} = \mathbf{p}_1 + \mathbf{p}_2$) is *zero*, we conclude that the *total* momentum, $\mathbf{p}_{tot}$, of the system must remain constant:

$$\mathbf{p}_{tot} = \mathbf{p}_1 + \mathbf{p}_2 = \text{constant} \tag{9.4}$$

or, equivalently,

$$\mathbf{p}_{1i} + \mathbf{p}_{2i} = \mathbf{p}_{1f} + \mathbf{p}_{2f} \tag{9.5}$$

where $\mathbf{p}_{1i}$ and $\mathbf{p}_{2i}$ are initial values and $\mathbf{p}_{1f}$ and $\mathbf{p}_{2f}$ are final values of the momentum during a time period, *dt*, over which the reaction pair interacts. Equation 9.5 in component form says that the total momenta in the *x, y,* and *z* directions are all independently conserved; that is,

$$p_{ix} = p_{fx} \qquad p_{iy} = p_{fy} \qquad p_{iz} = p_{fz} \tag{9.6}$$

This result is known as the **law of conservation of linear momentum.** It is considered as one of the most important laws of mechanics. We can state it as follows:

> **Whenever two isolated, uncharged particles interact with each other, their total momentum remains constant.**

That is, *the total momentum of an isolated system at all times equals its initial momentum.*

We can also describe the law of conservation of momentum in another way. Since we require that the system be isolated, no external forces are present, and the total momentum of the system remains constant.

Notice that we have made no statement concerning the nature of the forces acting on the system. The only requirement was that the forces must be *internal* to the system. Thus, momentum is constant for an isolated two-particle system *regardless* of the nature of the internal forces. One can use a similar and equivalent

The force from a nitrogen-propelled, hand-controlled device allows an astronaut to move about freely in space without restrictive tethers. *(Courtesy of NASA)*

Conservation of momentum

argument to show that the law of conservation of linear momentum also applies to an isolated system of many particles.

EXAMPLE 9.1   The Recoiling Pitching Machine

A baseball player uses a pitching machine to help him improve his batting average. He places the 50-kg machine on a frozen pond as in Figure 9.2. The machine fires a 0.15-kg baseball horizontally with a velocity of 36i m/s. What is the recoil velocity of the machine?

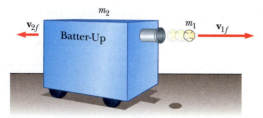

FIGURE 9.2 (Example 9.1) When the baseball is fired horizontally to the right, the pitching machine recoils to the left. The total momentum of the system before and after firing is zero.

Reasoning   We take the system to consist of the baseball and the pitching machine. Because of the force of gravity and the normal force, the system is not really isolated. However, both of these forces are directed perpendicularly to the mo-

tion of the system. Therefore, momentum is constant in the x direction because there are no external forces in this direction (assuming the surface is frictionless).

Solution   The total momentum of the system before firing is zero ($m_1 v_{1i} + m_2 v_{2i} = 0$). Therefore, the total momentum after firing must be zero; that is,

$$m_1 v_{1f} + m_2 v_{2f} = 0$$

With $m_1 = 0.15$ kg, $v_{1i} = 36i$ m/s, and $m_2 = 50$ kg, solving for v_{2f}, we find the recoil velocity of the pitching machine to be

$$v_{2f} = -\frac{m_2}{m_2} v_{1f} = -\left(\frac{0.15 \text{ kg}}{50 \text{ kg}}\right)(36i \text{ m/s}) = -0.11i \text{ m/s}$$

The negative sign for v_{2f} indicates that the pitching machine is moving to the left after firing, in the direction opposite the direction of motion of the cannon. In the words of Newton's third law, for every force (to the left) on the pitching machine, there is an equal but opposite force (to the right) on the ball. Because the pitching machine is much more massive than the ball, the acceleration and consequent speed of the pitching machine are much smaller than the acceleration and speed of the ball.

EXAMPLE 9.2   Decay of the Kaon at Rest

A meson is a nuclear particle that is more massive than an electron but less massive than a proton or neutron. One type of meson, called the neutral kaon (K^0), decays into a pair of charged pions (π^+ and π^-) that are oppositely charged but equal in mass, as in Figure 9.3. A pion is a particle associated with the strong nuclear force that binds the protons and neutrons together in the nucleus. Assuming the kaon is initially at rest, prove that after the decay, the two pions must have momenta that are equal in magnitude and opposite in direction.

Solution   The decay of the kaon, represented in Figure 9.3, can be written

$$K^0 \longrightarrow \pi^+ + \pi^-$$

If we let p^+ be the momentum of the positive pion and p^- be the momentum of the negative pion after the decay, then the final momentum of the system can be written

$$p_f = p^+ + p^-$$

Because the pion is at rest before the decay, we know that $p_i = 0$. Furthermore, because momentum is conserved, $p_i =$

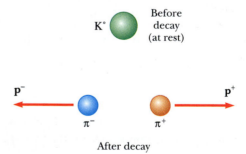

FIGURE 9.3 (Example 9.2) A kaon at rest decays spontaneously into a pair of oppositely charged pions. The pions move apart with momenta that are equal in magnitude but opposite in direction.

$p_f = 0$, so that $p^+ + p^- = 0$, or

$$p^+ = -p^-$$

Thus, we see that the two momentum vectors of the pions are equal in magnitude and opposite in direction.

9.2   IMPULSE AND MOMENTUM

As we have seen, the momentum of a particle changes if a net force acts on the particle. Let us assume that a single force **F** acts on a particle and that this force may vary with time. According to Newton's second law, $\mathbf{F} = d\mathbf{p}/dt$, or

$$d\mathbf{p} = \mathbf{F}\,dt \tag{9.7}$$

We can integrate this expression to find the change in the momentum of a particle. If the momentum of the particle changes from $\mathbf{p}_i$ at time t_i to $\mathbf{p}_f$ at time t_f, then integrating Equation 9.7 gives

$$\Delta\mathbf{p} = \mathbf{p}_f - \mathbf{p}_i = \int_{t_i}^{t_f} \mathbf{F}\,dt \tag{9.8}$$

The quantity on the right side of Equation 9.8 is called the *impulse* of the force **F** for the time interval $\Delta t = t_f - t_i$. Impulse is a vector defined by

$$\mathbf{I} \equiv \int_{t_i}^{t_f} \mathbf{F}\,dt = \Delta\mathbf{p} \tag{9.9}$$

Impulse of a force

That is, the impulse of the force F equals the change in the momentum of the particle.

Impulse-momentum theorem

This statement, known as the **impulse-momentum theorem**, is equivalent to Newton's second law. From this definition, we see that impulse is a vector quantity having a magnitude equal to the area under the force-time curve, as described in Figure 9.4a. In this figure, it is assumed that the force varies in time in the general manner shown and is nonzero in the time interval $\Delta t = t_f - t_i$. The direction of the impulse vector is the same as the direction of the change in momentum. Impulse has the dimensions of momentum, that is, ML/T. Note that impulse is *not* a property of the particle itself; rather, it is a measure of the degree to which an external force changes the momentum of the particle. Therefore, when we say that an impulse is given to a particle, it is implied that momentum is transferred from an external agent to that particle.

Since the force can generally vary in time as in Figure 9.4a, it is convenient to define a time-averaged force $\bar{\mathbf{F}}$ as

$$\bar{\mathbf{F}} \equiv \frac{1}{\Delta t}\int_{t_i}^{t_f} \mathbf{F}\,dt \tag{9.10}$$

where $\Delta t = t_f - t_i$. Therefore, we can express Equation 9.9 as

$$\mathbf{I} = \Delta\mathbf{p} = \bar{\mathbf{F}}\,\Delta t \tag{9.11}$$

This average force, described in Figure 9.4b, can be thought of as the constant force that would give the same impulse to the particle in the time interval Δt as the actual time-varying force gives over this same interval.

In principle, if **F** is known as a function of time, the impulse can be calculated from Equation 9.9. The calculation becomes especially simple if the force acting on the particle is constant. In this case, $\bar{\mathbf{F}} = \mathbf{F}$ and Equation 9.11 becomes

$$\mathbf{I} = \Delta\mathbf{p} = \mathbf{F}\,\Delta t \tag{9.12}$$

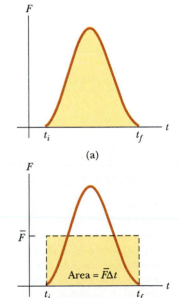

FIGURE 9.4   (a) A force acting on a particle may vary in time. The impulse is the area under the force versus time curve. (b) The average force (horizontal dashed line) gives the same impulse to the particle in the time Δt as the time-varying force described in (a).

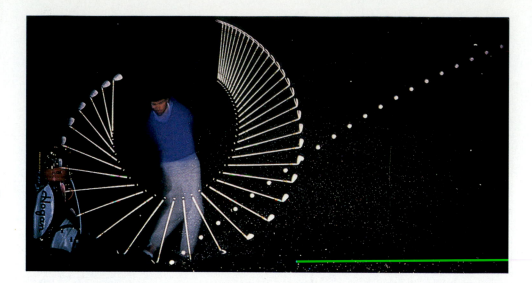

During the brief time the club is in contact with the ball, the ball gains momentum as a result of the collision, and the club loses the same amount of momentum. *(Courtesy Michel Hans/Photo Researchers)*

In many physical situations, we shall use the so-called **impulse approximation:** *Assume that one of the forces exerted on a particle acts for a short time but is much larger than any other force present.* This approximation is especially useful in treating collisions, where the duration of the collision is very short. When this approximation is made, we refer to the force as an *impulsive force*. For example, when a baseball is struck with a bat, the time of the collision is about 0.01 s, and the average force the bat exerts on the ball in this time is typically several thousand newtons. This is much greater than the force of gravity, so the impulse approximation is justified. When we use this approximation, it is important to remember that $\mathbf{p}_i$ and $\mathbf{p}_f$ represent the momenta *immediately* before and after the collision, respectively. Therefore, in the impulse approximation very little motion of the particle takes place during the collision.

EXAMPLE 9.3   Teeing Off

A golf ball of mass 50 g is struck with a club (Fig. 9.5). The force on the ball varies from zero when contact is made up to some maximum value (where the ball is deformed) back to zero when the ball leaves the club. Thus, the force-time curve is qualitatively described by Figure 9.4. Assuming that the ball travels 200 m, estimate the magnitude of the impulse due to the collision.

Solution   Neglecting air resistance, we can use Equation 4.18 for the range of a projectile:

$$R = \frac{v_0{}^2}{g} \sin 2\theta_0$$

Let us assume that the launch angle is 45°, the angle that provides the maximum range for any given launch speed. The initial speed of the ball is then estimated to be

$$v_0 = \sqrt{Rg} = \sqrt{(200 \text{ m})(9.80 \text{ m/s}^2)} = 44 \text{ m/s}$$

FIGURE 9.5   A golf ball being struck by a club. *(© Harold E. Edgerton. Courtesy of Palm Press, Inc.)*

Since $v_i = 0$ and $v_f = v_0$ for the ball, the magnitude of the impulse imparted to the ball is

$$I = \Delta p = mv_0 = (50 \times 10^{-3} \text{ kg})\left(44\ \frac{\text{m}}{\text{s}}\right) = \boxed{2.2 \text{ kg} \cdot \text{m/s}}$$

Exercise   If the club is in contact with the ball 4.5×10^{-4} s, estimate the magnitude of the average force exerted by the club on the ball.

Answer 4.9×10^3 N. This force is extremely large compared with the weight (gravity force) of the ball, which is only 0.49 N.

EXAMPLE 9.4   How Good Are the Bumpers?

In a particular crash test, an automobile of mass 1500 kg collides with a wall as in Figure 9.6a. The initial and final velocities of the automobile are $\mathbf{v}_i = -15.0\mathbf{i}$ m/s and $\mathbf{v}_f = 2.6\mathbf{i}$ m/s. If the collision lasts for 0.150 s, find the impulse due to the collision and the average force exerted on the automobile.

Solution   The initial and final momenta of the automobile are

$$\mathbf{p}_i = m\mathbf{v}_i = (1500 \text{ kg})(-15.0\mathbf{i} \text{ m/s})$$
$$= -2.25 \times 10^4\mathbf{i} \text{ kg} \cdot \text{m/s}$$

$$\mathbf{p}_f = m\mathbf{v}_f = (1500 \text{ kg})(2.6\mathbf{i} \text{ m/s})$$
$$= 0.39 \times 10^4\mathbf{i} \text{ kg} \cdot \text{m/s}$$

Hence, the impulse is

$$\mathbf{I} = \Delta\mathbf{p} = \mathbf{p}_f - \mathbf{p}_i = 0.39 \times 10^4\mathbf{i} \text{ kg} \cdot \text{m/s}$$
$$- (-2.25 \times 10^4\mathbf{i} \text{ kg} \cdot \text{m/s})$$

$$\mathbf{I} = \boxed{2.64 \times 10^4\mathbf{i} \text{ kg} \cdot \text{m/s}}$$

The average force exerted on the automobile is

$$\overline{\mathbf{F}} = \frac{\Delta\mathbf{p}}{\Delta t} = \frac{2.64 \times 10^4\mathbf{i} \text{ kg} \cdot \text{m/s}}{0.150 \text{ s}} = \boxed{1.76 \times 10^5\mathbf{i} \text{ N}}$$

Before

−15.0 m/s

After

2.6 m/s

FIGURE 9.6   (Example 9.4).

CONCEPTUAL EXAMPLE 9.5

A boxer wisely moves his head backward just before receiving a punch. How does this maneuver help reduce the force of impact?

Reasoning   As the boxer moves away from the moving fist, the time his head is in contact with the fist is increased. The impulse-momentum theorem tells us that the average force exerted on an object multiplied by the time during which the force acts equals the change in momentum of the object. The average force exerted on the boxer's head is reduced when he extends the time of contact by moving away from his opponent's fist. For the same reason, it is better to stop a truck out of control by hitting a haystack (long collision time and small force), rather than a brick wall (short collision time and large force).

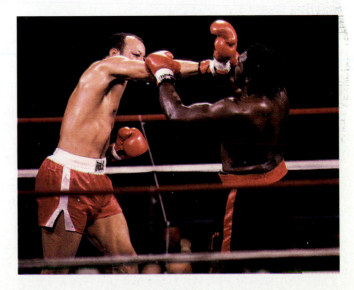

(Conceptual Example 9.5) Boxer moving away from a punch. *(Focus on Sports)*

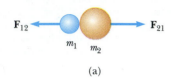

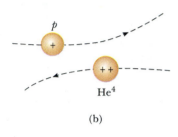

FIGURE 9.7   (a) The collision between two objects as the result of direct physical contact. (b) The "collision" between two charged particles which results in a change in direction for each particle. In this case the action-reaction forces are electrostatic, and the two particles do not experience physical contact.

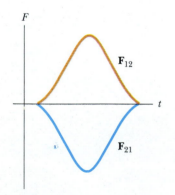

FIGURE 9.8   The impulse forces as a function of time for the two colliding particles described in Figure 9.7a. Note that $\mathbf{F}_{12} = -\mathbf{F}_{21}$.

CONCEPTUAL EXAMPLE 9.6

A karate expert is able to break a stack of boards with a swift blow with the side of his bare hand. How is this possible?

Reasoning   The arm and hand have a large momentum before the collision with the boards. This momentum is reduced quickly as the hand is in contact with the boards for a short time. Since the contact time is very small, the impact force (the force of the hand on the boards) is very large. Thus, a karate expert obtains the best result by delivering the blow in a short time.

9.3   COLLISIONS

In this section we use the law of conservation of linear momentum to describe what happens when two particles collide. We use the term **collision** to represent the event of two particles coming together for a short time and thereby producing impulsive forces on each other. *The force due to the collision is assumed to be much larger than any external forces present.*

A collision may entail physical contact between two macroscopic objects, as described in Figure 9.7a, but the notion of what we mean by collision must be generalized because "physical contact" on a submicroscopic scale is ill-defined and hence meaningless. To understand this, consider a collision on an atomic scale (Fig. 9.7b), such as the collision of a proton with an alpha particle (the nucleus of the helium atom). Because the two particles are positively charged, they never come into physical contact with each other; instead, they repel each other because of the strong electrostatic force between them at close separations. When two particles of masses m_1 and m_2 collide as in Figure 9.7, the impulse forces may vary in time in a complicated way, one of which is described in Figure 9.8. If $\mathbf{F}_{12}$ is the force exerted on m_1 by m_2, and if we assume that no external forces act on the particles, then the change in momentum of m_1 due to the collision is given by Equation 9.8:

$$\Delta \mathbf{p}_1 = \int_{t_i}^{t_f} \mathbf{F}_{12} \, dt$$

Likewise, if $\mathbf{F}_{21}$ is the force exerted on m_2 by m_1, the change in momentum of m_2 is

$$\Delta \mathbf{p}_2 = \int_{t_i}^{t_f} \mathbf{F}_{21} \, dt$$

Newton's third law states that the force exerted on m_1 by m_2 is equal to and opposite the force exerted on m_2 by m_1. Hence, we conclude that

$$\Delta \mathbf{p}_1 = -\Delta \mathbf{p}_2$$
$$\Delta \mathbf{p}_1 + \Delta \mathbf{p}_2 = 0$$

Since the total momentum of the system is $\mathbf{p}_{tot} = \mathbf{p}_1 + \mathbf{p}_2$, we conclude that the *change* in the momentum of the system due to the collision is zero:

$$\mathbf{p}_{tot} = \mathbf{p}_1 + \mathbf{p}_2 = \text{constant}$$

This is precisely what we expect as an overall result because there are no external forces acting on the system (Section 9.2). Since the impulsive forces due to the collision are internal, they do not change the total momentum of the system (only external forces can do that). Therefore, we conclude that *the total momentum of a system just before the collision equals the total momentum of a system just after the collision.*

Momentum is conserved for any collision

EXAMPLE 9.7 Carry Collision Insurance

An 1800-kg car stopped at a traffic light is struck from the rear by a 900-kg car and the two become entangled. If the smaller car was moving at 20 m/s before the collision, what is the speed of the entangled mass after the collision?

Reasoning The total momentum of the system (the two cars) before the collision equals the total momentum of the system after the collision because momentum is conserved for any type of collision.

Solution The magnitude of the total momentum of the system before the collision is equal to that of the smaller car because the larger car is initially at rest:

$$p_i = m_i v_i = (900 \text{ kg})(20 \text{ m/s}) = 1.80 \times 10^4 \text{ kg} \cdot \text{m/s}$$

After the collision, the mass that moves is the sum of the masses of the cars. The magnitude of the momentum of the combination is

$$p_f = (m_1 + m_2) v_f = (2700 \text{ kg}) v_f$$

Equating the momentum before to the momentum after and solving for v_f, the speed of the combined mass, we have

$$v_f = \frac{p_i}{m_1 + m_2} = \frac{1.80 \times 10^4 \text{ kg} \cdot \text{m/s}}{2700 \text{ kg}} = \boxed{6.67 \text{ m/s}}$$

CONCEPTUAL EXAMPLE 9.8

As a ball falls toward the Earth, its momentum increases. How would you reconcile this fact with the law of conservation of momentum?

Reasoning The momentum of the ball increases as it accelerates downward. A large change in its momentum occurs when it strikes the floor and reverses its direction of motion. The external forces exerted on the ball are the downward force of gravity and the upward normal force that acts while the ball is in contact with the floor to change its momentum. However, if we think of the ball and Earth together as our system, these forces are internal and do not change the total momentum of the system. That is, the total momentum of the system (ball and Earth) is constant. As the ball falls, the Earth moves up to meet it, on the order of 10^{25} times more slowly. Because the Earth's mass is so large, its upward motion is negligibly small.

CONCEPTUAL EXAMPLE 9.9

An open box slides across an icy (frictionless) surface of a frozen lake. What happens to the speed of the box as water from a rain shower collects in it, assuming that the rain falls vertically downward into the box? Explain.

Reasoning The rain initially has no horizontal component of momentum. The rain drops acquire a horizontal component of momentum as they collide with the box. Because the momentum of the system (box plus rain) must be conserved, the horizontal component of momentum of the box must decrease. Hence, the speed of the box decreases continuously as it collects water.

9.4 ELASTIC AND INELASTIC COLLISIONS IN ONE DIMENSION

As we have seen, momentum is conserved in any type of collision. Kinetic energy, however, is generally *not* constant in a collision because some of it is converted to thermal energy, to internal elastic potential energy when the objects are deformed, and to rotational energy.

Inelastic collision

We define various types of collisions on the basis of whether or not kinetic energy is constant. An **inelastic collision** is one in which *total kinetic energy is not constant (even though momentum is constant)*. The collision of a rubber ball with a hard surface is inelastic since some of the kinetic energy of the ball is lost when it is deformed while in contact with the surface. When two objects collide and stick together after the collision, some kinetic energy is lost, and the collision is called **perfectly inelastic**. For example, if two vehicles collide and become entangled, as in Example 9.7, they move with some common velocity after the perfectly inelastic collision. If a meteorite collides with the Earth, it becomes buried, and the collision is perfectly inelastic.

Elastic collision

An **elastic collision** is one in which *total kinetic energy is constant (as well as momentum)*. Billiard-ball collisions and the collisions of air molecules with the walls of a container at ordinary temperatures are highly elastic. Real collisions in the macroscopic world are only approximately elastic because some deformation and loss of kinetic energy take place. Collisions between atomic and subatomic particles may also be inelastic, but are often elastic on the average. Elastic and perfectly inelastic collisions are limiting cases; most collisions fall in a category between them.

Perfectly Inelastic Collisions

Consider two particles of masses m_1 and m_2 moving with initial velocities $\mathbf{v}_{1i}$ and $\mathbf{v}_{2i}$ along a straight line, as in Figure 9.9. If the two particles collide head-on, stick together, and move with some common velocity $\mathbf{v}_f$ after the collision, the collision is perfectly inelastic. Therefore, we can say that the total momentum before the collision equals the total momentum of the composite system after the collision:

$$m_1\mathbf{v}_{1i} + m_2\mathbf{v}_{2i} = (m_1 + m_2)\mathbf{v}_f \tag{9.13}$$

$$\mathbf{v}_f = \frac{m_1\mathbf{v}_{1i} + m_2\mathbf{v}_{2i}}{m_1 + m_2} \tag{9.14}$$

Elastic Collisions

Now consider two particles that undergo an elastic head-on collision (Fig. 9.10). In this case, both momentum and kinetic energy are constant; therefore, we can write

$$m_1\mathbf{v}_{1i} + m_2\mathbf{v}_{2i} = m_1\mathbf{v}_{1f} + m_2\mathbf{v}_{2f} \tag{9.15}$$

$$\tfrac{1}{2}m_1v_{1i}^2 + \tfrac{1}{2}m_2v_{2i}^2 = \tfrac{1}{2}m_1v_{1f}^2 + \tfrac{1}{2}m_2v_{2f}^2 \tag{9.16}$$

Because all velocities in Figure 9.10 are to the left or right, they can be represented by the corresponding speeds, where v is taken to be positive if a particle moves to the right and negative if it moves to the left.

Before collision

(a)

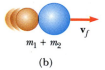

After collision

$m_1 + m_2$

(b)

FIGURE 9.9 Schematic representation of a perfectly inelastic head-on collision between two particles.

Before collision   After collision

m_1 (a) m_2 (b)

FIGURE 9.10 Schematic representation of an elastic head-on collision between two particles.

In a typical problem involving elastic collisions, there are two unknown quantities, and Equations 9.15 and 9.16 can be solved simultaneously to find these. However, an alternative approach, one that involves a little mathematical manipulation of Equation 9.16, often simplifies this process. To see this, let's cancel the factor of $\frac{1}{2}$ in Equation 9.16 and rewrite it as

$$m_1(v_{1i}^2 - v_{1f}^2) = m_2(v_{2f}^2 - v_{2i}^2)$$

Here we have moved the terms containing m_1 to one side of the equation and those containing m_2 to the other. Next, let us factor both sides:

$$m_1(v_{1i} - v_{1f})(v_{1i} + v_{1f}) = m_2(v_{2f} - v_{2i})(v_{2f} + v_{2i}) \qquad (9.17)$$

We now separate the terms containing m_1 and m_2 in the equation for the conservation of momentum (Eq. 9.15) to get

$$m_1(v_{1i} - v_{1f}) = m_2(v_{2f} - v_{2i}) \qquad (9.18)$$

To obtain our final result, we divide Equation 9.17 by Equation 9.18 and get

$$v_{1i} + v_{1f} = v_{2f} + v_{2i}$$

$$v_{1i} - v_{2i} = -(v_{1f} - v_{2f}) \qquad (9.19)$$

This equation, in combination with Equation 9.15, can be used to solve problems dealing with perfectly elastic collisions. According to Equation 9.19, the relative speed of the two objects before the collision, $v_{1i} - v_{2i}$, equals the negative of their relative speed after the collision, $-(v_{1f} - v_{2f})$.

Suppose that the masses and the initial velocities of both particles are known. Equations 9.15 and 9.16 can be solved for the final speeds in terms of the initial speeds, since there are two equations and two unknowns:

$$v_{1f} = \left(\frac{m_1 - m_2}{m_1 + m_2}\right) v_{1i} + \left(\frac{2m_2}{m_1 + m_2}\right) v_{2i} \qquad (9.20)$$

$$v_{2f} = \left(\frac{2m_1}{m_1 + m_2}\right) v_{1i} + \left(\frac{m_2 - m_1}{m_1 + m_2}\right) v_{2i} \qquad (9.21)$$

Elastic collision: relations between final and initial speeds

It is important to remember that the appropriate signs for v_{1i} and v_{2i} must be included in Equations 9.20 and 9.21. For example, if m_2 is moving to the left initially, as we will see in Figure 9.11, then v_{2i} is negative.

Let us consider some special cases: If $m_1 = m_2$, then $v_{1f} = v_{2i}$ and $v_{2f} = v_{1i}$. That is, the particles exchange speeds if they have equal masses. This is nearly what one observes in billiard ball collisions.

If m_2 is initially at rest, $v_{2i} = 0$, and Equations 9.20 and 9.21 become

$$v_{1f} = \left(\frac{m_1 - m_2}{m_1 + m_2}\right) v_{1i} \qquad (9.22)$$

$$v_{2f} = \left(\frac{2m_1}{m_1 + m_2}\right) v_{1i} \qquad (9.23)$$

If m_1 is very large compared with m_2, we see from Equations 9.22 and 9.23 that $v_{1f} \approx v_{1i}$ and $v_{2f} \approx 2v_{1i}$. That is, when a very heavy particle collides head-on with a very light one initially at rest, the heavy particle continues its motion unaltered after the collision, while the light particle rebounds with a speed equal to about twice the initial speed of the heavy particle. An example of such a collision would be that of a moving heavy atom, such as uranium, with a light atom, such as hydrogen.

If m_2 is much larger than m_1 and m_2 is initially at rest, then $v_{1f} \approx -v_{1i}$ and $v_{2f} \approx v_{2i} = 0$. That is, when a very light particle collides head-on with a very heavy particle initially at rest, the light particle has its velocity reversed, while the heavy particle remains approximately at rest. For example, imagine what happens when a marble hits a stationary bowling ball.

EXAMPLE 9.10   The Ballistic Pendulum

The ballistic pendulum (Fig. 9.11) is a system used to measure the speed of a fast-moving projectile, such as a bullet. The bullet is fired into a large block of wood suspended from some light wires. The bullet is stopped by the block, and the entire system swings through a height h. Because the collision is perfectly inelastic and momentum is conserved, Equation 9.14 gives the speed of the system right after the collision when we assume the impulse approximation. The kinetic energy right after the collision is

$$(1) \qquad K = \tfrac{1}{2}(m_1 + m_2)v_f^2$$

With $v_{2i} = 0$, Equation 9.14 becomes

$$(2) \qquad v_f = \frac{m_1 v_{1i}}{m_1 + m_2}$$

Substituting this value of v_f into (1) gives

$$K = \frac{m_1^2 v_{1i}^2}{2(m_1 + m_2)}$$

where v_{1i} is the initial speed of the bullet. Note that this kinetic energy is less than the initial kinetic energy of the bullet. In all the energy changes that take place after the collision, however, energy is constant; the kinetic energy at the bottom is transformed to potential energy at the height h:

$$\frac{m_1^2 v_{1i}^2}{2(m_1 + m_2)} = (m_1 + m_2)gh$$

$$v_{1i} = \left(\frac{m_1 + m_2}{m_1}\right)\sqrt{2gh}$$

Hence, it is possible to obtain the initial speed of the bullet by measuring h and the two masses. Why would it be incorrect to equate the initial kinetic energy of the incoming bullet to the final gravitational energy of the bullet-block combination?

Exercise   In a ballistic pendulum experiment, suppose that $h = 5.00$ cm, $m_1 = 5.00$ g, and $m_2 = 1.00$ kg. Find (a) the

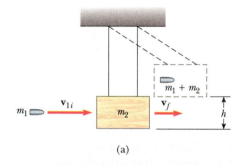

(a)

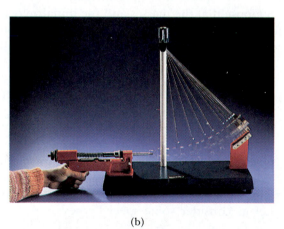

(b)

FIGURE 9.11   (Example 9.10) (a) Diagram of a ballistic pendulum. Note that $\mathbf{v}_f$ is the velocity of the system right after the perfectly inelastic collision. (b) Multiflash photo of a ballistic pendulum used in the laboratory.   *(Courtesy of Central Scientific Co.)*

initial speed of the projectile, and (b) the loss in energy due to the collision.

Answer   199 m/s; 98.5 J.

EXAMPLE 9.11 A Two-Body Collision with Spring

A block of mass $m_1 = 1.60$ kg initially moving to the right with a speed of 4.00 m/s on a frictionless horizontal track collides with a spring attached to a second block of mass $m_2 = 2.10$ kg moving to the left with a speed of 2.50 m/s, as in Figure 9.12a. The spring has a spring constant of 600 N/m. (a) At the instant when m_1 is moving to the right with a speed of 3.00 m/s, as in Figure 9.12b, determine the speed of m_2.

Solution First, note that the initial velocity of m_2 is -2.50 m/s because its direction is to the left. Since the total momentum is conserved, we have

$$m_1 v_{1i} + m_2 v_{2i} = m_1 v_{1f} + m_2 v_{2f}$$

$$(1.60 \text{ kg})(4.00 \text{ m/s}) + (2.10 \text{ kg})(-2.50 \text{ m/s})$$
$$= (1.60 \text{ kg})(3.00 \text{ m/s}) + (2.10 \text{ kg}) v_{2f}$$

$$v_{2f} = -1.74 \text{ m/s}$$

The negative value for v_{2f} means that m_2 is still moving to the left at the instant we are considering.

(b) Determine the distance the spring is compressed at that instant.

Solution To determine the compression in the spring, x, shown in Figure 9.12b, we use conservation of energy since there are no friction or other nonconservative forces acting on the system. Thus, we have

$$\tfrac{1}{2} m_1 v_{1i}^2 + \tfrac{1}{2} m_2 v_{2i}^2 = \tfrac{1}{2} m_1 v_{1f}^2 + \tfrac{1}{2} m_2 v_{2f}^2 + \tfrac{1}{2} kx^2$$

Substituting the given values and the result to part (a) into this expression gives

$$x = 0.173 \text{ m}$$

Exercise Find the velocity of m_1 and the compression in the spring at the instant that m_2 is at rest.

Answer 0.719 m/s to the right; 0.251 m.

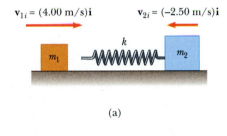

(a)

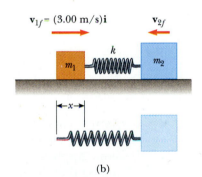

(b)

FIGURE 9.12 (Example 9.11).

EXAMPLE 9.12 Slowing Down Neutrons by Collisions

In a nuclear reactor, neutrons are produced when a $^{235}_{92}\text{U}$ atom splits in a process called fission. These neutrons are moving at about 10^7 m/s and must be slowed down to about 10^3 m/s before they take part in another fission event. They are slowed down by being passed through a solid or liquid material called a *moderator*. The slowing-down process involves elastic collisions. Let us show that a neutron can lose most of its kinetic energy if it collides elastically with a moderator containing light nuclei, such as deuterium (in "heavy water," D_2O) or carbon (in graphite).

Reasoning Because momentum and energy are constant, Equations 9.22 and 9.23 can be applied to the head-on collision of a neutron with the moderator nucleus.

Solution Let us assume that the moderator nucleus of mass m_m is at rest initially and that the neutron of mass m_n and initial speed v_{ni} collides head-on with it. The initial kinetic energy of the neutron is

$$K_{ni} = \tfrac{1}{2} m_n v_{ni}^2$$

After the collision, the neutron has a kinetic energy $\tfrac{1}{2} m_n v_{nf}^2$, where v_{nf} is given by Equation 9.22:

$$K_{nf} = \tfrac{1}{2} m_n v_{nf}^2 = \frac{m_n}{2} \left(\frac{m_n - m_m}{m_n + m_m} \right)^2 v_{ni}^2$$

Therefore, the fraction of the total kinetic energy possessed by the neutron after the collision is

$$(1) \quad f_n = \frac{K_{nf}}{K_{ni}} = \left(\frac{m_n - m_m}{m_n + m_m}\right)^2$$

From this result, we see that the final kinetic energy of the neutron is small when m_m is close to m_n and is zero when $m_n = m_m$.

We can calculate the kinetic energy of the moderator nucleus after the collision using Equation 9.23:

$$K_{mf} = \tfrac{1}{2} m_m v_{mf}^2 = \frac{2 m_n^2 m_m}{(m_n + m_m)^2} v_{ni}^2$$

Hence, the fraction of the total kinetic energy transferred to the moderator nucleus is

$$(2) \quad f_m = \frac{K_{mf}}{K_{ni}} = \frac{4 m_n m_m}{(m_n + m_m)^2}$$

Since the total energy is constant, (2) can also be obtained from (1) with the condition that $f_n + f_m = 1$, so that $f_m = 1 - f_n$.

Suppose that heavy water is used for the moderator. For collisions of the neutrons with deuterium nuclei in D_2O ($m_m = 2 m_n$), $f_n = 1/9$ and $f_m = 8/9$. That is, 89% of the neutron's kinetic energy is transferred to the deuterium nucleus. In practice, the moderator efficiency is reduced because head-on collisions are very unlikely to occur.

How do the results differ with graphite as the moderator?

9.5   TWO-DIMENSIONAL COLLISIONS

In Sections 9.1 and 9.3, we showed that the total momentum of a system of two particles is constant when the system is isolated. For a general collision of two particles, this result implies that the total momentum in each of the directions *x*, *y*, and *z* is constant. However, an important subset of collisions takes place in a plane. The game of billiards is a familiar example involving multiple collisions of objects moving on a two-dimensional surface. For such two-dimensional collisions, we obtain two component equations for conservation of momentum:

$$m_1 v_{1ix} + m_2 v_{2ix} = m_1 v_{1fx} + m_2 v_{2fx}$$

$$m_1 v_{1iy} + m_2 v_{2iy} = m_1 v_{1fy} + m_2 v_{2fy}$$

Let us consider a two-dimensional problem in which a particle of mass m_1 collides with a particle of mass m_2, where m_2 is initially at rest, as in Figure 9.13. After the collision, m_1 moves at an angle θ with respect to the horizontal and m_2 moves at an angle ϕ with respect to the horizontal. This is called a *glancing* collision. Applying the law of conservation of momentum in component form, and noting that the total *y* component of momentum is zero, we get

$$m_1 v_{1i} = m_1 v_{1f} \cos\theta + m_2 v_{2f} \cos\phi \qquad (9.24)$$

$$0 = m_1 v_{1f} \sin\theta - m_2 v_{2f} \sin\phi \qquad (9.25)$$

We now have two independent equations. So long as only two of the preceding quantities are unknown, we can solve the problem completely.

Because the collision is elastic, we can use Equation 9.16 (conservation of

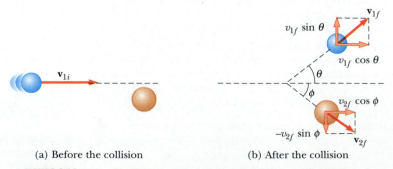

(a) Before the collision   (b) After the collision

FIGURE 9.13   An elastic glancing collision between two particles.

energy), with $v_{2i} = 0$, to give

$$\tfrac{1}{2}m_1v_1{}_i{}^2 = \tfrac{1}{2}m_1v_1{}_f{}^2 + \tfrac{1}{2}m_2v_2{}_f{}^2 \tag{9.26}$$

Knowing the initial speed of the moving particle, and the masses, we are left with four unknowns. Since we only have three equations, one of the four remaining quantities (v_{1f}, v_{2f}, θ, or ϕ) must be given to determine the motion after the collision from conservation principles alone.

If the collision is inelastic, kinetic energy is *not* constant, and Equation 9.26 does *not* apply.

Problem-Solving Strategy
Collisions

The following procedure is recommended when dealing with problems involving collisions between two objects.

- Set up a coordinate system and define your velocities with respect to that system. It is convenient to have the *x* axis coincide with one of the initial velocities.
- In your sketch of the coordinate system, draw and label all velocity vectors and include all the given information.
- Write expressions for the *x* and *y* components of the momentum of each object before and after the collision. Remember to include the appropriate signs for the components of the velocity vectors.

This simulator enables you to model and investigate a wide variety of collisions such as collisions between moving objects and fixed boundaries as well as collisions between two moving objects in one and two dimensions (elastic and inelastic). For example, you will be able to make an animated simulation of a 1200-kg car traveling at 30 m/s colliding with a 1000-kg sports car traveling at 50 m/s in the opposite direction.

Collisions

- Write expressions for the total momentum in the *x* direction *before* and *after* the collision and equate the two. Repeat this procedure for the total momentum in the *y* direction. These steps follow from the fact that, because the momentum of the *system* is constant in any collision, the total momentum along any direction must be constant. Remember, it is the momentum of the *system* that is constant, not the momenta of the individual objects.
- If the collision is inelastic, kinetic energy is not constant, and additional information is probably required. If the collision is totally inelastic, the final velocities of the two objects are equal. Proceed to solve the momentum equations for the unknown quantities.
- If the collision is elastic, kinetic energy is constant, and you can equate the total kinetic energy before the collision to the total kinetic energy after the collision to get an additional relationship between the velocities.

EXAMPLE 9.13   Collision at an Intersection

A 1500-kg car traveling east with a speed of 25.0 m/s collides at an intersection with a 2500-kg van traveling north at a speed of 20.0 m/s, as shown in Figure 9.14. Find the direction and magnitude of the velocity of the wreckage after the collision, assuming that the vehicles undergo a perfectly inelastic collision (that is, they stick together).

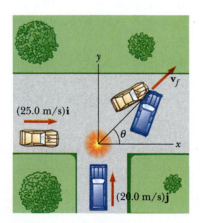

FIGURE 9.14   (Example 9.13) Top view of an eastbound car colliding with a northbound van.

Solution   Let us choose east to be along the positive *x* direction and north to be along the positive *y* direction, as in Figure 9.14. Before the collision, the only object having momentum in the *x* direction is the car. Thus, the magnitude of the total initial momentum of the system (car plus van) in the *x*

direction is

$$\sum p_{xi} = (1500 \text{ kg})(25.0 \text{ m/s}) = 3.75 \times 10^4 \text{ kg·m/s}$$

Now let us assume that the wreckage moves at an angle θ and speed v after the collision. The magnitude of the total momentum in the *x* direction after the collision is

$$\sum p_{xf} = (4000 \text{ kg})v \cos \theta$$

Because the total momentum in the *x* direction is constant, we can equate these two equations to get

$$(1) \qquad (3.75 \times 10^4) \text{ kg·m/s} = (4000 \text{ kg})v \cos \theta$$

Similarly, the total initial momentum of the system in the *y* direction is that of the van, whose magnitude is equal to $(2500 \text{ kg})(20.0 \text{ m/s})$. Applying conservation of momentum to the *y* direction, we have

$$\sum p_{yi} = \sum p_{yf}$$

$$(2500 \text{ kg})(20.0 \text{ m/s}) = (4000 \text{ kg})v \sin \theta$$

$$(2) \qquad 5.00 \times 10^4 \text{ kg·m/s} = (4000 \text{ kg})v \sin \theta$$

If we divide (2) by (1), we get

$$\tan \theta = \frac{5.00 \times 10^4}{3.75 \times 10^4} = 1.33$$

$$\theta = \boxed{53.1°}$$

When this angle is substituted into (2), the value of v is

$$v = \frac{5.00 \times 10^4 \text{ kg·m/s}}{(4000 \text{ kg}) \sin 53°} = \boxed{15.6 \text{ m/s}}$$

EXAMPLE 9.14   Proton-Proton Collision

A proton collides in a perfectly elastic fashion with another proton initially at rest. The incoming proton has an initial speed of 3.50×10^5 m/s and makes a glancing "collision" with the second proton, as in Figure 9.13. After the collision, one proton moves at an angle of 37.0° to the original direction of motion, and the second deflects at an angle ϕ to the same axis. Find the final speeds of the two protons and the angle ϕ.

Solution Since $m_1 = m_2$, $\theta = 37.0°$, and we are given $v_{1i} = 3.50 \times 10^5$ m/s, Equations 9.25 and 9.26 become

$$v_{1f} \cos 37.0° + v_{2f} \cos \phi = 3.50 \times 10^5$$

$$v_{1f} \sin 37.0° - v_{2f} \sin \phi = 0$$

$$v_{1f}^2 + v_{2f}^2 = (3.50 \times 10^5)^2$$

Solving these three equations with three unknowns simulta-

neously gives

$$v_{1f} = \boxed{2.80 \times 10^5 \text{ m/s}} \qquad v_{2f} = \boxed{2.11 \times 10^5 \text{ m/s}}$$

$$\phi = \boxed{53.0°}$$

Note that $\theta + \phi = 90°$. This result is not accidental. *Whenever two equal masses collide elastically in a glancing collision and one of them is initially at rest, their final velocities are always at right angles to each other.* The next example illustrates this point in more detail.

EXAMPLE 9.15 Billiard Ball Collision

In a game of billiards, a player wishes to sink the target ball in the corner pocket, as shown in Figure 9.15. If the angle to the corner pocket is 35°, at what angle θ is the cue ball deflected? Assume that friction and rotational motion ("English") are unimportant, and assume the collision is elastic.

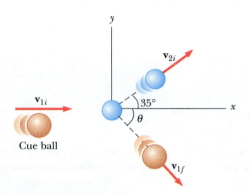

FIGURE 9.15 (Example 9.15).

Solution Since the target is initially at rest, $v_{2i} = 0$, conservation of energy (Eq. 9.16) gives

$$\tfrac{1}{2} m_1 v_{1i}^2 = \tfrac{1}{2} m_1 v_{1f}^2 + \tfrac{1}{2} m_2 v_{2f}^2$$

But $m_1 = m_2$, so that

$$(1) \qquad v_{1i}^2 = v_{1f}^2 + v_{2f}^2$$

Applying conservation of momentum to the two-dimensional collision gives

$$(2) \qquad \mathbf{v}_{1i} = \mathbf{v}_{1f} + \mathbf{v}_{2f}$$

If we square both sides of (2), we get

$$v_{1i}^2 = (\mathbf{v}_{1f} + \mathbf{v}_{2f}) \cdot (\mathbf{v}_{1f} + \mathbf{v}_{2f}) = v_{1f}^2 + v_{2f}^2 + 2\mathbf{v}_{1f} \cdot \mathbf{v}_{2f}$$

But $\mathbf{v}_{1f} \cdot \mathbf{v}_{2f} = v_{1f} v_{2f} \cos(\theta + 35°)$, and so

$$(3) \qquad v_{1i}^2 = v_{1f}^2 + v_{2f}^2 + 2 v_{1f} v_{2f} \cos(\theta + 35°)$$

Subtracting (1) from (3) gives

$$2 v_{1f} v_{2f} \cos(\theta + 35°) = 0$$

$$\cos(\theta + 35°) = 0$$

$$\theta + 35° = 90° \qquad \text{or} \qquad \theta = \boxed{55°}$$

Again, this result shows that whenever two equal masses undergo a glancing elastic collision and one of them is initially at rest, they move at right angles to each other after the collision.

9.6 THE CENTER OF MASS

In this section we describe the overall motion of a mechanical system in terms of a very special point called the **center of mass** of the system. The mechanical system can be either a system of particles, such as a collection of atoms in a container, or an extended object, such as a gymnast leaping through the air. We shall see that the mechanical system moves as if all its mass were concentrated at the center of mass. Furthermore, if the resultant external force on the system is $\mathbf{F}$ and the total mass of the system is M, the center of mass moves with an acceleration given by $\mathbf{a} = \mathbf{F}/M$. That is, the system moves as if the resultant external force were applied to a single particle of mass M located at the center of mass. This behavior is independent of other motion, such as rotation or vibration of the system. This result was implicitly assumed in earlier chapters since nearly all examples referred to the motion of extended objects.

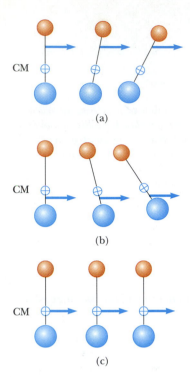

(a)

(b)

(c)

FIGURE 9.16 Two unequal masses are connected by a light, rigid rod. (a) The system rotates clockwise when a force is applied between the smaller mass and the center of mass. (b) The system rotates counterclockwise when the force is applied between the larger mass and the center of mass. (c) The system moves in the direction of the force without rotating when the force is applied at the center of mass.

Consider a mechanical system consisting of a pair of particles connected by a light, rigid rod (Fig. 9.16). The center of mass is located somewhere on the line joining the particles and is closer to the larger mass. If a single force is applied at some point on the rod somewhere between the center of mass and the smaller mass, the system rotates clockwise (Fig. 9.16a). If the force is applied at a point on the rod somewhere between the center of mass and the larger mass, the system rotates counterclockwise (Fig. 9.16b). If the force is applied at the center of mass, the system moves in the direction of F without rotating (Fig. 9.16c). Thus, the center of mass can be easily located.

One can describe the position of the center of mass of a system as being the weighted average position of the system's mass. For example, the center of mass of the pair of particles described in Figure 9.17 is located on the x axis and lies somewhere between the particles. Its x coordinate is

$$x_{CM} \equiv \frac{m_1 x_1 + m_2 x_2}{m_1 + m_2} \tag{9.27}$$

For example, if $x_1 = 0$, $x_2 = d$, and $m_2 = 2m_1$, we find that $x_{CM} = \frac{2}{3}d$. That is, the center of mass lies closer to the more massive particle. If the two masses are equal, the center of mass lies midway between the particles.

We can extend the center of mass concept to a system of many particles in

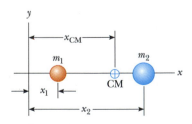

FIGURE 9.17 The center of mass of two particles on the x axis is located at x_{CM}, a point between the particles, closer to the larger mass.

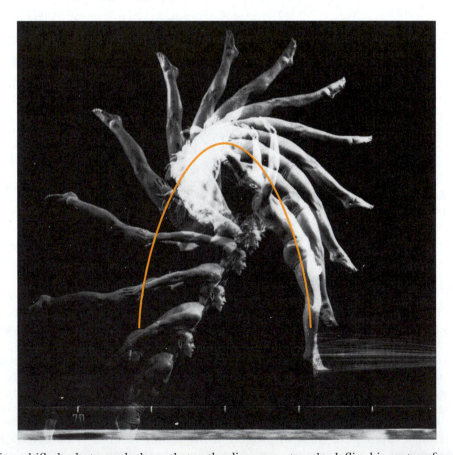

This multiflash photograph shows that as the diver executes a back flip, his center of mass follows a parabolic path, the same path that a particle would follow. (© *The Harold E. Edgerton 1992 Trust. Courtesy of Palm Press, Inc.*)

three dimensions. The x coordinate of the center of mass of n particles is defined to be

$$x_{\text{CM}} \equiv \frac{m_1 x_1 + m_2 x_3 + m_3 x_3 + \cdots + m_n x_n}{m_1 + m_2 + m_3 + \cdots + m_n} = \frac{\Sigma m_i x_i}{\Sigma m_i} \tag{9.28}$$

where x_i is the x coordinate of the ith particle and Σm_i is the *total mass* of the system. For convenience, we shall express the total mass as $M = \Sigma m_i$, where the sum runs over all n particles. The y and z coordinates of the center of mass are similarly defined by the equations

$$y_{\text{CM}} \equiv \frac{\Sigma m_i y_i}{M} \quad \text{and} \quad z_{\text{CM}} \equiv \frac{\Sigma m_i z_i}{M} \tag{9.29}$$

The center of mass can also be located by its position vector, $\mathbf{r}_{\text{CM}}$. The rectangular coordinates of this vector are x_{CM}, y_{CM}, and z_{CM}, defined in Equations 9.28 and 9.29. Therefore,

$$\begin{aligned}\mathbf{r}_{\text{CM}} &= x_{\text{CM}}\mathbf{i} + y_{\text{CM}}\mathbf{j} + z_{\text{CM}}\mathbf{k}\\ &= \frac{\Sigma m_i x_i \mathbf{i} + \Sigma m_i y_i \mathbf{j} + \Sigma m_i z_i \mathbf{k}}{M}\end{aligned}$$

$$\mathbf{r}_{\text{CM}} \equiv \frac{\Sigma m_i \mathbf{r}_i}{M} \tag{9.30}$$

where $\mathbf{r}_i$ is the position vector of the ith particle, defined by

$$\mathbf{r}_i \equiv x_i \mathbf{i} + y_i \mathbf{j} + z_i \mathbf{k}$$

Although locating the center of mass for an extended object is somewhat more cumbersome, the basic ideas we have discussed still apply. We can think of an extended object as a system of a large number of particles (Fig. 9.18). The particle separation is very small, and so the object can be considered to have a continuous mass distribution. By dividing the object into elements of mass Δm_i, with coordinates x_i, y_i, z_i, we see that the x coordinate of the center of mass is approximately

$$x_{\text{CM}} \approx \frac{\Sigma x_i \Delta m_i}{M}$$

with similar expressions for y_{CM} and z_{CM}. If we let the number of elements, n, approach infinity, then x_{CM} is given precisely. In this limit, we replace the sum by an integral and replace Δm_i by the differential element dm, so that

$$x_{\text{CM}} = \lim_{\Delta m_i \to 0} \frac{\Sigma x_i \Delta m_i}{M} = \frac{1}{M} \int x \, dm \tag{9.31}$$

Likewise, for y_{CM} and z_{CM} we get

$$y_{\text{CM}} = \frac{1}{M} \int y \, dm \quad \text{and} \quad z_{\text{CM}} = \frac{1}{M} \int z \, dm \tag{9.32}$$

We can express the vector position of the center of mass of an extended object in the form

$$\mathbf{r}_{\text{CM}} = \frac{1}{M} \int \mathbf{r} \, dm \tag{9.33}$$

which is equivalent to the three expressions given by Equations 9.31 and 9.32.

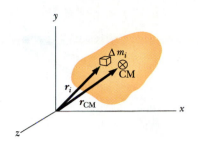

FIGURE 9.18 An extended object can be considered a distribution of many small elements of mass Δm_i. The center of mass is located at the vector position $\mathbf{r}_{\text{CM}}$, which has coordinates x_{CM}, y_{CM}, and z_{CM}.

Vector position of the center of mass for a system of particles

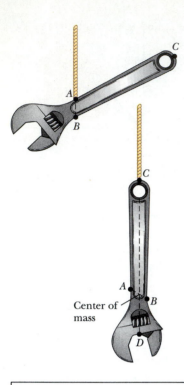

FIGURE 9.19 An experimental technique for determining the center of mass of a wrench. The wrench is hung freely from two different pivots, A and C. The intersection of the two vertical lines AB and CD locates the center of mass.

The center of mass of any symmetric object lies on an axis of symmetry and on any plane of symmetry.[3] For example, the center of mass of a rod lies in the rod, midway between its ends. The center of mass of a sphere or a cube lies at its geometric center.

One can determine the center of mass of an irregularly shaped object, such as a wrench, by suspending the wrench first from one point and then from another (Fig. 9.19). The wrench is first hung from point A, and a vertical line AB (which can be established with a plumb bob) is drawn when the wrench has stopped swinging. The wrench is then hung from point C, and a second vertical line, CD, is drawn. The center of mass coincides with the intersection of these two lines. In general, if the wrench is hung freely from any point, the vertical line through this point must pass through the center of mass.

Since an extended object is a continuous distribution of mass, each small mass element is acted upon by the force of gravity. The net effect of all of these forces is equivalent to the effect of a single force, $M\mathbf{g}$, acting through a special point, called the **center of gravity**. If $\mathbf{g}$ is constant over the mass distribution, then the center of gravity coincides with the center of mass. If an extended object is pivoted at its center of gravity, it balances in any orientation.

EXAMPLE 9.16   The Center of Mass of Three Particles

A system consists of three particles located at the corners of a right triangle as in Figure 9.20. Find the center of mass of the system.

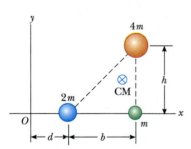

FIGURE 9.20   (Example 9.16) The center of mass of the three particles is located inside the triangle.

Solution   Using the basic defining equations for the coordinates of the center of mass, and noting that $z_{CM} = 0$, we get

$$x_{CM} = \frac{\sum m_i x_i}{M} = \frac{2md + m(d+b) + 4m(d+b)}{7m} = d + \frac{5}{7}b$$

$$y_{CM} = \frac{\sum m_i y_i}{M} = \frac{2m(0) + m(0) + 4mh}{7m} = \frac{4}{7}h$$

Therefore, we can express the position vector to the center of mass measured from the origin as

$$\mathbf{r}_{CM} = x_{CM}\mathbf{i} + y_{CM}\mathbf{j} = (d + \tfrac{5}{7}b)\mathbf{i} + \tfrac{4}{7}h\mathbf{j}$$

EXAMPLE 9.17   The Center of Mass of a Rod

(a) Show that the center of mass of a rod of mass M and length L lies midway between its ends, assuming the rod has a uniform mass per unit length (Fig. 9.21).

Solution   By symmetry, we see that $y_{CM} = z_{CM} = 0$ if the rod is placed along the x axis. Furthermore, if we call the mass per unit length λ (the linear mass density), then $\lambda = M/L$ for

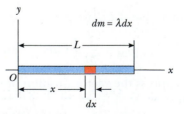

FIGURE 9.21   (Example 9.17) The center of mass of a uniform rod of length L is located at $x_{CM} = L/2$.

---

[3] This statement is valid only for objects that have a uniform mass-per-unit volume.

a uniform rod. If we divide the rod into elements of length dx, then the mass of each element is $dm = \lambda\, dx$. Since an arbitrary element is at a distance x from the origin, Equation 9.31 gives

$$x_{CM} = \frac{1}{M}\int_0^L x\, dm = \frac{1}{M}\int_0^L x\lambda\, dx = \frac{\lambda}{M}\frac{x^2}{2}\Big]_0^L = \frac{\lambda L^2}{2M}$$

Because $\lambda = M/L$, this reduces to

$$x_{CM} = \frac{L^2}{2M}\left(\frac{M}{L}\right) = \frac{L}{2}$$

One can also argue that by symmetry, $x_{CM} = L/2$.

(b) Suppose a rod is *nonuniform* such that its mass per unit length varies linearly with x according to the expression $\lambda = \alpha x$, where α is a constant. Find the x coordinate of the center of mass as a fraction of L.

Solution In this case, we replace dm by $\lambda\, dx$, where λ is not constant. Therefore, x_{CM} is

$$x_{CM} = \frac{1}{M}\int_0^L x\, dm = \frac{1}{M}\int_0^L x\lambda\, dx = \frac{\alpha}{M}\int_0^L x^2\, dx = \frac{\alpha L^3}{3M}$$

We can eliminate α by noting that the total mass of the rod is related to α through the relationship

$$M = \int dm = \int_0^L \lambda\, dx = \int_0^L \alpha x\, dx = \frac{\alpha L^2}{2}$$

Substituting this into the expression for x_{CM} gives

$$x_{CM} = \frac{\alpha L^3}{3\alpha L^2/2} = \frac{2}{3}L$$

EXAMPLE 9.18 The Center of Mass of a Right Triangle

An object of mass M is in the shape of a right triangle whose dimensions are shown in Figure 9.22. Locate the coordinates of the center of mass, assuming the object has a uniform mass per unit area.

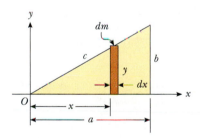

FIGURE 9.22 (Example 9.18).

Solution To evaluate the x coordinate of the center of mass, we divide the triangle into narrow strips of width dx and height y as in Figure 9.22. The mass dm of each strip is

$$dm = \frac{\text{total mass}}{\text{total area}} \times \text{area of strip}$$

$$= \frac{M}{\frac{1}{2}ab}\,(y\, dx) = \left(\frac{2M}{ab}\right) y\, dx$$

Therefore, the x coordinate of the center of mass is

$$x_{CM} = \frac{1}{M}\int x\, dm = \frac{1}{M}\int_0^a x\left(\frac{2M}{ab}\right) y\, dx = \frac{2}{ab}\int_0^a xy\, dx$$

In order to evaluate this integral, we must express the variable y in terms of the variable x. From similar triangles in Figure 9.22, we see that

$$\frac{y}{x} = \frac{b}{a} \qquad \text{or} \qquad y = \frac{b}{a}x$$

With this substitution, x_{CM} becomes

$$x_{CM} = \frac{2}{ab}\int_0^a x\left(\frac{b}{a}x\right) dx = \frac{2}{a^2}\int_0^a x^2\, dx = \frac{2}{a^2}\left[\frac{x^3}{3}\right]_0^a = \frac{2}{3}a$$

By a similar calculation, we get for the y coordinate of the center of mass

$$y_{CM} = \frac{1}{3}b$$

9.7 MOTION OF A SYSTEM OF PARTICLES

We can begin to understand the physical significance and utility of the center of mass concept by taking the time derivative of the position vector given by Equation 9.30. Assuming that M remains constant for a system of particles, that is, no particles enter or leave the system, we get the following expression for the **velocity of the** center of mass of the system:

$$v_{CM} = \frac{dr_{CM}}{dt} = \frac{1}{M} \sum m_i \frac{dr_i}{dt}$$

$$v_{CM} = \frac{\sum m_i v_i}{M} \tag{9.34}$$

where v_i is the velocity of the ith particle. Rearranging Equation 9.34 gives

$$M v_{CM} = \sum m_i v_i = \sum p_i = p_{tot} \tag{9.35}$$

Therefore, we conclude that the **total linear momentum of the system** equals the total mass multiplied by the velocity of the center of mass. In other words, the total linear momentum of the system is equal to that of a single particle of mass M moving with a velocity v_{CM}.

If we now differentiate Equation 9.35 with respect to time, we get the acceleration of the center of mass of the system:

$$a_{CM} = \frac{dv_{CM}}{dt} = \frac{1}{M} \sum m_i \frac{dv_i}{dt} = \frac{1}{M} \sum m_i a_i \tag{9.36}$$

Rearranging this expression and using Newton's second law, we get

$$M a_{CM} = \sum m_i a_i = \sum F_i \tag{9.37}$$

where F_i is the force on particle i.

The forces on any particle in the system may include both external forces (from outside the system) and internal forces (from within the system). However, by Newton's third law, the force exerted by particle 1 on particle 2, for example, is equal to and opposite the force exerted by particle 2 on particle 1. Thus, when we sum over all internal forces in Equation 9.37, they cancel in pairs and the net force on the system is due *only* to external forces. Thus, we can write Equation 9.37 in the form

$$\sum F_{ext} = M a_{CM} = \frac{dp_{tot}}{dt} \tag{9.38}$$

That is, the resultant external force on a system of particles equals the total mass of the system multiplied by the acceleration of the center of mass. If we compare this to Newton's second law for a single particle, we see that

the center of mass moves like an imaginary particle of mass M under the influence of the resultant external force on the system.

In the absence of external forces, the center of mass moves with uniform velocity, as in the case of the rotating wrench shown in Figure 9.23.

Finally, we see that if the resultant external force is zero, then from Equation 9.38 it follows that

$$\frac{dp_{tot}}{dt} = M a_{CM} = 0$$

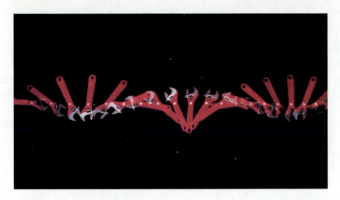

FIGURE 9.23   Multiflash photograph of a wrench moving on a horizontal surface. The center of mass of the wrench moves in a straight line as the wrench rotates about this point, shown by the white dot.   *(Education Development Center, Newton, Mass.)*

so that

$$\mathbf{p}_{tot} = M\mathbf{v}_{CM} = \text{constant} \qquad \left(\text{when } \sum \mathbf{F}_{ext} = 0\right) \qquad (9.39)$$

That is, the total linear momentum of a system of particles is constant if there are no external forces acting on the system. Therefore, it follows that for an isolated system of particles, both the total momentum and the velocity of the center of mass are constant in time. This is a generalization to a many-particle system of the law of conservation of momentum discussed in Section 9.1 for a two-particle system.

Suppose an isolated system consisting of two or more members is at rest. The center of mass of such a system remains at rest unless acted upon by an external force. For example, consider a system made up of a swimmer and a raft, with the system initially at rest. When the swimmer dives off the raft, the center of mass of the system remains at rest (if we neglect friction between raft and water). Furthermore, the linear momentum of the diver is equal in magnitude to that of the raft but opposite in direction.

As another example, suppose an unstable atom initially at rest suddenly decays into two fragments of masses M_1 and M_2, with velocities $\mathbf{v}_1$ and $\mathbf{v}_2$, respectively. Since the total momentum of the system before the decay is zero, the total momentum of the system after the decay must also be zero. Therefore, we see that $M_1\mathbf{v}_1 + M_2\mathbf{v}_2 = 0$. If the velocity of one of the fragments after the decay is known, the recoil velocity of the other fragment can be calculated.

CONCEPTUAL EXAMPLE 9.19

A boy stands at one end of a floating raft that is stationary relative to the shore. He then walks to the opposite end of the raft, away from the shore. Does the raft move? Explain.

Reasoning   Yes, the raft moves toward the shore. Neglecting friction between the raft and water, there are no horizontal forces acting on the system consisting of the boy and raft. Therefore, the center of mass of the system remains fixed relative to the shore (or any stationary point). As the boy moves away from the shore, the boat must move toward shore such that the center of mass of the system remains constant. An alternative explanation is that the momentum of the system remains constant if friction is neglected. As the boy acquires a momentum away from the shore, the boat must acquire an equal momentum toward the shore such that the total momentum of the system is always zero.

CONCEPTUAL EXAMPLE 9.20

Suppose you tranquilize a polar bear on a glacier as part of a research effort (Fig. 9.24). How might you be able to estimate the weight of the polar bear using a measuring tape, a rope, and a knowledge of your own weight?

Reasoning   Tie one end of the rope around the bear. Lay out the tape measure on the ice between the bear's original position and yours, as you hold the opposite end of the rope.

Take off your spiked shoes and pull on the rope. Both you and the bear will slide over the ice until you meet. From the tape observe how far you have slid, x_y, and how far the bear has slid, x_b. The point where you meet the bear is the constant location of the center of mass of the system (bear plus you), so you can determine the mass of the bear from $m_b x_b = m_y x_y$. (Unfortunately, if the bear now wakes up you cannot get back to your spiked shoes.)

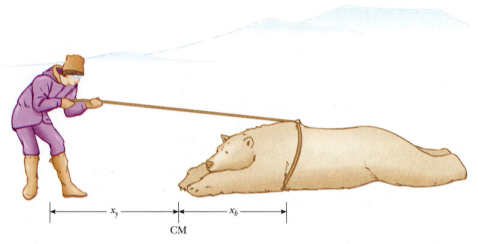

x_y x_b

CM

FIGURE 9.24   (Conceptual Example 9.20) The center of mass of an isolated system remains at rest unless acted on by an external force. How can you determine the weight of the polar bear?

CONCEPTUAL EXAMPLE 9.21   Exploding Projectile

A projectile is fired into the air and suddenly explodes into several fragments (Fig. 9.25). What can be said about the motion of the center of mass of the fragments after the collision?

Reasoning   Neglecting air resistance, the only external force on the projectile is the force of gravity. Thus, the projectile follows a parabolic path. If the projectile did not explode, it would continue to move along the parabolic path indicated by the broken line in Figure 9.25. Since the forces due to the explosion are internal, they do not affect the motion of the center of mass. Thus, after the explosion the center of mass of the fragments follows the same parabolic path the projectile would have followed if there had been no explosion.

FIGURE 9.25   (Conceptual Example 9.21) When a projectile explodes into several fragments, the center of mass of the fragments follows the same parabolic path the projectile would have taken had there been no explosion.

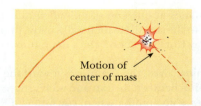

Motion of
center of mass

EXAMPLE 9.22   The Exploding Rocket

A rocket is fired vertically upward. At the instant it reaches an altitude of 1000 m and a speed of 300 m/s, it explodes into three equal fragments. One fragment continues to move upward with a speed of 450 m/s following the explosion. The second fragment has a speed of 240 m/s and is moving east right after the explosion. What is the velocity of the third fragment right after the explosion?

Solution   Let us call the total mass of the rocket M; hence, the mass of each fragment is $M/3$. The total momentum just before the explosion must equal the total momentum of the fragments right after the explosion since the forces of the explosion are internal to the system and cannot affect its total momentum.

Before the explosion:

$$\mathbf{p}_{tot} = M\mathbf{v}_0 = 300M\mathbf{j}$$

After the explosion:

$$\mathbf{p}_{tot} = 240\left(\frac{M}{3}\right)\mathbf{i} + 450\left(\frac{M}{3}\right)\mathbf{j} + \frac{M}{3}\mathbf{v}$$

where $\mathbf{v}$ is the unknown velocity of the third fragment. Equating these two expressions gives

$$M\frac{\mathbf{v}}{3} + 80M\mathbf{i} + 150M\mathbf{j} = 300M\mathbf{j}$$

$$\mathbf{v} = \boxed{(-240\mathbf{i} + 450\mathbf{j})\,\text{m/s}}$$

Exercise   Find the position of the center of mass relative to the ground 3.00 s after the explosion. (Assume the rocket engine is nonoperative after the explosion.)

Answer   The x coordinate of the center of mass doesn't change. The y coordinate of the center of mass at this instant is 1.86 km.

*9.8   ROCKET PROPULSION

When ordinary vehicles, such as automobiles and locomotives, are propelled, the driving force for the motion is one of friction. In the case of the automobile, the driving force is the force exerted by the road on the car. A locomotive "pushes" against the tracks; hence, the driving force is the force exerted by the tracks on the locomotive. However, a rocket moving in space has no air or tracks to push against. Therefore, the source of the propulsion of a rocket must be different. Figure 9.26

FIGURE 9.26   Lift-off of the space shuttle Columbia. Enormous thrust is generated by the shuttle's liquid-fueled engines, aided by the two solid-fuel boosters. Many physical principles from mechanics, thermodynamics, and electricity and magnetism are involved in such a launch.   *(Courtesy of NASA)*

is a dramatic photograph of a spacecraft at lift-off. *The operation of a rocket depends upon the law of conservation of linear momentum as applied to a system of particles, where the system is the rocket plus its ejected fuel.*

Rocket propulsion can be understood by first considering the mechanical system consisting of a machine gun mounted on a cart on wheels. As the gun is fired, each bullet receives a momentum mv in some direction, where v is measured with respect to a stationary Earth frame. For each bullet that is fired, the gun and cart must receive a compensating momentum in the opposite direction. That is, the reaction force exerted by the bullet on the gun accelerates the cart and gun. If there are n bullets fired each second, then the average force exerted on the gun is $\mathbf{F}_{av} = nm\mathbf{v}$.

In a similar manner, as a rocket moves in free space, *its linear momentum changes when some of its mass is released in the form of ejected gases. Since the ejected gases acquire some momentum, the rocket receives a compensating momentum in the opposite direction.* Therefore, *the rocket is accelerated as a result of the "push," or thrust, from the exhaust gases.* In free space, the center of mass of the system moves uniformly, independent of the propulsion process. It is interesting to note that the rocket and machine gun represent cases of the inverse of an inelastic collision; that is, momentum is conserved, but the kinetic energy of the system is increased (at the expense of internal energy).

Suppose that at some time t, the momentum of the rocket plus the fuel is $(M + \Delta m)v$ (Fig. 9.27a). At some short time later, Δt, the rocket ejects fuel of mass Δm and the rocket's speed therefore increases to $v + \Delta v$ (Fig. 9.27b). If the fuel is ejected with a speed v_e relative to the rocket, the speed of the fuel relative to a stationary frame of reference is $v - v_e$. Thus, if we equate the total initial momentum of the system to the total final momentum, we get

$$(M + \Delta m)v = M(v + \Delta v) + \Delta m(v - v_e)$$

Simplifying this expression gives

$$M \Delta v = \Delta m(v_e)$$

We also could have arrived at this result by considering the system in the center-of-mass frame of reference, that is, a frame whose velocity equals the center-of-mass velocity. In this frame, the total momentum is zero; therefore, if the rocket gains a momentum $M \Delta v$ by ejecting some fuel, the exhaust gases obtain a momentum $\Delta m(v_e)$ in the *opposite* direction, and so $M \Delta v - \Delta m(v_e) = 0$. If we now take the limit as Δt goes to zero, then $\Delta v \rightarrow dv$ and $\Delta m \rightarrow dm$. Furthermore, the increase in the exhaust mass, dm, corresponds to an equal decrease in the rocket mass, so that $dm = -dM$. Note that dM is given a negative sign because it represents a decrease in mass. Using this fact, we get

$$M \, dv = -v_e \, dM \tag{9.40}$$

Integrating this equation, and taking the initial mass of the rocket plus fuel to be M_i and the final mass of the rocket plus its remaining fuel to be M_f, we get

$$\int_{v_i}^{v_f} dv = -v_e \int_{M_i}^{M_f} \frac{dM}{M}$$

$$v_f - v_i = v_e \ln\left(\frac{M_i}{M_f}\right) \tag{9.41}$$

This is the basic expression of rocket propulsion. First, it tells us that the increase in speed is proportional to the exhaust speed, v_e. Therefore, the exhaust

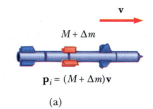

$M + \Delta m$

v

$\mathbf{p}_i = (M + \Delta m)\mathbf{v}$

(a)

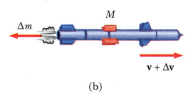

Δm

M

$\mathbf{v} + \Delta\mathbf{v}$

(b)

FIGURE 9.27 Rocket propulsion. (a) The initial mass of the rocket is $M + \Delta m$ at a time t, and its speed is v. (b) At a time $t + \Delta t$, the rocket's mass has reduced to M, and an amount of fuel Δm has been ejected. The rocket's speed increases by an amount Δv.

Expression for rocket propulsion

speed should be very high. Second, the increase in speed is proportional to the logarithm of the ratio M_i/M_f. Therefore, this ratio should be as large as possible, which means that the mass of the rocket without its fuel should be as small as possible and the rocket should carry as much fuel as possible.

The **thrust** on the rocket is the force exerted on the rocket by the ejected exhaust gases. We can obtain an expression for the thrust from Equation 9.40:

$$\text{Thrust} = M\frac{dv}{dt} = \left| v_e \frac{dM}{dt} \right| \qquad (9.42)$$

Here we see that the thrust increases as the exhaust speed increases and as the rate of change of mass (burn rate) increases.

EXAMPLE 9.23   A Rocket in Space

A rocket moving in free space has a speed of 3.0×10^3 m/s relative to Earth. Its engines are turned on, and fuel is ejected in a direction opposite the rocket's motion at a speed of 5.0×10^3 m/s relative to the rocket. (a) What is the speed of the rocket relative to Earth once its mass is reduced to one half its mass before ignition?

Solution   Applying Equation 9.41, we get

$$v_f = v_i + v_e \ln\left(\frac{M_i}{M_f}\right)$$

$$= 3.0 \times 10^3 + 5.0 \times 10^3 \ln\left(\frac{M_i}{0.5M_i}\right)$$

$$= 6.5 \times 10^3 \text{ m/s}$$

(b) What is the thrust on the rocket if it burns fuel at the rate of 50 kg/s?

Solution

$$\text{Thrust} = \left| v_e \frac{dM}{dt} \right| = \left(5.0 \times 10^3 \frac{m}{s} \right)\left(50 \frac{kg}{s} \right)$$

$$= 2.5 \times 10^5 \text{ N}$$

SUMMARY

The **linear momentum** of a particle of mass m moving with a velocity **v** is

$$\mathbf{p} \equiv m\mathbf{v} \qquad (9.1)$$

The law of **conservation of linear momentum** says that the total momentum of an isolated system is constant. If two particles form an isolated system, their total momentum is constant regardless of the nature of the force between them. Therefore, the total momentum of the system at all times equals its initial total momentum, or

$$\mathbf{p}_{1i} + \mathbf{p}_{2i} = \mathbf{p}_{1f} + \mathbf{p}_{2f} \qquad (9.5)$$

The **impulse** of a force **F** on a particle is equal to the change in the momentum of the particle:

$$\mathbf{I} \equiv \int_{t_i}^{t_f} \mathbf{F}\, dt = \Delta\mathbf{p} \qquad (9.9)$$

This is known as the **impulse-momentum theorem.**

Impulsive forces are often very strong compared with other forces on the system, and usually act for a very short time, as in the case of collisions.

When two particles collide, the total momentum of the system before the collision always equals the total momentum after the collision, regardless of the nature of the collision. An **inelastic collision** is one for which mechanical energy is not constant. A perfectly inelastic collision corresponds to the situation where the colliding bodies stick together after the collision. An **elastic collision** is one in which kinetic energy is constant.

In a two- or three-dimensional collision, the components of momentum in each of the three directions (x, y, and z) are conserved independently.

The **vector position of the center of mass of a system of particles** is defined as

$$\mathbf{r}_{CM} \equiv \frac{\Sigma m_i \mathbf{r}_i}{M} \tag{9.30}$$

where $M = \Sigma m_i$ is the total mass of the system and $\mathbf{r}_i$ is the vector position of the ith particle.

The **vector position of the center of mass of a rigid body** can be obtained from the integral expression

$$\mathbf{r}_{CM} = \frac{1}{M} \int \mathbf{r} \, dm \tag{9.33}$$

The **velocity of the center of mass for a system of particles** is

$$\mathbf{v}_{CM} = \frac{\Sigma m_i \mathbf{v}_i}{M} \tag{9.34}$$

The total momentum of a system of particles equals the total mass multiplied by the velocity of the center of mass, that is, $\mathbf{p}_{tot} = M\mathbf{v}_{CM}$.

Newton's second law applied to a system of particles is

$$\sum \mathbf{F}_{ext} = M\mathbf{a}_{CM} = \frac{d\mathbf{p}_{tot}}{dt} \tag{9.38}$$

where $\mathbf{a}_{CM}$ is the acceleration of the center of mass and the sum is over all external forces. Therefore, the center of mass moves like an imaginary particle of mass M under the influence of the resultant external force on the system. It follows from Equation 9.38 that the total momentum of the system is constant if there are no external forces acting on it.

QUESTIONS

1. If the kinetic energy of a particle is zero, what is its linear momentum? If the total energy of a particle is zero, is its linear momentum necessarily zero? Explain.
2. If the speed of a particle is doubled, by what factor is its momentum changed? By what factor is its kinetic energy changed?
3. If two particles have equal kinetic energies, are their momenta necessarily equal? Explain.
4. Does a large force always produce a larger impulse on a body than a small force? Explain.
5. An isolated system is initially at rest. Is it possible for parts of the system to be in motion at some later time? If so, explain how this might occur.
6. If two objects collide and one is initially at rest, is it possible for both to be at rest after the collision? Is it possible for one to be at rest after the collision? Explain.
7. Explain how linear momentum is conserved when a ball bounces from a floor.
8. Is it possible to have a collision in which all of the kinetic energy is lost? If so, cite an example.
9. In a perfectly elastic collision between two particles, does the kinetic energy of each particle change as a result of the collision?
10. When a ball rolls down an incline, its linear momentum increases. Does this imply that momentum is not conserved? Explain.
11. Consider a perfectly inelastic collision between a car and a large truck. Which vehicle loses more kinetic energy as a result of the collision?
12. Can the center of mass of a body lie outside the body? If so, give examples.
13. Three balls are thrown into the air simultaneously. What

is the acceleration of their center of mass while they are in motion?

14. A meter stick is balanced in a horizontal position with the index fingers of the right and left hands. If the two fingers are brought together, the stick remains balanced and the two fingers always meet at the 50-cm mark regardless of their original positions (try it!). Explain.

15. A sharpshooter fires a rifle while standing with the butt of the gun against his shoulder. If the forward momentum of a bullet is the same as the backward momentum of the gun, why isn't it as dangerous to be hit by the gun as by the bullet?

16. A piece of mud is thrown against a brick wall and sticks to the wall. What happens to the momentum of the mud? Is momentum conserved? Explain.

17. Early in this century, Robert Goddard proposed sending a rocket to the Moon. Critics took the position that in a vacuum, such as exists between Earth and Moon, the gases emitted by the rocket would have nothing to push against to propel the rocket. According to *Scientific American* (January 1975), Goddard placed a gun in a vacuum and fired a blank cartridge from it. (A blank cartridge fires only the wadding and hot gases of the burning gunpowder.) What happened when the gun was fired?

18. A pole-vaulter falls from a height of 6.0 m onto a foam rubber pad. Can you calculate his speed just before he reaches the pad? Can you calculate the force exerted on him due to the collision? Explain.

19. As a ball falls toward the Earth, the momentum of the ball increases. Reconcile this fact with the law of conservation of momentum.

20. Explain how you would use a balloon to demonstrate the mechanism responsible for rocket propulsion.

21. Explain the maneuver of decelerating a spacecraft. What other maneuvers are possible?

22. Does the center of mass of a rocket in free space accelerate? Explain. Can the speed of a rocket exceed the exhaust speed of the fuel? Explain.

23. A ball is dropped from a tall building. Identify the system for which linear momentum is constant.

24. A bomb, initially at rest, explodes into several pieces. Is linear momentum constant? (b) Is kinetic energy constant? Explain.

25. Is kinetic energy always lost in an inelastic collision? Explain.

26. A skater is standing still on a frictionless ice rink. Her friend throws a Frisbee straight at her. In which of the following cases is the largest momentum transferred to the skater? (a) She catches the Frisbee and holds it, (b) she catches it momentarily but drops it, (c) she catches it and then throws it back to her friend.

27. The Moon revolves around the Earth. Is the Moon's linear momentum constant? Is its kinetic energy constant? Assume the orbit is circular.

28. A large bedsheet is held vertically by two students. A third student, who happens to be the star pitcher on the baseball team, throws a raw egg at the sheet. Explain why the egg does not break when it hits the sheet, regardless of its initial speed. (If you try this one, make sure the pitcher hits the sheet near its center, and do not allow the egg to fall on the floor after being caught.)

29. A raw egg dropped to the floor breaks apart upon impact. However, a raw egg dropped onto a thick foam rubber cushion from a height of about 1 m rebounds without breaking. Why is this possible? (In this demonstration, be sure to catch the egg after the first bounce.)

PROBLEMS

Review Problem

A 60-kg person running at an initial speed of 4.0 m/s jumps onto a 120-kg cart initially at rest. The person slides on the cart's surface, and finally comes to rest relative to the cart. The coefficient of kinetic friction between the person and cart is 0.40 and the friction between the cart and ground can be neglected. (a) Find the final velocity of the person and cart relative to the Earth. (b) Find the frictional force exerted on the person while sliding on the surface of the cart. (c) How long does the frictional force act on the person? (d) Find the change in momentum of the person and the change in momentum of the cart. (e) Determine the displacement of the person relative to the Earth while sliding on the surface of the cart. (f) Determine the displacement of the cart relative to the Earth while the person slides on its surface. (g) Find the change in kinetic energy of the person. (i) Find the change in kinetic energy of the cart. (j) Explain why the answers to (g) and (i) differ. (What kind of collision is this, and what accounts for the loss in mechanical energy?)

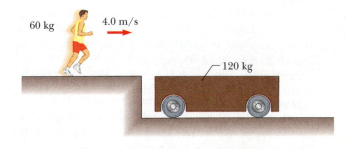

60 kg   4.0 m/s

120 kg

Section 9.1   Linear Momentum and Its Conservation
Section 9.2   Impulse and Momentum

1. A 3.0-kg particle has a velocity of $(3.0\mathbf{i} - 4.0\mathbf{j})$ m/s. Find its x and y components of momentum and the magnitude of its total momentum.

2. A 7.00-kg bowling ball moves in a straight line at 3.00 m/s. How fast must a 2.45-g Ping-Pong ball move in a straight line so that the two balls have the same momentum?

3. A child bounces a superball on the sidewalk. The linear impulse delivered by the sidewalk to the ball is 2.00 N·s during the 1/800 s of contact. What is the magnitude of the average force exerted on the ball by the sidewalk?

4. A superball with a mass of 60 g is dropped from a height of 2.0 m. It rebounds to a height of 1.8 m. What is the change in its linear momentum during the collision with the floor?

5. The force $\mathbf{F}_x$ acting on a 2.0-kg particle varies in time as shown in Figure P9.5. Find (a) the impulse of the force, (b) the final velocity of the particle if it is initially at rest, (c) its final velocity if it is initially moving along the x axis with a velocity of -2.0 m/s, and (d) the average force exerted on the particle for the time interval $t_i = 0$ to $t_f = 5.0$ s.

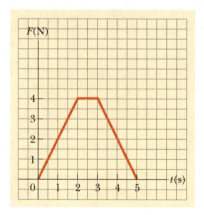

FIGURE P9.5

6. In a slow-pitch softball game, a 0.200-kg softball crossed the plate at 15.0 m/s at an angle of 45.0° below the horizontal. The ball was hit at 40.0 m/s, 30.0° above the horizontal. (a) Determine the impulse applied to the ball. (b) If the force on the ball increased linearly for 4.00 ms, held constant for 20.0 ms, then decreased to zero linearly in another 4.00 ms, find the maximum force on the ball.

7. An estimated force-time curve for a baseball struck by a bat is shown in Figure P9.7. From this curve, determine (a) the impulse delivered to the ball, (b) the average force exerted on the ball, and (c) the peak force exerted on the ball.

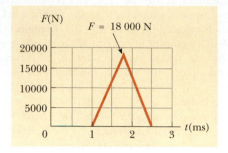

FIGURE P9.7

8. A garden hose is held in the manner shown in Figure P9.8. What force is necessary to hold the nozzle stationary if the discharge rate is 0.60 kg/s with a speed of 25 m/s?

FIGURE P9.8

9. A machine gun fires 35.0 g bullets at a speed of 750.0 m/s. If the gun can fire 200 bullets/min, what is the average force the shooter must exert to keep the gun from moving?

10. (a) If the momentum of an object is doubled in magnitude, what happens to its kinetic energy? (b) If the kinetic energy of an object is tripled, what happens to its momentum?

11. A 0.50-kg football is thrown with a speed of 15 m/s. A stationary receiver catches the ball and brings it to

☐ indicates problems that have full solutions available in the Student Solutions Manual and Study Guide.

rest in 0.020 s. (a) What is the impulse delivered to the ball? (b) What is the average force exerted on the receiver?

12. A car is stopped for a traffic signal. When the light turns green, the car accelerates, increasing its speed from zero to 5.20 m/s in 0.832 s. What linear impulse and average force does a 70.0-kg passenger in the car experience?

13. A 0.15-kg baseball is thrown with a speed of 40 m/s. It is hit straight back at the pitcher with a speed of 50 m/s. (a) What is the impulse delivered to the baseball? (b) Find the average force exerted by the bat on the ball if the two are in contact for 2.0×10^{-3} s. Compare this with the weight of the ball and determine whether or not the impulse approximation is valid in the situation.

14. A tennis player receives a shot with the ball (0.060 kg) traveling horizontally at 50 m/s and returns the shot with the ball traveling horizontally at 40 m/s in the opposite direction. What is the impulse delivered to the ball by the racket?

15. A 3.0-kg steel ball strikes a wall with a speed of 10 m/s at an angle of 60° with the surface. It bounces off with the same speed and angle (Fig. P9.15). If the ball is in contact with the wall for 0.20 s, what is the average force exerted on the ball by the wall?

15A. A steel ball of mass m strikes a wall with a speed of v at an angle θ with the surface. It bounces off with the same speed and angle (Fig. P9.15). If the ball is in contact with the wall for a time t, what is the average force exerted on the ball by the wall?

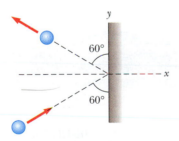

FIGURE P9.15

16. Water falls without splashing at a rate of 0.25 liter/s from a height of 60 m into a 0.75-kg bucket on a scale. If the bucket is originally empty, what does the scale read after 3.0 s?

Section 9.3 Collisions
Section 9.4 Elastic and Inelastic Collisions in One Dimension

17. A 79.5-kg man holding a 0.500-kg ball stands on a frozen pond next to a wall. He throws the ball at the wall with a speed of 10.0 m/s (relative to the ground) and then catches the ball after it rebounds from the

wall. (a) How fast is he moving after he catches the ball? (Ignore the projectile motion of the ball, and assume that it loses no energy in its collision with the wall.) (b) How many times does the man have to go through this process before his speed reaches 1.00 m/s relative to the ground?

18. Two blocks of masses M and $3M$ are placed on a horizontal, frictionless surface. A light spring is attached to one of them, and the blocks are pushed together with the spring between them (Fig. P9.18). A cord holding them together is burned, after which the block of mass $3M$ moves to the right with a speed of 2.00 m/s. What is the speed of the block of mass M?

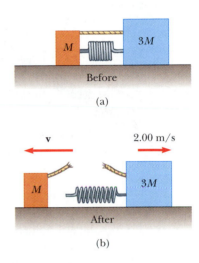

FIGURE P9.18

19. A 60.0-kg astronaut is on a space walk away from the shuttle when her tether line breaks. She is able to throw her 10.0-kg oxygen tank away from the shuttle with a speed of 12.0 m/s to propel herself back to the shuttle (Fig. P9.19). Assuming that she starts from rest (relative to the shuttle), determine the maximum distance she can be from the craft when the line breaks and still return within 60.0 s (the amount of time she can hold her breath).

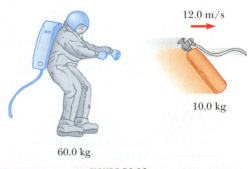

FIGURE P9.19

20. Identical air cars ($m = 200$ g) are equipped with identical springs ($k = 3000$ N/m). The cars move toward each other with speeds of 3.00 m/s on a horizontal air track and collide, compressing the springs (Fig. P9.20). Find the maximum compression of each spring.

20A. Identical air cars each of mass m are equipped with identical springs each having a force constant k. The cars move toward each other with speeds v on a horizontal air track and collide, compressing the springs (Fig. P9.20). Find the maximum compression of each spring.

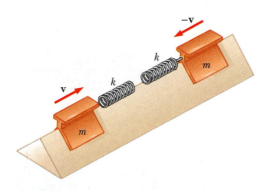

FIGURE P9.20

21. A 45.0-kg girl is standing on a plank that has a mass of 150 kg. The plank, originally at rest, is free to slide on a frozen lake, which is a flat, frictionless supporting surface. The girl begins to walk along the plank at a constant speed of 1.5 m/s relative to the plank. (a) What is her speed relative to the ice surface? (b) What is the speed of the plank relative to the ice surface?

22. A 7.00-kg bowling ball initially at rest is dropped from a height of 3.00 m. (a) What is the speed of the Earth coming up to meet the ball just before the ball hits the ground? Use 5.98×10^{24} kg as the mass of the Earth. (b) Use your answer to part (a) to justify ignoring the motion of the Earth when dealing with the motions of terrestrial objects.

23. A 2000-kg meteorite has a speed of 120 m/s just before colliding head-on with the Earth. Determine the recoil speed of the Earth (mass 5.98×10^{24} kg).

24. Gayle runs at a speed of 4.0 m/s and dives onto a sled that is initially at rest on the top of a frictionless snow-covered hill. After she and the sled have descended a vertical distance of 5.0 m, her brother, who is initially at rest, hops on her back and together they continue down the hill. What is their speed at the bottom if the total vertical drop is 15.0 m? Gayle's mass is 50.0 kg, the sled's is 5.0 kg, and her brother's is 30.0 kg.

25. A 10.0-g bullet is stopped in a block of wood ($m =$

5.00 kg). The speed of the bullet-plus-wood combination immediately after the collision is 0.600 m/s. What was the original speed of the bullet?

26. A 90-kg halfback running north with a speed of 10 m/s is tackled by a 120-kg opponent running south with a speed of 4.0 m/s. If the collision is perfectly inelastic and head-on, (a) calculate the speed and direction of the players just after the tackle and (b) determine the energy lost as a result of the collision. Account for the missing energy.

27. A 1200-kg car traveling initially with a speed of 25.0 m/s in an easterly direction crashes into the rear end of a 9000-kg truck moving in the same direction at 20.0 m/s (Fig. P9.27). The velocity of the car right after the collision is 18.0 m/s to the east. (a) What is the velocity of the truck right after the collision? (b) How much mechanical energy is lost in the collision? Account for this loss in energy.

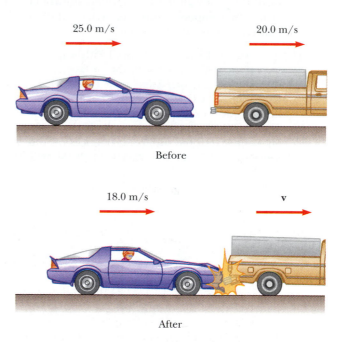

FIGURE P9.27

28. A railroad car of mass 2.5×10^4 kg moving with a speed of 4.0 m/s collides and couples with three other coupled railroad cars, each of the same mass as the single car and moving in the same direction with an initial speed of 2.0 m/s. (a) What is the speed of the four cars after the collision? (b) How much energy is lost in the collision?

29. A neutron in a reactor makes an elastic head-on collision with the nucleus of a carbon atom initially at rest. (a) What fraction of the neutron's kinetic energy is transferred to the carbon nucleus? (b) If the initial kinetic energy of the neutron is 1.6×10^{-13} J,

find its final kinetic energy and the kinetic energy of the carbon nucleus after the collision. (The mass of the carbon nucleus is about 12 times the mass of the neutron.)

30. A ball of mass m is suspended by a string of length L above a block standing on end, as shown in Figure P9.30. The ball is pulled back an angle θ and released. In trial A, it rebounds elastically off the block. In trial B, two-sided tape causes the ball to stick to the block in a totally inelastic collision. In which case is the ball more likely to knock the block over? Explain.

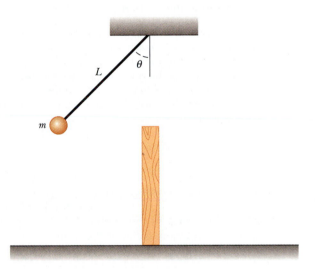

FIGURE P9.30

31. A 75-kg ice skater moving at 10 m/s crashes into a stationary skater of equal mass. After the collision, the two skaters move as a unit at 5.0 m/s. The average force a human skater can experience without breaking a bone is 4500 N. If the impact time is 0.10 s, does a bone break?

32. A 0.10-kg block is released from rest from the top of a 40.0° frictionless incline. When it has fallen a vertical distance of 1.5 m, a 0.015-kg bullet is fired into the block along a path parallel to the slope of the incline and momentarily brings the block to rest. (a) Find the speed of the bullet just before impact. (b) What bullet speed is needed to send the block up the incline to its initial position?

33. A 75.0-kg man stands in a 100.0-kg rowboat at rest in still water. He faces the back of the boat and throws a 5.00-kg rock out of the back at a speed of 20.0 m/s. The boat recoils forward and comes to rest 4.2 m from its original position. Calculate (a) the initial recoil speed of the boat, (b) the loss in mechanical energy due to the frictional force exerted by the water, and (c) the effective coefficient of friction between boat and water.

34. As shown in Figure P9.34, a bullet of mass m and speed v passes completely through a pendulum bob of mass M. The bullet emerges with a speed $v/2$. The pendulum bob is suspended by a stiff rod of length ℓ and negligible mass. What is the minimum value of v such that the pendulum bob will barely swing through a complete vertical circle?

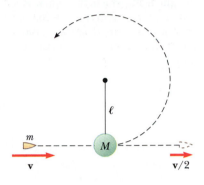

FIGURE P9.34

35. A 12-g bullet is fired into a 100-g wooden block initially at rest on a horizontal surface. After impact, the block slides 7.5 m before coming to rest. If the coefficient of friction between block and surface is 0.65, what was the speed of the bullet immediately before impact?

35A. A bullet of mass m_1 is fired into a wooden block of mass m_2 initially at rest on a horizontal surface. After impact, the block slides a distance d before coming to rest. If the coefficient of friction between block and surface is μ, what was the speed of the bullet immediately before impact?

36. A 7.00-g bullet fired into a 1.00-kg block of wood held in a vise penetrates to a depth of 8.00 cm. Then the vise is removed, the block of wood is placed on a frictionless horizontal surface, and another 7.00-g bullet is fired from the gun into it. To what depth does this second bullet penetrate?

37. Consider a frictionless track ABC as shown in Figure P9.37. A block of mass $m_1 = 5.001$ kg is released from A. It makes a head-on elastic collision with a block of mass $m_2 = 10.0$ kg at B, initially at rest. Cal-

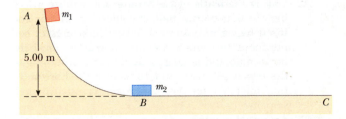

FIGURE P9.37

culate the maximum height to which m_1 rises after the collision.

38. A block of mass $m_1 = 2.0$ kg starts from rest on a surface inclined at 53° to the horizontal. The coefficient of kinetic friction between surface and block is $\mu_k = 0.25$. (a) If the speed of the block at the foot of the incline is 8.0 m/s to the right, determine the height from which the block was released. (b) Another block of mass $m_2 = 6.0$ kg is at rest on the smooth horizontal surface. Block m_1 collides with block m_2. After collision, the blocks stick together and move to the right. Determine the speed of the blocks after collision.

39. A 20.0-g bullet is fired horizontally into a 1.0-kg wooden block resting on a horizontal surface ($\mu_k = 0.25$). The bullet goes through the block and comes out with a speed of 250 m/s. If the block travels 5.0 m before coming to rest, what was the initial speed of the bullet?

40. Two blocks of mass $m_1 = 2.00$ kg and $m_2 = 4.00$ kg are released from a height of 5.00 m on a frictionless track as shown in Figure P9.40. The blocks undergo an elastic head-on collision. (a) Determine the two velocities just before collision. (b) Using Equations 9.20 and 9.21 determine the two velocities immediately after collision. (c) Determine the maximum height to which each block rises after collision.

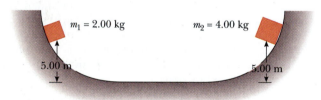

$m_1 = 2.00$ kg $m_2 = 4.00$ kg

5.00 m   5.00 m

FIGURE P9.40

Section 9.5   Two-Dimensional Collisions

41. A 3.00-kg mass with an initial velocity of $5.00\mathbf{i}$ m/s collides with and sticks to a 2.00-kg mass with an initial velocity of $-3.00\mathbf{j}$ m/s. Find the final velocity of the composite mass.

42. During the battle of Gettysburg, the gunfire was so intense that several bullets collided in midair and fused together. Assume a 5.0-g Union musket ball moving to the right at 250 m/s, and 20.0° above the horizontal, and a 3.0-g Confederate ball moving to the left at 280 m/s, and 15° above the horizontal. Immediately after they fuse together, what is their velocity?

43. An unstable nucleus of mass 17×10^{-27} kg initially at rest disintegrates into three particles. One of the particles, of mass 5.0×10^{-27} kg, moves along the y axis with a speed of 6.0×10^6 m/s. Another particle, of mass 8.4×10^{-27} kg, moves along the x axis with a speed of 4.0×10^6 m/s. Find (a) the velocity of the third particle and (b) the total energy given off in the process.

44. A 0.30-kg puck, initially at rest on a horizontal, frictionless surface, is struck by a 0.20-kg puck moving initially along the x axis with a speed of 2.0 m/s. After the collision, the 0.20-kg puck has a speed of 1.0 m/s at an angle of $\theta = 53°$ to the positive x axis (Fig. 9.13). (a) Determine the velocity of the 0.30-kg puck after the collision. (b) Find the fraction of kinetic energy lost in the collision.

45. Two shuffleboard disks of equal mass, one orange and the other yellow, are involved in a perfectly elastic, glancing collision. The yellow disk is initially at rest and is struck by the orange disk moving with a speed of 5.00 m/s. After the collision the orange disk moves along a direction that makes an angle of 37.0° with its initial direction of motion, and the velocity of the yellow disk is perpendicular to that of the orange disk (after the collision). Determine the final speed of each disk.

45A. Two shuffleboard disks of equal mass, one orange and the other yellow, are involved in a perfectly elastic, glancing collision. The yellow disk is initially at rest and is struck by the orange disk moving with a speed v_0. After the collision the orange disk moves along a direction that makes an angle θ with its initial direction of motion, and the velocity of the yellow disk is perpendicular to that of the orange disk (after the collision). Determine the final speed of each disk.

46. A block of mass m_1 moves east on a tabletop, traveling at a speed v_0 toward a block of mass m_2, which is at rest. After the collision, the first block moves south with a speed v. (a) Show that

$$v \le \sqrt{\frac{m_2 - m_1}{m_2 + m_1}}\, v_0$$

(*Hint:* Assume $K_{after} \le K_{before}$.) (b) What does this expression for v tell you about m_1 and m_2?

47. A billiard ball moving at 5.00 m/s strikes a stationary ball of the same mass. After the collision, the first ball moves at 4.33 m/s at an angle of 30.0° with respect to the original line of motion. Assuming an elastic collision (and ignoring friction and rotational motion), find the struck ball's velocity.

48. A 200-g cart moves on a horizontal, frictionless surface with a constant speed of 25.0 cm/s. A 50.0-g piece of modeling clay is dropped vertically onto the cart. (a) If the clay sticks to the cart, find the final speed of the system. (b) After the collision, the clay has no momentum in the vertical direction. Does this mean that the law of conservation of momentum is violated?

49. A particle of mass m, moving with speed v, collides obliquely with an identical particle initially at rest. Show that if the collision is elastic, the two particles move at 90° from each other after collision. (*Hint:* $(A + B)^2 = A^2 + B^2 + 2AB \cos \theta$.)

50. A mass m_1 having an initial velocity v_1 collides with a stationary mass m_2. After the collision, m_1 and m_2 are deflected as shown in Figure P9.50. The velocity of m_1 after collision is v_1'. Show that

$$\tan \theta_2 = \frac{v_1' \sin \theta_1}{v_1 - v_1 \cos \theta_2}$$

From the information given and the result you derived, can you make the assumption that the collision is perfectly elastic?

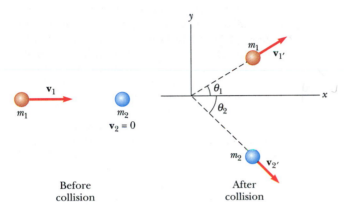

Before collision

After collision

FIGURE P9.50

Section 9.6   The Center of Mass

51. A 3.00-kg particle is located on the x axis at $x = -5.00$ m, and a 4.00-kg particle is on the x axis at $x = 3.00$ m. Find the center of mass of this two-particle system.

52. A water molecule consists of an oxygen atom with two hydrogen atoms bound to it (Fig. P9.52). The angle between the two bonds is 106°. If each bond is 0.100 nm long, where is the center of mass of the molecule?

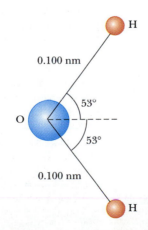

FIGURE P9.52

53. The mass of the Sun is 329 390 Earth masses, and the mean distance from the center of the Sun to the center of the Earth is 1.496×10^8 km. Treating the Earth and Sun as particles, with each mass concentrated at its respective geometric center, how far from the center of the Sun is the center of mass of the Earth-Sun system? Compare this distance with the mean radius of the Sun (6.960×10^5 km).

54. The separation between the hydrogen and chlorine atoms of the HCl molecule is about 1.30×10^{-10} m. Determine the location of the center of mass of the molecule as measured from the hydrogen atom. (Chlorine is 35 times more massive than hydrogen.)

55. Figure P9.55 shows three uniform objects—rod, right triangle, and square—with their masses given, along with their coordinates. Determine the center of mass for this three-object system.

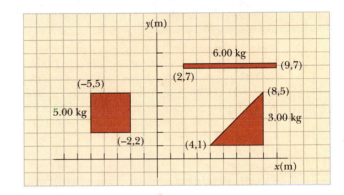

FIGURE P9.55

56. A circular hole of diameter a is cut out of a uniform square of sheet metal having sides $2a$, as in Figure P9.56. Where is the center of mass for the remaining portion?

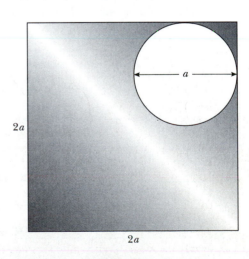

FIGURE P9.56

57. A uniform piece of sheet steel is shaped as in Figure P9.57. Compute the x and y coordinates of the center of mass of the piece.

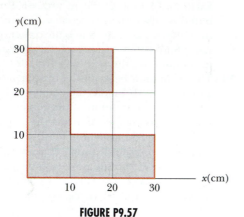

FIGURE P9.57

Section 9.7   Motion of a System of Particles

58. A block of mass $m = 2.0$ kg is placed on the top of an incline of mass $M = 8.0$ kg, height $h = 2.0$ m, and base length $L = 6.0$ m. If the block is released from rest (Fig. P9.58a), how far has the incline moved when the block reaches the bottom (Fig. P9.58b)? Assume all surfaces are frictionless. (*Hint:* The x coordinate for the center of mass of the block-incline system is fixed (why?) — see Example 9.18 for finding the center of mass of the incline.)

59. A 2.0-kg particle has a velocity $(2.0\mathbf{i} - 3.0\mathbf{j})$ m/s, and a 3.0-kg particle has a velocity $(1.0\mathbf{i} + 6.0\mathbf{j})$ m/s. Find (a) the velocity of the center of mass and (b) the total momentum of the system.

60. A 2.0-kg particle has a velocity of $\mathbf{v}_1 = (2.0\mathbf{i} - 10t\mathbf{j})$ m/s, where t is in seconds. A 3.0-kg particle moves with a constant velocity of $\mathbf{v}_2 = 4.0\mathbf{i}$ m/s. At $t = 0.50$ s, find (a) the velocity of the center of mass, (b) the acceleration of the center of mass, and (c) the total momentum of the system.

61. A 3.00-g particle is moving at 3.00 m/s toward a stationary 7.00-g particle. (a) With what speed does each approach the center of mass? (b) What is the momentum of each particle, relative to the center of mass?

61A. A particle of mass m_1 is moving at a speed v_1 toward a stationary particle of mass m_2. (a) With what speed does each approach the center of mass? (b) What is the momentum of each particle, relative to the center of mass?

62. Romeo entertains Juliet by playing his guitar from the rear of their boat in still water. After the serenade, Juliet carefully moves to the rear of the boat (away from shore) to plant a kiss on Romeo's cheek. If the 80-kg boat is facing shore and the 55-kg Juliet moves 2.7 m toward the 77-kg Romeo, how far does the boat move toward shore?

62A. Romeo entertains Juliet by playing his guitar from the rear of their boat in still water. After the serenade, Juliet carefully moves to the rear of the boat (away from shore) to plant a kiss on Romeo's cheek. If the boat (mass m_B) is facing shore and Juliet (mass m_J) moves a distance d toward Romeo (mass m_R), how far does the boat move toward shore?

*Section 9.8   Rocket Propulsion

63. A rocket engine consumes 80 kg of fuel per second. If the exhaust speed is 2.5×10^3 m/s, calculate the thrust on the rocket.

64. The first stage of a Saturn V space vehicle consumes fuel at the rate of 1.5×10^4 kg/s, with an exhaust speed of 2.6×10^3 m/s. (a) Calculate the thrust produced by these engines. (b) Find the initial acceleration of the vehicle on the launch pad if its initial mass is 3.0×10^6 kg. [*Hint:* You must include the force of gravity to solve part (b).]

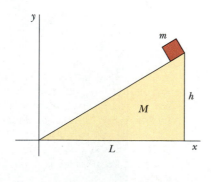

(a)

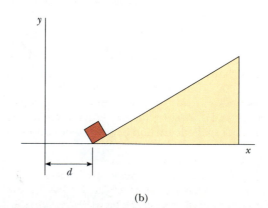

(b)

FIGURE P9.58

65. A 3000-kg rocket has 4000 kg of fuel on board. The rocket is coasting through space at 100.0 m/s and needs to boost its speed to 300.0 m/s. It does this by firing its engines and ejecting fuel at a relative speed of 650.0 m/s until the desired speed is reached. How much fuel is left on board after this maneuver?

66. A rocket for use in deep space is to have the capability of boosting a payload (plus the rocket frame and engine) of 3.0 metric tons to a speed of 10 000 m/s with an engine and fuel designed to produce an exhaust speed of 2000 m/s. (a) How much fuel plus oxidizer is required? (b) If a different fuel and engine design could give an exhaust speed of 5000 m/s, what amount of fuel and oxidizer would be required for the same task?

67. Fuel aboard a rocket has a density of 1.4×10^3 kg/m^3 and is ejected with a speed of 3.0×10^3 m/s. If the engine is to provide a thrust of 2.5×10^6 N, what volume of fuel must be burned per second?

67A. Fuel aboard a rocket has a density ρ and is ejected with a speed v. If the engine is to provide a thrust F, what volume of fuel must be burned per second?

68. A rocket with an initial total mass M_i is launched vertically from the Earth's surface. When the launch fuel has been completely burned, the rocket has reached an altitude small compared to the Earth's radius (so that the gravitational field strength may be considered constant during the burn). Show that the final speed is $v = -v_e \ln(M_i/M_f) - gt$, where the time of burn t_1 is $t_1 = (M_i - M_f)(dm/dt)^{-1}$. ($M_f$ is the final total mass of the rocket, v_e is the exhaust gas speed, and dm/dt is the constant rate of fuel consumption.)

ADDITIONAL PROBLEMS

69. A golf ball ($m = 46$ g) is struck a blow that makes an angle of 45° with the horizontal. The drive lands 200 m away on a flat fairway. If the golf club and ball are in contact for 7.0 ms, what is the average force of impact? (Neglect air resistance.)

70. A 12.0-g bullet is fired horizontally into a 100-g wooden block that is initially at rest on a rough horizontal surface and connected to a massless spring of constant 150 N/m. If the bullet-block system compresses the spring by 0.800 m, what was the speed of the bullet just as it enters the block? Assume that the coefficient of kinetic friction between block and surface is 0.60.

71. A 30-06 caliber hunting rifle fires a bullet of mass 0.012 kg with a muzzle velocity of 600 m/s to the right. The rifle has a mass of 4.0 kg. (a) What is the recoil velocity of the rifle as the bullet leaves the rifle? (b) If the rifle is stopped by the hunter's shoulder in a distance of 2.5 cm, what is the average force exerted on the shoulder by the rifle? (c) If the hunter's shoulder is partially restricted from recoiling, would the force exerted on the shoulder be the same as in part (b)? Explain.

72. An 8.00-g bullet is fired into a 2.50-kg block initially at rest at the edge of a frictionless table of height 1.00 m (Fig. P9.72). The bullet remains in the block, and after impact the block lands 2.00 m from the bottom of the table. Determine the initial speed of the bullet.

72A. A bullet of mass m is fired into a block of mass M initially at rest at the edge of a frictionless table of height h (Fig. P9.72). The bullet remains in the block, and after impact the block lands a distance d from the bottom of the table. Determine the initial speed of the bullet.

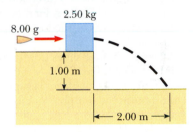

FIGURE P9.72

73. An attack helicopter is equipped with a 20-mm cannon that fires 130-g shells in the forward direction with a muzzle speed of 800 m/s. The fully loaded helicopter has a mass of 4000 kg. A burst of 160 shells is fired in a 4.0-s interval. What is the resulting average force on the helicopter and by what amount is its forward speed reduced?

74. A small block of mass $m_1 = 0.500$ kg is released from rest at the top of a curved frictionless wedge of mass $m_2 = 3.00$ kg, which sits on a frictionless horizontal surface as in Figure P9.74a. When it leaves the wedge, the block's velocity is 4.00 m/s to the right, as in Figure P9.74b. (a) What is the velocity of the wedge after the block reaches the horizontal surface? (b) What is the height h of the wedge?

74A. A small block of mass m_1 is released from rest at the top of a curved frictionless wedge of mass m_2, which sits on a frictionless horizontal surface as in Figure P9.74a. When it leaves the wedge, the block's velocity is v_1 to the right, as in Figure P9.74b. (a) What is the velocity of the wedge after the block reaches the horizontal surface? (b) What is the height h of the wedge?

75. A jet aircraft is traveling at 500 mi/h (223 m/s) in horizontal flight. The engine takes in air at a rate of

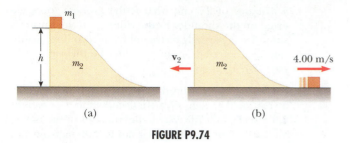

(a)   (b)

FIGURE P9.74

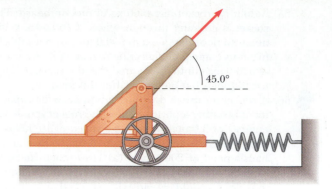

FIGURE P9.80

80.0 kg/s and burns fuel at a rate of 3.00 kg/s. If the exhaust gases are ejected at 600 m/s relative to the aircraft, find the thrust of the jet engine and the delivered horsepower.

76. Two ice skaters approach each other at right angles. Skater A has a mass of 50.0 kg and travels in the $+x$ direction at 2.00 m/s. Skater B has a mass of 70.0 kg and is moving in the $+y$ direction at 1.5 m/s. They collide and cling together. Find (a) the final velocity of the couple and (b) the loss in kinetic energy due to the collision.

77. A 75.0-kg firefighter slides down a pole while a constant frictional force of 300 N retards her motion. A horizontal 20.0-kg platform is supported by a spring at the bottom of the pole to cushion the fall. The firefighter starts from rest 4.00 m above the platform, and the spring constant is 4000 N/m. Find (a) the firefighter's speed just before she collides with the platform and (b) the maximum distance the spring is compressed. (Assume the frictional force acts during the entire motion.)

78. A 70.0-kg baseball player jumps vertically up to catch a baseball of mass 0.160 kg traveling horizontally with a speed of 35.0 m/s. If the vertical speed of the player at the instant of catching the ball is 0.200 m/s, determine the player's velocity just after the catch.

79. An 80.0-kg astronaut is working on the engines of his ship, which is drifting through space with a constant velocity. The astronaut, wishing to get a better view of the Universe, pushes against the ship and later finds himself 30.0 m behind the ship. Without a thruster, the only way to return to the ship is to throw his 0.500-kg wrench directly away from the ship. If he throws the wrench with a speed of 20.0 m/s, how long does it take the astronaut to reach the ship?

80. A cannon is rigidly attached to a carriage, which can move along horizontal rails but is connected to a post by a large spring of force constant $k = 2.00 \times 10^4$ N/m (Fig. P9.80). The cannon fires a 200-kg projectile at a velocity of 125 m/s directed 45.0° above the horizontal. (a) If the mass of the cannon and its carriage is 5000 kg, find the recoil speed of the cannon. (b) Determine the maximum extension of the spring. (c) Consider the system to consist of the cannon, carriage, and shell. Is the momentum of this system constant during the firing? Why or why not?

81. A chain of length L and total mass M is released from rest with its lower end just touching the top of a table, as in Figure P9.81a. Find the force exerted by the table on the chain after the chain has fallen through a distance x, as in Figure P9.81b. (Assume each link comes to rest the instant it reaches the table.)

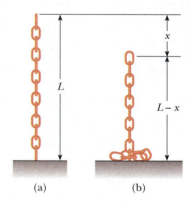

(a)   (b)

FIGURE P9.81

82. Two gliders are set in motion on an air track. A spring of force constant k is attached to the near side of one glider. The first glider of mass m_1 has velocity $\mathbf{v}_1$ and the second glider of mass m_2 has velocity $\mathbf{v}_2$, as in Figure P9.82 ($\mathbf{v}_1 > \mathbf{v}_2$). When m_1 collides with the spring attached to m_2 and compresses the spring to its maximum compression x_m, the velocity of the gliders is $\mathbf{v}$. In terms of $\mathbf{v}_1$, $\mathbf{v}_2$, m_1, m_2, and k, find

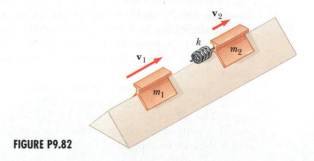

FIGURE P9.82

(a) the velocity **v** at maximum compression, (b) the maximum compression x_m, and (c) the velocities of each glider after m_1 has lost contact with the spring.

83. A 40.0-kg child stands at one end of a 70.0-kg boat that is 4.00 m in length (Fig. P9.83). The boat is initially 3.00 m from the pier. The child notices a turtle on a rock at the far end of the boat and proceeds to walk to that end to catch the turtle. Neglecting friction between boat and water, (a) describe the subsequent motion of the system (child plus boat). (b) Where is the child *relative to the pier* when he reaches the far end of the boat? (c) Will he catch the turtle? (Assume he can reach out 1.00 m from the end of the boat.)

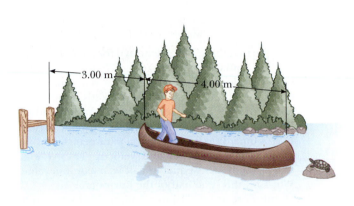

FIGURE P9.83

84. Two carts of equal mass, $m = 0.250$ kg, are placed on a frictionless track that has a light spring of force constant $k = 50.0$ N/m attached to one end of it, as in Figure P9.84. The blue cart is given an initial velocity of $\mathbf{v}_0 = 3.00$ m/s to the right, and the red cart is initially at rest. If the carts collide elastically, find (a) the velocity of the carts just after the first collision and (b) the maximum compression in the spring.

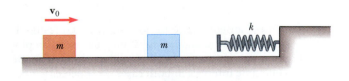

FIGURE P9.84

85. A 5.00-g bullet moving with an initial speed of 400 m/s is fired into and passes through a 1.00-kg block, as in Figure P9.85. The block, initially at rest on a

frictionless, horizontal surface, is connected to a spring of force constant 900 N/m. If the block moves 5.00 cm to the right after impact, find (a) the speed at which the bullet emerges from the block and (b) the energy lost in the collision.

85A. A bullet of mass m moving with an initial speed v_0 is fired into and passes through a block of mass M, as in Figure P9.85. The block, initially at rest on a frictionless, horizontal surface, is connected to a spring of force constant k. If the block moves a distance d to the right after impact, find (a) the speed at which the bullet emerges from the block and (b) the energy lost in the collision.

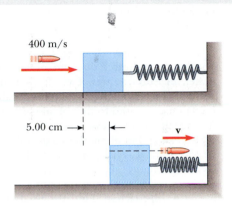

FIGURE P9.85

86. A student performs a ballistic pendulum experiment using an apparatus similar to that shown in Figure 9.11b. She obtains the following average data; $h = 8.68$ cm, $m_1 = 68.8$ g, and $m_2 = 263$ g. (a) Determine the initial speed v_{1i} of the projectile. (b) The second part of her experiment is to obtain v_{1i} by firing the same projectile horizontally (with the pendulum removed from the path), by measuring its horizontal displacement, x, and vertical displacement, y (Fig. P9.86). Show that the initial speed of the projectile is

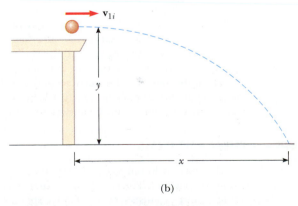

(b)

FIGURE P9.86

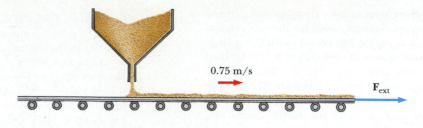

FIGURE P9.89

related to x and y through the relation

$$v_{1i} = \frac{x}{\sqrt{2y/g}}$$

What numerical value does she obtain for v_{1i} based on her measured values of $x = 257$ cm and $y = 85.3$ cm? What factors might account for the difference in this value compared to that obtained in part (a)?

87. Two particles, of masses m and $3m$, are moving toward each other along the x axis with the same initial speeds v_0. Mass m is traveling to the left, and mass $3m$ is traveling to the right. They undergo a head-on elastic collision, and each rebounds along the same line as it approached. Find the final speeds of the particles.

88. Two particles, of masses m and $3m$, are moving toward each other along the x axis with the same initial speeds v_0. Mass m is traveling to the left, and mass $3m$ is traveling to the right. They undergo an elastic glancing collision such that m is moving downward after the collision at a right angle to its initial direction. (a) Find the final speeds of the two masses. (b) What is the angle θ at which $3m$ is scattered?

89. Sand from a stationary hopper falls on a moving conveyor belt at the rate of 5.0 kg/s, as in Figure P9.89. The conveyor belt is supported by frictionless rollers and moves at 0.75 m/s under the action of a horizontal external force $\mathbf{F}_{ext}$ supplied by the motor that drives the belt. Find (a) the sand's rate of change of momentum in the horizontal direction, (b) the frictional force exerted by the belt on the sand, (c) the magnitude of $\mathbf{F}_{ext}$, (d) the work done by $\mathbf{F}_{ext}$ in one second, and (e) the kinetic energy acquired by the falling sand each second due to the change in its horizontal motion. (f) Why is the answer to (d) different from the answer to (e)?

SPREADSHEET PROBLEMS

S1. A rocket has an initial mass of 20 000 kg, of which 20% is the payload. The rocket burns fuel at a rate of 200 kg/s and exhausts gas at a relative speed of 2.00 km/s. Its acceleration dv/dt is determined from the

equation of motion

$$M\frac{dv}{dt} = v_e \left| \frac{dM}{dt} \right| + F_{ext}$$

Assume that there are no external forces and that the rocket's initial velocity is zero. When $F_{ext} = 0$, the rocket's velocity is $v(t) = v_e \ln(M_i/M)$, where M is the mass at time t and M_i is the rocket's initial mass. Spreadsheet 9.1 calculates the acceleration and velocity of the rocket and plots the velocity as a function of time. (a) Using the given parameters, find the maximum acceleration and velocity. (b) At what time is the velocity equal to half its maximum value? Why isn't this time half of the burn time?

S2. Modify Spreadsheet 9.1 to calculate the distance traveled by the rocket in Problem S9.1. Add a new column to the spreadsheet to find the new position x_{i+1}. Estimate the new position of the rocket by

$$x_{i+1} = x_i + \tfrac{1}{2}(v_{i+1} + v_i)\Delta t,$$

where x_i is the old position, v_i is the old velocity, and v_{i+1} is the new velocity.

S3. If the rocket in the previous problem burns its fuel twice as fast (400 kg/s) while maintaining the same exhaust speed, how far does it travel before it runs out of fuel? The velocity reached by the rocket is the same in both cases; that is

$$v_f - v_i = v_e \ln\left(\frac{M_i}{M_f}\right)$$

Is there an advantage in burning the fuel faster? Is there a disadvantage? *Hint:* It may be useful to plot the accelerations, velocities, and positions for the same time intervals in both cases. In which case does the rocket travel farther? In which case does the rocket have to withstand the largest acceleration? By how much?

S4. Applying conservation of momentum to the motion of a rocket, we find:

$$M\frac{dv}{dt} = v_e \left| \frac{dM}{dt} \right| + F_{ext},$$

where $v_e |dM/dt|$ is the thrust of the rocket and M is the mass of the rocket. If the rocket is fired vertically upward from Earth, we have $F_{ext} = -mg$. Dividing the equation of motion by M gives

$$\frac{dv}{dt} = \frac{v_e}{M}\left|\frac{dM}{dt}\right| - g$$

If the rocket gets far enough away from Earth during its "burn" so that the variation in the Earth's gravitational field must be taken into account, then the constant g in the equation above must be replaced by $g[R_E/(R_E + x)]^2$, where R_E is the radius of the Earth and x is the distance above the Earth. Write your own program or modify Spreadsheet 6.1 to calculate and plot the velocity and acceleration of a rocket during its burn time, taking into account the variation of gravity with the distance x from the Earth's surface. Use the following data: $M_i = 2.50 \times 10^6$ kg; burn rate $|dM/dt| = 1.35 \times 10^4$ kg/s; and thrust $= 25 \times 10^6$ N. The payload is 30% of the original mass. With these data, $v_e = 1.85$ km/s. How does the final speed of the rocket at burnout compare to the final speed if g were constant?

S5. To design a better golf club driver, a mechanical engineer equips a test club with a device that measures the impact forces generated when the driver strikes the ball. The force data as a function of time are recorded in the following table:

t (ms)	F (N)
0.45	0
0.50	20
0.55	210
0.60	1350
0.65	4350
0.70	12755
0.75	14430
0.80	11675
0.85	3005
0.90	1130
0.95	400
1.00	40
1.05	0
1.10	0

Using these data, calculate the impulse imparted to the ball by numerically integrating the force-versus-time data. Find the average force acting on the ball. (a) Use a simple rectangular integration approximation. Compare the impulse and average force you calculate with Example 9.3. (b) Repeat part (a) using a trapezoidal integration approximation.

Rotation of a Rigid
Object About a Fixed Axis

The pair of skaters in this figure-skating routine is executing the beautiful "death spiral." In this circular motion, the female skater is spun in a circular arc by her partner with her body nearly parallel to the ice. (© Duomo/Steven E. Sutton, 1994)

When an extended object, such as a wheel, rotates about its axis, the motion cannot be analyzed by treating the object as a particle, since at any given time different parts of the object have different velocities and accelerations. For this reason, it is convenient to consider an extended object as a large number of particles, each with its own velocity and acceleration.

In dealing with the rotation of an object, analysis is greatly simplified by assuming the object to be rigid. A **rigid object** is defined as one that is nondeformable or, to say the same thing another way, one in which the separations between all pairs of particles remain constant. All real bodies are deformable to some extent; however, our rigid-object model is useful in many situations where deformation is negligible. In this chapter, we treat the rotation of a rigid object about a fixed axis, commonly referred to as *pure rotational motion.*

When the rigid-body approximation is not adequate, it often suffices as a first approximation. Thus, a rotating galaxy is far from a rigid body, but for many purposes, the internal motions may be neglected when treating the entire galaxy. A rotating molecule is likely to expand because of the rotation, but may be adequately approximated as a rigid body with the longer bond lengths.

Rigid object

10.1 ANGULAR VELOCITY AND ANGULAR ACCELERATION

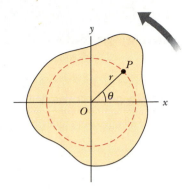

Figure 10.1 illustrates a planar rigid object of arbitrary shape confined to the *xy* plane and rotating about a fixed axis through *O* perpendicular to the plane of the figure. A particle on the object at *P* is at a fixed distance *r* from the origin and rotates in a circle of radius *r* about *O*. In fact, *every* particle on the object undergoes circular motion about *O*. It is convenient to represent the position of *P* with its polar coordinates (r, θ). In this representation, the only coordinate that changes in time is the angle θ; *r* remains constant. (In rectangular coordinates, both *x* and *y* vary in time.) As the particle moves along the circle from the positive *x* axis $(\theta = 0)$ to *P*, it moves through an arc length *s*, which is related to the angular position θ through the relationship

$$s = r\theta \tag{10.1a}$$

$$\theta = \frac{s}{r} \tag{10.1b}$$

It is important to note the units of θ in Equation 10.1b. The angle θ is the ratio of an arc length and the radius of the circle and, hence, is a pure number. However, we commonly give θ the artificial unit **radian** (rad), where

FIGURE 10.1 Rotation of a rigid object about a fixed axis through *O* perpendicular to the plane of the figure. (In other words, the axis of rotation is the *z* axis.) Note that a particle at *P* rotates in a circle of radius *r* centered at *O*.

> one radian is the angle subtended by an arc length equal to the radius of the arc.

Radian

Since the circumference of a circle is $2\pi r$, it follows from Equation 10.1b that 360° corresponds to an angle of $2\pi r/r$ rad or 2π rad (one revolution). Hence, 1 rad = $360°/2\pi \approx 57.3°$. To convert an angle in degrees to an angle in radians, we use the fact that 2π radians = 360°:

$$\theta \text{ (rad)} = \frac{\pi}{180°} \theta \text{ (deg)}$$

For example, 60° equals $\pi/3$ rad, and 45° equals $\pi/4$ rad.

As the particle in question on our rigid object travels from *P* to *Q* in a time Δt, the radius vector sweeps out an angle $\Delta\theta = \theta_2 - \theta_1$ (Fig. 10.2), which equals the **angular displacement**. We define the **average angular speed** $\bar{\omega}$ (omega) as the ratio of this angular displacement to the time interval Δt:

$$\bar{\omega} \equiv \frac{\theta_2 - \theta_1}{t_2 - t_1} = \frac{\Delta\theta}{\Delta t} \tag{10.2}$$

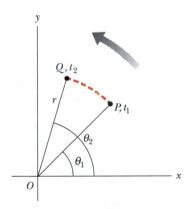

FIGURE 10.2 A particle on a rotating rigid object moves from *P* to *Q* along the arc of a circle. In the time interval $\Delta t = t_2 - t_1$, the radius vector sweeps out an angle $\Delta\theta = \theta_2 - \theta_1$.

In analogy to linear speed, the **instantaneous angular speed**, ω, is defined as the limit of the ratio in Equation 10.2 as Δt approaches zero:

$$\omega = \lim_{\Delta t \to 0} \frac{\Delta\theta}{\Delta t} = \frac{d\theta}{dt} \tag{10.3}$$

Instantaneous angular speed

Angular speed has units of radians per second, or s^{-1}, since radians are not dimensional. Let us adopt the convention that the fixed axis of rotation for the rigid object is the *z* axis, as in Figure 10.1. We take ω to be positive when θ is increasing (counterclockwise motion) and negative when θ is decreasing (clockwise motion).

If the instantaneous angular speed of an object changes from ω_1 to ω_2 in the time interval Δt, the object has an angular acceleration. The **average angular acceleration** $\bar{\alpha}$ (alpha) of a rotating object is defined as the ratio of the change in the angular speed to the time interval Δt:

Average angular acceleration

$$\bar{\alpha} \equiv \frac{\omega_2 - \omega_1}{t_2 - t_1} = \frac{\Delta \omega}{\Delta t} \tag{10.4}$$

In analogy to linear acceleration, the **instantaneous angular acceleration** is defined as the limit of the ratio $\Delta \omega / \Delta t$ as Δt approaches zero:

Instantaneous angular acceleration

$$\alpha \equiv \lim_{\Delta t \to 0} \frac{\Delta \omega}{\Delta t} = \frac{d\omega}{dt} \tag{10.5}$$

Angular acceleration has units of rad/s^2, or s^{-2}. Note that α is positive when ω is increasing in time and negative when ω is decreasing in time.

When rotating about a fixed axis, every particle on a rigid object has the same angular speed and the same angular acceleration. That is, the quantities ω and α characterize the rotational motion of the entire rigid object. Using these quantities, we can greatly simplify the analysis of rigid-body rotation. The angular displacement (θ), angular speed (ω), and angular acceleration (α) are analogous to linear displacement (x), linear speed (v), and linear acceleration (a), respectively, for the corresponding motion discussed in Chapter 2. The variables θ, ω, and α differ dimensionally from the variables x, v, and a, only by a length factor.

We have already indicated how the signs for ω and α are determined; however, we have not specified any direction in space associated with these vector quantities.[1] For rotation about a fixed axis, the only direction in space that uniquely specifies the rotational motion is the direction along the axis of rotation. Therefore, the directions $\boldsymbol{\omega}$ and $\boldsymbol{\alpha}$ are along this axis. If an object rotates in the xy plane as in Figure 10.1, the direction of $\boldsymbol{\omega}$ is out of the plane of the diagram when the rotation is counterclockwise and into the plane of the diagram when the rotation is clockwise. To illustrate this convention, it is convenient to use the *right-hand rule* shown in Figure 10.3a. The four fingers of the right hand are wrapped in the direction of the rotation. The extended right thumb points in the direction of $\boldsymbol{\omega}$. Figure 10.3b illustrates that $\boldsymbol{\omega}$ is also in the direction of advance of a similarly rotating right-handed screw. Finally, the sense of **a** follows from its definition as $d\boldsymbol{\omega}/dt$. It is the same as $\boldsymbol{\omega}$ if the angular speed (the magnitude of $\boldsymbol{\omega}$) is increasing in time and antiparallel to $\boldsymbol{\omega}$ if the angular speed is decreasing in time.

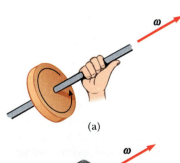

(a)

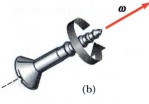

(b)

FIGURE 10.3 (a) The right-hand rule for determining the direction of the angular velocity vector. (b) The direction of $\boldsymbol{\omega}$ is in the direction of advance of a right-handed screw.

10.2 ROTATIONAL KINEMATICS: ROTATIONAL MOTION WITH CONSTANT ANGULAR ACCELERATION

In the study of linear motion, we found that the simplest form of accelerated motion to analyze is motion under constant linear acceleration (Chapter 2). Likewise, for rotational motion about a fixed axis, the simplest accelerated motion to analyze is motion under constant angular acceleration. Therefore, we next develop kinematic relationships for rotational motion under constant angular acceleration. If we write Equation 10.5 in the form $d\omega = \alpha \, dt$ and let $\omega = \omega_0$ at $t_0 = 0$,

---

[1] Although we do not verify it here, the instantaneous angular velocity and instantaneous angular acceleration are vector quantities, but the corresponding average values are not. This is because angular displacement is not a vector quantity for finite rotations.

TABLE 10.1   A Comparison of Kinematic Equations for Rotational and Linear Motion Under Constant Acceleration

Rotational Motion About a Fixed Axis with α = Constant Variables: θ and ω	Linear Motion with a = Constant Variables: x and v
$\omega = \omega_0 + \alpha t$	$v = v_0 + at$
$\theta = \theta_0 + \frac{1}{2}(\omega_0 + \omega)t$	$x = x_0 + \frac{1}{2}(v_0 + v)t$
$\theta = \theta_0 + \omega_0 t + \frac{1}{2}\alpha t^2$	$x = x_0 + v_0 t + \frac{1}{2}at^2$
$\omega^2 = \omega_0^2 + 2\alpha(\theta - \theta_0)$	$v^2 = v_0^2 + 2a(x - x_0)$

we can integrate this expression directly:

$$\omega = \omega_0 + \alpha t \qquad (\alpha = \text{constant}) \qquad (10.6)$$

> Rotational kinematic equations

Likewise, substituting Equation 10.6 into Equation 10.3 and integrating once more (with $\theta = \theta_0$ at $t_0 = 0$), we get

$$\theta = \theta_0 + \omega_0 t + \frac{1}{2}\alpha t^2 \qquad (10.7)$$

If we eliminate t from Equations 10.6 and 10.7, we get

$$\omega^2 = \omega_0^2 + 2\alpha(\theta - \theta_0) \qquad (10.8)$$

Notice that these kinematic expressions for rotational motion under constant angular acceleration are of the same form as those for linear motion under constant linear acceleration with the substitutions $x \rightarrow \theta$, $v \rightarrow \omega$, and $a \rightarrow \alpha$. Table 10.1 compares the kinematic equations for rotational and linear motion. Furthermore, the expressions are valid for both rigid-body rotation and particle motion about a *fixed* axis.

EXAMPLE 10.1   Rotating Wheel

A wheel rotates with a constant angular acceleration of 3.50 rad/s². If the angular speed of the wheel is 2.00 rad/s at $t_0 = 0$, (a) what angle does the wheel rotate through in 2.00 s?

Solution

$$\theta - \theta_0 = \omega_0 t + \frac{1}{2}\alpha t^2 = \left(2.00 \, \frac{\text{rad}}{\text{s}}\right)(2.00 \text{ s})$$

$$+ \frac{1}{2}\left(3.50 \, \frac{\text{rad}}{\text{s}^2}\right)(2.00 \text{ s})^2$$

$$= 11.0 \text{ rad} = 630° = 1.75 \text{ rev}$$

(b) What is the angular speed at $t = 2.00$ s?

$$\omega = \omega_0 + \alpha t = 2.00 \text{ rad/s} + \left(3.50 \, \frac{\text{rad}}{\text{s}^2}\right)(2.00 \text{ s})$$

$$= 9.00 \text{ rad/s}$$

We could also obtain this result using Equation 10.8 and the results of part (a). Try it!

Exercise   Find the angle that the wheel rotates through between $t = 2.00$ s and $t = 3.00$ s.

Answer   10.8 rad.

10.3   RELATIONSHIPS BETWEEN ANGULAR AND LINEAR QUANTITIES

In this section we derive some useful relationships between the angular speed and acceleration of a rotating rigid object and the linear speed and acceleration of an arbitrary point in the object. In order to do so, we must keep in mind that when a rigid object rotates about a fixed axis as in Figure 10.4, every particle of the object moves in a circle the center of which is the axis of rotation.

We can relate the angular speed of the rotating object to the tangential speed of a point P on the object. Since P moves in a circle, the linear velocity vector $\mathbf{v}$ is always tangent to the circular path, and hence the phrase *tangential velocity*. The magnitude of the tangential velocity of the point P is, by definition, ds/dt, where s is the distance traveled by this point measured along the circular path. Recalling that $s = r\theta$ (Eq. 10.1a) and noting that r is constant, we get

$$v = \frac{ds}{dt} = r\frac{d\theta}{dt}$$

$$v = r\omega \tag{10.9}$$

That is, the tangential speed of a point on a rotating rigid object equals the distance of that point from the axis of rotation multiplied by the angular speed. Therefore, although every point on the rigid object has the same *angular* speed, not every point has the same *linear* speed. In fact, Equation 10.9 shows that the linear speed of a point on the rotating object increases as one moves outward from the center of rotation, as you intuitively expect.

We can relate the angular acceleration of the rotating rigid object to the tangential acceleration of the point P by taking the time derivative of v:

$$a_t = \frac{dv}{dt} = r\frac{d\omega}{dt}$$

$$a_t = r\alpha \tag{10.10}$$

That is, the magnitude of the tangential component of the linear acceleration of a point on a rotating rigid object equals the distance of that point from the axis of rotation multiplied by the angular acceleration.

In Chapter 4 we found that a point rotating in a circular path undergoes a centripetal, or radial, acceleration of magnitude v^2/r directed toward the center of rotation (Fig. 10.5). Since $v = r\omega$ for the point P on the rotating object, we can

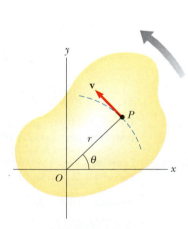

FIGURE 10.4   As a rigid object rotates about the fixed axis through O, the point P has a linear velocity $\mathbf{v}$ that is always tangent to the circular path of radius r.

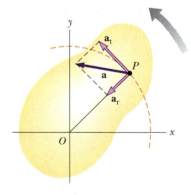

FIGURE 10.5   As a rigid object rotates about a fixed axis through O, the point P experiences a tangential component of linear acceleration, $\mathbf{a}_t$, and a centripetal component of linear acceleration, $\mathbf{a}_r$. The total linear acceleration of this point is $\mathbf{a} = \mathbf{a}_t + \mathbf{a}_r$.

express the magnitude of the centripetal acceleration as

$$a_r = \frac{v^2}{r} = r\omega^2 \qquad (10.11)$$

The total linear acceleration of the point is $\mathbf{a} = \mathbf{a}_t + \mathbf{a}_r$. Therefore, the magnitude of the total linear acceleration of the point P on the rotating rigid object is

$$a = \sqrt{a_t^2 + a_r^2} = \sqrt{r^2\alpha^2 + r^2\omega^4} = r\sqrt{\alpha^2 + \omega^4} \qquad (10.12)$$

CONCEPTUAL EXAMPLE 10.2

When a wheel of radius R rotates about a fixed axis as in Figure 10.3, do all points on the wheel have the same angular speed? Do they all have the same linear speed? If the angular speed is constant and equal to ω, describe the linear speeds and linear accelerations of the points located at $r = 0$, $r = R/2$, and $r = R$, where the points are measured from the center of the wheel.

Reasoning   Yes, all points on the wheel have the same anular speed. This is why we use angular quantities to describe

rotational motion. Not all points on the wheel have the same linear speed. The point at $r = 0$ has zero linear speed and zero linear acceleration; a point at $r = R/2$ has a linear speed $v = R\omega/2$ and a linear acceleration equal to the centripetal acceleration $v^2/(R/2) = R\omega^2/2$. (The tangential acceleration is zero at all points since ω is constant.) A point on the rim at $r = R$ has a linear speed $v = R\omega$ and a linear acceleration $R\omega^2$.

EXAMPLE 10.3   A Rotating Turntable

The turntable of a record player rotates initially at the rate 33 rev/min and takes 20.0 s to come to rest. (a) What is the angular acceleration of the turntable, assuming the acceleration is uniform?

Solution   Recalling that 1 rev $= 2\pi$ rad, we see that the initial angular speed is

$$\omega_0 = \left(33.0\, \frac{\text{rev}}{\text{min}}\right)\left(2\pi\, \frac{\text{rad}}{\text{rev}}\right)\left(\frac{1}{60}\, \frac{\text{min}}{\text{s}}\right) = 3.46\ \text{rad/s}$$

Using $\omega = \omega_0 + \alpha t$ and the fact that $\omega = 0$ at $t = 20.0$ s, we get

$$\alpha = -\frac{\omega_0}{t} = -\frac{3.46\ \text{rad/s}}{20.0\ \text{s}} = \boxed{-0.173\ \text{rad/s}^2}$$

where the negative sign indicates that ω is decreasing.

(b) How many rotations does the turntable make before coming to rest?

Solution   Using Equation 10.7, we find that the angular displacement in 20.0 s is

$$\Delta\theta = \theta - \theta_0 = \omega_0 t + \tfrac{1}{2}\alpha t^2$$

$$= [3.46(20.0) + \tfrac{1}{2}(-0.173)(20.0)^2]\text{rad} = \boxed{34.6\ \text{rad}}$$

This corresponds to $34.6/2\pi$ rev, or 5.50 rev.

(c) If the radius of the turntable is 14.0 cm, what are the magnitudes of the radial and tangential components of the linear acceleration of a point on the rim at $t = 0$?

Solution   We can use $a_t = r\alpha$ and $a_r = r\omega^2$, which give

$$a_t = r\alpha = (14.0\ \text{cm})\left(-0.173\, \frac{\text{rad}}{\text{s}^2}\right) = \boxed{-2.42\ \text{cm/s}^2}$$

$$a_r = r\omega_0^2 = (14.0\ \text{cm})\left(3.46\, \frac{\text{rad}}{\text{s}}\right)^2 = \boxed{168\ \text{cm/s}^2}$$

Exercise   What is the initial linear speed of a point on the rim of the turntable?

Answer   48.4 cm/s.

10.4   ROTATIONAL ENERGY

Let us consider a rigid object as a collection of small particles and let us assume that the object rotates about the fixed z axis with an angular speed ω (Fig. 10.6). Each particle has kinetic energy determined by its mass and speed. If the mass of

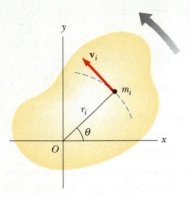

FIGURE 10.6 A rigid object rotating about the z axis with angular speed ω. The kinetic energy of the particle of mass m_i is $\frac{1}{2}m_i v_i^2$. The total energy of the object is $\frac{1}{2}I\omega^2$.

Rotational kinetic energy

the ith particle is m_i and its speed is v_i, its kinetic energy is

$$K_i = \tfrac{1}{2}m_i v_i^2$$

To proceed further, we must recall that although every particle in the rigid object has the same angular speed, ω, the individual linear speeds depend on the perpendicular distance r_i from the axis of rotation according to the expression $v_i = r_i \omega$ (Eq. 10.9). The *total* energy of the rotating rigid object is the sum of the kinetic energies of the individual particles:

$$K_R = \sum K_i = \sum \tfrac{1}{2}\, m_i v_i^2 = \tfrac{1}{2}\sum m_i r_i^2 \omega^2$$

$$K_R = \tfrac{1}{2}\left(\sum m_i r_i^2\right)\omega^2 \tag{10.13}$$

where we have factored ω^2 from the sum since it is common to every particle. The quantity in parentheses is called the **moment of inertia, I:**

$$I = \sum m_i r_i^2 \tag{10.14}$$

Using this notation, we can express the rotational energy of the rotating rigid object (Eq. 10.13) as

$$K_R = \tfrac{1}{2}\, I\omega^2 \tag{10.15}$$

From the definition of moment of inertia, we see that it has dimensions of ML^2 (kg·m² in SI units and g·cm² in cgs units). Although we commonly refer to the quantity $\frac{1}{2}I\omega^2$ as the **rotational energy**, it is not a new form of energy. It is ordinary kinetic energy, because it was derived from a sum over individual kinetic energies of the particles contained in the rigid object. However, the form of the energy given by Equation 10.15 is a convenient one when dealing with rotational motion, providing we know how to calculate I. It is important that you recognize the analogy between kinetic energy associated with linear motion, $\frac{1}{2}mv^2$, and rotational energy, $\frac{1}{2}I\omega^2$. The quantities I and ω in rotational motion are analogous to m and v in linear motion, respectively. (In fact, I takes the place of m every time we compare a linear-motion equation to its rotational counterpart.)

EXAMPLE 10.4   The Oxygen Molecule

Consider the diatomic molecule oxygen, O_2, which is rotating in the xy plane about the z axis passing through its center, perpendicular to its length. The mass of each oxygen atom is 2.66×10^{-26} kg, and at room temperature, the average separation between the two oxygen atoms is $d = 1.21 \times 10^{-10}$ m (the atoms are treated as point masses). (a) Calculate the moment of inertia of the molecule about the z axis.

Solution   Since the distance of each atom from the z axis is $d/2$, the moment of inertia about the z axis is

$$I = \sum m_i r_i^2 = m\left(\frac{d}{2}\right)^2 + m\left(\frac{d}{2}\right)^2 = \frac{md^2}{2}$$

$$= \left(\frac{2.66 \times 10^{-26}}{2}\text{ kg}\right)(1.21 \times 10^{-10}\text{ m})^2$$

$$= 1.95 \times 10^{-46}\text{ kg·m}^2$$

(b) If the angular speed of the molecule about the z axis is 4.60×10^{12} rad/s, what is its rotational kinetic energy?

Solution

$$K_R = \tfrac{1}{2}I\omega^2$$

$$= \tfrac{1}{2}(1.95 \times 10^{-46}\text{ kg·m}^2)\left(4.60 \times 10^{12}\,\frac{\text{rad}}{\text{s}}\right)^2$$

$$= 2.06 \times 10^{-21}\text{ J}$$

This is approximately equal to the average kinetic energy associated with the linear motion of the molecule at room temperature, which is about 6.2×10^{-21} J.

EXAMPLE 10.5   Four Rotating Masses

Four point masses are fastened to the corners of a frame of negligible mass lying in the *xy* plane (Fig. 10.7). (a) If the rotation of the system occurs about the *y* axis with an angular speed ω, find the moment of inertia about the *y* axis and the rotational kinetic energy about this axis.

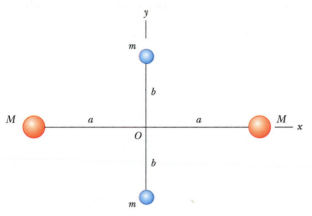

FIGURE 10.7   (Example 10.5) All four masses are at a fixed separation as shown. The moment of inertia depends on the axis about which it is evaluated.

Solution   First, note that the two masses *m* that lie on the *y* axis do not contribute to I_y (that is, $r_i = 0$ for these masses about this axis). Applying Equation 10.14, we get

$$I_y = \sum m_i r_i^2 = Ma^2 + Ma^2 = 2Ma^2$$

Therefore, the rotational energy about the *y* axis is

$$K_R = \tfrac{1}{2}I_y\omega^2 = \tfrac{1}{2}(2Ma^2)\,\omega^2 = \boxed{Ma^2\omega^2}$$

The fact that the masses *m* do not enter into this result makes sense, since they have no motion about the chosen axis of rotation; hence, they have no kinetic energy.

(b) Suppose the system rotates in the *xy* plane about an axis through *O* (the *z* axis). Calculate the moment of inertia about the *z* axis and the rotational energy about this axis.

Solution   Since r_i in Equation 10.14 is the *perpendicular* distance to the axis of rotation, we get

$$I_z = \sum m_i r_i^2 = Ma^2 + Ma^2 + mb^2 + mb^2 = \boxed{2Ma^2 + 2mb^2}$$

$$K_R = \tfrac{1}{2}I_z\omega^2 = \tfrac{1}{2}(2Ma^2 + 2mb^2)\,\omega^2 = \boxed{(Ma^2 + mb^2)\,\omega^2}$$

Comparing the results for (a) and (b), we conclude that the moment of inertia and, therefore, the rotational energy associated with a given angular speed depend on the axis of rotation. In (b), we expect the result to include all masses and distances because all four masses are in motion for rotation in the *xy* plane. Furthermore, the fact that the rotational energy in (a) is smaller than in (b) indicates that it would take less effort (work) to set the system into rotation about the *y* axis than about the *z* axis.

10.5   CALCULATION OF MOMENTS OF INERTIA

We can evaluate the moment of inertia of an extended object by imagining that the object is divided into many small volume elements, each of mass Δm. We use the definition $I = \sum r_i^2\,\Delta m_i$ and take the limit of this sum as $\Delta m \rightarrow 0$. In this limit, the sum becomes an integral over the whole object:

$$I = \lim_{\Delta m_i \to 0} \sum r_i^2\,\Delta m_i = \int r^2\,dm \qquad (10.16)$$

To evaluate the moment of inertia using Equation 10.16, it is necessary to express each volume element (of mass *dm*) in terms of its coordinates. It is common to define a mass density in various forms. For a three-dimensional object, it is appropriate to use the *volume density*, that is, *mass per unit volume*:

$$\rho = \lim_{\Delta V \to 0} \frac{\Delta m}{\Delta V} = \frac{dm}{dV}$$

$$dm = \rho\,dV$$

Therefore, the moment of inertia for a three-dimensional object can be expressed in the form

$$I = \int \rho r^2 \, dV$$

If the object is homogeneous, then ρ is constant and the integral can be evaluated for a known geometry. If ρ is not constant, then its variation with position must be specified.

When dealing with a sheet of uniform thickness t, rather than with a three-dimensional object, it is convenient to define a *surface density* $\sigma = \rho t$, which signifies *mass per unit area*.

Finally, when mass is distributed along a uniform rod of cross-sectional area A, we sometimes use *linear density*, $\lambda = \rho A$, where λ is defined as *mass per unit length*.

EXAMPLE 10.6   Uniform Hoop

Find the moment of inertia of a uniform hoop of mass M and radius R about an axis perpendicular to the plane of the hoop, through its center (Fig. 10.8).

Solution   All mass elements are the same distance $r = R$ from the axis, and so, applying Equation 10.16, we get for the moment of inertia about the z axis through O:

$$I_z = \int r^2 \, dm = R^2 \int dm = \boxed{MR^2}$$

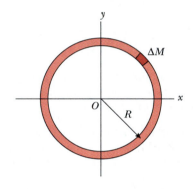

FIGURE 10.8   (Example 10.6) The mass elements of a uniform hoop are all the same distance from O.

EXAMPLE 10.7   Uniform Rigid Rod

Calculate the moment of inertia of a uniform rigid rod of length L and mass M (Fig. 10.9) about an axis perpendicular to the rod (the y axis) and passing through its center of mass.

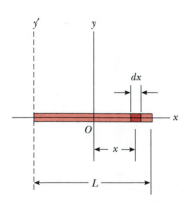

FIGURE 10.9   (Example 10.7) A uniform rigid rod of length L. The moment of inertia about the y axis is less than that about the y' axis.

Solution   The shaded length element dx has a mass dm equal to the mass per unit length multiplied by dx:

$$dm = \frac{M}{L} \, dx$$

Substituting this expression for dm into Equation 10.16, with $r = x$, we get

$$I_y = \int r^2 \, dm = \int_{-L/2}^{L/2} x^2 \frac{M}{L} \, dx = \frac{M}{L} \int_{-L/2}^{L/2} x^2 \, dx$$

$$= \frac{M}{L}\left[\frac{x^3}{3}\right]_{-L/2}^{L/2} = \boxed{\tfrac{1}{12} ML^2}$$

Exercise   Calculate the moment of inertia of a uniform rigid rod about an axis perpendicular to the rod and passing through one end (the y' axis). Note that the calculation requires that the limits of integration be from $x = 0$ to $x = L$.

Answer $\tfrac{1}{3} ML^2$.

EXAMPLE 10.8   Uniform Solid Cylinder

A uniform solid cylinder has a radius R, mass M, and length L. Calculate its moment of inertia about its central axis (the z axis in Fig. 10.10).

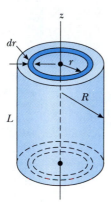

FIGURE 10.10   (Example 10.8) Calculating I about the z axis for a uniform solid cylinder.

Reasoning   It is convenient to divide the cylinder into many cylindrical shells each of radius r, thickness dr, and length L, as in Figure 10.10. (Cylindrical shells are chosen because we want all mass elements dm to have a single value for r, which makes the calculation of I more straightforward.) The volume of each shell is its cross-sectional area multiplied by the length, or $dV = dA \cdot L = (2\pi r\, dr)\, L$. If the mass per unit volume is ρ, then the mass of this differential volume element is $dm = \rho\, dV = \rho 2\pi rL\, dr$.

Solution   Substituting this expression for dm into Equation 10.16, we get

$$I_z = \int r^2\, dm = 2\pi\rho L \int_0^R r^3\, dr = \frac{\pi \rho L R^4}{2}$$

Because the total volume of the cylinder is $\pi R^2 L$, we see that $\rho = M/V = M/\pi R^2 L$. Substituting this value into the above result gives

$$I_z = \tfrac{1}{2}MR^2$$

As we saw in the previous examples, the moments of inertia of rigid bodies with simple geometry (high symmetry) are relatively easy to calculate provided the rotation axis coincides with an axis of symmetry. Table 10.2 gives the moments of inertia for a number of bodies about specific axes.[2]

The calculation of moments of inertia about an arbitrary axis can be somewhat cumbersome, even for a highly symmetric object. There is an important theorem, however, called the *parallel-axis theorem,* that often simplifies the calculation. Suppose the moment of inertia about any axis through the center of mass is I_{CM}. The parallel-axis theorem states that the moment of inertia about any axis that is parallel to and a distance D away from the axis that passes through the center of mass is

$$I = I_{CM} + MD^2 \qquad\qquad (10.17)$$

Parallel-axis theorem

*Proof of the Parallel-Axis Theorem   Suppose an object rotates in the xy plane about the z axis as in Figure 10.11 and the coordinates of the center of mass are x_{CM}, y_{CM}. Let the mass element Δm have coordinates x, y. Since this element is at a distance $r = \sqrt{x^2 + y^2}$ from the z axis, the moment of inertia about the z axis is

$$I = \int r^2\, dm = \int (x^2 + y^2)\, dm$$

However, we can relate the coordinates x, y of the mass element Δm to the coordinates of the mass element relative to the center of mass, x', y'. If the coordinates of

---

[2] Civil engineers use the moment of inertia concept to characterize the elastic properties (rigidity) of such structures as loaded beams. Hence, this concept is often useful even in a nonrotational context.

TABLE 10.2   Moments of Inertia of Some Rigid Objects

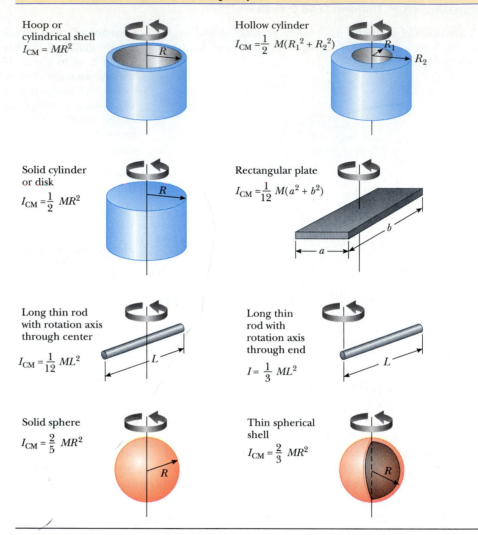

Hoop or cylindrical shell
$$I_{CM} = MR^2$$

Hollow cylinder
$$I_{CM} = \frac{1}{2} M(R_1{}^2 + R_2{}^2)$$

Solid cylinder or disk
$$I_{CM} = \frac{1}{2} MR^2$$

Rectangular plate
$$I_{CM} = \frac{1}{12} M(a^2 + b^2)$$

Long thin rod with rotation axis through center
$$I_{CM} = \frac{1}{12} ML^2$$

Long thin rod with rotation axis through end
$$I = \frac{1}{3} ML^2$$

Solid sphere
$$I_{CM} = \frac{2}{5} MR^2$$

Thin spherical shell
$$I_{CM} = \frac{2}{3} MR^2$$

the center of mass are x_{CM}, y_{CM}, then from Figure 10.11 we see that the relationships between the unprimed and primed coordinates are $x = x' + x_{CM}$ and $y = y' + y_{CM}$. Therefore,

$$I = \int [(x' + x_{CM})^2 + (y' + y_{CM})^2] \, dm$$

$$= \int [(x')^2 + (y')^2] \, dm + 2x_{CM} \int x' \, dm + 2y_{CM} \int y' \, dm + (x_{CM}{}^2 + y_{CM}{}^2) \int dm$$

The first term on the right is, by definition, the moment of inertia about an axis that is parallel to the z axis and passes through the center of mass. The second two terms on the right are zero because by definition of the center of mass, $\int x' \, dm = \int y' \, dm = 0$. Finally, the last term on the right is simply MD^2, since $\int dm = M$ and $D^2 = x_{CM}{}^2 + y_{CM}{}^2$. Therefore, we conclude that

$$I = I_{CM} + MD^2$$

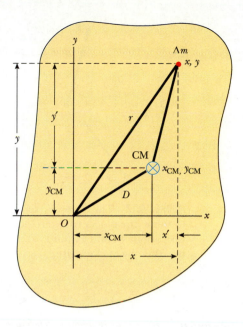

FIGURE 10.11 The parallel-axis theorem: If the moment of inertia about an axis perpendicular to the figure through the center of mass is I_{CM}, then the moment of inertia about the z axis is $I_z = I_{CM} + MD^2$.

EXAMPLE 10.9 Applying the Parallel-Axis Theorem

Consider once again a uniform rigid rod of mass M and length L as in Figure 10.9. Find the moment of inertia of the rod about an axis perpendicular to the rod through one end (the y' axis in Fig. 10.9).

Solution Since the moment of inertia of the rod about an axis perpendicular to the rod passing through its center of mass is $ML^2/12$ and the distance between this axis and the parallel axis through its end is $D = L/2$, the parallel-axis

theorem gives

$$I = I_{CM} + MD^2 = \tfrac{1}{12}ML^2 + M\left(\frac{L}{2}\right)^2 = \tfrac{1}{3}ML^2$$

Exercise Calculate the moment of inertia of the rod about a perpendicular axis through the point $x = L/4$.

Answer $I = \tfrac{7}{48}ML^2$.

10.6 TORQUE

When a force is exerted on a rigid object pivoted about an axis, the object tends to rotate about that axis. The tendency of a force to rotate an object about some axis is measured by a quantity called **torque** τ (Greek letter tau). Consider the wrench pivoted on the axis through O in Figure 10.12. The applied force **F** acts at a general angle ϕ (Greek letter phi) to the horizontal. We define the magnitude of the torque associated with the force **F** by the expression

$$\tau \equiv rF \sin \phi = Fd \tag{10.18}$$

Definition of torque

where r is the distance between the pivot point and the point of application of **F** and d is the perpendicular distance from the pivot point to the line of action of **F**. (The *line of action* of a force is an imaginary line extending out both ends of the vector representing the force. The blue dashed line extending from the tail of **F** in Figure 10.12 is part of the line of action of **F**.) From the right triangle in Figure 10.12, we see that $d = r \sin \phi$. This quantity d is called the **moment arm** (or *lever arm*) of **F**. The moment arm represents the perpendicular distance from the rotation axis to the line of action of **F**.

Moment arm

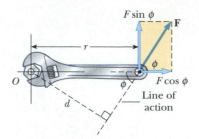

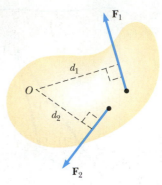

FIGURE 10.12   The force **F** has a greater rotating tendency about *O* as *F* increases and as the moment arm, *d*, increases. It is the component $F \sin \phi$ that tends to rotate the system about *O*.

FIGURE 10.13   The force $\mathbf{F}_1$ tends to rotate the object counterclockwise about *O*, and $\mathbf{F}_2$ tends to rotate the object clockwise.

It is very important that you recognize that *torque is defined only when a reference axis is specified*. Torque is the product of a force and the moment arm of that force, and moment arm is defined only in terms of the axis of rotation.

The only component of **F** that tends to cause rotation is $F \sin \phi$, the component perpendicular to *r*. The horizontal component, $F \cos \phi$, because it passes through *O*, has no tendency to produce rotation. From the definition of torque, we see that the rotating tendency increases as **F** increases and as *d* increases. For example, it is easier to close a door if we push at the doorknob rather than at a point close to the hinge.

If there are two or more forces acting on a rigid object, as in Figure 10.13, each has a tendency to produce rotation about the pivot at *O*. In this example, $\mathbf{F}_2$ has a tendency to rotate the object clockwise, and $\mathbf{F}_1$ has a tendency to rotate the object counterclockwise. We use the convention that the sign of the torque resulting from a force is positive if its turning tendency of the force is counterclockwise and negative if the turning tendency is clockwise. For example, in Figure 10.13, the torque resulting from $\mathbf{F}_1$, which has a moment arm of d_1, is positive and equal to $+F_1 d_1$; the torque from $\mathbf{F}_2$ is negative and equal to $-F_2 d_2$. Hence, the net torque about *O* is

$$\tau_{net} = \tau_1 + \tau_2 = F_1 d_1 - F_2 d_2$$

Torque should not be confused with force. Torque has units of force times length, newton-meters in SI units, and should be reported in these units.

The torque exerted on the nail by the hammer increases as the length of the handle (lever arm) increases.   *(© Richard Megna, 1991, Fundamental Photographs)*

EXAMPLE 10.10   The Net Torque on a Cylinder

A one-piece cylinder is shaped as in Figure 10.14, with a core section protruding from the larger drum. The cylinder is free to rotate around the central axis shown in the drawing. A rope wrapped around the drum, of radius R_1, exerts a force $\mathbf{F}_1$ to the right on the cylinder. A rope wrapped around the core, of radius R_2, exerts a force $\mathbf{F}_2$ downward on the cylinder. (a) What is the net torque acting on the cylinder about the rotation axis (which is the z axis in Fig. 10.14)?

Solution The torque due to F_1 is $-R_1F_1$ and is negative because it tends to produce a clockwise rotation. The torque due to F_2 is $+R_2F_2$ and is positive because it tends to produce a counterclockwise rotation. Therefore, the net torque about the rotation axis is

$$\tau_{net} = \tau_1 + \tau_2 = \boxed{R_2F_2 - R_1F_1}$$

(b) Suppose $F_1 = 5.0$ N, $R_1 = 1.0$ m, $F_2 = 6.0$ N, and $R_2 = 0.50$ m. What is the net torque about the rotation axis and which way does the cylinder rotate?

$$\tau_{net} = (6.0\text{ N})(0.50\text{ m}) - (5.0\text{ N})(1.0\text{ m}) = \boxed{-2.0\text{ N}\cdot\text{m}}$$

Since the net torque is negative, the cylinder rotates clockwise.

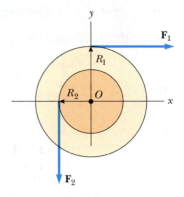

FIGURE 10.14 (Example 10.10) A solid cylinder pivoted about the z axis through O. The moment arm of F_1 is R_1, and the moment arm of F_2 is R_2.

10.7 RELATIONSHIP BETWEEN TORQUE AND ANGULAR ACCELERATION

In this section we show that the angular acceleration of a rigid object rotating about a fixed axis is proportional to the net torque acting about that axis. Before discussing the more complex case of rigid-body rotation, however, it is instructive first to discuss the case of a particle rotating about some fixed point under the influence of an external force.

Consider a particle of mass m rotating in a circle of radius r under the influence of a tangential force F_t as in Figure 10.15 and a central (radial) force F_c not shown in the figure. (The central force *must* be present to keep the particle moving in its circular path.) The tangential force provides a tangential acceleration a_t, and

$$F_t = ma_t$$

The torque about the center of the circle due to F_t is

$$\tau = F_t r = (ma_t)r$$

Since the tangential acceleration is related to the angular acceleration through the relation $a_t = r\alpha$ (Eq. 10.10), the torque can be expressed as

$$\tau = (mr\alpha)r = (mr^2)\alpha$$

Recall from Equation 10.14 that mr^2 is the moment of inertia of the rotating mass about the z axis passing through the origin, so that

$$\tau = I\alpha \qquad (10.19)$$

That is, *the torque acting on the particle is proportional to its angular acceleration,* and the proportionality constant is the moment of inertia. It is important to note that $\tau = I\alpha$ is the rotational analogue of Newton's second law of motion, $F = ma$.

Now let us extend this discussion to a rigid object of arbitrary shape rotating about a fixed axis as in Figure 10.16. The object can be regarded as an infinite number of mass elements dm of infinitesimal size. If we impose a cartesian coordinate system on the object, then each mass element rotates in a circle about the origin, and each has a tangential acceleration a_t produced by an external tangential force dF_t. For any given element, we know from Newton's second law that

$$dF_t = (dm)a_t$$

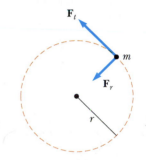

FIGURE 10.15 A particle rotating in a circle under the influence of a tangential force F_t. A central force F_c must also be present to maintain the circular motion.

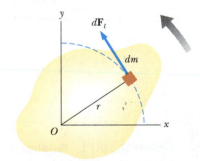

FIGURE 10.16 A rigid object rotating about an axis through O. Each mass element dm rotates about O with the same angular acceleration α, and the net torque on the object is proportional to α.

The torque $d\tau$ associated with the force $d\mathbf{F}_t$ acting about the origin is

$$d\tau = r\, dF_t = (r\, dm)\, a_t$$

Since $a_t = r\alpha$, the expression for $d\tau$ becomes

$$d\tau = (r\, dm)\, r\alpha = (r^2\, dm)\, \alpha$$

It is important to recognize that although each mass element of the rigid object may have a different linear acceleration $\mathbf{a}_t$, they all have the *same* angular acceleration, α. With this in mind, the above expression can be integrated to obtain the net torque of the external forces about O:

$$\tau_{net} = \int (r^2\, dm)\, \alpha = \alpha \int r^2\, dm$$

where α can be taken outside the integral because it is common to all mass elements. Since the moment of inertia of the object about the rotation axis through O is defined by $I = \int r^2\, dm$ (Eq. 10.16), the expression for τ_{net} becomes

$$\tau_{net} = I\alpha \tag{10.20}$$

Again we see that the net torque about the rotation axis is proportional to the angular acceleration of the object with the proportionality factor being I, a quantity that depends upon the axis of rotation and upon the size and shape of the object.

In view of the complex nature of the system, it is interesting to note that $\tau_{net} = I\alpha$ is strikingly simple and in complete agreement with experimental observations. The simplicity of the result lies in the manner in which the motion is described.

Although each point on a rigid object rotating about a fixed axis may not experience the same force, linear acceleration, or linear velocity, each point experiences the same angular acceleration and angular velocity at any instant. Therefore, at any instant the rotating rigid object as a whole is characterized by specific values for angular acceleration, net torque, and angular velocity.

Finally, note that the result $\tau_{net} = I\alpha$ also applies when the forces acting on the mass elements have radial components as well as tangential components. This is because the line of action of all radial components must pass through the axis of rotation, and hence all radial components produce zero torque about that axis.

EXAMPLE 10.11   Rotating Rod

A uniform rod of length L and mass M is free to rotate about a frictionless pivot at one end in a vertical plane, as in Figure 10.17. The rod is released from rest in the horizontal position. What is the initial angular acceleration of the rod and the initial linear acceleration of the right end of the rod?

Reasoning   We cannot use our kinematics equations to find α or a because they are not constant in this situation. We

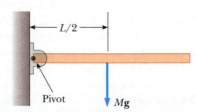

FIGURE 10.17   (Example 10.11) The uniform rod is pivoted at the left end.

have enough information to find the torque, however, which we can then use in the torque-angular acceleration relationship (Eq. 10.19) to find α and then a.

Solution   The only force contributing to torque about an axis through the pivot is the force of gravity $M\mathbf{g}$ exerted on the rod. (The force exerted by the pivot on the rod has zero torque about the pivot because its moment arm is zero.) To compute the torque on the rod, we can assume that the force of gravity is located at the geometric center and, thus, acts at the center of mass as shown in Figure 10.17. The magnitude of the torque due to this force about an axis through the pivot is

$$\tau = \frac{MgL}{2}$$

Since $\tau = I\alpha$, where $I = \frac{1}{3}ML^2$ for this axis of rotation (Table 10.2), we get

$$I\alpha = Mg\frac{L}{2}$$

$$\alpha = \frac{Mg(L/2)}{\frac{1}{3}ML^2} = \boxed{\frac{3g}{2L}}$$

This angular acceleration is common to all points on the rod.

To find the linear acceleration of the right end of the rod, we use the relationship $a_t = R\alpha$, with $R = L$. This gives

$$a_t = L\alpha = \boxed{\tfrac{3}{2}g}$$

This result, that $a_t > g$, is rather interesting. It means that if a coin were placed at the end of the rod, the end of the rod would fall faster than the coin when released.

Other points on the rod have a linear acceleration less than $\frac{3}{2}g$. For example, the middle of the rod has an acceleration $\frac{3}{4}g$.

Exercise   Show that in Figure 10.17, the force of gravity can be treated as a single force acting at the center of the rod.

EXAMPLE 10.12   Angular Acceleration of a Wheel

A wheel of radius R, mass M, and moment of inertia I is mounted on a frictionless, horizontal axle as in Figure 10.18. A light cord wrapped around the wheel supports an object of mass m. Calculate the linear acceleration of the object, the angular acceleration of the wheel, and the tension in the cord.

Reasoning   The torque acting on the wheel about its axis of rotation is $\tau = TR$, where T is the force exerted by the cord on the rim of the wheel. (The weight of the wheel and the normal force exerted by the axle on the wheel pass through the axis of rotation and produce no torque.) Since $\tau = I\alpha$, we get

$$\tau = I\alpha = TR$$

$$(1) \qquad \alpha = \frac{TR}{I}$$

Now let us apply Newton's second law to the motion of the suspended object, making use of the free-body diagram in Figure 10.18, taking the upward direction to be positive:

$$\sum F_y = T - mg = - ma$$

$$(2) \qquad a = \frac{mg - T}{m}$$

The linear acceleration of the suspended object is equal to the tangential acceleration of a point on the rim of the wheel. Therefore, the angular acceleration of the wheel and this linear acceleration are related by $a = R\alpha$. Using this fact to-

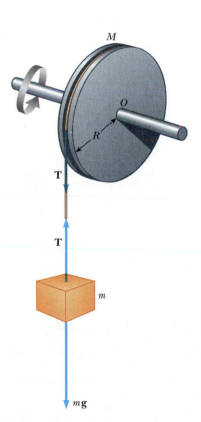

FIGURE 10.18   (Example 10.12) The cord attached to an object of mass m is wrapped around the wheel, and the tension in the cord produces a torque about the axle through O.

gether with (1) and (2) gives

$$(3) \qquad a = R\alpha = \frac{TR^2}{I} = \frac{mg - T}{m}$$

$$(4) \qquad T = \frac{mg}{1 + mR^2/I}$$

Substituting (4) into (3), and solving for a and α gives

$$a = \frac{g}{1 + I/mR^2}$$

$$\alpha = \frac{a}{R} = \frac{g}{R + I/mR}$$

Exercise The wheel in Figure 10.18 is a solid disk of $M = 2.00$ kg, $R = 30.0$ cm, and $I = 0.0900$ kg·m². The suspended object has a mass of $m = 0.500$ kg. Find the tension in the cord and the angular acceleration of the wheel.

Answer 3.27 N; 10.9 rad/s².

EXAMPLE 10.13   The Atwood Machine Revisited

Two masses m_1 and m_2 are connected to each other by a light cord that passes over two identical pulleys, each having a moment of inertia I as in Figure 10.19a. Find the acceleration of each mass and the tensions T_1, T_2, and T_3 in the cord. (Assume no slipping between cord and pulleys.)

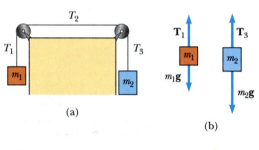

(a)

(b)

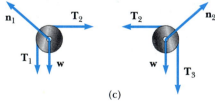

(c)

FIGURE 10.19   (Example 10.13).

Reasoning First, let us write Newton's second law of motion as applied to each mass, taking $m_2 > m_1$. (The free-body diagrams for the two masses are shown in Figure 10.19b, and the upward direction is taken to be positive.)

$$T_1 - m_1 g = m_1 a \qquad (1)$$

$$T_3 - m_2 g = -m_2 a \qquad (2)$$

Next, we must include the effect of the pulleys on the motion. Free-body diagrams for the pulleys are shown in Fig-

ure 10.19c. The net torque about the axle for the pulley on the left is $(T_2 - T_1)R$, while the net torque for the pulley on the right is $(T_3 - T_2)R$. Using the relation $\tau_{net} = I\alpha$ for each pulley, and noting that each pulley has the same α, we get

$$(T_2 - T_1)R = I\alpha \qquad (3)$$

$$(T_3 - T_2)R = I\alpha \qquad (4)$$

We now have four equations with four unknowns: a, T_1, T_2, and T_3. These can be solved simultaneously. Adding Equations (3) and (4) gives

$$(T_3 - T_1)R = 2I\alpha \qquad (5)$$

Subtracting Equation (2) from Equation (1) gives

$$T_1 - T_3 + m_2 g - m_1 g = (m_1 + m_2)a$$

or

$$T_3 - T_1 = (m_2 - m_1)g - (m_1 + m_2)a \qquad (6)$$

Substituting Equation (6) into Equation (5), we have

$$[(m_2 - m_1)g - (m_1 + m_2)a]R = 2I\alpha$$

Since $\alpha = a/R$, this can be simplified as follows:

$$(m_2 - m_1)g - (m_1 + m_2)a = 2I\frac{a}{R^2}$$

or

$$a = \frac{(m_2 - m_1)g}{m_1 + m_2 + 2\dfrac{I}{R^2}} \qquad (7)$$

This value of a can then be substituted into Equations (1) and (2) to give T_1 and T_3. Finally, T_2 can be found from Equation (3) or Equation (4).

10.8 WORK, POWER, AND ENERGY IN ROTATIONAL MOTION

The description of a rotating rigid object would not be complete without a discussion of the rotational kinetic energy and how its change is related to the work done by external forces. In this section, we shall see that the important relationship $\tau_{\text{net}} = I\alpha$ can also be obtained by considering the rate at which energy is changing with time.

We again restrict our discussion to rotation about a fixed axis. Consider a rigid object pivoted at O in Figure 10.20. Suppose a single external force $\mathbf{F}$ is applied at P. The work done by $\mathbf{F}$ as the object rotates through an infinitesimal distance $ds = r\, d\theta$ in a time dt is

$$dW = \mathbf{F} \cdot d\mathbf{s} = (F \sin \phi)\, r\, d\theta$$

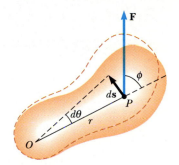

FIGURE 10.20 A rigid object rotates about an axis through O under the action of an external force F applied at P.

where $F \sin \phi$ is the tangential component of $\mathbf{F}$, or, in other words, the component of the force along the displacement. Note from Figure 10.20 that *the radial component of* $\mathbf{F}$ *does no work because it is perpendicular to the displacement.*

Since the magnitude of the torque due to $\mathbf{F}$ about O is defined to be $rF \sin \phi$ by Equation 10.18, we can write the work done for the infinitesimal rotation as

$$dW = \tau\, d\theta \qquad (10.21)$$

The rate at which work is being done by $\mathbf{F}$ as the object rotates about the fixed axis is

$$\frac{dW}{dt} = \tau \frac{d\theta}{dt}$$

Since dW/dt is the instantaneous power P (Section 7.5) delivered by the force, and since $d\theta/dt = \omega$, this expression reduces to

$$P = \frac{dW}{dt} = \tau\omega \qquad (10.22)$$

Power delivered to a rigid object rotating about a fixed axis

This expression is analogous to $P = Fv$ in the case of linear motion, and the expression $dW = \tau\, d\theta$ is analogous to $dW = F_x\, dx$.

The Work and Energy in Rotational Motion

In linear motion, we found the energy concept, and in particular the work-energy theorem, extremely useful in describing the motion of a system. The energy concept can be equally useful in describing rotational motion. From what we learned of linear motion, we expect that for rotation of a symmetric object about a fixed axis, the work done by external forces equals the change in the rotational energy.

To show that this is in fact the case, let us begin with $\tau = I\alpha$. Using the chain rule from the calculus, we can express the torque as

$$\tau = I\alpha = I\frac{d\omega}{dt} = I\frac{d\omega}{d\theta}\frac{d\theta}{dt} = I\frac{d\omega}{d\theta}\omega$$

Rearranging this expression and noting that $\tau\, d\theta = dW$, we get

$$\tau\, d\theta = dW = I\omega\, d\omega$$

Integrating this expression, we get for the total work done

Work-energy relation for
rotational motion

$$W = \int_{\theta_0}^{\theta} \tau \, d\theta = \int_{\omega_0}^{\omega} I\omega \, d\omega = \tfrac{1}{2} I\omega^2 - \tfrac{1}{2} I\omega_0^2 \qquad (10.23)$$

where the angular speed changes from ω_0 to ω as the angular displacement changes from θ_0 to θ. That is,

> the net work done by external forces in rotating a symmetric rigid object about a fixed axis equals the change in the object's rotational energy.

Table 10.3 lists the various equations we have discussed pertaining to rotational motion, together with the analogous expressions for linear motion. The last two equations in Table 10.3, involving angular momentum L, are discussed in Chapter 11 and are included here only for the sake of completeness.

TABLE 10.3   A Comparison of Useful Equations in Rotational and Translational Motion

Rotational Motion About a Fixed Axis	Linear Motion
Angular speed $\omega = d\theta/dt$	Linear speed $v = dx/dt$
Angular acceleration $\alpha = d\omega/dt$	Linear acceleration $a = dv/dt$
Resultant torque $\Sigma\tau = I\alpha$	Resultant force $\Sigma F = Ma$
If $\alpha = $ constant $\begin{cases} \omega = \omega_0 + \alpha t \\ \theta - \theta_0 = \omega_0 t + \tfrac{1}{2}\alpha t^2 \\ \omega^2 = \omega_0^2 + 2\alpha(\theta - \theta_0) \end{cases}$	If $a = $ constant $\begin{cases} v = v_0 + at \\ x - x_0 = v_0 t + \tfrac{1}{2}at^2 \\ v^2 = v_0^2 + 2a(x - x_0) \end{cases}$
Work $W = \int_{\theta_0}^{\theta} \tau \, d\theta$	Work $W = \int_{x_0}^{x} F_x \, dx$
Rotational energy $K_R = \tfrac{1}{2}I\omega^2$	Translational energy $K = \tfrac{1}{2}mv^2$
Power $P = \tau\omega$	Power $P = Fv$
Angular momentum $L = I\omega$	Linear momentum $p = mv$
Resultant torque $\tau = dL/dt$	Resultant force $F = dp/dt$

CONCEPTUAL EXAMPLE 10.14

Consider an object in the shape of a hoop lying in the xy plane with all of its mass concentrated on its rim. In two separate experiments, the hoop is rotated by an external agent from rest to an angular speed ω. In one experiment, the rotation occurs about the z axis passing through the center of the hoop. In the other experiment, the rotation occurs about an axis parallel to z passing through a point P on the rim of the hoop. Which rotation requires more work?

Reasoning   Rotation about the axis through P requires more work. The moment of inertia of the hoop about the axis

through the center of the hoop is $I_{CM} = MR^2$, whereas by the parallel axis theorem, the moment of inertia about the axis through P is $I_P = I_{CM} + MR^2 = MR^2 + MR^2 = 2MR^2$. That is, $I_P = 2I_{CM}$. Using Equation 10.23, taking the initial angular speed to be zero, we see that the work required by the external agent for rotation about the axis through P is twice the work required for rotation about the axis through the center of the hoop.

EXAMPLE 10.15   Rotating Rod Revisited

A uniform rod of length L and mass M is free to rotate on a frictionless pin through one end (Fig. 10.21). The rod is released from rest in the horizontal position. (a) What is the angular speed of the rod at its lowest position?

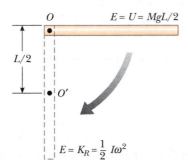

FIGURE 10.21   (Example 10.15) A uniform rigid rod pivoted at O rotates in a vertical plane under the action of gravity.

Reasoning and Solution   The question can be answered by considering the mechanical energy of the system. When the rod is horizontal, it has no rotational energy. The potential energy relative to the lowest position of its center of mass (O') is $MgL/2$. When it reaches its lowest position, the energy is entirely rotational energy, $\frac{1}{2}I\omega^2$, where I is the moment of inertia about the pivot. Since $I = \frac{1}{3}ML^2$ (Table 10.2) and since mechanical energy is constant, we have

$$\tfrac{1}{2}MgL = \tfrac{1}{2}I\omega^2 = \tfrac{1}{2}\left(\tfrac{1}{3}ML^2\right)\omega^2$$

$$\omega = \sqrt{\frac{3g}{L}}$$

(b) Determine the linear speed of the center of mass and the linear speed of the lowest point on the rod in the vertical position.

Solution

$$v_{\mathrm{CM}} = r\omega = \frac{L}{2}\,\omega = \tfrac{1}{2}\sqrt{3gL}$$

The lowest point on the rod has a linear speed equal to $2v_{\mathrm{CM}} = \sqrt{3gL}$.

EXAMPLE 10.16   Connected Masses

Consider two masses connected by a string passing over a pulley having a moment of inertia I about its axis of rotation, as in Figure 10.22. The string does not slip on the pulley, and the system is released from rest. Find the linear speeds of the masses after m_2 descends through a distance h, and the angular speed of the pulley at this time.

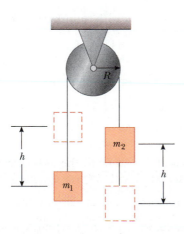

FIGURE 10.22   (Example 10.16).

Reasoning and Solution   If we neglect friction in the system, then mechanical energy is constant and we can state that the increase in kinetic energy equals the decrease in potential energy. Since $K_i = 0$ (the system is initially at rest), we have

$$\Delta K = K_f - K_i = \tfrac{1}{2}m_1v^2 + \tfrac{1}{2}m_2v^2 + \tfrac{1}{2}I\omega^2$$

where m_1 and m_2 have a common speed. Because $v = R\omega$, this expression becomes

$$\Delta K = \tfrac{1}{2}\left(m_1 + m_2 + \frac{I}{R^2}\right)v^2$$

From Figure 10.22, we see that m_2 loses potential energy while m_1 gains potential energy. That is, $\Delta U_2 = -m_2gh$ and $\Delta U_1 = m_1gh$. Applying the principle of conservation of energy in the form $\Delta K + \Delta U_1 + \Delta U_2 = 0$ gives

$$\tfrac{1}{2}\left(m_1 + m_2 + \frac{I}{R^2}\right)v^2 + m_1gh - m_2gh = 0$$

$$v = \left[\frac{2(m_2 - m_1)gh}{\left(m_1 + m_2 + \dfrac{I}{R^2}\right)}\right]^{1/2}$$

Since $v = R\omega$, the angular speed of the pulley at this instant is $\omega = v/R$.

Exercise   Repeat the calculation of v in Example 10.16 using $\tau_{\mathrm{net}} = I\alpha$ applied to the pulley and Newton's second law applied to m_1 and m_2. Make use of the procedure presented in Examples 10.12 and 10.13.

SUMMARY

If a particle rotates in a circle of radius r through an angle θ (measured in radians), the arc length it moves through is $s = r\theta$.

The **instantaneous angular speed** of a particle rotating in a circle or of a rigid object rotating about a fixed axis is

$$\omega = \frac{d\theta}{dt} \tag{10.3}$$

where ω is in rad/s, or s^{-1}.

The **instantaneous angular acceleration** of a rotating object is

$$\alpha = \frac{d\omega}{dt} \tag{10.5}$$

and has units of rad/s^2, or s^{-2}. When a rigid object rotates about a fixed axis, every part of the object has the same angular speed and the same angular acceleration.

If a particle or object rotates about a fixed axis under constant angular acceleration α, one can apply equations of kinematics in analogy with kinematic equations for linear motion under constant linear acceleration:

$$\omega = \omega_0 + \alpha t \tag{10.6}$$

$$\theta = \theta_0 + \omega_0 t + \tfrac{1}{2}\alpha t^2 \tag{10.7}$$

$$\omega^2 = \omega_0{}^2 + 2\alpha(\theta - \theta_0) \tag{10.8}$$

When a rigid object rotates about a fixed axis, the angular speed and angular acceleration are related to the linear speed and tangential linear acceleration through the relationships

$$v = r\omega \tag{10.9}$$

$$a_t = r\alpha \tag{10.10}$$

The **moment of inertia of a system of particles** is

$$I = \sum m_i r_i^2 \tag{10.14}$$

If a rigid object rotates about a fixed axis with angular speed ω, its **rotational energy** can be written

$$K_R = \tfrac{1}{2}I\omega^2 \tag{10.15}$$

where I is the moment of inertia about the axis of rotation.

The **moment of inertia of a rigid object** is

$$I = \int r^2 \, dm \tag{10.16}$$

where r is the distance from the mass element dm to the axis of rotation.

The magnitude of the **torque** associated with a force $\mathbf{F}$ acting on an object is

$$\tau = Fd \tag{10.18}$$

where d is the moment arm of the force, which is the perpendicular distance from some origin to the line of action of the force. Torque is a measure of the tendency of the force to rotate the object about some axis.

If a rigid object free to rotate about a fixed axis has a **net external torque** acting on it, the object undergoes an angular acceleration α, where

$$\tau_{\text{net}} = I\alpha \tag{10.20}$$

The rate at which work is done by external forces in rotating a rigid object about a fixed axis, or the **power** delivered, is

$$P = \tau\omega \qquad\qquad (10.22)$$

The net work done by external forces in rotating a rigid object about a fixed axis equals the change in the rotational energy of the object:

$$W = \tfrac{1}{2}I\omega^2 - \tfrac{1}{2}I\omega_0^2 \quad . \qquad\qquad (10.23)$$

QUESTIONS

1. What is the magnitude of the angular velocity, $\boldsymbol{\omega}$, of the second hand of a clock? What is the direction of $\boldsymbol{\omega}$ as you view a clock hanging vertically? What is the magnitude of the angular acceleration $\boldsymbol{\alpha}$ of the second hand?

2. A wheel rotates counterclockwise in the xy plane. What is the direction of $\boldsymbol{\omega}$? What is the direction of $\boldsymbol{\alpha}$ if the angular velocity is decreasing in time?

3. Are the kinematic expressions for θ, ω, and α valid when the angular displacement is measured in degrees instead of in radians?

4. A turntable rotates at a constant rate of 45 rev/min. What is its angular speed in radians per second? What is the magnitude of its angular acceleration?

5. Suppose $a = b$ and $M > m$ for the system of particles described in Figure 10.7. About what axis (x, y, or z) does the moment of inertia have the smallest value? the largest value?

6. Suppose the rod in Figure 10.9 has a nonuniform mass distribution. In general, would the moment of inertia about the y axis still equal $\frac{1}{12}ML^2$? If not, could the moment of inertia be calculated without knowledge of the manner in which the mass is distributed?

7. Suppose that only two external forces act on a rigid body, and the two forces are equal in magnitude but opposite in direction. Under what conditions does the body rotate?

8. Explain how you might use the apparatus described in Example 10.12 to determine the moment of inertia of the wheel. (If the wheel does not have a uniform mass density, the moment of inertia is not necessarily equal to $\frac{1}{2}MR^2$.)

9. Using the results from Example 10.12, how would you

calculate the angular speed of the wheel and the linear speed of the suspended mass at $t = 2$ s, if the system is released from rest at $t = 0$? Is the expression $v = R\omega$ valid in this situation?

10. If a small sphere of mass M were placed at the end of the rod in Figure 10.21, would the result for ω be greater than, less than, or equal to the value obtained in Example 10.15?

11. Explain why changing the axis of rotation of an object changes its moment of inertia.

12. Is it possible to change the translational kinetic energy of an object without changing its rotational energy?

13. Two cylinders having the same dimensions are set into rotation about their long axes with the same angular speed. One is hollow, and the other if filled with water. Which cylinder will be easier to stop rotating?

14. Must an object be rotating to have a nonzero moment of inertia?

15. If you see an object rotating, is there necessarily a net torque acting on it?

16. Can a (momentarily) stationary object have a nonzero angular acceleration?

17. The polar diameter of the Earth is slightly less than the equatorial diameter. How would the moment of inertia of the Earth change if some mass from near the equator were removed and transferred to the polar regions to make the Earth a perfect sphere?

18. During a wrecking operation, a tall chimney is toppled by an explosive charge at its base. The chimney ruptures on its lower half while toppling, so that the bottom part reaches the ground before the top part. Explain why this occurs.

PROBLEMS

Section 10.2   Rotational Kinematics: Rotational Motion with Constant Angular Acceleration

1. A wheel starts from rest and rotates with constant angular acceleration to an angular speed of 12.0

rad/s in 3.00 s. Find (a) the magnitude of the angular acceleration of the wheel and (b) the angle in radians through which it rotates in this time.

2. The turntable of a record player rotates at a rate of $33\frac{1}{3}$ rev/min and takes 60.0 s to come to rest when

☐ indicates problems that have full solutions available in the Student Solutions Manual and Study Guide.

switched off. Calculate (a) the magnitude of its angular acceleration and (b) the number of revolutions it makes before coming to rest.

3. What is the angular speed in radians per second of (a) the Earth in its orbit about the Sun and (b) the Moon in its orbit about the Earth?

4. (a) The hour and minute hands on a clock coincide at 12 o'clock. Determine all the other times (up to the second) when these two hands coincide. (b) If the clock has a second hand, determine all the times when the three hands coincide, given that they all coincide at 12 o'clock.

5. An electric motor rotating a grinding wheel at 100 rev/min is switched off. Assuming constant negative angular acceleration of magnitude 2.00 rad/s², (a) how long does it take the wheel to stop? (b) Through how many radians does it turn during the time found in (a)?

6. The angular position of a point on a wheel is described by $\theta = 5.0 + 10t + 2.0t^2$ rad. Determine the angular position, speed, and acceleration of the point at $t = 0$ and $t = 3.0$ s.

7. A rotating wheel requires 3.0 s to rotate 37 rev. Its angular speed at the end of the 3.0 s interval is 98 rad/s. What is its constant angular acceleration?

8. A car accelerates uniformly from rest and reaches a speed of 22 m/s in 9 s. If the diameter of a tire is 58 cm, find (a) the number of revolutions the tire makes during this motion, assuming no slipping. (b) What is the final rotational speed of a tire in revolutions per second?

Section 10.3   Relationships Between Angular and Linear Quantities

9. A racing car travels on a circular track of radius 250 m. If the car moves with a constant linear speed of 45.0 m/s, find (a) its angular speed and (b) the magnitude and direction of its acceleration.

9A. A racing car travels on a circular track of radius R. If the car moves with a constant linear speed v, find (a) its angular speed and (b) the magnitude and direction of its acceleration.

10. A car traveling on a flat (unbanked) circular track accelerates uniformly from rest with a tangential acceleration of 1.70 m/s². The car makes it one quarter of the way around the circle before skidding off the track. Determine the coefficient of static friction between car and track.

11. A wheel 2.00 m in diameter rotates with a constant angular acceleration of 4.00 rad/s². The wheel starts at rest at $t = 0$, and the radius vector at point P on the rim makes an angle of 57.3° with the horizontal at this time. At $t = 2.00$ s, find (a) the angular speed of the wheel, (b) the linear speed and acceleration of the point P, and (c) the position of the point P.

12. A discus thrower accelerates a discus from rest to a speed of 25.0 m/s by whirling it through 1.25 rev. Assume the discus moves on the arc of a circle 1.00 m in radius. (a) Calculate the final angular speed of the discus. (b) Determine the magnitude of the angular acceleration of the discus, assuming it to be constant. (c) Calculate the acceleration time.

13. A disk 8.00 cm in radius rotates at a constant rate of 1200 rev/min about its central axis. Determine (a) its angular speed, (b) the linear speed at a point 3.00 cm from its center, (c) the radial acceleration of a point on the rim, and (d) the total distance a point on the rim moves in 2.00 s.

14. A car is traveling at 36 km/h on a straight road. The radius of the tires is 25 cm. Find the angular speed of one of the tires with its axle taken as the axis of rotation.

15. A 6.00-kg block is released from A on a frictionless track shown in Figure P10.15. Determine the radial and tangential components of acceleration for the block at P.

15A. A block of mass m is released from A on a frictionless track shown in Figure P10.15. Determine the radial and tangential components of acceleration for the block at P.

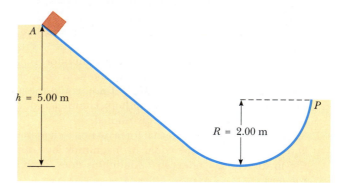

FIGURE P10.15

Section 10.4   Rotational Energy

16. A can of soup has a mass 215 g, height 10.8 cm, and diameter 6.38 cm. It is placed at rest on the top of an incline that is 3.00 m long and at 25.0° to the horizontal. Using energy methods, calculate the moment of inertia of the can if it takes 1.50 s to reach the bottom of the incline.

17. The four particles in Figure P10.17 are connected by rigid rods of negligible mass. The origin is at the center of the rectangle. If the system rotates in the xy plane about the z axis with an angular speed of 6.00 rad/s, calculate (a) the moment of inertia of the system about the z axis and (b) the rotational energy of the system.

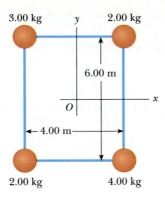

3.00 kg   *y*   2.00 kg

6.00 m

O ——— *x*

←—4.00 m——→

2.00 kg   4.00 kg

FIGURE P10.17

18. The center of mass of a pitched baseball (radius = 3.8 cm) moves at 38 m/s. The ball spins about an axis through its center of mass with an angular speed of 125 rad/s. Calculate the ratio of the rotational energy to the translational kinetic energy. Treat the ball as a uniform sphere.

18A. The center of mass of a pitched baseball of radius R moves at a speed v. The ball spins about an axis through its center of mass with an angular speed ω. Calculate the ratio of the rotational energy to the translational kinetic energy. Treat the ball as a uniform sphere.

19. Three particles are connected by rigid rods of negligible mass lying along the y axis (Fig. P10.19). If the system rotates about the x axis with an angular speed of 2.00 rad/s, find (a) the moment of inertia about the x axis and the total rotational energy evaluated from $\frac{1}{2}I\omega^2$ and (b) the linear speed of each particle and the total energy evaluated from $\Sigma\frac{1}{2}m_i v_i^2$.

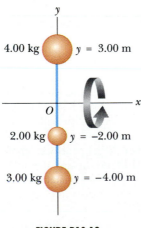

y

4.00 kg $y = 3.00$ m

O ——————— *x*

2.00 kg $y = -2.00$ m

3.00 kg $y = -4.00$ m

FIGURE P10.19

20. The hour and minute hands of Big Ben in London are 2.7 m and 4.5 m long and have masses of 60 kg

and 100 kg, respectively. Calculate the total rotational kinetic energy of the two hands about the axis of rotation. (Model the hands as long, thin rods.)

21. Two masses M and m are connected by a rigid rod of length L and negligible mass as in Figure P10.21. For an axis perpendicular to the rod, show that the system has the minimum moment of inertia when the axis passes through the center of mass. Show that this moment of inertia is $I = \mu L^2$, where $\mu = mM/(mM)$.

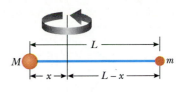

M ←——— L ———→ m

←x→←——$L-x$——→

FIGURE P10.21

Section 10.5   Calculation of Moments of Inertia

22. Three identical thin rods of length L and mass m are placed perpendicular to each other as shown in Figure P10.22. The setup is rotated about an axis that passes through the end of one rod and is parallel to another. Determine the moment of inertia of this arrangement.

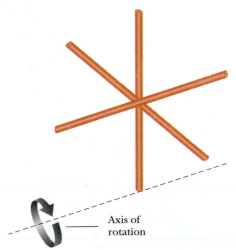

Axis of rotation

FIGURE P10.22

23. Use the parallel-axis theorem and Table 10.2 to find the moments of inertia of (a) a solid cylinder about an axis parallel to the center-of-mass axis and passing through the edge of the cylinder and (b) a solid sphere about an axis tangent to its surface.

Section 10.6   Torque

24. Using the lengths and masses given in Problem 20, (a) determine the total torque due to the weight of Big Ben's hands about the axis of rotation when the

time reads (i) 3:00, (ii) 5:15, (iii) 6:00, (iv) 8:20, (v) 9:45. (Model the hands as long, thin rods.) (b) Determine all times (up to the second) when the total torque about the axis of rotation is zero.

25. Find the net torque on the wheel in Figure P10.25 about the axle through O if $a = 10$ cm and $b = 25$ cm.

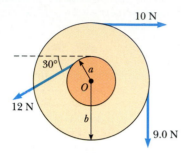

FIGURE P10.25

26. Find the mass m needed to balance the 1500-kg truck on the incline shown in Figure P10.26. Assume all pulleys are frictionless and massless.

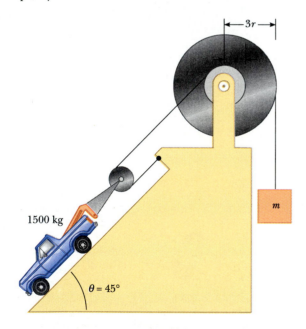

FIGURE P10.26

27. A flywheel in the shape of a solid cylinder of radius $R = 0.60$ m and mass $M = 15$ kg can be brought to an angular speed of 12 rad/s in 0.60 s by a motor exerting a constant torque. After the motor is turned off, the flywheel makes 20 rev before coming to rest because of friction (assumed constant during rotation). What percentage of the power generated by the motor is used to overcome friction?

Section 10.7   Relationship Between Torque and Angular Acceleration

28. A bicycle wheel has a diameter of 64.0 cm and a mass of 1.80 kg. The bicycle is placed on a stationary stand on rollers, and a resistive force of 120 N is applied to the rim of the tire. Assume all the mass of the wheel is concentrated on the outside radius. To give the wheel an acceleration of 4.5 rad/s^2, what force must be applied by a chain passing over (a) a 9.0-cm-diameter sprocket and (b) a 5.6-cm-diameter sprocket?

29. A block of mass $m_1 = 2.00$ kg and one of mass $m_2 = 6.00$ kg are connected by a massless string over a pulley that is in the shape of a disk having radius $R = 0.25$ m and mass $M = 10.0$ kg. In addition, the blocks are allowed to move on a fixed block-wedge of angle $\theta = 30.0°$ as in Figure P10.29. The coefficient of kinetic friction is 0.36 for both blocks. Determine (a) the acceleration of the two blocks and (b) the tensions in the string on both sides of the pulley.

29A. A block of mass m_1 and one of mass m_2 are connected by a massless string over a pulley that is in the shape of a disk having radius R and mass M. In addition, the blocks are allowed to move on a fixed block-wedge of angle θ as in Figure P10.29. Determine (a) the acceleration of the two blocks and (b) the tensions in the string on both sides of the pulley.

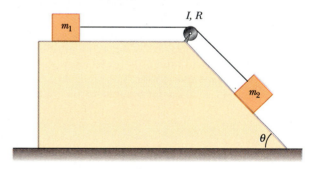

FIGURE P10.29

30. A model airplane whose mass is 0.75 kg is tethered by a wire so that it flies in a circle 30 m in radius. The airplane engine provides a net thrust of 0.80 N perpendicular to the tethering wire. (a) Find the torque the net thrust produces about the center of the circle. (b) Find the angular acceleration of the airplane when it is in level flight. (c) Find the linear acceleration of the airplane tangent to its flight path.

Section 10.8   Work, Power, and Energy in Rotational Motion

31. A cylindrical rod 24 cm long has mass 1.2 kg and radius 1.5 cm. A 20-kg ball of diameter 8.0 cm is attached to one end. The arrangement is originally ver-

tical with the ball at the top and is free to pivot about the other end. After the ball-rod system falls a quarter turn, what are (a) its rotational kinetic energy, (b) its angular speed, and (c) the linear speed of the ball? (d) How does this linear ball speed compare with the speed if the ball had fallen freely from a distance equal to the radius (28 cm)?

32. A 15-kg mass and a 10-kg mass are suspended by a pulley that has a radius of 10 cm and a mass of 3.0 kg (Fig. P10.32). The cord has a negligible mass and causes the pulley to rotate without slipping. The pulley rotates without friction. The masses start from rest 3.0 m apart. Treat the pulley as a uniform disk, and determine the speeds of the two masses as they pass each other.

32A. A mass m_1 and a mass m_2 are suspended by a pulley that has a radius R and a mass m_3 (Fig. P10.32). The cord has a negligible mass and causes the pulley to rotate without slipping. The pulley rotates without friction. The masses start from rest a distance d apart. Treat the pulley as a uniform disk, and determine the speeds of the two masses as they pass each other.

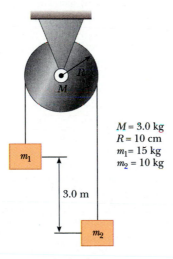

$M = 3.0$ kg
$R = 10$ cm
$m_1 = 15$ kg
$m_2 = 10$ kg

3.0 m

FIGURE P10.32

33. (a) A uniform solid disk of radius R and mass M is free to rotate on a frictionless pivot through a point on its rim (Fig. P10.33). If the disk is released from rest in the position shown by the green circle, what is the speed of its center of mass when the disk reaches the position indicated by the dashed circle? (b) What is the speed of the lowest point on the disk in the dashed position? (c) Repeat part (a) for a uniform hoop.

34. A potter's wheel, a thick stone disk of radius 0.50 m and mass 100 kg, is freely rotating at 50 rev/min. The potter can stop the wheel in 6.0 s by pressing a

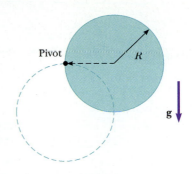

Pivot R

g

FIGURE P10.33

wet rag against the rim and exerting a radially inward force of 70 N. Find the effective coefficient of kinetic friction between wheel and wet rag.

35. A weight of 50.0 N is attached to the free end of a light string wrapped around a pulley of radius 0.250 m and mass 3.00 kg. The pulley is free to rotate in a vertical plane about the horizontal axis passing through its center. The weight is released 6.00 m above the floor. (a) Determine the tension in the string, the acceleration of the mass, and the speed with which the weight hits the floor. (b) Find the speed calculated in part (a) by using the principle of conservation of energy.

36. A bus is designed to draw its power from a rotating flywheel that is brought up to its maximum speed (3000 rpm) by an electric motor. The flywheel is a solid cylinder of mass 1000 kg and diameter 1.00 m. If the bus requires an average power of 10.0 kW, how long does the flywheel rotate?

ADDITIONAL PROBLEMS

37. A grinding wheel is in the form of a uniform solid disk of radius 7.00 cm and mass 2.00 kg. It starts from rest and accelerates uniformly under the action of the constant torque of 0.600 N·m that the motor exerts on the wheel. (a) How long does the wheel take to reach its final speed of 1200 rev/min? (b) Through how many revolutions does it turn while accelerating?

38. Two identical disks having moment of inertia 0.0080 kg·m² and radius 10 cm are free to pivot on frictionless axles. One disk has a 2.0-N weight attached to it by a cord wrapped around the circumference, while the other has a force of 2.0 N applied to its cord. Which disk is turning faster after 1.0 m of cord has been unwrapped? Explain.

39. The fishing pole in Figure P10.39 makes an angle of 20° with the horizontal. What is the torque exerted by the fish about an axis perpendicular to the page and passing through the fisher's hand.

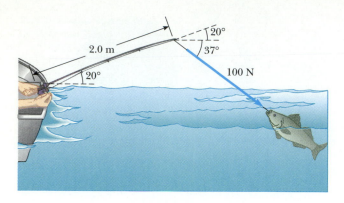

FIGURE P10.39

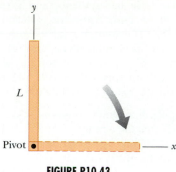

FIGURE P10.43

40. The density of the Earth, at any distance r from its center, is approximately

$$\rho = \left[14.2 - 11.6 \left(\frac{r}{R} \right) \right] \times 10^3 \text{ kg/m}^3$$

where R is the radius of the Earth. Show that this density leads to a moment of inertia $I = 0.330 MR^2$ about an axis through the center, where M is the mass of the Earth.

41. A 4.00-m length of light nylon cord is wound around a uniform cylindrical spool of radius 0.500 m and mass 1.00 kg. The spool is mounted on a frictionless axle and is initially at rest. The cord is pulled from the spool with a constant acceleration of magnitude 2.50 m/s². (a) How much work has been done on the spool when it reaches an angular speed of 8.00 rad/s? (b) Assuming there is enough cord on the spool, how long does it take the spool to reach this angular speed? (c) Is there enough cord on the spool?

42. A flywheel in the form of a heavy circular disk of diameter 0.600 m and mass 200 kg is mounted on a frictionless bearing. A motor connected to the flywheel accelerates it from rest to 1000 rev/min. (a) What is the moment of inertia of the flywheel? (b) How much work is done on it during this acceleration? (c) After 1000 rev/min is achieved, the motor is disengaged. A friction brake is used to slow the rotational rate to 500 rev/min. How much energy is dissipated as heat in the friction brake?

43. A long uniform rod of length L and mass M is pivoted about a horizontal, frictionless pin through one end. The rod is released from rest in a vertical position as in Figure P10.43. At the instant the rod is horizontal, find (a) its angular speed, (b) the magnitude of its angular acceleration, (c) the x and y components of the acceleration of its center of mass, and (d) the components of the reaction force at the pivot.

44. A bicycle is turned upside down while its owner repairs a flat tire. A friend spins the other wheel of radius 0.381 m and observes that drops of water fly off tangentially. She measures the height reached by drops moving vertically (Fig. P10.44). A drop that breaks loose from the tire on one turn rises $h = 54.0$ cm above the tangent point. A drop that breaks loose on the next turn rises 51.0 cm above the tangent point. The height to which the drops rise decreases because the angular speed of the wheel decreases. From this information, determine the magnitude of the average angular acceleration of the wheel.

44A. A bicycle is turned upside down while its owner repairs a flat tire. A friend spins the other wheel of radius R and observes that drops of water fly off tangentially. She measures the height reached by drops moving vertically (Fig. P10.44). A drop that breaks loose from the tire on one turn rises a distance h_1 above the tangent point. A drop that breaks loose on the next turn rises a distance $h_2 < h_1$ above the tangent point. The height to which the drops rise decreases because the angular speed of the wheel decreases. From this information, determine the magnitude of the average angular acceleration of the wheel.

45. *Radius of gyration:* For any given rotational axis, the radius of gyration, K, of a rigid body is defined by the expression $K^2 = I/M$, where M is the total mass of the body and I is the moment of inertia about the given axis. In other words, the radius of gyration is the distance between an imaginary point mass M, and the axis of rotation with I for the point mass about that axis is the same as for the rigid body. Find the radius of gyration of (a) a solid disk of radius R, (b) a uniform rod of length L, and (c) a solid sphere of radius R, all three rotating about a central axis.

46. A bright physics student purchases a wind vane for her father's garage. The vane consists of a rooster sitting on top of an arrow. The vane is fixed to a vertical shaft of radius r and mass m that is free to

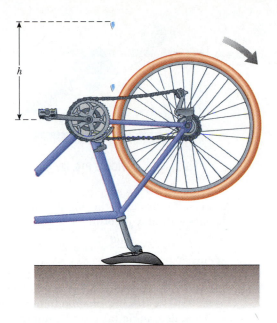

FIGURE P10.44

turn in its roof mount as shown in Figure P10.46. The student sets up an experiment to measure the rotational inertia of the rooster and arrow. String wound about the shaft passes over a pulley and is connected to a mass M hanging over the edge of the roof. When the mass M is released, the student determines the time t that the mass takes to fall through a distance h. From these data, she is able to find the rotational inertia I of the rooster and arrow. Find the expression for I in terms of m, M, r, g, h, and t.

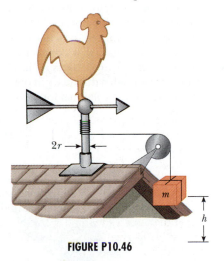

FIGURE P10.46

47. The top in Figure P10.47 has moment of inertia 4.00×10^{-4} kg·m^2 and is initially at rest. It is free to rotate about the stationary axis AA'. A string wrapped around a peg along the axis of the top is

pulled in such a manner as to maintain a constant tension of 5.57 N. If the string does not slip while being unwound, what is the angular speed of the top after 80.0 cm of string has been pulled off the peg?

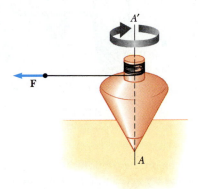

FIGURE P10.47

48. A cord is wrapped around a pulley of mass m and radius r. The free end of the cord is connected to a block of mass M. The block starts from rest and then slides down an incline that makes an angle θ with the horizontal. The coefficient of kinetic friction between block and incline is μ. (a) Use energy methods to show that the block's speed as a function of displacement d down the incline is

$$v = \left[4gd\left(\frac{M}{m + 2M} \right)(\sin\theta - \mu\cos\theta) \right]^{1/2}$$

(b) Find the magnitude of the acceleration of the block in terms of μ, m, M, g, and θ.

49. (a) What is the rotational energy of the Earth about its spin axis? The radius of the Earth is 6370 km and its mass is 5.98×10^{24} kg. Treat the Earth as a sphere of moment of inertia $\frac{2}{5}MR^2$. (b) The rotational energy of the Earth is decreasing steadily because of tidal friction. Estimate the change in rotational energy in one day, given that the rotational period decreases by about 10 μs each year.

50. The speed of a moving bullet can be determined by allowing the bullet to pass through two rotating paper disks mounted a distance d apart on the same axle (Fig. P10.50). From the angular displacement $\Delta\theta$ of the two bullet holes in the disks and the rotational speed of the disks, we can determine the speed v of the bullet. Find the bullet speed for the following data: $d = 80$ cm, $\omega = 900$ rev/min, and $\Delta\theta = 31°$.

51. The blocks shown in Figure 10.51 are connected by a string of negligible mass passing over a pulley of radius $R = 0.250$ m and moment of inertia I. The

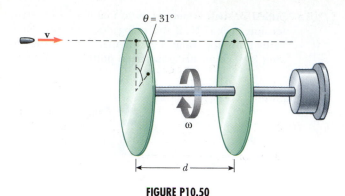

$\theta = 31°$

FIGURE P10.50

block on the frictionless incline is moving up with a constant acceleration of magnitude $a = 2.00 \text{ m/s}^2$. (a) Determine T_1 and T_2, the tensions in the two parts of the string, and (b) find the moment of inertia of the pulley.

51A. The blocks shown in Figure 10.51 are connected by a string of negligible mass passing over a pulley of radius R and moment of inertia I. The block on the incline is moving up with a constant acceleration of magnitude a. (a) Determine T_1 and T_2, the tensions in the two parts of the string, and (b) find the moment of inertia of the pulley.

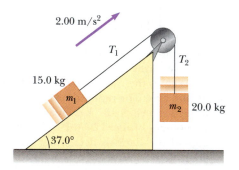

2.00 m/s²

T_1

T_2

15.0 kg

m_1

m_2 20.0 kg

37.0°

FIGURE P10.51

52. The pulley shown in Figure P10.52 has a radius R and moment of inertia I. One end of the mass m is connected to a spring of force constant k, and the other end is fastened to a cord wrapped around the pulley. The pulley axle and the incline are frictionless. If the pulley is wound counterclockwise so as to stretch the spring a distance d from its unstretched position and then released from rest, find (a) the angular speed of the pulley when the spring is again unstretched and (b) a numerical value for the angular speed at this point if $I = 1.0 \text{ kg} \cdot \text{m}^2$, $R = 0.30$ m, $k = 50$ N/m, $m = 0.50$ kg, $d = 0.20$ m, and $\theta = 37°$.

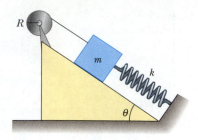

R

m

k

θ

FIGURE P10.52

53. As a result of friction, the angular speed of a wheel changes with time according to

$$\frac{d\theta}{dt} = \omega_0 e^{-\sigma t}$$

where ω_0 and σ are constants. The angular speed changes from 3.50 rad/s at $t = 0$ to 2.00 rad/s at $t = 9.30$ s. Use this information to determine σ and ω_0. Then determine (a) the magnitude of the angular acceleration at $t = 3.00$ s, (b) the number of revolutions the wheel makes in the first 2.50 s, and (c) the number of revolutions it makes before coming to rest.

54. A wheel is formed from a hoop and n equally spaced spokes. The mass of the hoop is M and its radius (and hence length of each spoke) is R. If the mass of each spoke is m, determine the moment of inertia of the wheel (a) about an axis through its center and perpendicular to the plane of the wheel and (b) about an axis through the hoop and perpendicular to the plane of the wheel.

55. A uniform, hollow, cylindrical spool has inside radius $R/2$, outside radius R, and mass M (Fig. P10.55). It is mounted so as to rotate on a fixed horizontal axle. A mass m is connected to the end of a string wound around the spool. The mass m falls from rest through a distance y in time t. Show that the torque due to the frictional forces between spool and axle is

$$\tau_f = R\left[m\left(g - \frac{2y}{t^2} \right) - \frac{5}{4} M\left(\frac{y}{t^2} \right) \right]$$

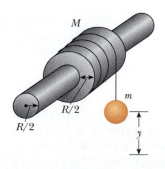

M

$R/2$

$R/2$

m

y

FIGURE P10.55

56. An electric motor can accelerate a Ferris wheel of moment of inertia $I = 20\ 000$ kg·m^2 from rest to 10 rev/min in 12 s. When the motor is turned off, friction causes the wheel to slow down from 10 to 8.0 rev/min in 10 s. Determine (a) the torque generated by the motor to bring the wheel to 10 rev/min and (b) the power needed to maintain this rotational speed.

SPREADSHEET PROBLEM

S1. A disk having a moment of inertia of 100 kg·m^2 is free to rotate about a fixed axis through its center as in Figure 10.18. A tangential force whose magnitude ranges from $F = 0$ to $F = 50.0$ N can be applied at any distance ranging from $R = 0$ to $R = 3.00$ m measured from the axis of rotation. Using Spreadsheet 10.1, find values of F and R that causes the disk to complete 2 revolutions in 10.0 s. Are there unique values for F and R?

CHAPTER 11

Rolling Motion, Angular Momentum, and Torque

Derek Swinson, professor of physics at the University of New Mexico, demonstrating the "gyro-ski" technique. The skier initiates a turn by lifting the axle of the rotating bicycle wheel. The direction of the turn depends on whether the left or right hand is used to lift the axle from the horizontal. Ignoring friction and gravity, the angular momentum of the system (the skier and the bicycle wheel) remains constant. *(Courtesy of Derek Swinson)*

I n the previous chapter we learned how to treat the rotation of a rigid body about a fixed axis. This chapter deals in part with the more general case, where the axis of rotation is not fixed in space. We begin by describing the rolling motion of an object. Next, we define a vector product, a convenient mathematical tool for expressing such quantities as torque and angular momentum. The central point of this chapter is to develop the concept of the angular momentum of a system of particles, a quantity that plays a key role in rotational dynamics. In analogy to the conservation of linear momentum, we find that angular momentum is always conserved. Like the law of conservation of linear momentum, the conservation of angular momentum is a fundamental law of physics, equally valid for relativistic and quantum systems.

11.1 ROLLING MOTION OF A RIGID BODY

In this section we treat the motion of a rigid body that is rotating about a moving axis. The general motion of a rigid body in space is very complex. However, we can simplify matters by restricting our discussion to a homogeneous rigid body having a high degree of symmetry, such as a cylinder, sphere, or hoop. Furthermore, we assume that the body undergoes rolling motion in a plane.

Suppose a cylinder is rolling on a straight path as in Figure 11.1. The center of mass moves in a straight line, while a point on the rim moves in a more complex path, which corresponds to the path of a cycloid. As we shall see later in this chapter, it is convenient to view the rolling motion as a combination of rotation about the center of mass and translation of the center of mass.

Now consider a uniform cylinder of radius R rolling without slipping on a horizontal surface (Fig. 11.2). As the cylinder rotates through an angle θ, its center of mass moves a distance $s = R\theta$. Therefore, the speed and acceleration magnitude of the center of mass for *pure rolling motion* are

$$v_{CM} = \frac{ds}{dt} = R\frac{d\theta}{dt} = R\omega \tag{11.1}$$

$$a_{CM} = \frac{dv_{CM}}{dt} = R\frac{d\omega}{dt} = R\alpha \tag{11.2}$$

The linear velocities of various points on the rolling cylinder are illustrated in Figure 11.3. Note that the linear velocity of any point is in a direction perpendicular to the line from that point to the contact point. At any instant, the point P is at rest relative to the surface since sliding does not occur.

A general point on the cylinder, such as Q, has both horizontal and vertical components of velocity. However, the points P and P' and the point at the center of mass are unique and of special interest. Relative to the surface on which the cylinder is moving, the center of mass moves with a speed $v_{CM} = R\omega$, whereas the contact point P has zero speed. The point P' has a speed $2v_{CM} = 2R\omega$, since all points on the cylinder have the same angular speed.

We can express the total energy of the rolling cylinder as

$$K = \tfrac{1}{2}I_P\omega^2 \tag{11.3}$$

where I_P is the moment of inertia about the axis through P. Applying the parallel-

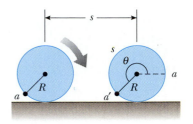

FIGURE 11.2 For pure rolling motion, as the cylinder rotates through an angle θ, its center moves a distance $s = R\theta$.

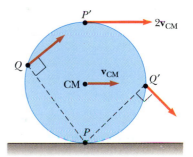

FIGURE 11.3 All points on a rolling body move in a direction perpendicular to an axis through the contact point P. The center of the body moves with a velocity $\mathbf{v}_{CM}$, while the point P' moves with a velocity $2\mathbf{v}_{CM}$.

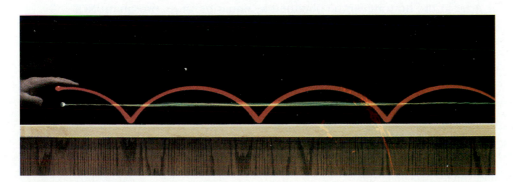

FIGURE 11.1 Light sources at the center and rim of a rolling cylinder illustrate the different paths these points take. The center moves in a straight line, as indicated by the green line, while a point on the rim moves in the path of a cycloid, as indicated by the red curve. *(Courtesy of Henry Leap and Jim Lehman)*

axis theorem, we can substitute $I_P = I_{CM} + MR^2$ into Equation 11.3 to get

$$K = \tfrac{1}{2}I_{CM}\omega^2 + \tfrac{1}{2}MR^2\omega^2$$

$$K = \tfrac{1}{2}I_{CM}\omega^2 + \tfrac{1}{2}Mv_{CM}^2 \qquad (11.4)$$

Total kinetic energy of a rolling body

where we have used the fact that $v_{CM} = R\omega$.

We can think of Equation 11.4 as follows: The term $\tfrac{1}{2}I_{CM}\omega^2$ represents the rotational kinetic energy about the center of mass, and the term $\tfrac{1}{2}Mv_{CM}^2$ represents the kinetic energy the cylinder would have if it were just translating through space without rotating. Thus, we can say that

the total kinetic energy of an object undergoing rolling motion is the sum of the rotational kinetic energy about the center of mass and the translational kinetic energy of the center of mass.

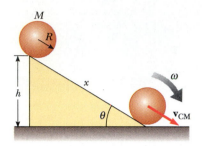

FIGURE 11.4 A round object rolling down an incline. Mechanical energy is conserved if no slipping occurs and there is no rolling friction.

We can use energy methods to treat a class of problems concerning the rolling motion of a rigid object down a rough incline. We assume that the object in Figure 11.4 does not slip and is released from rest at the top of the incline. Note that rolling motion is possible only if a frictional force is present between the object and the incline to produce a net torque about the center of mass. Despite the presence of friction, there is no loss of mechanical energy since the contact point is at rest relative to the surface at any instant. On the other hand, if the object were to slide, mechanical energy would be lost as motion progressed.

Using the fact that $v_{CM} = R\omega$ for pure rolling motion, we can express Equation 11.4 as

$$K = \tfrac{1}{2}I_{CM}\left(\frac{v_{CM}}{R}\right)^2 + \tfrac{1}{2}Mv_{CM}^2$$

$$K = \tfrac{1}{2}\left(\frac{I_{CM}}{R^2} + M\right)v_{CM}^2 \qquad (11.5)$$

When the rolling object reaches the bottom of the incline, work equal to Mgh has been done on it by the gravitational field, where h is the height of the incline. Because the body starts from rest at the top, its kinetic energy at the bottom, given by Equation 11.5, must equal this energy gain. Therefore, the speed of the center of mass at the bottom can be obtained by equating these two quantities:

$$\tfrac{1}{2}\left(\frac{I_{CM}}{R^2} + M\right)v_{CM}^2 = Mgh$$

$$v_{CM} = \left(\frac{2gh}{1 + I_{CM}/MR^2}\right)^{1/2} \qquad (11.6)$$

EXAMPLE 11.1   Sphere Rolling Down an Incline

If the object in Figure 11.4 is a solid sphere, calculate the speed of its center of mass at the bottom and determine the magnitude of the linear acceleration of the center of mass.

$$v_{CM} = \left(\frac{2gh}{1 + \frac{\frac{2}{5}MR^2}{MR^2}}\right)^{1/2} = \left(\frac{10}{7}gh\right)^{1/2}$$

Solution   For a uniform solid sphere, $I_{CM} = \tfrac{2}{5}MR^2$, and therefore Equation 11.6 gives

The vertical displacement is related to the displacement x along the incline through the relationship $h = x \sin\theta$.

Hence, after squaring both sides, we can express the equation above as

$$v_{CM}^2 = \frac{10}{7} gx \sin \theta$$

Comparing this with the familiar expression from kinematics, $v_{CM}^2 = 2a_{CM}x$, we see that the acceleration of the center of mass is

$$a_{CM} = \boxed{\tfrac{5}{7}g \sin \theta}$$

These results are quite interesting in that both the speed and the acceleration of the center of mass are *independent* of the mass and radius of the sphere! That is, *all homogeneous solid spheres experience the same speed and acceleration on a given incline.*

If we repeated the calculations for a hollow sphere, a solid cylinder, or a hoop, we would obtain similar results. The constant factors that appear in the expressions for v_c and a_c depend only on the moment of inertia about the center of mass for the specific body. In all cases, the acceleration of the center of mass is *less* than $g \sin \theta$, the value it would have if the plane were frictionless and no rolling occurred.

EXAMPLE 11.2   Another Look at the Rolling Sphere

In this example, let us consider the solid sphere rolling down an incline and verify the results of Example 11.1 using dynamic methods. The free-body diagram for the sphere is illustrated in Figure 11.5.

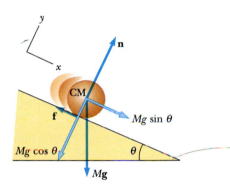

FIGURE 11.5   (Example 11.2) Free-body diagram for a solid sphere rolling down an incline.

Solution   Newton's second law applied to the motion of the center of mass gives

$$(1) \qquad \sum F_x = Mg \sin \theta - f = Ma_{CM}$$
$$\sum F_y = n - Mg \cos \theta = 0$$

where x is measured downward along the inclined plane. Now let us write an expression for the torque acting on the sphere. A convenient axis to choose is an axis through the center of the sphere, perpendicular to the plane of the figure.[1] Since n and Mg go through this origin, they have zero moment arms and do not contribute to the torque. However, the force of friction produces a torque about this axis equal to fR in the clockwise direction; therefore

$$\tau_{CM} = fR = I_{CM}\alpha$$

Since $I_{CM} = \tfrac{2}{5}MR^2$ and $\alpha = \alpha_{CM}/R$, we get

$$(2) \qquad f = \frac{I_{CM}\alpha}{R} = \left(\frac{\tfrac{2}{5}MR^2}{R}\right)\frac{a_{CM}}{R} = \tfrac{2}{5}Ma_{CM}$$

Substituting (2) into (1) gives

$$a_{CM} = \boxed{\tfrac{5}{7}g \sin \theta}$$

which agrees with the result of Example 11.1. Note that $F_{net} = ma$ applies rigorously if F_{net} is the net force and a is the acceleration of the center of mass. Hence, in this case, although the frictional force does no work, it contributes to F_{net} and thus decreases the acceleration of the center of mass.

---

[1] You should note that although the point at the center of mass is not an inertial frame, the expression $\tau_{CM} = I\alpha$ still applies in the center-of-mass frame.

11.2   THE VECTOR PRODUCT AND TORQUE

Consider a force **F** acting on a rigid body at the vector position **r** (Fig. 11.6). *The origin O is assumed to be in an inertial frame, so that Newton's second law is valid.* The *magnitude* of the torque due to this force relative to the origin is, by definition, $rF \sin \phi$, where ϕ is the angle between **r** and **F**. The axis about which **F** tends to produce rotation is perpendicular to the plane formed by **r** and **F**. If the force lies in the xy plane as in Figure 11.6, then the torque, **τ**, is represented by a vector parallel to the z axis. The force in Figure 11.6 creates a torque that tends to rotate the body counterclockwise looking down the z axis, and so the sense of **τ** is toward

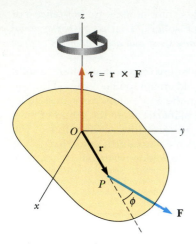

FIGURE 11.6   The torque vector τ lies in a direction perpendicular to the plane formed by the position vector **r** and the applied force **F**.

increasing z and τ is therefore in the positive z direction. If we reversed the direction of **F** in Figure 11.6, τ would then be in the negative z direction. The torque involves two vectors, **r** and **F**, and is in fact defined to be equal to the *vector product*, or *cross product,* of **r** and **F**:

$$\tau \equiv \mathbf{r} \times \mathbf{F} \tag{11.7}$$

We must now give a formal definition of the vector product. Given any two vectors **A** and **B**, the vector product $\mathbf{A} \times \mathbf{B}$ is defined as a third vector **C**, the *magnitude* of which is $AB \sin \theta$, where θ is the angle included between **A** and **B**. That is, if **C** is given by

$$\mathbf{C} = \mathbf{A} \times \mathbf{B} \tag{11.8}$$

then its magnitude is

$$C \equiv AB \sin \theta \tag{11.9}$$

Note that the quantity $AB \sin \theta$ is equal to the area of the parallelogram formed by **A** and **B**, as shown in Figure 11.7. The *direction* of $\mathbf{A} \times \mathbf{B}$ is perpendicular to the plane formed by **A** and **B**, as in Figure 11.7, and its sense is determined by the advance of a right-handed screw when turned from **A** to **B** through the angle θ. A more convenient rule to use for the direction of $\mathbf{A} \times \mathbf{B}$ is the right-hand rule illustrated in Figure 11.7. The four fingers of the right hand are pointed along **A** and then "wrapped" into **B** through the angle θ. The direction of the erect right thumb is the direction of $\mathbf{A} \times \mathbf{B}$. Because of the notation, $\mathbf{A} \times \mathbf{B}$ is often read "A cross B"; hence the term *cross product.*

Some properties of the vector product that follow from its definition are as follows:

Properties of the vector product

1. Unlike the scalar product, the order in which the two vectors are multiplied in a cross product is important, that is,

$$\mathbf{A} \times \mathbf{B} = -(\mathbf{B} \times \mathbf{A}) \tag{11.10}$$

Therefore, if you change the order of the cross product, you must change the sign. You could easily verify this relationship with the right-hand rule (Fig. 11.7).

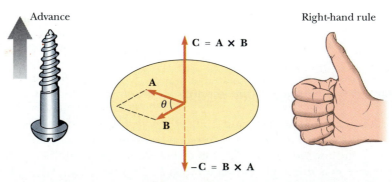

FIGURE 11.7   The vector product $\mathbf{A} \times \mathbf{B}$ is a third vector **C** having a magnitude $AB \sin \theta$ equal to the area of the parallelogram shown. The direction of **C** is perpendicular to the plane formed by **A** and **B**, and its sense is determined by the right-hand rule.

2. If **A** is parallel to **B** ($\theta = 0°$ or $180°$), then $\mathbf{A} \times \mathbf{B} = 0$; therefore, it follows that $\mathbf{A} \times \mathbf{A} = 0$.
3. If **A** is perpendicular to **B**, then $|\mathbf{A} \times \mathbf{B}| = AB$.
4. The vector product obeys the *distributive law,* that is,

$$\mathbf{A} \times (\mathbf{B} + \mathbf{C}) = \mathbf{A} \times \mathbf{B} + \mathbf{A} \times \mathbf{C} \qquad (11.11)$$

5. The derivative of the cross product with respect to some variable such as t is

$$\frac{d}{dt}(\mathbf{A} \times \mathbf{B}) = \mathbf{A} \times \frac{d\mathbf{B}}{dt} + \frac{d\mathbf{A}}{dt} \times \mathbf{B} \qquad (11.12)$$

where it is important to preserve the multiplicative order of **A** and **B**, in view of Equation 11.10.

It is left as an exercise to show from Equations 11.8 and 11.9 and the definition of unit vectors that the cross products of the rectangular unit vectors **i**, **j**, and **k** obey the following expressions:

$$\mathbf{i} \times \mathbf{i} = \mathbf{j} \times \mathbf{j} = \mathbf{k} \times \mathbf{k} = 0 \qquad (11.13a)$$

$$\mathbf{i} \times \mathbf{j} = -\mathbf{j} \times \mathbf{i} = \mathbf{k} \qquad (11.13b)$$

$$\mathbf{j} \times \mathbf{k} = -\mathbf{k} \times \mathbf{j} = \mathbf{i} \qquad (11.13c)$$

$$\mathbf{k} \times \mathbf{i} = -\mathbf{i} \times \mathbf{k} = \mathbf{j} \qquad (11.13d)$$

Cross products of unit vectors

Signs are interchangeable. For example, $\mathbf{i} \times (-\mathbf{j}) = -\mathbf{i} \times \mathbf{j} = -\mathbf{k}$.

The cross product of *any* two vectors **A** and **B** can be expressed in the following determinant form:

$$\mathbf{A} \times \mathbf{B} = \begin{vmatrix} \mathbf{i} & \mathbf{j} & \mathbf{k} \\ A_x & A_y & A_z \\ B_x & B_y & B_z \end{vmatrix}$$

Expanding this determinant gives the result

$$\mathbf{A} \times \mathbf{B} = (A_y B_z - A_z B_y)\mathbf{i} + (A_z B_x - A_x B_z)\mathbf{j} + (A_x B_y - A_y B_x)\mathbf{k} \qquad (11.14)$$

EXAMPLE 11.3 The Cross Product

Two vectors lying in the xy plane are given by the equations $\mathbf{A} = 2\mathbf{i} + 3\mathbf{j}$ and $\mathbf{B} = -\mathbf{i} + 2\mathbf{j}$. Find $\mathbf{A} \times \mathbf{B}$, and verify explicitly that $\mathbf{A} \times \mathbf{B} = -\mathbf{B} \times \mathbf{A}$.

Solution Using Equations 11.13a through 11.13d for the cross product of unit vectors gives

$$\mathbf{A} \times \mathbf{B} = (2\mathbf{i} + 3\mathbf{j}) \times (-\mathbf{i} + 2\mathbf{j})$$

$$= 2\mathbf{i} \times 2\mathbf{j} + 3\mathbf{j} \times (-\mathbf{i}) = 4\mathbf{k} + 3\mathbf{k} = \boxed{7\mathbf{k}}$$

(We have omitted the terms with $\mathbf{i} \times \mathbf{i}$ and $\mathbf{j} \times \mathbf{j}$, because they are zero.)

$$\mathbf{B} \times \mathbf{A} = (-\mathbf{i} + 2\mathbf{j}) \times (2\mathbf{i} + 3\mathbf{j})$$

$$= -\mathbf{i} \times 3\mathbf{j} + 2\mathbf{j} \times 2\mathbf{i} = -3\mathbf{k} - 4\mathbf{k} = \boxed{-7\mathbf{k}}$$

Therefore, $\mathbf{A} \times \mathbf{B} = -\mathbf{B} \times \mathbf{A}$.

As an alternative method for finding $\mathbf{A} \times \mathbf{B}$, we could use Equation 11.8, with $A_x = 2$, $A_y = 3$, $A_z = 0$ and $B_x = -1$, $B_y = 2$, $B_z = 0$:

$$\mathbf{A} \times \mathbf{B} = (0)\mathbf{i} + (0)\mathbf{j} + [2 \times 2 - 3 \times (-1)]\mathbf{k} = 7\mathbf{k}$$

Exercise Use the results to this example and Equation 11.9 to find the angle between **A** and **B**.

Answer 60.3°.

11.3   ANGULAR MOMENTUM OF A PARTICLE

A particle of mass m, located at the vector position $\mathbf{r}$, moves with a velocity $\mathbf{v}$ (Fig. 11.8).

> The instantaneous angular momentum $\mathbf{L}$ of the particle relative to the origin O is defined by the cross product of the instantaneous vector position of the particle and its instantaneous linear momentum $\mathbf{p}$:

Angular momentum of a particle

$$\mathbf{L} \equiv \mathbf{r} \times \mathbf{p} \tag{11.15}$$

The SI units of angular momentum are $kg \cdot m^2/s$. It is important to note that both the magnitude and direction of $\mathbf{L}$ depend on the choice of origin. The direction of $\mathbf{L}$ is perpendicular to the plane formed by $\mathbf{r}$ and $\mathbf{p}$, and its sense is governed by the right-hand rule. For example, in Figure 11.8, $\mathbf{r}$ and $\mathbf{p}$ are in the xy plane, so that $\mathbf{L}$ points in the z direction. Since $\mathbf{p} = m\mathbf{v}$, the magnitude of $\mathbf{L}$ is

$$L = mvr \sin \phi \tag{11.16}$$

where ϕ is the angle between $\mathbf{r}$ and $\mathbf{p}$. It follows that L is zero when $\mathbf{r}$ is parallel to $\mathbf{p}$ ($\phi = 0$ or $180°$). In other words, when the particle moves along a line that passes through the origin, it has zero angular momentum with respect to the origin. On the other hand, if $\mathbf{r}$ is perpendicular to $\mathbf{p}$ ($\phi = 90°$), then $L = mvr$. At that instant the particle moves exactly as though it were on the rim of a wheel rotating about the origin in a plane defined by $\mathbf{r}$ and $\mathbf{p}$.

Alternatively, one should note that a particle has nonzero angular momentum about some point if its position measured from that point is nonzero and the velocity vector would not carry it through that point, even if the position vector does not rotate about the point. On the other hand, if the position vector simply increases or decreases in length, the particle moves along a line passing through the origin and, therefore, has zero angular momentum with respect to that origin.

In linear motion, we found that the resultant force on a particle equals the time rate of change of its linear momentum (Eq. 9.3). We now show that the resultant torque acting on a particle equals the time rate of change of its angular momentum. Let us start by writing the torque on the particle in the form

$$\boldsymbol{\tau} = \mathbf{r} \times \mathbf{F} = \mathbf{r} \times \frac{d\mathbf{p}}{dt} \tag{11.17}$$

where we have used Newton's second law in the form $\mathbf{F} = d\mathbf{p}/dt$. Now let us differentiate Equation 11.15 with respect to time using the rule given by Equation 11.12:

$$\frac{d\mathbf{L}}{dt} = \frac{d}{dt}(\mathbf{r} \times \mathbf{p}) = \mathbf{r} \times \frac{d\mathbf{p}}{dt} + \frac{d\mathbf{r}}{dt} \times \mathbf{p}$$

It is important to adhere to the order of terms since $\mathbf{A} \times \mathbf{B} = -\mathbf{B} \times \mathbf{A}$.

The last term on the right in the above equation is zero, because $\mathbf{v} = d\mathbf{r}/dt$ is parallel to $\mathbf{p}$ (property 2 of the vector product). Therefore,

$$\frac{d\mathbf{L}}{dt} = \mathbf{r} \times \frac{d\mathbf{p}}{dt} \tag{11.18}$$

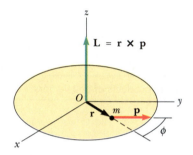

FIGURE 11.8   The angular momentum $\mathbf{L}$ of a particle of mass m and momentum $\mathbf{p}$ located at the position $\mathbf{r}$ is a vector given by $\mathbf{L} = \mathbf{r} \times \mathbf{p}$. The value of $\mathbf{L}$ depends on the origin and is a vector perpendicular to both $\mathbf{r}$ and $\mathbf{p}$.

Comparing Equations 11.17 and 11.18, we see that

$$\boldsymbol{\tau} = \frac{d\mathbf{L}}{dt} \tag{11.19}$$

Torque equals time rate of
change of angular momentum

which is the rotational analog of Newton's second law, $\mathbf{F} = d\mathbf{p}/dt$. This rotational
result says that

> the torque acting on a particle is equal to the time rate of change of the
> particle's angular momentum.

It is important to note that Equation 11.19 is valid only if the origins of $\boldsymbol{\tau}$ and $\mathbf{L}$ are
the *same*. Equation 11.19 is also valid when there are several forces acting on the
particle, in which case $\boldsymbol{\tau}$ is the net torque on the particle. *Furthermore, the expression is
valid for any origin fixed in an inertial frame.* Of course, the same origin must be used
in calculating all torques as well as the angular momentum.

A System of Particles

The total angular momentum, $\mathbf{L}$, of a system of particles about some point is
defined as the vector sum of the angular momenta of the individual particles:

$$\mathbf{L} = \mathbf{L}_1 + \mathbf{L}_2 + \cdots + \mathbf{L}_n = \sum \mathbf{L}_i$$

where the vector sum is over all n particles in the system.

Since the individual momenta of the particles may change in time, the total
angular momentum may do so also. In fact, from Equations 11.17 and 11.18, we
find that the time rate of change of the total angular momentum equals the vector
sum of all torques, both those associated with internal forces between particles and
those associated with external forces. However, the net torque associated with
internal forces is zero. To understand this, recall that Newton's third law tells us
that the internal forces occur in equal and opposite pairs. If we assume that these
forces lie along the line of separation of each pair of particles, then the torque due
to each action-reaction force pair is zero. By summation, we see that the net internal torque vanishes. Finally, we conclude that the total angular momentum can
vary with time only if there is a net external torque on the system, so that we have

$$\sum \boldsymbol{\tau}_{\text{ext}} = \sum \frac{d\mathbf{L}_i}{dt} = \frac{d}{dt} \sum \mathbf{L}_i = \frac{d\mathbf{L}}{dt} \tag{11.20}$$

That is,

> the time rate of change of the total angular momentum of the system about
> some origin in an inertial frame equals the net external torque acting on
> the system about that origin.

Note that Equation 11.20 is the rotational analog of Equation 9.3, $\mathbf{F}_{\text{ext}} = d\mathbf{p}/dt$, for
a system of particles.

CONCEPTUAL EXAMPLE 11.4

Can a particle moving in a straight line have nonzero angular momentum?

Reasoning   A particle has angular momentum about an origin when moving in a straight line as long as its line of motion does not pass through the origin. Its angular momentum is zero only if the line of motion passes through the origin, in which case **r** is parallel to **v**, and $L = r \times p = r \times mv = 0$.

EXAMPLE 11.5   Linear Motion

A particle of mass m moves in the xy plane with a velocity v along a straight line (Fig. 11.9). What are the magnitude and direction of its angular momentum (a) with respect to the origin O and (b) with respect to the origin O'?

Solution   (a) From the definition of angular momentum, $L = r \times p = rmv \sin \phi(- k)$. Therefore the magnitude of **L** is

$$L = mvr \sin \phi = \boxed{mvd}$$

where $d = r \sin \phi$ is the distance of closest approach of the particle from the origin. The direction of **L** from the right-hand rule is into the diagram, and we can write the vector expression $L = -(mvd)k$.

(b) Because the direction of **v** passes through O', the angular momentum relative to O' is zero.

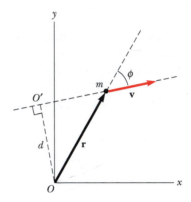

FIGURE 11.9   (Example 11.5) A particle moving in a straight line with a velocity **v** has an angular momentum equal in magnitude to mvd relative to O, where $d = r \sin \phi$ is the distance of closest approach to the origin. The vector $L = r \times p$ points *into* the diagram in this case.

EXAMPLE 11.6   Circular Motion

A particle moves in the xy plane in a circular path of radius r, as in Figure 11.10. (a) Find the magnitude and direction of its angular momentum relative to O when its velocity is **v**.

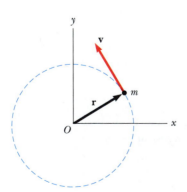

FIGURE 11.10   (Example 11.6) A particle moving in a circle of radius r has an angular momentum equal in magnitude to mvr relative to the center. The vector $L = r \times p$ points *out* of the diagram.

Solution   Since **r** is perpendicular to **v**, $\phi = 90°$ and the magnitude of **L** is simply

$$L = mvr \sin 90° = \boxed{mvr} \qquad \text{(for r perpendicular to v)}$$

The direction of **L** is perpendicular to the plane of the circle, and its sense depends on the direction of **v**. If the sense of the rotation is counterclockwise, as in Figure 11.10, then by the right-hand rule, the direction of $L = r \times p$ is *out* of the paper. Hence, we can write the vector expression $L = (mvr)k$. If the particle were to move clockwise, **L** would point into the paper.

(b) Find an alternative expression for L in terms of the angular speed, ω.

Solution   Since $v = r\omega$ for a particle rotating in a circle, we can express L as

$$L = mvr = mr^2\omega = \boxed{I\omega}$$

where I is the moment of inertia of the particle about the z axis through O. In this case the angular momentum is in the *same* direction as the angular velocity vector, ω (see Section 10.1), and so we can write $\mathbf{L} = I\boldsymbol{\omega} = I\omega\mathbf{k}$.

Exercise A car of mass 1500 kg moves on a circular race track of radius 50 m with a speed of 40 m/s. What is the magnitude of its angular momentum relative to the center of the track?

Answer 3.0×10^6 kg·m²/s.

11.4 ROTATION OF A RIGID BODY ABOUT A FIXED AXIS

Consider a rigid body rotating about an axis that is fixed in direction. We assume that the z axis coincides with the axis of rotation, as in Figure 11.11. Each particle of the rigid body rotates in the xy plane about the z axis with an angular speed ω. The magnitude of the angular momentum of the particle of mass m_i is $m_i v_i r_i$ about the origin O. Because $v_i = r_i\omega$, we can express the magnitude of the angular momentum of the ith particle as

$$L_i = m_i r_i^2 \omega$$

The vector $\mathbf{L}_i$ is directed along the z axis, corresponding to the direction of ω.

We can now find the z component of the angular momentum of the rigid body by taking the sum of L_i over all particles of the body:

$$L_z = \sum m_i r_i^2 \omega = \left(\sum m_i r_i^2\right)\omega$$

or

$$L_z = I\omega \tag{11.21}$$

where I is the moment of inertia of the rigid body about the z axis.

Now let us differentiate Equation 11.21 with respect to time, noting that I is constant for a rigid body:

$$\frac{dL_z}{dt} = I\frac{d\omega}{dt} = I\alpha \tag{11.22}$$

where α is the angular acceleration relative to the axis of rotation. Because dL_z/dt is equal to the net torque (Eq. 11.20), we can express Equation 11.22 as

$$\sum \tau_{\text{ext}} = \frac{dL_z}{dt} = I\alpha \tag{11.23}$$

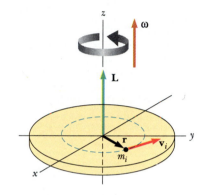

FIGURE 11.11 When a rigid body rotates about an axis, the angular momentum $\mathbf{L}$ is in the same direction as the angular velocity ω, according to the expression $\mathbf{L} = I\boldsymbol{\omega}$.

That is, the net external torque acting on an object rotating about a fixed axis equals the moment of inertia about the axis of rotation multiplied by the angular acceleration of the object relative to that axis.

You should note that if a symmetrical object rotates about a fixed axis passing through its center of mass, you can write Equation 11.21 in vector form, $\mathbf{L} = I\boldsymbol{\omega}$, where $\mathbf{L}$ is the total angular momentum of the object measured with respect to the axis of rotation. Furthermore, the expression is valid for any object, regardless of its symmetry, if **L** stands for the component of angular momentum along the axis of rotation.[2]

EXAMPLE 11.7   Rotating Sphere

A uniform solid sphere of radius $R = 0.50$ m and mass 15 kg rotates about the z axis through its center, as in Figure 11.12. Find the magnitude of its angular momentum when the angular speed is 3.0 rad/s.

Solution   The moment of inertia of the sphere about an axis through its center is, from Table 10.2,

$$I = \tfrac{2}{5}MR^2 = \tfrac{2}{5}(15 \text{ kg})(0.50 \text{ m})^2 = 1.5 \text{ kg} \cdot \text{m}^2$$

Therefore, the magnitude of the angular momentum is

$$L = I\omega = (1.5 \text{ kg} \cdot \text{m}^2)(3.0 \text{ rad/s}) = \boxed{4.5 \text{ kg} \cdot \text{m}^2/\text{s}}$$

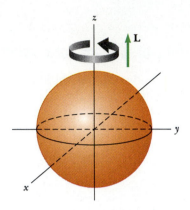

FIGURE 11.12   (Example 11.7) A sphere that rotates about the z axis in the direction shown has an angular momentum **L** in the positive z direction. If the direction of rotation is reversed, **L** will point in the negative z direction.

EXAMPLE 11.8   Rotating Rod

A rigid rod of mass M and length ℓ rotates in a vertical plane about a frictionless pivot through its center (Fig. 11.13). Particles of masses m_1 and m_2 are attached at the ends of the rod. (a) Determine the magnitude of the angular momentum of the system when the angular speed is ω.

Solution   The moment of inertia of the system equals the sum of the moments of inertia of the three components: the rod, m_1, and m_2. Using Table 10.2, we find that the total moment of inertia about the z axis through O is

$$I = \frac{1}{12} M\ell^2 + m_1\left(\frac{\ell}{2}\right)^2 + m_2\left(\frac{\ell}{2}\right)^2 = \frac{\ell^2}{4}\left(\frac{M}{3} + m_1 + m_2\right)$$

Therefore, when the angular speed is ω, the magnitude of the angular momentum is

$$L = I\omega = \frac{\ell^2}{4}\left(\frac{M}{3} + m_1 + m_2\right)\omega$$

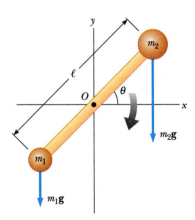

FIGURE 11.13   (Example 11.8) Since gravitational forces act on the system rotating in a vertical plane, there is in general a net nonzero torque about O when $m_1 \neq m_2$, which in turn produces an angular acceleration according to $\tau_{\text{net}} = I\alpha$.

---

[2] In general, the expression $\mathbf{L} = I\boldsymbol{\omega}$ is not always valid. If a rigid body rotates about an arbitrary axis, **L** and $\boldsymbol{\omega}$ may point in different directions. In fact, in this case, the moment of inertia cannot be treated as a scalar. Strictly speaking, $\mathbf{L} = I\boldsymbol{\omega}$ applies only to rigid bodies of any shape that rotate about one of three mutually perpendicular axes (called *principal axes*) through the center of mass. This is discussed in more advanced texts on mechanics.

(b) Determine the magnitude of the angular acceleration of the system when the rod makes an angle θ with the horizontal.

Solution The torque due to the force m_1g about the pivot is

$$\tau_1 = m_1 g \frac{\ell}{2} \cos \theta \qquad (\tau_1 \text{ is out of the plane})$$

The torque due to the force m_2g about the pivot is

$$\tau_2 = -m_2 g \frac{\ell}{2} \cos \theta \qquad (\tau_2 \text{ is into the plane})$$

Hence, the net torque about O is

$$\tau_{net} = \tau_1 + \tau_2 = \tfrac{1}{2}(m_1 - m_2)g\ell \cos \theta$$

You should note that the direction of τ_{net} is out of the plane if $m_1 > m_2$ and is into the plane if $m_1 > m_2$.

To find α, we use $\tau_{net} = I\alpha$, where I was obtained in (a):

$$\alpha = \frac{\tau_{net}}{I} = \frac{2(m_1 - m_2)g \cos \theta}{\ell \left(\dfrac{M}{3} + m_1 + m_2 \right)}$$

Note that α is zero when θ is $\pi/2$ or $-\pi/2$ (vertical position) and a maximum when θ is 0 or π (horizontal position). Furthermore, the angular speed of the system changes because α is not zero.

Exercise If $m_1 > m_2$, at what value of θ is ω a maximum? Knowing the angular speed at some instant, how would you calculate the linear speed of m_1 and m_2?

EXAMPLE 11.9 **Two Connected Masses**

Two masses, m_1 and m_2, are connected by a light cord that passes over a pulley of radius R and moment of inertia I about its axle, as in Figure 11.14. The mass m_2 slides on a frictionless, horizontal surface. Determine the acceleration of the two masses using the concepts of angular momentum and torque.

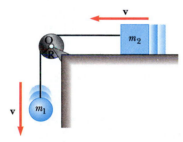

FIGURE 11.14 (Example 11.9).

Reasoning and Solution First, calculate the angular momentum of the system, which consists of the two masses plus the pulley. Then calculate the torque about an axis along the axle of the pulley through O. At the instant m_1 and m_2 have a speed v, the angular momentum of m_1 is $m_1 vR$, and that of m_2 is $m_2 vR$. At the same instant, the angular momentum of the pulley is $I\omega = Iv/R$. Therefore, the total angular momentum of the system is

$$(1) \qquad L = m_1 vR + m_2 vR + I \frac{v}{R}$$

Now let us evaluate the total external torque on the system about the axle. Because it has zero moment arm, the force exerted by the axle on the pulley does not contribute to the torque. Furthermore, the normal force acting on m_2 is balanced by its weight m_2g, and so these forces do not contribute to the torque. The external force m_1g produces a torque about the axle equal in magnitude to m_1gR, where R is the moment arm of the force about the axle. This is the total external torque about O; that is, $\tau_{ext} = m_1gR$. Using this result, together with (1) and Equation 11.23 gives

$$\tau_{ext} = \frac{dL}{dt}$$

$$m_1 gR = \frac{d}{dt} \left[(m_1 + m_2)Rv + I \frac{v}{R} \right]$$

$$(2) \qquad m_1 gR = (m_1 + m_2)R \frac{dv}{dt} + \frac{I}{R} \frac{dv}{dt}$$

Because $dv/dt = a$, we can solve this for a to get

$$a = \frac{m_1 g}{(m_1 + m_2) + I/R^2}$$

You may wonder why we did not include the forces that the cord exerts on the objects in evaluating the net torque about the axle. The reason is that these forces are internal to the system under consideration. Only the external torques contribute to the change in angular momentum.

11.5 CONSERVATION OF ANGULAR MOMENTUM

In Chapter 9 we found that the total linear momentum of a system of particles remains constant when the resultant external force acting on the system is zero. We have an analogous conservation law in rotational motion:

Conservation of angular momentum

> The total angular momentum of a system is constant if the resultant external torque acting on the system is zero.

This follows directly from Equation 11.20, where we see that if

$$\sum \tau_{\text{ext}} = \frac{d\mathbf{L}}{dt} = 0 \qquad (11.24)$$

then

$$\mathbf{L} = \text{constant} \qquad (11.25)$$

For a system of particles, we write this conservation law as $\sum \mathbf{L}_n =$ constant. If a body undergoes a redistribution of its mass, then its moment of inertia changes and we express this conservation of angular momentum in the form

$$\mathbf{L}_i = \mathbf{L}_f = \text{constant} \qquad (11.26)$$

If the system is a body rotating about a *fixed* axis, such as the z axis, then we can write $L_z = I\omega$, where L_z is the component of $\mathbf{L}$ along the axis of rotation and I is the moment of inertia about this axis. In this case, we can express the conservation of angular momentum as

$$I_i \omega_i = I_f \omega_f = \text{constant} \qquad (11.27)$$

This expression is valid for rotations either about a fixed axis or about an axis through the center of mass of the system as long as the axis remains parallel to itself. We require only that the net external torque be zero.

Although we do not prove it here, there is an important theorem concerning the angular momentum relative to the center of mass:

> The resultant torque acting on a body about an axis through the center of mass equals the time rate of change of angular momentum regardless of the motion of the center of mass.

This theorem applies even if the center of mass is accelerating, provided τ and $\mathbf{L}$ are evaluated relative to the center of mass.

In Equation 11.27 we have a third conservation law to add to our list. We can now state that the energy, linear momentum, and angular momentum of an isolated system all remain constant.

There are many examples that demonstrate conservation of angular momentum. You may have observed a figure skater undergoing a spin motion in the finale of an act. The angular speed of the skater increases upon pulling his or her hands and feet close to the body. Neglecting friction between skates and ice, we see that there are no external torques on the skater. The change in angular speed is due to the fact that, because angular momentum is conserved, the product $I\omega$ remains constant and a decrease in the moment of inertia of the skater causes an increase in the angular speed. Similarly, when divers (or acrobats) wish to make several somersaults, they pull their hands and feet close to their bodies in order to rotate at a higher rate. In these cases, the external force due to gravity acts through the center of mass and, hence, exerts no torque about this point. Therefore, the

Nancy Kerrigan, winner of a Silver Medal in the 1994 Winter Olympics. In this spin her angular speed increases when she pulls her arms in close to her body, demonstrating that angular momentum is conserved.
(© Duomo/Steven E. Sutton, 1994)

angular momentum about the center of mass must be constant, or $I_i\omega_i = I_f\omega_f$. For example, when divers wish to double their angular speed, they must reduce their moment of inertia to half its initial value.

CONCEPTUAL EXAMPLE 11.10

A particle moves in a straight line, and you are told that the net torque acting on it is zero about some unspecified origin. Does this necessarily imply that the net force on the particle is zero? Can you conclude that its velocity is constant?

Reasoning   The net force is not necessarily zero. If the line of action of the net force passes through the origin, the net torque about an axis passing through that origin is zero, even though the net force is not zero. Because the net force is not necessarily zero, you cannot conclude that its velocity is constant.

EXAMPLE 11.11   Formation of a Neutron Star

A star of radius 1.0×10^4 km rotates about its axis with a period of 30 days. The star undergoes a supernova explosion, whereby its core collapses into a neutron star of radius 3.0 km. Estimate the period of the neutron star.

Solution   Let us assume that during the collapse of the star, (1) no torque acts on it, (2) it remains spherical, and (3) its mass remains constant. Since I is proportional to r^2, and $\omega = 2\pi/T$, conservation of angular momentum (Eq. 11.27) gives $T_f = T_i(r_f/r_i)^2 = (30 \text{ days})(3.0 \text{ km}/1.0 \times 10^4 \text{ km})^2 = 2.7 \times 10^{-6}$ days $= 0.23$ s. Thus the neutron star rotates about four times each second.

EXAMPLE 11.12   A Projectile–Cylinder Collision

A projectile of mass m and velocity v_0 is fired at a solid cylinder of mass M and radius R (Fig. 11.15). The cylinder is initially at rest and is mounted on a fixed horizontal axle that runs through the center of mass. The line of motion of the projectile is perpendicular to the axle and at a distance $d < R$ from the center. Find the angular speed of the system after the projectile strikes and adheres to the surface of the cylinder.

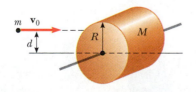

FIGURE 11.15   (Example 11.12) The angular momentum of the system before the collision equals the angular momentum right after the collision with respect to the center of mass if we take the center of mass to be along the cylinder axis (an approximation).

Reasoning   Let us evaluate the angular momentum of the system (projectile + cylinder) about the axle of the cylinder.

The net external torque on the system is zero about this axle. Hence, the angular momentum of the system is the same before and after the collision.

Solution   Before the collision, only the projectile has angular momentum with respect to a point on the axle. The magnitude of this angular momentum is $mv_0 d$, and it is directed along the axle into the paper. After the collision, the total angular momentum of the system is $I\omega$, where I is the total moment of inertia about the axle (projectile + cylinder). Since the total angular momentum is constant, we get

$$mv_0 d = I\omega = (\tfrac{1}{2}MR^2 + mR^2)\omega$$

$$\omega = \frac{mv_0 d}{\tfrac{1}{2}MR^2 + mR^2}$$

This suggests another technique for measuring the speed of a bullet.

Exercise   In this example, mechanical energy is not conserved since the collision is inelastic. Show that $\tfrac{1}{2}I\omega^2 < \tfrac{1}{2}mv_0^2$. What do you suppose accounts for the energy loss?

EXAMPLE 11.13   The Merry-Go-Round

A horizontal platform in the shape of a circular disk rotates in a horizontal plane about a frictionless vertical axle (Fig. 11.16). The platform has a mass $M = 100$ kg and a radius $R = 2.0$ m. A student whose mass is $m = 60$ kg walks slowly from the rim of the platform toward the center. If the angular speed of the system is 2.0 rad/s when the student is at the rim, (a) calculate the angular speed when the student has reached a point 0.50 m from the center.

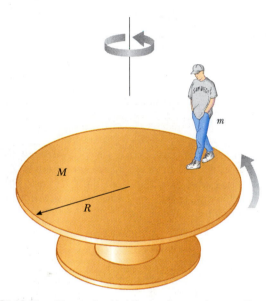

FIGURE 11.16 (Example 11.13) As the student walks toward the center of the rotating platform, the angular speed of the system increases since the angular momentum must remain constant.

Solution   Let us call the moment of inertia of the platform I_p and the moment of inertia of the student I_s. Treating the

student as a point mass m, we can write the initial moment of inertia of the system about the axis of rotation

$$I_i = I_p + I_s = \tfrac{1}{2}MR^2 + mR^2$$

When the student has walked to the position $r < R$, the moment of inertia of the system reduces to

$$I_f = \tfrac{1}{2}MR^2 + mr^2$$

Since there are no external torques on the system (student + platform) about the axis of rotation, we can apply the law of conservation of angular momentum:

$$I_i\omega_i = I_f\omega_f$$

$$(\tfrac{1}{2}MR^2 + mR^2)\,\omega_i = (\tfrac{1}{2}MR^2 + mr^2)\,\omega_f$$

$$\omega_f = \left(\frac{\tfrac{1}{2}MR^2 + mR^2}{\tfrac{1}{2}MR^2 + mr^2}\right)\omega_i$$

$$\omega_f = \left(\frac{200 + 240}{200 + 15}\right)(2.0 \text{ rad/s})$$

$$= \boxed{4.1 \text{ rad/s}}$$

(b) Calculate the initial and final rotational energies of the system.

Solution

$$K_i = \tfrac{1}{2}I_i\omega_i^2 = \tfrac{1}{2}(440 \text{ kg}\cdot\text{m}^2)\left(2.0\,\frac{\text{rad}}{\text{s}}\right)^2 = \boxed{880 \text{ J}}$$

$$K_f = \tfrac{1}{2}I_f\omega_f^2 = \tfrac{1}{2}(215 \text{ kg}\cdot\text{m}^2)\left(4.1\,\frac{\text{rad}}{\text{s}}\right)^2 = \boxed{1.8 \times 10^3 \text{ J}}$$

Note that the rotational energy of the system increases! What accounts for this increase in energy?

CONCEPTUAL EXAMPLE 11.14

A student sits on a pivoted stool while holding a pair of weights, as in Figure 11.17. The stool is free to rotate about a vertical axis with negligible friction. The student is set in rotating motion with the weights outstretched. Why does the angular speed of the system increase as the weights are pulled inward?

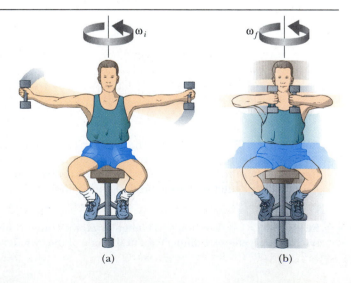

FIGURE 11.17 (Conceptual Example 11.14) (a) The student is given an initial angular speed while holding two masses as shown. (b) When the masses are pulled in close to the body, the angular speed of the system increases. Why?

(a)   (b)

Reasoning The initial angular momentum of the system (student + weights + stool) is $I_i\omega_i$. After the weights are pulled in, the angular momentum of the system is $I_f\omega_f$. Note that $I_f < I_i$ because the weights are now closer to the axis of rotation, reducing the moment of inertia. Since the net external torque on the system is zero, angular momentum is constant, and so $I_i\omega_i = I_f\omega_f$. Therefore, $\omega_f > \omega_i$.

As in the previous example, the kinetic energy of the system increases as the weights are pulled inward. The increase in rotational energy arises from the fact that the student does work in pulling the weights toward the axis of rotation.

EXAMPLE 11.15 The Spinning Bicycle Wheel

In another favorite classroom demonstration, a student holds the axle of a spinning bicycle wheel while seated on a pivoted stool (Fig. 11.18). The student and stool are initially at rest while the wheel is spinning in a horizontal plane with an initial angular momentum $\mathbf{L}_0$ pointing upward. Explain what happens when the wheel is inverted about its center by 180°.

FIGURE 11.18 (Example 11.15) The wheel is initially spinning when the student is at rest. What happens when the wheel is inverted?

Solution In this situation, the system consists of the student, wheel, and stool. Initially, the total angular momentum of the system is $\mathbf{L}_0$, corresponding to the contribution from the spinning wheel. As the wheel is inverted, a torque is supplied by the student, but this is internal to the system. There is no external torque acting on the system about the vertical axis. Therefore, *the angular momentum of the system about the vertical axis is constant.*

Initially, we have

$$\mathbf{L}_{\text{system}} = \mathbf{L}_0 \qquad \text{(upward)}$$

After the wheel is inverted,

$$\mathbf{L}_{\text{system}} = \mathbf{L}_0 = \mathbf{L}_{\text{student + stool}} + \mathbf{L}_{\text{wheel}}$$

In this case, $\mathbf{L}_{\text{wheel}} = -\mathbf{L}_0$ because the wheel is now rotating in the opposite sense. Therefore

$$\mathbf{L}_0 = \mathbf{L}_{\text{student + stool}} - \mathbf{L}_0$$

$$\mathbf{L}_{\text{student + stool}} = 2\mathbf{L}_0$$

This result tells us that, as the wheel is inverted, the student and stool will start to turn, acquiring an angular momentum having a magnitude twice that of the spinning wheel and directed upward.

This student is holding the axle of a spinning bicycle wheel while seated on a pivoted stool. The student and stool are initially at rest while the wheel is spinning in a horizontal plane. When the wheel is inverted about its center by 180°, the student and stool begin to rotate because angular momentum is conserved. *(Courtesy of Central Scientific Co.)*

---

*11.6 THE MOTION OF GYROSCOPES AND TOPS

A very unusual and fascinating type of motion that you probably have observed is that of a top spinning about its axis of symmetry as in Fig. 11.19a. If the top spins about its axis very rapidly, the axis will rotate about the vertical direction as indicated, thereby sweeping out a cone. The motion of the axis of the top about the

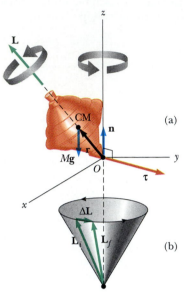

(a)

(b)

FIGURE 11.19 Precessional motion of a top spinning about its axis of symmetry. The only external forces acting on the top are the normal force **n**, and the force of gravity, *M***g**. The direction of the angular momentum, **L**, is along the axis of symmetry.

vertical, known as **precessional motion,** is usually slow compared with the spin motion of the top. It is quite natural to wonder why the top doesn't fall over. Since the center of mass is not directly above the pivot point *O*, there is clearly a net torque acting on the top about *O* due to the weight force *M***g**. From this description, it is easy to see that the top would certainly fall if it were not spinning. However, because the top is spinning, it has an angular momentum **L** directed along its axis of symmetry. As we shall show, the motion of the rotation axis about the *z* axis (the precessional motion) arises from the fact that the torque produces a change in the *direction* of the rotation axis. This is an excellent example of the importance of the directional nature of angular momentum.

The two forces acting on the top are the downward force of gravity, *M***g**, and the normal force, **n**, acting upward at the pivot point *O*. The normal force produces no torque about the pivot since its moment arm is zero. However, the force of gravity produces a torque $\boldsymbol{\tau} = \mathbf{r} \times M\mathbf{g}$ about *O*, where the direction of $\boldsymbol{\tau}$ is perpendicular to the plane formed by **r** and *M***g**. By necessity, the vector $\boldsymbol{\tau}$ lies in a horizontal plane perpendicular to the angular momentum vector. The net torque and angular momentum of the body are related through Equation 11.19:

$$\boldsymbol{\tau} = \frac{d\mathbf{L}}{dt}$$

From this expression, we see that the nonzero torque produces a change in angular momentum $d\mathbf{L}$, which is in the same direction as $\boldsymbol{\tau}$. Therefore, like the torque vector, $d\mathbf{L}$ must also be at right angles to **L**. Figure 11.19b illustrates the resulting precessional motion of the axis of the top. In a time Δt, the change in angular momentum $\Delta \mathbf{L} = \mathbf{L}_f - \mathbf{L}_i = \boldsymbol{\tau} \Delta t$. Because $\Delta \mathbf{L}$ is perpendicular to **L**, the magnitude of **L** doesn't change ($|\mathbf{L}_i| = |\mathbf{L}_f|$). Rather, what is changing is the *direction* of **L**. Since the change in angular momentum is in the direction of $\boldsymbol{\tau}$, which lies in the *xy* plane, the top undergoes precessional motion. Thus, the effect of the torque is to deflect the angular momentum of the top in a direction perpendicular to its spin axis.

The essential features of precessional motion can be illustrated by considering the simple gyroscope shown in Fig. 11.20a. This device consists of a wheel free to spin about an axle that is pivoted at a distance *h* from the center of mass of the wheel. When given an angular velocity $\boldsymbol{\omega}$ about its axis, the wheel will have a spin angular momentum $\mathbf{L} = I\boldsymbol{\omega}$ directed along the axle as shown. Let us consider the torque acting on the wheel about the pivot *O*. Again, the force **n** of the support on the axle produces no torque about *O*. On the other hand, the weight *M***g** produces a torque of magnitude *Mgh* about *O*. The direction of this torque is perpendicular to the axle (and perpendicular to **L**), as shown in Figure 11.20. This torque causes the angular momentum to change in the direction perpendicular to the axle. Hence, the axle moves in the direction of the torque, that is, in the horizontal plane.

There is an assumption we must make in order to simplify the description of the system. The total angular momentum of the precessing wheel is the sum of the spin angular momentum, $I\boldsymbol{\omega}$, and the angular momentum due to the motion of the center of mass about the pivot. In our treatment, we shall neglect the contribution from the center-of-mass motion and take the total angular momentum to be just $I\boldsymbol{\omega}$. In practice, this is a good approximation if $\boldsymbol{\omega}$ is made very large.

In a time *dt* the torque due to the weight force adds to the system angular momentum equal to $dL = \tau \, dt = (Mgh) \, dt$. When added vectorially to the original total angular momentum, $I\boldsymbol{\omega}$, this additional angular momentum causes a shift in the direction of the total angular momentum.

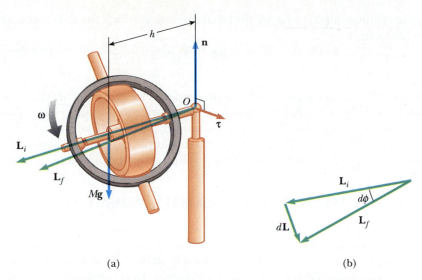

(a)   (b)

FIGURE 11.20   (a) The motion of a simple gyroscope pivoted a distance h from its center of mass. Note that the weight Mg produces about the pivot a torque that is perpendicular to the axle. (b) This torque results in a change in angular momentum $d\mathbf{L}$ in the direction perpendicular to the axle. The axle sweeps out an angle $d\phi$ in a time dt.

The vector diagram in Fig. 11.20b shows that in the time dt, the angular momentum vector rotates through an angle $d\phi$, which is also the angle through which the axle rotates. From the vector triangle formed by the vectors $\mathbf{L}_i$, $\mathbf{L}_f$, and $d\mathbf{L}$, we see that

$$d\phi = \frac{dL}{L} = \frac{(Mgh)\,dt}{L}$$

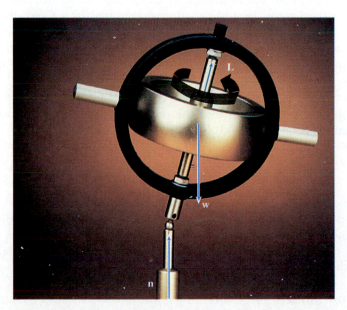

This toy gyroscope undergoes precessional motion about the vertical axis as it spins about its axis of symmetry. The only forces acting on it are the force of gravity, w, and the upward force of the pivot, n. The direction of its angular momentum, L, is along the axis of symmetry.   *(Courtesy of Central Scientific Company)*

Using the relationship $L = I\omega$, we find that the rate at which the axle rotates about the vertical axis is

$$\omega_p = \frac{d\phi}{dt} = \frac{Mgh}{I\omega} \qquad (11.28)$$

The angular frequency ω_p is called the **precessional frequency**. This result is valid only when $\omega_p \ll \omega$. Otherwise, a much more complicated motion is involved. As you can see from Equation 11.28, the condition that $\omega_p \ll \omega$ is met when $I\omega$ is large compared with Mgh. Furthermore, note that the precessional frequency decreases as ω increases, that is, as the wheel spins faster about its axis of symmetry.

*11.7 ANGULAR MOMENTUM AS A FUNDAMENTAL QUANTITY

We have seen that the concept of angular momentum is very useful for describing the motion of macroscopic systems. However, the concept is also valid on a submicroscopic scale and has been used extensively in the development of modern theories of atomic, molecular, and nuclear physics. In these developments, it was found that the angular momentum of a system is a fundamental quantity. The word *fundamental* in this context implies that angular momentum is an intrinsic property of atoms, molecules, and their constituents.

In order to explain the results of a variety of experiments on atomic and molecular systems, it is necessary to assign discrete values to the angular momentum. These discrete values are some multiple of a fundamental unit of angular momentum, which equals $\hbar = h/2\pi$, where h is called Planck's constant:

$$\text{Fundamental unit of angular momentum} = \hbar = 1.054 \times 10^{-34} \frac{\text{kg} \cdot \text{m}^2}{\text{s}^2}$$

Let us accept this postulate without proof for the time being and show how it can be used to estimate the frequency of a diatomic molecule. Consider the O_2 molecule as a rigid rotor, that is, two atoms separated by a fixed distance d and rotating about the center of mass (Fig. 11.21). Equating the rotational angular momentum to the fundamental unit $\hbar$, we can estimate the lowest frequency:

$$I_{CM}\omega \approx \hbar \qquad \text{or} \qquad \omega \approx \frac{\hbar}{I_{CM}}$$

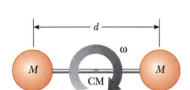

FIGURE 11.21 The rigid-rotor model of the diatomic molecule. The rotation occurs about the center of mass in the plane of the diagram.

In Example 10.4, we found that the moment of inertia of the O_2 molecule about this axis of rotation is 2.03×10^{-46} kg·m². Therefore,

$$\omega \approx \frac{\hbar}{I_{CM}} = \frac{1.054 \times 10^{-34} \text{ kg} \cdot \text{m}^2/\text{s}}{1.95 \times 10^{-46} \text{ kg} \cdot \text{m}^2} = 5.41 \times 10^{11} \text{ rad/s}$$

Actual frequencies are a multiple of this smallest possible value.

This simple example shows that certain classical concepts and models might be useful in describing some features of atomic and molecular systems. However, a wide variety of phenomena on the submicroscopic scale can be explained only if we assume discrete values of the angular momentum associated with a particular type of motion.

The Danish physicist Niels Bohr (1885–1962) accepted and adapted this radical idea of discrete momentum values in his theory of the hydrogen atom. Strictly classical models were unsuccessful in describing many properties of the hydrogen atom. Bohr postulated that the electron could occupy only those circular orbits about the proton for which the orbital angular momentum was equal to $n\hbar$, where

n is an integer. That is, he made the bold assumption that the orbital angular momentum is quantized. From this simple model, the rotational frequencies of the electron in the various orbits can be estimated (Problem 35).

SUMMARY

The **total kinetic energy** of a rigid body, such as a cylinder, that is rolling on a rough surface without slipping equals the rotational kinetic energy about its center of mass, $\frac{1}{2}I_{CM}\omega^2$, plus the translational kinetic energy of the center of mass, $\frac{1}{2}Mv_{CM}{}^2$:

$$K = \frac{1}{2}I_{CM}\omega^2 + \frac{1}{2}Mv_{CM}{}^2 \tag{11.4}$$

In this expression, v_{CM} is the velocity of the center of mass and $v_{CM} = R\omega$ for pure rolling motion.

The **torque** τ due to a force $\mathbf{F}$ about an origin in an inertial frame is defined to be

$$\tau \equiv \mathbf{r} \times \mathbf{F} \tag{11.7}$$

Given two vectors $\mathbf{A}$ and $\mathbf{B}$, their **cross product** $\mathbf{A} \times \mathbf{B}$ is a vector $\mathbf{C}$ having a magnitude

$$C \equiv AB \sin \theta \tag{11.9}$$

where θ is the angle included between $\mathbf{A}$ and $\mathbf{B}$. The direction of the vector $\mathbf{C} = \mathbf{A} \times \mathbf{B}$ is perpendicular to the plane formed by $\mathbf{A}$ and $\mathbf{B}$, and its sense is determined by the right-hand rule. Some properties of the cross product include the facts that $\mathbf{A} \times \mathbf{B} = -\mathbf{B} \times \mathbf{A}$ and $\mathbf{A} \times \mathbf{A} = 0$.

The **angular momentum** $\mathbf{L}$ of a particle of linear momentum $\mathbf{p} = m\mathbf{v}$ is

$$\mathbf{L} \equiv \mathbf{r} \times \mathbf{p} = m\mathbf{r} \times \mathbf{v} \tag{11.15}$$

where $\mathbf{r}$ is the vector position of the particle relative to an origin in an inertial frame.

The **net external torque** acting on a particle or rigid body is equal to the time rate of change of its angular momentum:

$$\sum \tau_{ext} = \frac{d\mathbf{L}}{dt} \tag{11.20}$$

The z *component* of **angular momentum of a rigid body rotating about a fixed axis** (the z axis) is

$$L_z = I\omega \tag{11.21}$$

where I is the moment of inertia about the axis of rotation, and ω is its angular velocity.

The **net external torque** acting on a rigid body equals the product of its moment of inertia about the axis of rotation and its angular acceleration:

$$\sum \tau_{ext} = I\alpha \tag{11.23}$$

If the net external torque acting on a system is zero, the total angular momentum of the system is constant. Applying this **conservation of angular momentum** law to a body whose moment of inertia changes gives

$$I_i\omega_i = I_f\omega_f = \text{constant} \tag{11.27}$$

QUESTIONS

1. Is it possible to calculate the torque acting on a rigid body without specifying a center of rotation? Is the torque independent of the location of the center of rotation?
2. Is the triple product defined by $\mathbf{A} \cdot (\mathbf{B} \times \mathbf{C})$ a scalar or vector quantity? Explain why the operation $(\mathbf{A} \cdot \mathbf{B}) \times \mathbf{C}$ has no meaning.
3. In the expression for torque, $\boldsymbol{\tau} = \mathbf{r} \times \mathbf{F}$, is $\mathbf{r}$ equal to the moment arm? Explain.
4. If the torque acting on a particle about an arbitrary origin is zero, what can you say about its angular momentum about that origin?
5. Suppose that the velocity vector of a particle is completely specified. What can you conclude about the direction of its angular momentum vector with respect to the direction of motion?
6. If a single force acts on an object, and the torque due to that force is nonzero about some point, is there any other point about which the torque is zero?
7. If a system of particles is in motion, is it possible for the total angular momentum to be zero about some origin? Explain.
8. A ball is thrown in such a way that it does not spin about its own axis. Does this mean that the angular momentum is zero about an arbitrary origin? Explain.
9. Why is it easier to keep your balance on a moving bicycle than on a bicycle at rest?
10. A scientist at a hotel sought assistance from a bellhop to carry a mysterious suitcase. When the unaware bellhop rounded a corner carrying the suitcase, it suddenly moved away from him for some unknown reason. At this point, the alarmed bellhop dropped the suitcase and ran off. What do you suppose might have been in the suitcase?
11. When a cylinder rolls on a horizontal surface as in Figure 11.4, are there any points on the cylinder that have only a vertical component of velocity at some instant? If so, where are they?
12. Three objects of uniform density—a solid sphere, a solid cylinder, and a hollow cylinder—are placed at the top of an incline (Fig. 11.22). If they all are released from rest at the same elevation and roll without slipping, which reaches the bottom first? Which reaches last? You should try this at home and note that the result is independent of the masses and radii.

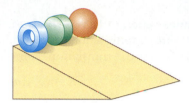

FIGURE 11.22

13. A mouse is initially at rest on a horizontal turntable mounted on a frictionless vertical axle. If the mouse begins to walk around the perimeter, what happens to the turntable? Explain.
14. Stars originate as large bodies of slowly rotating gas. Because of gravity, these regions of gas slowly decrease in size. What happens to the angular velocity of a star as it shrinks? Explain.
15. Often when a high diver wants to turn a flip in midair, she will draw her legs up against her chest. Why does this make her rotate faster? What should she do when she wants to come out of her flip?
16. As a tether ball winds around a pole, what happens to its angular velocity? Explain.
17. For a particle undergoing uniform circular motion, how are its linear momentum $\mathbf{p}$ and the angular momentum $\mathbf{L}$ oriented with respect to each other?
18. Why do tightrope walkers carry a long pole to help balance themselves?
19. Two balls have the same size and mass. One is hollow while the other is solid. How could you decide which is which without breaking them apart?
20. A particle is moving in a circle with constant speed. Locate one point about which the particle's angular momentum is constant and another about which it changes with time.

PROBLEMS

Review Problem

A rigid, massless rod has three equal masses attached to it as shown in the figure. The rod is free to rotate about a frictionless axle perpendicular to the rod through the point P and is released from rest in the horizontal position at $t = 0$. Assuming m and d are known, find (a) the moment of inertia of the system (rod plus masses) about the pivot, (b) the

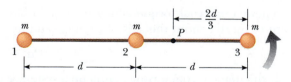

torque acting on the system at $t = 0$, (c) the angular acceleration of the system at $t = 0$, (d) the linear acceleration of the mass labeled 3 at $t = 0$, (e) the maximum kinetic energy of the system, (f) the maximum angular speed reached by the rod, (g) the maximum angular momentum of the system, and (h) the maximum speed reached by the mass labeled 2.

Section 11.1 Rolling Motion of a Rigid Body

1. A cylinder of mass 10.0 kg rolls without slipping on a horizontal surface. At the instant its center of mass has a speed of 10.0 m/s, determine (a) the translational kinetic energy of its center of mass, (b) the rotational energy about its center of mass, and (c) its total energy.

2. A solid sphere has a radius of 0.200 m and a mass of 150 kg. How much work is required to get the sphere rolling with an angular speed of 50.0 rad/s on a horizontal surface? (Assume the sphere starts from rest and rolls without slipping.)

3. (a) Determine the acceleration of the center of mass of a uniform solid disk rolling down an incline and compare this acceleration with that of a uniform hoop. (b) What is the minimum coefficient of friction required to maintain pure rolling motion for the disk?

4. A uniform solid disk and a uniform hoop are placed side by side at the top of an incline of height h. If they are released from rest and roll without slipping, determine their speeds when they reach the bottom. Which object reaches the bottom first?

5. A bowling ball has a mass of 4.0 kg, a moment of inertia of $1.6 \times 10^{-2} \text{ kg} \cdot \text{m}^2$, and a radius of 0.10 m. If it rolls down the lane without slipping at a linear speed of 4.0 m/s, what is its total energy?

5A. A bowling ball has a mass M, radius R, and a moment of inertia of $\frac{2}{5}MR^2$. If it rolls down the lane without slipping at a linear speed v, what is its total energy in terms of M and v?

6. A ring of mass 2.4 kg, inner radius 6.0 cm, and outer radius 8.0 cm is rolling (without slipping) up an inclined plane that makes an angle of $\theta = 36.9°$ with the horizontal (Fig. P11.6). At the moment the ring is $x = 2.0$ m up the plane its speed is 2.8 m/s. The ring continues up the plane for some additional distance and then rolls back down. Assuming that the plane is long enough so that the ring does not roll off the top end, how far up the plane does it go?

Section 11.2 The Vector Product and Torque

7. Two vectors are given by $\mathbf{A} = -3\mathbf{i} + 4\mathbf{j}$ and $\mathbf{B} = 2\mathbf{i} + 3\mathbf{j}$. Find (a) $\mathbf{A} \times \mathbf{B}$ and (b) the angle between $\mathbf{A}$ and $\mathbf{B}$.

8. A student claims to have found a vector $\mathbf{A}$ such that $(2\mathbf{i} - 3\mathbf{j} + 4\mathbf{k}) \times \mathbf{A} = (4\mathbf{i} + 3\mathbf{j} - \mathbf{k})$. Do you believe this claim? Explain.

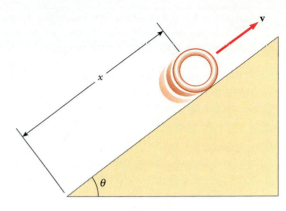

FIGURE P11.6

9. Vector $\mathbf{A}$ is in the negative y direction, and vector $\mathbf{B}$ is in the negative x direction. What are the directions of (a) $\mathbf{A} \times \mathbf{B}$ and (b) $\mathbf{B} \times \mathbf{A}$?

10. A particle is located at the vector position $\mathbf{r} = (\mathbf{i} + 3\mathbf{j})$ m, and the force acting on it equals $(3\mathbf{i} + 2\mathbf{j})$ N. What is the torque about (a) the origin and (b) the point having coordinates $(0, 6)$ m?

11. If $|\mathbf{A} \times \mathbf{B}| = \mathbf{A} \cdot \mathbf{B}$, what is the angle between $\mathbf{A}$ and $\mathbf{B}$?

12. Verify Equation 11.14 and show that the cross product may be written

$$\mathbf{A} \times \mathbf{B} = \begin{vmatrix} \mathbf{i} & \mathbf{j} & \mathbf{k} \\ A_x & A_y & A_z \\ B_x & B_y & B_z \end{vmatrix}$$

13. Two forces $\mathbf{F}_1$ and $\mathbf{F}_2$ act along the two sides of an equilateral triangle as shown in Figure P11.13. Find a third force $\mathbf{F}_3$ to be applied at B and along BC that will make the net torque about the point of intersection of the altitudes zero. Will the net torque change if $\mathbf{F}_3$ is applied not at B but at any other point along BC?

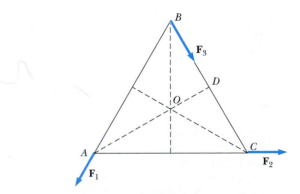

FIGURE P11.13

14. A force $\mathbf{F} = (2.0\mathbf{i} + 3.0\mathbf{j})$ N is applied to an object that is pivoted about a fixed axis aligned along the z coordinate axis. If the force is applied at the point $\mathbf{r} = (4.0\mathbf{i} + 5.0\mathbf{j} + 0\mathbf{k})$ m, find (a) the magnitude of the net torque about the z axis and (b) the direction of the torque vector $\boldsymbol{\tau}$.

Section 11.3   Angular Momentum of a Particle

15. A light rigid rod 1.00 m in length rotates in the xy plane about a pivot through the rod's center. Two particles of masses 4.00 kg and 3.00 kg are connected to its ends (Fig. P11.15). Determine the angular momentum of the system about the origin at the instant the speed of each particle is 5.00 m/s.

15A. A light rigid rod of length d rotates in the xy plane about a pivot through the rod's center. Two particles of masses m_1 and m_2 are connected to its ends (Fig. P11.15). Determine the angular momentum of the system about the origin at the instant the speed of each particle is v.

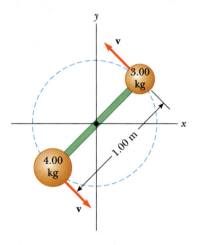

FIGURE P11.15

16. At a certain instant the position of a stone in a sling is given by $\mathbf{r} = (1.7\mathbf{i})$ m. The linear momentum $\mathbf{p}$ of the stone is $(12\mathbf{j})$ kg·m/s. Calculate its angular momentum $\mathbf{L} = \mathbf{r} \times \mathbf{p}$.

17. The position vector of a particle of mass 2.0 kg is given as a function of time by $\mathbf{r} = (6.0\mathbf{i} + 5.0t\mathbf{j})$ m. Determine the angular momentum of the particle as a function of time.

18. A conical pendulum consists of a bob of mass m moving in a circular path in a horizontal plane as shown in Figure P11.18. During the motion, the supporting wire of length ℓ maintains a constant angle θ with the vertical. Show that the magnitude of the angular momentum of the bob about the support point is

$$L = \sqrt{\frac{m^2 g \ell^3 \sin^4 \theta}{\cos \theta}}$$

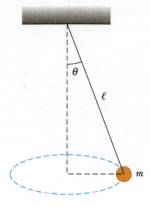

FIGURE P11.18

19. A particle of mass m moves in a circle of radius R at a constant speed v, as shown in Figure P11.19. If the motion begins at point Q, determine the angular momentum of the particle about point P as a function of time.

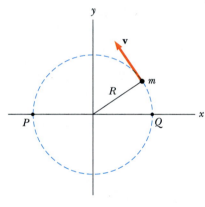

FIGURE P11.19

20. An airplane of mass 12 000 kg flies level to the ground at an altitude of 10.0 km with a constant speed of 175 m/s relative to the Earth. (a) What is the magnitude of the airplane's angular momentum relative to a ground observer directly below the airplane? (b) Does this value change as the airplane continues its motion along a straight line?

21. A ball having mass m is fastened at the end of a flagpole that is connected to the side of a tall building at point P shown in Figure P11.21. The length of the flagpole is ℓ, and it makes an angle θ with the horizontal. If the ball becomes loose and starts to fall, determine its angular momentum as a function of time about P. Neglect air resistance.

22. A 4.0-kg mass is attached to a light cord, which is wound around a pulley (Fig. 10.18). The pulley is a uniform solid cylinder of radius 8.0 cm and mass 2.0 kg. (a) What is the net torque on the system

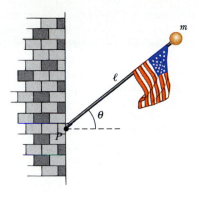

FIGURE P11.21

about the point O? (b) When the mass has a speed v, the pulley has an angular speed $\omega = v/R$. Determine the total angular momentum of the system about O. (c) Using the fact that $\tau = dL/dt$ and your result from (b), calculate the acceleration of the mass.

23. A particle of mass m is shot with an initial velocity $\mathbf{v}_0$ making an angle θ with the horizontal as shown in Figure P11.23. The particle moves in the gravitational field of the Earth. Find the angular momentum of the particle about the origin when the particle is (a) at the origin, (b) at the highest point of its trajectory, and (c) just before it hits the ground. (d) What torque causes its angular momentum to change?

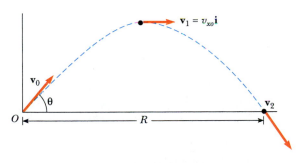

FIGURE P11.23

Section 11.4 Rotation of a Rigid Body About a Fixed Axis

24. A uniform solid disk of mass 3.00 kg and radius 0.200 m rotates about a fixed axis perpendicular to its face. If the angular frequency of rotation is 6.00 rad/s, calculate the angular momentum of the disk when the axis of rotation (a) passes through its center of mass and (b) passes through a point midway between the center and the rim.

24A. A uniform solid disk of mass M and radius R rotates about a fixed axis perpendicular to its face. If the angular frequency of rotation is ω, calculate the angular momentum of the disk when the axis of

rotation (a) passes through its center of mass and (b) passes through a point midway between the center and the rim.

25. A particle of mass 0.400 kg is attached to the 100-cm mark of a meter stick of mass 0.100 kg. The meter stick rotates on a horizontal, frictionless table with an angular speed of 4.00 rad/s. Calculate the angular momentum of the system when the stick is pivoted about an axis (a) perpendicular to the table through the 50.0-cm mark and (b) perpendicular to the table through the 0-cm mark.

26. The hour and minute hands of Big Ben in London are 2.7 m and 4.5 m long and have masses of 60 kg and 100 kg, respectively. Calculate their total angular momentum about the center point. Treat the hands as long, thin rods.

Section 11.5 Conservation of Angular Momentum

27. A cylinder for which the moment of inertia is I_1 rotates about a vertical, frictionless axle with angular velocity ω_0. A second cylinder, this one having moment of inertia I_2 and initially not rotating, drops onto the first cylinder (Fig. P11.27). Since the surfaces are not frictionless, the two eventually reach the same angular velocity ω. (a) Calculate ω. (b) Show that energy is lost in this situation and calculate the ratio of the final to the initial rotational energy.

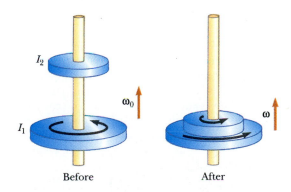

Before   After

FIGURE P11.27

28. A merry-go-round of radius $R = 2.0$ m has a moment of inertia $I = 250$ kg·m² and is rotating at 10 rev/min. A 25-kg child jumps onto the edge of the merry-go-round. What is the new angular speed of the merry-go-round?

29. A 60-kg woman stands at the rim of a horizontal turntable having a moment of inertia of 500 kg·m² and a radius of 2.00 m. The turntable is initially at rest and is free to rotate about a frictionless, vertical axle through its center. The woman then starts walking around the rim clockwise (as viewed from above the system) at a constant speed of 1.50 m/s relative to

the Earth. (a) In what direction and with what angular speed does the turntable rotate? (b) How much work does the woman do to set the turntable into motion?

30. A uniform rod of mass 100 g and length 50.0 cm rotates in a horizontal plane about a fixed, vertical, frictionless pin through its center. Two small beads, each of mass 30.0 g, are mounted on the rod such that they are able to slide without friction along its length. Initially the beads are held by catches at positions 10.0 cm on each side of center, at which time the system rotates at an angular speed of 20.0 rad/s. Suddenly, the catches are released and the small beads slide outward along the rod. Find (a) the angular speed of the system at the instant the beads reach the ends of the rod and (b) the angular speed of the rod after the beads fly off the ends.

30A. A uniform rod of mass M and length d rotates in a horizontal plane about a fixed, vertical, frictionless pin through its center. Two small beads, each of mass m, are mounted on the rod such that they are able to slide without friction along its length. Initially the beads are held by catches at positions x (where $x < d/2$) on each side of center, at which time the system rotates at an angular speed ω. Suddenly, the catches are released and the small beads slide outward along the rod. Find (a) the angular speed of the system at the instant the beads reach the ends of the rod and (b) the angular speed of the rod after the beads fly off the ends.

31. The student in Figure 11.17 holds two weights, each of mass 10.0 kg. When his arms are extended horizontally, the weights are 1.00 m from the axis of rotation and he rotates with an angular speed of 2.00 rad/s. The moment of inertia of student plus stool is 8.00 kg·m² and is assumed to be constant. If the student pulls the weights horizontally to 0.250 m from the rotation axis, calculate (a) the final angular speed of the system and (b) the change in the mechanical energy of the system.

32. A puck of mass 80.0 g and radius 4.00 cm slides along an air table at 1.5 m/s as shown in Figure P11.32a. It makes a glancing collision with a second puck of radius 6.00 cm and mass 120 g (initially at

rest) such that their rims just touch. The pucks stick together and spin after the collision (Fig. P11.32b). What are (a) the angular momentum of the system relative to the center of mass and (b) its angular speed about the center of mass?

33. A wooden block of mass M resting on a frictionless horizontal surface is attached to a rigid rod of length ℓ and of negligible mass (Fig. P11.33). The rod is pivoted at the other end. A bullet of mass m traveling parallel to the horizontal surface and normal to the rod with speed v hits the block and gets embedded in it. (a) What is the angular momentum of the bullet-block system? (b) What fraction of the original kinetic energy is lost in the collision?

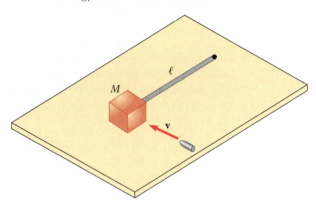

FIGURE P11.33

34. A wheel-shaped space station has a radius of 100 m and a moment of inertia of 5.00×10^8 kg·m². A crew of 150 live on the rim, and the station is rotated so that they experience an apparent gravity of g (Fig. P11.34). If 100 people move to the center, the angular rotation speed changes. What apparent gravity is experienced by those remaining at the rim? Assume an average mass of 65.0 kg per crew member.

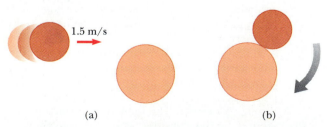

(a)   (b)

FIGURE P11.32

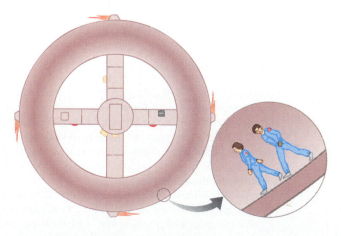

FIGURE P11.34

*Section 11.7   Angular Momentum as a Fundamental Quantity

35. In the Bohr model of the hydrogen atom, the electron moves in a circular orbit of radius 0.529×10^{-10} m around the proton. Assuming the orbital angular momentum of the electron is equal to h, calculate (a) the orbital speed of the electron, (b) the kinetic energy of the electron, and (c) the angular frequency of the electron's motion.

ADDITIONAL PROBLEMS

36. A uniform solid sphere of radius r is placed on the inside surface of a hemispherical bowl of radius R. The sphere is released from rest at an angle θ to the vertical and rolls without slipping (Fig. P11.36). Determine the angular speed of the sphere when it reaches the bottom of the bowl.

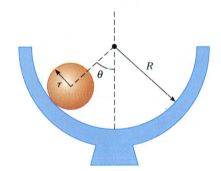

FIGURE P11.36

37. Halley's comet moves about the Sun in an elliptical orbit, with its closest approach to the Sun being 0.59 AU and its greatest distance from the Sun being 35 AU (1 AU = the Earth-Sun distance). If the comet's speed at closest approach is 54 km/s, what is its speed when it is farthest from the Sun, assuming that its angular momentum about the Sun is constant?

38. A thin uniform cylindrical turntable of radius 2.00 m and mass 30.0 kg rotates in a horizontal plane with an initial angular speed of 4π rad/s. The turntable bearing is frictionless. A small clump of clay of mass 0.250 kg is dropped onto the turntable and sticks at a point 1.80 m from the center of rotation. (a) Find the final angular speed of the clay and turntable. (Treat the clay as a point mass.) (b) Is mechanical energy constant in this collision? Explain and use numerical results to verify your answer.

39. A string is wound around a uniform disk of radius R and mass M. The disk is released from rest with the string vertical and its top end tied to a fixed support (Fig. P11.39). As the disk descends, show that (a) the tension in the string is one-third the weight of the disk, (b) the magnitude of the acceleration of the

center of mass is $2g/3$, and (c) the speed of the center of mass is $(4gh/3)^{1/2}$. Verify your answer to (c) using the energy approach.

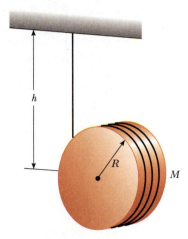

FIGURE P11.39

40. A constant horizontal force **F** is applied to a lawn roller in the form of a uniform solid cylinder of radius R and mass M (Fig. P11.40). If the roller rolls without slipping on the horizontal surface, show that (a) the acceleration of the center of mass is $2F/3M$ and (b) the minimum coefficient of friction necessary to prevent slipping is $F/3Mg$. (*Hint:* Take the torque with respect to the center of mass.)

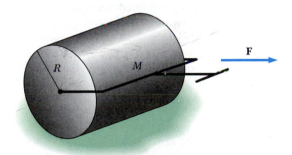

FIGURE P11.40

41. A light rope passes over a light, frictionless pulley. One end is fastened to a bunch of bananas of mass M, and a monkey of mass M clings to the other end (Fig. P11.41). The monkey climbs the rope in an attempt to reach the bananas. (a) Treating the system as consisting of the monkey, bananas, rope, and pulley, evaluate the net torque about the pulley axis. (b) Using the results to (a), determine the total angular momentum about the pulley axis and describe the motion of the system. Will the monkey reach the bananas?

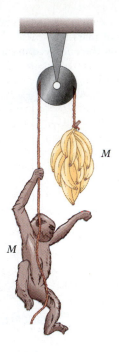

FIGURE P11.41

42. A small, solid sphere of mass m and radius r rolls without slipping along the track shown in Figure P11.42. If it starts from rest at the top of the track at a height h, where h is large compared to r, (a) what is the minimum value of h (in terms of the radius of the loop R) such that the sphere completes the loop? (b) What are the force components on the sphere at the point P if $h = 3R$?

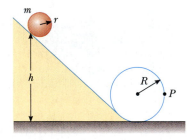

FIGURE P11.42

43. This problem describes a method of determining the moment of inertia of an irregularly shaped object such as the payload for a satellite. Figure P11.43 shows one method of determining I experimentally. A mass m is suspended by a cord wound around the inner shaft (radius r) of a turntable supporting the object. When the mass is released from rest, it descends uniformly a distance h, acquiring a speed v. Show that the moment of inertia I of the equipment (including the turntable) is $mr^2(2gh/v^2 - 1)$.

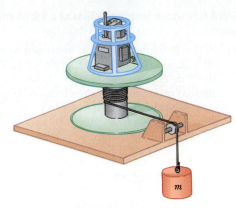

FIGURE P11.43

44. Consider the problem of the solid sphere rolling down an incline as described in Example 11.1. (a) Choose the instantaneous axis through the contact point P as the axis of the origin for the torque equation and show that the acceleration of the center of mass is $a_{\mathrm{CM}} = \frac{5}{7}g \sin \theta$. (b) Show that the minimum coefficient of friction such that the sphere rolls without slipping is $\mu_{\min} = \frac{2}{7} \tan \theta$.

45. A common physics demonstration (Fig. P11.45) consists of a ball resting r meters away from the hinged end of a board of length ℓ that is elevated at an angle θ with the horizontal. A cup attached to the board at r_c is to catch the ball when the support stick is suddenly removed. Show that (a) the ball, placed at the very end, will lag behind the stick in falling when $\theta < 35.3°$ and (b) the cup should be placed at

$$r_c = \frac{2\ell}{3 \cos \theta}$$

for this limiting angle. (c) If a ball is at the end of a 1.0-m stick at this critical angle, show that the cup must be 18.4 cm from the hinged end.

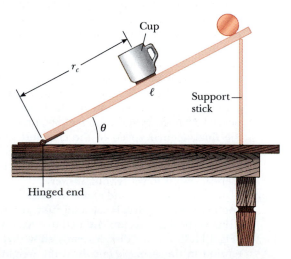

FIGURE P11.45

46. A bowling ball is both sliding and spinning on a horizontal surface such that its rotational kinetic energy equals its translational kinetic energy. What is the ratio of the ball's center-of-mass speed to the tangential speed of a point on its surface?

47. A constant torque of 25.0 N·m is applied to a grindstone for which the moment of inertia is 0.130 kg·m². Using energy principles, find the angular speed after the grindstone has made 15.0 rev. (Neglect friction.)

48. A projectile of mass m moves to the right with speed v_0 (Fig. P11.48a). The projectile strikes and sticks to the end of a stationary rod of mass M and length d that is pivoted about a frictionless axle through its center (Fig. P11.48b). (a) Find the angular speed of the system right after the collision. (b) Determine the fractional loss in mechanical energy due to the collision.

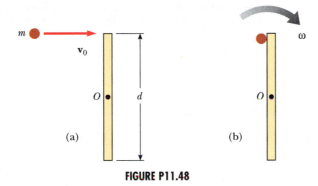

FIGURE P11.48

49. A mass m is attached to a cord passing through a small hole in a frictionless, horizontal surface (Fig. P11.49). The mass is initially orbiting with speed v_0 in a circle of radius r_0. The cord is then slowly pulled from below, decreasing the radius of the circle to r. (a) What is the speed of the mass when the radius is r? (b) Find the tension in the cord as a function of r. (c) How much work W is done in moving m from r_0 to r? (*Note:* The tension depends on r.) (d) Obtain nu-

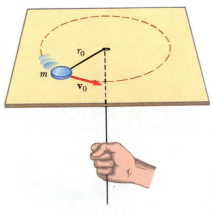

FIGURE P11.49

merical values for v, T, and W when $r = 0.100$ m, $m = 50.0$ g, $r_0 = 0.300$ m, and $v_0 = 1.50$ m/s.

50. A bowling ball is given an initial speed v_0 on an alley such that it initially slides without rolling. The coefficient of friction between ball and alley is μ. Show that at the time pure rolling motion occurs, (a) the speed of the ball's center of mass is $5v_0/7$ and (b) the distance it has traveled is $12v_0^2/49\mu g$. (*Hint:* When pure rolling motion occurs, $v_{CM} = R\omega$. Since the frictional force provides the deceleration, from Newton's second law it follows that $a_{CM} = \mu g$.)

51. A trailer with loaded weight w is being pulled by a vehicle with a force **F**, as in Figure P11.51. The trailer is loaded such that its center of mass is located as shown. Neglect the force of rolling friction and assume the trailer has an acceleration of magnitude a. (a) Find the vertical component of **F** in terms of the given parameters. (b) If $a = 2.00$ m/s² and $h = 1.50$ m, what must be the value of d in order that $F_y = 0$ (no vertical load on the vehicle)? (c) Find F_x and F_y given that $w = 1500$ N, $d = 0.800$ m, $L = 3.00$ m, $h = 1.50$ m, and $a = -2.00$ m/s².

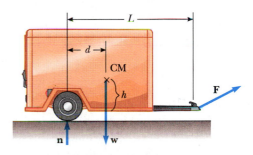

FIGURE P11.51

52. (a) A thin rod of length h and mass M is held vertically with its lower end resting on a frictionless horizontal surface. The rod is then let go to fall freely. Determine the speed of its center of mass just before it hits the horizontal surface. (b) Suppose the rod were pivoted at its lower end. Determine the speed of the rod's center of mass just before it hits the surface.

53. Two astronauts (Fig. P11.53), each having a mass of 75 kg, are connected by a 10-m rope of negligible mass. They are isolated in space, orbiting their center of mass at speeds of 5.0 m/s. Calculate (a) the magnitude of the angular momentum of the system by treating the astronauts as particles and (b) the rotational energy of the system. By pulling on the rope, the astronauts shorten the distance between them to 5.0 m. (c) What is the new angular momentum of the system? (d) What are their new speeds? (e) What is the new rotational energy of the system? (f) How much work is done by the astronauts in shortening the rope?

53A. Two astronauts (Fig. P11.53), each having a mass M, are connected by a rope of length d having negligible mass. They are isolated in space, orbiting their center of mass at speeds v. Calculate (a) the magnitude of the angular momentum of the system by treating the astronauts as particles and (b) the rotational energy of the system. By pulling on the rope, the astronauts shorten the distance between them to $d/2$. (c) What is the new angular momentum of the system? (d) What are their new speeds? (e) What is the new rotational energy of the system? (f) How much work is done by the astronauts in shortening the rope?

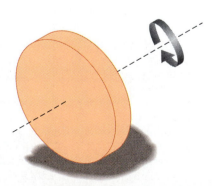

FIGURE P11.53

54. A solid cube of wood of side $2a$ and mass M is resting on a horizontal suface. The cube is constrained to rotate about an axis AB (Fig. P11.54). A bullet of mass m and speed v is shot at the face opposite $ABCD$ at a height of $4a/3$. The bullet gets embedded in the cube. Find the minimum value of v required to tip the cube so that it falls on face $ABCD$. Assume $m < M$.

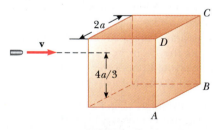

FIGURE P11.54

55. Toppling chimneys often break apart in mid-fall because the mortar between the bricks cannot withstand much tension force. As the chimney falls, this tension supplies the centripetal forces on the topmost segments that they need to keep them traveling in an arc. For simplicity, let us model the chimney as a uniform rod of length ℓ pivoted at the lower end. The rod starts at rest in a vertical position (with the pivot at the bottom) and falls over under the influ-

ence of gravity. What fraction of the length of the rod has a tangential acceleration greater than $g \sin \theta$, where θ is the angle the chimney makes with the vertical axis?

56. A solid sphere is positioned at the top of an incline that makes an angle θ with the horizontal. This initial position of the sphere is a vertical distance h above the ground. The sphere is released and moves down the incline. Calculate the speed of the sphere when it reaches the bottom of the incline in the case where (a) it rolls without slipping and (b) it slips frictionlessly without rolling. Compare the times required to reach the bottom in cases (a) and (b).

57. A spool of wire of mass M and radius R is unwound under a constant force **F** (Fig. P11.57). Assuming the spool is a uniform solid cylinder that doesn't slip, show that (a) the acceleration of the center of mass is $4F/3M$ and (b) the force of friction is to the *right* and equal in magnitude to $F/3$. (c) If the cylinder starts from rest and rolls without slipping, what is the speed of its center of mass after it has rolled through a distance d?

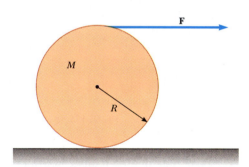

FIGURE P11.57

58. A uniform solid disk is set into rotation with an angular speed ω_0 about an axis through its center. While still rotating at this speed, the disk is placed into contact with a horizontal surface and released as in Figure P11.58. (a) What is the angular speed of the disk

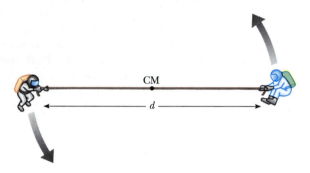

FIGURE P11.58

once pure rolling takes place? (b) Find the fractional loss in kinetic energy from the time the disk is released until pure rolling occurs. (*Hint:* Consider torques about the center of mass.)

59. Suppose a solid disk of radius R is given an angular speed ω_0 about an axis through its center and then lowered to a horizontal surface and released, as in Problem 58 (Fig. P11.58). Furthermore, assume that the coefficient of friction between disk and surface is μ. (a) Show that the time it takes pure rolling motion to occur is $R\omega_0/3\mu g$. (b) Show that the distance the disk travels before pure rolling occurs is $R^2\omega_0^2/18\mu g$.

60. A large, cylindrical roll of tissue paper of initial radius R lies on a long, horizontal surface with the open end of the paper nailed to the surface. The roll is given a slight shove ($v_0 \approx 0$) and commences to unroll. (a) Determine the speed of the center of mass of the roll when its radius has diminished to r. (b) Calculate a numerical value for this speed at $r = 1.0$ mm, assuming $R = 6.0$ m. (c) What happens to the energy of the system when the paper is completely unrolled? (*Hint:* Assume the roll has a uniform density and apply energy methods.)

61. A solid cube of side $2a$ and mass M is sliding on a frictionless surface with uniform velocity $\mathbf{v}_0$ as in Figure P11.61a. It hits a small obstacle at the end of the table, which causes the cube to tilt as in Figure P11.61b. Find the minimum value of $\mathbf{v}_0$ such that the cube falls off the table. Note that the moment of inertia of the cube about an axis along one of its edges is $8Ma^2/3$. (*Hint:* The cube undergoes an inelastic collision at the edge.)

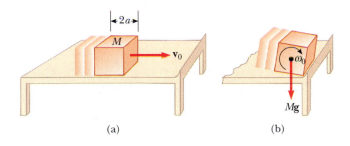

(a)   (b)

FIGURE P11.61

 62. In a demonstration known as the ballistics cart, a ball is projected vertically upward from a cart moving with constant velocity along the horizontal direction. The ball lands in the catching cup of the cart because both the cart and ball have the same horizontal component of velocity. Consider a ballistics cart moving on an incline making an angle θ with the horizontal as in Figure P11.62. The cart (including wheels) has a mass M and the moment of inertia of each of the two wheels is $mR^2/2$. (a) Using conservation of energy (assuming no friction between cart and axle),

and assuming pure rolling motion (no slipping), show that the acceleration of the cart along the incline is

$$a_x = \left(\frac{M}{M + 2m}\right) g\sin\theta$$

(b) Note that the x component of acceleration of the ball released by the cart is $g\sin\theta$. Thus, the x component of the cart's acceleration is *smaller* than that of the ball by the factor $M/(M + 2m)$. Use this fact and kinematic equations to show that the ball overshoots the cart by an amount Δx, where

$$\Delta x = \left(\frac{4m}{M + 2m}\right)\left(\frac{\sin\theta}{\cos^2\theta}\right)\frac{v_{y0}^2}{g}$$

and v_{y0} is the initial speed of the ball imparted to it by the spring in the cart. (c) Show that the distance d that the ball travels measured along the incline is

$$d = \frac{2v_{y0}^2}{g}\frac{\sin\theta}{\cos^2\theta}$$

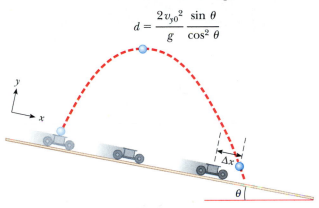

FIGURE P11.62

63. A spool of wire rests on a horizontal surface as in Figure P11.63, and, when pulled, does not slip at the contact point P. The spool is pulled in the directions indicated by the vectors $\mathbf{F}_1$, $\mathbf{F}_2$, $\mathbf{F}_3$, and $\mathbf{F}_4$. For each force, determine the direction the spool rolls. Note that the line of action of $\mathbf{F}_2$ passes through P.

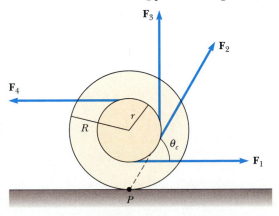

FIGURE P11.63

64. The spool shown in Figure P11.63 has inner radius r and outer radius R. The angle θ between the applied force and the horizontal can be varied. Show that the critical angle for which the spool does not roll and remains stationary is given by $\cos \theta_c = r/R$. (*Hint:* At the critical angle, the line of action of the applied force passes through the contact point.)

65. A plank having mass $M = 6.0$ kg rides on top of two identical solid cylindrical rollers each having radius $R = 5.0$ cm and mass $m = 2.0$ kg (Fig. P11.65). The plank is pulled by a constant horizontal force $F = 6.0$ N applied to its end and perpendicular to the axes of the cylinders (which are parallel). The cylin-ders roll without slipping on a flat surface. There is also no slipping between cylinders and plank. (a) Find the acceleration of the plank and of the rollers. (b) What frictional forces are acting?

FIGURE P11.65

Static Equilibrium and Elasticity

The Chinese acrobats in this difficult formation represent a balanced system. The external forces acting on the system, as shown by the blue vectors, are the weights of the acrobats, $\mathbf{w}_1$ and $\mathbf{w}_2$, and the upward force of the support, $\mathbf{n}$, on the lower acrobat. The vector sum of these external forces must be zero in such a balanced system. The net external torque acting on such a balanced system must also be zero. *(J.P. Lafont, Sygma)*

I n Chapters 10 and 11 we studied the dynamics of a rigid object, that is, one whose parts remain at a fixed separation with respect to each other when subjected to external forces. Part of this chapter is concerned with the conditions under which a rigid object is in equilibrium. The term *equilibrium* implies either that the object is at rest or that its center of mass moves with constant velocity. We deal here only with the former, which are referred to as objects in *static equilibrium*. Static equilibrium represents a common situation in engineering practice, and the principles involved are of special interest to civil engineers, architects, and mechanical engineers. Those of you who are engineering students will undoubtedly take an intensified course in statics in the future.

In Chapter 5 we stated that one necessary condition for equilibrium is that the net force on an object be zero. If the object is treated as a particle, this is the only condition that must be satisfied for equilibrium. That is, if the net force on the particle is zero, the particle remains at rest (if originally at rest) or moves with constant velocity (if originally in motion).

The situation with real (extended) objects is more complex when the objects cannot be treated as particles. In order for an extended object to be in static equilibrium, the net force on it must be zero *and* the net torque about any origin must be zero. In order to establish whether or not an object is in equilibrium, we must know its size and shape, the forces acting on different parts of it, and the points of application of the various forces.

The last section of this chapter deals with the realistic situation of objects that deform under load conditions. Such deformations are usually elastic in nature and do not affect the conditions of equilibrium. By *elastic* we mean that when the deforming forces are removed, the object returns to its original shape. Several elastic constants are defined, each corresponding to a different type of deformation.

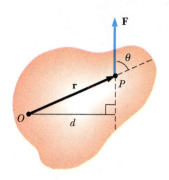

FIGURE 12.1   A single force **F** acts on a rigid object at the point *P*. The moment arm of **F** relative to *O* is the perpendicular distance *d* from **O** to the line of action of **F**.

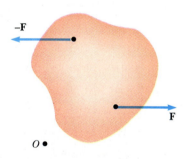

FIGURE 12.2   The two forces acting on the object are equal in magnitude and opposite in direction, yet the object is not in equilibrium.

Equivalent forces

12.1   THE CONDITIONS OF EQUILIBRIUM OF A RIGID OBJECT

Consider a single force **F** acting on a rigid object as in Figure 12.1. The effect of the force depends on its point of application, *P*. If **r** is the position vector of this point relative to *O*, the torque associated with the force **F** about *O* is given by Equation 11.7:

$$\boldsymbol{\tau} = \mathbf{r} \times \mathbf{F}$$

Recall from Section 11.2 that the vector $\boldsymbol{\tau}$ is perpendicular to the plane formed by **r** and **F**. Furthermore, the sense of $\boldsymbol{\tau}$ is determined by the sense of the rotation that **F** tends to give to the object. The right-hand rule can be used to determine the direction of $\boldsymbol{\tau}$: Close your right hand such that your four fingers wrap in the direction of rotation that **F** tends to give the object; your thumb then points in the direction of $\boldsymbol{\tau}$. Hence, in Figure 12.1, $\boldsymbol{\tau}$ is directed out of the paper.

As you can see from Figure 12.1, the tendency of **F** to make the object rotate about an axis through *O* depends on the moment arm *d* as well as on the magnitude of **F**. By definition, the magnitude of $\boldsymbol{\tau}$ is *Fd*.

Now suppose either of two forces, $\mathbf{F}_1$ and $\mathbf{F}_2$, act on a rigid object. The two forces will have the same effect on the object only if they have the same magnitude, the same direction, and the same line of action. In other words,

two forces $\mathbf{F}_1$ and $\mathbf{F}_2$ are equivalent if and only if $F_1 = F_2$ and if the two produce the same torque about any given point.

Two equal and opposite forces that are *not* equivalent are shown in Figure 12.2. The force directed to the right tends to rotate the object clockwise about an axis perpendicular to the diagram through *O*, whereas the force directed to the left tends to rotate it counterclockwise about that axis.

When pivoted about an axis through its center of mass, an object undergoes an angular acceleration about this axis if there is a nonzero torque acting. As an example, suppose an object is pivoted about an axis through its center of mass as in

Figure 12.3. Two equal and opposite forces act in the directions shown, such that their lines of action do not pass through the center of mass. A pair of forces acting in this manner form what is called a **couple**. (The two forces shown in Figure 12.2 also form a couple.) Since each force produces the same torque, *Fd*, the net torque has a magnitude 2*Fd*. Clearly, the object rotates clockwise and undergoes an angular acceleration about the axis. This is a nonequilibrium situation as far as the rotational motion is concerned. That is, the "unbalanced," or net, torque on the object gives rise to an angular acceleration α according to the relationship $\tau_{net} = 2Fd = I\alpha$ (Eq. 10.20).

In general, an object is in rotational equilibrium only if its angular acceleration $\alpha = 0$. Since $\tau_{net} = I\alpha$ for rotation about a fixed axis, a necessary condition of equilibrium is that *the net torque about any origin must be zero*. We now have *two necessary conditions for equilibrium of an object*, which can be stated as follows:

- The resultant external force must equal zero.

$$\sum F = 0 \tag{12.1}$$

- The resultant external torque must be zero about *any* origin.

$$\sum \tau = 0 \tag{12.2}$$

The first condition is a statement of translational equilibrium; it tells us that the linear acceleration of the center of mass of the object must be zero when viewed from an inertial reference frame. The second condition is a statement of rotational equilibrium and tells us that the angular acceleration about any axis must be zero. In the special case of **static equilibrium**, which is the main subject of this chapter, the object is at rest so that it has no linear or angular speed (that is, $v_{CM} = 0$ and $\omega = 0$).

The two vector expressions given by Equations 12.1 and 12.2 are equivalent, in general, to six scalar equations, three from the first condition of equilibrium, and three from the second (corresponding to x, y, and z components). Hence, in a complex system involving several forces acting in various directions, you would be faced with solving a set of equations with many unknowns. Here, we restrict our discussion to situations in which all the forces lie in the *xy* plane. (Forces whose vector representations are in the same plane are said to be *coplanar*.) With this restriction, we shall have to deal with only three scalar equations. Two of these come from balancing the forces in the x and y directions. The third comes from the torque equation, namely, that the net torque about *any* point in the *xy* plane must be zero. Hence, the two conditions of equilibrium provide the equations

$$\sum F_x = 0 \qquad \sum F_y = 0 \qquad \sum \tau_z = 0 \tag{12.3}$$

where the axis of the torque equation is arbitrary, as we show later.

There are two cases of equilibrium that are often encountered. The first deals with an object subjected to only two forces, and the second is concerned with an object subjected to three forces.

Case I *An object subjected to two forces is in equilibrium if and only if the two forces are equal in magnitude, opposite in direction, and have the same line of action.* Figure 12.4a shows a situation in which the object is not in equilibrium because the two forces are not along the same line. Note that the torque about any axis, such as one through *P*, is not zero, which violates the second condition of equilibrium. In Figure 12.4b, the object is in equilibrium because the forces have the same line of action. In this situation, it is easy to see that the net torque about any axis is zero.

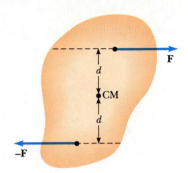

FIGURE 12.3 Two equal and opposite forces acting on the object form a couple. In this case, the object rotates clockwise. The net torque about the center of mass is 2*Fd*.

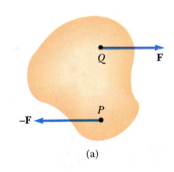

(a)

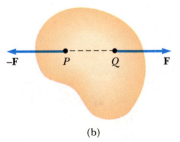

(b)

FIGURE 12.4 (a) The object is not in equilibrium because the two forces do not have the same line of action. (b) The object is in equilibrium because the two forces act along the same line.

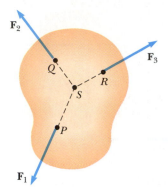

FIGURE 12.5 If three forces act on an object in equilibrium, their lines of action must intersect at a point S (or they must be parallel to each other).

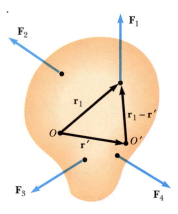

FIGURE 12.6 Construction for showing that if the net torque about origin O is zero, the net torque about any other origin, such as O', is also zero.

Case II *If an object subjected to three forces is in equilibrium, the lines of action of the three forces must intersect at a common point.* That is, the forces must be *concurrent.* (One exception to this rule is the situation in which none of the lines of action intersect. In this situation, the forces must all be parallel to each other and lie in the same plane.) Figure 12.5 illustrates the general rule. The lines of action of the three forces pass through the point S. The conditions of equilibrium require that $F_1 + F_2 + F_3 = 0$ and that the net torque about any axis be zero. Note that as long as the forces are concurrent, the net torque about an axis through S must be zero.

We can easily show that, regardless of the number of forces acting, if an object is in translational equilibrium and if the net torque is zero with respect to one point, it must also be zero about any other point. The point can be inside or outside the boundaries of the object. Consider an object under the action of several forces such that the resultant force $\Sigma F = F_1 + F_2 + F_3 + \cdots = 0$. Figure 12.6 describes this situation (for clarity, only four forces are shown). The point of application of F_1 relative to O is specified by the position vector r_1. Similarly, the points of application of $F_2, F_3, \ldots$ are specified by $r_2, r_3, \ldots$ (not shown). The net torque about O is

$$\sum \tau_0 = r_1 \times F_1 + r_2 \times F_2 + r_3 \times F_3 + \cdots$$

Now consider another arbitrary point, O', having a position vector r' relative to O. The point of application of F_1 relative to O' is identified by the vector $r_1 - r'$. Likewise, the point of application of F_2 relative to O' is $r_2 - r'$, and so forth. Therefore, the torque about O' is

$$\Sigma \tau_{O'} = (r_1 - r') \times F_1 + (r_2 - r') \times F_2 + (r_3 - r') \times F_3 + \cdots$$
$$= r_1 \times F_1 + r_2 \times F_2 + r_3 \times F_3 + \cdots - r' \times (F_1 + F_2 + F_3 + \cdots)$$

Since the net force is assumed to be zero, the last term in this last expression vanishes and we see that the torque about O' is equal to the torque about O. Hence,

> if an object is in translational equilibrium and the net torque is zero about one point, it is also zero about any other point.

12.2 MORE ON THE CENTER OF GRAVITY

Whenever we deal with rigid objects, one of the forces we must consider is the weight of the object, that is, the force of gravity acting on it. In order to compute the torque due to the weight force, all of the weight can be considered as being concentrated at a single point called the *center of gravity.* As we shall see, the center of gravity of an object coincides with its center of mass if the object is in a uniform gravitational field.

Consider an object of arbitrary shape lying in the xy plane, as in Figure 12.7. Suppose the object is divided into a large number of very small particles of masses $m_1, m_2, m_3, \ldots$ having coordinates $(x_1, y_1), (x_2, y_2), (x_3, y_3), \ldots$ In Chapter 9 we defined the x coordinate of the center of mass of such an object to be

$$x_{CM} = \frac{m_1 x_1 + m_2 x_2 + m_3 x_3 + \cdots}{m_1 + m_2 + m_3 + \cdots} = \frac{\Sigma m_i x_i}{\Sigma m_i}$$

The y coordinate of the center of mass is similar to this, with x_{CM} replaced by y_{CM}.

Let us now examine the situation from another point of view by considering the weight of each part of the object, as in Figure 12.8. Each particle contributes a torque about the origin equal to the particle's weight multiplied by its moment arm. For example, the torque due to the weight $m_1 g_1$ is $m_1 g_1 x_1$, and so forth. We wish to locate the center of gravity, the one position of the single force w (the total weight of the object) whose effect on the rotation of the object is the same as that of the individual particles. Equating the torque exerted by w acting at the center of gravity to the sum of the torques acting on the individual particles gives

$$(m_1 g_1 + m_2 g_2 + m_3 g_3 + \cdots)x_{cg} = m_1 g_1 x_1 + m_2 g_2 x_2 + m_3 g_3 x_3 + \cdots$$

This expression accounts for the fact that the gravitational field strength, g, can in general vary over the object. If we assume that g is uniform over the object (as is usually the case), then the g terms cancel and we get

$$x_{cg} = \frac{m_1 x_1 + m_2 x_2 + m_3 x_3 + \cdots}{m_1 + m_2 + m_3 + \cdots} \qquad (12.4)$$

In other words, *the center of gravity is located at the center of mass as long as the object is in a uniform gravitational field.*

In several examples presented in the next section, we shall be concerned with homogeneous, symmetric objects for which the center of gravity coincides with the geometric center of the object. A rigid object in a uniform gravitational field can be balanced by a single force equal in magnitude to the weight of the object, as long as the force is directed upward through the center of gravity.

12.3 EXAMPLES OF RIGID OBJECTS IN STATIC EQUILIBRIUM

In working static equilibrium problems, it is important to recognize external forces acting on the object. Failure to do so will result in an incorrect analysis. The following procedure is recommended when analyzing an object in equilibrium under the action of several external forces:

Problem-Solving Strategy
Objects in Equilibrium

- Make a sketch of the object under consideration.
- Draw a free-body diagram and label all external forces acting on the object. Try to guess the correct direction for each force. If you select a direction that leads to a negative sign in your solution for a force, do not be alarmed; this merely means that the direction of the force is the opposite of what you guessed.
- Resolve all forces into rectangular components, choosing a convenient coordinate system. Then apply the first condition for equilibrium. Remember to keep track of the signs of the various force components.
- Choose a convenient axis for calculating the net torque on the object. Remember that the choice of the origin for the torque equation is arbitrary; therefore, choose an origin that will simplify your calculation as much as possible. Becoming adept at this is a matter of practice.
- The first and second conditions of equilibrium give a set of linear equations with several unknowns. All that is left is to solve the simultaneous equations for the unknowns in terms of the known quantities.

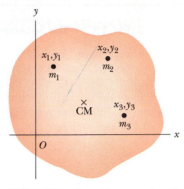

FIGURE 12.7 An object can be divided into many small particles each having a specific mass and coordinates. These particles can be used to locate the center of mass.

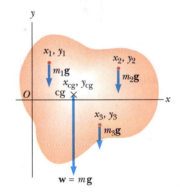

FIGURE 12.8 The center of gravity of the object is located at the center of mass if the value of **g** is constant over the object.

EXAMPLE 12.1   The Seesaw

A uniform board of weight 40.0 N supports two children weighing 500 N and 350 N, respectively, as shown in Figure 12.9. If the support (called the *fulcrum*) is under the center of gravity of the board and if the 500-N child is 1.50 m from the center, (a) determine the upward force **n** exerted on the board by the support.

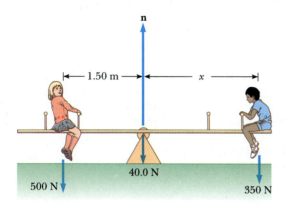

FIGURE 12.9   (Example 12.1) A balanced system.

Solution   First note that, in addition to **n**, the external forces acting on the board are the weights of the children and

the weight of the board, all of which act downward. We can assume that the center of gravity of the board is at its geometric center because we were told that the board is uniform. Since the system is in equilibrium, the upward force **n** must balance all the downward forces. From $\Sigma F_y = 0$, we have

$$n - 500 \text{ N} - 350 \text{ N} - 40.0 \text{ N} = 0 \quad \text{or} \quad n = \boxed{890 \text{ N}}$$

(The equation $\Sigma F_x = 0$ also applies to this situation, but it is unnecessary to consider this equation because we have no forces acting horizontally on the board.)

(b) Determine where the 350-N child should sit to balance the system.

Solution   To find this position, we must invoke the second condition for equilibrium. Taking the center of gravity of the board as the axis for our torque equation, we see from $\Sigma \tau = 0$ that

$$(500 \text{ N})(1.50 \text{ m}) - (350 \text{ N})x = 0$$

$$x = \boxed{2.14 \text{ m}}$$

Exercise   If the fulcrum did not lie under the center of gravity of the board, what other information would you need to solve the problem?

EXAMPLE 12.2   A Weighted Hand

A 50.0-N weight is held in the hand with the forearm in the horizontal position, as in Figure 12.10a. The biceps muscle is attached 3.00 cm from the joint, and the weight is 35.0 cm from the joint. Find the upward force that the biceps exerts on the forearm and the downward force exerted by the upper arm on the forearm and acting at the joint. Neglect the weight of the forearm.

Solution   The forces acting on the forearm are equivalent to those acting on a bar as shown in Figure 12.10b, where **F** is the upward force exerted by the biceps and **R** is the downward force exerted by the upper arm at the joint. From the first condition for equilibrium, we have

$$(1) \qquad \Sigma F_y = F - R - 50.0 \text{ N} = 0$$

From the second condition for equilibrium, we know that the sum of the torques about any point must be zero. With the joint O as the axis, we have

$$Fd - w\ell = 0$$

$$F(3.00 \text{ cm}) - (50.0 \text{ N})(35.0 \text{ cm}) = 0$$

$$F = \boxed{583 \text{ N}}$$

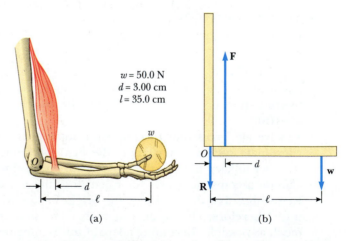

FIGURE 12.10   (Example 12.2) (a) In this skeletal view, the bicep muscle is pulling upward with a force **F** (essentially) at right angles to the forearm. (b) The mechanical model for the system described in (a).

This value for F can be substituted into (1) to give $R = 533$ N. Hence, the forces at joints and in muscles can be extremely large.

Exercise In reality, the biceps makes an angle of 15.0° with the vertical, so that $\mathbf{F}$ has both a vertical and a horizontal component. Find the value of $\mathbf{F}$ and the components of $\mathbf{R}$ including this fact in your analysis.

Answer $F = 604$ N, $R_x = 156$ N, $R_y = 533$ N.

EXAMPLE 12.3 Standing on a Horizontal Beam

A uniform horizontal beam of length 8.00 m and weight 200 N is attached to a wall by a pin connection. Its far end is supported by a cable that makes an angle of 53.0° with the horizontal (Fig. 12.11a). If a 600-N person stands 2.00 m from the wall, find the tension in the cable and the force exerted by the wall on the beam.

Solution First we must identify all the external forces acting on the beam. These are its weight, the force $\mathbf{T}$ exerted by the cable, the force $\mathbf{R}$ exerted by the wall at the pivot (the direction of this force is unknown), and the weight of the person on the beam. These are all indicated in the free-body diagram for the beam (Fig. 12.11b). If we resolve $\mathbf{T}$ and $\mathbf{R}$ into horizontal and vertical components and apply the first condition for equilibrium, we get

(1) $\qquad \sum F_x = R\cos\theta - T\cos 53.0° = 0$

(2) $\qquad \sum F_y = R\sin\theta + T\sin 53.0° - 600\text{ N} - 200\text{ N} = 0$

Because R, T, and θ are all unknown, we cannot obtain a solution from these expressions alone. (The number of simultaneous equations must equal the number of unknowns in order for us to be able to solve for the unknowns.)

Now let us invoke the condition for rotational equilibrium. A convenient axis to choose for our torque equation is the one that passes through the pivot at O. The feature that makes this point so convenient is that the force $\mathbf{R}$ and the horizontal component of $\mathbf{T}$ both have a lever arm of zero and, hence, zero torque about this pivot. Recalling our convention for the sign of the torque about an axis and noting that the lever arms of the 600-N, 200-N, and $T\sin 53°$ forces are 2.00 m, 4.00 m, and 8.00 m, respectively, we get

$$\sum \tau_O = (T\sin 53.0°)(8.00\text{ m}) - (600\text{ N})(2.00\text{ m})$$
$$- (200\text{ N})(4.00\text{ m}) = 0$$

$$T = \boxed{313\text{ N}}$$

Thus the torque equation with this axis gives us one of the unknowns directly! This value is substituted into (1) and (2) to give

$$R\cos\theta = 188\text{ N}$$

$$R\sin\theta = 550\text{ N}$$

We divide these two equations and recall the trigonometric

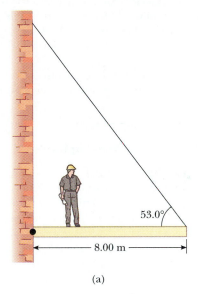

(a)

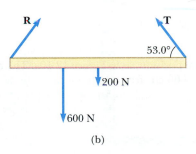

(b)

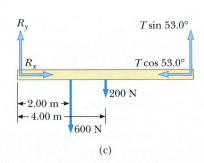

(c)

FIGURE 12.11 (Example 12.3) (a) A uniform beam supported by a cable. (b) The free-body diagram for the beam.

identity $\sin \theta / \cos \theta = \tan \theta$ to get

$$\tan \theta = \frac{550 \text{ N}}{188 \text{ N}} = 2.93$$

$$\theta = 71.1°$$

Finally,

$$R = \frac{188 \text{ N}}{\cos \theta} = \frac{188 \text{ N}}{\cos 71.1°} = \boxed{581 \text{ N}}$$

If we had selected some other axis for the torque equation, the solution would have been the same. For example, if we had chosen to have the axis pass through the center of gravity of the beam, the torque equation would involve both T and R. However, this equation, coupled with (1) and (2), could still be solved for the unknowns. Try it!

When many forces are involved in a problem of this nature, it is convenient to set up a table of forces, lever arms, and torques. For instance, in the example just given, we would construct the following table. Setting the sum of the terms in the last column equal to zero represents the condition of rotational equilibrium.

Force Component	Lever Arm Relative to O (m)	Torque About O (N·m)
$T \sin 53.0°$	8.00	$(8.00) T \sin 53°$
$T \cos 53.0°$	0	0
200 N	4.00	$-(4.00)(200)$
600 N	2.00	$-(2.00)(600)$
$R \sin \theta$	0	0
$R \cos \theta$	0	0

EXAMPLE 12.4   The Leaning Ladder

A uniform ladder of length ℓ and weight $w = 50$ N rests against a smooth, vertical wall (Fig. 12.12a). If the coefficient of static friction between ladder and ground is $\mu_s = 0.40$, find the minimum angle θ_{min} such that the ladder does not slip.

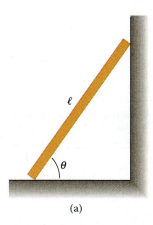

(a)

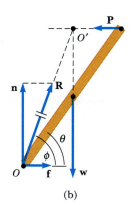

(b)

FIGURE 12.12   (Example 12.4) (a) A uniform ladder at rest, leaning against a smooth wall. The floor is rough. (b) The free-body diagram for the ladder. Note that the forces **R**, **w**, and **P** pass through a common point O'.

Solution   The free-body diagram showing all the external forces acting on the ladder is illustrated in Figure 12.12b. The reaction **R** exerted by the ground on the ladder is the vector sum of a normal force, **n**, and the force of friction, **f**. The reaction force **P** exerted by the wall on the ladder is horizontal, since the wall is frictionless. From the first condition of equilibrium applied to the ladder, we have

$$\sum F_x = f - P = 0$$

$$\sum F_y = n - w = 0$$

Since $w = 50$ N, we see from the second equation here that $n = w = 50$ N. Furthermore, *when the ladder is on the verge of slipping, the force of friction must be a maximum*, given by $f_{max} = \mu_s n = 0.40(50 \text{ N}) = 20$ N. (Recall Eq. 5.9: $f_s \leq \mu_s n$.) Thus, at this angle, $P = 20$ N.

To find θ, we must use the second condition of equilibrium. When the torques are taken about the origin O at the bottom of the ladder, we get

$$\sum \tau_O = P\ell \sin \theta - w \frac{\ell}{2} \cos \theta = 0$$

But $P = 20$ N when the ladder is about to slip and $w = 50$ N, so that this expression gives

$$\tan \theta_{min} = \frac{w}{2P} = \frac{50 \text{ N}}{40 \text{ N}} = 1.25$$

$$\theta_{min} = \boxed{51°}$$

It is interesting to note that the result does not depend on ℓ or w. The answer depends only on μ_s.

An alternative approach to analyzing this problem is to consider the intersection O' of the lines of action of forces **w** and **P**. Since the torque about any origin must be zero, the torque about O' must be zero. This requires that the line of action of **R** (the resultant of **n** and **f**) pass through O'. That is, since three forces act on this stationary object, the forces must be concurrent. With this condition, one could then obtain the angle ϕ that **R** makes with the horizontal (where ϕ is greater than θ), assuming the length of the ladder is known.

Exercise   For the angles labeled in Figure 12.12, show that $\tan \phi = 2 \tan \theta$.

EXAMPLE 12.5   Raising a Cylinder

A cylinder of weight w and radius R is to be raised onto a step of height h as shown in Figure 12.13. A rope is wrapped around the cylinder and pulled horizontally. Assuming the cylinder doesn't slip on the step, find the minimum force $\mathbf{F}$ necessary to raise the cylinder and the reaction force at P exerted by the step on the cylinder.

Solution   When the cylinder is just about to be raised, the reaction force at Q goes to zero. Hence, at this time there are only three forces on the cylinder, as shown in Figure 12.13b. From the dotted triangle in Figure 12.13a, we see that the moment arm d of the weight relative to the point P is

$$d = \sqrt{R^2 - (R - h)^2} = \sqrt{2Rh - h^2}$$

The moment arm of $\mathbf{F}$ relative to P is $2R - h$. Therefore, the net torque acting on the cylinder about P is

$$wd - F(2R - h) = 0$$

$$w\sqrt{2Rh - h^2} - F(2R - h) = 0$$

$$\boxed{F = \frac{w\sqrt{2Rh - h^2}}{2R - h}}$$

We can determine the components of $\mathbf{n}$ by using the first condition of equilibrium:

$$\sum F_x = F - n\cos\theta = 0$$

$$\sum F_y = n\sin\theta - w = 0$$

Dividing gives

$$(1) \qquad \tan\theta = \frac{w}{F}$$

and solving for n gives

$$(2) \qquad \boxed{n = \sqrt{w^2 + F^2}}$$

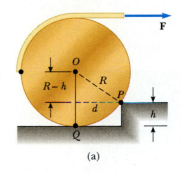

(a)

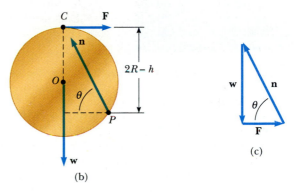

(b)

(c)

FIGURE 12.13   (Example 12.5) (a) A cylinder of weight w being pulled by a force $\mathbf{F}$ over a step. (b) The free-body diagram for the cylinder when it is just about to be raised. (c) The vector sum of the three external forces is zero.

Exercise   Solve this problem by noting that the three forces acting on the cylinder are concurrent and therefore all pass through the point C. The three forces form the sides of the triangle shown in Figure 12.13c.

12.4   ELASTIC PROPERTIES OF SOLIDS

In our study of mechanics thus far, we have assumed that objects remain undeformed when external forces act on them. In reality, all objects are deformable. That is, it is possible to change the shape or size of an object (or both) by applying external forces. Although these changes are observed as large-scale deformations, the internal forces that resist the deformation are due to short-range forces between atoms.

We shall discuss the deformation of solids in terms of the concepts of stress and strain. **Stress** is a quantity that is proportional to the force causing a deformation; more specifically, stress is the external force per unit cross-sectional area acting on an object. **Strain** is a measure of the degree of deformation. It is found that, for sufficiently small stresses, the stress is proportional to the strain; the constant of proportionality depends on the material being deformed and on the nature of the deformation. We call this proportionality constant the **elastic modulus.** The elastic modulus is therefore the ratio of stress to strain:

Elastic modulus

$$\text{Elastic modulus} \equiv \frac{\text{stress}}{\text{strain}} \qquad (12.5)$$

We consider three types of deformation and define an elastic modulus for each:

- **Young's modulus,** which measures the resistance of a solid to a change in its length.
- **Shear modulus,** which measures the resistance to motion of the planes of a solid sliding past each other
- **Bulk modulus,** which measures the resistance that solids or liquids offer to changes in their volume

A plastic model of an arch structure under load conditions observed between two crossed polarizers. The stress pattern is produced in those regions where the stresses are greatest. Such patterns and models are useful in finding the optimum design for architectural components. *(Peter Aprahamian/Science Photo Library)*

Young's Modulus: Elasticity in Length

Consider a long bar of cross-sectional area A and length L_0 that is clamped at one end (Fig. 12.14). When an external force **F** is applied along the bar and perpendicular to the cross-section, internal forces in the bar resist distortion (''stretching''), but the bar attains an equilibrium in which its length is greater than L_0 and in which the external force is exactly balanced by internal forces. In such a situation, the bar is said to be stressed. We define the **tensile stress** as the ratio of the magnitude of the external force F to the cross-sectional area A. The **tensile strain** in this case is defined as the ratio of the change in length, ΔL, to the original length, L_0, and is therefore a dimensionless quantity. Thus, we can use Equation 12.6 to define **Young's modulus:**

Young's modulus

$$Y \equiv \frac{\text{tensile stress}}{\text{tensile strain}} = \frac{F/A}{\Delta L/L_0} \qquad (12.6)$$

This quantity is typically used to characterize a rod or wire stressed under either tension or compression. Note that because strain is a dimensionless quantity, Y has units of force per unit area. Typical values are given in Table 12.1. Experiments

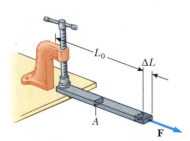

FIGURE 12.14 A long bar clamped at one end is stretched by an amount ΔL under the action of a force **F**.

TABLE 12.1   Typical Values for Elastic Modulus

Substance	Young's Modulus (N/m²)	Shear Modulus (N/m²)	Bulk Modulus (N/m²)
Aluminum	7.0×10^{10}	2.5×10^{10}	7.0×10^{10}
Brass	9.1×10^{10}	3.5×10^{10}	6.1×10^{10}
Copper	11×10^{10}	4.2×10^{10}	14×10^{10}
Steel	20×10^{10}	8.4×10^{10}	16×10^{10}
Tungsten	35×10^{10}	14×10^{10}	20×10^{10}
Glass	$6.5-7.8 \times 10^{10}$	$2.6-3.2 \times 10^{10}$	$5.0-5.5 \times 10^{10}$
Quartz	5.6×10^{10}	2.6×10^{10}	2.7×10^{10}
Water	—	—	0.21×10^{10}
Mercury	—	—	2.8×10^{10}

show that (a) for a fixed applied force, the change in length is proportional to the original length and (b) the force necessary to produce a given strain is proportional to the cross-sectional area. Both of these observations are in accord with Equation 12.6.

The **elastic limit** of a substance is defined as the maximum stress that can be applied to the substance before it becomes permanently deformed. It is possible to exceed the elastic limit of a substance by applying a sufficiently large stress (Fig. 12.15). When the stress exceeds the elastic limit, the object is permanently distorted and does not return to its original shape after the stress is removed. Hence, the shape of the object is permanently changed. As the stress is increased even further, the material will ultimately break.

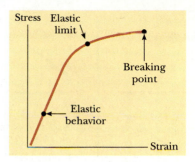

FIGURE 12.15 Stress-versus-strain curve for an elastic solid.

Shear Modulus: Elasticity of Shape

Another type of deformation occurs when an object is subjected to a force **F** tangential to one of its faces while the opposite face is held fixed by a force such as the force of friction, **f**$_s$ (Fig. 12.16a). The stress in this case is called a shear stress. If the object is originally a rectangular block, a shear stress results in a shape whose cross-section is a parallelogram. As the stress increases, the stress-strain curve is no longer a straight line. A book pushed sideways as in Figure 12.16b is an example of an object under a shear stress. There is no change in volume under this deformation to a first approximation (for small distortions).

We define the **shear stress** as F/A, the ratio of the tangential force to the area, A, of the face being sheared. The **shear strain** is defined as the ratio $\Delta x/h$, where Δx is the horizontal distance the sheared face moves and h is the height of the object. In terms of these quantities, the **shear modulus** is

$$S \equiv \frac{\text{shear stress}}{\text{shear strain}} = \frac{F/A}{\Delta x/h} \tag{12.7}$$

Values of the shear modulus for some representative materials are given in Table 12.1. The units of shear modulus are force per unit area.

Bulk Modulus: Volume Elasticity

Bulk modulus characterizes the response of a substance to uniform squeezing. Suppose that the external forces acting on an object are at right angles to all its faces (Fig. 12.17) and distributed uniformly over all the faces. As we shall see in Chapter 15, such uniformly distributed forces occur when an object is immersed in a fluid. An object subject to this type of deformation undergoes a change in volume but no change in shape. The **volume stress**, ΔP, is defined as the ratio of the magnitude of the normal force, F, to the area, A. The quantity $\Delta P = F/A$ is called the **pressure**. The volume strain is equal to the change in volume, ΔV, divided by the original volume, V. Thus, from Equation 12.6 we can characterize a volume compression in terms of the **bulk modulus**, defined as

$$B \equiv \frac{\text{volume stress}}{\text{volume strain}} = -\frac{F/A}{\Delta V/V} = -\frac{\Delta P}{\Delta V/V} \tag{12.8}$$

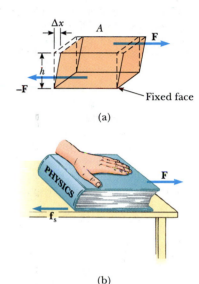

FIGURE 12.16 (a) A shear deformation in which a rectangular block is distorted by two forces of equal magnitude but opposite directions applied to two parallel faces. (b) A book under shear stress.

Bulk modulus

A negative sign is inserted in this defining equation so that B is a positive number. This maneuver is necessary because an increase in pressure (positive ΔP) causes a decrease in volume (negative ΔV) and vice versa.

Table 12.1 lists bulk moduli for some materials. If you look up such values in a different source, you will often find that the reciprocal of the bulk modulus is

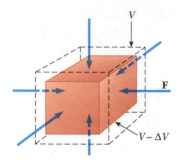

FIGURE 12.17 When a solid is under uniform pressure, it undergoes a change in volume but no change in shape. This cube is compressed on all sides by forces normal to its six faces.

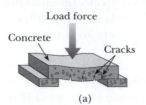
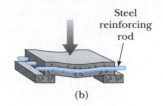
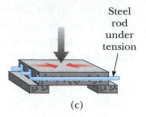

FIGURE 12.18 (a) A concrete slab with no reinforcement tends to crack under a heavy load. (b) The strength of the concrete is increased by using steel tensile reinforcement rods. (c) The concrete is further strengthened by prestressing it with steel rods under tension.

listed. The reciprocal of the bulk modulus is called the **compressibility** of the material.

Note from Table 12.1 that both solids and liquids have a bulk modulus. However, there is no shear modulus and no Young's modulus for liquids because a liquid does not sustain a shearing stress or a tensile stress (it flows instead).

Prestressed Concrete

If the stress on a solid object exceeds a certain value, the object will fracture. The maximum stress that can be applied before fracture occurs depends on the nature of the material and the type of applied stress. For example, concrete has a tensile strength of about 2×10^6 N/m², a compressive strength of 20×10^6 N/m², and a shear strength of 2×10^6 N/m². If the applied stress exceeds these values, the concrete fractures. It is common practice to use large safety factors to prevent failure in concrete structures.

Concrete is normally very brittle when cast in thin sections. Thus, concrete slabs tend to sag and crack at unsupported areas, as in Figure 12.18a. The slab can be strengthened by using steel rods to reinforce the concrete, as in Figure 12.18b. Because concrete is much stronger under compression than under tension, vertical columns of concrete that are under compression can support very heavy loads, whereas horizontal beams of concrete tend to sag and crack because of their smaller shear strength. A significant increase in shear strength is achieved, however, by prestressing the reinforced concrete, as in Figure 12.18c. As the concrete is being poured, the steel rods are held under tension by external forces. The external forces are released after the concrete cures, which results in a permanent tension in the steel and hence a compressive stress on the concrete. This enables the concrete slab to support a much heavier load.

EXAMPLE 12.6   Measuring Young's Modulus

A load of 102 kg is supported by a wire of length 2.0 m and cross-sectional area 0.10 cm². The wire is stretched by 0.22 cm. Find the tensile stress, tensile strain, and Young's modulus for the wire.

Solution

$$\text{Tensile stress} = \frac{F}{A} = \frac{Mg}{A} = \frac{(102 \text{ kg})(9.80 \text{ m/s}^2)}{0.10 \times 10^{-4} \text{ m}^2}$$

$$= 1.0 \times 10^8 \text{ N/m}^2$$

$$\text{Tensile strain} = \frac{\Delta L}{L_0} = \frac{0.22 \times 10^{-2} \text{ m}}{2.0 \text{ m}}$$

$$= 0.11 \times 10^{-2}$$

$$Y = \frac{\text{tensile stress}}{\text{tensile strain}} = \frac{1.0 \times 10^8 \text{ N/m}^2}{0.11 \times 10^{-2}}$$

$$= 9.1 \times 10^{10} \text{ N/m}^2$$

Comparing this value for Y with the values in Table 12.1, we conclude that the wire is probably made of brass.

EXAMPLE 12.7 Squeezing a Lead Sphere

A solid lead sphere of volume 0.50 m³ is lowered to a depth in the ocean where the water pressure is equal to 2.0×10^7 N/m². The bulk modulus of lead is equal to 7.7×10^9 N/m². What is the change in volume of the sphere?

Solution From the definition of bulk modulus, we have

$$B = -\frac{\Delta P}{\Delta V / V}$$

$$\Delta V = -\frac{V \Delta P}{B}$$

In this case, the change in pressure, ΔP, is 2.0×10^7 N/m². (This is large relative to atmospheric pressure, 1.01×10^5 N/m².) Therefore,

$$\Delta V = -\frac{(0.50 \text{ m}^3)(2.0 \times 10^7 \text{ N/m}^2)}{7.7 \times 10^9 \text{ N/m}^2} = -1.3 \times 10^{-3} \text{ m}^3$$

The negative sign indicates a decrease in volume.

SUMMARY

A rigid object is in **equilibrium** if and only if *the resultant external force on it is zero* and *the resultant external torque on it is zero about any origin:*

$$\sum \mathbf{F} = 0 \tag{12.1}$$

$$\sum \boldsymbol{\tau} = 0 \tag{12.2}$$

The first condition is the *condition of translational equilibrium,* and the second is the *condition of rotational equilibrium.*

If two forces act on a rigid object, the object is in equilibrium if and only if the forces are equal in magnitude and opposite in direction and have the same line of action.

When three forces act on a rigid object that is in equilibrium, the three forces must be concurrent, that is, their lines of action must intersect at a common point.

The force of gravity exerted on an object can be considered to act at a single point called the **center of gravity.** The center of gravity of an object coincides with the center of mass if the object is in a uniform gravitational field.

The elastic properties of a substance can be described using the concepts of stress and strain. **Stress** is a quantity proportional to the force producing a deformation; **strain** is a measure of the degree of deformation. Stress is proportional to strain, and the constant of proportionality is the **elastic modulus:**

$$\text{Elastic modulus} \equiv \frac{\text{stress}}{\text{strain}} \tag{12.5}$$

Three common types of deformation are (1) the resistance of a solid to elongation under a load, characterized by **Young's modulus,** Y; (2) the resistance of a solid to the motion of internal planes sliding past each other, characterized by the **shear modulus,** S; (3) the resistance of a solid (or a liquid) to a volume change, characterized by the **bulk modulus,** B.

A large balanced rock at the Garden of the Gods in Colorado Springs, Colorado, an example of stable equilibrium. *(Photo by David Serway)*

QUESTIONS

1. Can a body be in equilibrium if only one external force acts on it? Explain.
2. Can a body be in equilibrium if it is in motion? Explain.
3. Locate the center of gravity for the following uniform objects: (a) sphere, (b) cube, (c) right circular cylinder.
4. The center of gravity of an object may be located outside the object. Give a few examples for which this is the case.
5. You are given an arbitrarily shaped piece of plywood, together with a hammer, nail, and plumb bob. How could

you use these items to locate the center of gravity of the plywood? (*Hint:* Use the nail to suspend the plywood.)

6. In order for a chair to be balanced on one leg, where must the center of gravity of the chair be located?

7. Give an example in which the net torque acting on an object is zero and yet the net force is nonzero.

8. Give an example in which the net force acting on an object is zero and yet the net torque is nonzero.

9. Can an object be in equilibrium if the only torques acting on it produce clockwise rotation?

10. A tall crate and a short crate of equal mass are placed side by side on an incline (without touching each other). As the incline angle is increased, which crate will topple first? Explain.

11. When lifting a heavy object, why is it recommended to keep the back as vertical as possible, lifting from the knees, rather than bending over and lifting from the waist?

12. Would you expect the center of gravity and the center of mass of the Empire State Building to coincide precisely? Explain.

13. Give several examples where several forces are acting on a system in such a way that their sum is zero but the system is not in equilibrium.

14. If you measure the net torque and the net force on a system to be zero, (a) could the system still be rotating with respect to you? (b) Could it be translating with respect to you?

15. A ladder is resting inclined against a wall. Would you feel safer climbing up the ladder if you were told that the ground is frictionless but the wall is rough or that the wall is frictionless but the ground is rough? Justify your answer.

16. What kind of deformation does a cube of Jello exhibit when it "jiggles"?

PROBLEMS

Section 12.1   The Conditions of Equilibrium of a Rigid Object

1. A baseball player holds a 36-oz bat (weight = 10.0 N) with one hand at the point O (Fig. P12.1). The bat is in equilibrium. The weight of the bat acts along a line 60 cm to the right of O. Determine the force and the torque exerted on the bat by the player.

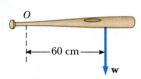

— 60 cm —

w

FIGURE P12.1

2. Write the necessary conditions of equilibrium for the body shown in Figure P12.2. Take the origin of the torque equation at the point O.

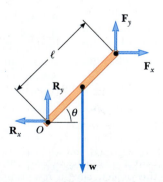

FIGURE P12.2

[3.] A uniform beam of weight w and length ℓ has weights w_1 and w_2 at two positions, as in Figure P12.3. The beam is resting at two points. For what value of x will the beam be balanced at P such that the normal force at O is zero?

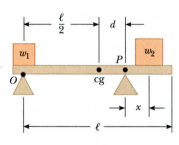

FIGURE P12.3

4. A letter "A" is formed from two uniform pieces of metal each of weight 26.0 N and length 1.00 m, hinged at the top and held together by a horizontal wire of length 1.20 m (Fig. P12.4). The structure rests on a frictionless surface. If the wire is connected at points a distance of 0.65 m from the top of the letter, determine the tension in the wire.

[5.] A ladder of weight 400 N and length 10.0 m is placed against a frictionless vertical wall. A person weighing 800 N stands on the ladder 2.00 m from the bottom as measured along the ladder. The foot of the ladder is 8.00 m from the bottom of the wall. Calculate the

☐ indicates problems that have full solutions available in the Student Solutions Manual and Study Guide.

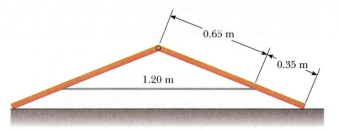

FIGURE P12.4

force exerted by the wall, and the normal force exerted by the floor on the ladder.

5A. A ladder of weight w_1 and length L is placed against a frictionless vertical wall. A person weighing w_2 stands on the ladder a distance x from the bottom as measured along the ladder. The foot of the ladder is a distance d from the bottom of the wall. Find expressions for the force exerted by the wall, and the normal force exerted by the floor on the ladder.

6. A student gets his car stuck in a snow drift. Not at a loss, having studied physics, he attaches one end of a stout rope to the vehicle and the other end to the trunk of a nearby tree, allowing for a small amount of slack. The student then exerts a force **F** on the center of the rope in the direction perpendicular to the car-tree line, as shown in Figure P12.6. If the rope is inextensible and if the magnitude of the applied force is 500 N, what is the force on the car? (Assume equilibrium conditions.)

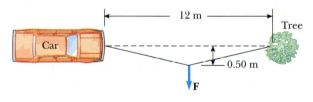

FIGURE P12.6

Section 12.2   More on the Center of Gravity

7. A circular pizza of radius R has a circular piece of radius $R/2$ removed from one side as shown in Figure P12.7. Clearly the center of gravity has moved from C to C' along the x axis. Show that the distance

FIGURE P12.7

from C to C' is $R/6$. (Assume the thickness and density of the pizza are uniform throughout.)

8. A carpenter's square has the shape of an "L," as in Figure P12.8. Locate its center of gravity.

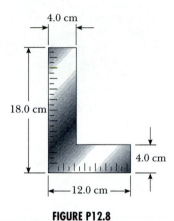

FIGURE P12.8

9. Consider the following mass distribution: 5.0 kg at (0, 0) m, 3.0 kg at (0, 4.0) m, and 4.0 kg at (3.0, 0) m. Where should a fourth mass of 8.0 kg be placed so that the center of gravity of the four-mass arrangement will be at (0, 0)?

10. Pat builds a track for his model car out of wood (Fig. P12.10). The track is 5.0 cm wide, 1.0 m high, and 3.0 m long. The runway is cut such that it forms a parabola, $y = (x - 3)^2/9$. Locate the horizontal position of the center of gravity of this track.

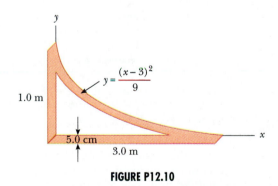

FIGURE P12.10

Section 12.3   Examples of Rigid Objects in Static Equilibrium

11. Chris is pushing his sister Nicole in a wheelbarrow when it is stopped by a brick 8.0 cm high (Fig. P12.11). The handles make an angle of 15.0° with the horizontal. A downward force of 400 N is exerted on the wheel, which has a radius of 20.0 cm. (a) What force must Chris apply along the handles in order to just start the wheel over the brick? (b) What is the force (magnitude and direction) that the brick exerts on the wheel just as the wheel begins to lift

FIGURE P12.11

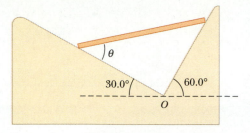

FIGURE P12.16

over the brick? Assume in both parts that the brick remains fixed and does not slide along the ground.

12. Two pans of a balance are 50.0 cm apart. The fulcrum of the balance has been shifted 1.0 cm away from the center by a dishonest shopkeeper. By what percentage is the true weight of the goods being marked up by the shopkeeper? (Assume the balance has negligible mass.)

13. A ladder having a uniform density and a mass m rests against a frictionless vertical wall at an angle of 60°. The lower end rests on a flat surface where the coefficient of static friction is $\mu_s = 0.40$. A student with a mass $M = 2m$ attempts to climb the ladder. What fraction of the length L of the ladder will the student have reached when the ladder begins to slip?

14. A uniform plank of length 6.0 m and mass 30 kg rests horizontally on a scaffold, with 1.5 m of the plank hanging over one end of the scaffold. How far can a painter of mass 70 kg walk on the overhanging part of the plank before it tips?

14A. A uniform plank of length L and mass m_1 rests horizontally on a scaffold, with a length d of the plank hanging over one end of the scaffold. How far can a painter of mass m_2 walk on the overhanging part of the plank before it tips?

15. A 1500-kg automobile has a wheel base (the distance between the axles) of 3.0 m. The center of mass of the automobile is on the center line at a point 1.2 m behind the front axle. Find the force exerted by the ground on each wheel.

16. A uniform rod of weight w and length L is supported at its ends by a frictionless trough as shown in Figure P12.16. (a) Show that the center of gravity of the rod is directly over point O when the rod is in equilibrium. (b) Determine the equilibrium value of the angle θ.

17. A cue stick strikes a cue ball and delivers a horizontal impulse in such a way that the ball rolls without slipping as it starts to move. At what height above the ball's center (in terms of the radius of the ball) was the blow struck?

18. A flexible chain weighing 40 N hangs between two hooks located at the same height (Fig. P12.18). At each hook, the tangent to the chain makes an angle $\theta = 42°$ with the horizontal. Find (a) the magnitude of the force each hook exerts on the chain and (b) the tension in the chain at its midpoint. (*Hint:* For part (b), make a free-body diagram for half the chain.)

FIGURE P12.18

19. A hemispherical sign 1.0 m in diameter and of uniform mass density is supported by two strings as shown in Figure P12.19. What fraction of the sign's weight is supported by each string?

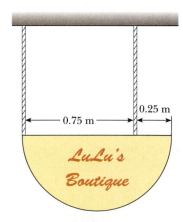

FIGURE P12.19

20. Sir Lost dons his armor and sets out from the castle on his trusty steed in his quest to rescue fair damsels from dragons (Fig. P12.20). Unfortunately his aide lowered the drawbridge too far and finally stopped it 20.0° below the horizontal. Sir Lost and his steed

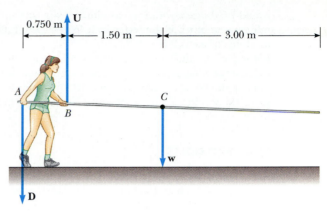

FIGURE P12.22

FIGURE P12.20

stop when their combined center of mass is 1.0 m from the end of the bridge. The bridge is 8.0 m long and has a mass of 2000 kg; the lift cable is attached to the bridge 5.0 m from the castle end and to a point 12.0 m above the bridge. Sir Lost's mass combined with his armor and steed is 1000 kg. (a) Determine the tension in the cable and (b) the horizontal and vertical force components acting on the bridge at the castle end.

21. Two identical uniform bricks of length L are placed in a stack over the edge of a horizontal surface with the maximum overhang possible without falling, as in Figure P12.21. Find the distance x.

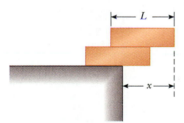

FIGURE P12.21

22. A vaulter holds a 29.4-N pole in equilibrium by exerting an upward force, **U**, with her leading hand, and a downward force, **D**, with her trailing hand, as shown in Figure P12.22. If we assume that the weight of the pole acts at its midpoint, what are the magnitudes of **U** and **D**?

Section 12.4 Elastic Properties of Solids

23. A 200-kg load is hung on a wire of length 4.0 m, cross-sectional area 0.20×10^{-4} m², and Young's modulus 8.0×10^{10} N/m². What is its increase in length?

24. A steel piano wire 1.12 m long has a cross-sectional area of 6.0×10^{-3} cm². When under a tension of 115 N, how much does it stretch?

25. Assume that Young's modulus for bone is 1.5×10^{10} N/m² and that the bone will fracture if a shear stress of more than 1.5×10^{8} N/m² is exerted. (a) What is the maximum force that can be exerted on the femur bone in the leg if it has a minimum effective diameter of 2.5 cm? (b) If this much force is applied compressively, by how much does the 25.0-cm-long bone shorten?

26. If the elastic limit of copper is 1.5×10^{8} N/m², determine the minimum diameter a copper wire can have under a load of 10 kg if its elastic limit is not to be exceeded.

27. A 2.0-m-long cylindrical steel wire with a cross-sectional diameter of 4.0 mm is placed over a frictionless pulley, with one end of the wire connected to a 5.00-kg mass and the other end connected to a 3.00-kg mass. By how much does the wire stretch while the masses are in motion?

27A. A cylindrical steel wire of length L with a cross-sectional diameter d is placed over a frictionless pulley, with one end of the wire connected to a mass m_1 and the other end connected to a mass m_2. By how much does the wire stretch while the masses are in motion?

28. Calculate the density of sea water at a depth of 1000 m where the hydraulic pressure is approximately 1.000×10^{7} N/m². (The density of sea water at the surface is 1.030×10^{3} kg/m³.)

29. If the shear stress in steel exceeds about 4.0×10^{8} N/m², the steel ruptures. Determine the shearing force necessary to (a) shear a steel bolt 1.0 cm in diameter and (b) punch a 1.0-cm-diameter hole in a 0.50-cm-thick steel plate.

30. (a) Find the minimum diameter of a steel wire 18 m long that will elongate no more than 9.0 mm when a load of 380 kg is hung on the lower end. (b) If the elastic limit for this steel is 3.0×10^{8} N/m², will permanent deformation occur with this load?

31. When water freezes, it expands about 9%. What

would be the pressure increase inside your automobile engine block if the water in it froze? (The bulk modulus of ice is $2.0 \times 10^9 \ N/m^2$.)

32. For safety in climbing, a mountaineer uses a 50-m nylon rope that is 10 mm in diameter. When supporting the 90-kg climber on one end, the rope elongates 1.6 m. Find Young's modulus for the rope material.

ADDITIONAL PROBLEMS

33. A bridge of length 50 m and mass 8.0×10^4 kg is supported at each end as in Figure P12.33. A truck of mass 3.0×10^4 kg is located 15 m from one end. What are the forces on the bridge at the points of support?

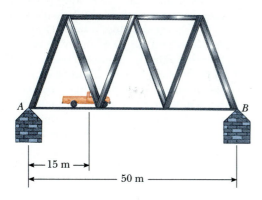

- 15 m
- 50 m

FIGURE P12.33

34. A solid sphere of radius R and mass M is placed in a wedge as shown in Figure P12.34. The inner surfaces of the wedge are frictionless. Determine the forces exerted by the wedge on the sphere at the two contact points.

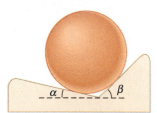

α β

FIGURE P12.34

35. A 10-kg monkey climbs up a 120-N uniform ladder of length L, as in Figure P12.35. The upper and lower ends of the ladder rest on frictionless surfaces. The lower end is fastened to the wall by a horizontal rope that can support a maximum tension of 110 N. (a) Draw a free-body diagram for the ladder. (b) Find the tension in the rope when the monkey is one third the way up the ladder. (c) Find the maximum distance d the monkey can walk up the ladder

before the rope breaks, expressing your answer as a fraction of L.

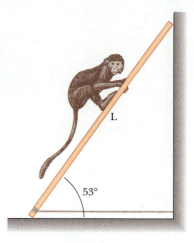

L

$53°$

FIGURE P12.35

36. A hungry bear weighing 700 N walks out on a beam in an attempt to retrieve a basket of food hanging at the end of the beam (Fig. P12.36). The beam is uniform, weighs 200 N, and is 6.00 m long; the basket weighs 80.0 N. (a) Draw a free-body diagram for the beam. (b) When the bear is at $x = 1.00$ m, find the tension in the wire and the components of the force exerted by the wall on the left end of the beam. (c) If the wire can withstand a maximum tension of 900 N, what is the maximum distance the bear can walk before the wire breaks?

x

$60.0°$

Goodies

FIGURE P12.36

37. Old MacDonald had a farm, and on that farm he had a gate (Fig. P12.37). The gate is 3.0 m long and 1.8 m tall with hinges attached to the top and bottom. The guy wire makes an angle of 30.0° with the top of the gate and is tightened by turnbuckle to a tension of 200 N. The mass of the gate is 40.0 kg. (a) Determine the horizontal force exerted on the gate by the bot-

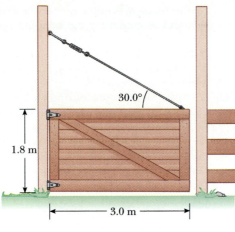

FIGURE P12.37

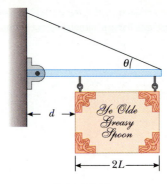

FIGURE P12.39

tom hinge. (b) Find the horizontal force exerted by the upper hinge. (c) Determine the combined vertical force exerted by both hinges. (d) What must be the tension in the guy wire so that the horizontal force exerted by the upper hinge is zero?

38. A 1200-N uniform boom is supported by a cable as in Figure P12.38. The boom is pivoted at the bottom, and a 2000-N object hangs from its top. Find the tension in the cable and the components of the reaction force on the boom by the floor.

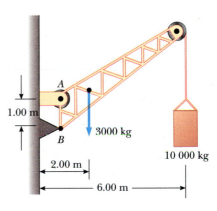

FIGURE P12.40

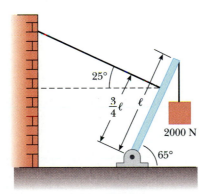

FIGURE P12.38

39. A uniform sign of weight w and width $2L$ hangs from a light, horizontal beam, hinged at the wall and supported by a cable (Fig. P12.39). Determine (a) the tension in the cable and (b) the components of the reaction force exerted by the wall on the beam in terms of w, d, L, and θ.

40. A crane of mass 3000 kg supports a load of 10 000 kg as in Figure P12.40. The crane is pivoted with a smooth pin at A and rests against a smooth support at B. Find the reaction forces at A and B.

41. A 15-m uniform ladder weighing 500 N rests against a frictionless wall. The ladder makes a 60.0° angle with the horizontal. (a) Find the horizontal and vertical forces the ground exerts on the base of the ladder when an 800-N firefighter is 4.00 m from the bottom. (b) If the ladder is just on the verge of slipping when the firefighter is 9.00 m up, what is the coefficient of static friction between ladder and ground?

41A. A uniform ladder of length L and mass m_1 rests against a frictionless wall. The ladder makes an angle θ with the horizontal. (a) Find the horizontal and vertical forces the ground exerts on the base of the ladder when a firefighter of mass m_2 is a distance x from the bottom. (b) If the ladder is just on the verge of slipping when the firefighter is a distance d from the bottom, what is the coefficient of static friction between ladder and ground?

42. A uniform ladder weighing 200 N is leaning against a wall (Fig. 12.12). The ladder slips when θ is 60°. Assuming the coefficients of static friction at the wall and the ground are the same, obtain a value for μ_s.

43. A 10 000-N shark is supported by a cable attached to a 4.00-m rod that can pivot at the base. Calculate the cable tension needed to hold the system in the posi-

tion shown in Figure P12.43. Find the horizontal and vertical forces exerted on the base of the rod. (Neglect the weight of the rod.)

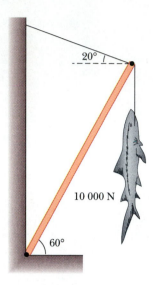

FIGURE P12.43

44. When a person stands on tiptoe (a strenuous position), the position of the foot is as shown in Figure P12.44a. The total weight of the body w is supported

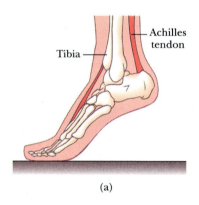

(a)

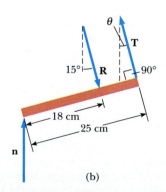

(b)

FIGURE P12.44

by the force **n** exerted by the floor on the toe. A mechanical model for the situation is shown in Figure P12.44b, where **T** is the force exerted by the Achilles tendon on the foot and **R** is the force exerted by the tibia on the foot. Find the values of T, R, and θ when $w = 700$ N.

45. A person bends over and lifts a 200-N object as in Figure P12.45a, with the back in the horizontal position (a terrible way to lift an object). The back muscle attached at a point two thirds up the spine maintains the position of the back, where the angle between the spine and this muscle is 12.0°. Using the mechanical model shown in Figure P12.45b and taking the weight of the upper body to be 350 N, find the tension in the back muscle and the compressional force in the spine.

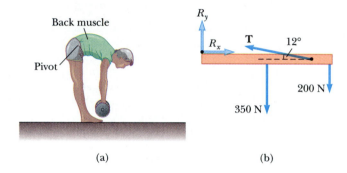

(a)   (b)

FIGURE P12.45

46. Two 200-N traffic lights are suspended from a single cable as shown in Figure P12.46. Neglect the cable weight and (a) prove that if $\theta_1 = \theta_2$, then $T_1 = T_2$. (b) Determine the three tensions if $\theta_1 = \theta_2 = 8.0°$.

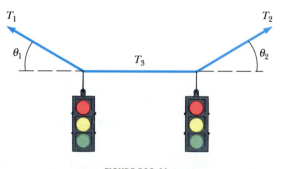

FIGURE P12.46

47. A force acts on a rectangular block weighing 400 N as in Figure P12.47. (a) If the block slides with constant speed when $F = 200$ N and $h = 0.400$ m, find the coefficient of sliding friction and the position of the resultant normal force. (b) If $F = 300$ N, find the value of h for which the block just begins to tip.

47A. A force F acts on a rectangular block of mass m as in Figure P12.47. (a) If the block slides with constant speed, find the coefficient of sliding friction and the position of the resultant normal force. (b) Find the value of h for which the block just begins to tip.

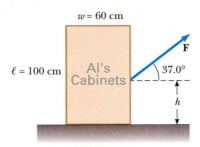

$w = 60$ cm

$\ell = 100$ cm

Al's Cabinets

F

37.0°

h

FIGURE P12.47

48. Consider the rectangular block of Problem 47. A force **F** is applied horizontally at the upper edge. (a) What is the minimum force required to start to tip the block? (b) What is the minimum coefficient of static friction required for the block to tip with the application of a force of this magnitude? (c) Find the magnitude and direction of the minimum force required to tip the block if the point of application can be chosen anywhere on the block.

49. A uniform beam of weight w is inclined at an angle θ to the horizontal with its upper end supported by a horizontal rope tied to a wall and its lower end resting on a rough floor (Fig. P12.49). (a) If the coefficient of static friction between beam and floor is μ_s, determine an expression for the maximum weight W that can be suspended from the top before the beam slips. (b) Determine the magnitude of the reaction force at the floor and the magnitude of the force exerted by the beam on the rope at P in terms of w, W, and μ_s.

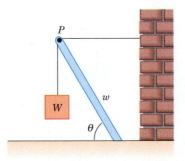

P

W

w

θ

FIGURE P12.49

50. Figure P12.50 shows a truss that supports a downward force of 1000 N applied at the point B. Neglect-

ing the weight of the truss, apply the conditions of equilibrium to prove that $n_A = 366$ N and $n_C = 634$ N.

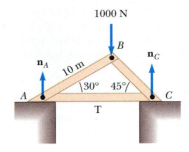

1000 N

B

n_A

n_C

10 m

30° 45°

A

C

T

FIGURE P12.50

51. A stepladder of negligible weight is constructed as shown in Figure P12.51. A painter of mass 70.0 kg stands on the ladder 3.00 m from the bottom. Assuming the floor is frictionless, find (a) the tension in the horizontal bar connecting the two halves of the ladder, (b) the normal forces at A and B, and (c) the components of the reaction force at the hinge C that the left half of the ladder exerts on the right half. (*Hint:* Treat each half of the ladder separately.)

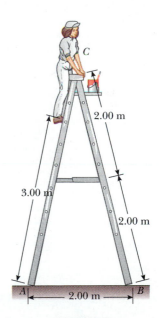

C

2.00 m

3.00 m

2.00 m

A 2.00 m B

FIGURE P12.51

52. A flat dance floor of dimensions 20.0 m by 20.0 m has a mass of 1000 kg. Three dance couples, each of mass 125 kg, start in the top left, top right, and bottom left corners. (a) Where is the initial center of gravity? (b) The couple in the bottom left corner moves 10.0 m to the right. Where is the new center of gravity? (c) What was the speed of the center of

gravity if it took that couple 8.00 s to change positions?

53. A shelf bracket is mounted on a vertical wall by a single screw, as shown in Figure P12.53. Neglecting the weight of the bracket, find the horizontal component of the force that the screw exerts on the bracket when an 80.0-N vertical force is applied as shown. (*Hint:* Imagine that the bracket is slightly loose.)

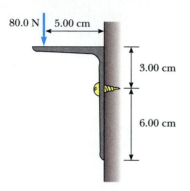

80.0 N   5.00 cm

3.00 cm

6.00 cm

FIGURE P12.53

54. Figure P12.54 shows a claw hammer as it pulls a nail out of a horizontal surface. If a force of 150 N is exerted horizontally as shown, find (a) the force exerted by the hammer claws on the nail and (b) the force exerted by the surface on the point of contact with the hammer head. Assume that the force the hammer exerts on the nail is parallel to the nail.

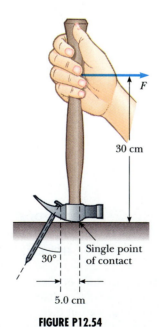

F

30 cm

Single point of contact

30°

5.0 cm

FIGURE P12.54

55. Figure P12.55 shows a vertical force applied tangentially to a uniform cylinder of weight *w*. The coeffi-

cient of static friction between the cylinder and all surfaces is 0.50. Find, in terms of *w*, the maximum force F that can be applied without causing the cylinder to rotate. (*Hint:* When the cylinder is on the verge of slipping, both friction forces are at their maximum values. Why?)

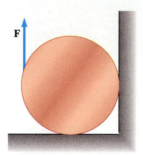

F

FIGURE P12.55

56. A wire of length *L*, Young's modulus *Y*, and cross-sectional area *A* is stretched elastically by an amount ΔL. By Hooke's law, the restoring force is $-k\,\Delta L$. (a) Show that $k = YA/L$. (b) Show that the work done in stretching the wire by an amount ΔL is

$$\text{Work} = \frac{1}{2}\,\frac{YA}{L}\,(\Delta L)^2$$

57. Two racket balls are placed in a glass as shown in Figure P12.57. Their centers and the point *A* lie on a straight line. (a) Assume that the walls are frictionless, and determine P_1, P_2, and P_3. (b) Determine the magnitude of the force exerted on the right ball by the left ball. Assume each ball has a mass of 170 g.

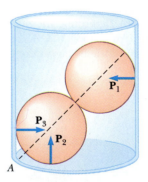

P_1

P_3

P_2

A

FIGURE P12.57

58. In Figure P12.58, the scales read $w_1 = 380$ N and $w_2 = 320$ N. Neglecting the weight of the supporting plank, how far from the woman's feet is her center of mass, given that her height is 2.0 m?

59. (a) Estimate the force with which a karate master strikes a board if the hand's speed at time of impact is 10.0 m/s, decreasing to 1.0 m/s during a 0.0020 s

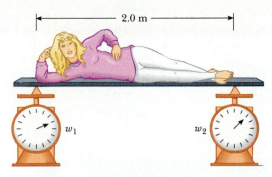

2.0 m

w_1 w_2

FIGURE P12.58

time-of-contact with the board. The mass of coordinated hand and arm is 1.0 kg. (b) Estimate the shear stress if this force is exerted on a 1.0-cm-thick pine board that is 10 cm wide. (c) If the maximum shear stress a pine board can receive before breaking is 3.6×10^6 N/m^2, will the board break?

60. A steel cable 3.0 cm^2 in cross-sectional area has a mass of 2.4 kg per meter of length. If 500 m of the cable is hung over a vertical cliff, how much does the cable stretch under its own weight? $Y_{steel} = 2.0 \times 10^{11}$ N/m^2.

61. The bottom and top of a bucket have radii of 25.0 cm and 35.0 cm, respectively. The bucket is 30.0 cm high and filled with water. Where is the center of gravity? (Ignore the weight of the bucket itself.)

Oscillatory Motion

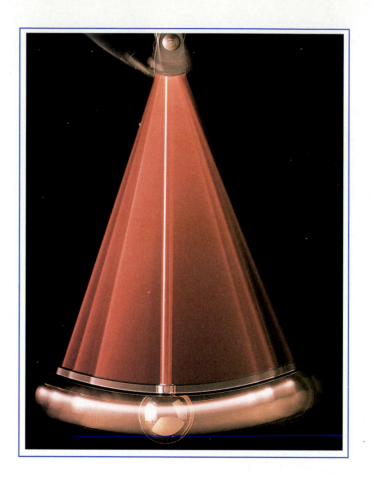

Time exposure of a simple pendulum, which consists of a small metal sphere suspended by a light string. The time it takes the pendulum to undergo one complete oscillation is the period of its motion, while the maximum displacement of the pendulum from the vertical position is called the amplitude. As we shall see in this chapter, the period of its motion for small amplitudes depends only on the length of the pendulum and the value of the free-fall acceleration. *(James Stevenson/SPL/Photo Researchers)*

A very special kind of motion occurs when the force on a body is proportional to the displacement of the body from equilibrium. If this force always acts toward the equilibrium position of the body, there is a repetitive back-and-forth motion about this position. Such motion is an example of what is called *periodic* or *oscillatory* motion.

You are most likely familiar with several examples of periodic motion, such as the oscillations of a mass on a spring, the motion of a pendulum, and the vibrations of a stringed musical instrument. Numerous systems exhibit oscillatory motion. For example, the molecules in a solid oscillate about their equilibrium positions; electromagnetic waves, such as light waves, radar, and radio waves, are characterized by oscillating electric and magnetic field vectors; and in alternating-current circuits, voltage, current, and electrical charge vary periodically with time.

Most of the material in this chapter deals with *simple harmonic motion*. In this type of motion, an object oscillates between two spatial positions for an indefinite period of time with no loss in mechanical energy. In real mechanical systems, retarding (frictional) forces are always present and these forces are considered in an optional section at the end of the chapter.

13.1   SIMPLE HARMONIC MOTION

A particle moving along the x axis is said to exhibit **simple harmonic motion** when x, its displacement from equilibrium, varies in time according to the relationship

$$x = A \cos(\omega t + \phi) \qquad (13.1)$$

Displacement versus time for simple harmonic motion

where A, ω, and ϕ are constants of the motion. In order to give physical significance to these constants, it is convenient to plot x as a function of t, as in Figure 13.1. First, note that A, called the **amplitude** of the motion, is the maximum displacement of the particle in either the positive or negative x direction. The constant angle ϕ is called the **phase constant** (or phase angle) and along with the amplitude A is determined uniquely by the initial displacement and initial velocity of the particle. The constants ϕ and A tell us what the displacement was at time $t = 0$. The quantity $(\omega t + \phi)$ is called the **phase** of the motion and is useful in comparing the motions of two systems of particles. Note that the function x is periodic and repeats itself when ωt increases by 2π rad.

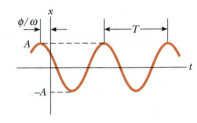

FIGURE 13.1   Displacement versus time for a particle undergoing simple harmonic motion. The amplitude of the motion is A, and the period is T.

The **period**, T, of the motion is the time it takes the particle to go through one full cycle. That is, the value of x at time t equals the value of x at time $t + T$. We can show that $T = 2\pi/\omega$ by using the fact that the phase increases by 2π rad in a time T:

$$\omega t + \phi + 2\pi = \omega(t + T) + \phi$$

Hence, $\omega T = 2\pi$, or

$$T = \frac{2\pi}{\omega} \qquad (13.2)$$

Period

The inverse of the period is called the **frequency** of the motion, f. The frequency represents the *number of oscillations the particle makes per unit time*:

$$f = \frac{1}{T} = \frac{\omega}{2\pi} \qquad (13.3)$$

Frequency

The units of f are cycles/s, or hertz (Hz).

Rearranging Equation 13.3 gives

$$\omega = 2\pi f = \frac{2\pi}{T} \qquad (13.4)$$

Angular frequency

The constant ω is called the **angular frequency** and has units of radians per second. We shall discuss the geometric significance of ω in Section 13.4.

We can obtain the speed of a particle undergoing simple harmonic motion by differentiating Equation 13.1 with respect to time:

$$v = \frac{dx}{dt} = -\omega A \sin(\omega t + \phi) \qquad (13.5)$$

Speed in simple harmonic motion

The acceleration of the particle is

Acceleration in simple harmonic motion

$$a = \frac{dv}{dt} = -\omega^2 A \cos(\omega t + \phi) \tag{13.6}$$

Since $x = A\cos(\omega t + \phi)$, we can express Equation 13.6 in the form

$$a = -\omega^2 x \tag{13.7}$$

From Equation 13.5 we see that since the sine and cosine functions oscillate between ± 1, the extreme values of v are $\pm \omega A$. Equation 13.6 tells us that the extreme values of the acceleration are $\pm \omega^2 A$. Therefore, the maximum values of the speed and acceleration are

Maximum values of speed and acceleration in simple harmonic motion

$$v_{\max} = \omega A \tag{13.8}$$

$$a_{\max} = \omega^2 A \tag{13.9}$$

Figure 13.2a represents the displacement versus time for an arbitrary value of the phase constant. The velocity and acceleration versus time curves are illustrated in Figures 13.2b and 13.2c. These curves show that the phase of the velocity differs from the phase of the displacement by $\pi/2$ rad, or 90°. That is, when x is a maximum or a minimum, the velocity is zero. Likewise, when x is zero, the speed is a maximum. Furthermore, note that the phase of the acceleration differs from the phase of the displacement by π rad, or 180°. That is, when x is a maximum, a is a maximum in the opposite direction.

The phase constant ϕ is important when comparing the motion of two or more oscillating particles. Suppose that the initial position x_0 and initial speed v_0 of a single oscillator are given, that is, at $t = 0$, $x = x_0$ and $v = v_0$. Under these conditions, Equations 13.1 and 13.5 give

$$x_0 = A\cos\phi \quad \text{and} \quad v_0 = -\omega A \sin\phi \tag{13.10}$$

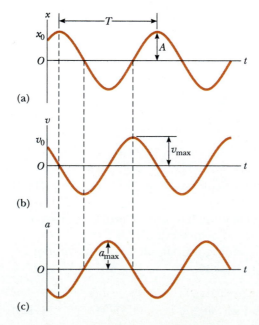

FIGURE 13.2 Graphical representation of simple harmonic motion: (a) displacement versus time, (b) velocity versus time, and (c) acceleration versus time. Note that at any specified time the velocity is 90° out of phase with the displacement and the acceleration is 180° out of phase with the displacement.

Dividing the second of these equations by the first eliminates A, giving $v_0/x_0 = -\omega \tan \phi$, or

$$\tan \phi = -\frac{v_0}{\omega x_0} \qquad (13.11)$$

Furthermore, if we square Equations 13.10 and add terms, we get $x_0{}^2 + \left(\dfrac{v_0}{\omega}\right)^2 = A^2 \cos^2 \phi + A^2 \sin^2 \phi$. Solving for A, we find

$$A = \sqrt{x_0{}^2 + \left(\frac{v_0}{\omega}\right)^2} \qquad (13.12)$$

> The phase angle ϕ and amplitude A can be obtained from the initial conditions

Thus, we see that ϕ and A are known if x_0, ω, and v_0 are given.

The following are important properties of a particle moving in simple harmonic motion:

- The displacement, velocity, and acceleration all vary sinusoidally with time but are not in phase, as shown in Figure 13.2.
- The acceleration of the particle is proportional to the displacement but in the opposite direction.
- The frequency and the period of the motion are independent of the amplitude.

> Properties of simple harmonic motion

CONCEPTUAL EXAMPLE 13.1

Does the acceleration of a simple harmonic oscillator remain constant during its motion? Is the acceleration ever zero? Explain.

Reasoning In simple harmonic motion, the acceleration is not constant. The acceleration is zero whenever the object passes through its equilibrium position, to the right whenever the object is to the left of its equilibrium position, and to the left whenever the object is to the right of its equilibrium position. In general, the acceleration is proportional to the displacement but oppositely directed.

EXAMPLE 13.2 An Oscillating Body

A body oscillates with simple harmonic motion along the x axis. Its displacement varies with time according to the equation

$$x = (4.00 \text{ m}) \cos\left(\pi t + \frac{\pi}{4}\right)$$

where t is in seconds and the angles in the second parentheses are in radians. (a) Determine the amplitude, frequency, and period of the motion.

Solution By comparing this equation with the general equation for simple harmonic motion, $x = A \cos(\omega t + \phi)$, we see that $A = 4.00$ m and $\omega = \pi$ rad/s; therefore we find $f = \omega/2\pi = \pi/2\pi = 0.500 \text{ s}^{-1}$ and $T = 1/f = 2.00$ s.

(b) Calculate the velocity and acceleration of the body at any time t.

Solution

$$v = \frac{dx}{dt} = -(4.00 \text{ m/s}) \sin\left(\pi t + \frac{\pi}{4}\right) \frac{d}{dt}(\pi t)$$

$$= -(4.00\pi \text{ m/s}) \sin\left(\pi t + \frac{\pi}{4}\right)$$

$$a = \frac{dv}{dt} = -(4.00\pi \text{ m/s}^2) \cos\left(\pi t + \frac{\pi}{4}\right) \frac{d}{dt}(\pi t)$$

$$= -(4.00\pi^2 \text{ m/s}^2) \cos\left(\pi t + \frac{\pi}{4}\right)$$

(c) Using the results to part (b), determine the position, velocity, and acceleration of the body at $t = 1.00$ s.

Solution   Noting that the angles in the trigonometric functions are in radians, we get at $t = 1.00$ s

$$x = (4.00 \text{ m}) \cos\left(\pi + \frac{\pi}{4}\right) = (4.00 \text{ m}) \cos\left(\frac{5\pi}{4}\right)$$

$$= (4.00 \text{ m})(-0.707) = \boxed{-2.83 \text{ m}}$$

$$v = -(4.00\pi \text{ m/s}) \sin\left(\frac{5\pi}{4}\right)$$

$$= -(4.00\pi \text{ m/s})(-0.707) = \boxed{8.89 \text{ m/s}}$$

$$a = -(4.00\pi^2 \text{ m/s}^2) \cos\left(\frac{5\pi}{4}\right)$$

$$= -(4.00\pi^2 \text{ m/s}^2)(-0.707) = \boxed{27.9 \text{ m/s}^2}$$

(d) Determine the maximum speed and maximum acceleration of the body.

Solution   From the general expressions for v and a found in part (b), we see that the maximum values of the sine and cosine functions are unity. Therefore, v varies between $\pm 4.00\pi$ m/s, and a varies between $\pm 4.00\pi^2$ m/s^2. Thus,

$v_{max} = 4.00\pi$ m/s and $a_{max} = 4.00\pi^2$ m/s^2. The same results are obtained using $v_{max} = \omega A$ and $a_{max} = \omega^2 A$, where $A = 4.00$ m and $\omega = \pi$ rad/s.

(e) Find the displacement of the body between $t = 0$ and $t = 1.00$ s.

Solution   The x coordinate at $t = 0$ is

$$x_0 = (4.00 \text{ m}) \cos\left(0 + \frac{\pi}{4}\right) = (4.00 \text{ m})(0.707) = 2.83 \text{ m}$$

In part (c), we found that the coordinate at $t = 1.00$ s is -2.83 m; therefore, the displacement between $t = 0$ and $t = 1.00$ s is

$$\Delta x = x - x_0 = -2.83 \text{ m} - 2.83 \text{ m} = \boxed{-5.66 \text{ m}}$$

Because the particle's velocity changes sign during the first second, the magnitude of Δx is not the same as the distance traveled in the first second.

Exercise   What is the phase of the motion at $t = 2.00$ s?

Answer $9\pi/4$ rad.

13.2 MASS ATTACHED TO A SPRING

Consider a physical system consisting of a mass attached to the end of a spring, where the mass is free to move on a horizontal, frictionless track (Fig. 13.3). When the spring is neither stretched nor compressed, the mass is at the position $x = 0$, called the *equilibrium position* of the system. We know from experience that such a system will oscillate back and forth if disturbed from the equilibrium position. Because the surface is frictionless, the mass moves in simple harmonic motion. An experimental arrangement that clearly demonstrates that such a system moves in simple harmonic motion is illustrated in Figure 13.4, in which a mass oscillating vertically on a spring has a pen attached to it. While the mass is in motion, a sheet of paper is moved horizontally, and the pen traces out a sinusoidal pattern.

We can understand this motion qualitatively by first recalling that, when the mass is displaced a small distance x from equilibrium, the spring exerts a force on m given by Hooke's law, as expressed by Equation 7.9:

$$F = -kx$$

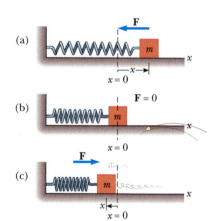

(a)

(b)

(c)

FIGURE 13.3   A mass attached to a spring on a frictionless track moves in simple harmonic motion. (a) When the mass is displaced to the right of equilibrium, the displacement is positive and the acceleration is negative. (b) At the equilibrium position, $x = 0$, the acceleration is zero but the speed is a maximum. (c) When the displacement is negative, the acceleration is positive.

As we learned in Section 7.3, we call this a linear restoring force because it is linearly proportional to the displacement and is always directed toward the equilibrium position and therefore *opposite* the displacement. That is, when the mass is displaced to the right in Figure 13.3, x is positive and the restoring force is to the left. When the mass is displaced to the left of $x = 0$, then x is negative and $\mathbf{F}$ is to the right.

If we apply Newton's second law to the motion of the mass in the x direction, we get

$$F = -kx = ma$$

$$a = -\frac{k}{m}x \tag{13.13}$$

That is, *the acceleration is proportional to the displacement of the mass from equilibrium and is in the opposite direction.* If the mass is displaced a maximum distance $x = A$ at some initial time and released from rest, its initial acceleration is $-kA/m$ (its extreme negative value). When the mass passes through the equilibrium position, $x = 0$ and its acceleration is zero. At this instant, its speed is a maximum. The mass then continues to travel to the left of equilibrium and finally reaches $x = -A$, at which time its acceleration is kA/m (maximum positive) and its speed is again zero. Thus, we see that the mass oscillates between the turning points $x = \pm A$. In one full cycle of its motion, the mass travels a distance $4A$.

We now describe the motion in a quantitative fashion. Recall that $a = dv/dt = d^2x/dt^2$, and so we can express Equation 13.13 as

$$\frac{d^2x}{dt^2} = -\frac{k}{m}x$$

If we denote the ratio k/m by the symbol ω^2, this equation becomes

$$\frac{d^2x}{dt^2} = -\omega^2x \qquad (13.14)$$

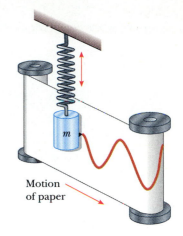

FIGURE 13.4 An experimental apparatus for demonstrating simple harmonic motion. A pen attached to the oscillating mass traces out a sine wave on the moving chart paper.

What we now require is a solution to Equation 13.14—that is, a function $x(t)$ that satisfies this second-order differential equation. Since Equations 13.14 and 13.7 are equivalent, we see that the solution must be that of simple harmonic motion:

$$x(t) = A\cos(\omega t + \phi)$$

To see this explicitly, note that if $x(t) = A\cos(\omega t + \phi)$ then

$$\frac{dx}{dt} = A\frac{d}{dt}\cos(\omega t + \phi) = -\omega A\sin(\omega t + \phi)$$

$$\frac{d^2x}{dt^2} = -\omega A\frac{d}{dt}\sin(\omega t + \phi) = -\omega^2 A\cos(\omega t + \phi)$$

Comparing the expressions for x and d^2x/dt^2, we see that $d^2x/dt^2 = -\omega^2x$ and Equation 13.14 is satisfied.

The following general statement can be made based on the foregoing discussion:

> Whenever the force acting on a particle is linearly proportional to the displacement and in the opposite direction, the particle moves in simple harmonic motion.

Since the period is $T = 2\pi/\omega$ and the frequency is the inverse of the period, we can express the period and frequency of the motion for this system as

$$T = \frac{2\pi}{\omega} = 2\pi\sqrt{\frac{m}{k}} \qquad (13.15)$$

Period and frequency for mass-spring system

$$f = \frac{1}{T} = \frac{1}{2\pi}\sqrt{\frac{k}{m}} \qquad (13.16)$$

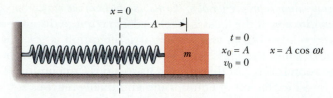

FIGURE 13.5 A mass-spring system that starts from rest at $x_0 = A$. In this case, $\phi = 0$, and so $x = A \cos \omega t$.

That is, the period and frequency depend only on the mass and on the force constant of the spring. As we might expect, the frequency is larger for a stiffer spring (the stiffer the spring, the higher the value of k) and decreases with increasing mass.

Special Case I In order to better understand the physical significance of our solution of the equation of motion, let us consider the following special case. Suppose we pull the mass a distance A from equilibrium and release it from rest from this stretched position, as in Figure 13.5. We must then require that our solution for $x(t)$ obey the initial conditions that at $t = 0$, $x_0 = A$, and $v_0 = 0$. These conditions are met if we choose $\phi = 0$, giving $x = A \cos \omega t$ as our solution. To check this solution, we note that it satisfies the condition that $x_0 = A$ at $t = 0$, since $\cos 0 = 1$. Thus, we see that A and ϕ contain the information on initial conditions.

Now let us investigate the behavior of the velocity and acceleration for this special case. Since $x = A \cos \omega t$,

$$v = \frac{dx}{dt} = -\omega A \sin \omega t$$

and

$$a = \frac{dv}{dt} = -\omega^2 A \cos \omega t$$

From the preceding velocity expression, we see that at $t = 0$, $v_0 = 0$, as we require. The expression for the acceleration tells us that at $t = 0$, $a = -\omega^2 A$. Physically this negative acceleration makes sense, because the force on the mass is directed to the left when the displacement is positive. In fact, at the extreme position shown in Figure 13.5, $F = -kA$ (to the left), and the initial acceleration is $-kA/m$.

We could also use a more formal approach to show that $x = A \cos \omega t$ is the correct solution by using the relationship $\tan \phi = -v_0/\omega x_0$ (Eq. 13.11). Since $v_0 = 0$ at $t = 0$, $\tan \phi = 0$ and so $\phi = 0$ or π. Only $\phi = 0$ yields the correct sign for x_0.

The displacement, velocity, and acceleration versus time are plotted in Figure 13.6 for this special case. Note that the acceleration reaches extreme values of $\pm \omega^2 A$ when the displacement has extreme values of $\mp A$. Furthermore, the velocity has extreme values of $\pm \omega A$, which both occur at $x = 0$. Hence, the quantitative solution agrees with our qualitative description of this system.

Special Case II Now suppose that the mass is given an initial velocity v_0 to the right while the system is in the equilibrium position, so that at $t = 0$, $x_0 = 0$, and $v = v_0$ (Fig. 13.7). Our solution must now satisfy these initial conditions. Since the mass is moving toward positive x values at $t = 0$ and since $x_0 = 0$ at $t = 0$, the solution has the form $x = A \sin \omega t$.

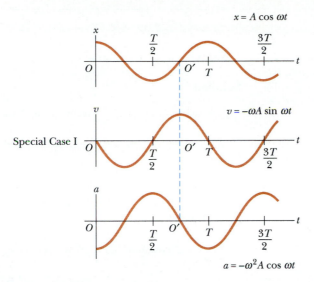

$$x = A \cos \omega t$$

$$v = -\omega A \sin \omega t$$

Special Case I

$$a = -\omega^2 A \cos \omega t$$

FIGURE 13.6   Displacement, velocity, and acceleration versus time for a particle undergoing simple harmonic motion under the initial conditions that at $t = 0$, $x = A$, and $v = 0$.

Applying Equation 13.11 and the initial condition that $x_0 = 0$ at $t = 0$ gives $\tan \phi = -\infty$ and $\phi = -\pi/2$. Hence, the solution is $x = A \cos(\omega t - \pi/2)$, which can be written $x = A \sin \omega t$. Furthermore, from Equation 13.11 we see that $A = v_0/\omega$; therefore, we can express our solution as

$$x = \frac{v_0}{\omega} \sin \omega t$$

The velocity and acceleration in this case are

$$v = \frac{dx}{dt} = v_0 \cos \omega t$$

$$a = \frac{dv}{dt} = -\omega v_0 \sin \omega t$$

These results are consistent with the fact that the mass always has a maximum speed at $x = 0$, and the force and acceleration are zero at this position. The graphs of these functions versus time in Figure 13.6 correspond to the origin at O'. What is the solution for x if the mass is initially moving to the left in Figure 13.7?

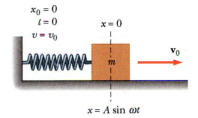

$x_0 = 0$
$t = 0$
$v = v_0$

$x = 0$

$x = A \sin \omega t$

FIGURE 13.7   The mass-spring system starts its motion at the equilibrium position, $x = 0$ at $t = 0$. If its initial velocity is v_0 to the right, its x coordinate varies as

$$x = \frac{v_0}{\omega} \sin \omega t.$$

EXAMPLE 13.3   **Watch Out for Potholes**

A car of mass 1300 kg is constructed using a frame supported by four springs. Each spring has a force constant of 20 000 N/m. If two people riding in the car have a combined mass of 160 kg, find the frequency of vibration of the car when it is driven over a pothole in the road.

Solution   We assume the weight is evenly distributed. Thus, each spring supports one fourth of the load. The total mass supported by the springs is 1460 kg, and therefore each

spring supports 365 kg. Hence, the frequency of vibration is, from Equation 13.16,

$$f = \frac{1}{2\pi} \sqrt{\frac{k}{m}} = \frac{1}{2\pi} \sqrt{\frac{20\ 000\ \text{N/m}}{365\ \text{kg}}} = 1.18\ \text{Hz}$$

Exercise   How long does it take the car to execute two complete vibrations?

Answer   1.70 s.

EXAMPLE 13.4   A Mass-Spring System

A mass of 200 g is connected to a light spring of force constant 5.00 N/m and is free to oscillate on a horizontal, frictionless track. If the mass is displaced 5.00 cm from equilibrium and released from rest, as in Figure 13.5, (a) find the period of its motion.

Solution   This situation corresponds to Special Case I, where $x = A \cos \omega t$ and $A = 5.00 \times 10^{-2}$ m. Therefore,

$$\omega = \sqrt{\frac{k}{m}} = \sqrt{\frac{5.00 \text{ N/m}}{200 \times 10^{-3} \text{ kg}}} = 5.00 \text{ rad/s}$$

and

$$T = \frac{2\pi}{\omega} = \frac{2\pi}{5} = \boxed{1.26 \text{ s}}$$

(b) Determine the maximum speed of the mass.

Solution

$$v_{\text{max}} = \omega A = (5.00 \text{ rad/s})(5.00 \times 10^{-2} \text{ m}) = \boxed{0.250 \text{ m/s}}$$

(c) What is the maximum acceleration of the mass?

Solution

$$a_{\text{max}} = \omega^2 A = (5.00 \text{ rad/s})^2(5.00 \times 10^{-2} \text{ m}) = \boxed{1.25 \text{ m/s}^2}$$

(d) Express the displacement, speed, and acceleration as functions of time.

Solution   The expression $x = A \cos \omega t$ is our solution for Special Case I, and so we can use the results from (a), (b), and (c) to get

$$x = A \cos \omega t = \boxed{(0.0500 \text{ m}) \cos 5.00t}$$

$$v = -\omega A \sin \omega t = \boxed{-(0.250 \text{ m/s}) \sin 5.00t}$$

$$a = -\omega^2 A \cos \omega t = \boxed{-(1.25 \text{ m/s}^2) \cos 5.00t}$$

13.3   ENERGY OF THE SIMPLE HARMONIC OSCILLATOR

Let us examine the mechanical energy of the mass-spring system described in Figure 13.5. Since the surface is frictionless, we expect that the total mechanical energy is constant, as was shown in Chapter 8. We can use Equation 13.5 to express the kinetic energy as

Kinetic energy of a simple harmonic oscillator

$$K = \tfrac{1}{2}mv^2 = \tfrac{1}{2}m\omega^2 A^2 \sin^2(\omega t + \phi) \tag{13.17}$$

The elastic potential energy stored in the spring for any elongation x is given by $\tfrac{1}{2}kx^2$. Using Equation 13.1, we get

Potential energy of a simple harmonic oscillator

$$U = \tfrac{1}{2}kx^2 = \tfrac{1}{2}kA^2 \cos^2(\omega t + \phi) \tag{13.18}$$

We see that K and U are *always* positive quantities. Since $\omega^2 = k/m$, we can express the total energy of the simple harmonic oscillator as

$$E = K + U = \tfrac{1}{2}kA^2[\sin^2(\omega t + \phi) + \cos^2(\omega t + \phi)]$$

But $\sin^2\theta + \cos^2\theta = 1$, where $\theta = \omega t + \phi$; therefore, this equation reduces to

Total energy of a simple harmonic oscillator

$$E = \tfrac{1}{2}kA^2 \tag{13.19}$$

That is, the energy of a simple harmonic oscillator is a constant of the motion and proportional to the square of the amplitude. In fact, the total mechanical energy is equal to the maximum potential energy stored in the spring when $x = \pm A$. At these points, $v = 0$ and there is no kinetic energy. At the equilibrium position, $x = 0$ and $U = 0$, so that the total energy is all in the form of kinetic energy. That is, at $x = 0$, $E = \tfrac{1}{2}mv_{\text{max}}^2 = \tfrac{1}{2}m\omega^2 A^2$.

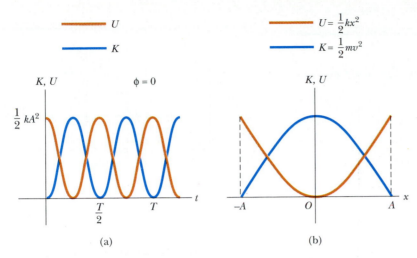

FIGURE 13.8 (a) Kinetic energy and potential energy versus time for a simple harmonic oscillator with $\phi = 0$. (b) Kinetic energy and potential energy versus displacement for a simple harmonic oscillator. In either plot, note that $K + U =$ constant.

Plots of the kinetic and potential energies versus time are shown in Figure 13.8a, where we have taken $\phi = 0$. As mentioned above, both K and U are always positive and their sum at all times is a constant equal to $\frac{1}{2}kA^2$, the total energy of the system. The variations of K and U with displacement are plotted in Figure 13.8b. Energy is continuously being transformed between potential energy stored in the spring and the kinetic energy of the mass.

EXAMPLE 13.5   Oscillations on a Horizontal Surface

A 0.500-kg mass connected to a light spring of force constant 20.0 N/m oscillates on a horizontal, frictionless track. (a) Calculate the total energy of the system and the maximum speed of the mass if the amplitude of the motion is 3.00 cm.

Solution   Using Equation 13.19, we get

$$E = \tfrac{1}{2}kA^2 = \tfrac{1}{2}\left(20.0\,\frac{\text{N}}{\text{m}}\right)(3.00 \times 10^{-2}\,\text{m})^2 = 9.00 \times 10^{-3}\,\text{J}$$

When the mass is at $x = 0$, $U = 0$ and $E = \frac{1}{2}mv_{\max}^2$; therefore,

$$\tfrac{1}{2}mv_{\max}^2 = \boxed{9.00 \times 10^{-3}\,\text{J}}$$

$$v_{\max} = \sqrt{\frac{18.0 \times 10^{-3}\,\text{J}}{0.500\,\text{kg}}} = \boxed{0.190\ \text{m/s}}$$

(b) What is the velocity of the mass when the displacement is equal to 2.00 cm?

Solution   We can apply Equation 13.20 directly:

$$v = \pm\sqrt{\frac{k}{m}\,(A^2 - x^2)} = \pm\sqrt{\frac{20.0}{0.500}\,(3.00^2 - 2.00^2) \times 10^{-4}}$$

$$= \boxed{\pm 0.141\ \text{m/s}}$$

The positive and negative signs indicate that the mass could be moving to the right or left at this instant.

(c) Compute the kinetic and potential energies of the system when the displacement equals 2.00 cm.

Solution   Using the result to part (b), we get

$$K = \tfrac{1}{2}mv^2 = \tfrac{1}{2}(0.500\ \text{kg})(0.141\ \text{m/s})^2$$

$$= \boxed{4.97 \times 10^{-3}\,\text{J}}$$

$$U = \tfrac{1}{2}kx^2 = \tfrac{1}{2}\left(20.0\,\frac{\text{N}}{\text{m}}\right)(2.00 \times 10^{-2}\,\text{m})^2$$

$$= \boxed{4.00 \times 10^{-3}\,\text{J}}$$

Note that the sum $K + U$ equals the total energy, E.

Exercise   For what values of x does the speed of the mass equal 0.100 m/s?

Answer ± 2.55 cm.

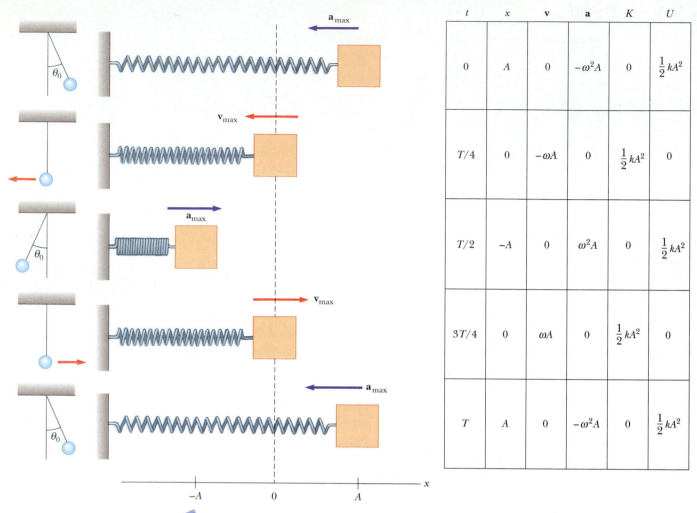

t	x	v	a	K	U
0	A	0	$-\omega^2 A$	0	$\frac{1}{2}kA^2$
$T/4$	0	$-\omega A$	0	$\frac{1}{2}kA^2$	0
$T/2$	$-A$	0	$\omega^2 A$	0	$\frac{1}{2}kA^2$
$3T/4$	0	ωA	0	$\frac{1}{2}kA^2$	0
T	A	0	$-\omega^2 A$	0	$\frac{1}{2}kA^2$

FIGURE 13.9 Simple harmonic motion for a mass-spring system and its analogy to the motion of a simple pendulum. The parameters in the table at the right refer to the mass-spring system, assuming that at $t = 0$, $x = A$ so that $x = A \cos \omega t$ (Special Case I).

Figure 13.9 illustrates the position, velocity, acceleration, kinetic energy, and potential energy of the mass-spring system for one full period of the motion. Most of the ideas discussed so far are incorporated in this important figure. Study it carefully.

Finally, we can use energy conservation to obtain the velocity for an arbitrary displacement x by expressing the total energy at some arbitrary position as

$$E = K + U = \tfrac{1}{2}mv^2 + \tfrac{1}{2}kx^2 = \tfrac{1}{2}kA^2$$

| Velocity as a function of position for a simple harmonic oscillator

$$v = \pm\sqrt{\frac{k}{m}\,(A^2 - x^2)} = \pm\omega\sqrt{A^2 - x^2} \qquad (13.20)$$

Again, this expression substantiates the fact that the speed is a maximum at $x = 0$ and is zero at the turning points, $x = \pm A$.

13.4   THE PENDULUM

The **simple pendulum** is another mechanical system that moves in oscillatory motion. It consists of a point mass m suspended by a light string of length L, where the upper end of the string is fixed as in Figure 13.10. The motion occurs in a vertical plane and is driven by the gravitational force. We shall show that, provided the angle θ is small, the motion is that of a simple harmonic oscillator. The forces acting on the mass are the force exerted by the string **T** and the gravitational force $m\mathbf{g}$. The tangential component of the gravitational force, $mg \sin \theta$, always acts toward $\theta = 0$, opposite the displacement. Therefore, the tangential force is a restoring force, and we can write the equation of motion in the tangential direction:

$$F_t = - mg \sin \theta = m \frac{d^2 s}{dt^2}$$

where s is the displacement measured along the arc and the minus sign indicates that F_t acts toward the equilibrium position. Since $s = L\theta$ and L is constant, this equation reduces to

$$\frac{d^2 \theta}{dt^2} = - \frac{g}{L} \sin \theta$$

The right side is proportional to $\sin \theta$ rather than to θ; hence, we conclude that the motion is not simple harmonic motion since it is not of the form of Equation 13.14. However, if we assume that θ is small, we can use the approximation

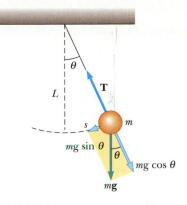

FIGURE 13.10   When θ is small, the simple pendulum oscillates in simple harmonic motion about the equilibrium position ($\theta = 0$). The restoring force is $mg \sin \theta$, the component of the weight tangent to the circle.

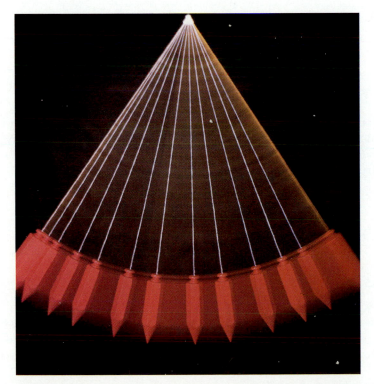

The motion of a simple pendulum captured with multiflash photography. Is the motion simple harmonic in this case?   *(Paul Silverman/Fundamental Photographs)*

The Foucault pendulum at the Smithsonian Institution in Washington, D.C. This type of pendulum was first used by the French physicist Jean Foucault to verify the Earth's rotation experimentally. During its swinging motion, the pendulum's plane of oscillation appears to rotate, as the bob successively knocks over the red indicators arranged in a horizontal circle.   *(Courtesy of the Smithsonian Institution)*

$\sin \theta \approx \theta$, where θ is measured in radians.[1] Therefore, the equation of motion becomes

Equation of motion for the simple pendulum (small θ)

$$\frac{d^2\theta}{dt^2} = -\frac{g}{L}\,\theta \tag{13.21}$$

Now we have an expression that is of exactly the same form as Equation 13.14, and so we conclude that the motion is simple harmonic motion. Therefore, θ can be written as $\theta = \theta_0 \cos(\omega t + \phi)$, where θ_0 is the *maximum angular displacement* and the angular frequency ω is

Angular frequency of motion for the simple pendulum

$$\omega = \sqrt{\frac{g}{L}} \tag{13.22}$$

The period of the motion is

Period of motion for the simple pendulum

$$T = \frac{2\pi}{\omega} = 2\pi\sqrt{\frac{L}{g}} \tag{13.23}$$

In other words, *the period and frequency of a simple pendulum depend only on the length of the string and the value of g.* Since the period is independent of the mass, we conclude that all simple pendulums of equal length at the same location oscillate with

---

[1] This approximation can be understood by examining the series expansion for $\sin \theta$, which is $\sin \theta = \theta - \theta^3/3! + \cdots$. For small values of θ, we see that $\sin \theta \approx \theta$. The difference between θ (in radians) and $\sin \theta$ for $\theta = 15°$ is only about 1%.

equal periods.[2] The analogy between the motion of a simple pendulum and the mass-spring system is illustrated in Figure 13.9.

The simple pendulum can be used as a timekeeper. It is also a convenient device for making precise measurements of the free-fall acceleration. Such measurements are important since variations in local values of **g** can provide information on the location of oil and other valuable underground resources.

CONCEPTUAL EXAMPLE 13.6

A pendulum bob is made with a sphere filled with water. What would happen to the frequency of vibration of this pendulum if there were a hole in the sphere that allowed the water to leak out slowly? Neglect both the mass of the string and air resistance.

Reasoning   The frequency of a pendulum is equal to the inverse of its period. From Equation 13.23, we see that $f = (1/2\pi)\sqrt{g/L}$. Because the frequency depends only on the length of the pendulum and the free-fall acceleration, and

not on the mass, the frequency will not change appreciably if the sphere is small compared to the length of the suspension. However, as the water leaks out of the sphere, the frequency first decreases (as the distance from the pivot to the center of mass of the sphere increases). After the water level in the sphere reaches the half-way point, the frequency begins to increase again until the sphere is empty. At that point, the frequency is the same as it was when the sphere was completely filled with water.

EXAMPLE 13.7   A Measure of Height

A man enters a tall tower, needing to know its height. He notes that a long pendulum extends from the ceiling almost to the floor and that its period is 12.0 s. How tall is the tower?

Solution   If we use $T = 2\pi\sqrt{L/g}$ and solve for L, we get

$$L = \frac{gT^2}{4\pi^2} = \frac{(9.80 \text{ m/s}^2)(12.0 \text{ s})^2}{4\pi^2} = 35.7 \text{ m}$$

Exercise   If the pendulum described in this example is taken to the Moon, where the free-fall acceleration is 1.67 m/s², what is the period there?

Answer   29.1 s.

The Physical Pendulum

If a hanging object oscillates about a fixed axis that does not pass through its center of mass, and the object cannot be accurately approximated as a point mass, then it must be treated as a physical, or compound, pendulum. Consider a rigid object pivoted at a point O that is a distance d from the center of mass (Fig. 13.11). The torque about O is provided by the weight of the object, and the magnitude of the torque is $mgd \sin \theta$. Using the fact that $\tau = I\alpha$, where I is the moment of inertia about the axis through O, we get

$$- mgd \sin \theta = I\frac{d^2\theta}{dt^2}$$

The minus sign indicates that the torque about O tends to decrease θ. That is, the weight of the object produces a restoring torque.

---

[2] The period of oscillation for the simple pendulum with arbitrary amplitude is

$$T = 2\pi\sqrt{\frac{L}{g}}\left(1 + \frac{1}{4}\sin^2\frac{\theta_0}{2} + \frac{9}{64}\sin^4\frac{\theta_0}{2} + \cdots\right)$$

where θ_0 is the maximum angular displacement in radians.

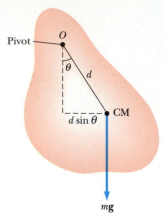

FIGURE 13.11 The physical pendulum consists of a rigid body pivoted at the point *O*, which is not at the center of mass. At equilibrium, the weight vector passes through *O*, corresponding to $\theta = 0$. The restoring torque about *O* when the system is displaced through an angle θ is $mgd \sin \theta$.

If we again assume that θ is small, then the approximation $\sin \theta \approx \theta$ is valid and the equation of motion reduces to

$$\frac{d^2\theta}{dt^2} = -\left(\frac{mgd}{I}\right)\theta = -\omega^2\theta \tag{13.24}$$

This equation is of the same form as Equation 13.14, and so the motion is simple harmonic motion. That is, the solution of Equation 13.24 is $\theta = \theta_0 \cos(\omega t + \phi)$, where θ_0 is the maximum angular displacement and

$$\omega = \sqrt{\frac{mgd}{I}}$$

The period is

$$T = \frac{2\pi}{\omega} = 2\pi\sqrt{\frac{I}{mgd}} \tag{13.25}$$

You can use this result to measure the moment of inertia of a planar rigid body. If the location of the center of mass and hence of *d* is known, the moment of inertia can be obtained by measuring the period. Finally, note that Equation 13.25 reduces to the period of a simple pendulum (Eq. 13.23) when $I = md^2$, that is, when all the mass is concentrated at the center of mass.

EXAMPLE 13.8   A Swinging Rod

A uniform rod of mass *M* and length *L* is pivoted about one end and oscillates in a vertical plane (Fig. 13.12). Find the period of oscillation if the amplitude of the motion is small.

Solution   In Chapter 10 we found that the moment of inertia of a uniform rod about an axis through one end is $\frac{1}{3}ML^2$. The distance *d* from the pivot to the center of mass is $L/2$. Substituting these quantities into Equation 13.25 gives

$$T = 2\pi\sqrt{\frac{\frac{1}{3}ML^2}{Mg\frac{L}{2}}} = 2\pi\sqrt{\frac{2L}{3g}}$$

Comment   In one of the Moon landings, an astronaut walking on the Moon's surface had a belt hanging from his space suit, and the belt oscillated as a compound pendulum. A scientist on Earth observed this motion on TV and from it was able to estimate the free-fall acceleration on the Moon. How do you suppose this calculation was done?

Exercise   Calculate the period of a meter stick pivoted about one end and oscillating in a vertical plane as in Figure 13.12.

Answer   1.64 s.

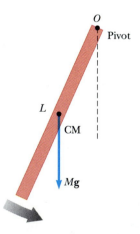

FIGURE 13.12   (Example 13.8) A rigid rod oscillating about a pivot through one end is a physical pendulum with $d = L/2$ and, from Table 10.2, $I_0 = \frac{1}{3}ML^2$.

Torsional Pendulum

Figure 13.13 shows a rigid body suspended by a wire attached at the top of a fixed support. When the body is twisted through some small angle θ, the twisted wire exerts a restoring torque on the body proportional to the angular displacement.

That is,

$$\tau = -\kappa\theta$$

where κ (the Greek letter kappa) is called the *torsion constant* of the support wire. The value of κ can be obtained by applying a known torque to twist the wire through a measurable angle θ. Applying Newton's second law for rotational motion gives

$$\tau = -\kappa\theta = I\frac{d^2\theta}{dt^2}$$

$$\frac{d^2\theta}{dt^2} = -\frac{\kappa}{I}\theta \qquad (13.26)$$

Again, this is the equation of motion for a simple harmonic oscillator, with $\omega = \sqrt{\kappa/I}$ and a period

$$T = 2\pi\sqrt{\frac{I}{\kappa}} \qquad (13.27)$$

This system is called a *torsional pendulum*. There is no small-angle restriction in this situation, as long as the response of the wire remains linear. The balance wheel of a watch oscillates as a torsional pendulum, energized by the mainspring. Torsional pendulums are also used in laboratory galvanometers and the Cavendish torsional balance.

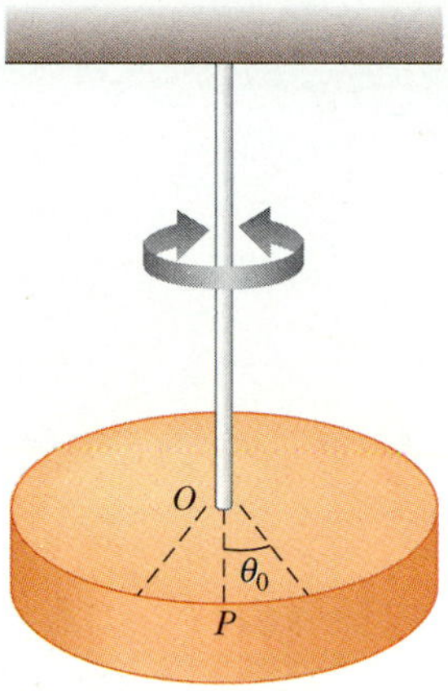

FIGURE 13.13 A torsional pendulum consists of a rigid body suspended by a wire attached to a rigid support. The body oscillates about the line *OP* with an amplitude θ_0.

*13.5 COMPARING SIMPLE HARMONIC MOTION WITH UNIFORM CIRCULAR MOTION

We can better understand and visualize many aspects of simple harmonic motion by studying its relationship to uniform circular motion. Figure 13.14 shows an experimental arrangement useful for developing this concept. Shadows are cast on a screen by a peg attached to the rim of a rotating disk and a mass attached to a spring. The two shadows move together when the period of the rotating disk equals the period of the oscillating mass-spring system.

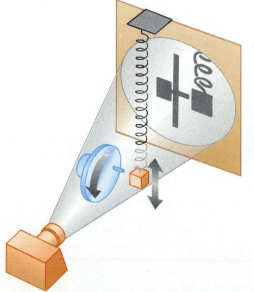

FIGURE 13.14 Experimental setup for demonstrating the connection between simple harmonic motion and uniform circular motion. The shadow of a peg on a turntable is projected on the screen along with the shadow of a mass on a spring. When the period of rotation of the turntable equals the period of oscillation of the mass, the shadows move together.

Consider a particle at point P moving in a circle of radius A with constant angular speed ω (Fig. 13.15a). We refer to this circle as the *reference circle* for the motion. As the particle rotates, its position vector rotates about the origin, O. Consider a particle located on the circumference of the circle of radius A, as in Figure 13.15a, with the line OP making an angle ϕ with the x axis at $t = 0$. We call this circle a reference circle for comparison of simple harmonic motion and uniform circular motion and take the position of P at $t = 0$ as our reference position. If the particle moves along the circle with constant angular speed ω until OP makes an angle θ with the x axis as in Figure 13.15b, then at some time $t > 0$, the angle between OP and the x axis is $\theta = \omega t + \phi$. As the particle rotates on the reference circle, the angle that OP makes with the x axis changes with time. Furthermore, the projection of P onto the x axis, labeled point Q, moves back and forth along a line superimposed on the horizontal diameter of the reference circle, between the limits $x = \pm A$.

Note that points P and Q always have the same x coordinate. From the right triangle OPQ, we see that this x coordinate is

$$x = A \cos(\omega t + \phi) \tag{13.28}$$

This expression shows that the point Q moves with simple harmonic motion along the x axis. Therefore, we conclude that

> simple harmonic motion along a straight line can be represented by the projection of uniform circular motion along a diameter of a reference circle.

By a similar argument, you can see from Figure 13.15b that the projection of P along the y axis also exhibits harmonic motion. Therefore, *uniform circular motion can be considered a combination of two simple harmonic motions,* one along x and one along y, where the two differ in phase by $90°$.

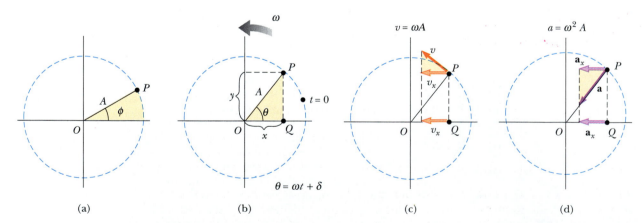

FIGURE 13.15 Relationship between the uniform circular motion of a point P and the simple harmonic motion of the point Q. A particle at P moves in a circle of radius A with constant angular frequency ω. (a) Reference circle showing position of P at $t = 0$. (b) The x components of the points P and Q are equal and vary in time as $x = A \cos(\omega t + \phi)$. (c) The x component of velocity of P equals the velocity of Q. (d) The x component of the acceleration of P equals the acceleration of Q.

This geometric interpretation shows that the time for one complete revolution of the point P on the reference circle is equal to the period of motion, T, for simple harmonic motion between $x = \pm A$. That is, the angular speed of P is the same as the angular frequency, ω, of simple harmonic motion along the x axis. The phase constant ϕ for simple harmonic motion corresponds to the initial angle that OP makes with the x axis. The radius of the reference circle, A, equals the amplitude of the simple harmonic motion.

Since the relationship between linear and angular speed for circular motion is $v = r\omega$ (Eq. 10.9), the particle moving on the reference circle of radius A has a velocity of magnitude ωA. From the geometry in Figure 13.15c, we see that the x component of this velocity is $-\omega A \sin(\omega t + \phi)$. By definition, the point Q has a velocity given by dx/dt. Differentiating Equation 13.28 with respect to time, we find that the velocity of Q is the same as the x component of the velocity of P.

The acceleration of P on the reference circle is directed radially inward toward O and has a magnitude $v^2/A = \omega^2 A$. From the geometry in Figure 13.15d, we see that the x component of this acceleration is $-\omega^2 A \cos(\omega t + \phi)$. This value is also the acceleration of the projected point Q along the x axis, as you can verify from Equation 13.28.

EXAMPLE 13.9   Circular Motion with Constant Speed

A particle rotates counterclockwise in a circle of radius 3.00 m with a constant angular speed of 8.00 rad/s. At $t = 0$, the particle has an x coordinate of 2.00 m. (a) Determine the x coordinate as a function of time.

Solution Since the amplitude of the particle's motion equals the radius of the circle and $\omega = 8.00$ rad/s, we have

$$x = A \cos(\omega t + \phi) = (3.00 \text{ m}) \cos(8.00t + \phi)$$

We can evaluate ϕ using the initial condition that $x = 2.00$ m at $t = 0$:

$$2.00 \text{ m} = (3.00 \text{ m}) \cos(0 + \phi)$$

$$\phi = \cos^{-1}\left(\tfrac{2.00}{3.00}\right) = 48.2° = 0.841 \text{ rad}$$

Therefore, the x coordinate versus time is of the form

$$x = \ (3.00 \text{ m}) \cos(8.00t + 0.841)$$

Note that the angles in the cosine function are in radians.

(b) Find the x components of the particle's velocity and acceleration at any time t.

Solution

$$v_x = \frac{dx}{dt} = (-3.00)(8.00) \sin(8.00t + 0.841)$$

$$= -(24.0 \text{ m/s}) \sin(8.00t + 0.841)$$

$$a_x = \frac{dv_x}{dt} = (-24.0)(8.00) \cos(8.00t + 0.841)$$

$$= -(192 \text{ m/s}^2) \cos(8.00t + 0.841)$$

From these results, we conclude that $v_{max} = 24$ m/s and $a_{max} = 192$ m/s^2. Note that these values also equal the tangential velocity, ωA, and centripetal acceleration, $\omega^2 A$.

---

*13.6   DAMPED OSCILLATIONS

The oscillatory motions we have considered so far have been for ideal systems, that is, systems that oscillate indefinitely under the action of a linear restoring force. In real systems, dissipative forces, such as friction, are present and retard the motion. Consequently, the mechanical energy of the system diminishes in time, and the motion is said to be *damped*.

One common type of retarding force, which we discussed in Chapter 6, is proportional to the speed and acts in the direction opposite the motion. This retarding force is often observed when an object moves through a gas. Because the

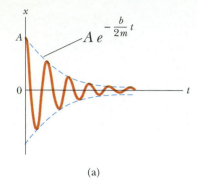

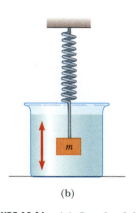

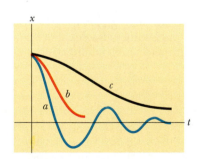

FIGURE 13.16   (a) Graph of the displacement versus time for a damped oscillator. Note the decrease in amplitude with time. (b) One example of a damped oscillator is a mass on a spring submersed in a liquid.

retarding force can be expressed as $\mathbf{R} = -b\mathbf{v}$, where b is a constant, and the restoring force of the system is $-kx$, we can write Newton's second law as

$$\sum F_x = -kx - bv = ma_x$$

$$-kx - b\frac{dx}{dt} = m\frac{d^2x}{dt^2} \tag{13.29}$$

The solution of this equation requires mathematics that may not be familiar to you as yet, and so we simply state it here without proof. When the retarding force is small compared with kx—that is, when b is small—the solution to Equation 13.29 is

$$x = Ae^{-\frac{b}{2m}t}\cos(\omega t + \phi) \tag{13.30}$$

where the angular frequency of motion is

$$\omega = \sqrt{\frac{k}{m} - \left(\frac{b}{2m}\right)^2} \tag{13.31}$$

This result can be verified by substituting Equation 13.30 into Equation 13.29. Figure 13.16a shows the displacement as a function of time in this case. We see that *when the retarding force is small compared with the restoring force, the oscillatory character of the motion is preserved but the amplitude decreases in time,* and the motion ultimately ceases. Any system that behaves in this way is known as a **damped oscillator.** The dashed blue lines in Figure 13.16a, which define what is called the *envelope* of the oscillatory curve, represent the exponential factor that appears in Equation 13.30. This envelope shows that *the amplitude decays exponentially with time.* For motion with a given spring constant and particle mass, the oscillations dampen more rapidly as the maximum value of the retarding force approaches the maximum value of the restoring force. One example of a damped harmonic oscillator is a mass immersed in a fluid as in Figure 13.16b.

It is convenient to express the angular frequency of a damped oscillator in the form

$$\omega = \sqrt{\omega_0^2 - \left(\frac{b}{2m}\right)^2}$$

where $\omega_0 = \sqrt{k/m}$ represents the angular frequency in the absence of a retarding force (the undamped oscillator). In other words, when $b = 0$, the retarding force is zero and the system oscillates with its natural frequency, ω_0. As the magnitude of the retarding force approaches the magnitude of the restoring force in the spring, the oscillations dampen more rapidly. When b reaches a critical value b_c such that $b_c/2m = \omega_0$, the system does not oscillate and is said to be **critically damped.** In this case, once released from rest at some nonequilibrium position, the system returns to equilibrium and then stays there. The displacement versus time graph for this case is the red curve in Figure 13.17.

If the medium is so viscous that the retarding force is greater than the restoring force—that is, if $b/2m > \omega_0$—the system is **overdamped.** Again, the displaced system does not oscillate but simply returns to its equilibrium position. As the damping increases, the time it takes the displacement to reach equilibrium also increases, as indicated in Figure 13.17. In any case, when friction is present, the energy of the oscillator eventually falls to zero. The lost mechanical energy dissipates into thermal energy in the retarding medium.

FIGURE 13.17   Plots of displacement versus time for an underdamped oscillator (a), a critically damped oscillator (b), and an overdamped oscillator (c).

*13.7 FORCED OSCILLATIONS

It is possible to compensate for the energy loss in a damped system by applying an external force that does positive work on the system. At any instant, energy can be put into the system by an applied force that acts in the direction of motion of the oscillator. For example, a child on a swing can be kept in motion by appropriately timed pushes. The amplitude of motion remains constant if the energy input per cycle of motion exactly equals the energy lost as a result of friction.

A common example of a forced oscillator is a damped oscillator driven by an external force that varies periodically, such as $F = F_0 \cos \omega t$, where ω is the angular frequency of the force and F_0 is a constant. Adding this driving force to the left side of Equation 13.29 gives

$$F_0 \cos \omega t - b\frac{dx}{dt} - kx = m\frac{d^2 x}{dt^2} \tag{13.32}$$

As earlier, the solution of this equation is not presented. After a sufficiently long period of time, when the energy input per cycle equals the energy lost per cycle, a steady-state condition is reached in which the oscillations proceed with constant amplitude. At this time, when the system is in a steady state, the solution of Equation 13.32 is

$$x = A \cos(\omega t + \phi) \tag{13.33}$$

where

$$A = \frac{F_0/m}{\sqrt{(\omega^2 - \omega_0{}^2)^2 + \left(\dfrac{b\omega}{m}\right)^2}} \tag{13.34}$$

and where $\omega_0 = \sqrt{k/m}$ is the angular frequency of the undamped oscillator $(b = 0)$. One can argue that, in steady state, the oscillator must physically have the same frequency as the driving force, and so the solution given by Equation 13.33 is expected. In fact, when this solution is substituted into Equation 13.32, one finds that it is indeed a solution, provided the amplitude is given by Equation 13.34.

Equation 13.34 shows that the motion of the forced oscillator is not damped since it is being driven by an external force. That is, the external agent provides the necessary energy to overcome the losses due to the retarding force. Note that the mass oscillates at the angular frequency of the driving force, ω. For small damping, the amplitude becomes large when the frequency of the driving force is near the natural frequency of oscillation. The dramatic increase in amplitude near the natural frequency ω_0 is called **resonance**, and ω_0 is called the **resonance frequency** of the system.

Physically, the reason for large-amplitude oscillations at the resonance frequency is that energy is being transferred to the system under the most favorable conditions. This can be better understood by taking the first time derivative of x, which gives an expression for the velocity of the oscillator. In doing so, one finds that v is proportional to $\sin(\omega t + \phi)$. When the applied force is in phase with the velocity, the rate at which work is done on the oscillator by the force **F** (or the power) equals Fv. Since the quantity Fv is always positive when **F** and **v** are in phase, we conclude that *at resonance the applied force is in phase with the velocity and the power transferred to the oscillator is a maximum.*

Figure 13.18 is a graph of the amplitude as a function of frequency for the forced oscillator, with and without a retarding force. Note that the amplitude

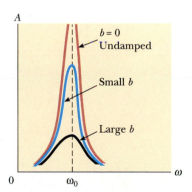

FIGURE 13.18 Graph of amplitude versus frequency for a damped oscillator when a periodic driving force is present. When the frequency of the driving force equals the natural frequency, ω_0, resonance occurs. Note that the shape of the resonance curve depends on the size of the damping coefficient, b.

FIGURE 13.19   If pendulum *P* is set into oscillation, pendulum *Q* will eventually oscillate with the greatest amplitude because its length is equal to that of *P* and so they have the same natural frequency of vibration. The pendulums are oscillating in a direction perpendicular to the plane formed by the stationary strings.

increases with decreasing damping ($b \rightarrow 0$) and that the resonance curve broadens as the damping increases. Under steady-state conditions and at any driving frequency, the energy transferred into the system equals the energy lost because of the damping force; hence, the average total energy of the oscillator remains constant. In the absence of a damping force ($b = 0$), we see from Equation 13.35 that the steady-state amplitude approaches infinity as $\omega \rightarrow \omega_0$. In other words, if there are no losses in the system and if we continue to drive an initially motionless oscillator with a periodic force that is in phase with the velocity, the amplitude of motion builds without limit (Fig. 13.18). This limitless building does not occur in practice because some damping is always present. That is, at resonance the amplitude is large but finite for small damping.

One experiment that demonstrates a resonance phenomenon is illustrated in Figure 13.19. Several pendulums of different lengths are suspended from a stretched string. If one of them, such as *P*, is set in sideways motion, the others begin to oscillate because they are coupled by the stretched string. Pendulum *Q*, whose length is the same as that of *P* (and hence the two pendulums have the same natural frequency), oscillates with the greatest amplitude.

Later in the text we shall see that resonance appears in other areas of physics. For example, certain electrical circuits have natural frequencies. A bridge has natural frequencies that can be set into resonance by an appropriate driving force. A striking example of such resonance occurred in 1940, when the Tacoma Narrows bridge in Washington was destroyed by resonant vibrations. Although the winds were not particularly strong on that occasion, the bridge ultimately collapsed because turbulences generated by the wind blowing through the bridge structure occurred at a frequency that matched the natural frequency of the structure.

Many other examples of resonant vibrations can be cited. A resonant vibration you may have experienced is the "singing" of telephone wires in the wind. Machines often break if one vibrating part is at resonance with some other moving part. Finally, soldiers marching in cadence across bridges have been known to set up resonant vibrations in the structure, causing it to collapse. In one famous accident, which occurred in France in 1850, a collapsed suspension bridge resulted in the death of 226 soldiers.

SUMMARY

The position of a simple harmonic oscillator varies periodically in time according to the expression

$$x = A \cos(\omega t + \phi) \tag{13.1}$$

where A is the amplitude of the motion, ω is the angular frequency, and ϕ is the phase constant. The value of ϕ depends on the initial position and velocity of the oscillator.

The time T for one complete vibration is called the **period** of the motion:

$$T = \frac{2\pi}{\omega} \tag{13.2}$$

The inverse of the period is the **frequency** of the motion, which equals the number of oscillations per second.

The **velocity** and **acceleration** of a simple harmonic oscillator are

$$v = \frac{dx}{dt} = -\omega A \sin(\omega t + \phi) \qquad (13.5)$$

$$a = \frac{dv}{dt} = -\omega^2 A \cos(\omega t + \phi) \qquad (13.6)$$

Thus, the maximum speed is ωA, and the maximum acceleration is $\omega^2 A$. The speed is zero when the oscillator is at its turning points, $x = \pm A$, and a maximum at the equilibrium position, $x = 0$. The magnitude of the acceleration is a maximum at the turning points and zero at the equilibrium position.

A mass-spring system moves in simple harmonic motion on a frictionless surface, with a period

$$T = \frac{2\pi}{\omega} = 2\pi\sqrt{\frac{m}{k}} \qquad (13.15)$$

The **kinetic energy** and **potential energy** for a simple harmonic oscillator vary with time and are given respectively by

$$K = \tfrac{1}{2}mv^2 = \tfrac{1}{2}m\omega^2 A^2 \sin^2(\omega t + \phi) \qquad (13.17)$$

$$U = \tfrac{1}{2}kx^2 = \tfrac{1}{2}kA^2 \cos^2(\omega t + \phi) \qquad (13.18)$$

The **total energy** of a simple harmonic oscillator is a constant of the motion and is

$$E = \tfrac{1}{2}kA^2 \qquad (13.19)$$

The potential energy of a simple harmonic oscillator is a maximum when the particle is at its turning points (maximum displacement from equilibrium) and zero at the equilibrium position. The kinetic energy is zero at the turning points and a maximum at the equilibrium position.

A **simple pendulum** of length L moves in simple harmonic motion for small angular displacements from the vertical, with a period

$$T = 2\pi\sqrt{\frac{L}{g}} \qquad (13.23)$$

A **physical pendulum** moves in simple harmonic motion about a pivot that does not go through the center of mass. The period of this motion is

$$T = 2\pi\sqrt{\frac{I}{mgd}} \qquad (13.25)$$

where I is the moment of inertia about an axis through the pivot and d is the distance from the pivot to the center of mass.

QUESTIONS

1. What is the total distance traveled by a body executing simple harmonic motion in a time equal to its period if its amplitude is A?

2. If the coordinate of a particle varies as $x = -A \cos \omega t$, what is the phase constant in Equation 13.1? At what position does the particle begin its motion?

3. Does the displacement of an oscillating particle between $t = 0$ and a later time t necessarily equal the position of the particle at time t? Explain.

4. Determine whether or not the following quantities can be in the same direction for a simple harmonic oscillator: (a) displacement and velocity, (b) velocity and acceleration, (c) displacement and acceleration.

5. Can the amplitude A and phase constant ϕ be deter-

mined for an oscillator if only the position is specified at $t = 0$? Explain.

6. Describe qualitatively the motion of a mass-spring system when the mass of the spring is not neglected.

7. If a mass-spring system is hung vertically and set into oscillation, why does the motion eventually stop?

8. Explain why the kinetic and potential energies of a mass-spring system can never be negative.

9. A mass-spring system undergoes simple harmonic motion with an amplitude A. Does the total energy change if the mass is doubled but the amplitude is not changed? Do the kinetic and potential energies depend on the mass? Explain.

10. What happens to the period of a simple pendulum if the pendulum's length is doubled? What happens to the period if the mass that is suspended is doubled?

11. A simple pendulum is suspended from the ceiling of a stationary elevator, and the period is determined. Describe the changes, if any, in the period when the elevator (a) accelerates upward, (b) accelerates downward, and (c) moves with constant velocity.

12. A simple pendulum undergoes simple harmonic motion when θ is small. Is the motion periodic when θ is large? How does the period of motion change as θ increases?

13. Give a few examples of damped oscillations that are commonly observed.

14. Will damped oscillations occur for any values of b and k? Explain.

15. Is it possible to have damped oscillations when a system is at resonance? Explain.

16. At resonance, what does the phase constant ϕ equal in Equation 13.33? (*Hint:* Compare this equation with the expression for the driving force, which must be in phase with the velocity at resonance.)

17. A platoon of soldiers marches in step along a road. Why are they ordered to break step when crossing a bridge?

18. Give as many examples as you can in the workings of an automobile where the motion is simple harmonic or damped.

19. If a grandfather clock were running slow, how could we adjust the length of the pendulum to correct the time?

PROBLEMS

Section 13.1   Simple Harmonic Motion

1. The displacement of a particle at $t = 0.25$ s is given by the expression $x = (4.0 \text{ m}) \cos(3.0\pi t + \pi)$, where x is in meters and t is in seconds. Determine (a) the frequency and period of the motion, (b) the amplitude of the motion, (c) the phase constant, and (d) the displacement of the particle at $t = 0.25$ s.

2. A small ball is set in horizontal motion by rolling it with a speed of 3.00 m/s across a room 12.0 m long, between two walls. Assume that the collisions made with each wall are perfectly elastic and that the motion is perpendicular to the two walls. (a) Show that the motion is periodic, and (b) determine its period. Is this motion simple harmonic? Explain.

3. A ball dropped from a height of 4.00 m makes a perfectly elastic collision with the ground. Assuming no energy lost due to air resistance, (a) show that the motion is periodic and (b) determine the period of the motion. (c) Is the motion simple harmonic? Explain.

4. If the initial position and velocity of an object moving in simple harmonic motion are x_0, v_0, and a_0, and if the angular frequency of oscillation is ω, (a) show that the position and speed of the object for all time can be written as

$$x(t) = x_0 \cos \omega t + \left(\frac{v_0}{\omega}\right) \sin \omega t$$

$$v(t) = -x_0 \omega \sin \omega t + v_0 \cos \omega t$$

(b) If the amplitude of the motion is A, show that

$$v^2 - ax = v_0^2 - a_0 x_0 = A^2 \omega^2$$

5. The displacement of an object is $x = (8.0 \text{ cm}) \cos(2.0t + \pi/3)$, where x is in centimeters and t is in seconds. Calculate (a) the speed and acceleration at $t = \pi/2$ s, (b) the maximum speed and the earliest time $(t > 0)$ at which the particle has this speed, and (c) the maximum acceleration and the earliest time $(t > 0)$ at which the particle has this acceleration.

6. A 20-g particle moves in simple harmonic motion with a frequency of 3.0 oscillations/s and an amplitude of 5.0 cm. (a) Through what total distance does the particle move during one cycle of its motion? (b) What is its maximum speed? Where does this occur? (c) Find the maximum acceleration of the particle. Where in the motion does the maximum acceleration occur?

7. A particle moving along the x axis in simple harmonic motion starts from the origin at $t = 0$ and moves to the right. If the amplitude of its motion is 2.00 cm and the frequency is 1.50 Hz, (a) show that its displacement is given by $x = (2.00 \text{ cm}) \sin(3.00t)$. Determine (b) the maximum speed and the earliest time $(t > 0)$ at which the particle has this speed, (c) the maximum acceleration and the earliest time $(t > 0)$ at which the particle has this acceleration, and (d) the total distance traveled between $t = 0$ and $t = 1.00$ s.

☐ indicates problems that have full solutions available in the Student Solutions Manual and Study Guide.

8. A piston in an automobile engine is in simple harmonic motion. If its amplitude of oscillation from centerline is ± 5.0 cm and its mass is 2.0 kg, find the maximum velocity and acceleration of the piston when the auto engine is running at the rate of 3600 rev/min.

Section 13.2   Mass Attached to a Spring

(*Note:* Neglect the mass of the spring in all of these problems.)

9. A block of unknown mass is attached to a spring of spring constant 6.50 N/m and undergoes simple harmonic motion with an amplitude of 10.0 cm. When the mass is halfway between its equilibrium position and the endpoint, its speed is measured to be + 30.0 cm/s. Calculate (a) the mass of the block, (b) the period of the motion, and (c) the maximum acceleration of the block.

10. A spring stretches by 3.9 cm when a 10-g mass is hung from it. If a 25-g mass attached to this spring oscillates in simple harmonic motion, calculate the period of motion.

11. A 7.00-kg mass is hung from the bottom end of a vertical spring fastened to an overhead beam. The mass is set into vertical oscillations having a period of 2.60 s. Find the force constant of the spring.

12. A 1.0-kg mass attached to a spring of force constant 25 N/m oscillates on a horizontal, frictionless track. At $t = 0$, the mass is released from rest at $x = -3.0$ cm. (That is, the spring is compressed by 3.0 cm.) Find (a) the period of its motion, (b) the maximum values of its speed and acceleration, and (c) the displacement, velocity, and acceleration as functions of time.

13. A simple harmonic oscillator takes 12.0 s to undergo five complete vibrations. Find (a) the period of its motion, (b) the frequency in Hz, and (c) the angular frequency in rad/s.

14. A 1.0-kg mass is attached to a horizontal spring. The spring is initially stretched by 0.10 m and the mass is released from rest there. After 0.50 s, the speed of the mass is zero. What is the maximum speed of the mass?

15. A 0.50-kg mass attached to a spring of force constant 8.0 N/m vibrates in simple harmonic motion with an amplitude of 10 cm. Calculate (a) the maximum value of its speed and acceleration, (b) the speed and acceleration when the mass is 6.0 cm from the equilibrium position, and (c) the time it takes the mass to move from $x = 0$ to $x = 8.0$ cm.

16. (a) A 100-g block is placed on top of a 200-g block as shown in Figure P13.16. The coefficient of static friction between the blocks is 0.20. The lower block is now moved back and forth horizontally in simple harmonic motion having an amplitude of 6.0 cm. Keeping the amplitude constant, what is the highest frequency for which the upper block will not slip relative to the lower block? (b) Suppose the lower block is moved vertically in simple harmonic motion rather than horizontally. The frequency is held constant at 2.0 oscillations/s while the amplitude is gradually increased. Determine the amplitude at which the upper block will no longer maintain contact with the lower block.

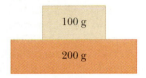

FIGURE P13.16

17. A particle that hangs from a spring oscillates with an angular frequency of 2.00 rad/s. The spring is suspended from the ceiling of an elevator car and hangs motionless (relative to the elevator car) as the car descends at a constant speed of 1.50 m/s. The car then stops suddenly. (a) With what amplitude does the particle oscillate? (b) What is the equation of motion for the particle? (Choose the upward direction to be positive.)

17A. A particle that hangs from a spring oscillates with an angular frequency ω. The spring is suspended from the ceiling of an elevator car and hangs motionless (relative to the elevator car) as the car descends at a constant speed v. The car then stops suddenly. (a) With what amplitude does the particle oscillate? (b) What is the equation of motion for the particle? (Choose the upward direction to be positive.)

Section 13.3   Energy of the Simple Harmonic Oscillator

(*Note:* Neglect the mass of the spring in all these problems.)

18. A 1.5-kg block at rest on a tabletop is attached to a horizontal spring having constant 19.6 N/m. The spring is initially unstretched. A constant 20.0-N horizontal force is applied to the object causing the spring to stretch. (a) Determine the speed of the block after it has moved 0.30 m from equilibrium if the surface between the block and tabletop is frictionless. (b) Answer part (a) if the coefficient of kinetic friction between block and tabletop is 0.20.

18A. A block of mass m, at rest on a tabletop is attached to a horizontal spring having constant k. The spring is initially unstretched. A constant horizontal force F is applied to the object causing the spring to stretch. (a) Determine the speed of the block after it has moved a distance d from equilibrium if the surface between the block and tabletop is frictionless. (b) Answer part (a) if the coefficient of kinetic friction between block and tabletop is μ.

19. An automobile having a mass of 1000 kg is driven into a brick wall in a safety test. The bumper behaves

like a spring of constant 5.0×10^6 N/m and compresses 3.16 cm as the car is brought to rest. What was the speed of the car before impact, assuming no energy is lost during impact with the wall?

20. A bullet of mass 10.0 g is fired into and embeds in a 2.00-kg block attached to a spring with constant 19.6 N/m. (a) How far will the spring be compressed if the speed of the bullet just before striking the block is 300 m/s and the block slides along a frictionless track? (b) Answer part (a) if the coefficient of kinetic friction between track and block is 0.200.

21. The amplitude of a system moving in simple harmonic motion is doubled. Determine the change in (a) the total energy, (b) the maximum speed, (c) the maximum acceleration, and (d) the period.

22. A 50-g mass connected to a spring of force constant 35 N/m oscillates on a horizontal, frictionless surface with an amplitude of 4.0 cm. Find (a) the total energy of the system and (b) the speed of the mass when the displacement is 1.0 cm. When the displacement is 3.0 cm, find (c) the kinetic energy and (d) the potential energy.

23. A particle executes simple harmonic motion with an amplitude of 3.00 cm. At what displacement from the midpoint of its motion does its speed equal one half of its maximum speed?

24. A 2.00-kg mass is attached to a spring and placed on a horizontal smooth surface. A horizontal force of 20.0 N is required to hold the mass at rest when it is pulled 0.200 m from its equilibrium position (the origin of the x axis). The mass is now released from rest with an initial displacement of $x_0 = 0.200$ m, and it subsequently undergoes simple harmonic oscillations. Find (a) the force constant of the spring, (b) the frequency of the oscillations, and (c) the maximum speed of the mass. Where does this maximum speed occur? (d) Find the maximum acceleration of the mass. Where does it occur? (e) Find the total energy of the oscillating system. When the displacement equals one-third the maximum value, find (f) the speed and (g) the acceleration.

Section 13.4   The Pendulum

25. A simple pendulum has a period of 2.50 s. (a) What is its length? (b) What would its period be on the Moon, where $g_{Moon} = 1.67$ m/s^2?

26. A light rod extends rigidly 0.50 m out from one end of a meterstick. The stick is suspended at the far end of the rod and set into oscillation. (a) Determine the period of oscillation. (b) By what percentage does this differ from a 1.0-m-long simple pendulum?

27. A simple pendulum is 5.0 m long. (a) What is the period of simple harmonic motion for this pendulum if it is located in an elevator accelerating upward at 5.0 m/s^2? (b) What is the answer to part (a) if the elevator is accelerating downward at 5.0 m/s^2? (c) What is the period of simple harmonic motion for this pendulum if it is placed in a truck that is accelerating horizontally at 5.0 m/s^2?

27A. A simple pendulum has a length L. (a) What is the period of simple harmonic motion for this pendulum if it is located in an elevator accelerating upward with an acceleration a? (b) What is the answer to part (a) if the elevator is accelerating downward with an acceleration a? (c) What is the period of simple harmonic motion for this pendulum if it is placed in a truck that is accelerating horizontally with an acceleration a?

28. A mass is attached to the end of a string to form a simple pendulum. The period of its harmonic motion is measured for small angular displacements and three lengths, each time clocking the motion with a stopwatch for 50 oscillations. For lengths of 1.00 m, 0.75 m, and 0.50 m, total times of 99.8 s, 86.6 s, and 71.1 s are measured for 50 oscillations. (a) Determine the period of motion for each length. (b) Determine the mean value of g obtained from these three independent measurements, and compare it with the accepted value. (c) Plot T^2 versus L, and obtain a value for g from the slope of your best-fit straight-line graph. Compare this value with that obtained in part (b).

29. A simple pendulum has a mass of 0.250 kg and a length of 1.00 m. It is displaced through an angle of 15.0° and then released. What are (a) the maximum speed? (b) the maximum angular acceleration? (c) the maximum restoring force?

30. Consider the physical pendulum of Figure 13.11. (a) If its moment of inertia about an axis passing through its center of mass and parallel to the axis passing through its pivot point is I_{CM}, show that its period is

$$T = 2\pi\sqrt{\frac{I_{CM} + md^2}{mgd}}$$

where d is the distance between the pivot point and center of mass. (b) Show that the period has a minimum value when d satisfies $md^2 = I_{CM}$.

31. A simple pendulum has a length of 3.00 m. Determine the change in its period if it is taken from a point where $g = 9.80$ m/s^2 to an elevation where the free-fall acceleration decreases to 9.79 m/s^2.

32. A horizontal rod 1.0 m long, with masss 2.0 kg, is suspended from a wire at its center, to form a torsional pendulum. If the resulting period is 3.0 minutes, what is the torsion constant for the wire?

33. A physical pendulum in the form of a planar body moves in simple harmonic motion with a frequency of 0.450 Hz. If the pendulum has a mass of 2.20 kg and the pivot is located 0.350 m from the center of mass, determine the moment of inertia of the pendulum.

34. The angular displacement of a pendulum is represented by the equation $\theta = 0.32 \cos \omega t$, where θ is in radians and $\omega = 4.43$ rad/s. Determine the period and length of the pendulum.

35. A clock balance wheel has a period of oscillation of 0.250 s. The wheel is constructed so that 20.0 g of mass is concentrated around a rim of radius 0.500 cm. What are (a) the wheel's moment of inertia and (b) the torsion constant of the attached spring?

*Section 13.5 Comparing Simple Harmonic Motion with Uniform Circular Motion

36. While riding behind a car traveling at 3.0 m/s, you notice that one of the car's tires has a small hemispherical boss on its rim, as in Figure P13.36. (a) Explain why the boss, from your viewpoint behind the car, executes simple harmonic motion. (b) If the radii of the car's tires are 0.30 m, what is the boss's period of oscillation?

Boss

FIGURE P13.36

37. Consider the simplified single-piston engine in Figure P13.37. If the wheel rotates at a constant angular speed ω, explain why the piston rod oscillates in simple harmonic motion.

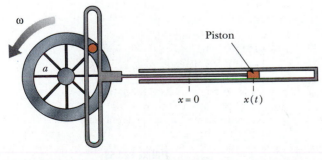

FIGURE P13.37

*Section 13.6 Damped Oscillations

38. Show that the time rate of change of mechanical energy for a damped, undriven oscillator is given by $dE/dt = -bv^2$ and hence is always negative. (*Hint:* Differentiate the expression for the mechanical energy of an oscillator, $E = \frac{1}{2}mv^2 + \frac{1}{2}kx^2$, and use Eq. 13.29.)

39. A pendulum of length 1.00 m is released from an initial angle of 15.0°. After 1000 s, its amplitude is reduced by friction to 5.5°. What is the value of $b/2m$?

*Section 13.7 Forced Oscillations

40. A 2.00-kg mass attached to a spring is driven by an external force $F = (3.00 \text{ N}) \cos(2\pi t)$. If the force constant of the spring is 20.0 N/m, determine (a) the period and (b) the amplitude of the motion. (*Hint:* Assume there is no damping, that is, $b = 0$, and use Eq. 13.34.)

41. Calculate the resonant frequencies of (a) a 3.00-kg mass attached to a spring of force constant 240 N/m and (b) a simple pendulum 1.50 m in length.

42. Consider an undamped forced oscillator at resonance so that $\omega = \omega_0 = \sqrt{k/m}$. The equation of motion is

$$\frac{d^2 x}{dt^2} + \omega^2 x = \left(\frac{F_0}{m}\right) \cos \omega t$$

Show by direct substitution that the solution of this equation is

$$x(t) = x_0 \cos \omega t + \left(\frac{v_0}{\omega}\right) \sin \omega t + \left(\frac{F_0}{2m\omega}\right) t \sin \omega t$$

where x_0 and v_0 are its initial position and velocity.

43. A weight of 40.0 N is suspended from a spring that has a force constant of 200 N/m. The system is undamped and is subjected to a harmonic force of frequency 10.0 Hz, resulting in a forced-motion amplitude of 2.00 cm. Determine the maximum value of the force.

ADDITIONAL PROBLEMS

44. A car with bad shock absorbers bounces up and down with a period of 1.5 s after hitting a bump. The car has a mass of 1500 kg and is supported by four springs of equal force constant k. Determine a value for k.

45. A large passenger of mass 150 kg sits in the car of Problem 44. What is the new period of oscillation?

46. A block rests on a flat plate that executes vertical simple harmonic motion with a period of 1.2 s. What is the maximum amplitude of the motion for which the block does not separate from the plate?

47. When the simple pendulum illustrated in Figure

P13.47 makes an angle with the vertical, its speed is v.
(a) Calculate the total mechanical energy of the pendulum as a function of v and θ. (b) Show that when θ is small, the potential energy can be expressed as $\frac{1}{2} mgL\theta^2 = \frac{1}{2} m\omega^2 s^2$. (*Hint:* In part (b), approximate $\cos\theta$ by $\cos\theta \approx 1 - \theta^2/2$.)

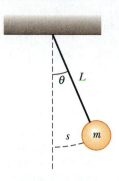

FIGURE P13.47

48. A horizontal platform vibrates with simple harmonic motion in the horizontal direction with a period of 2.0 s. A body on the platform starts to slide when the amplitude of vibration reaches 0.30 m. Find the coefficient of static friction between body and platform.

49. A particle of mass m slides inside a hemispherical bowl of radius R. Show that, for small displacements from equilibrium, the particle moves in simple harmonic motion with an angular frequency equal to that of a simple pendulum of length R. That is, $\omega = \sqrt{g/R}$.

50. A horizontal plank of mass m and length L is pivoted at one end, and the opposite end is attached to a spring of force constant k (Fig. P13.50). The moment of inertia of the plank about the pivot is $\frac{1}{3} mL^2$. When the plank is displaced a small angle θ from the horizontal and released, show that it moves with simple harmonic motion with an angular frequency $\omega = \sqrt{3k/m}$.

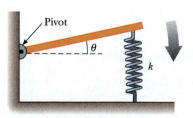

FIGURE P13.50

51. A mass M is attached to the end of a uniform rod of mass M and length L that is pivoted at the top (Fig. P13.51). (a) Determine the tensions in the rod at the pivot and at the point P when the system is stationary. (b) Calculate the period of oscillation for small dis-

placements from equilibrium, and determine this period for $L = 2.00$ m. (*Hint:* Assume the mass at the end of the rod is a point mass and use Eq. 13.25.)

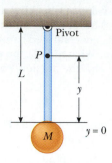

FIGURE P13.51

52. Consider a damped oscillator as illustrated in Figure 13.16. Assume the mass is 375 g, the spring constant is 100 N/m, and $b = 0.100$ kg/s. (a) How long does it take for the amplitude to drop to half its initial value? (b) How long does it take for the mechanical energy to drop to half its initial value? (c) Show that, in general, the rate at which the amplitude decreases in a damped harmonic oscillator is one-half the rate at which the mechanical energy decreases.

53. A small, thin disk of radius r and mass m is attached rigidly to the face of a second thin disk of radius R and mass M as shown in Figure P13.53. The center of the small disk is located at the edge of the large disk. The large disk is mounted at its center on a frictionless axle. The assembly is rotated through an angle θ from its equilibrium position and released. (a) Show that the speed of the center of the small disk as it passes through the equilibrium position is

$$v = 2\left[\frac{Rg(1 - \cos\theta)}{(M/m) + (r/R)^2 + 2} \right]^{1/2}$$

(b) Show that the period of the motion is

$$T = 2\pi\left[\frac{(M + 2m)R^2 + mr^2}{2mgR} \right]^{1/2}$$

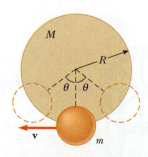

FIGURE P13.53

54. A mass m is connected to two springs of force constants k_1 and k_2 as in Figures P13.54a and P13.54b. In each case, the mass moves on a frictionless table and is displaced from equilibrium and released. Show that in each case the mass exhibits simple harmonic motion with periods

(a) $\quad T = 2\pi\sqrt{\dfrac{m(k_1 + k_2)}{k_1 k_2}}$

(b) $\quad T = 2\pi\sqrt{\dfrac{m}{k_1 + k_2}}$

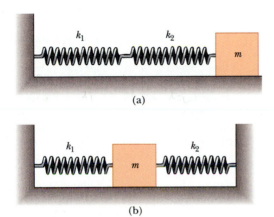

(a)

(b)

FIGURE P13.54

55. A pendulum of length L and mass M has a spring of force constant k connected to it at a distance h below its point of suspension (Fig. P13.55). Find the frequency of vibration of the system for small values of the amplitude (small θ). (Assume the vertical suspension of length L is rigid, but neglect its mass.)

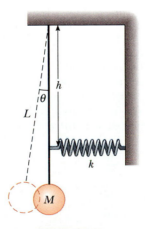

FIGURE P13.55

56. A mass m is oscillating freely on a vertical spring (Fig. P13.56). When $m = 0.810$ kg, the period is 0.910 s. An unknown mass on the same spring has a period of 1.16 s. Determine (a) the spring constant k and (b) the unknown mass.

56A. A mass m is oscillating freely on a vertical spring with a period T (Fig. P13.56). An unknown mass m' on the same spring oscillates with a period T'. Determine (a) the spring constant k and (b) the unknown mass m'.

FIGURE P13.56

57. A large block P executes horizontal simple harmonic motion by sliding across a frictionless surface with a frequency $f = 1.5$ Hz. Block B rests on it, as shown in Figure P13.57, and the coefficient of static friction between the two is $\mu_s = 0.60$. What maximum amplitude of oscillation can the system have if the block is not to slip?

57A. A large block P executes horizontal simple harmonic motion by sliding across a frictionless surface with a frequency f. Block B rests on it, as shown in Figure P13.57, and the coefficient of static friction between the two is μ_s. What maximum amplitude of oscillation can the system have if the block is not to slip?

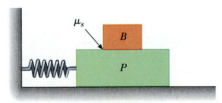

FIGURE P13.57

58. A long, thin rod of mass M and length L oscillates about its center on a cylinder of radius R (Fig. P13.58). Show that small displacements give rise to simple harmonic motion with a period $\pi L/\sqrt{3gR}$.

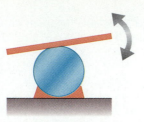

FIGURE P13.58

59. A simple pendulum having a length of 2.23 m and a mass of 6.74 kg is given an initial speed of 2.06 m/s at its equilibrium position. Assume it undergoes simple harmonic motion and determine its (a) period, (b) total energy, and (c) maximum angular displacement.

60. A mass, $m_1 = 9.0$ kg, is in equilibrium while connected to a light spring of constant $k = 100$ N/m that is fastened to a wall as in Figure P13.60a. A second mass, $m_2 = 7.0$ kg, is slowly pushed up against mass m_1, compressing the spring by the amount $A = 0.2$ m, as shown in Figure P13.60b. The system is then released, causing both masses to start moving to the right on the frictionless surface. (a) When m_1 reaches the equilibrium point, m_2 loses contact with

m_1 (Fig. P13.60c) and moves to the right with speed v. Determine the value of v. (b) How far apart are the masses when the spring is fully stretched for the first time (D in Fig. P13.60d)? (*Hint:* First determine the period of oscillation and the amplitude of the m_1-spring system after m_2 loses contact with m_1.)

61. The mass of the deuterium molecule (D_2) is twice that of the hydrogen molecule (H_2). If the vibrational frequency of H_2 is 1.30×10^{14} Hz, what is the vibrational frequency of D_2, assuming that the "spring constant" of attracting forces is the same for the two molecules?

62. Show that if a torsional pendulum is twisted through an angle and then held the potential energy is $U = \frac{1}{2}\kappa\theta^2$.

63. A 2.00-kg block hangs without vibrating at the end of a spring ($k = 500$ N/m) that is attached to the ceiling of an elevator car. The car is rising with an upward acceleration of $g/3$ when the acceleration suddenly ceases (at $t = 0$). (a) What is the angular frequency of oscillation of the block after the acceleration ceases? (b) By what amount is the spring stretched during the time that the elevator car is accelerating? (c) What are the amplitude of the oscillation and the initial phase angle observed by a rider in the car? Take the upward direction to be positive.

64. A solid sphere (radius = R) rolls without slipping in a cylindrical trough (radius = $5R$) as shown in Figure P13.64. Show that, for small displacements from equilibrium perpendicular to the length of the trough, the sphere executes simple harmonic motion with a period $T = 2\pi\sqrt{28R/5g}$.

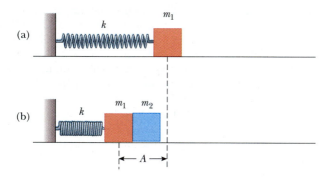

(a)

(b)

(c)

(d)

FIGURE P13.60

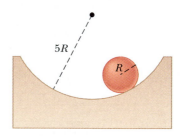

FIGURE P13.64

65. A mass m is connected to two rubber bands of length L, each under tension T, as in Figure P13.65. The mass is displaced vertically by a small distance y. Assuming the tension does not change, show that (a) the restoring force is $-(2T/L)y$ and (b) the sys-

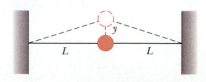

FIGURE P13.65

tem exhibits simple harmonic motion with an angular frequency $\omega = \sqrt{2T/mL}$.

66. A light, cubical container of volume a^3 is initially filled with a liquid of mass density ρ. The cube is initially supported by a light string to form a pendulum of length L_0 measured from the center of mass of the filled container. The liquid is allowed to flow from the bottom of the container at a constant rate (dM/dt). At any time t, the level of the fluid in the container is h and the length of the pendulum is L (measured relative to the instantaneous center of mass). (a) Sketch the apparatus and label the dimensions a, h, L_0, and L. (b) Find the time rate of change of the period as a function of time t. (c) Find the period as a function of time.

67. When a mass M connected to the end of a spring of mass $m_s = 7.40$ g and force constant k is set into simple harmonic motion, the period of its motion is

$$T = 2\pi\sqrt{\frac{M + (m_s/3)}{k}}$$

A two-part experiment is conducted using various masses suspended vertically from the spring. (a) Displacements of 17.0, 29.3, 35.3, 41.3, 47.1, and 49.3 cm are measured for M values of 20.0, 40.0, 50.0, 60.0, 70.0, and 80.0 g, respectively. Construct a graph of Mg versus x, and perform a linear least-squares fit to the data. From the slope of your graph, determine a value for k for this spring. (b) The system is now set into simple harmonic motion, and periods are measured with a stopwatch. With $M = 80.0$ g, the total time for 10 oscillations is measured to be 13.41 s. The experiment is repeated with M values of 70.0, 60.0, 50.0, 40.0, and 20.0 g, with corresponding times for 10 oscillations of 12.52, 11.67, 10.67, 9.62, and 7.03 s. Obtain experimental values for T for each of these M values. Plot a graph of T^2 versus M and determine a value for k from the slope of the linear least-squares fit through the data points. Compare this value of k with that obtained in part (a). (c) Obtain a value for m_s from your graph and compare it with the given value of 7.40 g.

SPREADSHEET PROBLEMS

S1. Use a spreadsheet to plot the position x of an object undergoing simple harmonic motion as a function of time. (a) Use $A = 2.00$ m, $\omega = 5.00$ rad/s, and $\phi = 0$. On the same graph, plot the function for $\phi = \pi/4$, $\pi/2$, and π. Explain how the different phase constants affect the position of the object. (b) Repeat part (a) with $\omega = 7.50$ rad/s. How does the change in ω affect the plots?

S2. The potential and kinetic energies of a particle of mass m oscillating on a spring are given by Equations 13.17 and 13.18. Use a spreadsheet to plot $U(t)$,

$K(t)$, and their sum as functions of time on the same graph. The input variables should be k, A, m, and ϕ. What happens to the sum of $U(t)$ and $K(t)$ as the input variables are changed?

S3. Consider a simple pendulum of length L that is displaced from the vertical by an angle θ_0 and released. In this problem, you are to determine the angular displacement for any value of the initial angular displacement. Modify Spreadsheet 6.1 or write your own program to integrate the equation of motion of a simple pendulum:

$$\frac{d^2\theta}{dt^2} = -\frac{g}{L}\sin\theta$$

Take the initial conditions to be $\theta = \theta_0$ and $d\theta/dt = 0$ at $t = 0$. Choose values for θ_0 and L and find the angular displacement θ as a function of time. Using the same values of θ_0, compare your results for θ with those obtained from $\theta(t) = \theta_0 \cos \omega_0 t$ where $\omega_0 = \sqrt{g/L}$. Repeat for different values of θ_0. Be sure to include large initial angular displacements. Does the period change when the initial angular displacement is changed? How do the periods for large values of θ_0 compare to those for small values of θ_0? (*Note:* Using the modified Euler's method to solve this differential equation, you may find that the amplitude tends to increase with time. The fourth-order Runge-Kutta method would be a better choice to solve the differential equation. However, if you chose dt small enough, the solution using the modified Euler's method will still be good.)

S4. In the real world friction is always present. Modify your spreadsheet for Problem S3 to include the force of air resistance: $F_{air} = -bv$. In this situation, the equation of motion is

$$\frac{d^2\theta}{dt^2} = -\frac{g}{L}\sin\theta - \frac{b}{m}\frac{d\theta}{dt}$$

where m is the mass of the pendulum and b is the drag coefficient. Integrate this equation and calculate the sum of the kinetic and potential energies as a function of time using your calculated values of θ and $d\theta/dt$. Since energy is not constant, when in the pendulum's cycle is energy dissipated the fastest?

S5. The differential equation for the position x of a driven, damped harmonic oscillator of mass m is

$$m\frac{d^2x}{dt^2} + b\frac{dx}{dt} + m\omega_0^2 x = F_0 \cos \omega t$$

where ω_0 and b are constants, F_0 is the amplitude of the driving force, and ω is the driving angular frequency. For the case of a mass on a spring, $\omega_0 = \sqrt{k/m}$, where k is the force constant. Write a computer program or spreadsheet to integrate this equation of motion. Your input variables are the damping coefficient b, the mass m, ω_0 (the natural frequency of the system), and F_0. Choose values for the initial velocity and position and calculate the position x as a function of time. Vary the driving angular frequency ω. Pick some values near ω_0. Does resonance occur? (*Note*: Your program may "crash" or blow up near resonance. The amplitude of the motion may have very large values when $\omega = \omega_0$. Also, the Euler method usually tends to overestimate the solution. Other methods such as the 4th-order Runge-Kutta method discussed in the Supplement give a more accurate solution.)

CHAPTER 14

The Law of Gravity

Astronauts F. Story Musgrave and Jeffrey A. Hoffman complete the final of five space walks needed to repair the Hubble Space Telescope in December 1993. *(Courtesy NASA)*

Prior to 1686, a large amount of data had been collected on the motions of the Moon and the planets but a clear understanding of the forces that caused these celestial bodies to move the way they did was not available. In that year, however, Newton provided the key that unlocked the secrets of the heavens. He knew, from his first law, that a net force had to be acting on the Moon because without such a force the Moon would move in a straight-line path rather than in its almost circular orbit. Newton reasoned that this force was the gravitational attraction that the Earth exerts on the Moon. He also concluded that there could be nothing special about the Earth-Moon system or about the Sun and its planets that would cause gravitational forces to act on them alone. In other words, he saw that the same force of attraction that causes the Moon to follow its path also causes an apple to fall to Earth from a tree. He wrote, "I deduced that the forces which keep the planets in their orbs must be reciprocally as the squares of their distances from the centers about which they revolve; and thereby compared the force requisite to keep the Moon in her orb with force of gravity at the surface of the Earth; and found them answer pretty nearly.''

In this chapter we study the law of gravity. Emphasis is placed on describing the motion of the planets, because astronomical data provide an important test of the validity of the law of gravity. We show that the laws of planetary motion developed by Johannes Kepler follow from the law of gravity and the concept of the conservation of angular momentum. A general expression for the gravitational potential energy is derived, and the energetics of planetary and satellite motion are treated. The law of gravity is also used to determine the force between a particle and an extended body.

14.1   NEWTON'S LAW OF GRAVITY

It has been said that Newton was struck on the head by a falling apple while napping under a tree (or some variation of this legend). This accident supposedly prompted him to imagine that perhaps all bodies in the Universe are attracted to each other in the same way the apple was attracted to the Earth. Newton then analyzed astronomical data on the motion of the Moon around the Earth. From this analysis, he made the bold statement that the force law governing the motion of planets has the *same* mathematical form as the force law that attracts a falling apple to the Earth.

In 1686 Newton published his work on the law of gravity in his *Mathematical Principles of Natural Philosophy*. **Newton's law of gravity** states that

> every particle in the Universe attracts every other particle with a force that is directly proportional to the product of their masses and inversely proportional to the square of the distance between them.

If the particles have masses m_1 and m_2 and are separated by a distance r, the magnitude of this gravitational force is

The law of gravity

$$F_g = G \frac{m_1 m_2}{r^2} \tag{14.1}$$

where G is a universal constant called the *universal gravitational constant,* which has been measured experimentally. Its value in SI units is

$$G = 6.672 \times 10^{-11} \frac{\text{N} \cdot \text{m}^2}{\text{kg}^2} \tag{14.2}$$

The force law given by Equation 14.1 is often referred to as an **inverse-square law** because the magnitude of the force varies as the inverse square of the separation of the particles. We can express this force in vector form by defining a unit vector $\hat{\mathbf{r}}_{12}$ (Fig. 14.1). Because this unit vector is in the direction of the displacement vector $\mathbf{r}_{12}$ directed from m_1 to m_2, the force exerted on m_2 by m_1 is

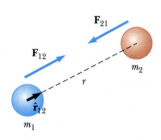

FIGURE 14.1   The gravitational force between two particles is attractive. The unit vector $\hat{\mathbf{r}}_{12}$ is directed from m_1 to m_2. Note that $\mathbf{F}_{12} = -\mathbf{F}_{21}$.

$$\mathbf{F}_{21} = -G \frac{m_1 m_2}{r_{12}^2} \hat{\mathbf{r}}_{12} \tag{14.3}$$

where the minus sign indicates that m_2 is attracted to m_1, and so the force must be directed toward m_1. Likewise, by Newton's third law the force exerted on m_1 by m_2, designated $\mathbf{F}_{12}$, is equal in magnitude to $\mathbf{F}_{21}$ and in the opposite direction. That is, these forces form an action-reaction pair, and $\mathbf{F}_{12} = -\mathbf{F}_{21}$.

There are several features of the inverse-square law that deserve some attention. The gravitational force is a field force that always exists between two particles, regardless of the medium that separates them. The force varies as the inverse square of the distance between the particles and therefore decreases rapidly with increasing separation. Finally, the gravitational force is proportional to the mass of each particle.

Properties of the gravitational force

Another important fact is that *the gravitational force exerted by a finite-size, spherically symmetric mass distribution on a particle outside the sphere is the same as if the entire mass of the sphere were concentrated at its center.* For example, the force exerted by the Earth on a particle of mass m at the Earth's surface has the magnitude

$$F_g = G\frac{M_E m}{R_E^2}$$

where M_E is the Earth's mass and R_E is the Earth's radius. This force is directed toward the center of the Earth.

14.2   MEASUREMENT OF THE GRAVITATIONAL CONSTANT

The universal gravitational constant, G, was measured in an important experiment by Henry Cavendish in 1798. The Cavendish apparatus consists of two small spheres each of mass m fixed to the ends of a light horizontal rod suspended by a fine fiber or thin metal wire, as in Figure 14.2. Two large spheres each of mass M are then placed near the smaller spheres. The attractive force between the smaller and larger spheres causes the rod to rotate and twist the wire suspension. If the system is oriented as shown in Figure 14.2, the rod rotates clockwise when viewed from above. The angle through which it rotates is measured by the deflection of a light beam reflected from a mirror attached to the vertical suspension. The deflected spot of light is an effective technique for amplifying the motion. The experiment is carefully repeated with different masses at various separations. In addition to providing a value for G, the results show that the force is attractive, proportional to the product mM, and inversely proportional to the square of the distance r.

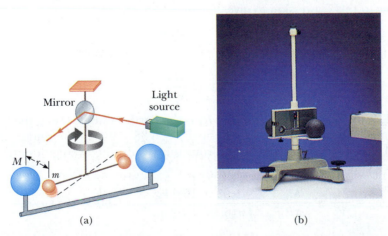

(a)   (b)

FIGURE 14.2   (a) Schematic diagram of the Cavendish apparatus for measuring G. The smaller spheres of mass m are attracted to the large spheres of mass M, and the rod rotates through a small angle. A light beam reflected from a mirror on the rotating apparatus measures the angle of rotation. The dashed line represents the original position of the rod. (b) Photograph of a student Cavendish apparatus.   *(Courtesy of PASCO Scientific)*

EXAMPLE 14.1 Three Interacting Masses

Three uniform spheres of mass 2.00 kg, 4.00 kg, and 6.00 kg are placed at the corners of a right triangle as in Figure 14.3. Calculate the resultant gravitational force on the 4.00-kg mass, assuming the spheres are isolated from the rest of the Universe.

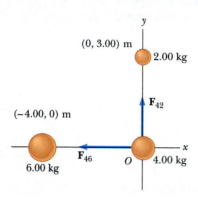

FIGURE 14.3 (Example 14.1) The resultant gravitational force on the 4.00-kg mass is the vector sum $\mathbf{F}_{46} + \mathbf{F}_{42}$.

Reasoning First we calculate the individual forces on the 4.00-kg mass due to the 2.00-kg and 6.00-kg masses separately, and then we find the vector sum to get the resultant force.

Solution The force exerted on the 4.00-kg mass by the 2.00-kg mass is directed upward and given by

$$\mathbf{F}_{42} = G \frac{m_4 m_2}{r_{42}^2} \mathbf{j}$$

$$= \left(6.67 \times 10^{-11} \frac{\text{N} \cdot \text{m}^2}{\text{kg}^2} \right) \frac{(4.00 \text{ kg})(2.00 \text{ kg})}{(3.00 \text{ m})^2} \mathbf{j}$$

$$= 5.93 \times 10^{-11} \mathbf{j} \text{ N}$$

The force exerted on the 4.00-kg mass by the 6.00-kg mass is directed to the left:

$$\mathbf{F}_{46} = G \frac{m_4 m_6}{r_{46}^2} (-\mathbf{i})$$

$$= \left(-6.67 \times 10^{-11} \frac{\text{N} \cdot \text{m}^2}{\text{kg}^2} \right) \frac{(4.00 \text{ kg})(6.00 \text{ kg})}{(4.00 \text{ m})^2} \mathbf{i}$$

$$= -10.0 \times 10^{-11} \mathbf{i} \text{ N}$$

Therefore, the resultant force on the 4.00-kg mass is

$$\mathbf{F}_4 = \mathbf{F}_{42} + \mathbf{F}_{46} = (-10.0\mathbf{i} + 5.93\mathbf{j}) \times 10^{-11} \text{ N}$$

Exercise Find the magnitude and direction of the force $\mathbf{F}_4$.

Answer 11.6×10^{-11} N at an angle of 149° with the positive *x* axis.

14.3 WEIGHT AND GRAVITATIONAL FORCE

In Chapter 5 when defining the weight of a body of mass *m* as *mg*, we referred to *g* simply as the magnitude of the free-fall acceleration. Now we are in a position to obtain a more fundamental description of *g*. Since the force on a freely falling body of mass *m* near the surface of the Earth is given by Equation 14.1, we can equate *mg* to this expression to give

$$mg = G \frac{M_E m}{R_E^2}$$

Free-fall acceleration near the Earth's surface

$$g = G \frac{M_E}{R_E^2} \tag{14.4}$$

where M_E is the mass of the Earth and R_E is the Earth's radius. Using the facts that $g = 9.80$ m/s² at the Earth's surface and $R_E \approx 6.38 \times 10^6$ m, we find from Equation 14.4 that $M_E = 5.98 \times 10^{24}$ kg. From this result, the average density of the Earth is calculated to be

$$\rho_E = \frac{M_E}{V_E} = \frac{M_E}{\frac{4}{3}\pi R_E^3} = \frac{5.98 \times 10^{24} \text{ kg}}{\frac{4}{3}\pi (6.38 \times 10^6 \text{ m})^3} = 5.50 \times 10^3 \text{ kg/m}^3$$

Since this value is about twice the density of most rocks at the Earth's surface, we conclude that the inner core of the Earth has a density much higher than the average value.

Now consider a body of mass *m* a distance *h* above the Earth's surface, or a distance *r* from the Earth's center, where $r = R_E + h$. The magnitude of the gravitational force acting on this mass is

$$F_g = G \frac{M_E m}{r^2} = G \frac{M_E m}{(R_E + h)^2}$$

If the body is in free-fall, then $F_g = mg'$ and we see that g', the free-fall acceleration experienced by an object at the altitude h, is

$$g' = \frac{GM_E}{r^2} = \frac{GM_E}{(R_E + h)^2} \qquad (14.5)$$

Variation of g with altitude

Thus, it follows that g' *decreases* with *increasing altitude*. Since the true weight of a body is mg', we see that as $r \rightarrow \infty$, the true weight approaches zero.

EXAMPLE 14.2   Variation of g with Altitude h

Determine the magnitude of the free-fall acceleration at an altitude of 500 km. By what percentage is the weight of a body reduced at this altitude?

Solution   Using Equation 14.5 with $h = 500$ km, $R_E = 6.38 \times 10^6$ m, and $M_E = 5.98 \times 10^{24}$ kg gives

$$g' = \frac{GM_E}{(R_E + h)^2}$$

$$= \frac{(6.67 \times 10^{-11} \text{ N} \cdot \text{m}^2/\text{kg}^2)(5.98 \times 10^{24} \text{ kg})}{(6.38 \times 10^6 + 0.500 + 10^6)^2 \text{ m}^2}$$

$$= \boxed{8.43 \text{ m/s}^2}$$

Since $g'/g = 8.43/9.8 = 0.86$, we conclude that the weight of a body is reduced by about 14% at an altitude of 500 km. Values of g' at other altitudes are listed in Table 14.1.

TABLE 14.1   Free-fall Acceleration, g', at Various Altitudes Above the Earth's Surface

Altitude h (km)	g' (m/s²)
1000	7.33
2000	5.68
3000	4.53
4000	3.70
5000	3.08
6000	2.60
7000	2.23
8000	1.93
9000	1.69
10 000	1.49
50 000	0.13
∞	0

14.4   KEPLER'S LAWS

The movements of the planets, stars, and other celestial bodies have been observed by people for thousands of years. In early history, scientists regarded the Earth as the center of the Universe. This so-called geocentric model was elaborated and formalized by the Greek astronomer Claudius Ptolemy (100–170) in the second century A.D. and was accepted for the next 1400 years. In 1543, the Polish astronomer Nicolaus Copernicus (1473–1543) suggested that the Earth and the other planets revolve in circular orbits about the Sun (the heliocentric model).

The Danish astronomer Tycho Brahe (1546–1601) made accurate astronomical measurements over a period of 20 years and provided a rigorous test of the alternate models of the Solar System. It is interesting to note that these precise observations, made on the planets and 777 stars visible to the naked eye, were carried out with a large sextant and compass without a telescope, which had not yet been invented.

The German astronomer Johannes Kepler, who was Brahe's assistant, acquired Brahe's astronomical data and spent about 16 years trying to deduce a mathemati-

Johannes Kepler (1571–1630), a German astronomer, is best known for developing the laws of planetary motion based on the careful observations of Tycho Brahe. Throughout his life, Kepler was sidetracked by mystic ideas dating back to the ancient Greeks. For example, he believed in the motion of the "music of the spheres" proposed by Pythagoras, in which each planet in its motion sounds out an exact musical note. After spending several years trying to work out a "regular-solid theory" of the planets, he concluded that the Copernican view of circular planetary orbits had to be abandoned for the view that the planetary orbits are ellipses with the Sun always at one of the foci.   *(Art Resource)*

cal model for the motion of the planets. After many laborious calculations, he found that Brahe's precise data on the revolution of Mars about the Sun provided the answer. Such data are difficult to sort out because the Earth is also in motion about the Sun. Kepler's analysis first showed that the concept of circular orbits about the Sun had to be abandoned. He eventually discovered that the orbit of Mars could be accurately described by an ellipse with the Sun at one focal point. He then generalized this analysis to include the motion of all planets. The complete analysis is summarized in three statements, known as **Kepler's laws:**

1. All planets move in elliptical orbits with the Sun at one of the focal points.
2. The radius vector drawn from the Sun to a planet sweeps out equal areas in equal time intervals.
3. The square of the orbital period of any planet is proportional to the cube of the semimajor axis of the elliptical orbit.

Half a century later, Newton demonstrated that these laws are the consequence of a simple force that exists between any two masses. Newton's law of gravity, together with his development of the laws of motion, provides the basis for a full mathematical solution to the motion of planets and satellites. More important, Newton's law of gravity correctly describes the gravitational attractive force between *any* two masses.

14.5 THE LAW OF GRAVITY AND THE MOTION OF PLANETS

In formulating his law of gravity, Newton used the following observation, which supports the assumption that the gravitational force is proportional to the inverse square of the separation. Let us compare the acceleration of the Moon in its orbit with the acceleration of an object falling near the Earth's surface, such as the legendary apple (Fig. 14.4). Assume that both accelerations have the same cause, namely, the gravitational attraction of the Earth. From the inverse-square law, Newton found that the acceleration of the Moon toward the Earth (centripetal acceleration) should be proportional to $1/r_M^2$, where r_M is the Earth-Moon separation. Furthermore, the acceleration of the apple toward the Earth should vary as $1/R_E^2$, where R_E is the radius of the Earth. Using the values $r_M = 3.84 \times 10^8$ m and $R_E = 6.37 \times 10^6$ m, we predict the ratio of the Moon's acceleration, a_M, to the

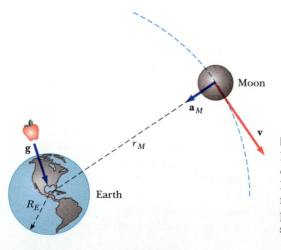

FIGURE 14.4 As it revolves about the Earth, the Moon experiences a centripetal acceleration a_M directed toward the Earth. An object near the Earth's surface, such as the apple shown here, experiences an acceleration g. (Dimensions are not to scale.)

apple's acceleration, g, to be

$$\frac{a_M}{g} = \frac{(1/r_M)^2}{(1/R_E)^2} = \left(\frac{R_E}{r_M}\right)^2 = \left(\frac{6.37 \times 10^6 \text{ m}}{3.84 \times 10^8 \text{ m}}\right)^2 = 2.75 \times 10^{-4}$$

Therefore

$$a_M = (2.75 \times 10^{-4})(9.80 \text{ m/s}^2) = 2.70 \times 10^{-3} \text{ m/s}^2$$

The centripetal acceleration of the Moon can also be calculated from a knowledge of its mean distance from the Earth and its orbital period, $T = 27.32$ days $= 2.36 \times 10^6$ s. In a time T, the Moon travels a distance $2\pi r_M$, which equals the circumference of its orbit. Therefore, its orbital speed is $2\pi r_M/T$, and its centripetal acceleration is

$$a_M = \frac{v^2}{r_M} = \frac{(2\pi r_M/T)^2}{r_M} = \frac{4\pi^2 r_M}{T^2} = \frac{4\pi^2(3.84 \times 10^8 \text{ m})}{(2.36 \times 10^6 \text{ s})^2} = 2.72 \times 10^{-3} \text{ m/s}^2$$

The agreement between this value and the value obtained above using g provides strong evidence that the inverse-square law of force is correct.

Although these results must have been very encouraging to Newton, he was deeply troubled by an assumption made in the analysis. In order to evaluate the acceleration of an object at the Earth's surface, the Earth was treated as if its mass were all concentrated at its center. That is, Newton assumed that the Earth acts as a particle as far as its influence on an exterior object is concerned. Several years later in 1686, and based on his pioneering work in the development of the calculus, Newton proved this point.

Kepler's Third Law

It is informative to show that Kepler's third law can be predicted from the inverse-square law for circular orbits.[1] Consider a planet of mass M_p moving about the Sun of mass M_S in a circular orbit, as in Figure 14.5. Since the gravitational force exerted on the planet by the Sun is equal to the central force needed to keep the planet moving in a circle,

$$\frac{GM_S M_p}{r^2} = \frac{M_p v^2}{r}$$

But the orbital speed of the planet is simply $2\pi r/T$, where T is its period; therefore, the above expression becomes

$$\frac{GM_S}{r^2} = \frac{(2\pi r/T)^2}{r}$$

$$T^2 = \left(\frac{4\pi^2}{GM_S}\right) r^3 = K_S r^3 \qquad (14.6)$$

where K_S is a constant given by

$$K_S = \frac{4\pi^2}{GM_S} = 2.97 \times 10^{-19} \text{ s}^2/\text{m}^3$$

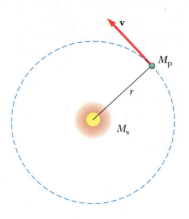

FIGURE 14.5 A planet of mass M_p moving in a circular orbit about the Sun. The orbits of all planets except Mars, Mercury, and Pluto are nearly circular.

Kepler's third law

---

[1] The orbits of all planets except Mars, Mercury, and Pluto are very close to being circular. For example, the ratio of the semiminor to the semimajor axis for the Earth is $b/a = 0.999\,86$.

(Left) The Hubble Space Telescope's Faint Object Camera has obtained the clearest image ever of Pluto and its moon, Charon. Pluto is the bright object at the center of the frame; Charon is the fainter object in the lower left. Charon's orbit around Pluto is a circle seen nearly edge-on from Earth. *(Right)* Separate views of Jupiter and of Periodic Comet Shoemaker-Levy 9, both taken with the Hubble Space Telescope about two months before Jupiter and the comet collided in July 1994, were put together in a computer. Their relative sizes and distance were altered. The black spot on Jupiter is the shadow of its moon Io.   *(Both photos courtesy NASA)*

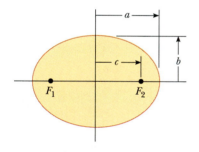

FIGURE 14.6 Plot of an ellipse. The semimajor axis has a length a, and the semiminor axis has a length b. The focal points are located at a distance c from the center, where $a^2 = b^2 + c^2$, and the eccentricity is defined as $e = c/a$.

Equation 14.6 is Kepler's third law. The law is also valid for elliptical orbits if we replace r by the length of the semimajor axis, a (Fig. 14.6). Note that the constant of proportionality, K_S, is independent of the mass of the planet. Therefore, Equation 14.6 is valid for *any* planet. If we were to consider the orbit of a satellite about the Earth, such as the Moon, then the constant would have a different value, with the Sun's mass replaced by the Earth's mass. In this case, the proportionality constant equals $4\pi^2/GM_E$.

A collection of useful planetary data is given in Table 14.2. The last column of this table verifies that T^2/r^3 is a constant whose value is $K_S = 4\pi^2/GM_S = 2.97 \times 10^{-19}$ s^2/m^3.

TABLE 14.2   Useful Planetary Data

Body	Mass (kg)	Mean Radius (m)	Period (s)	Distance from Sun (m)	$\dfrac{T^2}{r^3}\left(\dfrac{\text{s}^2}{\text{m}^3}\right)$
Mercury	3.18×10^{23}	2.43×10^6	7.60×10^6	5.79×10^{10}	2.97×10^{-19}
Venus	4.88×10^{24}	6.06×10^6	1.94×10^7	1.08×10^{11}	2.99×10^{-19}
Earth	5.98×10^{24}	6.37×10^6	3.156×10^7	1.496×10^{11}	2.97×10^{-19}
Mars	6.42×10^{23}	3.37×10^6	5.94×10^7	2.28×10^{11}	2.98×10^{-19}
Jupiter	1.90×10^{27}	6.99×10^7	3.74×10^8	7.78×10^{11}	2.97×10^{-19}
Saturn	5.68×10^{26}	5.85×10^7	9.35×10^8	1.43×10^{12}	2.99×10^{-19}
Uranus	8.68×10^{25}	2.33×10^7	2.64×10^9	2.87×10^{12}	2.95×10^{-19}
Neptune	1.03×10^{26}	2.21×10^7	5.22×10^9	4.50×10^{12}	2.99×10^{-19}
Pluto	$\approx 1.4 \times 10^{22}$	$\approx 1.5 \times 10^6$	7.82×10^9	5.91×10^{12}	2.96×10^{-19}
Moon	7.36×10^{22}	1.74×10^6	—	—	—
Sun	1.991×10^{30}	6.96×10^8	—	—	—

EXAMPLE 14.3   The Mass of the Sun

Calculate the mass of the Sun using the fact that the period of the Earth is 3.156×10^7 s and its distance from the Sun is 1.496×10^{11} m.

Solution   Using Equation 14.6, we get

$$M_S = \frac{4\pi^2 r^3}{GT^2}$$

$$= \frac{4\pi^2(1.496 \times 10^{11} \text{ m})^3}{\left(6.67 \times 10^{-11} \dfrac{\text{N}\cdot\text{m}^2}{\text{kg}^2}\right)(3.156 \times 10^7 \text{ s})^2}$$

$$= 1.99 \times 10^{30} \text{ kg}$$

Note that the Sun is 333 000 times as massive as the Earth!

Kepler's Second Law and Conservation of Angular Momentum

Consider a planet of mass M_p moving about the Sun in an elliptical orbit (Fig. 14.7). The gravitational force acting on the planet is always along the radius vector, directed toward the Sun. Such a force directed toward or away from a fixed point (that is, one that is a function of r only) is called a **central force**. The torque acting on the planet due to this central force is clearly zero since F is parallel to r. That is,

$$\boldsymbol{\tau} = \mathbf{r} \times \mathbf{F} = \mathbf{r} \times F(r)\hat{\mathbf{r}} = 0$$

But recall from Equation 11.19 that the torque equals the time rate of change of angular momentum, or $\boldsymbol{\tau} = d\mathbf{L}/dt$. Therefore,

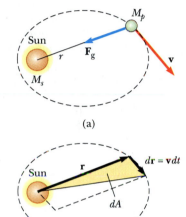

because $\boldsymbol{\tau} = 0$, the angular momentum L of the planet is a constant of the motion:

$$\mathbf{L} = \mathbf{r} \times \mathbf{p} = M_p\mathbf{r} \times \mathbf{v} = \text{constant}$$

FIGURE 14.7   (a) The gravitational force acting on a planet is directed toward the Sun, along the radius vector. (b) As a planet orbits the Sun, the area swept out by the radius vector in a time dt is equal to one half the area of the parallelogram formed by the vectors r and $d\mathbf{r} = $ v dt.

Since L is a constant of the motion, we see that the planet's motion at any instant is restricted to the plane formed by r and v.

We can relate this result to the following geometric consideration. The radius vector r in Figure 14.7b sweeps out an area dA in a time dt. This area equals one-half the area $|\mathbf{r} \times d\mathbf{r}|$ of the parallelogram formed by the vectors r and $d\mathbf{r}$. Since the displacement of the planet in a time dt is $d\mathbf{r} = $ v dt, we get

$$dA = \tfrac{1}{2}|\mathbf{r} \times d\mathbf{r}| = \tfrac{1}{2}|\mathbf{r} \times \mathbf{v}\, dt| = \frac{L}{2M_p}\, dt$$

$$\frac{dA}{dt} = \frac{L}{2M_p} = \text{constant} \qquad (14.7)$$

Kepler's second law

where L and M_p are both constants of the motion. Thus, we conclude that

the radius vector from the Sun to a planet sweeps out equal areas in equal times.

It is important to recognize that this result, which is Kepler's second law, is a consequence of the fact that the force of gravity is a central force, which in turn implies that angular momentum remains constant. Therefore, the second law applies to *any* situation that involves a central force, whether inverse-square or not.

The inverse-square nature of the force of gravity is not revealed by Kepler's second law. Although we do not prove it here, Kepler's first law is a direct consequence of the fact that the gravitational force varies as $1/r^2$. That is, under an inverse-square force law, the orbits of the planets can be shown to be ellipses with the Sun at one focus.

EXAMPLE 14.4   Motion in an Elliptical Orbit

A satellite of mass m moves in an elliptical orbit about the Earth (Fig. 14.8). The minimum and maximum distances of the planet from the Earth are called the *perihelion* (indicated by p in Fig. 14.8) and *aphelion* (indicated by a), respectively. If the speed of the satellite at p is v_p, what is its speed at a?

Solution   The angular momentum of the planet relative to the Earth is $M_p\mathbf{r} \times \mathbf{v}$. At the points a and p, $\mathbf{v}$ is perpendicular to $\mathbf{r}$. Therefore, the magnitude of the angular momentum at these positions is $L_a = M_p v_a r_a$ and $L_p = M_p v_p r_p$. Because angular momentum is constant, we see that

$$M_p v_a r_a = M_p v_p r_p$$

$$v_a = \frac{r_p}{r_a} v_p$$

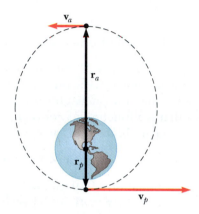

FIGURE 14.8   (Example 14.4) As a satellite moves about the Earth in an elliptical orbit, its angular momentum is constant. Therefore, $M_a v_a r_a = M_p v_p r_p$, where the subscripts a and p represent aphelion and perihelion, respectively.

14.6   THE GRAVITATIONAL FIELD

When Newton first published his theory of gravitation, his contemporaries found it difficult to accept the concept of a field force that could act through a distance. They asked how it was possible for two masses to interact even though they were not in contact with each other. Although Newton himself could not answer this question, his theory was considered a success because it satisfactorily explained the motion of the planets.

An alternative approach in describing the gravitational interaction, therefore, is to introduce the concept of a **gravitational field** that covers every point in space. When a particle of mass m is placed at a point where the field is the vector $\mathbf{g}$, the particle experiences a force $\mathbf{F}_g = m\mathbf{g}$. In other words, the field exerts a force on the particle. Hence, the gravitational field is defined by

Gravitational field

$$\mathbf{g} \equiv \frac{\mathbf{F}_g}{m} \tag{14.8}$$

That is, the gravitational field at a point in space equals the gravitational force experienced by a test mass placed at that point divided by the test mass. As an example, consider an object of mass m near the Earth's surface. The gravitational force on the object is directed toward the center of the Earth and has a magnitude mg. Since the gravitational force on the object has a magnitude $GM_E m/r^2$ (where

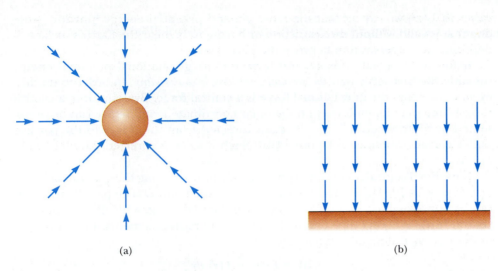

(a)   (b)

FIGURE 14.9 (a) The gravitational field vectors in the vicinity of a uniform spherical mass such as the Earth vary in both direction and magnitude. (b) The gravitational field vectors in a small region near the Earth's surface are uniform; that is, they match in both direction and magnitude.

M_E is the mass of the Earth), the field **g** at a distance r from the center of the Earth is

$$\mathbf{g} = \frac{\mathbf{F}_g}{m} = -\frac{GM_E}{r^2}\hat{\mathbf{r}} \qquad (14.9)$$

where $\hat{\mathbf{r}}$ is a unit vector pointing radially outward from the Earth, and the minus sign indicates that the field points toward the center of the Earth, as in Figure 14.9a. Note that the field vectors at different points surrounding the Earth vary in both direction and magnitude. In a small region near the Earth's surface, the downward field g is approximately constant and uniform, as indicated in Figure 14.9b. Equation 14.9 is valid at all points *outside* the Earth's surface, assuming that the Earth is spherical. At the Earth's surface, where $r = R_E$, **g** has a magnitude of 9.80 N/kg.

CONCEPTUAL EXAMPLE 14.5

How would you explain the fact that Saturn and Jupiter have periods much greater than one year?

Reasoning Kepler's third law (Eq. 14.6), which applies to all the planets, tells us that the period of a planet is propor-

tional to $r^{3/2}$. Because Saturn and Jupiter are farther than Earth from the Sun, they have longer periods. The Sun's gravitational field (whose magnitude is GM_S/r^2) is much weaker at a distant Jovian planet. Thus, an outer planet experiences much smaller centripetal acceleration than Earth, and a correspondingly longer period.

14.7 GRAVITATIONAL POTENTIAL ENERGY

In Chapter 8 we introduced the concept of gravitational potential energy, that is, the energy associated with the position of a particle. We emphasized the fact that the gravitational potential energy function, $U = mgy$, is valid only when the particle is near the Earth's surface. Since the gravitational force between two particles

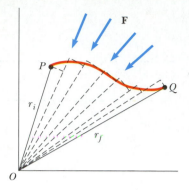

FIGURE 14.10 A particle moves from P to Q while under the action of a central force **F**, which is in the radial direction. The path is broken into a series of radial and circular segments. Since the work done along the circular segments is zero, the work done is independent of the path.

Work done by a central force

varies as $1/r^2$, we expect that the more general potential energy function—the one that is valid without the restriction of having to be near the Earth's surface—depends on the separation between the particles.

Before we calculate this general form for the gravitational potential energy function, we first verify that *the gravitational force is conservative*. In order to do this, we first note that the gravitational force is a central force. By definition, a central force is one that depends only on the polar coordinate r, and hence can be represented by $F(r)\hat{\mathbf{r}}$, where $\hat{\mathbf{r}}$ is a unit vector directed from the origin to the particle under consideration as in Figure 14.10. Such a force is directed parallel to the radius vector.

Consider a central force acting on a particle moving along the general path P to Q in Figure 14.10. The central force acts toward the point O. The path from P to Q can be approximated by a series of radial and circular segments. By definition, a central force is always directed along one of the radial segments; therefore, the work done by **F** along any *radial segment* is

$$dW = \mathbf{F} \cdot d\mathbf{r} = F(r)\ dr$$

You should recall that, by definition, the work done by a force that is perpendicular to the displacement is zero. Hence, the work done along any circular segment is zero because **F** is perpendicular to the displacement along these segments. Therefore, the total work done by **F** is the sum of the contributions along the radial segments:

$$W = \int_{r_i}^{r_f} F(r)\ dr$$

where the subscripts i and f refer to the initial and final positions. This result applies to *any* path from P to Q. Therefore, we conclude that *any central force is conservative*. We are now assured that a potential energy function can be obtained once the form of the central force is specified. You should recall from Chapter 8 that the change in the gravitational potential energy associated with a given displacement is defined as the negative of the work done by the gravitational force during that displacement, or

$$\Delta U = U_f - U_i = -\int_{r_i}^{r_f} F(r)\ dr \tag{14.10}$$

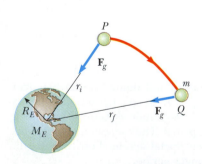

FIGURE 14.11 As a particle of mass m moves from P to Q above the Earth's surface, the gravitational potential energy changes according to Equation 14.10.

Change in gravitational potential energy

We can use this result to evaluate the gravitational potential energy function. Consider a particle of mass m moving between two points P and Q above the Earth's surface (Fig. 14.11). The particle is subject to the gravitational force given by Equation 14.1. We can express this force as

$$\mathbf{F}_g = -\frac{GM_E m}{r^2}\hat{\mathbf{r}}$$

where $\hat{\mathbf{r}}$ is a unit vector directed from the Earth to the particle and the negative sign indicates that the force is attractive. Substituting this expression for $\mathbf{F}_g$ into Equation 14.10, we can compute the change in the gravitational potential energy function:

$$U_f - U_i = GM_E m \int_{r_i}^{r_f} \frac{dr}{r^2} = GM_E m\left[-\frac{1}{r}\right]_{r_i}^{r_f}$$

$$U_f - U_i = -GM_E m\left(\frac{1}{r_f} - \frac{1}{r_i}\right) \tag{14.11}$$

As always, the choice of a reference point for the potential energy is completely arbitrary. It is customary to choose the reference point where the force is zero. Taking $U_i = 0$ at $r_i = \infty$, we obtain the important result

$$U(r) = -\frac{GM_E m}{r} \qquad (14.12)$$

Gravitational potential energy
$r > R_E$

This expression applies to the Earth-particle system where the two masses are separated by a distance r, provided that $r \geq R_E$. The result is not valid for particles moving inside the Earth, where $r < R_E$. (The situation where $r < R_E$ is treated in Section 14.10.) Because of our choice of U_i, the function $U(r)$ is always negative (Fig. 14.12).

Although Equation 14.12 was derived for the particle-Earth system, it can be applied to any two particles. That is, the gravitational potential energy associated with any pair of particles of masses m_1 and m_2 separated by a distance r is

$$U = -\frac{Gm_1 m_2}{r} \qquad (14.13)$$

This expression shows that the gravitational potential energy for any pair of particles varies as $1/r$, whereas the force between them varies as $1/r^2$. Furthermore, the potential energy is negative because the force is attractive and we have taken the potential energy as zero when the particle separation is infinity. Because the force between the particles is attractive, we know that an external agent must do positive work to increase the separation between them. The work done by the external agent produces an increase in the potential energy as the two particles are separated. That is, U becomes less negative as r increases.[2]

When two particles are separated by a distance r, an external agent has to supply an energy at least equal to $+ Gm_1 m_2 / r$ in order to separate the particles by an infinite distance. It is convenient to think of the absolute value of the potential energy as the *binding energy* of the system. If the external agent supplies an energy greater than the binding energy, $Gm_1 m_2 / r$, the additional energy of the system will be in the form of kinetic energy when the particles are at an infinite separation.

We can extend this concept to three or more particles. In this case, the total potential energy of the system is the sum over all pairs of particles.[3] Each pair contributes a term of the form given by Equation 14.13. For example, if the system contains three particles as in Figure 14.13, we find that

$$U_{\text{total}} = U_{12} + U_{13} + U_{23} = -G\left(\frac{m_1 m_2}{r_{12}} + \frac{m_1 m_3}{r_{13}} + \frac{m_2 m_3}{r_{23}}\right) \qquad (14.14)$$

The absolute value of U_{total} represents the work needed to separate the particles by an infinite distance. If the system consists of four particles, there are six terms in the sum, corresponding to the six distinct pairs of interaction forces.

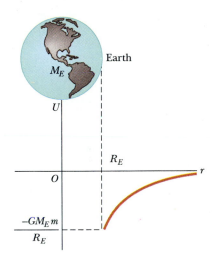

FIGURE 14.12 Graph of the gravitational potential energy, U, versus r for a particle above the Earth's surface. The potential energy goes to zero as r approaches ∞.

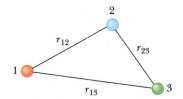

FIGURE 14.13 Diagram of three interacting particles.

---

[2] Note that part of the work done can also produce a change in kinetic energy of the system. That is, if the work done in separating the particles exceeds the increase in potential energy, the excess energy is accounted for by the increase in kinetic energy of the system.

[3] The fact that potential energy terms can be added for all pairs of particles stems from the experimental fact that gravitational forces obey the superposition principle. That is, if $\Sigma F = F_{12} + F_{13} + F_{23} + \cdots$ then there exists a potential energy term for each interaction F_{ij}.

EXAMPLE 14.6   **The Change in Potential Energy**

A particle of mass m is displaced through a small vertical distance Δy near the Earth's surface. Let us show that the general expression for the change in gravitational potential energy given by Equation 14.11 reduces to the familiar relationship $\Delta U = mg\,\Delta y$.

Solution   We can express Equation 14.11 in the form

$$\Delta U = -GM_E m\left(\frac{1}{r_f} - \frac{1}{r_i}\right) = GM_E m\left(\frac{r_f - r_i}{r_i r_f}\right)$$

If both the initial and the final position of the particle are close to the Earth's surface, then $r_f - r_i = \Delta y$ and $r_i r_f \approx R_E^2$. (Recall that r is measured from the center of the Earth.) Therefore, the change in potential energy becomes

$$\Delta U \approx \frac{GM_E m}{R_E^2}\,\Delta y = mg\,\Delta y$$

where we have used the fact that $g = GM_E/R_E^2$. Keep in mind that the reference point is arbitrary because it is the change in potential energy that is meaningful.

14.8   ENERGY CONSIDERATIONS IN PLANETARY AND SATELLITE MOTION

Consider a body of mass m moving with a speed v in the vicinity of a massive body of mass M, where $M \gg m$. The system might be a planet moving around the Sun or a satellite in orbit around the Earth. If we assume that M is at rest in an inertial reference frame, then the total energy E of the two-body system when the bodies are separated by a distance r is the sum of the kinetic energy of m and the potential energy of the system, given by Equation 14.13[4]:

$$E = K + U$$

$$E = \tfrac{1}{2}mv^2 - \frac{GMm}{r} \tag{14.15}$$

Furthermore, the total energy is constant if we assume the system is isolated. Therefore as m moves from P to Q in Figure 14.11, the total energy remains constant and Equation 14.15 gives

$$E = \tfrac{1}{2}mv_i^2 - \frac{GMm}{r_i} = \tfrac{1}{2}mv_f^2 - \frac{GMm}{r_f} \tag{14.16}$$

This result shows that E may be positive, negative, or zero, depending on the speed of m. However, for a bound system, such as the Earth and Sun, E is necessarily *less than zero*. We can easily establish that $E < 0$ for the system consisting of a mass m moving in a circular orbit about a body of mass M, where $M \gg m$ (Fig. 14.14). Newton's second law applied to m gives

$$\frac{GMm}{r^2} = \frac{mv^2}{r}$$

Multiplying both sides by r and dividing by 2 gives

$$\tfrac{1}{2}mv^2 = \frac{GMm}{2r} \tag{14.17}$$

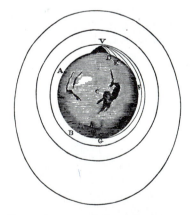

"... the greater the velocity ... with which (a stone) is projected, the farther it goes before it falls to the Earth. We may therefore suppose the velocity to be so increased, that it would describe an arc of 1, 2, 5, 10, 100, 1000 miles before it arrived at the Earth, till at last, exceeding the limits of the Earth, it should pass into space without touching."—Newton, *System of the World.*

---

[4] You might recognize that we have ignored the acceleration and kinetic energy of the larger mass. To see that this simplification is reasonable, consider an object of mass m falling toward the Earth. Since the center of mass of the object-Earth system is stationary, it follows that $mv = M_E v_E$. Thus, the Earth acquires a kinetic energy equal to

$$\tfrac{1}{2}M_E v_E^2 = \tfrac{1}{2}\frac{m^2}{M_E}v^2 = \frac{m}{M_E}K$$

where K is the kinetic energy of the object. Since $M_E \gg m$, the kinetic energy of the Earth is negligible.

Substituting this into Equation 14.15, we obtain

$$E = \frac{GMm}{2r} - \frac{GMm}{r}$$

$$E = -\frac{GMm}{2r} \qquad (14.18)$$

This clearly shows that *the total energy must be negative in the case of circular orbits.* Note that *the kinetic energy is positive and equal to one half the magnitude of the potential energy.* The absolute value of E is also equal to the binding energy of the system.

The total mechanical energy is also negative in the case of elliptical orbits. The expression for E for elliptical orbits is the same as Equation 14.18 with r replaced by the semimajor axis length, a.

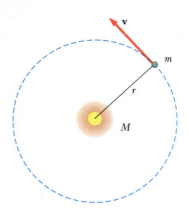

FIGURE 14.14   A body of mass m moving in a circular orbit about a much larger body of mass M.

Both the total energy and the total angular momentum of a planet-Sun system are constants of the motion.

EXAMPLE 14.7   Changing the Orbit of a Satellite

Calculate the work required to move an Earth satellite of mass m from a circular orbit of radius $2R_E$ to one of radius $3R_E$.

Solution   Applying Equation 14.18, we get for the total initial and final energies

$$E_i = -\frac{GM_E m}{4R_E} \qquad E_f = -\frac{GM_E m}{6R_E}$$

Therefore, the work required to increase the energy of the system is

$$W = E_f - E_i = -\frac{GM_E m}{6R_E} - \left(-\frac{GM_E m}{4R_E} \right) = \boxed{\frac{GM_E m}{12R_E}}$$

For example, if we take $m = 10^3$ kg, we find that the work required is $W = 5.2 \times 10^9$ J, which is the energy equivalent of 39 gal of gasoline.

If we wish to determine how the energy is distributed after work is done on the system, we find from Equation 14.17 that the change in kinetic energy is $\Delta K = -GM_E m/12R_E$ (it decreases), while the corresponding change in potential energy is $\Delta U = GM_E m/6R_E$ (it increases). Thus, the work done on the system is $W = \Delta K + \Delta U = GM_E m/12R_E$, as we calculated above. In other words, part of the work done goes into increasing the potential energy and part goes into decreasing the kinetic energy.

Escape Speed

Suppose an object of mass m is projected vertically upward from the Earth's surface with an initial speed v_i, as in Figure 14.15. We can use energy considerations to find the minimum value of the initial speed such that the object will escape the Earth's gravitational field. Equation 14.16 gives the total energy of the object at any point when its speed and distance from the center of the Earth are known. At the surface of the Earth, $v_i = v$ and $r_i = R_E$. When the object reaches its maximum altitude, $v_f = 0$ and $r_f = r_{max}$. Because the total energy of the system is constant, substitution of these conditions into Equation 14.16 gives

$$\tfrac{1}{2}mv_i^2 - \frac{GM_E m}{R_E} = -\frac{GM_E m}{r_{max}}$$

Solving for v_i^2 gives

$$v_i^2 = 2GM_E \left(\frac{1}{R_E} - \frac{1}{r_{max}} \right) \qquad (14.19)$$

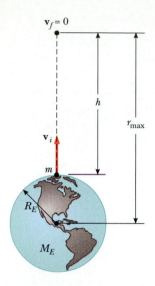

FIGURE 14.15   An object of mass *m* projected upward from the Earth's surface with an initial speed v_i reaches a maximum altitude *h*.

Therefore, if the initial speed is known, this expression can be used to calculate the maximum altitude *h*, because we know that $h = r_{max} - R_E$.

We are now in a position to calculate the minimum speed the object must have at the Earth's surface in order to escape from the influence of the Earth's gravitational field. Traveling at this minimum speed, the object can *just* reach infinity with a final speed of zero. Setting $r_{max} = \infty$ in Equation 14.19 and taking $v_i = v_{esc}$ (the escape speed), we get

$$v_{esc} = \sqrt{\frac{2GM_E}{R_E}} \tag{14.20}$$

Note that this expression for v_{esc} is independent of the mass of the object. In other words, a spacecraft has the same escape speed as a molecule. Furthermore, the result is independent of the direction of the velocity, provided the trajectory does not intersect the Earth.

If the object is given an initial speed equal to v_{esc}, its total energy is equal to zero. This can be seen by noting that when $r = \infty$, the object's kinetic energy and its potential energy are both zero. If v_i is greater than v_{esc}, the total energy is greater than zero and the object has some residual kinetic energy at $r = \infty$.

EXAMPLE 14.8   Escape Speed of a Rocket

Calculate the escape speed from the Earth for a 5000-kg spacecraft, and determine the kinetic energy it must have at the Earth's surface in order to escape the Earth's gravitational field.

Solution   Using Equation 14.20 with $M_E = 5.98 \times 10^{24}$ kg and $R_E = 6.37 \times 10^6$ m gives

$$v_{esc} = \sqrt{\frac{2GM_E}{R_E}}$$

$$= \sqrt{\frac{2(6.67 \times 10^{-11}\,\text{N·m}^2/\text{kg}^2)(5.98 \times 10^{24}\,\text{kg})}{6.37 \times 10^6\,\text{m}}}$$

$$= \boxed{1.12 \times 10^4\,\text{m/s}}$$

This corresponds to about 25 000 mi/h.
The kinetic energy of the spacecraft is

$$K = \tfrac{1}{2}mv_{esc}^2 = \tfrac{1}{2}(5.00 \times 10^3\,\text{kg})(1.12 \times 10^4\,\text{m/s})^2$$

$$= \boxed{3.14 \times 10^{11}\,\text{J}}$$

Finally, you should note that Equations 14.19 and 14.20 can be applied to objects projected from any planet. That is, in general, the escape speed from the surface of any planet of mass *M* and radius *R* is

Escape speed

$$v_{esc} = \sqrt{\frac{2GM}{R}}$$

A list of escape speeds for the planets, the Moon, and the Sun is given in Table 14.3. Note that the values vary from 1.1 km/s for Pluto to about 618 km/s for the Sun. These results, together with some ideas from the kinetic theory of gases (Chapter 21), explain why some planets have atmospheres and others do not. As we shall see later, a gas molecule has an average kinetic energy that depends on its temperature. Hence, lighter molecules, such as hydrogen and helium, have a higher average speed than the heavier species at the same temperature. When the speed of the lighter molecules is not much less than the escape speed, a significant fraction of them have a chance to escape from the planet.

This mechanism also explains why the Earth does not retain hydrogen and helium molecules in its atmosphere while much heavier molecules, such as oxygen and nitrogen, do not escape. On the other hand, the very large escape speed for Jupiter enables that planet to retain hydrogen, the primary constituent of its atmosphere. Similarly, the very large mass of the Sun allows it to retain hydrogen and helium, as well as heavier gases. In contrast, Mercury, which is small and hot, has no atmosphere.

TABLE 14.3	Escape Speeds from the Surfaces of the Planets, the Moon, and the Sun
Planet	v_{esc} **(km/s)**
Mercury	4.3
Venus	10.3
Earth	11.2
Mars	5.0
Jupiter	60
Saturn	36
Uranus	22
Neptune	24
Pluto	1.1
Moon	2.3
Sun	618

*14.9   THE GRAVITATIONAL FORCE BETWEEN AN EXTENDED OBJECT AND A PARTICLE

We have emphasized that the law of universal gravitation given by Equation 14.3 is valid only if the interacting objects are considered as particles. In view of this, how can we calculate the force between a particle and an object having finite dimensions? This is accomplished by treating the extended object as a collection of particles and making use of integral calculus. We first evaluate the potential energy function, from which the force can be calculated.

The potential energy associated with a system consisting of a point mass m and an extended body of mass M is obtained by dividing the body into segments of mass ΔM_i (Fig. 14.16). The potential energy associated with this element and with the particle of mass m is $-Gm\,\Delta M_i/r_i$, where r_i is the distance from the particle to the element ΔM_i. The total potential energy of the system is obtained by taking the sum over all segments as $\Delta M_i \rightarrow 0$. In this limit, we can express U in integral form as

$$U = -Gm \int \frac{dM}{r} \qquad (14.21)$$

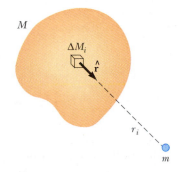

FIGURE 14.16   A particle of mass m interacting with an extended object of mass M. The total gravitational force exerted on the particle by the object can be obtained by taking a vector sum over all forces due to each segment of the object.

Once U has been evaluated, the force exerted on m can be obtained by taking the negative derivative of this scalar function (see Section 8.6). If the extended body has spherical symmetry, the function U depends only on r and the force is given by $-dU/dr$. We treat this situation in Section 14.10. In principle, one can evaluate U for any geometry; however, the integration can be cumbersome.

An alternative approach to evaluating the gravitational force between a particle and an extended body is to perform a vector sum over all segments of the body. Using the procedure outlined in evaluating U and the law of gravity (Eq. 14.3), we obtain for the total force on the particle

$$\mathbf{F}_g = -Gm \int \frac{dM}{r^2}\,\hat{\mathbf{r}} \qquad (14.22)$$

Force on a particle due to a spherical shell

where $\hat{\mathbf{r}}$ is a unit vector directed from the element dM toward the particle (see Fig. 14.16). This procedure is not always recommended, because working with a vector function is more difficult than working with the scalar potential energy function. However, if the geometry is simple, as in the following example, the evaluation of $\mathbf{F}$ can be straightforward.

EXAMPLE 14.9   Gravitational Force Between a Mass and a Bar

A homogeneous bar of length ℓ and mass M is at a distance h from a point mass m (Fig. 14.17). Calculate the total gravitational force exerted on m by the bar.

Solution   The segment of the bar that has a length dx has a mass dM. Since the mass per unit length is a constant, it then follows that the ratio of masses, dM/M, is equal to the ratio of

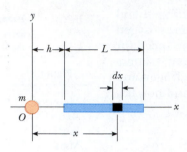

FIGURE 14.17 (Example 14.9) The gravitational force exerted on a particle at the origin by the bar is directed to the right. Note that the bar is *not* equivalent to a particle of mass M located at the center of mass of the bar.

lengths, dx/ℓ, and so $dM = (M/\ell)\,dx$. The variable r in Equation 14.22 is x in our case, and the force on m is to the right; therefore, we get

$$\mathbf{F}_g = Gm \int_{h}^{\ell+h} \frac{M}{\ell} \frac{dx}{x^2} \,\mathbf{i}$$

$$\mathbf{F}_g = \frac{GmM}{\ell} \left[-\frac{1}{x} \right]_{h}^{\ell+h} \mathbf{i} = \frac{GmM}{h(\ell+h)} \,\mathbf{i}$$

We see that the force on m is in the positive x direction, as expected, since the gravitational force is attractive.

Note that in the limit $\ell \to 0$, the force varies as $1/h^2$, which is what is expected for the force between two point masses. Furthermore, if $h \gg \ell$, the force also varies as $1/h^2$. This can be seen by noting that the denominator of the expression for $\mathbf{F}_g$ can be expressed in the form $h^2\left(1 + \dfrac{\ell}{h}\right)$, which is approximately equal to h^2. Thus, when bodies are separated by distances that are large compared with their characteristic dimensions, they behave like particles.

*14.10 GRAVITATIONAL FORCE BETWEEN A PARTICLE AND A SPHERICAL MASS

In this section we describe the gravitational force between a particle and a spherically symmetric mass distribution. We have already stated that a large sphere attracts a particle outside it as if the total mass of the sphere were concentrated at its center. Let us describe the nature of the force on a particle when the extended body is either a spherical shell or a solid sphere, and then apply these facts to some interesting systems.

Spherical Shell

Case 1. If a particle of mass m is located outside a spherical shell of mass M (say, point P in Fig. 14.18), the shell attracts the particle as though the mass of the shell were concentrated at its center.

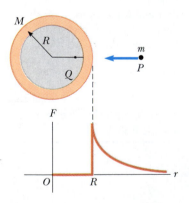

FIGURE 14.18 The gravitational force on a particle when it is outside the spherical shell is GMm/r^2 and acts toward the center. The force on the particle is zero everywhere inside the shell.

Case 2. If the particle is located inside the shell (point Q in Fig. 14.18), the force on it is zero. We can express these two important results in the following way:

$$\mathbf{F}_g = -\frac{GMm}{r^2}\,\hat{\mathbf{r}} \qquad \text{for } r \geq R \qquad (14.23a)$$

$$\mathbf{F}_g = 0 \qquad \text{for } r < R \qquad (14.23b)$$

The force as a function of the distance r is plotted in Figure 14.18. Note that the shell does not act as a gravitational shield. Even when inside the shell, the particle may experience forces due to other masses outside the shell.

Solid Sphere

Case 1. If a particle of mass m is located outside a homogeneous solid sphere of mass M (at point P in Fig. 14.19), the sphere attracts the particle as though the mass of the sphere were concentrated at its center. That is, Equation 14.23a applies in this situation. This follows from Case 1 above, since a solid sphere can be considered a collection of concentric spherical shells.

Case 2. If a particle of mass m is located inside a homogeneous solid sphere of mass M (at point Q in Fig. 14.19), the force on m is due *only* to the mass M' contained within the sphere of radius $r < R$, represented by the dashed circle in Figure 14.19. In other words,

$$\mathbf{F}_g = -\frac{GmM}{r^2}\hat{\mathbf{r}} \qquad \text{for } r \geqslant R \qquad (14.24a)$$

$$\mathbf{F}_g = -\frac{GmM'}{r^2}\hat{\mathbf{r}} \qquad \text{for } r < R \qquad (14.24b)$$

Force on a particle due to a solid sphere

Because the sphere is assumed to have a uniform density, it follows that the ratio of masses M'/M is equal to the ratio of volumes V/V', where V is the total volume of the sphere and V' is the volume within the dotted surface. That is,

$$\frac{M'}{M} = \frac{V'}{V} = \frac{\frac{4}{3}\pi r^3}{\frac{4}{3}\pi R^3} = \frac{r^3}{R^3}$$

Solving this equation for M' and substituting the value obtained into Equation 14.24b, we get

$$\mathbf{F}_g = -\frac{GmM}{R^3}r\hat{\mathbf{r}} \qquad \text{for } r < R \qquad (14.25)$$

That is, the force goes to zero at the center of the sphere, as we would intuitively expect. The force as a function of r is plotted in Figure 14.19.

Case 3. If a particle is located inside a solid sphere having a density ρ that is spherically symmetric but not uniform, then M' in Equation 14.24b is given by an integral of the form $M' = \int \rho\, dV$, where the integration is taken over the volume contained within the dashed circle in Figure 14.19. This integral can be evaluated if the radial variation of ρ is given. The integral is easily evaluated if the mass distribution has spherical symmetry, that is, if ρ is a function of r only. In this case, we take the volume element dV as the volume of a spherical shell of radius r and thickness dr, so that $dV = 4\pi r^2\, dr$. For example, if $\rho(r) = Ar$, where A is a constant, it is left as a problem (Problem 63) to show that $M' = \pi A r^4$. Hence we see from Equation 14.24b that F is proportional to r^2 in this case and is zero at the center.

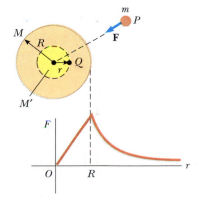

FIGURE 14.19 The gravitational force on a particle when it is outside a uniform solid sphere is GMm/r^2 and is directed toward the center. The force on the particle when it is inside such a sphere is proportional to r and goes to zero at the center.

CONCEPTUAL EXAMPLE 14.10

A particle is projected through a small hole into the interior of a large spherical shell. Describe the subsequent motion of the particle in the interior of the shell.

Reasoning The gravitational force (and field) is zero inside the spherical shell (Eq. 14.23b). Because the force on the particle is zero once inside the shell, it moves with constant velocity in the direction of its original motion until it hits the inside wall of the shell. Its path thereafter depends on the nature of the collision of the object with the wall and the direction in which it was projected as it passed through the hole.

EXAMPLE 14.11   A Free Ride, Thanks to Gravity

An object moves in a smooth, straight tunnel dug between two points on the Earth's surface (Fig. 14.20). Show that the object moves with simple harmonic motion and find the period of its motion. Assume that the Earth's density is uniform throughout its volume.

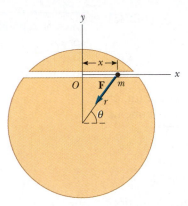

FIGURE 14.20   (Example 14.11) A particle moves along a tunnel dug through the Earth. The component of the gravitational force F_g along the x axis is the driving force for the motion. Note that this component always acts toward the origin O.

Solution   When the object is in the tunnel, the gravitational force exerted on the object acts toward the Earth's center and is given by Equation 14.25:

$$F_g = -\frac{GmM_E}{R_E^{\,3}}\,r$$

The y component of this force is balanced by the normal force exerted by the tunnel wall, and the x component is

$$F_x = -\frac{GmM_E}{R_E^{\,3}}\,r\cos\theta$$

Since the x coordinate of the object is $x = r\cos\theta$, we can write

$$F_x = -\frac{GmM_E}{R_E^{\,3}}\,x$$

Applying Newton's second law to the motion along x gives

$$F_x = -\frac{GmM_E}{R_E^{\,3}}\,x = ma$$

$$a = -\frac{GM_E}{R_E^{\,3}}\,x = -\omega^2 x$$

But this is the equation of simple harmonic motion with angular speed ω (Chapter 13), where

$$\omega = \sqrt{\frac{GM_E}{R_E^{\,3}}}$$

The period is calculated using the data in Table 14.2 and the above result:

$$T = \frac{2\pi}{\omega} = 2\pi\sqrt{\frac{R_E^{\,3}}{GM_E}}$$

$$= 2\pi\sqrt{\frac{(6.37 \times 10^6)^3}{(6.67 \times 10^{-11})(5.98 \times 10^{24})}}$$

$$= 5.06 \times 10^3 \text{ s} = \boxed{84.3 \text{ min}}$$

This period is the same as that of a satellite in a circular orbit just above the Earth's surface. Note that the result is independent of the length of the tunnel.

It has been proposed to operate a mass-transit system between any two cities using this principle. A one-way trip would take about 42 min. A more precise calculation of the motion must account for the fact that the Earth's density is not uniform as we have assumed. More important, there are many practical problems to consider. For instance, it would be impossible to achieve a frictionless tunnel, and so some auxiliary power source would be acquired. Can you think of other problems?

SUMMARY

Newton's law of gravity states that the gravitational force of attraction between any two particles of masses m_1 and m_2 separated by a distance r has the magnitude

$$F_g = G\frac{m_1 m_2}{r^2} \tag{14.1}$$

where G is the universal gravitational constant, which has the value 6.672×10^{-11} N·m²/kg².

An object at a distance h above the Earth's surface experiences a gravitational force of magnitude mg', where g' is the **free-fall acceleration** at that elevation:

$$g' = \frac{GM_E}{r^2} = \frac{GM_E}{(R_E + h)^2} \tag{14.5}$$

In this expression, M_E is the mass of the Earth and R_E is the radius of the Earth. Thus, the weight of an object decreases as it moves away from the Earth's surface.

Kepler's laws of planetary motion state that

1. All planets move in elliptical orbits with the Sun at one of the focal points.
2. The radius vector drawn from the Sun to a planet sweeps out equal areas in equal time intervals.
3. The square of the orbital period of any planet is proportional to the cube of the semimajor axis for the elliptical orbit.

Kepler's second law is a consequence of the fact that the force of gravity is a *central force,* that is, one that is directed toward a fixed point. This implies that the angular momentum of the planet-Sun system is a constant of the motion.

Kepler's third law is consistent with the inverse-square nature of the law of universal gravitation. Newton's second law, together with the force law given by Equation 14.1, verifies that the period T and radius r of the orbit of a planet about the Sun are related by

$$T^2 = \left(\frac{4\pi^2}{GM_S}\right)r^3 \tag{14.6}$$

where M_S is the mass of the Sun.

Most planets have nearly circular orbits about the Sun. For elliptical orbits, Equation 14.6 is valid if r is replaced by the semimajor axis, a.

The gravitational force is conservative, and therefore a potential energy function can be defined. The **gravitational potential energy** associated with two particles separated by a distance r is

$$U = -\frac{Gm_1m_2}{r} \tag{14.13}$$

where U is taken to be zero at $r = \infty$. The total potential energy for a system of particles is the sum of energies for all pairs of particles, with each pair represented by a term of the form given by Equation 14.13.

If an isolated system consists of a particle of mass m moving with a speed v in the vicinity of a massive body of mass M, the *total energy E* of the system is the sum of the kinetic and potential energies:

$$E = \tfrac{1}{2}mv^2 - \frac{GMm}{r} \tag{14.15}$$

The total energy is a constant of the motion.

If m moves in a circular orbit of radius r about M, where $M \gg m$, the **total energy of the system** is

$$E = -\frac{GMm}{2r} \tag{14.18}$$

The total energy is negative for any bound system, that is, one in which the orbit is closed, such as an elliptical orbit.

QUESTIONS

1. Estimate the gravitational force between you and a person 2 m away from you.
2. Use Kepler's second law to convince yourself that the Earth must move faster in its orbit during December, when it is closest to the Sun, than during June, when it is farthest from the Sun.
3. If a system consists of five particles, how many terms appear in the expression for the total potential energy?
4. Is it possible to calculate the potential energy function associated with a particle and an extended body without knowing the geometry or mass distribution of the extended body?
5. Does the escape speed of a rocket depend on its mass? Explain.
6. Compare the energies required to reach the Moon for a 10^5-kg spacecraft and a 10^3-kg satellite.
7. Explain why it takes more fuel for a spacecraft to travel from the Earth to the Moon than for the return trip. Estimate the difference.
8. Is the potential energy associated with the Earth-Moon system greater than, less than, or equal to the kinetic energy of the Moon relative to the Earth?
9. Explain why there is no work done on a planet as it moves in a circular orbit around the Sun, even though a gravitational force is acting on the planet. What is the net work done on a planet during each revolution as it moves around the Sun in an elliptical orbit?
10. Explain why the force exerted on a particle by a uniform sphere must be directed toward the center of the sphere.

Would this be the case if the mass distribution of the sphere were not spherically symmetric?
11. Neglecting the density variation of the Earth, what would be the period of a particle moving in a smooth hole dug through the Earth's center?
12. At what position in its elliptical orbit is the speed of a planet a maximum? At what position is the speed a minimum?
13. If you are given the mass and radius of planet X, how would you calculate the free-fall acceleration on the surface of this planet?
14. If a hole could be dug to the center of the Earth, do you think that the force on a mass m would still obey Equation 14.1 there? What do you think the force on m would be at the center of the Earth?
15. In his 1798 experiment, Cavendish was said to have "weighed the Earth." Explain this statement.
16. The *Voyager* spacecraft was accelerated toward escape speed from the Sun by the gravitational force exerted on the spacecraft by Jupiter. How is this possible?
17. How would you find the mass of the Moon?
18. The *Apollo 13* spaceship developed trouble in the oxygen system about halfway to the Moon. Why did the mission continue on around the Moon, and then return home, rather than immediately turn back to Earth?
19. By how much is the free-fall acceleration at the Earth's equator reduced because of the rotation of the Earth? How does this effect vary with latitude?

PROBLEMS

Review Problem

A satellite of mass m is in an orbit of radius R around a planet of mass M in the equatorial plane of the planet. The satellite remains above the same point on the planet at all times. If the free-fall acceleration on the surface of the planet is g, find (a) the speed of the satellite, (b) the period of the satellite, (c) the kinetic energy of the satellite, (d) the potential energy of the satellite, (e) the radius of the planet, (f) the minimum possible period for the satellite, (g) the maximum kinetic energy of the satellite, (h) the minimum potential energy of the satellite, and (i) the escape speed of the satellite from this orbit.

Section 14.1 through Section 14.3

1. On the way to the Moon the Apollo astronauts reach a point where the Moon's gravitational pull is stronger than that of Earth's. (a) Determine the distance of this point from the center of the Earth. (b) What is the acceleration due to the Earth's gravity at this point?
2. A 200-kg mass and a 500-kg mass are separated by 0.400 m. (a) Find the net gravitational force exerted by these masses on a 50.0-kg mass placed midway between them. (b) At what position (other than infinitely remote ones) does the 50.0-kg mass experience a net force of zero?
3. A student proposes to measure the gravitational con-

□ indicates problems that have full solutions available in the Student Solutions Manual and Study Guide.

stant, G, by suspending two spherical masses from the ceiling of a tall cathedral and measuring the deflection from the vertical. If two 100.0-kg masses are suspended at the end of 45.00-m-long cables, and the cables are attached to the ceiling 1.000 m apart, what is the separation of the masses?

4. (a) Determine the change and fractional change in gravitational force that the Sun exerts on a 50.0-kg woman standing on the equator at noon and midnight. (*Hint:* Since Δr is so small, use differentials.) (b) By what percent does the weight of the 50.0-kg woman decrease during a total eclipse of the Sun (Fig. P14.4)?

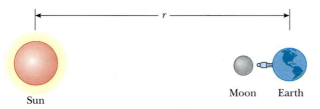

FIGURE P14.4

5. Three equal masses are located at three corners of a square of edge length ℓ as in Figure P14.5. Find the gravitational field **g** at the fourth corner due to these masses.

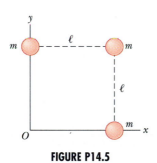

FIGURE P14.5

6. When a falling meteor is at a distance $d = 3R_E$ above the Earth's surface, what is its free-fall acceleration?

7. An astronaut weighs 140 N on the Moon's surface. When he is in a circular orbit about the Moon at an altitude $h = R_M$, what gravitational force does the Moon exert on him?

8. Two objects attract each other with a gravitational force of magnitude 1.0×10^{-8} N when separated by 20 cm. If the total mass of the two objects is 5.0 kg, what is the mass of each?

9. If the mass of Mars is $0.107 M_E$ and its radius is $0.53 R_E$, estimate the gravitational field g at the surface of Mars.

10. The free-fall acceleration on the surface of the Moon is about one-sixth that on the surface of the Earth. If the radius of the Moon is about $0.25 R_E$, find the ratio of their densities, ρ_{Moon}/ρ_{Earth}.

11. Plaskett's binary system consists of two stars that revolve in a circular orbit about a center of gravity midway between them. This means that the masses of the two stars are equal (Fig. P14.11). If the orbital speed of each star is 220 km/s and the orbital period of each is 14.4 days, find the mass M of each star. (For comparison, the mass of our Sun is 2×10^{30} kg.)

11A. Plaskett's binary system consists of two stars that revolve in a circular orbit about a center of gravity midway between them. This means that the masses of the two stars are equal (Fig. P14.11). If the orbital speed of each star is v and the orbital period of each is T, find the mass M of each star.

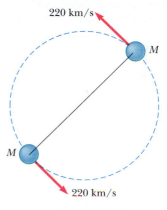

FIGURE P14.11

12. The Moon is 384 400 km distant from the Earth's center, and it completes an orbit in 27.3 days. (a) Determine the Moon's orbital speed. (b) How far does the Moon "fall" toward Earth in 1.00 s?

Section 14.4 Kepler's Laws
Section 14.5 The Law of Gravity and the Motion of Planets

13. During a solar eclipse, the Moon, Earth, and Sun all lie on the same line, with the Moon between the Earth and the Sun. (a) What force is exerted on the Moon by the Sun? (b) What force is exerted on the Moon by the Earth? (c) What force is exerted on the Earth by the Sun?

14. The *Explorer VIII* satellite, placed into orbit November 3, 1960, to investigate the ionosphere, had the following orbit parameters: perigee 459 km and apogee 2289 km (both distances above the Earth's surface); period 112.7 min. Find the ratio v_p/v_a.

15. Io, a small Moon of Jupiter, has an orbital period of 1.77 days and an orbital radius of 4.22×10^5 km. From these data, determine the mass of Jupiter.

16. A particle of mass m moves along a straight line with constant speed in the x direction a distance b from the x axis (Fig. P14.16). Show that Kepler's second law is satisfied by showing that the two shaded triangles in the figure have the same area when $t_4 - t_3 = t_2 - t_1$.

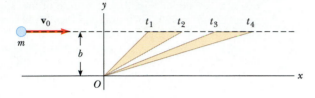

FIGURE P14.16

17. Two planets X and Y travel counterclockwise in circular orbits about a star as in Figure P14.17. The radii of their orbits are in the ratio $3:1$. At some time, they are aligned as in Figure P14.17a, making a straight line with the star. Five years later, planet X has rotated through 90° as in Figure P14.17b. Where is planet Y at this time?

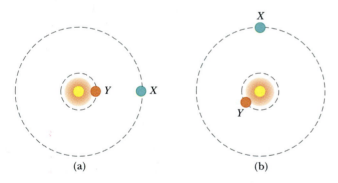

(a)   (b)

FIGURE P14.17

18. Geosynchronous satellites orbit the Earth 42 000 km from the Earth's center. Their angular speed at this height is the same as the rotational speed of the Earth, and so they appear stationary in the sky. What is the force acting on a 1000-kg satellite at this height?

19. A synchronous satellite, which always remains above the same point on a planet's equator, is put in orbit around Jupiter to study the famous red spot. Jupiter rotates once every 9.9 h. Use the data of Table 14.2 to find the altitude of the satellite.

20. Halley's comet approaches the Sun to within 0.57 A.U., and its orbital period is 75.6 years. (A.U. is the abbreviation for astronomical unit, where 1 A.U. = 1.50×10^6 km is the mean Earth-Sun distance.) How far from the Sun will Halley's comet travel before it starts its return journey? (Fig. P14.20.)

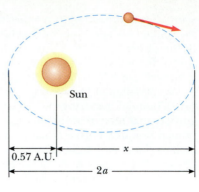

FIGURE P14.20

Section 14.6   The Gravitational Field

21. Compute the magnitude and direction of the gravitational field at a point P on the perpendicular bisector of two equal masses separated by a distance $2a$ as shown in Figure P14.21.

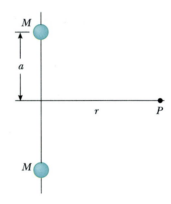

FIGURE P14.21

22. Find the gravitational field at a distance r along the axis of a thin ring of mass M and radius a.

23. At what point along the line connecting the Earth and the Moon is there zero gravitational force on an object? (Ignore the presence of the Sun and the other planets.)

Section 14.7   Gravitational Potential Energy

(*Note:* Assume $U = 0$ at $r = \infty$.)

24. A satellite of the Earth has a mass of 100 kg and is at an altitude of 2.00×10^6 m. (a) What is the potential energy of the satellite-Earth system? (b) What is the magnitude of the gravitational force exerted by the Earth on the satellite?

25. How much work is done by the Moon's gravitational field as a 1000-kg meteor comes in from outer space and impacts on the Moon's surface?

26. How much energy is required to move a 1000-kg mass from the Earth's surface to an altitude $h = 2R_E$?

26A. How much energy is required to move a mass m from the Earth's surface to an altitude h?

27. After it exhausts its nuclear fuel, the ultimate fate of our Sun is possibly to collapse to a *white dwarf*, which is a star that has approximately the mass of the Sun, but the radius of the Earth. Calculate (a) the average density of the white dwarf, (b) the free-fall acceleration at its surface, and (c) the gravitational potential energy of a 1.00-kg object at its surface.

Section 14.8 Energy Considerations in Planetary and Satellite Motion

28. Determine the escape speed for a rocket on the far side of Ganymede, the largest of Jupiter's moons. The radius of Ganymede is 2.64×10^6 m, and its mass is 1.495×10^{23} kg. The mass of Jupiter is 1.90×10^{27} kg, and the distance between Jupiter and Ganymede is 1.071×10^9 m. Be sure to include the gravitational effect due to Jupiter, but you may ignore the motion of Jupiter and Ganymede as they revolve about their center of mass (Fig. P14.28).

Jupiter

Ganymede

FIGURE P14.28

29. A spaceship is fired from the Earth's surface with an initial speed of 2.00×10^4 m/s. What will its speed be when it is very far from the Earth? (Neglect friction.)

30. A 1000-kg satellite orbits the Earth at an altitude of 100 km. It is desired to increase the altitude of the orbit to 200 km. How much energy must be added to the system to effect this change in altitude?

30A. A satellite of mass m orbits the Earth at an altitude h_1. It is desired to increase the altitude of the orbit to h_2. How much energy must be added to the system to effect this change in altitude?

31. In Robert Heinlein's *The Moon Is a Harsh Mistress,* the colonial inhabitants of the Moon threaten to launch rocks down onto the Earth if they are not given independence (or at least representation). Assuming that a rail gun could launch a rock of mass m at twice the lunar escape speed, calculate the speed of the rock as it enters the Earth's atmosphere.

32. A rocket is fired vertically, ejecting sufficient mass to move upward at a constant acceleration of $2g$. After 40.0 s, the rocket motors are turned off, and the rocket subsequently moves under the action of gravity alone, with negligible air resistance. Ignoring the variation of **g** with altitude, find (a) the maximum height the rocket reaches and (b) the total flight time from launch until the rocket returns to Earth. (c) Sketch a freehand (qualitative) graph of speed versus time for the flight.

33. A satellite moves in a circular orbit just above the surface of a planet. Show that the orbital speed v and escape speed of the satellite are related by the expression $v_{esc} = \sqrt{2}\,v$.

34. A satellite moves in an elliptical orbit about the Earth such that, at perigee and apogee positions, the distances from the Earth's center are, respectively, D and $4D$. Find the ratios (a) v_p/v_a and (b) E_p/E_a.

35. A 500-kg satellite is in a circular orbit at an altitude of 500 km above the Earth's surface. Because of air friction, the satellite eventually is brought to the Earth's surface, and it hits the Earth with a speed of 2.00 km/s. How much energy was absorbed by the atmosphere through friction?

35A. A satellite of mass m is in a circular orbit at an altitude h above the Earth's surface. Because of air friction, the satellite eventually is brought to the Earth's surface, and it hits the Earth with a speed v. How much energy was absorbed by the atmosphere through friction?

36. An artificial Earth satellite is "parked" in an equatorial circular orbit at an altitude of 1.00×10^3 km. What is the minimum additional speed that must be imparted to the satellite if it is to escape from Earth's gravitational attraction? How does this compare with the minimum escape speed for leaving from the Earth's surface?

37. (a) What is the minimum speed necessary for a spacecraft to escape the Solar System, starting at the Earth's orbit? (b) *Voyager 1* achieved a maximum speed of 125 000 km/h on its way to photograph Jupiter. Beyond what distance from the Sun is this speed sufficient to escape the Solar System?

*Section 14.9 The Gravitational Force Between an Extended Object and a Particle

38. A uniform rod of mass M is in the shape of a semicircle of radius R (Fig. P14.38). Calculate the force on a point mass m placed at the center of the semicircle.

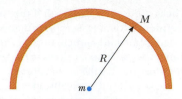

FIGURE P14.38

39. A spacecraft in the shape of a long cylinder has a length of 100 m and its mass with occupants is 1000 kg. It has strayed in too close to a 1.0-km radius black hole having a mass 100 times that of the Sun (Fig. P14.39). If the nose of the spacecraft points toward the center of the black hole, and if distance between the nose of the spaceship and the black hole's center is 10 km, (a) determine the average acceleration of the spaceship. (b) What is the difference in the acceleration experienced by the occupants in the nose of the ship and those in the rear of the ship farthest from the black hole?

39A. A spacecraft in the shape of a long cylinder has a length ℓ and its mass with occupants is m. It has strayed in too close to a black hole of radius R and mass M. If the nose of the spacecraft points toward the center of the black hole, and if distance between the nose of the spaceship and the black hole's center is d, (a) determine the average acceleration of the spaceship. (b) What is the difference in the acceleration experienced by the occupants in the nose of the ship and those in the rear of the ship farthest from the black hole?

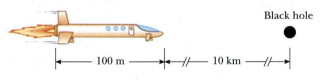

Black hole

|← 100 m →|←//— 10 km —//→|

FIGURE P14.39

*Section 14.10   Gravitational Force Between a Particle and a Spherical Mass

40. (a) Show that the period calculated in Example 14.11 can be written as

$$T = 2\pi\sqrt{\frac{R_E}{g}}$$

where g is the free-fall acceleration. (b) What would this period be if tunnels were made through the Moon? (c) What practical problem regarding these tunnels on Earth would be removed if they were built on the Moon?

41. A 500-kg uniform solid sphere has a radius of 0.400 m. Find the magnitude of the gravitational force exerted by the sphere on a 50.0-g particle located (a) 1.50 m from the center of the sphere, (b) at

the surface of the sphere, and (c) 0.200 m from the center of the sphere.

42. A uniform solid sphere of mass m_1 and radius R_1 is inside and concentric with a spherical shell of mass m_2 and radius R_2 (Fig. P14.42). Find the gravitational force exerted by the sphere on a particle of mass m located at (a) $r = a$, (b) $r = b$, (c) $r = c$, where r is measured from the center of the spheres.

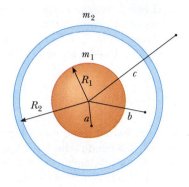

FIGURE P14.42

ADDITIONAL PROBLEMS

43. Calculate the fractional difference $\Delta g/g$ in the free-fall acceleration at points on the Earth's surface nearest to and farthest from the Moon, taking into account the gravitational effect of the Moon. (This difference is responsible for the occurrence of the *lunar tides* on the Earth.)

44. *Voyagers 1* and *2* surveyed the surface of Jupiter's moon Io and photographed active volcanoes spewing liquid sulfur to heights of 70 km above the surface of this moon. Estimate the speed with which the liquid sulfur left the volcano. Io's mass is 8.9×10^{22} kg, and its radius is 1820 km.

45. A cylindrical habitat in space 6.0 km in diameter and 30 km long has been proposed (by G. K. O'Neill, 1974). Such a habitat would have cities, land, and lakes on the inside surface and air and clouds in the center. This would all be held in place by rotation of the cylinder about its long axis. How fast would the cylinder have to rotate to imitate the Earth's gravitational field at the walls of the cylinder?

45A. A cylindrical habitat in space having diameter d and length L has been proposed (by G. K. O'Neill, 1974). Such a habitat would have cities, land, and lakes on the inside surface and air and clouds in the center. This would all be held in place by rotation of the cylinder about its long axis. How fast would the cylinder have to rotate to imitate the Earth's gravitational field at the walls of the cylinder?

46. Two spheres having masses M and $2M$ and radii R and $3R$, respectively, are released from rest when the

distance between their centers is $12R$. How fast will each sphere be moving when they collide? Assume that the two spheres interact only with each other.

47. In introductory physics laboratories, a typical Cavendish balance for measuring the gravitational constant G uses lead spheres of masses 1.50 kg and 15.0 g whose centers are separated by 4.50 cm. Calculate the gravitational force between these spheres, treating each as a point mass located at the center of the sphere.

48. Consider two identical uniform rods of length L and mass m lying along the same line and having their closest points separated by a distance d (Fig. P14.48). Show that the mutual gravitational force between these rods has a magnitude

$$F = \frac{Gm^2}{L^2} \ln\left(\frac{(L+d)^2}{d(2L+d)}\right)$$

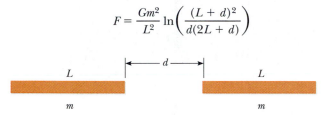

FIGURE P14.48

49. An object of mass m moves in a smooth straight tunnel of length L dug through a chord of the Earth as discussed in Example 14.11 (Fig. 14.20). (a) Determine the effective force constant of the harmonic motion and the amplitude of the motion. (b) Using energy considerations, find the maximum speed of the object. Where does this maximum speed occur? (c) Obtain a numerical value for the maximum speed if $L = 2500$ km.

50. For any body orbiting the Sun, Kepler's third law may be written $T^2 = kr^3$ where T is the orbital period and r is the semimajor axis of the orbit. (a) What is the value of k if T is measured in years and r is measured in A.U.? (See Problem 20.) (b) Use your value of k to find the orbital period of Jupiter if its mean radius from the Sun is 5.2 A.U.

51. Three point objects having masses m, $2m$, and $3m$ are fixed at the corners of a square of side length a such that the lighter object is at the upper left-hand corner, the heavier object is at the lower left-hand corner, and the remaining object is at the upper right-hand corner. Determine the magnitude and direction of the resulting gravitational field at the center of the square.

52. An airplane in a wide "outside" loop can create an apparent weight of zero inside the aircraft cabin. What must be the radius of curvature of the flight path for an aircraft moving at 480 km/h to create a condition of weightlessness inside the aircraft?

53. What angular speed (in revolutions per minute) is needed for a centrifuge to produce an acceleration of $1000g$ at a radius of 10.0 cm?

54. The Earth-Sun distance is 1.521×10^{11} m at the aphelion and 1.471×10^{11} m at the perihelion. If the Earth's orbital speed at the perihelion is 3.027×10^4 m/s, determine (a) its orbital speed at the aphelion, (b) the kinetic and potential energy at the perihelion, and (c) the kinetic and potential energy at the aphelion. Is the total energy constant? (Neglect the effect of the Moon and other planets.)

55. Two hypothetical planets of masses m_1 and m_2 and radii r_1 and r_2, respectively, are at rest when they are an infinite distance apart. Because of their gravitational attraction, they head toward each other on a collision course. (a) When their center-to-center separation is d, find the speed of each planet and their relative velocity. (b) Find the kinetic energy of each planet just before they collide if $m_1 = 2.0 \times 10^{24}$ kg, $m_2 = 8.0 \times 10^{24}$ kg, $r_1 = 3.0 \times 10^6$ m, and $r_2 = 5.0 \times 10^6$ m. (*Hint:* Both energy and momentum are conserved.)

56. After a supernova explosion, a star may undergo a gravitational collapse to an extremely dense state known as a neutron star, in which all the electrons and protons are squeezed together to form neutrons. A neutron star having a mass about equal to that of the Sun would have a radius of about 10 km. Find (a) the free-fall acceleration at its surface, (b) the weight of a 70-kg person at its surface, and (c) the energy required to remove a neutron of mass 1.67×10^{-27} kg from its surface to infinity.

57. When it orbited the Moon, the *Apollo 11* spacecraft's mass was 9.979×10^3 kg, its period was 119 min, and its mean distance from the Moon's center was 1.849×10^6 m. Assuming its orbit was circular and the Moon to be a uniform sphere, find (a) the mass of the Moon, (b) the orbital speed of the spacecraft, and (c) the minimum energy required for the craft to leave the orbit and escape the Moon's gravitational field.

58. Studies of the relationship of the Sun to its galaxy—the Milky Way—have revealed that the Sun is located near the outer edge of the galactic disc, about 30 000 light years from the center. Furthermore, it has been found that the Sun has an orbital speed of approximately 250 km/s around the galactic center. (a) What is the period of the Sun's galactic motion? (b) What is the approximate mass of the Milky Way galaxy? Using the fact that the Sun is a typical star, estimate the number of stars in the Milky Way.

59. X-ray pulses from Cygnus X-1, a celestial x-ray source, have been recorded during high-altitude rocket flights. The signals can be interpreted as originating when a blob of ionized matter orbits a black hole with a period of 5.0 ms. If the blob were in a circular orbit about a black hole whose mass is $20M_{Sun}$, what is the orbit radius?

60. *Vanguard I*, launched March 3, 1958, is the oldest man-made satellite still in orbit. Its initial orbit had

an apogee of 3970 km and a perigee of 650 km. Its maximum speed was 8.23 km/s and it had a mass of 1.60 kg. (a) Determine the period of the orbit. (Use the semimajor axis.) (b) Determine the speeds at apogee and perigee. (c) Find the total energy of the satellite.

61. A 200-kg satellite is placed in Earth orbit 200 km above the surface. (a) Assuming a circular orbit, how long does the satellite take to complete one orbit? (b) What is the satellite's speed? (c) What is the minimum energy necessary to place this satellite in orbit (assuming no air friction)?

61A. A satellite of mass m is placed in Earth orbit at an altitude h. (a) Assuming a circular orbit, how long does the satellite take to complete one orbit? (b) What is the satellite's speed? (c) What is the minimum energy necessary to place this satellite in orbit (assuming no air friction)?

62. In Larry Niven's science fiction novel *Ringworld,* a ring of material rotates about a star (Fig. P14.62). The rotational speed of the ring is 1.25×10^6 m/s and its radius is 1.53×10^{11} m. The inhabitants of this ringworld experience a normal contact force **n**. Acting alone, this normal force would produce an inward acceleration of 9.90 m/s². Additionally, the star at the center of the ring exerts a gravitational force on the ring and its inhabitants. (a) Show that the total centripetal acceleration of the inhabitants is 10.2 m/s². (b) The difference between the total acceleration and the acceleration provided by the normal force is due to the gravitational attraction of the central star. Show that the mass of the star is approximately 10^{32} kg.

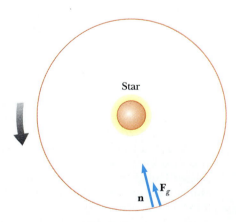

FIGURE P14.62

63. A sphere of mass M and radius R has a nonuniform density that varies with r, the distance from its center, according to the expression $\rho = Ar$, for $0 \le r \le R$. (a) What is the constant A in terms of M and R? (b) Determine the force on a particle of mass m placed outside the sphere. (c) Determine the force on the particle if it is inside the sphere. (*Hint:* See Section 14.10 and note that the distribution is spherically symmetric.)

64. Two stars of masses M and m, separated by a distance d, revolve in circular orbits about their center of mass (Fig. P14.64). Show that each star has a period given by

$$T^2 = \frac{4\pi^2}{G(M + m)} d^3$$

(*Hint:* Apply Newton's second law to each star, and note that the center-of-mass condition requires that $Mr_2 = mr_1$, where $r_1 + r_2 = d$.)

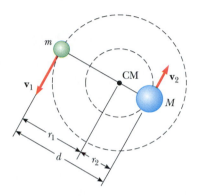

FIGURE P14.64

65. In 1978, astronomers at the U.S. Naval Observatory discovered that Pluto has a Moon, called Charon, that eclipses the planet every 6.4 days. Given that the center-to-center separation between Pluto and Charon is 19 700 km, find the total mass $(M + m)$ of the two bodies. (*Hint:* See Problem 64.)

66. In an effort to explain large meteor collisions with the Earth, scientists have postulated the existence of a companion star that is extremely dim and extremely far from the Sun. If this star (which some astronomers call Nemesis) has an orbital period of 3.0×10^7 years around the Sun-Nemesis center of mass, and a mass of $0.20M_{\text{Sun}}$, determine the average distance of this star from the Sun. $M_{\text{Sun}} = 2.0 \times 10^{30}$ kg.

67. A satellite is in a circular orbit about a planet of radius R. If the altitude of the satellite is h and its period is T, (a) show that the density of the planet is

$$\rho = \frac{3\pi}{GT^2}\left(1 + \frac{h}{R}\right)^3$$

(b) Calculate the average density of the planet if the period is 200 min and the satellite's orbit is close to the planet's surface.

68. It is claimed that a commercially available, portable gravity meter is sensitive enough to detect changes in

g to 1 part in 10^{11}. At the Earth's surface, what change in elevation would produce this variation? Assume the radius of the Earth is 6.0×10^6 m.

69. A particle of mass m is located inside a uniform solid sphere of radius R and mass M. If the particle is at a distance r from the center of the sphere, (a) show that the gravitational potential energy of the system is $U = (GmM/2R^3)r^2 - 3GmM/2R$. (b) How much work is done by the gravitational force in bringing the particle from the surface of the sphere to its center?

SPREADSHEET PROBLEMS

S1. Four point masses are fixed as shown in Figure PS14.1. The gravitational potential energy for a test particle of mass m moving along the x axis in the gravitational field of the fixed masses can be written as

$$U(x) = -\frac{GM_1 m}{r_1} - \frac{GM_2 m}{r_2} - \frac{GM_3 m}{r_3} - \frac{GM_4 m}{r_4}$$

where $r_1 = r_4 = [b^2 + (x + a^2)]^{1/2}$ and $r_2 = r_3 = [b^2 + (x - a)^2]^{1/2}$. Spreadsheet 14.1 calculates $U(x)$ and the force $F(x)$ exerted on the test particle $= -dU(x)/dx$. The quantities M_1, M_2, M_3, M_4, a, and b are input parameters. The mass of the test particle is 1.00 kg. Use $a = 0$, $b = 1.00$ m, $M_1 = M_3 = 0$, $M_2 = M_4 = 1.00$ kg. (This is the case of two particles on the y axis at $y = +1.00$ m and $y = -1.00$ m.) (a) Plot $U(x)$ and $F(x)$ versus x for $x = -4.00$ m to $+4.00$ m. (b) If the test particle has an energy of -7.50×10^{-11} J, describe the particle's motion. (c) How does the force on the particle vary as the particle moves along the x axis?

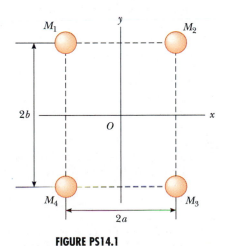

FIGURE PS14.1

S2. In Spreadsheet 14.1, let $a = 1.00$ m, $b = 0.50$ m, and $M_1 = M_2 = M_3 = M_4 = 1.00$ kg. The mass of the test

particle is 1.00 kg. (a) Plot $U(x)$ and $F(x)$ versus x. (b) If the test particle has an energy of -2.00×10^{-10} J, describe its motion. (c) If the particle has an energy of -3.00×10^{-10} J, what motion is possible?

S3. A projectile is fired straight up from the surface of the Earth with an initial speed v_0. If v_0 is less than the escape speed $v_{esc} = \sqrt{2gR_E} = 11.2$ km/s, the projectile will rise to a maximum height h above the Earth's surface and return. Neglecting air resistance and using conservation of energy, show that

$$h = \frac{R_E v_0^2}{2gR_E - v_0}$$

where R_E is the Earth's radius. Write your own spreadsheet to calculate h for $v_0 = 1, 2, 3, \ldots,$ 11 km/s. Plot h/R_E versus v_0. What happens as v_0 approaches the escape speed v_{esc}?

S4. The acceleration of an object moving in the gravitational field of the Earth is

$$a = -GM_E \frac{\mathbf{r}}{r^3}$$

where r is the position vector directed from the center of the Earth to the object. Choosing the origin at the center of the Earth, and assuming the object is moving in the xy-plane, the acceleration has rectangular components

$$a_x = -\frac{GM_E x}{(x^2 + y^2)^{3/2}}$$

$$a_y = -\frac{GM_E y}{(x^2 + y^2)^{3/2}}$$

Write your own spreadsheet (or computer program) to find the position of the object as a function of time. (Use the techniques developed in Spreadsheet 6.1 as a model for your program. Assume that the initial position of the object is at $x = 0$ and $y = 2R_E$, where R_E is the radius of the Earth, and give the object an initial velocity of 5 km/s in the x direction. The time increment should be made as small as practical. Try 5 s. Plot the x and y coordinates of the object as functions of time. Does the object hit the Earth? Vary the initial velocity until a circular orbit is found.

S5. Modify your spreadsheet (or program) from Problem S4 to calculate the kinetic, potential, and total energies as functions of time. How do they vary with time for the various cases you tried?

S6. Modify your spreadsheet (or program) from Problem S14.3 to calculate the angular momentum of the object. In rectangular coordinates $L = m(xv_y - yv_x)$. How does the angular momentum vary with time?

Kepler's second law of planetary motion implies that L remains constant. Is your calculated L constant?

S7. Imagine that you are an intrepid space traveler and you discover a Universe where your spaceship no longer behaves normally. You suspect that the law of gravity does not have the inverse square dependence that it should. Modify the spreadsheet (or program) in Problem S4 to allow for a gravitational law of any inverse power. Investigate the shape of the orbits for different inverse powers. Is energy constant for orbital motion in this strange Universe? Are Kepler's laws still valid?

Fluid Mechanics

A scuba diver plays with an octopus in Poor Knights Island, New Zealand. As the diver descends to greater depths, the water pressure increases above atmospheric pressure, and the internal pressure of the body also increases accordingly to maintain equilibrium. Scuba divers have been able to swim at depths of over 1000 ft. *(Darryl Torckler/Tony Stone Images)*

Matter is normally classified as being in one of three states: solid, liquid, or gaseous. Everyday experience tells us that a solid has a definite volume and shape. A brick maintains its familiar shape and size day in and day out. We also know that a liquid has a definite volume but no definite shape. Finally, a gas has neither definite volume nor definite shape. These definitions help us to picture the states of matter, but they are somewhat artificial. For example, asphalt and plastics are normally considered solids, but over long periods of time they tend to flow like liquids. Likewise, most substances can be a solid, liquid, or gas (or combinations of these), depending on the temperature and pressure. In general, the time it takes a particular substance to change its shape in response to an external force determines whether we treat the substance as a solid, liquid, or gas.

A **fluid** is a collection of molecules that are randomly arranged and held together by weak cohesive forces and forces exerted by the walls of a container. Both liquids and gases are fluids.

In our treatment of the mechanics of fluids, we shall see that no new physical principles are needed to explain such effects as the buoyant force on a submerged object and the dynamic lift on an airplane wing. First, we consider a fluid at rest and derive an expression for the pressure exerted by the fluid as a function of its density and depth. We then treat fluids in motion, an area of study called fluid dynamics. A fluid in motion can be described by a model in which certain simplifying assumptions are made. We use this model to analyze some situations of practical importance. An underlying principle known as the *Bernoulli principle* enables us to determine relationships between the pressure, density, and velocity at every point in a fluid. We conclude the chapter with a brief discussion of internal friction in a fluid and turbulent motion.

15.1 PRESSURE

The study of fluid mechanics involves the density of a substance, defined as its mass per unit volume. For this reason, Table 15.1 lists the densities of various substances. These values vary slightly with temperature, since the volume of a substance is temperature dependent (as we shall see in Chapter 19). Note that under standard conditions (0°C and atmospheric pressure) the densities of gases are about 1/1000 the densities of solids and liquids. This difference implies that the average molecular spacing in a gas under these conditions is about ten times greater than in a solid or liquid.

Fluids do not sustain shearing stresses, and thus the only stress that can exist on an object submerged in a fluid is one that tends to compress the object. The force exerted by the fluid on the object is always perpendicular to the surfaces of the object, as shown in Figure 15.1.

The pressure at a specific point in a fluid can be measured with the device pictured in Figure 15.2. The device consists of an evacuated cylinder enclosing a light piston connected to a spring. As the device is submerged in a fluid, the fluid presses down on the top of the piston and compresses the spring until the inward force of the fluid is balanced by the outward force of the spring. The fluid pressure can be measured directly if the spring is calibrated in advance. This is accomplished by applying a known force to the spring to compress it a given distance.

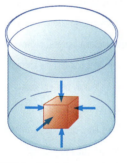

FIGURE 15.1 The force of the fluid on a submerged object at any point is perpendicular to the surface of the object. The force of the fluid on the walls of the container is perpendicular to the walls at all points.

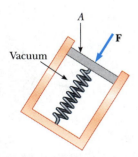

FIGURE 15.2 A simple device for measuring pressure in a fluid.

TABLE 15.1	**Densities of Some Common Substances**		
Substance	ρ (kg/m³)ᵃ	Substance	ρ (kg/m³)ᵃ
Ice	0.917×10^3	Water	1.00×10^3
Aluminum	2.70×10^3	Sea water	1.03×10^3
Iron	7.86×10^3	Ethyl alcohol	0.806×10^3
Copper	8.92×10^3	Benzene	0.879×10^3
Silver	10.5×10^3	Mercury	13.6×10^3
Lead	11.3×10^3	Air	1.29
Gold	19.3×10^3	Oxygen	1.43
Platinum	21.4×10^3	Hydrogen	8.99×10^{-2}
Glycerine	1.26×10^3	Helium	1.79×10^{-1}

ᵃ All values are at standard atmospheric pressure and temperature (STP), that is, atmospheric pressure and 0°C. To convert to grams per cubic centimeter, multiply by 10^{-3}.

If *F* is the magnitude of the normal force on the piston and *A* is the surface area of the piston, then the pressure, *P*, of the fluid at the level to which the device has been submerged is defined as the ratio of force to area:

$$P \equiv \frac{F}{A} \qquad (15.1)$$

Definition of pressure

To define the pressure at a specific point, consider a fluid enclosed as shown in Figure 15.2. If the normal force exerted by the fluid is *F* over a surface element of area δA that contains the point in question, then the pressure at that point is

$$P = \lim_{\delta A \to 0} \frac{F}{\delta A} = \frac{dF}{dA} \qquad (15.2)$$

As we see in the next section, the pressure in a fluid varies with depth. Therefore, to get the total force on a flat wall of a container, we have to integrate Equation 15.2 over the surface.

Since pressure is force per unit area, it has units of N/m^2 in the SI system. Another name for the SI unit of pressure is **pascal** (Pa).

$$1 \text{ Pa} \equiv 1 \text{ N/m}^2 \qquad (15.3)$$

CONCEPTUAL EXAMPLE 15.1

A woman wearing high-heeled shoes is invited into a home in which the kitchen has vinyl floor covering. Why should the homeowner be concerned?

Reasoning   She can exert enough pressure on the floor to dent or puncture the floor covering. The large pressure is caused by the fact that her weight is distributed over the very small cross-sectional area of her high heels. She should be asked to remove her high heels and put on some slippers.

CONCEPTUAL EXAMPLE 15.2

The daring physics professor, after a long lecture, stretches out for a nap on a bed of nails as in the photograph. How is this possible?

Reasoning   If you try to support your entire weight on a single nail, the pressure on your body is your weight divided by the very small area of the nail. This pressure is sufficiently large to penetrate the skin. However, if you distribute your weight over several hundred nails, as the professor is doing, the pressure is considerably reduced because the area that supports your weight is the total area of all nails in contact with your body. (Note that lying on a bed of nails is much more comfortable than sitting on the bed. Standing on the bed without shoes is not recommended.)

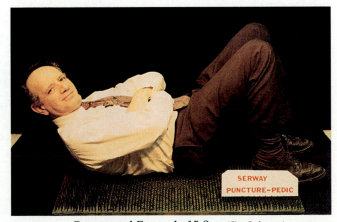

Conceptual Example 15.2.   (*Jim Lehman*)

Snowshoes prevent the person from sinking into the soft snow because the person's weight is spread over a larger area, which reduces the pressure on the snow's surface.   *(Earl Young/FPG)*

Variation of pressure with depth

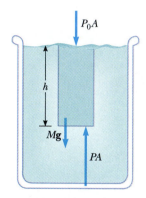

FIGURE 15.3   The variation of pressure with depth in a fluid. The net force on the volume of water within the darker region must be zero.

15.2   VARIATION OF PRESSURE WITH DEPTH

As divers well know, the pressure in the sea or a lake increases as they dive to greater depths. Likewise, atmospheric pressure decreases with increasing altitude. For this reason, aircraft flying at high altitudes must have pressurized cabins.

We now show how the pressure in a liquid increases linearly with depth. Consider a liquid of density ρ at rest and open to the atmosphere as in Figure 15.3. Let us select a sample of the liquid contained within an imaginary cylinder of cross-sectional area A extending from the surface of the liquid to a depth h. The pressure exerted by the fluid on the bottom face is P, and the pressure on the top face of the cylinder is atmospheric pressure, P_0. Therefore, the upward force exerted by the liquid on the bottom of the cylinder is PA, and the downward force exerted by the atmosphere on the top is P_0A. Because the mass of liquid in the cylinder is $\rho V = \rho Ah$, the weight of the fluid in the cylinder is $w = \rho gV = \rho gAh$. Because the cylinder is in equilibrium, the upward force at the bottom must be greater than the downward force at the top of the sample to support its weight:

$$PA - P_0A = \rho hgA$$

or

$$P = P_0 + \rho gh \qquad (15.4)$$

where we usually take atmospheric pressure to be $P_0 = 1.00 \text{ atm} \approx 1.01 \times 10^5$ Pa. In other words,

> the absolute pressure P at a depth h below the surface of a liquid open to the atmosphere is *greater* than atmospheric pressure by an amount ρgh.

This result also verifies that the pressure is the same at all points having the same depth, independent of the shape of the container.

In view of the fact that the pressure in a liquid depends only upon depth, any increase in pressure at the surface must be transmitted to every point in the fluid. This was first recognized by the French scientist Blaise Pascal (1623–1662) and is called **Pascal's law:**

> A change in the pressure applied to an enclosed liquid is transmitted undiminished to every point of the liquid and to the walls of the container.

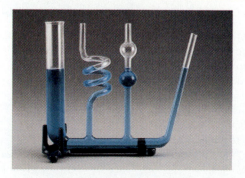

This photograph illustrates that the pressure in a liquid is the same at all points having the same elevation. Note that the shape of the vessel does not affect the pressure.   *(Courtesy of Central Scientific Co.)*

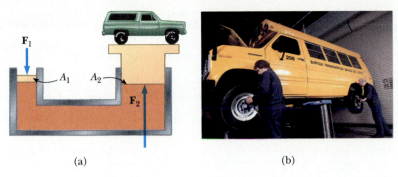

(a)   (b)

FIGURE 15.4   (a) Diagram of a hydraulic press. Since the increase in pressure is the same at the left and right sides, a small force F_1 at the left produces a much larger force F_2 at the right. (b) A bus under repair is supported by a hydraulic lift in a garage.   *(Superstock)*

An important application of Pascal's law is the hydraulic press illustrated by Figure 15.4. A force F_1 is applied to a small piston of area A_1. The pressure is transmitted through a liquid to a larger piston of area A_2. Since the pressure is the same on both sides, we see that $P = F_1/A_1 = F_2/A_2$. Therefore, the force F_2 is larger than F_1 by the multiplying factor A_2/A_1. Hydraulic brakes, car lifts, hydraulic jacks, and forklifts all make use of this principle.

CONCEPTUAL EXAMPLE 15.3

A typical silo on a farm has many bands wrapped around its perimeter as shown in the photograph. Why is the spacing between successive bands smaller at the lower portions of the silo?

Reasoning   If you think of the grain stored in the silo as a fluid, the pressure the grain exerts on the walls of the silo increases with increasing depth just as water pressure in a lake increases with increasing depth. Thus, the spacing between bands is made smaller at the lower portions to overcome the larger outward forces on the walls in these regions.

Conceptual Example 15.3.   *(Henry Leap)*

EXAMPLE 15.4   The Car Lift

In a car lift used in a service station, compressed air exerts a force on a small piston of circular cross-section having a radius of 5.00 cm. This pressure is transmitted by a liquid to a second piston of radius 15.0 cm. What force must the compressed air exert in order to lift a car weighing 13 300 N? What air pressure will produce this force?

Solution   Because the pressure exerted by the compressed air is transmitted undiminished throughout the fluid, we have

$$F_1 = \left(\frac{A_1}{A_2}\right)F_2 = \frac{\pi(5.00 \times 10^{-2}\,\text{m})^2}{\pi(15.0 \times 10^{-2}\,\text{m})^2}\,(1.33 \times 10^4\,\text{N})$$

$$= 1.48 \times 10^3\,\text{N}$$

The air pressure that will produce this force is

$$P = \frac{F_1}{A_1} = \frac{1.48 \times 10^3\,\text{N}}{\pi(5.00 \times 10^{-2}\,\text{m})^2} = 1.88 \times 10^5\,\text{Pa}$$

This pressure is approximately twice atmospheric pressure.

The input work (the work done by F_1) is equal to the output work (the work done by F_2), so that energy is conserved.

EXAMPLE 15.5   The Water Bed

A water bed is 2.00 m on a side and 30.0 cm deep. (a) Find its weight.

Solution   Since the density of water is 1000 kg/m³ (Table 15.1), the mass of the bed is

$$M = \rho V = (1000\,\text{kg/m}^3)(1.20\,\text{m}^3) = 1.20 \times 10^3\,\text{kg}$$

and its weight is

$$w = Mg = (1.20 \times 10^3\,\text{kg})(9.80\,\text{m/s}^2) = 1.18 \times 10^4\,\text{N}$$

This is equivalent to approximately 2640 lb (as compared with a regular bed that weighs approximately 300 lb). In order to support such a heavy load, you would be well advised to keep your water bed in the basement or on a sturdy, well-supported floor.

(b) Find the pressure exerted on the floor when the bed rests in its normal position. Assume that the entire lower surface of the bed makes contact with the floor.

Solution   The weight of the bed is 1.18×10^4 N. The cross-sectional area is 4.00 m² when the bed is in its normal position. This gives a pressure exerted on the floor of

$$P = \frac{1.18 \times 10^4\,\text{N}}{4.00\,\text{m}^2} = 2.95 \times 10^3\,\text{Pa}$$

Exercise   Calculate the pressure exerted on the floor when the bed rests on its side.

Answer   Since the area of its side is 0.600 m², the pressure is 1.96×10^4 Pa.

EXAMPLE 15.6   Pressure in the Ocean

Calculate the pressure at an ocean depth of 1000 m. Assume the density of sea water is 1.024×10^3 kg/m³ and take $P_0 = 1.01 \times 10^5$ Pa.

Solution

$$P = P_0 + \rho g h$$
$$= 1.01 \times 10^5\,\text{Pa} + (1.024 \times 10^3\,\text{kg/m}^3)$$
$$\times (9.80\,\text{m/s}^2)(1.00 \times 10^3\,\text{m})$$

$$P = 1.01 \times 10^7\,\text{Pa}$$

This is 100 times greater than atmospheric pressure! Obviously, the design and construction of vessels that will withstand such enormous pressures are not trivial matters.

Exercise   Calculate the total force exerted on the outside of a circular submarine window of diameter 30.0 cm at this depth.

Answer 7.00×10^5 N.

EXAMPLE 15.7   The Force on a Dam

Water is filled to a height H behind a dam of width w (Fig. 15.5). Determine the resultant force on the dam.

Reasoning   We cannot calculate the force on the dam by simply multiplying the area times the pressure, because the pressure varies with depth. The problem can be solved by finding the force dF on a narrow horizontal strip at depth h,

and then integrate the expression to find the total force on the dam.

Solution   The pressure at the depth h beneath the surface at the shaded portion is

$$P = \rho g h = \rho g(H - y)$$

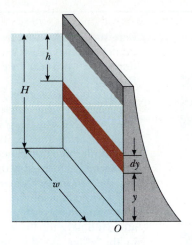

FIGURE 15.5   (Example 15.7) The total force on a dam must be obtained from the expression $F = \int P\, dA$, where dA is the area of the dark strip.

(We have left out atmospheric pressure because it acts on both sides of the dam.) Using Equation 15.2, we find the force on the shaded strip of area $dA = w\,dy$ to be

$$dF = P\,dA = \rho g(H - y)\,w\,dy$$

Therefore, the total force on the dam is

$$F = \int P\, dA = \int_0^H \rho g(H - y)\,w\,dy = \tfrac{1}{2}\rho g w H^2$$

Note that because the pressure increases with depth, the dam is designed such that its thickness increases with depth, as in Figure 15.5.

Exercise   Use the fact that the pressure increases linearly with depth to find the average pressure on the dam and the total force on the dam.

15.3   PRESSURE MEASUREMENTS

One simple device for measuring pressure is the open-tube manometer illustrated in Figure 15.6a. One end of a U-shaped tube containing a liquid is open to the atmosphere, and the other end is connected to a system of unknown pressure P. The difference in pressure $P - P_0$ is equal to ρgh. Therefore, we see that $P = P_0 + \rho gh$. The pressure P is called the **absolute pressure**, while the difference $P - P_0$ is called the **gauge pressure**. For example, the pressure you measure in your bicycle tire is gauge pressure.

Another instrument used to measure pressure is the common barometer, invented by Evangelista Torricelli (1608–1647). A long tube closed at one end is filled with mercury and then inverted into a dish of mercury (Fig. 15.6b). The closed end of the tube is nearly a vacuum, so its pressure can be taken as zero. Therefore, it follows that $P_0 = \rho_{Hg}gh$, where ρ_{Hg} is the density of the mercury and h is the height of the mercury column. One atmosphere (1 atm) of pressure is defined to be the pressure equivalent of a column of mercury that is exactly 0.7600 m in height at 0°C, with $g = 9.80665$ m/s². At this temperature, mercury has a density of 13.595×10^3 kg/m³; therefore,

$$P_0 = \rho_{Hg}gh = (13.595 \times 10^3 \text{ kg/m}^3)(9.80665 \text{ m/s}^2)(0.7600 \text{ m})$$
$$= 1.013 \times 10^5 \text{ Pa}$$

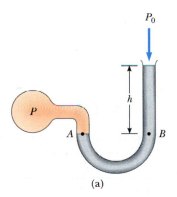

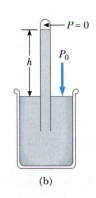

FIGURE 15.6   Two devices for measuring pressure: (a) an open-tube manometer and (b) a mercury barometer.

15.4   BUOYANT FORCES AND ARCHIMEDES' PRINCIPLE

Archimedes' principle can be stated as follows:

> Any body completely or partially submerged in a fluid is buoyed up by a force equal to the weight of the fluid displaced by the body.

Everyone has experienced Archimedes' principle. As an example of a common experience, recall that it is relatively easy to lift someone if the person is in a

Archimedes

| 2 8 7 – 2 1 2 B.C. |

rchimedes, a Greek mathematician, physicist, and engineer, was perhaps the greatest scientist of antiquity. He was the first to compute accurately the ratio of a circle's circumference to its diameter and also showed how to calculate the volume and surface area of spheres, cylinders, and other geometric shapes. He is well known for discovering the nature of the buoyant force acting on objects and was also a gifted inventor. One of his practical inventions, still in use today, is the Archimedes' screw, an inclined rotating coiled tube used originally to lift water from the holds of ships. He also invented the catapult and devised systems of levers, pulleys, and weights for raising heavy loads. Such inventions were successfully used by the soldiers to defend his native city, Syracuse, during a two-year siege by the Romans.

According to legend, Archimedes was asked by King Hieron to determine whether the king's crown was made of pure gold or had been alloyed with some other metal. The task was to be performed without damaging the crown. Archimedes presumably arrived at a solution while taking a bath, noting a partial loss of weight after submerging his arms and legs in the water. As the story goes, he was so excited about his great discovery that he ran through the streets of Syracuse naked shouting "Eureka!," which is Greek for "I have found it."

swimming pool whereas lifting that same individual on dry land is much harder. Evidently, water provides partial support to any object placed in it. The upward force that the fluid exerts on an object submerged in it is called the **buoyant force**. According to Archimedes' principle,

> the magnitude of the buoyant force always equals the weight of the fluid displaced by the object.

The buoyant force acts vertically upward through what was the center of gravity of the displaced fluid.

Archimedes' principle can be verified in the following manner. Suppose we focus our attention on the indicated cube of fluid in the container of Figure 15.7. This cube of fluid is in equilibrium under the action of the forces on it. One of these forces is its weight. What cancels this downward force? Apparently, the rest of the fluid inside the container is holding it in equilibrium. Thus, the buoyant force **B** on the cube of fluid is exactly equal in magnitude to the weight of the fluid inside the cube:

$$B = w$$

Now, imagine that the cube of fluid is replaced by a cube of steel of the same dimensions. What is the buoyant force on the steel? The fluid surrounding a cube behaves in the same way whether a cube of fluid or a cube of steel is being buoyed up; therefore, *the buoyant force acting on the steel is the same as the buoyant force acting on a cube of fluid of the same dimensions.* This result applies for a submerged object of any shape, size, or density.

Let us show explicitly that the buoyant force is equal in magnitude to the weight of the displaced fluid. The pressure at the bottom of the cube in Figure 15.7

FIGURE 15.7 The external forces on the cube of water are the force of gravity w and the buoyancy force **B**. Under equilibrium conditions, $B = w$.

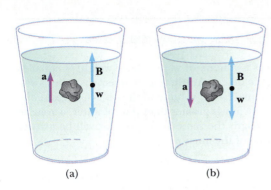

(a) (b)

FIGURE 15.8 (a) A totally submerged object that is less dense than the fluid in which it is submerged will experience a net upward force. (b) A totally submerged object that is denser than the fluid sinks.

Hot-air balloons over Albuquerque, New Mexico. Since hot air is less dense than cold air, there is a net upward buoyant force on the balloons. *(William Moriarity/ Rainbow)*

is greater than the pressure at the top by an amount $\rho_f g h$, where ρ_f is the density of the fluid and h is the height of the cube. Since the pressure difference, ΔP, is equal to the buoyant force per unit area, that is, $\Delta P = B/A$, we see that $B = (\Delta P)A = (\rho_f g h)A = \rho_f g V$, where V is the volume of the cube. Since the mass of the fluid in the cube is $M = \rho_f V$, we see that

$B = w = \rho_f$ where w is the weight of the displaced fluid.

Before proceeding with a few examples, it is instructive to compare the forces acting on a totally submerged object with those acting on a floating object.

Case I. A Totally Submerged Object   When an object is totally submerged in a fluid of density ρ_f, the upward buoyant force is given by $B = \rho_f V_0 g$, where V_0 is the volume of the object. If the object has a density ρ_0, its weight is equal to $w = Mg = \rho_0 V_0 g$, and the net force on it is $B - w = (\rho_f - \rho_0)V_0 g$. Hence, if the density of the object is less than the density of the fluid as in Figure 15.8a, the unsupported object will accelerate upward. If the density of the object is greater than the density of the fluid as in Figure 15.8b, the unsupported object will sink.

Case II. A Floating Object   Now consider an object in static equilibrium floating on a fluid, that is, an object that is only partially submerged. In this case, the upward buoyant force is balanced by the downward weight of the object. If V is the volume of the fluid displaced by the object (which corresponds to that volume of the object beneath the fluid level), then the buoyant force has a magnitude $B = \rho_f V g$. Since the weight of the object is $w = Mg = \rho_0 V_0 g$, and $w = B$, we see that $\rho_f V g = \rho_0 V_0 g$, or

$$\frac{\rho_0}{\rho_f} = \frac{V}{V_0} \tag{15.6}$$

Under normal conditions, the average density of a fish is slightly greater than the density of water. This being the case, a fish would sink if it did not have some mechanism for adjusting its density. The fish accomplishes this by internally regulating the size of its swim bladder. In this manner, fish are able to swim to various depths.

CONCEPTUAL EXAMPLE 15.8

Steel is much denser than water. How, then, do ships made of steel float?

Reasoning   The hull of the ship is full of air, and the density of air is about one thousandth the density of water. Hence, the total weight of the ship is less than the weight of an equal volume of water.

CONCEPTUAL EXAMPLE 15.9

A person in a boat floating in a small pond throws an anchor overboard. Does the level of the pond rise, fall, or remain the same?

Reasoning The level of the pond falls. This is because the anchor displaces more water while in the boat. A floating object displaces a volume of water whose weight is equal to the weight of the object. A submerged object displaces a volume of water equal to the volume of the object. Because the density of the anchor is greater than that of water, a volume of water that weighs the same as the anchor will be greater than the volume of the anchor.

EXAMPLE 15.10   A Submerged Object

A piece of aluminum is suspended from a string and then completely immersed in a container of water (Fig. 15.9). The mass of the aluminum is 1.0 kg, and its density is 2.7×10^3 kg/m³. Calculate the tension in the string before and after the aluminum is immersed.

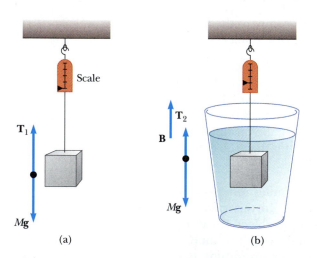

FIGURE 15.9 (Example 15.10) (a) When the aluminum is suspended in air, the scale reads the true weight, Mg (neglecting the buoyancy of air). (b) When the aluminum is immersed in water, the buoyant force **B** reduces the scale reading to $T_2 = Mg - B$.

Solution When the aluminum is suspended in air, as in Figure 15.9a, the tension in the string, T_1 (the reading on the scale), is equal to the weight, Mg, of the aluminum, assuming that the buoyant force of air can be neglected:

$$T_1 = Mg = (1.0 \text{ kg})(9.80 \text{ m/s}^2) = \boxed{9.8 \text{ N}}$$

When immersed in water, the aluminum experiences an upward buoyant force **B**, as in Figure 15.9b, which reduces the tension in the string. Since the system is in equilibrium,

$$T_2 + B - Mg = 0$$
$$T_2 = Mg - B = 9.8 \text{ N} - B$$

In order to calculate B, we must first calculate the volume of the aluminum:

$$V_{Al} = \frac{M}{\rho_{Al}} = \frac{1.0 \text{ kg}}{2.7 \times 10^3 \text{ kg/m}^3} = 3.7 \times 10^{-4} \text{ m}^3$$

Since the buoyant force equals the weight of the water displaced, we have

$$B = M_w g = \rho_w V_{Al} g$$
$$= (1.0 \times 10^3 \text{ kg/m}^3)(3.7 \times 10^{-4} \text{ m}^3)(9.80 \text{ m/s}^2)$$
$$= 3.6 \text{ N}$$

Therefore,

$$T_2 = 9.8 \text{ N} - B = 9.8 \text{ N} - 3.6 \text{ N} = \boxed{6.2 \text{ N}}$$

EXAMPLE 15.11   The Floating Ice Cube

An ice cube floats in a glass of water as in Figure 15.10. What fraction of the cube lies above the water level?

Reasoning and Solution This problem corresponds to Case II described in the text. The weight of the ice cube is $w = \rho_i V_i g$, where $\rho_i = 917$ kg/m³ and V_i is the volume of the whole ice cube. The upward buoyant force equals the weight of the displaced water; that is, $B = \rho_w V g$, where V is the volume of the ice cube beneath the water (the shaded region in

FIGURE 15.10 (Example 15.11).

Fig. 15.10) and ρ_w is the density of the water, $\rho_w = 1000$ kg/m³. Since $\rho_i V_i g = \rho_w V g$, the fraction of ice beneath water is $V/V_i = \rho_i/\rho_w$. Hence, the fraction of ice above the water level is

$$f = 1 - \frac{\rho_i}{\rho_f} = 1 - \frac{917 \text{ kg/m}^3}{1000 \text{ kg/m}^3} = \boxed{0.083 \quad \text{or} \quad 8.3\%}$$

Exercise What fraction of the volume of an iceberg lies below the level of the sea? The density of sea water is 1024 kg/m³.

Answer 0.896 or 89.6%.

15.5 FLUID DYNAMICS

Thus far, our study of fluids has been restricted to fluids at rest. We now turn our attention to fluid dynamics, that is, fluids in motion. Instead of trying to study the motion of each particle of the fluid as a function of time, we describe the properties of the fluid at each point as a function of time.

Flow Characteristics

When fluid is in motion, its flow can be characterized as being one of two main types. The flow is said to be **steady** or **laminar** if each particle of the fluid follows a smooth path, so that the paths of different particles never cross each other, as in Figure 15.11. Thus, in steady flow, the velocity of the fluid at any point remains constant in time.

Above a certain critical speed, fluid flow becomes **nonsteady** or **turbulent**. Turbulent flow is an irregular flow characterized by small whirlpool-like regions as in Figure 15.12. As an example, the flow of water in a stream becomes turbulent in regions where rocks and other obstructions are encountered, often forming "white water" rapids.

The term **viscosity** is commonly used in fluid flow to characterize the degree of internal friction in the fluid. This internal friction or viscous force is associated

FIGURE 15.11   A car undergoes an aerodynamic test in a wind tunnel. The streamlines in the airflow are made visible by smoke particles. What can you conclude about the speed of the airflow from this photograph?   *(Andy Sacks/Tony Stone Images)*

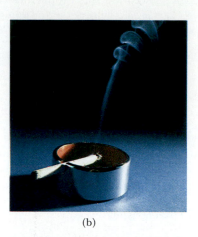

(a)   (b)

FIGURE 15.12   (a) Turbulent flow: the tip of a rotating blade (the dark region) forms a vortex in air that is being heated by an alcohol lamp (the wick is at the bottom). Note the air turbulence on both sides of the rotating blade.   *(© 1973 Kim Vandiver and Harold Edgerton, Palm Press, Inc.)* (b) Hot gases from a cigarette made visible by smoke particles. The smoke first moves in streamline flow at the bottom and then in turbulent flow above.   *(Werner Wolff/Black Star)*

High-speed photograph of secondary drop formation after the impact of a falling drop on a pool of water. After cratering, the surface collapsed to form a column of water. At the top of the column, water surface tension effects create a secondary drop. This can be seen to contain a very high proportion of red dye, so little mixing has taken place. The water column itself is beginning to collapse. *(Dr. David Gorham and Dr. Ian Hutchings/SPL/Photo Researchers)*

with the resistance to two adjacent layers of the fluid to move relative to each other. Because of viscosity, part of the kinetic energy of a fluid is converted to thermal energy. This is similar to the mechanism by which an object sliding on a rough horizontal surface loses kinetic energy.

Because the motion of a real fluid is complex and not yet fully understood, we make some simplifying assumptions in our approach. As we shall see, many features of real fluids in motion can be understood by considering the behavior of an ideal fluid. In our model of an **ideal fluid,** we make four assumptions:

- **Nonviscous fluid.**   In a nonviscous fluid, internal friction is neglected. An object moving through the fluid experiences no viscous force.
- **Steady flow.**   In steady flow, we assume that the velocity of the fluid at each point remains constant in time.
- **Incompressible fluid.**   The density of an incompressible fluid is assumed to remain constant in time.
- **Irrotational flow.**   Fluid flow is irrotational if there is no angular momentum of the fluid about any point. If a small wheel placed anywhere in the fluid does not rotate about the wheel's center of mass, the flow is irrotational. (If the wheel were to rotate, as it would if turbulence were present, the flow would be rotational.)

Properties of an ideal fluid

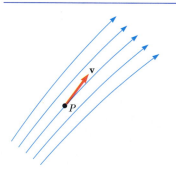

FIGURE 15.13   This diagram represents a set of streamlines (blue lines). A particle at *P* follows one of these streamlines, and its velocity is tangent to the streamline at each point along its path.

15.6   STREAMLINES AND THE EQUATION OF CONTINUITY

The path taken by a fluid particle under steady flow is called a **streamline**. The velocity of the fluid particle is always tangent to the streamline, as shown in Figure 15.13. No two streamlines can cross each other, for if they did, a fluid particle could move either way at the crossover point and then the flow would not be steady. A set of streamlines as shown in Figure 15.13 forms what is called a *tube of flow*. Note that fluid particles cannot flow into or out of the sides of this tube, since if they did, the streamlines would be crossing each other.

Consider an ideal fluid flowing through a pipe of nonuniform size as in Figure 15.14. The particles in the fluid move along the streamlines in steady flow. At all points the velocity of any particle is tangent to the streamline along which it moves.

In a small time interval Δt, the fluid at the bottom end of the pipe moves a distance $\Delta x_1 = v_1 \Delta t$. If A_1 is the cross-sectional area in this region, then the mass contained in the shaded region is $\Delta m_1 = \rho A_1 \Delta x_1 = \rho A_1 v_1 \Delta t$. Similarly, the fluid that moves through the upper end of the pipe in the time Δt has a mass $\Delta m_2 = \rho A_2 v_2 \Delta t$. However, since *mass is conserved* and because the flow is steady, the mass that crosses A_1 in a time Δt must equal the mass that crosses A_2 in a time Δt. That is, $\Delta m_1 = \Delta m_2$, or $\rho A_1 v_1 = \rho A_2 v_2$. If the density is common to both sides of this expression, we get

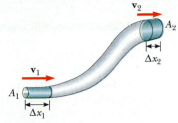

FIGURE 15.14 A fluid moving with streamline flow through a pipe of varying cross-sectional area. The volume of fluid flowing through A_1 in a time interval Δt must equal the volume flowing through A_2 in the same time interval. Therefore, $A_1 v_1 = A_2 v_2$.

$$A_1 v_1 = A_2 v_2 = \text{constant} \qquad (15.7)$$

This expression is called the **equation of continuity**. It says that

> the product of the area and the fluid speed at all points along the pipe is a constant for an incompressible fluid.

Therefore, as one would expect, the speed is high where the tube is constricted and low where the tube is wide. The product Av, which has the dimensions of volume/time, is called the *volume flux,* or flow rate. The condition $Av = \text{constant}$ is equivalent to the fact that the amount of fluid that enters one end of the tube in a given time interval equals the amount leaving in the same time interval, assuming no leaks.

EXAMPLE 15.12 Filling a Water Bucket

A water hose 2.00 cm in diameter is used to fill a 20.0-liter bucket. If it takes 1.00 min to fill the bucket, what is the speed v at which the water leaves the hose? (1 L = 10^3 cm^3)

Solution The cross-sectional area of the hose is

$$A = \pi r^2 = \pi \frac{d^2}{4} = \pi \left(\frac{2.00^2}{4} \right) \text{cm}^2 = \pi \text{ cm}^2$$

According to the data given, the flow rate is equal to 20.0 liters/min. Equating this to the product Av gives

$$Av = 20.0 \, \frac{\text{L}}{\text{min}} = \frac{20.0 \times 10^3 \text{ cm}^3}{60.0 \text{ s}}$$

$$v = \frac{20.0 \times 10^3 \text{ cm}^3}{(\pi \text{ cm}^2)(60.0 \text{ s})} = \boxed{106 \text{ cm/s}}$$

Exercise If the diameter of the hose is reduced to 1.00 cm, what will the speed of the water be as it leaves the hose, assuming the same flow rate?

Answer 424 cm/s.

15.7 BERNOULLI'S EQUATION

As a fluid moves through a pipe of varying cross-section and elevation, the pressure changes along the pipe. In 1738 the Swiss physicist Daniel Bernoulli first derived an expression that relates the pressure to fluid speed and elevation.

Consider the flow of an ideal fluid through a nonuniform pipe in a time Δt, as illustrated in Figure 15.15. The force on the lower end of the fluid is $P_1 A_1$, where P_1 is the pressure in section 1. The work done by this force is $W_1 = F_1 \Delta x_1 =$

Daniel Bernoulli was a Swiss physicist and mathematician who made important discoveries in hydrodynamics. Born into a family of mathematicians on February 8, 1700, he was the only member of the family to make a mark in physics. He was educated and received his doctorate in Basel, Switzerland.

Bernoulli's most famous work, *Hydrodynamica*, was published in 1738; it is both a theoretical and a practical study of equilibrium, pressure, and velocity of fluids. He showed that as the velocity of fluid flow increases, its

Daniel Bernoulli

| 1 7 0 0 – 1 7 8 2 |

pressure decreases. Referred to as "Bernoulli's principle," his work is used to produce a vacuum in chemical laboratories by connecting a vessel to a tube through which water is running rapidly.

Bernoulli's *Hydrodynamica* also attempted the first explanation of the behavior of gases with changing pressure and temperature; this was the beginning of the kinetic theory of gases.

(Drawing courtesy of The Bettmann Archive)

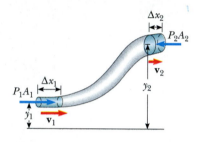

FIGURE 15.15 A fluid flowing through a constricted pipe with streamline flow. The fluid in the section of length Δx_1 moves to the section of length Δx_2. The volumes of fluid in the two sections are equal.

$P_1 A_1 \Delta x_1 = P_1 \Delta V$, where ΔV is the volume of section 1. In a similar manner, the work done on the fluid at the upper end in the time Δt is $W_2 = -P_2 A_2 \Delta x_2 = -P_2 \Delta V$. (The volume that passes through section 1 in a time Δt equals the volume that passes through section 2 in the same time interval.) This work is negative because the fluid force opposes the displacement. Thus the net work done by these forces in the time Δt is

$$W = (P_1 - P_2) \Delta V$$

Part of this work goes into changing the kinetic energy of the fluid, and part goes into changing the gravitational potential energy. If Δm is the mass passing through the pipe in the time Δt, then the change in its kinetic energy is

$$\Delta K = \tfrac{1}{2} (\Delta m) v_2^2 - \tfrac{1}{2} (\Delta m) v_1^2$$

The change in gravitational potential energy is

$$\Delta U = \Delta mgy_2 - \Delta mgy_1$$

We can apply the work-energy theorem in the form $W = \Delta K + \Delta U$ to this volume of fluid to give

$$(P_1 - P_2) \Delta V = \tfrac{1}{2} (\Delta m) v_2^2 - \tfrac{1}{2} (\Delta m) v_1^2 + \Delta mgy_2 - \Delta mgy_1$$

If we divide each term by ΔV and recall that $\rho = \Delta m / \Delta V$, the above expression reduces to

$$P_1 - P_2 = \tfrac{1}{2} \rho v_2^2 - \tfrac{1}{2} \rho v_1^2 + \rho gy_2 - \rho gy_1$$

Rearranging terms, we get

$$P_1 + \tfrac{1}{2} \rho v_1^2 + \rho gy_1 = P_2 + \tfrac{1}{2} \rho v_2^2 + \rho gy_2 \qquad (15.8)$$

This is **Bernoulli's equation** as applied to an ideal fluid. It is often expressed as

$$P + \tfrac{1}{2} \rho v^2 + \rho gy = \text{constant} \qquad (15.9)$$

Bernoulli's equation

Bernoulli's equation says that the sum of the pressure, (P), the kinetic energy per unit volume $(\tfrac{1}{2} \rho v^2)$, and gravitational potential energy per unit volume (ρgy) has the same value at all points along a streamline.

Note that Bernoulli's equation is *not* a sum of energy density terms, because P is pressure, not energy density.

When the fluid is at rest, $v_1 = v_2 = 0$ and Equation 15.8 becomes

$$P_1 - P_2 = \rho g(y_2 - y_1) = \rho g h$$

which agrees with Equation 15.4.

EXAMPLE 15.13   The Venturi Tube

The horizontal constricted pipe illustrated in Figure 15.16, known as a *Venturi tube,* can be used to measure flow speeds in an incompressible fluid. Let us determine the flow speed at point 2 if the pressure difference $P_1 - P_2$ is known.

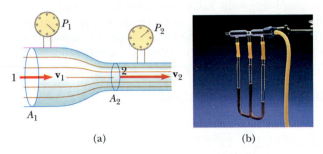

(a)   (b)

FIGURE 15.16   (a) (Example 15.13) The pressure P_1 is greater than the pressure P_2 since $v_1 < v_2$. This device can be used to measure the speed of fluid flow. (b) A Venturi tube.   *(Courtesy of Central Scientific Company)*

Solution   Since the pipe is horizontal, $y_1 = y_2$ and Equation 15.8 applied to points 1 and 2 gives

$$P_1 + \tfrac{1}{2}\rho v_1^2 = P_2 + \tfrac{1}{2}\rho v_2^2$$

From the equation of continuity (Eq. 15.7), we see that $A_1 v_1 = A_2 v_2$ or

$$v_1 = \frac{A_2}{A_1} v_2$$

Substituting this expression into the previous equation gives

$$P_1 + \tfrac{1}{2}\rho \left(\frac{A_2}{A_1}\right)^2 v_2^2 = P_2 + \tfrac{1}{2}\rho v_2^2$$

$$v_2 = A_1 \sqrt{\frac{2(P_1 - P_2)}{\rho(A_1^2 - A_2^2)}}$$

We can also obtain an expression for v_1 using this result and the continuity equation. Note that since $A_2 < A_1$, it follows that $P_1 > P_2$. In other words, the pressure is reduced in the constricted part of the pipe. This result is somewhat analogous to the following situation: Consider a very crowded room, where people are squeezed together. As soon as a door is opened and people begin to exit, the squeezing (pressure) is least near the door where the motion (flow) is greatest.

EXAMPLE 15.14   Torricelli's Law (Speed of Efflux)

A tank containing a liquid of density ρ has a hole in its side at a distance y_1 from the bottom (Fig. 15.17). The diameter of the hole is small compared to the diameter of the tank. The air above the liquid is maintained at a pressure P. Determine the speed at which the fluid leaves the hole when the liquid level is a distance h above the hole.

Solution   Because $A_2 \gg A_1$, the fluid is approximately at rest at the top, point 2. Applying Bernoulli's equation to points 1 and 2 and noting that at the hole $P_1 = P_0$, we get

$$P_0 + \tfrac{1}{2}\rho v_1^2 + \rho g y_1 = P + \rho g y_2$$

But $y_2 - y_1 = h$, and so this reduces to

$$v_1 = \sqrt{\frac{2(P - P_0)}{\rho} + 2gh}$$

The flow rate from the hole is $A_1 v_1$. When P is large compared with atmospheric pressure P_0 (and therefore the term

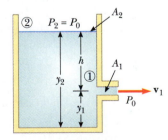

FIGURE 15.17   (Example 15.14) The water speed, v_1, from the hole in the side of the container is given by $v_1 = \sqrt{2gh}$.

$2gh$ can be neglected), the speed of efflux is mainly a function of P. Finally, if the tank is open to the atmosphere, then $P = P_0$ and $v_1 = \sqrt{2gh}$. In other words, the speed of efflux for an open tank is equal to that acquired by a body falling freely through a vertical distance h. This is known as **Torricelli's law.**

An instructor at the Massachusetts Institute of Technology tests his invention, a dimpled baseball bat, against a regulation bat in a wind tunnel. Just as dimples on jet aircraft and golf balls allow air to flow over them more efficiently, the dimpled bat also moved more efficiently and thus could strike a ball with more energy.   *(National Geographic Society)*

*15.8   OTHER APPLICATIONS OF BERNOULLI'S EQUATION

Consider the streamlines that flow around an airplane wing as shown in Figure 15.18. Let us assume that the airstream approaches the wing horizontally from the right with a velocity v_1. The tilt of the wing causes the airstream to be deflected downward with a velocity v_2. Because the airstream is deflected by the wing, the wing must exert a force on the airstream. According to Newton's third law, the airstream must exert an equal and opposite force **F** on the wing. This force has a vertical component called the **lift** (or aerodynamic lift) and a horizontal component called **drag**. The lift depends on several factors, such as the speed of the airplane, the area of the wing, its curvature, and the angle between the wing and the horizontal. As this angle increases, turbulent flow can set in above the wing to reduce the lift.

The lift on the wing is consistent with Bernoulli's equation. The speed of the airstream is greater above the wing, hence the air pressure above the wing is less than the pressure below the wing, resulting in a net upward force.

In general, an object experiences lift by any effect that causes the fluid to change its direction as it flows past the object. Some factors that influence the lift

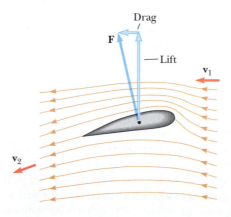

FIGURE 15.18   Streamline flow around a moving airplane wing. The air approaching from the right with a velocity v_1 is deflected downward by the wing, leaving the trailing edge of the wing with a velocity v_2. Because the airstream is deflected, it exerts a force **F** on the wing that has a vertical and horizontal component.

are the shape of the object, its orientation with respect to the fluid flow, spinning motion (for example, a spinning baseball), and the texture of the object's surface.

A number of devices operate in the manner described in Figure 15.19. A stream of air passing over an open tube reduces the pressure above the tube. This reduction in pressure causes the liquid to rise into the airstream. The liquid is then dispersed into a fine spray of droplets. You might recognize that this so-called atomizer is used in perfume bottles and paint sprayers. The same principle is used in the carburetor of a gasoline engine. In this case, the low-pressure region in the carburetor is produced by air drawn in by the piston through the air filter. The gasoline vaporizes, mixes with the air, and enters the cylinder of the engine for combustion.

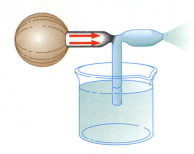

FIGURE 15.19 A stream of air passing over a tube dipped into a liquid will cause the liquid to rise in the tube as shown.

Bernoulli's equation explains one symptom called vascular flutter in a person with advanced arteriosclerosis. The artery is constricted as a result of an accumulation of plaque on its inner walls (Fig. 15.20). In order to maintain a constant flow rate through such a constricted artery, the driving pressure must increase. Such an increase in pressure requires a greater demand on the heart muscle. If the blood speed is sufficiently high in the constricted region, the artery may collapse under external pressure, causing a momentary interruption in blood flow. At this point, Bernoulli's equation does not apply, and the vessel reopens under arterial pressure. As the blood rushes through the constricted artery, the internal pressure drops and again the artery closes. Such variations in blood flow can be heard with a stethoscope. If the plaque becomes dislodged and ends up in a smaller vessel that delivers blood to the heart, the person can suffer a heart attack.

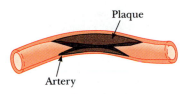

FIGURE 15.20 Blood must travel faster than normal through a constricted artery.

CONCEPTUAL EXAMPLE 15.15

A tornado or hurricane will often lift the roof of a house. Use Bernoulli's equation to explain how this occurs.

Reasoning The rapidly moving air characteristic of a tornado or hurricane causes the external pressure to fall below atmospheric pressure. However, the stationary air inside the house remains at normal atmospheric pressure. The pressure difference between the inside and outside results in an upward force on the roof, which can be large enough to lift it off the house.

CONCEPTUAL EXAMPLE 15.16

When a fast-moving train passes a train at rest, the two tend to be drawn together. How does Bernoulli's equation explain this phenomenon?

Reasoning As air is displaced by the moving train, the air passing between the trains has a higher relative speed than the air on the outside, which is free to expand. Thus, the air pressure is lower between the trains than on the outside, resulting in a net force that draws the trains closer to each other.

---

*15.9 ENERGY FROM THE WIND

Although the wind is a large potential source of energy (about 5 kW per acre in the United States), it has been harnessed only on a small scale. It has been estimated that, on a global scale, the winds account for a total available power of 2×10^{10} kW (about three times the current world power consumption). Therefore, if only a small percentage of the available power could be harnessed, wind power would represent a significant fraction of our energy needs. As with all indirect energy resources, wind power systems have some disadvantages, which in this case arise mainly from the variability of wind velocities.

We can use some of the ideas developed in this chapter to estimate wind power. Any wind-energy machine involves the conversion of the kinetic energy of moving air to the mechanical energy of an object, usually by means of a rotating shaft. The kinetic energy per unit volume of a moving column of air is

$$\frac{KE}{\text{volume}} = \tfrac{1}{2}\rho v^2$$

where ρ is the density of air and v is its speed. The rate of flow of air through a column of cross-sectional area A is Av (Fig. 15.21). This can be considered as the volume of air crossing a given surface area each second. In the working machine, A is the cross-sectional area of the wind-collecting system, such as a set of rotating propeller blades. Multiplying the kinetic energy per unit volume by the flow rate gives the rate at which energy is transferred, or, in other words, the power:

$$\text{Power} = \frac{KE}{\text{volume}} \times \frac{\text{volume}}{\text{time}} = (\tfrac{1}{2}\rho v^2)(Av) = \tfrac{1}{2}\rho v^3 A \qquad (15.10)$$

Therefore, the available power per unit area is

$$\frac{\text{Power}}{A} = \tfrac{1}{2}\rho v^3 \qquad (15.11)$$

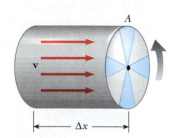

FIGURE 15.21   Wind moving through a cylindrical column of cross-sectional area A with a speed v.

According to this result, if the moving air column could be brought to rest, a power of $\tfrac{1}{2}\rho v^3$ would be available for each square meter that was intercepted. For example, if we assume a moderate speed of 12 m/s (27 mi/h) and take $\rho = 1.3$ kg/m³, we find that

$$\frac{\text{Power}}{A} = \tfrac{1}{2}(1.3 \text{ kg/m}^3)(12 \text{ m/s})^3 \approx 1100 \text{ W/m}^2 = 1.1 \text{ kW/m}^2$$

Because the power per unit area varies as the cube of the speed, its value doubles if v increases by only 26%. Conversely, the power output is halved if the speed decreases by 26%.

This calculation is based on ideal conditions and assumes that all of the kinetic energy is available for power. In reality, the airstream emerges from the wind generator with some residual speed, and more refined calculations show that, at

(a)

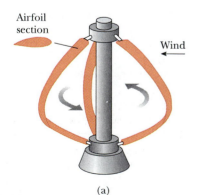

(b)

(c)

FIGURE 15.22   (a) A vertical-axis wind generator. (b) A horizontal-axis wind generator. (c) Photograph of a vertical-axis wind generator.   *(Courtesy of U.S. Department of Energy)*

best, one can extract only 59.3% of this quantity. The expression for the maximum available power per unit area for the ideal wind generator is

$$\frac{\text{Maximum power}}{A} = \frac{8}{27}\rho v^3 \qquad (15.12)$$

In a real wind machine, further losses resulting from the nonideal nature of the propeller, gearing, and generator reduce the total available power to around 15% of the value predicted by Equation 15.11. Sketches of two types of wind turbines are shown in Figure 15.22.

EXAMPLE 15.17   Power Output of a Windmill

Calculate the power output of a wind generator having a blade diameter of 80 m, assuming a wind speed of 10 m/s and an overall efficiency of 15%.

Solution   Since the radius of the blade is 40 m, the cross-sectional area of the propellers is

$$A = \pi r^2 = \pi (40 \text{ m})^2 = 5.0 \times 10^3 \text{ m}^2$$

If 100% of the available wind energy could be extracted, the maximum available power would be

$$\text{Maximum power} = \tfrac{1}{2}\rho A v^3$$
$$= \tfrac{1}{2}(1.2 \text{ kg/m}^3)(5.0 \times 10^3 \text{ m}^2)(10 \text{ m/s})^3$$

$$= 3.0 \times 10^6 \text{ W} = 3.0 \text{ MW}$$

Since the overall efficiency is 15%, the output power is

$$\text{Power} = 0.15(\text{maximum power}) = \boxed{0.45 \text{ MW}}$$

In comparison, a large steam-turbine plant has a power output of about 1 GW. Hence, 2200 such wind generators would be required to equal this output under these conditions. The large number of generators required for reasonable output power is clearly a major disadvantage of wind power. (See Problem 52.)

---

*15.10   VISCOSITY

A fluid does not support a shearing stress. However, fluids do offer some degree of resistance to shearing motion. This resistance to shearing motion is a form of internal friction called *viscosity*. The viscosity arises because of a frictional force between adjacent layers of the fluid as they slide past one another. The degree of viscosity of a fluid can be understood with the following example. If two plates of glass are separated by a layer of fluid such as oil, with one plate fixed in position, it is easy to slide one plate over the other (Fig. 15.23). However, if the fluid separating the plates is tar, the task of sliding one plate over the other becomes much more difficult. Thus, we would conclude that tar has a higher viscosity than oil. In Figure 15.23, note that the speed of successive layers of fluid increases linearly from 0 to v as one moves from a layer adjacent to the fixed plate to a layer adjacent to the moving plate.

Recall that in a solid a shearing stress gives rise to a relative displacement of adjacent layers (Section 12.4). In an analogous fashion, adjacent layers of a fluid under shear stress are set into relative motion. Again, consider two parallel layers, one fixed and one moving to the right under the action of an external force **F** as in Figure 15.23. Because of this motion, a portion of the fluid is distorted from its original shape, *ABCD*, at one instant to the shape *AEFD* after a short time interval. If you refer to Section 12.4, you will recognize that the fluid has undergone a shear strain. By definition, the shear stress on the fluid is equal to the ratio F/A, while the shear strain is defined by the ratio $\Delta x/\ell$:

$$\text{Shear stress} = \frac{F}{A} \qquad \text{Shear strain} = \frac{\Delta x}{\ell}$$

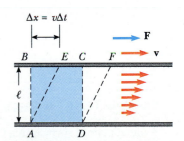

FIGURE 15.23   A layer of liquid between two solid surfaces in which the lower surface is fixed and the upper surface moves to the right with a velocity **v**.

TABLE 15.2   Coefficients of Viscosity of Various Fluids

Fluid	T(°C)	η (N·s/m²)
Water	20	1.0×10^{-3}
Water	100	0.3×10^{-3}
Whole blood	37	2.7×10^{-3}
Glycerine	20	830×10^{-3}
Motor oil (SAE 10)	30	250×10^{-3}
Air	20	1.8×10^{-5}

The upper plate moves with a speed v, and the fluid adjacent to this plate has the same speed. Thus, in a time Δt, the fluid at the moving plate travels a distance $\Delta x = v \Delta t$, and we can express the shear strain per unit time as

$$\frac{\text{Shear strain}}{\Delta t} = \frac{\Delta x/\ell}{\Delta t} = \frac{v}{\ell}$$

This equation states that the rate of change of shearing strain is v/ℓ.

The **coefficient of viscosity**, η, for the fluid is defined as the ratio of the shearing stress to the rate of change of the shear strain:

Coefficient of viscosity

$$\eta \equiv \frac{F/A}{v/\ell} = \frac{F\ell}{Av} \tag{15.13}$$

The SI unit of the coefficient of viscosity is N·s/m². The coefficients of viscosity for some fluids are given in Table 15.2.

The expression for η given by Equation 15.13 is valid only if the fluid speed varies linearly with position. In this case, it is common to say that the speed gradient, v/ℓ, is uniform. If the speed gradient is *not* uniform, we must express η in the general form

$$\eta \equiv \frac{F/A}{dv/dy} \tag{15.14}$$

where the speed gradient dv/dy is the change in speed with position as measured perpendicular to the direction of velocity.

EXAMPLE 15.18   Measuring the Coefficient of Viscosity

A metal plate of surface area 0.15 m² is connected to an 8.0-g mass via a string that passes over an ideal pulley (massless and frictionless), as in Figure 15.24. A lubricant having a film thickness of 0.30 mm is placed between plate and surface. When released, the plate moves to the right with a constant speed of 0.085 m/s. Find the coefficient of viscosity of the lubricant.

Solution   Because the plate moves with constant speed, its acceleration is zero. It moves to the right under the action of the force T exerted by the string and the frictional force f

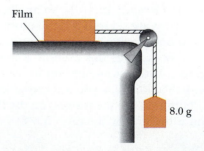

FIGURE 15.24   (Example 15.18).

associated with the viscous fluid. In this case, the tension is equal in magnitude to the suspended weight; therefore,

$$f = T = mg = (8.0 \times 10^{-3} \text{ kg})(9.80 \text{ m/s}^2) = 7.8 \times 10^{-2} \text{ N}$$

The lubricant in contact with the horizontal surface is at rest, while the layer in contact with the plate moves at the speed of the plate. Assuming the speed gradient is uniform, we have

$$\eta = \frac{F\ell}{Av} = \frac{(7.8 \times 10^{-2} \text{ N})(0.30 \times 10^{-3} \text{ m})}{(0.15 \text{ m}^2)(0.085 \text{ m/s})}$$

$$= 5.5 \times 10^{-3} \text{ N} \cdot \text{s/m}^2$$

SUMMARY

The **pressure**, P, in a fluid is the force per unit area that the fluid exerts on any surface:

$$P \equiv \frac{F}{A} \tag{15.1}$$

In the SI system, pressure has units of N/m^2, and 1 N/m^2 = 1 pascal (Pa).

The pressure in a fluid varies with depth h according to the expression

$$P = P_0 + \rho g h \tag{15.4}$$

where P_0 is atmospheric pressure (= 1.013×10^5 N/m^2) and ρ is the density of the fluid, assumed uniform.

Pascal's law states that when pressure is applied to an enclosed fluid, the pressure is transmitted undiminished to every point in the fluid and to every point on the walls of the container.

When an object is partially or fully submerged in a fluid, the fluid exerts an upward force on the object called the **buoyant force.** According to **Archimedes' principle,** the buoyant force is equal to the weight of the fluid displaced by the body.

Various aspects of fluid dynamics can be understood by assuming that the fluid is nonviscous and incompressible and that the fluid motion is a steady flow with no turbulence.

Using these assumptions, two important results regarding fluid flow through a pipe of nonuniform size can be obtained:

- The flow rate through the pipe is a constant, which is equivalent to stating that the product of the cross-sectional area, A, and the speed, v, at any point is a constant:

$$A_1 v_1 = A_2 v_2 = \text{constant} \tag{15.7}$$

- The sum of the pressure, kinetic energy per unit volume, and gravitational potential energy per unit volume has the same value at all points along a streamline:

$$P + \tfrac{1}{2}\rho v^2 + \rho g y = \text{constant} \tag{15.9}$$

QUESTIONS

1. Two drinking glasses having equal weights but different shapes and different cross-sectional areas are filled to the same level with water. According to the expression $P =$ $P_0 + \rho g h$, the pressure is the same at the bottom of both glasses. In view of this, why does one weigh more than the other?

2. If the top of your head has an area of 100 cm², what is the weight of the air above your head?

3. When you drink a liquid through a straw, you reduce the pressure in your mouth and let the atmosphere move the liquid. Explain how this works. Could you use a straw to sip a drink on the Moon?

4. Pascal used a barometer with water as the working fluid. Why is it impractical to use water for a typical barometer?

5. A helium-filled balloon rises until its density becomes the same as that of the air. If a sealed submarine begins to sink, will it go all the way to the bottom of the ocean or will it stop when its density becomes the same as that of the surrounding water?

6. A fish rests on the bottom of a bucket of water while the bucket is being weighed. When the fish begins to swim around, does the weight change?

7. Will a ship ride higher in the water of an inland lake or in the ocean? Why?

8. If 1 000 000 N of weight were placed on the deck of the World War II battleship North Carolina, the ship would sink only 2.5 cm lower in the water. What is the cross-sectional area of the ship at water level?

9. Lead has a greater density than iron, and both are denser than water. Is the buoyant force on a lead object greater than, less than, or equal to the buoyant force on an iron object of the same volume?

10. An ice cube is placed in a glass of water. What happens to the level of the water as the ice melts?

11. The water supply for a city is often provided from reservoirs built on high ground. Water flows from the reservoir, through pipes, and into your home when you turn the tap on your faucet. Why is the water flow more rapid out of a faucet on the first floor of a building than in an apartment on a higher floor?

12. Smoke rises in a chimney faster when a breeze is blowing. Use Bernoulli's equation to explain this phenomenon.

13. Why do many trailer trucks use wind deflectors on the top of their cabs? How do such devices reduce fuel consumption?

14. If you suddenly turn on your shower water at full speed, why is the shower curtain pushed inward?

15. If air from a hair dryer is blown over the top of a Ping-Pong ball, the ball can be suspended in air. Explain.

16. When ski-jumpers are airborne, why do they bend their bodies forward and keep their hands at their sides?

17. When an object is immersed in a liquid at rest, why is the net force on the object in the horizontal direction equal to zero?

18. Explain why a sealed bottle partially filled with a liquid can float.

19. When is the buoyant force on a swimmer greater—after the swimmer is exhaling or after inhaling?

20. A piece of unpainted wood is partially submerged in a container filled with water. If the container is sealed and pressurized above atmospheric pressure, does the wood rise, fall, or remain at the same level? (*Hint:* Wood is porous.)

21. A flat plate is immersed in a liquid at rest. For what orientation of the plate is the pressure on its flat surface uniform?

22. Because atmospheric pressure is about 10^5 N/m² and the area of a person's chest is about 0.13 m², the force of the atmosphere on one's chest is around 13 000 N. In view of this enormous force, why don't our bodies collapse?

23. How would you determine the density of an irregularly shaped rock?

24. Why do airplane pilots prefer to take off into the wind?

25. If you release a ball while inside a freely falling elevator, the ball remains in front of you rather than falling to the floor, because the ball, the elevator, and you all experience the same downward acceleration, g. What happens if you repeat this experiment with a helium-filled balloon? (This one is tricky.)

26. Two identical ships set out to sea. One is loaded with a cargo of Styrofoam, and the other is empty. Which ship is more submerged?

27. A small piece of steel is tied to a block of wood. When the wood is placed in a tub of water with the steel on top, half of the block is submerged. If the block is inverted so that the steel is under water, does the amount of the block submerged increase, decrease, or remain the same? What happens to the water level in the tub when the block is inverted?

28. Prairie dogs ventilate their burrows by building a mound over one entrance, which is open to a stream of air. A second entrance at ground level is open to almost stagnant air. How does this construction create an air flow through the burrow?

29. An unopened can of diet cola floats when placed in a tank of water, whereas a can of regular cola of the same brand sinks in the tank. What could explain this behavior?

30. The photograph (top left, p. 443) shows a glass cylinder containing four liquids of different densities. From top to bottom, the liquids are oil (orange), water (yellow), salt water (green), and mercury (silver). The cylinder also contains, from top to bottom, a Ping-Pong ball, a piece of wood, an egg, and a steel ball. (a) Which of these liquids has the lowest density, and which has the greatest? (b) What can you conclude about the density of each object?

31. In the photograph (top right, p. 443), an air stream moves from right to left through a tube that is constricted at the middle. Three Ping-Pong balls are levitated in

Question 16.   *(Galen Powell/Peter Arnold, Inc.)*

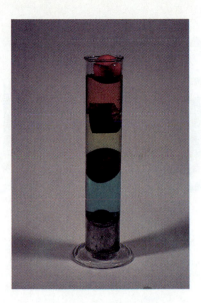

Question 30. *(Henry Leap and Jim Lehman)*

equilibrium above the vertical columns through which the air escapes. (a) Why is the ball at the right higher than the one in the middle? (b) Why is the ball at the left lower than the ball at the right even though the horizontal tube has the same dimensions at these two points?

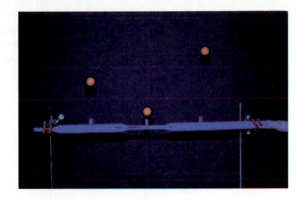

Question 31. *(Henry Leap and Jim Lehman)*

PROBLEMS

Section 15.1   Pressure

1. Calculate the mass of a solid iron sphere that has a diameter of 3.0 cm.

2. A small ingot of shiny grey metal has a volume of 25 cm^3 and a mass of 535 g. What is the metal? (See Table 15.1.)

3. Estimate the density of the *nucleus* of an atom. What does this result suggest concerning the structure of matter? (Use the fact that the mass of a proton is 1.67×10^{-27} kg and its radius is approximately 10^{-15} m.)

4. A king orders a gold crown having a mass of 0.5 kg. When it arrives from the metalsmith, the volume of the crown is found to be 185 cm^3. Is the crown made of solid gold?

5. A 50-kg woman balances on one heel of a pair of high-heeled shoes. If the heel is circular with radius 0.5 cm, what pressure does she exert on the floor?

6. What is the total mass of the Earth's atmosphere? (The radius of the Earth is 6.37×10^6 m, and atmospheric pressure at the surface is 1.01×10^5 N/m^2.)

7. Estimate the density of a neutron star. Such an object is thought to have a radius of only 10 km and a mass equal to that of the Sun. ($M_{Sun} = 1.99 \times 10^{30}$ kg.)

Section 15.2   Variation of Pressure with Depth

8. Determine the absolute pressure at the bottom of a lake that is 30 m deep.

9. A cubic sealed vessel with edge L is placed on a cart, which is moving horizontally with an acceleration a as in Figure P15.9. The cube is filled with a fluid having density ρ. Determine the pressure P at the center of the cube.

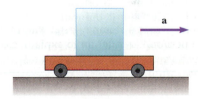

FIGURE P15.9

10. The small piston of a hydraulic lift has a cross-sectional area of 3.00 cm^2, and the large piston has an area of 200 cm^2 (Fig. 15.4). What force must be applied to the small piston to raise a load of 15.0 kN? (In service stations this is usually accomplished with compressed air.)

11. The spring of the pressure gauge shown in Figure 15.2 has a force constant of 1000 N/m, and the piston has a diameter of 2.0 cm. Find the depth in water for which the spring compresses by 0.50 cm.

12. A swimming pool has dimensions 30 m × 10 m and a flat bottom. When the pool is filled to a depth of 2.0 m with fresh water, what is the total force due to

□ indicates problems that have full solutions available in the Student Solutions Manual and Study Guide.

the water on the bottom? On each end? On each side?

13. What must be the contact area between a suction cup (completely exhausted) and a ceiling in order to support the weight of an 80.0-kg student?

14. The tank in Figure P15.14 is filled 2.0-m deep with water. At the bottom of the tank there is a rectangular hatch 1.0 m high and 2.0 m wide that is hinged at the top of the hatch. (a) Determine the net force on the hatch. (b) Find the torque exerted about the hinges.

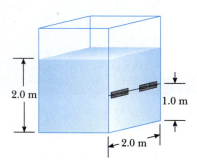

FIGURE P15.14

15. A solid copper ball with a diameter of 3.00 m at sea level is placed at the bottom of the ocean, at a depth of 10.000 km. If the density of seawater is 1030 kg/m³, by how much (approximately) does the diameter of the ball decrease when it reaches bottom? The bulk modulus of copper is 14×10^{10} N/m².

16. What is the hydrostatic force on the back of Grand Coulee Dam if the water in the reservoir is 150 m deep and the width of the dam is 1200 m?

17. In some places, the Greenland ice sheet is 1.0 km thick. Estimate the pressure on the ground underneath the ice. ($\rho_{ice} = 920$ kg/m³.)

Section 15.3   Pressure Measurements

18. Mercury is poured into a U-tube as in Figure P15.18a. The left arm of the tube has a cross-sectional area of $A_1 = 10.0$ cm², and the right arm has a cross-sectional area of $A_2 = 5.00$ cm². One hundred grams of water are then poured into the right arm as in Figure P15.18b. (a) Determine the length of the water column in the right arm of the U-tube. (b) Given that the density of mercury is 13.6 g/cm³, what distance, h, does the mercury rise in the left arm?

19. A U-tube of constant cross-sectional area, open to the atmosphere, is partially filled with mercury. Water is then poured into both arms. If the equilibrium configuration of the tube is as shown in Figure P15.19, with $h_2 = 1.00$ cm, determine the value of h_1.

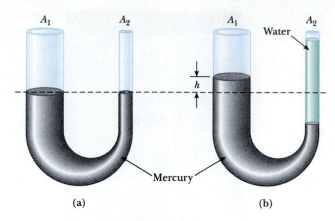

FIGURE P15.18

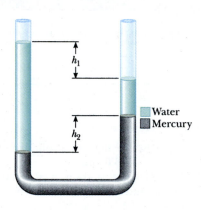

FIGURE P15.19

20. The open vertical tube in Figure P15.20 contains two fluids of densities ρ_1 and ρ_2, which do not mix. Show that the pressure at the depth $h_1 + h_2$ is given by the expression $P = P_0 + \rho_1 g h_1 + \rho_2 g h_2$.

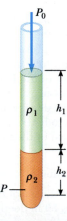

FIGURE P15.20

21. Blaise Pascal duplicated Torricelli's barometer using (as a Frenchman would) a red Bordeaux wine as the working liquid (Fig. P15.21). The density of the wine he used was 0.984×10^3 kg/m³. What was the height h of the wine column for normal atmospheric pressure? Would you expect the vacuum above the column to be as good as for mercury?

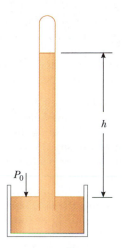

FIGURE P15.21

22. Normal atmospheric pressure is 1.013×10^5 Pa. The approach of a storm causes the height of a mercury barometer to drop by 20 mm from the normal height. What is the atmospheric pressure? (The density of mercury is 13.59 g/cm³.)

23. A simple U-tube that is open at both ends is partially filled with water (Fig. P15.23). Kerosene ($\rho_k = 0.82 \times 10^3$ kg/m³) is then poured into one arm of the tube, forming a column 6.0 cm in height, as shown in the diagram. What is the difference h in the heights of the two liquid surfaces?

23A. A simple U-tube that is open at both ends is partially filled with water (Fig. P15.23). Kerosene of density ρ_k is then poured into one arm of the tube, forming a

column of height h_k, as shown in the diagram. What is the difference h in the heights of the two liquid surfaces?

Section 15.4   Buoyant Forces and Archimedes' Principle

24. Helium balloons having mass 5.0 g when deflated and radii 20.0 cm each are used by a 20.0-kg boy in order to lift himself off the ground. How many balloons are needed if the density of helium is 0.18 kg/m³, and the density of air is 1.29 kg/m³.

25. What is the true weight of one cubic meter of balsa wood that has a specific gravity of 0.15? Note the true weight of an object is its weight in vacuum.

26. A long cylindrical tube of radius r is weighted on one end so that it floats upright in a fluid having density ρ. It is pushed down a distance x from its equilibrium position and released. Show that it will execute simple harmonic motion if the resistive effects of the water are neglected, and determine the period of the oscillations.

27. A cube of wood 20 cm on a side and having a density of 0.65×10^3 kg/m³ floats on water. (a) What is the distance from the top face of the cube to the water level? (b) How much lead weight has to be placed on top of the cube so that its top is just level with the water? (Assume its top face remains parallel to the water's surface.)

28. A balloon is filled with 400 m³ of helium. How big a payload can the balloon lift? (The density of air is 1.29 kg/m³; the density of helium is 0.180 kg/m³.)

29. A plastic sphere floats in water with 50% of its volume submerged. This same sphere floats in oil with 40% of its volume submerged. Determine the densities of the oil and the sphere.

30. A 10-kg block of metal measuring 12 cm × 10 cm × 10 cm is suspended from a scale and immersed in water as in Figure 15.9b. The 12-cm dimension is vertical, and the top of the block is 5.0 cm from the surface of the water. (a) What are the forces on the top and bottom of the block? (Take $P_0 = 1.0130 \times 10^5$ N/m².) (b) What is the reading of the spring scale? (c) Show that the buoyant force equals the difference between the forces at the top and bottom of the block.

31. A frog in a hemispherical pod (Fig. P15.31) finds that he just floats without sinking in a blue-green sea

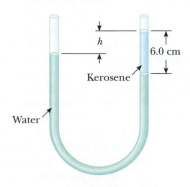

FIGURE P15.23

FIGURE P15.31

(density 1.35 g/cm³). If the pod has a radius of 6.0 cm and has negligible mass, what is the mass of the frog?

32. A light balloon filled with helium of density 0.180 kg/m³ is tied to a light string of length $L = 3.00$ m. The string is tied to the ground, forming an "inverted" simple pendulum as in Figure P15.32a. If the balloon is displaced slightly from equilibrium, as in Figure P15.32b, (a) show that the motion is simple harmonic and (b) determine the period of the motion. Take the density of air to be 1.29 kg/m³, and ignore any energy lost due to air friction.

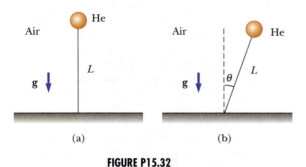

FIGURE P15.32

33. A Styrofoam slab has a thickness of 10 cm and a density of 300 kg/m³. What is the area of the slab if it floats just awash in fresh water when a 75-kg swimmer is aboard?

33A. A Styrofoam slab has a thickness h and a density ρ_s. What is the area of the slab if it floats just awash in fresh water when a swimmer of mass m is aboard?

34. A balloon is used to suspend a 0.020 m³ block of aluminum in water by filling it with air. (a) What volume of air is necessary to just suspend this with the top of the balloon at the water's surface? (b) If instead of being solid, the aluminum had a 0.0060 m³ hollow cavity in it, what fraction of the balloon would be above the water? Ignore the mass of the air in the balloon.

35. How many cubic meters of helium are required to lift a balloon with a 400-kg payload to a height of 8000 m? ($\rho_{He} = 0.18$ kg/m³.) Assume the balloon maintains a constant volume and that the density of air decreases with altitude z according to the expression $\rho_{air} = \rho_0 e^{-z/8000}$, where z is in meters, and ρ_0 ($= 1.29$ kg/m³) is the density of air at sea level.

Sections 15.5–15.7   Fluid Dynamics and Bernoulli's Equation

36. The rate of flow of water through a horizontal pipe is 2.00 m³/min. Determine the speed of flow at a point where the diameter of the pipe is (a) 10.0 cm, (b) 5.0 cm.

37. A large storage tank filled with water develops a small hole in its side at a point 16 m below the water level. If the rate of flow from the leak is 2.5 × 10⁻³ m³/min, determine (a) the speed at which the water leaves the hole and (b) the diameter of the hole.

37A. A large storage tank filled with water develops a small hole in its side at a point a distance h below the water level. If the rate of flow from the leak is R m³/min, determine (a) the speed at which the water leaves the hole and (b) the diameter of the hole.

38. The legendary Dutch boy who saved Holland by placing his finger in the hole of a dike had a finger 1.2 cm in diameter. Assuming the hole was 2.0 m below the surface of the sea (density 1030 kg/m³), (a) what was the force on his finger? (b) If he removed his finger from the hole, how long would it take the released water to fill one acre of land to a depth of one foot assuming the hole remained constant in size. (A typical U.S. family of four uses this much water in one year.)

39. A horizontal pipe 10.0 cm in diameter has a smooth reduction to a pipe 5.0 cm in diameter. If the pressure of the water in the larger pipe is 8.0 × 10⁴ Pa and the pressure in the smaller pipe is 6.0 × 10⁴ Pa, at what rate does water flow through the pipes?

40. Water is pumped from the Colorado River to Grand Canyon Village through a 15.0-cm-diameter pipe. The river is at 564 m elevation and the village is at 2096 m. (a) What is the minimum pressure the water must be pumped at to arrive at the village? (b) If 4500 m³ are pumped per day, what is the speed of the water in the pipe? (c) What additional pressure is necessary to deliver this flow? (*Note:* You may assume that the gravitational field strength and the density of air are constant over this range of elevations.)

41. Water flows through a fire hose of diameter 6.35 cm at a rate of 0.0120 m³/s. The fire hose ends in a nozzle of inner diameter 2.20 cm. What is the speed with which the water exits the nozzle?

42. The Garfield Thomas water tunnel at Pennsylvania State University has a circular cross-section that constricts from a diameter of 3.6 m to the test section, which is 1.2 m in diameter. If the speed of flow is 3.0 m/s in the larger-diameter pipe, determine the speed of flow in the test section.

43. Old Faithful Geyser in Yellowstone Park erupts at approximately 1-hour intervals, and the height of the fountain reaches 40 m. (a) With what speed does the water leave the ground? (b) What is the pressure (above atmospheric) in the heated underground chamber if its depth is 175 m?

*Section 15.8 Other Applications of Bernoulli's Equation

44. An airplane has a mass of 1.6×10^4 kg, and each wing has an area of 40.0 m². During level flight, the pressure on the lower wing surface is 7.0×10^4 Pa. Determine the pressure on the upper wing surface.

45. A pump is designed as a horizontal cylinder with a cross-sectional area of A and the open hole of cross-sectional area a with a very much smaller than A. A fluid having density ρ is forced out of the pump by a piston moving at a constant speed of v by an applied constant force F (Fig. P15.45). Determine the speed u of the jet of fluid.

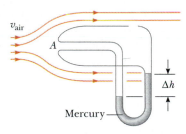

FIGURE P15.45

46. A Venturi tube may be used as a fluid flow meter (Fig. 15.16). If the difference in pressure $P_1 - P_2 = 21$ kPa, find the fluid flow rate in m³/s given that the radius of the outlet tube is 1.0 cm, the radius of the inlet tube is 2.0 cm, and the fluid is gasoline ($\rho = 700$ kg/m³).

47. A Pitot tube can be used to determine the velocity of air flow by measuring the difference between the total pressure and the static pressure (Fig. P15.47). If the fluid in the tube is mercury, density $\rho_{Hg} = 13\ 600$ kg/m³, and $\Delta h = 5.00$ cm, find the speed of air flow. (Assume that the air is stagnant at point A and take $\rho_{air} = 1.25$ kg/m³.)

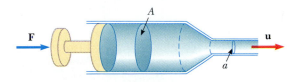

FIGURE P15.47

48. A siphon is used to drain water from a tank, as indicated in Figure P15.48. The siphon has a uniform diameter. Assume steady flow. (a) If the distance $h = 1.00$ m, find the speed of outflow at the end of the siphon. (b) What is the limitation on the height of the top of the siphon above the water surface? (In order to have a continuous flow of liquid, the pressure must not drop below the vapor pressure of the liquid.)

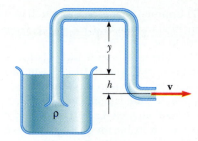

FIGURE P15.48

49. A large storage tank is filled to a height h_0. If the tank is punctured at a height h from the bottom of the tank (Fig. P15.49), how far from the tank will the stream land?

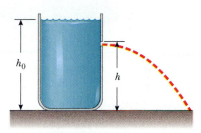

FIGURE P15.49

50. A hole is punched in the side of a 20-cm-tall container, full of water as shown in Figure P15.49. If the water is to shoot as far as possible horizontally, (a) how far from the bottom of the container should the hole be punched? (b) Neglecting friction losses, how far (initially) from the side of the container will the water land?

50A. A hole is punched in the side of a container of height h_0, full of water as shown in Figure P15.49. If the water is to shoot as far as possible horizontally, (a) how far from the bottom of the container should the hole be punched? (b) Neglecting friction losses, how far (initially) from the side of the container will the water land?

*Section 15.9 Energy from the Wind

51. Calculate the power output of a windmill having blades 10.0 m in diameter if the wind speed is 8.0 m/s. Assume that the efficiency of the system is 20%.

52. According to one rather ambitious plan, it would take 50 000 windmills, each 800 ft in diameter, to obtain an average output of 200 GW. These would be strategically located through the Great Plains, along the Aleutian Islands, and on floating platforms along the Atlantic and Gulf coasts and on the Great Lakes.

The annual energy consumption in the United States in 1980 was about 8.3×10^{19} J. What fraction of this could be supplied by the array of windmills?

53. Consider a windmill with blades of cross-sectional area A, as in Figure P15.53, and assume the mill is facing directly into the wind. If the wind speed is v, show that the kinetic energy of the air passing freely through an area A in a time Δt is $K = \frac{1}{2}\rho A v^3 \Delta t$. Why is it not possible for the blades to extract this amount of kinetic energy?

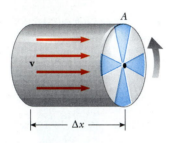

FIGURE P15.53

ADDITIONAL PROBLEMS

54. The water supply of a building is fed through a main 6.0-cm-diameter pipe. A 2.0-cm-diameter faucet tap located 2.0 m above the main pipe is observed to fill a 25-liter container in 30.0 s. (a) What is the speed at which the water leaves the faucet? (b) What is the *gauge pressure* in the 6.0-cm main pipe? (Assume the faucet is the only "leak" in the building.)

55. A Ping-Pong ball has a diameter of 3.8 cm and average density of 0.084 g/cm³. What force is required to hold it completely submerged under water?

56. Figure P15.56 shows a water tank with a valve at the bottom. If this valve is opened, what is the maximum height attained by the water stream coming out of the right side of the tank? Assume that $h = 10$ m, $L = 2.0$ m, and $\theta = 30°$ and that the cross-sectional area at A is very large compared with that at B.

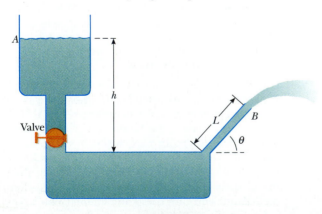

FIGURE P15.56

57. A helium-filled balloon is tied to a 2.0-m-long, 0.050-kg uniform string. The balloon is spherical with a radius of 0.40 m. When released, it lifts a length, h, of string and then remains in equilibrium, as in Figure P15.57. Determine the value of h. When deflated, the balloon has a mass of 0.25 kg.

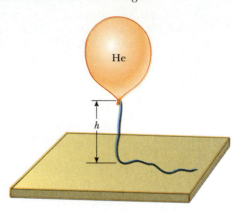

FIGURE P15.57

58. A tube of uniform cross-sectional area is open to the atmosphere and has the shape shown in Figure P15.58, with $\theta = 30.0°$. It is initially filled with water as in Figure P15.58a, and then oil (density = 750 kg/m³) is poured into the left arm, forming a column 0.80 m long (slant height, s) as in Figure P15.58b. By what distance, h, does the water column rise in the right arm?

58A. A tube of uniform cross-sectional area is open to the atmosphere and has the shape shown in Figure P15.58. It is initially filled with water as in Figure P15.58a, and then oil of density ρ_0 is poured into the left arm, forming a column of slant height s, as in Figure P15.58b. By what distance, h, does the water column rise in the right arm?

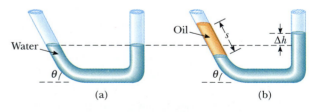

FIGURE P15.58

59. Water is forced out of a fire extinguisher by air pressure, as shown in Figure P15.59. How much gauge air pressure (above atmospheric) is required for a water jet to have a speed of 30 m/s when the water level is 0.50 m below the nozzle?

60. Torricelli was the first to realize that we live at the bottom of an ocean of air. He correctly surmised that

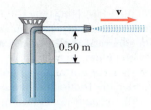

FIGURE P15.59

the pressure of our atmosphere is due to the weight of the air. The density of air at the Earth's surface is 1.29 kg/m³. The density decreases with increasing altitude as the atmosphere thins out. If we assume that the density is constant (1.29 kg/m³) up to some altitude h, and zero above, then h would represent the thickness of our atmosphere. Use this model to determine the value of h that gives a pressure of 1.0 atm at the surface of the Earth. Would the peak of Mt. Everest rise above the surface of such an atmosphere?

61. The true weight of a body is its weight when measured in a vacuum where there are no buoyant forces. A body of volume V is weighed in air on a balance using weights of density ρ. If the density of air is ρ_a and the balance reads w', show that the true weight w is

$$w = w' + \left(V - \frac{w'}{\rho g}\right)\rho_a g$$

62. A wooden dowel has a diameter of 1.20 cm. It floats in water with 0.40 cm of its diameter above water (Fig. P15.62). Determine the density of the dowel.

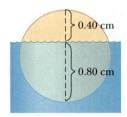

0.40 cm

0.80 cm

FIGURE P15.62

63. A pipe carrying water has a diameter of 2.5 cm. Estimate the maximum flow speed if the flow is to be laminar. Assume the temperature is 20°C.

64. A 1.0-kg beaker containing 2.0 kg of oil (density = 916.0 kg/m³) rests on a scale. A 2.0-kg block of iron is suspended from a spring scale and completely submerged in the oil as in Figure P15.64. Determine the equilibrium readings of both scales.

64A. A beaker of mass m_b containing oil of mass m_o (density = ρ_o) rests on a scale. A block of iron of

mass m_i is suspended from a spring scale and completely submerged in the oil as in Figure P15.64. Determine the equilibrium readings of both scales.

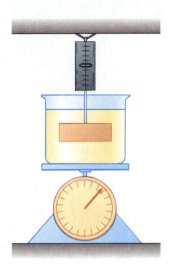

FIGURE P15.64

65. If a 1-megaton nuclear weapon is exploded at ground level, the peak overpressure (that is, the pressure increase above normal atmospheric pressure) will be 0.20 atm at a distance of 6.0 km. What force due to such an explosion will be exerted on the side of a house with dimensions 4.5 m × 22 m?

66. A light spring of constant $k = 90.0$ N/m rests vertically on a table (Fig. P15.66a). A 2.00-g balloon is filled with helium (density = 0.180 kg/m³) to a volume of 5.00 m³ and connected to the spring, causing it to expand as in Figure P15.66b. Determine the expansion length L when the balloon is in equilibrium.

66A. A light spring of constant k rests vertically on a table (Fig. P15.66a). A balloon of mass m is filled with helium (density = ρ_{He}) to a volume V and connected to the spring, causing it to expand as in Figure P15.66b. Determine the expansion length L when the balloon is in equilibrium.

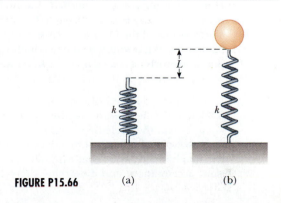

FIGURE P15.66   (a)   (b)

67. With reference to Figure 15.5, show that the total torque exerted by the water behind the dam about an axis through O is $\frac{1}{6}\rho gwH^3$. Show that the effective line of action of the total force exerted by the water is at a distance $\frac{1}{3}H$ above O.

68. In 1657 Otto von Guericke, inventor of the air pump, evacuated a sphere made of two brass hemispheres. Two teams of eight horses each could pull the hemispheres apart only on some trials, and then "with greatest difficulty" (Fig. P15.68). (a) Show that the force F required to pull the evacuated hemispheres apart is $\pi R^2 (P_0 - P)$, where R is the radius of the hemispheres and P is the pressure inside the hemispheres, which is much less than P_0. (b) Determine the force if $P = 0.10P_0$ and $R = 0.30$ m.

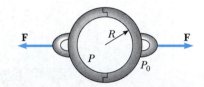

FIGURE P15.68   *(Henry Leap and Jim Lehman)*

69. In 1983, the United States began coining the cent piece out of copper-clad zinc rather than pure copper. If the mass of the old copper cent is 3.083 g while that of the new cent is 2.517 g, calculate the percent of zinc (by volume) in the new cent. The density of copper is 8.960 g/cm³ and that of zinc is 7.133 g/cm³. The new and old coins have the same volume.

70. How much air must be pushed downward at 40.0 m/s in order to keep an 800-kg helicopter aloft?

71. The flow rate of the Columbia River is approximately 3200 m³/s. What would be the maximum power output of the turbines in a dam if the water were to fall a vertical distance of 160 m?

72. A thin spherical shell of mass 4.0 kg and diameter 0.20 m is filled with helium (density = 0.180 kg/m³). It is then released from rest on the bottom of a pool of water that is 4.0 m deep. (a) Neglecting frictional effects, show that the shell rises with constant acceleration, and determine the value of that acceleration. (b) How long will it take for the top of the shell to reach the water surface?

72A. A thin spherical shell of mass m and diameter d is filled with helium (density = ρ_{He}). It is then released from rest on the bottom of a pool of water of depth h. (a) Neglecting frictional effects, show that the shell rises with constant acceleration, and determine the value of that acceleration. (b) How long will it take for the top of the shell to reach the water surface?

73. An incompressible, nonviscous fluid initially rests in the vertical portion of the pipe shown in Figure P15.73a, where $L = 2.0$ m. When the valve is opened, the fluid flows into the horizontal section of the pipe. What is the speed of the fluid when it is entirely in the horizontal section as in Figure P15.73b? Assume the cross-sectional area of the entire pipe is constant.

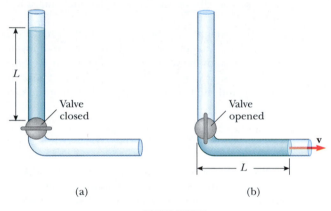

(a)   (b)

FIGURE P15.73

74. Water falls over a dam of height h meters at a rate of R kg/s. (a) Show that the power available from the water is

$$P = \tfrac{16}{27}Rgh$$

where g is the free-fall acceleration. (b) Each hydroelectric unit at the Grand Coulee Dam discharges water at a rate of 8.5×10^5 kg/s from a height of 87 m. The power developed by the falling water is converted to electric power with an efficiency of 85%. How much electric power is produced by each hydroelectric unit?

75. A U-tube open at both ends is partially filled with water (Fig. P15.75a). Oil (density = 750 kg/m³) is then poured into the right arm and forms a column $L = 5.00$ cm high (Fig. P15.75b). (a) Determine the difference, h, in the heights of the two liquid surfaces. Assume that the density of air is 1.29 kg/m³, but be sure to include differences in the atmospheric pressure due to differences in altitude. (b) The right arm is then shielded from any air motion while air is

blown across the top of the left arm until the surfaces of the two liquids are at the same height (Fig. P15.75c). Determine the speed of the air being blown across the left arm.

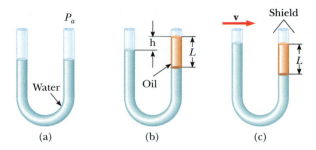

FIGURE P15.75

76. Show that the variation of atmospheric pressure with altitude is given by $P = P_0 e^{-\alpha h}$, where $\alpha = \rho_0 g / P_0$, P_0 is atmospheric pressure at some reference level, and ρ_0 is the atmospheric density at this level. Assume that the decrease in atmospheric pressure with increasing altitude is given by Equation 15.4 (so that $dP/dy = -\rho g$) and that the density of air is proportional to the pressure.

77. A cube of ice whose edge is 20 mm is floating in a glass of ice-cold water with one of its faces parallel to the water surface. (a) How far below the water surface is the bottom face of the block? (b) Ice-cold ethyl alcohol is gently poured onto the water surface to form a layer 5 mm thick above the water. When the ice cube attains hydrostatic equilibrium again, what will be the distance from the top of the water to the bottom face of the block? (c) Additional cold ethyl alcohol is poured onto the water surface until the top surface of the alcohol coincides with the top surface of the ice cube (in hydrostatic equilibrium). How thick is the required layer of ethyl alcohol?

78. The *spirit-in-glass thermometer*, invented in Florence, Italy, around 1654, consists of a tube of liquid (the spirit) containing a number of submerged glass spheres with slightly different masses (Fig. P15.78). At sufficiently low temperatures all the spheres float, but as the temperature rises, the spheres sink one after the other. The device is a crude but interesting

FIGURE P15.78   *(Courtesy Jeanne Maier)*

tool for measuring temperature. Suppose the tube is filled with ethyl alcohol, whose density is 0.78945 g/cm³ at 20.0°C and decreases to 0.78097 g/cm³ at 30.0°C. (a) If one of the spheres has a radius of 1.000 cm and is in equilibrium halfway up the tube at 20.0°C, determine its mass. (b) When the temperature increases to 30.0°C, what mass must a second sphere of the same radius have in order to be in equilibrium at the halfway point? (c) At 30.0°C the first sphere has fallen to the bottom of the tube. What upward force does the bottom of the tube exert on this sphere?

Aerial view of windsurfer riding on the crest of a wave. This dramatic photograph illustrates many physical principles that are described in the text. For example, the water wave carries energy and momentum as it travels from one location to another. The surfer and the surfboard move in a complex path under the action of several forces, including gravity, wind resistance, and the force of water on the surfboard. *(© Darrell Wong/Tony Stone Images)*

PART II

Mechanical Waves

The impetus is much quicker than the water, for it often happens that the wave flees the place of its creation, while the water does not; like the waves made in a field of grain by the wind, where we see the waves running across the field while the grain remains in place.

LEONARDO DA VINCI

As we look around us, we find many examples of objects that vibrate: a pendulum, the strings of a guitar, an object suspended on a spring, the piston of an engine, the head of a drum, the reed of a saxophone. Most elastic objects vibrate when an impulse is applied to them. That is, once they are distorted, their shape tends to be restored to some equilibrium configuration. Even at the atomic level, the atoms in a solid vibrate about some position as if they were connected to their neighbors by imaginary springs.

Wave motion is closely related to the phenomenon of vibration.

Sound waves, earthquake waves, waves on stretched strings, and water waves are all produced by some source of vibration. As a sound wave travels through some medium, such as air, the molecules of the medium vibrate back and forth; as a water wave travels across a pond, the water molecules vibrate up and down and backward and forward. As waves travel through a medium, the particles of the medium move in repetitive cycles. Therefore, the motion of the particles bears a strong resemblance to the periodic motion of a vibrating pendulum or a mass attached to a spring.

There are many other phenomena in nature whose explanation re-

quires us to understand the concepts of vibrations and waves. For instance, although many large structures, such as skyscrapers and bridges, appear to be rigid, they actually vibrate, a fact that must be taken into account by the architects and engineers who design and build them. To understand how radio and television work, we must understand the origin and nature of electromagnetic waves and how they propagate through space. Finally, much of what scientists have learned about atomic structure has come from information carried by waves. Therefore, we must first study waves and vibrations in order to understand the concepts and theories of atomic physics.

CHAPTER 16

Wave Motion

Large waves sometimes travel great distances over the surface of the ocean, yet the water does not flow with the wave. The crests and troughs of the wave often form repetitive patterns.   *(Superstock)*

Most of us experienced waves as children when we dropped a pebble into a pond. At the point where the pebble hits the water surface, waves are created by the impact. These waves move outward from the creation point in expanding circles until they finally reach the shore. If you were to examine carefully the motion of a leaf floating on the disturbance, you would see that the leaf moves up, down, and sideways about its original position but does not undergo any net displacement away from or toward the point where the pebble hits the water. The water molecules just beneath the leaf, as well as all the other water molecules on the pond surface, behave in the same way. That is, the water wave moves from one place to another, and yet the water is not carried with it.

An excerpt from a book by Einstein and Infeld gives the following remarks concerning wave phenomena.[1]

---

[1] A. Einstein and L. Infeld, *The Evolution of Physics,* New York, Simon and Schuster, 1961. Excerpt from "What is a Wave?".

A bit of gossip starting in Washington reaches New York very quickly, even though not a single individual who takes part in spreading it travels between these two cities. There are two quite different motions involved, that of the rumor, Washington to New York, and that of the persons who spread the rumor. The wind, passing over a field of grain, sets up a wave which spreads out across the whole field. Here again we must distinguish between the motion of the wave and the motion of the separate plants, which undergo only small oscillations. . . . The particles constituting the medium perform only small vibrations, but the whole motion is that of a progressive wave. The essentially new thing here is that for the first time we consider the motion of something which is not matter, but energy propagated through matter.

Water waves and the waves across a grainfield are only two examples of physical phenomena that have wavelike characteristics. The world is full of waves, the two main types of which are mechanical waves and electromagnetic waves. We have already mentioned examples of mechanical waves: sound waves, water waves, and "grain waves." In each case, there is some physical medium being disturbed—air molecules, water molecules, and stalks of grain in our three particular examples. Electromagnetic waves are a special class of waves that do not require a medium in order to propagate, some examples being visible light, radio waves, television signals, and x-rays. Here in Part II of this book, we shall study only mechanical waves.

The wave concept is abstract. When we observe what we call a water wave, what we see is a rearrangement of the water's surface. Without the water, there would be no wave. A wave traveling on a string would not exist without the string. Sound waves travel through air as a result of pressure variations from point to point. In such cases involving mechanical waves, what we interpret as a wave corresponds to the disturbance of a body or medium. Therefore, we can consider a wave to be the *motion of a disturbance.*

The mathematics used to describe wave phenomena is common to all waves. In general, we shall find that mechanical wave motion is described by specifying the positions of all points of the disturbed medium as a function of time.

16.1 INTRODUCTION

The mechanical waves discussed in this chapter require (1) some source of disturbance, (2) a medium that can be disturbed, and (3) some physical connection through which adjacent portions of the medium can influence each other. We shall find that all waves carry energy. The amount of energy transmitted through a medium and the mechanism responsible for that transport of energy differ from case to case. For instance, the power of ocean waves during a storm is much greater than the power of sound waves generated by a single human voice.

Three physical characteristics are important in characterizing waves: wavelength, frequency, and wave speed. One **wavelength** is the *minimum distance between any two points on a wave that behave identically,* as shown in Figure 16.1.

Most waves are periodic, and the **frequency** of such periodic waves is *the time rate at which the disturbance repeats itself.*

Waves travel with a specific speed, which depends on the properties of the medium being disturbed. For instance, sound waves travel through air at 20°C with a speed of about 344 m/s (781 mi/h), whereas the speed of sound in most solids is higher than 344 m/s. Electromagnetic waves travel very swiftly through a vacuum with a speed of approximately 3.00×10^8 m/s (186 000 mi/s).

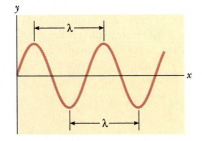

FIGURE 16.1 The wavelength λ of a wave is the distance between adjacent crests or adjacent troughs.

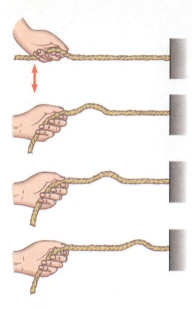

FIGURE 16.2 A wave pulse traveling down a stretched rope. The shape of the pulse is approximately unchanged as it travels along the rope.

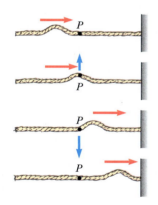

FIGURE 16.3 A pulse traveling on a stretched rope is a transverse wave. That is, any element *P* on the rope moves (blue arrows) in a direction perpendicular to the wave motion (red arrows).

16.2 TYPES OF WAVES

One way to demonstrate wave motion is to flick one end of a long rope that is under tension and has its opposite end fixed, as in Figure 16.2. In this manner, a single wave bump (called a wave pulse) is formed and travels (to the right in Fig. 16.2) with a definite speed. This type of disturbance is called a **traveling wave,** and Figure 16.2 represents four consecutive "snapshots" of the traveling wave. As we shall see later, the speed of the wave depends on the tension in the rope and on the properties of the rope. The rope is the medium through which the wave travels. The shape of the wave pulse changes very little as it travels along the rope.[2]

As the wave pulse travels, *each segment of the rope that is disturbed moves in a direction perpendicular to the wave motion.* Figure 16.3 illustrates this point for one particular segment, labeled *P*. Note that no part of the rope ever moves in the direction of the wave.

> A traveling wave that causes the particles of the disturbed medium to move perpendicular to the wave motion is called a **transverse wave.**

> A traveling wave that causes the particles of the medium to move parallel to the direction of wave motion is called a **longitudinal wave.**

Sound waves, which we discuss in Chapter 17, are one example of longitudinal waves. Sound waves in air are a series of high- and low-pressure regions, or disturbances, traveling in the same direction as the displacements. A longitudinal pulse can be easily produced in a stretched spring, as in Figure 16.4. The left end of the spring is given a sudden movement (consisting of a brief push to the right and equally brief pull to the left) along the length of the spring; this movement creates a sudden compression of the coils. The compressed region *C* (pulse) travels along the spring, and so we see that the disturbance is parallel to the wave motion. The compressed region is followed by a region where the coils are stretched.

Some waves in nature are neither transverse nor longitudinal, but a combination of the two. Surface water waves are a good example. When a water wave travels on the surface of deep water, water molecules at the surface move in nearly circular paths, as shown in Figure 16.5, where the water surface is drawn as a series of crests and troughs. Note that the disturbance has both transverse and longitudinal components. As the wave passes, water molecules at the crests move in the direction of

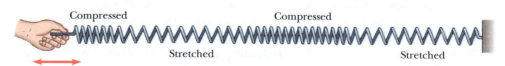

Compressed   Compressed

Stretched   Stretched

FIGURE 16.4 A longitudinal pulse along a stretched spring. The displacement of the coils is in the direction of the wave motion. For the starting motion described in the text, the compressed region is followed by a stretched region.

---

[2] Strictly speaking, the pulse will change its shape and gradually spread out during the motion. This effect is called *dispersion* and is common to many mechanical waves.

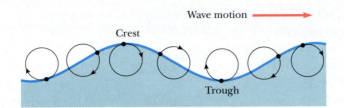

FIGURE 16.5   Wave motion on the surface of water. The molecules at the water's surface move in nearly circular paths. Each molecule is displaced horizontally and vertically from its equilibrium position, represented by circles.

the wave, and molecules at the troughs move in the opposite direction. Since the molecule at the crest in Figure 16.5 will soon be at a trough, its movement in the direction of the wave will soon be canceled by its movement in the opposite direction. Since this argument holds for every disturbed water molecule, we conclude that there is no net displacement of any water molecule.

16.3   ONE-DIMENSIONAL TRAVELING WAVES

Let us now give a mathematical description of a one-dimensional traveling wave. Consider again a wave pulse traveling to the right with constant speed v on a long taut string, as in Figure 16.6. The pulse moves along the x axis (the axis of the string), and the transverse displacement of the string (the medium) is measured with the coordinate y.

Figure 16.6a represents the shape and position of the pulse at time $t = 0$. At this time, the shape of the pulse, whatever it may be, can be represented as $y = f(x)$. That is, y is some definite function of x. The *maximum displacement of the string*, A, is called the **amplitude** of the wave. Since the speed of the wave pulse is v, it travels to the right a distance vt in a time t (Fig. 16.6b).

If the shape of the wave pulse doesn't change with time, we can represent the string displacement y for all later times measured in a stationary frame having the origin at 0 as

$$y = f(x - vt) \tag{16.1}$$

Wave traveling to the right

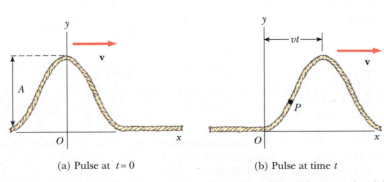

(a) Pulse at $t = 0$ (b) Pulse at time t

FIGURE 16.6   A one-dimensional wave pulse traveling to the right with a speed v. (a) At $t = 0$, the shape of the pulse is given by $y = f(x)$. (b) At some later time t, the shape remains unchanged and the vertical displacement of any point P of the medium is given by $y = f(x - vt)$.

If the wave pulse travels to the *left,* the string displacement is

$$y = f(x + vt) \qquad (16.2)$$

The displacement y, sometimes called the *wave function,* depends on the two variables x and t. For this reason, it is often written $y(x, t)$, which is read "y as a function of x and t."

It is important to understand the meaning of y. Consider a particular point P on the string, identified by a particular value of its coordinates. As the wave passes P, the y coordinate of this point increases, reaches a maximum, and then decreases to zero. Therefore, the **wave function** y represents *the y coordinate of any medium point P at any time t.* Furthermore, if t is fixed, then the wave function y as a function of x *defines a curve representing the shape of the pulse at this time.* This curve is equivalent to a "snapshot" of the wave at this time.

For a pulse that moves without changing shape, the speed of the pulse is the same as that of any feature along the pulse, such as the crest in Figure 16.6b. To find the speed of the pulse, we can calculate how far the crest moves in a short time and then divide this distance by the time interval. In order to follow the motion of the crest, some particular value, say x_0, must be substituted in Equation 16.1 for $x - vt$. Regardless of how x and t change individually, we must require that $x - vt = x_0$ in order to stay with the crest. This expression, therefore, represents the equation of motion of the crest. At $t = 0$, the crest is at $x = x_0$; at a time dt later, the crest is at $x = x_0 + v\,dt$. Therefore, in a time dt, the crest has moved a distance $dx = (x_0 + v\,dt) - x_0 = v\,dt$. Hence, the wave speed is

$$v = \frac{dx}{dt} \qquad (16.3)$$

As noted above, the wave velocity must not be confused with the transverse velocity (which is in the y direction) of a particle in the medium (nor with the longitudinal velocity for a longitudinal wave).

EXAMPLE 16.1   A Pulse Moving to the Right

A wave pulse moving to the right along the x axis is represented by the wave function

$$y(x, t) = \frac{2}{(x - 3.0t)^2 + 1}$$

where x and y are measured in centimeters and t is in seconds. Let us plot the waveform at $t = 0$, $t = 1.0$ s, and $t = 2.0$ s.

Solution   First, note that this function is of the form $y = f(x - vt)$. By inspection, we see that the speed of the wave is $v = 3.0$ cm/s. Furthermore, the wave amplitude (the maximum value of y) is given by $A = 2.0$ cm. At times $t = 0$, $t = 1.0$ s, and $t = 2.0$ s, the wave function expressions are

$$y(x, 0) = \frac{2}{x^2 + 1} \qquad \text{at } t = 0$$

$$y(x, 1.0) = \frac{2}{(x - 3.0)^2 + 1} \qquad \text{at } t = 1.0 \text{ s}$$

$$y(x, 2.0) = \frac{2}{(x - 6.0)^2 + 1} \qquad \text{at } t = 2.0 \text{ s}$$

We can now use these expressions to plot the wave function versus x at these times. For example, let us evaluate $y(x, 0)$ at $x = 0.50$ cm:

$$y(0.50, 0) = \frac{2}{(0.50)^2 + 1} = 1.6 \text{ cm}$$

Likewise, $y(1.0, 0) = 1.0$ cm, $y(2.0, 0) = 0.40$ cm, and so on. A continuation of this procedure for other values of x yields the waveform shown in Figure 16.7a. In a similar manner, one obtains the graphs of $y(x, 1.0)$ and $y(x, 2.0)$, shown in

Figures 16.7b and 16.7c, respectively. These snapshots show that the wave pulse moves to the right without changing its shape and has a constant speed of 3.0 cm/s.

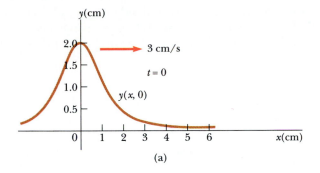

(a)

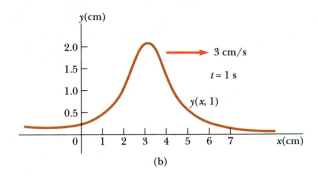

(b)

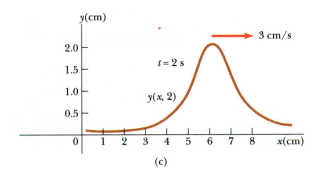

(c)

FIGURE 16.7   (Example 16.1) Graphs of the function $y(x, t) = 2/[(x - 3t)^2 + 1]$. (a) $t = 0$, (b) $t = 1$ s, and (c) $t = 2$ s.

16.4 SUPERPOSITION AND INTERFERENCE OF WAVES

Many interesting wave phenomena in nature cannot be described by a single moving pulse. Instead, one must analyze complex waveforms in terms of a combination of many traveling waves. To analyze such wave combinations, one can make use of the **superposition principle:**

> If two or more traveling waves are moving through a medium, the resultant wave function at any point is the algebraic sum of the wave functions of the individual waves.

Linear waves obey the superposition principle

Waves that obey this principle are called *linear waves,* and they are generally characterized by small wave amplitudes. Waves that violate the superposition principle are called *nonlinear waves* and are often characterized by large amplitudes. In this book, we deal only with linear waves.

One consequence of the superposition principle is that *two traveling waves can pass through each other without being destroyed or even altered.* For instance, when two pebbles are thrown into a pond and hit the surface at two places, the expanding circular surface waves do not destroy each other. In fact, they pass right through each other. The complex pattern that is observed can be viewed as two independent sets of expanding circles. Likewise, when sound waves from two sources move

Wave Motion

This simulator allows you to model wave motion involving one or two traveling waves. For the case of a single wave, you will be able to specify the velocity and shape of the wave by selecting values from a given list, or define your own wave shape and velocity. For a wave traveling on a string, you can specify whether an end of the string is fixed or free and examine how the wave is reflected at this end in each case. When studying two waves traveling on a string, you can investigate their superposition as they move through each other.

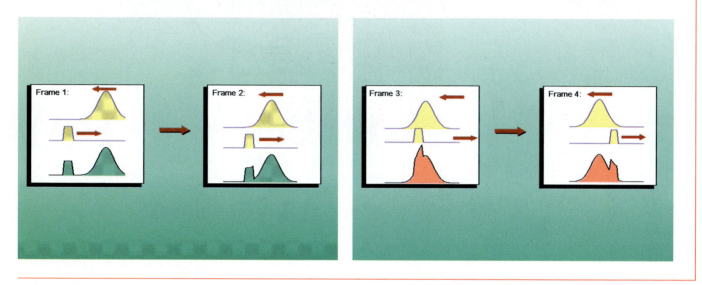

through air, they also pass through each other. The resulting sound one hears at a given point is the resultant of both disturbances.

A simple pictorial representation of the superposition principle is obtained by considering two pulses traveling in opposite directions on a taut string, as in Figure 16.8. The wave function for the pulse moving to the right is y_1, and the wave function for the pulse moving to the left is y_2. The pulses have the same speed but different shapes. Each pulse is assumed to be symmetric, and the displacement of the medium is in the positive y direction for both pulses. (Note that the superposition principle applies even if the two pulses are not symmetric and even when they travel at different speeds.) When the waves begin to overlap (Fig. 16.8b), the resulting complex waveform is given by $y_1 + y_2$. When the crests of the pulses coincide (Fig. 16.8c), the resulting waveform $y_1 + y_2$ is symmetric. The two pulses finally separate and continue moving in their original directions (Fig. 16.8d). Note that the final waveforms remain unchanged, as if the two pulses had never met!

The combination of separate waves in the same region of space to produce a resultant wave is called **interference**. For the two pulses shown in Figure 16.8, the displacement of the medium is in the positive y direction for both pulses, and the resultant waveform (when the pulses overlap) exhibits a displacement greater than those of the individual pulses. Since the displacements caused by the two pulses are in the same direction, we refer to their superposition as **constructive interference**.

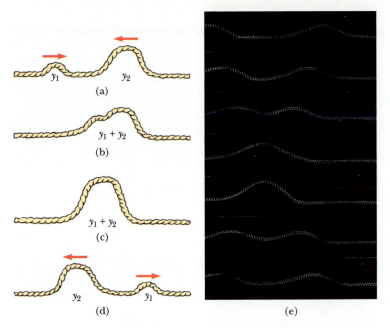

FIGURE 16.8 *(Left)* Two wave pulses traveling on a stretched string in opposite directions pass through each other. When the pulses overlap, as in (b) and (c), the net displacement of the string equals the sum of the displacements produced by each pulse. Since each pulse produces positive displacements of the string, we refer to their superposition as *constructive interference.* *(Right)* Photograph of superposition of two equal and symmetric pulses traveling in opposite directions on a stretched spring. *(Photo, Education Development Center, Newton, Mass.)*

Now consider two pulses traveling in opposite directions on a taut string, where now one is inverted relative to the other, as in Figure 16.9. In this case, when the pulses begin to overlap, the resultant waveform is given by $y_1 - y_2$. Again the two pulses pass through each other as indicated. Since the displacements caused by the two pulses are in opposite directions, we refer to their superposition as **destructive interference.**

Interference patterns produced by outward spreading waves from several drops of water falling into a pond. *(Martin Dohrn/SPL/Photo Researchers)*

Interference of water waves produced in a ripple tank. The sources of the waves are two objects that vibrate perpendicularly to the surface of the tank. *(Courtesy of Central Scientific CO.)*

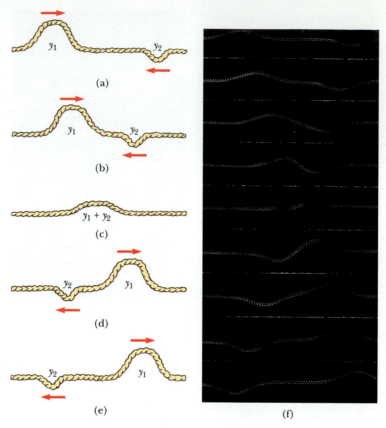

FIGURE 16.9 *(Left)* Two wave pulses traveling in opposite directions with displacements that are inverted relative to each other. When the two overlap as in (c), their displacements subtract from each other. *(Right)* Photograph of superposition of two symmetric pulses traveling in opposite directions, where one is inverted relative to the other.   *(Photo, Education Development Center, Newton, Mass.)*

16.5   THE SPEED OF WAVES ON STRINGS

The speed of linear mechanical waves depends only on the properties of the medium through which the wave travels. In this section, we focus on determining the speed of a transverse pulse traveling on a taut string. If the tension in the string is F and its mass per unit length is μ, then as we shall show, the wave speed is

Speed of a wave on a stretched string

$$v = \sqrt{\frac{F}{\mu}} \tag{16.4}$$

First, let us verify that this expression is dimensionally correct. The dimensions of F are $\mathrm{MLT^{-2}}$, and the dimensions of μ are $\mathrm{ML^{-1}}$. Therefore, the dimensions of F/μ are $\mathrm{L^2/T^2}$; hence the dimensions of $\sqrt{F/\mu}$ are $\mathrm{L/T}$, which are indeed the dimensions of speed. No other combination of F and μ is dimensionally correct if we assume that they are the only variables relevant to the situation.

Now let us use a mechanical analysis to derive the above expression. Consider a pulse moving to the right with a uniform speed v, measured relative to a stationary frame of reference. Instead of staying in this frame, it is more convenient to choose as our reference frame one that moves along with the pulse with the same speed, so that the pulse is at rest in this frame, as in Figure 16.10a. This change of reference frame is permitted because Newton's laws are valid in either a stationary frame or one that moves with constant velocity.

A small segment of the string of length Δs forms an approximate arc of a circle of radius R, as shown in Figure 16.10a and magnified in Figure 16.10b. In the pulse's frame of reference (which is moving to the right along with the pulse), the shaded segment is moving down with a speed v. This small segment has a centripetal acceleration equal to v^2/R, which is supplied by the force of tension $\mathbf{F}$ in the string. The force $\mathbf{F}$ acts on each side of the segment, tangent to the arc, as in Figure 16.10b. The horizontal components of $\mathbf{F}$ cancel, and each vertical component $F \sin \theta$ acts radially inward toward the center of the arc. Hence, the total radial force is $2F \sin \theta$. Since the segment is small, θ is small and we can use the familiar small-angle approximation $\sin \theta \approx \theta$. Therefore, the total radial force can be expressed as

$$F_r = 2F \sin \theta \approx 2F\theta$$

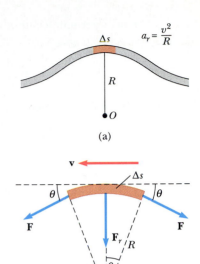

(a)

(b)

FIGURE 16.10 (a) To obtain the speed v of a wave on a stretched string, it is convenient to describe the motion of a small segment of the string in a moving frame of reference. (b) The net force on a small segment of length Δs is in the radial direction. The horizontal components of the tension force cancel.

The small segment has a mass $m = \mu \, \Delta s$. Since the segment forms part of a circle and subtends an angle 2θ at the center, $\Delta s = R(2\theta)$, and hence

$$m = \mu \, \Delta s = 2\mu R\theta$$

If we apply Newton's second law to this segment, the radial component of motion gives

$$F_r = \frac{mv^2}{R} \quad \text{or} \quad 2F\theta = \frac{2\mu R\theta v^2}{R}$$

where F_r is the force that supplies the centripetal acceleration of the segment and maintains the curvature at this point. Solving for v gives Equation 16.4. Notice that this derivation is based on the assumption that the pulse height is small relative to the length of the string. Using this assumption, we were able to use the approximation that $\sin \theta \approx \theta$. Furthermore, the model assumes that the tension F is not affected by the presence of the pulse, so that F is the same at all points on the string. Finally, this proof does *not* assume any particular shape for the pulse. Therefore, we conclude that a pulse of *any shape* will travel on the string with speed $v = \sqrt{F/\mu}$ without any change in pulse shape.

EXAMPLE 16.2 The Speed of a Pulse on a Cord

A uniform cord has a mass of 0.300 kg and a length of 6.00 m (Fig. 16.11). Find the speed of a pulse on this cord.

Solution The tension F in the cord is equal to the weight of the suspended 2.00-kg mass:

$$F = mg = (2.00 \text{ kg})(9.80 \text{ m/s}^2) = 19.6 \text{ N}$$

(This calculation of the tension neglects the small mass of the cord. Strictly speaking, the cord can never be exactly horizontal, and therefore the tension is not uniform.)

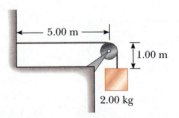

FIGURE 16.11 (Example 16.2) The tension F in the cord is maintained by the suspended mass. The wave speed is given by the expression $v = \sqrt{F/\mu}$.

The mass per unit length μ is

$$\mu = \frac{m}{\ell} = \frac{0.300 \text{ kg}}{6.00 \text{ m}} = 0.0500 \text{ kg/m}$$

Therefore, the wave speed is

$$v = \sqrt{\frac{F}{\mu}} = \sqrt{\frac{19.6 \text{ N}}{0.0500 \text{ kg/m}}} = \boxed{19.8 \text{ m/s}}$$

Exercise   Find the time it takes the pulse to travel from the wall to the pulley.

Answer   0.253 s.

16.6   REFLECTION AND TRANSMISSION OF WAVES

Whenever a traveling wave reaches a boundary, part or all of the wave is reflected. For example, consider a pulse traveling on a string fixed at one end (Fig. 16.12). When the pulse reaches the wall, it is reflected. Because the support attaching the string to the wall is rigid, the pulse does not transmit any part of the disturbance to the wall and its amplitude does not change.

Note that the reflected pulse is inverted. This can be explained as follows. When the pulse reaches the fixed end of the string, the string produces an upward force on the support. By Newton's third law, the support must then exert an equal and opposite (downward) reaction force on the string. This downward force causes the pulse to invert upon reflection.

Now consider another case, where this time the pulse arrives at the end of a string that is free to move vertically, as in Figure 16.13. The tension at the free end is maintained by tying the string to a ring of negligible mass that is free to slide vertically on a smooth post. Again the pulse is reflected, but this time it is not inverted. As the pulse reaches the post, it exerts a force on the free end of the string, causing the ring to accelerate upward. In the process, the ring would overshoot the height of the incoming pulse except that it is pulled back by the downward component of the tension force. This ring movement produces a reflected pulse that is not inverted, and whose amplitude is the same as that of the incoming pulse.

Finally, we may have a situation in which the boundary is intermediate between these two extreme cases, that is, one in which the boundary is neither rigid nor free. In this case, part of the incident pulse is transmitted and part is reflected. For instance, suppose a light string is attached to a heavier string as in Figure 16.14. When a pulse traveling on the light string reaches the boundary between the two,

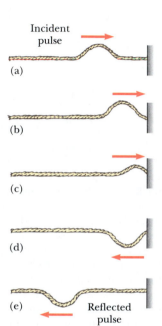

FIGURE 16.12   The reflection of a traveling wave pulse at the fixed end of a stretched string. The reflected pulse is inverted, but its shape remains the same.

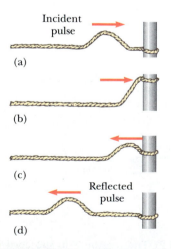

FIGURE 16.13   The reflection of a traveling wave pulse at the free end of a stretched string. The reflected pulse is not inverted.

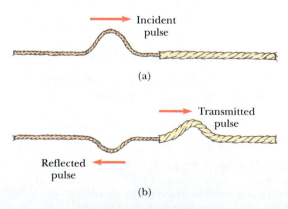

FIGURE 16.14   (a) A pulse traveling to the right on a light string attached to a heavier string. (b) Part of the incident pulse is reflected (and inverted), and part is transmitted to the heavier string. (Note that the change in pulse width is not shown.)

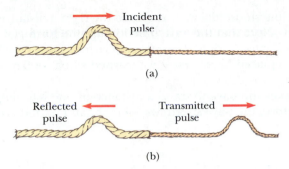

(a)

Reflected pulse ← ... → Transmitted pulse

(b)

FIGURE 16.15   (a) A pulse traveling to the right on a heavy string attached to a lighter string. (b) The incident pulse is partially reflected and partially transmitted. In this case, the reflected pulse is not inverted. (Note that the change in pulse width is not shown.)

part of the pulse is reflected and inverted and part is transmitted to the heavier string. As one would expect, the reflected pulse has a smaller amplitude than the incident pulse, because part of the incident energy is transferred to the pulse in the heavier string. The reflected pulse is inverted for the same reasons described above for the fixed-support case.

When a pulse traveling on a heavy string strikes the boundary between that string and a lighter one, as in Figure 16.15, again part is reflected and part is transmitted. In this case, however, the reflected pulse is not inverted.

In either case, the relative heights of the reflected and transmitted pulses depend on the relative densities of the two strings. If the strings are identical, there is no discontinuity at the boundary, and hence no reflection takes place.

In the previous section, we found that the speed of a wave on a string increases as the mass per unit length of the string decreases. In other words, a pulse travels more slowly on a heavy string than on a light string if both are under the same tension. The following general rules apply to reflected waves: *When a wave pulse travels from medium A to medium B and $v_A > v_B$ (that is, when B is denser than A), the pulse is inverted upon reflection. When a wave pulse travels from medium A to medium B and $v_A < v_B$ (A is denser than B), the pulse is not inverted upon reflection.*

16.7 SINUSOIDAL WAVES

In this section, we introduce an important waveform known as a **sinusoidal wave** whose shape is shown in Figure 16.16. The red curve represents a snapshot of the traveling sinusoidal wave at $t = 0$, and the blue curve represents a snapshot of the wave at some later time t. At $t = 0$, the vertical displacement of the curve can be written

$$y = A \sin\left(\frac{2\pi}{\lambda} x\right) \tag{16.5}$$

where the constant A represents the wave amplitude and the constant λ is the symbol we use for the wavelength. Thus, we see that the vertical displacement repeats itself whenever x is increased by an integral multiple of λ. If the wave moves to the right with a speed v, the wave function at some later time t is

$$y = A \sin\left[\frac{2\pi}{\lambda}(x - vt)\right] \tag{16.6}$$

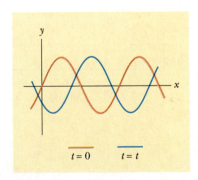

FIGURE 16.16   A one-dimensional sinusoidal wave traveling to the right with a speed v. The red curve represents a snapshot of the wave at $t = 0$, and the blue curve represents a snapshot at some later time t.

That is, the traveling sinusoidal wave moves to the right a distance vt in the time t, as in Figure 16.16. Note that the wave function has the form $f(x - vt)$ and represents a wave traveling to the right. If the wave were traveling to the left, the quantity $x - vt$ would be replaced by $x + vt$, as we learned when we developed Equations 16.1 and 16.2.

The time it takes the wave to travel a distance of one wavelength is called the **period**, T. Therefore, the wave speed, wavelength, and period are related by

$$v = \frac{\lambda}{T} \tag{16.7}$$

Substituting this into Equation 16.6, we find that

$$y = A \sin\left[2\pi\left(\frac{x}{\lambda} - \frac{t}{T}\right) \right] \tag{16.8}$$

This form of the wave function clearly shows the *periodic* nature of y. That is, at any given time t (a snapshot of the wave), y has the *same* value at the positions x, $x + \lambda$, $x + 2\lambda$, and so on. Furthermore, at any given position x, the value of y at times t, $t + T$, $t + 2T$, and so on is the same.

We can express the wave function in a convenient form by defining two other quantities, called the **angular wave number** k and the **angular frequency** ω:

Angular wave number

$$k \equiv \frac{2\pi}{\lambda} \tag{16.9}$$

Angular frequency

$$\omega \equiv \frac{2\pi}{T} \tag{16.10}$$

Using these definitions, we see that Equation 16.8 can be written in the more compact form

Wave function for a sinusoidal wave

$$y = A \sin(kx - \omega t) \tag{16.11}$$

The frequency of a sinusoidal wave, which we denote by the symbol f, is related to the period by the relationship

Frequency

$$f = \frac{1}{T} \tag{16.12}$$

The most common unit for frequency, as we learned in Chapter 13, is s^{-1}, or hertz (Hz). The corresponding unit for T is seconds.

Using Equations 16.9, 16.10, and 16.12, we can express the wave speed v in the alternative forms

$$v = \frac{\omega}{k} \tag{16.13}$$

Speed of a sinusoidal wave

$$v = \lambda f \tag{16.14}$$

The wave function given by Equation 16.11 assumes that the vertical displacement y is zero at $x = 0$ and $t = 0$. This need not be the case. If the vertical displacement is not zero at $x = 0$ and $t = 0$, we generally express the wave function in the form

$$y = A \sin(kx - \omega t - \phi) \qquad (16.15)$$

General relation for a sinusoidal wave

where ϕ is again called the **phase constant,** just as it was in our study of periodic motion in Chapter 13. This constant can be determined from the initial conditions.

EXAMPLE 16.3   A Traveling Sinusoidal Wave

A sinusoidal wave traveling in the positive x direction has an amplitude of 15.0 cm, a wavelength of 40.0 cm, and a frequency of 8.00 Hz. The vertical displacement of the medium at $t = 0$ and $x = 0$ is also 15.0 cm, as shown in Figure 16.17. (a) Find the angular wave number, period, angular frequency, and speed of the wave.

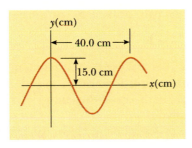

FIGURE 16.17   (Example 16.3) A sinusoidal wave of wavelength $\lambda = 40.0$ cm and amplitude $A = 15.0$ cm. The wave function can be written in the form $y = A \cos(kx - \omega t)$.

Solution   Using Equations 16.9, 16.10, 16.12, and 16.14, we find the following:

$$k = \frac{2\pi}{\lambda} = \frac{2\pi \text{ rad}}{40.0 \text{ cm}} = \boxed{0.157 \text{ rad/cm}}$$

$$T = \frac{1}{f} = \frac{1}{8.00 \text{ s}^{-1}} = \boxed{0.125 \text{ s}}$$

$$\omega = 2\pi f = 2\pi(8.00 \text{ s}^{-1}) = \boxed{50.3 \text{ rad/s}}$$

$$v = f\lambda = (8.00 \text{ s}^{-1})(40.0 \text{ cm}) = \boxed{320 \text{ cm/s}}$$

(b) Determine the phase constant ϕ, and write a general expression for the wave function.

Solution   Since $A = 15.0$ cm and since it is given that $y = 15.0$ cm at $x = 0$ and $t = 0$, substitution into Equation 16.15 gives

$$15 = 15 \sin(-\phi) \qquad \text{or} \qquad \sin(-\phi) = 1$$

Since $\sin(-\phi) = -\sin\phi$, we see that $\phi = -\pi/2$ rad (or $-90°$). Hence, the wave function is of the form

$$y = A \sin\left(kx - \omega t + \frac{\pi}{2}\right) = A \cos(kx - \omega t)$$

That the wave function must have this form can be seen by inspection, noting that the cosine argument is displaced by $90°$ from the sine function. Substituting the values for A, k, and ω into this expression gives

$$y = (15.0 \text{ cm}) \cos(0.157x - 50.3t)$$

Sinusoidal Waves on Strings

In Figure 16.2 we described how to create a wave pulse by jerking a taut string up and down once. To create a train of such pulses, normally referred to as a "wave train" or just plain "wave," we can replace the hand with a vibrating blade. Figure 16.18 represents snapshots of the wave created in this way at intervals of one quarter of a period. Note that because the blade vibrates in simple harmonic motion, *each particle of the string, such as P, also oscillates vertically with simple harmonic motion.* This must be the case because each particle follows the simple harmonic motion of the blade. Therefore, every segment of the string can be treated as a simple harmonic oscillator vibrating with a frequency equal to the frequency of

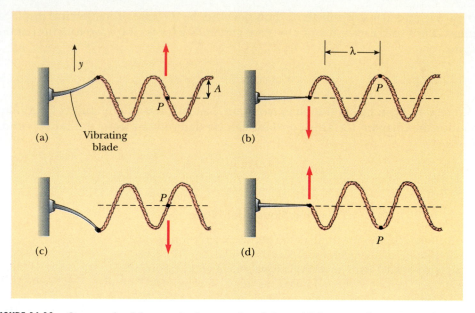

FIGURE 16.18   One method for producing a train of sinusoidal wave pulses on a continuous string. The left end of the string is connected to a blade that is set into vibration. Every segment of the string, such as the point *P*, oscillates with simple harmonic motion in the vertical direction.

vibration of the blade.[3] Note that although each segment oscillates in the *y* direction, the wave travels in the *x* direction with a speed *v*. Of course, this is the definition of a transverse wave. In this case, the energy carried by the traveling wave is supplied by the vibrating blade.

If the waveform at $t = 0$ is as described in Figure 16.18b, then the wave function can be written

$$y = A \sin(kx - \omega t)$$

We can use this expression to describe the motion of any point on the string. The point *P* (or any other point on the string) moves only vertically, and so *its x coordinate remains constant.* Therefore, the *transverse speed, v_y* (not to be confused with the wave speed *v*), and *transverse acceleration, a_y*, are

$$v_y = \frac{dy}{dt}\bigg]_{x=\text{constant}} = \frac{\partial y}{\partial t} = -\omega A \cos(kx - \omega t) \qquad (16.16)$$

$$a_y = \frac{dv_y}{dt}\bigg]_{x=\text{constant}} = \frac{\partial v_y}{\partial t} = -\omega^2 A \sin(kx - \omega t) \qquad (16.17)$$

The maximum values of these quantities are simply the absolute values of the coefficients of the cosine and sine functions:

$$(v_y)_{\text{max}} = \omega A \qquad (16.18)$$

$$(a_y)_{\text{max}} = \omega^2 A \qquad (16.19)$$

---

[3] In this arrangement, we are assuming that the mass always oscillates in a vertical line. The tension in the string would vary if the mass were allowed to move sideways. Such a motion would make the analysis very complex.

You should recognize that the transverse speed and transverse acceleration do not reach their maximum values simultaneously. In fact, the transverse speed reaches its maximum value (ωA) when $y = 0$, whereas the transverse acceleration reaches its maximum value ($\omega^2 A$) when $y = \pm A$. Finally, Equations 16.18 and 16.19 are identical to the corresponding equations for simple harmonic motion.

EXAMPLE 16.4   A Sinusoidally Driven String

The string shown in Figure 16.18 is driven at a frequency of 5.00 Hz. The amplitude of the motion is 12.0 cm, and the wave speed is 20.0 m/s. Determine the angular frequency and wave number for this wave, and write an expression for the wave function.

Solution   Using Equations 16.10, 16.12, and 16.13 gives

$$\omega = \frac{2\pi}{T} = 2\pi f = 2\pi(5.00 \text{ Hz}) = \boxed{31.4 \text{ rad/s}}$$

$$k = \frac{\omega}{v} = \frac{31.4 \text{ rad/s}}{20.0 \text{ m/s}} = \boxed{1.57 \text{ rad/m}}$$

Since $A = 12.0$ cm $= 0.120$ m, we have

$$y = A \sin(kx - \omega t) = \boxed{(0.120 \text{ m}) \sin(1.57x - 31.4t)}$$

Exercise   Calculate the maximum values for the transverse speed and transverse acceleration of any point on the string.

Answer   3.77 m/s; 118 m/s².

16.8   ENERGY TRANSMITTED BY SINUSOIDAL WAVES ON STRINGS

As waves propagate through a medium, they transport energy. This is easily demonstrated by hanging a mass on a taut string and then sending a pulse down the string, as in Figure 16.19. When the pulse meets the suspended mass, the mass is momentarily displaced, as in Figure 16.19b. In the process, energy is transferred to the mass since work must be done in moving it upward.

In this section, we describe the rate at which energy is transported along a string by a one-dimensional sinusoidal wave. Later, we shall extend these ideas to three-dimensional waves.

Consider a sinusoidal wave traveling on a string (Fig. 16.20). The source of the energy is some external agent at the left end of the string, which does work in producing the train of wave pulses. Let us focus on an element of the string of length Δx and mass Δm. Each such segment moves vertically with simple harmonic motion. Furthermore, all segments have the same angular frequency, ω, and the same amplitude, A. As we found in Chapter 13, the total energy E associated with a particle moving with simple harmonic motion is $E = \frac{1}{2}kA^2 = \frac{1}{2}m\omega^2 A^2$, where k is the equivalent force constant of the restoring force. If we apply this to the element of length Δx, we see that the total energy of this element is

$$\Delta E = \frac{1}{2}(\Delta m)\omega^2 A^2$$

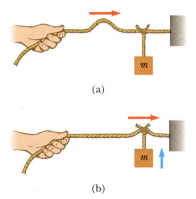

FIGURE 16.19   (a) A pulse traveling to the right on a stretched string on which a mass has been suspended. (b) Energy is transmitted to the suspended mass when the pulse arrives.

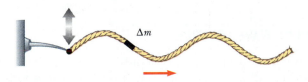

FIGURE 16.20   A sinusoidal wave traveling along the x axis on a stretched string. Every segment moves vertically, and each segment has the same total energy. The power transmitted by the wave equals the energy contained in one wavelength divided by the period of the wave.

If μ is the mass per unit length of the string, then the element of length Δx has a mass $\Delta m = \mu \, \Delta x$. Hence, we can express the energy ΔE as

$$\Delta E = \tfrac{1}{2}(\mu \, \Delta x) \omega^2 A^2 \qquad (16.20)$$

If the wave travels from left to right as in Figure 16.20, the energy ΔE arises from the work done on the element Δm by the string element to the left of Δm. Similarly, the element Δm does work on the element to its right, and so we see that energy is transmitted to the right. The rate at which energy is transmitted along the string —in other words, the power—is dE/dt. If we let Δx approach 0, Equation 16.20 gives

$$\text{Power} = \frac{dE}{dt} = \tfrac{1}{2}\left(\mu \, \frac{dx}{dt}\right) \omega^2 A^2$$

Since dx/dt is equal to the wave speed, v, we have

Power

$$\text{Power} = \tfrac{1}{2} \mu \omega^2 A^2 v \qquad (16.21)$$

This expression is generally valid, so that we can say *the power transmitted by any sinusoidal wave is proportional to the square of the frequency and to the square of the amplitude.*

Thus, we see that a wave traveling through a medium corresponds to energy transport through the medium, with no net transfer of matter. An oscillating source provides the energy and produces a sinusoidal disturbance in the medium. The disturbance is able to propagate through the medium as the result of the interaction between adjacent particles. In order to verify Equation 16.20 by direct experiment, one would have to design some device at the far end of the string to extract the energy of the wave without producing any reflections.

EXAMPLE 16.5   Power Supplied to a Vibrating String

A taut string having mass per unit length of $\mu = 5.00 \times 10^{-2}$ kg/m is under a tension of 80.0 N. How much power must be supplied to the rope to generate sinusoidal waves at a frequency of 60.0 Hz and an amplitude of 6.00 cm?

Solution   The wave speed on the string is

$$v = \sqrt{\frac{F}{\mu}} = \left(\frac{80.0 \text{ N}}{5.00 \times 10^{-2} \text{ kg/m}}\right)^{1/2} = 40.0 \text{ m/s}$$

Since $f = 60.0$ Hz, the angular frequency ω of the sinusoidal

waves on the string has the value

$$\omega = 2\pi f = 2\pi(60.0 \text{ Hz}) = 377 \text{ s}^{-1}$$

Using these values in Equation 16.21 for the power, with $A = 6.00 \times 10^{-2}$ m, gives

$$\begin{aligned} \text{Power} &= \tfrac{1}{2} \mu \omega^2 A^2 v \\ &= \tfrac{1}{2}(5.00 \times 10^{-2} \text{ kg/m})(377 \text{ s}^{-1})^2 \\ &\quad \times (6.00 \times 10^{-2} \text{ m})^2 (40.0 \text{ m/s}) \\ &= \boxed{512 \text{ W}} \end{aligned}$$

---

*16.9   THE LINEAR WAVE EQUATION

In Section 16.3 we introduced the concept of the wave function to represent waves traveling on a string. All wave functions $y(x, t)$ represent solutions of an equation called the *linear wave equation*. This equation gives a complete description of the wave motion, and from it one can derive an expression for the wave speed. Furthermore, the wave equation is basic to many forms of wave motion. In this section, we derive the wave equation as applied to waves on strings.

Consider a small segment of a string of length Δx and tension F, on which a traveling wave is propagating (Fig. 16.21). Let us assume that the ends of the segment make small angles θ_A and θ_B with the x axis. The net force on the segment in the vertical direction is

$$\sum F_y = F \sin \theta_B - F \sin \theta_A = F(\sin \theta_B - \sin \theta_A)$$

Since we have assumed that the angles are small, we can use the small-angle approximation $\sin \theta \approx \tan \theta$ and express the net force as

$$\sum F_y \approx F(\tan \theta_B - \tan \theta_A)$$

However, the tangents of the angles at A and B are defined as the slope of the curve at these points. Since the slope of a curve is given by $\partial y / \partial x$, we have[4]

$$\sum F_y \approx F\left[\left(\frac{\partial y}{\partial x}\right)_B - \left(\frac{\partial y}{\partial x}\right)_A \right] \tag{16.22}$$

We now apply Newton's second law to the segment, with the mass of the segment given by $m = \mu \, \Delta x$, where μ is the mass per unit length of the string:

$$\sum F_y = ma_y = \mu \, \Delta x \left(\frac{\partial^2 y}{\partial t^2}\right) \tag{16.23}$$

Equating Equation 16.23 to Equation 16.22 gives

$$\mu \, \Delta x \left(\frac{\partial^2 y}{\partial t^2}\right) = F\left[\left(\frac{\partial y}{\partial x}\right)_B - \left(\frac{\partial y}{\partial x}\right)_A \right]$$

$$\frac{\mu}{F} \frac{\partial^2 y}{\partial t^2} = \frac{(\partial y / \partial x)_B - (\partial y / \partial x)_A}{\Delta x} \tag{16.24}$$

The right side of this equation can be expressed in a different form if we note that the partial derivative of any function is defined as

$$\frac{\partial f}{\partial x} \equiv \lim_{\Delta x \to 0} \frac{f(x + \Delta x) - f(x)}{\Delta x}$$

If we associate $f(x + \Delta x)$ with $(\partial y / \partial x)_B$ and $f(x)$ with $(\partial y / \partial x)_A$, we see that in the limit $\Delta x \to 0$, Equation 16.24 becomes

$$\frac{\mu}{F} \frac{\partial^2 y}{\partial t^2} = \frac{\partial^2 y}{\partial x^2} \tag{16.25}$$

Linear wave equation

This is the linear wave equation as it applies to waves on a string.

We now show that the sinusoidal wave function represents a solution of this wave equation. If we take the sinusoidal wave function to be of the form $y(x, t) = A \sin(kx - \omega t)$, the appropriate derivatives are

$$\frac{\partial^2 y}{\partial t^2} = -\omega^2 A \sin(kx - \omega t)$$

$$\frac{\partial^2 y}{\partial x^2} = -k^2 A \sin(kx - \omega t)$$

Substituting these expressions into Equation 16.25 gives

$$k^2 = (\mu / F)\omega^2$$

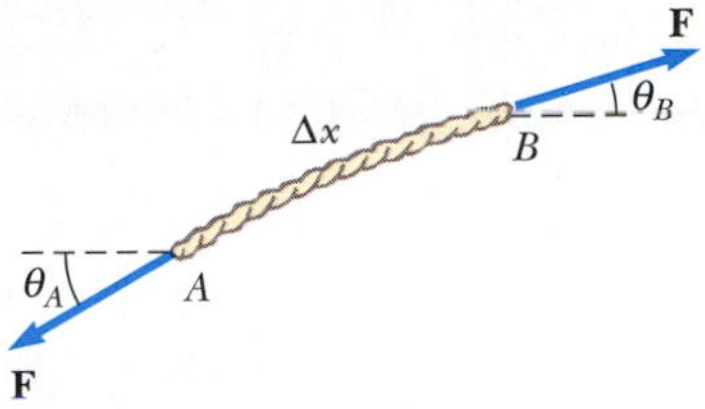

FIGURE 16.21   A segment of a string under tension F. Note that the slopes at points A and B are given by $\tan \theta_A$ and $\tan \theta_B$, respectively.

---

[4] It is necessary to use partial derivatives because y depends on both x and t.

Using the relationship $v = \omega/k$ in the above expression, we see that

$$v^2 = \frac{\omega^2}{k^2} = \frac{F}{\mu}$$

$$v = \sqrt{\frac{F}{\mu}}$$

which is Equation 16.4. This derivation represents another proof of the expression for the wave speed on a taut string.

The linear wave equation is often written in the form

Linear wave equation in general

$$\frac{\partial^2 y}{\partial x^2} = \frac{1}{v^2}\frac{\partial^2 y}{\partial t^2} \tag{16.26}$$

This expression applies in general to various types of traveling waves. For waves on strings, y represents the vertical displacement of the string. For sound waves, y corresponds to variations in the pressure or density of a gas. In the case of electromagnetic waves, y corresponds to electric or magnetic field components.

We have shown that the sinusoidal wave function is one solution of the linear wave equation. Although we do not prove it here, the linear wave equation is satisfied by *any* wave function having the form $y = f(x \pm vt)$. Furthermore, we have seen that the wave equation is a direct consequence of Newton's second law applied to any segment of the string.

SUMMARY

A **transverse wave** is one in which the particles of the medium move in a direction *perpendicular* to the direction of the wave velocity. An example is a wave on a taut string.

A **longitudinal wave** is one in which the particles of the medium move in a direction *parallel* to the direction of the wave velocity. Sound waves in fluids are longitudinal.

Any one-dimensional wave traveling with a speed v in the x direction can be represented by a wave function of the form

$$y = f(x \pm vt) \tag{16.1, 16.2}$$

where the $+$ sign applies to a wave traveling in the negative x direction and the $-$ sign applies to a wave traveling in the positive x direction. The shape of the wave at any instant (a snapshot of the wave) is obtained by holding t constant.

The **superposition principle** says that when two or more waves move through a medium, the resultant wave function equals the algebraic sum of the individual wave functions. Waves that obey this principle are said to be *linear*. When two waves combine in space, they interfere to produce a resultant wave. The **interference** may be **constructive** (when the individual displacements are in the same direction) or **destructive** (when the displacements are in opposite directions).

The **speed** of a wave traveling on a taut string of mass per unit length μ and tension F is

$$v = \sqrt{\frac{F}{\mu}} \tag{16.4}$$

When a **pulse** traveling on a string meets a fixed end, the pulse is reflected and inverted. If the pulse reaches a free end, it is reflected but not inverted.

The **wave function** for a one-dimensional sinusoidal wave traveling to the right can be expressed as

$$y = A \sin\left[\frac{2\pi}{\lambda}(x - vt)\right] = A \sin(kx - \omega t) \qquad \text{(16.6, 16.11)}$$

where A is the amplitude, λ is the wavelength, k is the angular wave number, and ω is the angular frequency. If T is the period (the time it takes the wave to travel a distance equal to one wavelength) and f is the frequency, then v, k, and ω can be written

$$v = \frac{\lambda}{T} = \lambda f \qquad \text{(16.7, 16.14)}$$

$$k \equiv \frac{2\pi}{\lambda} \qquad \text{(16.9)}$$

$$\omega = \frac{2\pi}{T} = 2\pi f \qquad \text{(16.10, 16.12)}$$

The **power** transmitted by a sinusoidal wave on a stretched string is

$$\text{Power} = \tfrac{1}{2}\mu\omega^2 A^2 v \qquad \text{(16.21)}$$

QUESTIONS

1. Why is a wave pulse traveling on a string considered a transverse wave?

2. How would you set up a longitudinal wave in a stretched spring? Would it be possible to set up a transverse wave in a spring?

3. By what factor would you have to increase the tension in a taut string in order to double the wave speed?

4. When traveling on a taut string, does a wave pulse always invert upon reflection? Explain.

5. Can two pulses traveling in opposite directions on the same string reflect from each other? Explain.

6. Does the vertical speed of a segment of a horizontal taut string through which a wave is traveling depend on the wave speed?

7. If you were to shake one end of a taut rope periodically three times each second, what would be the period of the sinusoidal waves set up in the rope?

8. A vibrating source generates a sinusoidal wave on a string under constant tension. If the power delivered to the string is doubled, by what factor does the amplitude change? Does the wave speed change under these circumstances?

9. Consider a wave traveling on a taut rope. What is the difference, if any, between the speed of the wave and the speed of a small section of the rope?

10. If a long rope is hung from a ceiling and waves are sent up the rope from its lower end, they do not ascend with constant speed. Explain.

11. What happens to the wavelength of a wave on a string when the frequency is doubled? Assume the tension in the string remains the same.

12. What happens to the speed of a wave on a taut string when the frequency is doubled? Assume the tension in the string remains the same.

13. How do transverse waves differ from longitudinal waves?

14. When all the strings on a guitar are stretched to the same tension, will the speed of a wave along the more massive bass strings be faster or slower than the speed of a wave on the lighter strings?

15. If you stretch a rubber hose and pluck it, you can observe a pulse traveling up and down the hose. What happens to the speed if you stretch the hose tighter? If you fill the hose with water?

16. In a longitudinal wave in a spring, the coils move back and forth in the direction of wave motion. Does the speed of the wave depend on the maximum speed of each coil?

17. When two waves interfere, can the amplitude of the resultant wave be larger than either of the two original waves? Under what conditions?

18. A solid can transport both longitudinal waves and transverse waves, but a fluid can transport only longitudinal waves. Why?

19. In an earthquake both S (transverse) and P (longitudinal) waves are sent out. The S waves travel through the Earth more slowly than the P waves (4.5 km/s versus 7.8 km/s). By detecting the time of arrival of the waves, how can one determine how far away the epicenter of the quake was? How many detection centers are necessary to pinpoint the location of the epicenter?

PROBLEMS

Section 16.3   One-Dimensional Traveling Waves

1. At $t = 0$, a transverse wave pulse in a wire is described by the function

$$y = \frac{6}{x^2 + 3}$$

where x and y are in meters. Write the function $y(x, t)$ that describes this wave if it is traveling in the positive x direction with a speed of 4.5 m/s.

2. Two wave pulses **A** and **B** are moving in opposite directions along a taut string with a speed of 2 cm/s. The amplitude of **A** is twice the amplitude of **B**. The pulses are shown in Figure P16.2 at $t = 0$. Sketch the shape of the string at $t = 1, 1.5, 2, 2.5,$ and 3 s.

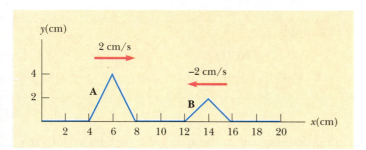

FIGURE P16.2

3. Two points, A and B, on the Earth are at the same longitude and 60.0° apart in latitude. An earthquake at point A sends two waves toward B. A transverse wave travels along the surface of the Earth at 4.50 km/s, and a longitudinal wave travels through the body of the Earth at 7.80 km/s. (a) Which wave arrives at B first? (b) What is the time difference between the arrivals of the two waves at B? Take the radius of the Earth to be 6370 km.

4. A wave moving along the x axis is described by

$$y(x, t) = 5.0e^{-(x+5.0t)^2}$$

where x is in meters and t is in seconds. Determine (a) the direction of the wave motion and (b) the speed of the wave.

5. Ocean waves with a crest-to-crest distance of 10 m can be described by

$$y(x, t) = (0.80 \text{ m}) \sin[0.63(x - vt)]$$

where $v = 1.2$ m/s. (a) Sketch $y(x, t)$ at $t = 0$. (b) Sketch $y(x, t)$ at $t = 2.0$ s. Note how the entire wave form has shifted 2.4 m in the positive x direction in this time interval.

Section 16.4   Superposition and Interference of Waves

6. Two waves in one string are described by the relationships

$$y_1 = 3.0 \cos(4.0x - 5.0t)$$
$$y_2 = 4.0 \sin(5.0x - 2.0t)$$

where y and x are in centimeters and t is in seconds. Find the superposition of the waves $y_1 + y_2$ at the points (a) $x = 1.0$, $t = 1.0$, (b) $x = 1.0$, $t = 0.50$, (c) $x = 0.50$, $t = 0$. (Remember that the arguments of the trigonometric functions are in radians.)

7. Two sinusoidal waves in a string are defined by the functions

$$y_1 = (2.0 \text{ cm}) \sin(20x - 30t)$$
$$y_2 = (2.0 \text{ cm}) \sin(25x - 40t)$$

where y and x are in centimeters and t is in seconds. (a) What is the phase difference between these two waves at the point $x = 5.0$ cm at $t = 2.0$ s? (b) What is the positive x value closest to the origin for which the two phases differ by $\pm \pi$ at $t = 2.0$ s? (This is where the two waves add to zero.)

8. Two waves are traveling in the same direction along a stretched string. Each has an amplitude of 4.0 cm, and they are 90° out of phase. Find the amplitude of the resultant wave.

8A. Two waves are traveling in the same direction along a stretched string. Each has an amplitude A, and they are out of phase by an angle ϕ. Find the amplitude of the resultant wave.

9. Two pulses traveling on the same string are described by

$$y_1 = \frac{5}{(3x - 4t)^2 + 2}$$

and

$$y_2 = \frac{-5}{(3x + 4t - 6)^2 + 2}$$

(a) In which direction does each pulse travel? (b) At what time do the two cancel? (c) At what point do the two waves always cancel?

Section 16.5   The Speed of Waves on Strings

10. Transverse waves with a speed of 50 m/s are to be produced in a taut string. A 5.0-m length of string with a total mass of 0.060 kg is used. What is the required tension?

□ indicates problems that have full solutions available in the Student Solutions Manual and Study Guide.

11. A piano string of mass per unit length 5.00×10^{-3} kg/m is under a tension of 1350 N. Find the speed with which a wave travels on this string.

12. An astronaut on the Moon wishes to measure the local value of g by timing pulses traveling down a wire that has a large mass suspended from it. Assume a wire of mass 4.00 g is 1.60 m long and has a 3.00-kg mass suspended from it. A pulse requires 36.1 ms to traverse the length of the wire. Calculate g from these data. (You may neglect the mass of the wire when calculating the tension in it.)

12A. An astronaut on the Moon wishes to measure the local value of g by timing pulses traveling down a wire that has a large mass suspended from it. Assume a wire of mass m and length L has a mass M suspended from it. A pulse requires a time t to traverse the length of the wire. Calculate g from these data. (You may neglect the mass of the wire when finding the tension in it.)

13. Transverse waves travel with a speed of 20.0 m/s in a string under a tension of 6.00 N. What tension is required for a wave speed of 30.0 m/s in the same string?

14. A simple pendulum consists of a ball of mass M hanging from a uniform string of mass m and length L, with $m \ll M$. If the period of oscillation for the pendulum is T, determine the speed of a transverse wave in the string when the pendulum hangs vertically.

15. The elastic limit of a piece of steel wire is 2.7×10^9 Pa. What is the maximum speed at which transverse wave pulses can propagate along this wire without exceeding this stress? (The density of steel is 7.86×10^3 kg/m³.)

16. A light string of mass 10.0 g and length $L = 3.00$ m has its ends tied to two walls that are separated by the distance $D = 2.00$ m. Two masses, each of mass $M = 2.00$ kg, are suspended from the string as in Figure P16.16. If a wave pulse is sent from point A, how long does it take to travel to point B?

16A. A light string of mass m and length L has its ends tied to two walls that are separated by the distance D. Two masses, each of mass M, are suspended from the string as in Figure P16.16. If a wave pulse is sent from point A, how long does it take to travel to point B?

17. A 30.0-m steel wire and a 20.0-m copper wire, both with 1.00-mm diameters, are connected end to end and stretched to a tension of 150 N. How long does it take a transverse wave to travel the entire length of the two wires?

18. A light string of mass per unit length 8.00 g/m has its ends tied to two walls separated by a distance equal to three fourths the length of the string (Fig. P16.18). A mass m is suspended from the center of the string,

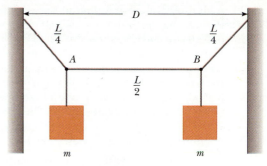

FIGURE P16.16

putting a tension in the string. (a) Find an expression for the transverse wave speed in the string as a function of the hanging mass. (b) How much mass should be suspended from the string to have a wave speed of 60.0 m/s?

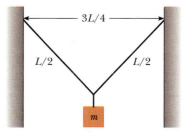

FIGURE P16.18

Section 16.7 Sinusoidal Waves

19. (a) Plot y versus t at $x = 0$ for a sinusoidal wave of the form $y = (15.0 \text{ cm}) \cos(0.157x - 50.3t)$, where x and y are in centimeters and t is in seconds. (b) Determine the period of vibration from this plot and compare your result with the value found in Example 16.3.

20. For a certain transverse wave, the distance between two successive maxima is 1.2 m and eight maxima pass a given point along the direction of travel every 12 s. Calculate the wave speed.

20A. For a certain transverse wave, the distance between two successive maxima is λ and N maxima pass a given point along the direction of travel every t s. Find an expression for the wave speed.

21. A sinusoidal wave is traveling along a rope. The oscillator that generates the wave completes 40.0 vibrations in 30.0 s. Also, a given maximum travels 425 cm along the rope in 10.0 s. What is the wavelength?

22. When a particular wire is vibrating with a frequency of 4.00 Hz, a transverse wave of wavelength 60.0 cm is produced. Determine the speed of wave pulses along the wire.

23. A sinusoidal wave traveling in the $-x$ direction (to the left) has an amplitude of 20.0 cm, a wavelength of 35.0 cm, and a frequency of 12.0 Hz. The displacement of the wave at $t = 0$, $x = 0$ is $y = -3.00$ cm, and the wave has a positive velocity here. (a) Sketch the wave at $t = 0$. (b) Find the angular wave number, period, angular frequency, and phase velocity of the wave. (c) Write an expression for the wave function $y(x, t)$.

24. A sinusoidal wave train is described by

$$y = (0.25 \text{ m}) \sin(0.30x - 40t)$$

where x and y are in meters and t is in seconds. Determine for this wave the (a) amplitude, (b) angular frequency, (c) angular wave number, (d) wavelength, (e) wave speed, and (f) direction of motion.

25. Two waves are described by

$$y_1(x, t) = 5.0 \sin(2.0x - 10t)$$

and

$$y_2(x, t) = 10 \cos(2.0x - 10t),$$

where x is in meters and t is in seconds. Show that the resulting wave is sinusoidal, and determine the amplitude and phase of this sinusoidal wave.

26. A bat can detect small objects such as an insect whose size is approximately equal to one wavelength of the sound the bat makes. If bats emit a chirp at a frequency of 60.0 kHz, and if the speed of sound in air is 340 m/s, what is the smallest insect a bat can detect?

27. (a) Write the expression for y as a function of x and t for a sinusoidal wave traveling along a rope in the *negative x* direction with the following characteristics: $A = 8.00$ cm, $\lambda = 80.0$ cm, $f = 3.00$ Hz, and $y(0, t) = 0$ at $t = 0$. (b) Write the expression for y as a function of x for the wave in part (a) assuming that $y(x, 0) = 0$ at the point $x = 10.0$ cm.

28. A transverse wave on a string is described by

$$y = (0.12 \text{ m}) \sin \pi(x/8 + 4t)$$

(a) Determine the transverse speed and acceleration of the string at $t = 0.20$ s for the point on the string located at $x = 1.6$ m. (b) What are the wavelength, period, and speed of propagation of this wave?

29. A transverse sinusoidal wave on a string has a period $T = 25.0$ ms and travels in the negative x direction with a speed of 30.0 m/s. At $t = 0$, a particle on the string at $x = 0$ has a displacement of 2.00 cm and travels to the left with a speed of 2.0 m/s. (a) What is the amplitude of the wave? (b) What is the initial phase angle? (c) What is the maximum transverse speed of the string? (d) Write the wave function for the wave.

30. A sinusoidal wave of wavelength 2.0 m and amplitude 0.10 m travels with a speed of 1.0 m/s on a string. Initially, the left end of the string is at the

origin and the wave moves from left to right. Find (a) the frequency and angular frequency, (b) the angular wave number, and (c) the wave function for this wave. Determine the equation of motion for (d) the left end of the string and (e) the point on the string at $x = 1.5$ m to the right of the left end. (f) What is the maximum speed of any point on the string?

31. A wave is described by $y = (2.0 \text{ cm}) \sin(kx - \omega t)$, where $k = 2.11$ rad/m, $\omega = 3.62$ rad/s, x is in meters, and t is in seconds. Determine the amplitude, wavelength, frequency, and speed of the wave.

32. A sinusoidal wave on a string is described by

$$y = (0.51 \text{ cm}) \sin(kx - \omega t)$$

where $k = 3.1$ rad/cm and $\omega = 9.3$ rad/s. How far does a wave crest move in 10 s? Does it move in the positive or negative x direction?

33. A transverse traveling wave on a taut wire has an amplitude of 0.200 mm and a frequency of 500 Hz and travels with a speed of 196 m/s. (a) Write an equation in SI units of the form $y = A \sin(kx - \omega t)$ for this wave. (b) The mass per unit length of this wire is 4.10 g/m. Find the tension in the wire.

34. A wave on a string is described by the wave function $y = (0.10 \text{ m}) \sin(0.50x - 20t)$. (a) Show that a particle in the string at $x = 2.0$ m executes harmonic motion. (b) Determine the frequency of oscillation of this particular point.

Section 16.8 Energy Transmitted by Sinusoidal Waves on Strings

35. A taut rope has a mass of 0.18 kg and a length of 3.6 m. What power must be supplied to the rope in order to generate sinusoidal waves having an amplitude of 0.10 m and a wavelength of 0.50 m and traveling with a speed of 30 m/s?

35A. A taut rope has a mass M and length L. What power must be supplied to the rope in order to generate sinusoidal waves having an amplitude A and wavelength λ and traveling with a speed v?

36. A two-dimensional water wave spreads in circular wavefronts. Show that the amplitude A at a distance r from the initial disturbance is proportional to $1/\sqrt{r}$. (*Hint:* Consider the energy concentrated in the outward-moving ripple.)

37. Transverse waves are being generated on a rope under constant tension. By what factor is the required power increased or decreased if (a) the length of the rope is doubled and the angular frequency remains constant, (b) the amplitude is doubled and the angular frequency is halved, (c) both the wavelength and the amplitude are doubled, and (d) both the length of the rope and the wavelength are halved?

38. Sinusoidal waves 5.00 cm in amplitude are to be transmitted along a string that has a linear density of 4.00×10^{-2} kg/m. If the maximum power delivered by the source is 300 W and the string is under a tension of 100 N, what is the highest vibrational frequency at which the source can operate?

39. A sinusoidal wave on a string is described by the equation

$$y = (0.15 \text{ m}) \sin(0.80x - 50t)$$

where x and y are in meters and t is in seconds. If the mass per unit length of this string is 12 g/m, determine (a) the speed of the wave, (b) the wavelength, (c) the frequency, and (d) the power transmitted to the wave.

40. A horizontal string can transmit a maximum power of P (without breaking) if a wave with amplitude A and angular frequency ω is traveling along it. In order to increase this maximum power, a student folds the string and uses this "double string" as a transmitter. Determine the maximum power that can be transmitted along the "double string."

*Section 16.9 The Linear Wave Equation

41. Show that the wave function $y = \ln[b(x - vt)]$ is a solution to Equation 16.26, where b is a constant.

42. Show that the wave function $y = e^{b(x - vt)}$ is a solution of the wave equation (Eq. 16.26), where b is a constant.

43. (a) Show that the function $y(x, t) = x^2 + v^2 t^2$ is a solution to the wave equation. (b) Show that the function above can be written as $f(x + vt) + g(x - vt)$, and determine the functional forms for f and g. (c) Repeat parts (a) and (b) for the function, $y(x, t) = \sin(x) \cos(vt)$.

ADDITIONAL PROBLEMS

44. A traveling wave propagates according to the expression $y = (4.0 \text{ cm}) \sin(2.0x - 3.0t)$ where x is in centimeters and t is in seconds. Determine (a) the amplitude, (b) the wavelength, (c) the frequency, (d) the period, and (e) the direction of travel of the wave.

45. The wave function for a linearly polarized wave on a taut string is (in SI units)

$$y(x, t) = (0.35 \text{ m}) \sin(10\pi t - 3\pi x + \pi/4)$$

(a) What are the speed and direction of travel of the wave? (b) What is the vertical displacement of the string at $t = 0$, $x = 0.10$ m? (c) What are the wavelength and frequency of the wave? (d) What is the maximum magnitude of the transverse speed of the string?

46. A block of mass $M = 2.0$ kg, supported by a string, rests on an incline making an angle of $\theta = 45°$ with the horizontal (Fig. P16.46). The length of the string is $L = 0.5$ m and its mass is $m = 2.0$ g, and hence much less than M. Determine the time it takes a transverse wave to travel from one end of the string to the other.

46A. A block of mass M, supported by a string, rests on an incline making an angle of θ with the horizontal (Fig. P16.46). The length of the string is L and its mass is $m \ll M$. Determine the time it takes a transverse wave to travel from one end of the string to the other.

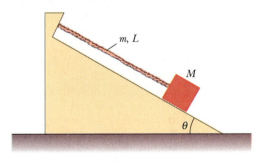

m, L

M

θ

FIGURE P16.46

47. (a) Determine the speed of transverse waves on a string under a tension of 80.0 N if the string has a length of 2.00 m and a mass of 5.00 g. (b) Calculate the power required to generate these waves if they have a wavelength of 16.0 cm and an amplitude of 4.00 cm.

48. A sinusoidal wave in a rope is described by the wave function $y = (0.20 \text{ m}) \sin[\pi(0.75x + 18t)]$ where x and y are in meters and t is in seconds. The rope has a linear mass density of 0.25 kg/m. If the tension in the rope is provided by an arrangement like the one illustrated in Figure 16.11, what is the value of the suspended mass?

49. A 2.0-kg block hangs from a rubber string, being supported so that the string is not stretched. The unstretched length of the string is 0.5 m and its mass is 5.0 g. The "spring constant" for the string is 100.0 N/m. The block is released and stops at the lowest point. (a) Determine the tension in the string when the block is at this lowest point. (b) What is the length of the string in this "stretched" position? (c) Find the speed of a transverse wave in the string if the block is held in this lowest position.

49A. A block of mass M hangs from a rubber string, being supported so that the string is not stretched. The unstretched length of the string is L_0 and its mass is m. The "spring constant" for the string is k. The block is released and stops at the lowest point. (a) Determine the tension in the string when the block is at this lowest point. (b) What is the length of the string in this "stretched" position? (c) Find the speed of a transverse wave in the string if the block is held in this lowest position.

50. A wire of density ρ is tapered so that its cross-sectional area varies with x, according to

$$A = (1.0 \times 10^{-3} x + 0.010) \text{ cm}^2$$

(a) If the wire is subject to a tension F, derive a relationship for the speed of a wave as a function of position. (b) If the wire is aluminum and is subject to a tension of 24 N, determine the speed at the origin and at $x = 10$ m.

51. Determine the speed and direction of propagation of each of the following sinusoidal waves, assuming that x is measured in meters and t in seconds.
(a) $y = 0.60 \cos(3.0x - 15t + 2)$
(b) $y = 0.40 \cos(3.0x + 15t - 2)$
(c) $y = 1.2 \sin(15t + 2.0x)$
(d) $y = 0.20 \sin(12t - x/2 + \pi)$

52. A rope of total mass m and length L is suspended vertically. Show that a transverse wave pulse will travel the length of the rope in a time $t = 2\sqrt{L/g}$. (*Hint:* First find an expression for the wave speed at any point a distance x from the lower end by considering the tension in the rope as resulting from the weight of the segment below that point.)

53. If mass M is suspended from the bottom of the rope in Problem 52, (a) show that the time for a transverse wave to travel the length of the rope is

$$t = 2\sqrt{\frac{L}{g}\left(\frac{\sqrt{M+m} - \sqrt{M}}{\sqrt{m}}\right)}$$

(b) Show that this reduces to the result of Problem 52 when $M = 0$. (c) Show that for $m \ll M$, the expression in part (a) reduces to

$$t = \sqrt{\frac{mL}{Mg}}$$

54. As a sound wave travels through the air, it produces pressure variations (above and below atmospheric pressure) given by $P = 1.27 \sin \pi(x - 340t)$ in SI units. Find (a) the amplitude of the pressure variations, (b) the frequency, (c) the wavelength in air, and (d) the speed of the sound wave.

55. An aluminum wire is clamped at each end under zero tension at room temperature (22°C). The tension in the wire is increased by reducing the temperature, which results in a decrease in the wire's equilibrium length. What strain ($\Delta L/L$) results in a transverse wave speed of 100 m/s? Take the cross-sectional area of the wire to be 5.0×10^{-6} m², the density to be 2.7×10^3 kg/m³, and Young's modulus to be 7.0×10^{10} N/m².

56. (a) Show that the speed of longitudinal waves along a spring of force constant k is $v = \sqrt{kL/\mu}$, where L is the unstretched length of the spring and μ is the mass per unit length. (b) A spring of mass 0.40 kg has an unstretched length of 2.0 m and a force constant of 100 N/m. Using the results to part (a), determine the speed of longitudinal waves along this spring.

57. It is stated in Problem 52 that a wave pulse travels from the bottom to the top of a rope of length L in a time $t = 2\sqrt{L/g}$. Use this result to answer the following questions. (It is *not* necessary to set up any new integrations.) (a) How long does it take for a wave pulse to travel halfway up the rope? (Give your answer as a fraction of the quantity $2\sqrt{L/g}$.) (b) A pulse starts traveling up the rope. How far has it traveled after a time $\sqrt{L/g}$?

58. A string of length L consists of two sections. The left half has mass per unit length $\mu = \mu_0/2$, while the right has a mass per unit length $\mu' = 3\mu = 3\mu_0/2$. Tension in the string is F_0. Notice from the data given that this string has the same total mass as a uniform string of length L and mass per unit length μ_0. (a) Find the speeds v and v' at which transverse wave pulses travel in the two sections. Express the speeds in terms of F_0 and μ_0, and also as multiples of the speed $v_0 = \sqrt{F_0/\mu_0}$. (b) Find the time required for a pulse to travel from one end of the string to the other. Give your result as a multiple of $T_0 = L/v_0$.

59. A wave pulse traveling along a string of linear mass density μ is described by the relationship

$$y = [A_0 e^{-bx}] \sin(kx - \omega t)$$

where the factors in brackets before the sine are said to be the amplitude. (a) What is the power $P(x)$ carried by this wave at a point x? (b) What is the power carried by this wave at the origin? (c) Compute the ratio $P(x)/P(0)$.

SPREADSHEET PROBLEM

S1. Two transverse wave pulses traveling in opposite directions along the x axis are represented by the following wave functions:

$$y_1(x, t) = \frac{6}{(x - 3t)^2} \qquad y_2(x, t) = -\frac{3}{(x + 3t)^2}$$

where x and y are measured in centimeters and t is in seconds. Write a spreadsheet or program to add the two pulses and obtain the shape of the composite waveform $y_{\text{tot}} = y_1 + y_2$ as a function of time. Plot y_{tot} versus x. Make separate plots for $t = 0$, 0.5, 1, 1.5, 2, 2.5, and 3.0 s.

CHAPTER 17

Sound Waves

Fennec foxes are small animals with very large ears, about four inches long. The sensitivity of their auditory system is enchanced by these large ears, which enable them to hear very faint sounds as they collect a larger cross-section of sound waves. *(Tom McHugh/Photo Researchers)*

Sound waves are the most important example of longitudinal waves. They can travel through any material medium with a speed that depends on the properties of the medium. As the waves travel, the particles in the medium vibrate to produce density and pressure changes along the direction of motion of the wave. These changes result in a series of high- and low-pressure regions called *condensations* and *rarefactions,* respectively. If the source of the sound waves vibrates sinusoidally, the pressure variations are also sinusoidal. We shall find that the mathematical description of harmonic sound waves is identical to that of harmonic string waves discussed in the previous chapter.

There are three categories of mechanical waves that cover different ranges of frequency: (1) *Audible waves* (usually called **sound waves**) are waves that lie within the range of sensitivity of the human ear, typically, 20 Hz to 20 000 Hz. They can be generated in a variety of ways, such as by musical instruments, human vocal cords, and loudspeakers. (2) *Infrasonic waves* are waves having frequencies below the audible range. Earthquake waves are an example. (3) *Ultrasonic waves* are waves having frequencies above the audible range. For example, they can be generated by inducing vibrations in a quartz crystal with an applied alternating electric field. All may be longitudinal or transverse in solids but only longitudinal in fluids.

Any device that transforms one form of power into another is called a *transducer.* In addition to the loudspeaker (which transforms electric power to power in audible waves) and the quartz crystal (electric power to ultrasonic power), ceramic and magnetic phonograph pickups are common examples of sound transducers.

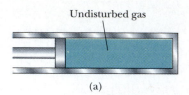

Undisturbed gas

(a)

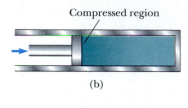

Compressed region

(b)

(c)

(d)

FIGURE 17.1   Motion of a longitudinal pulse through a compressible medium. The compression (darker region) is produced by the moving piston.

17.1   SPEED OF SOUND WAVES

The speed of sound waves depends on the compressibility and the inertia of the medium. If the medium has a bulk modulus B (Section 12.4) and an equilibrium density ρ, the speed of sound waves in that medium is

$$v = \sqrt{\frac{B}{\rho}} \qquad (17.1)$$

It is interesting to compare Equation 17.1 with the expression for the speed of transverse waves on a string, $v = \sqrt{F/\mu}$, discussed in the previous chapter. In both cases, the wave speed depends on an elastic property of the medium (B or F) and on an inertial property of the medium (ρ or μ). In fact, the speed of *all mechanical waves* follows an expression of the general form

$$v = \sqrt{\frac{\text{elastic property}}{\text{inertial property}}}$$

Let us describe pictorially the motion of a one-dimensional longitudinal pulse moving through a long tube containing a compressible gas or liquid (Fig. 17.1). A piston at the left end can be moved to the right to compress the fluid and create the pulse. Before the piston is moved, the medium is undisturbed and of uniform density, as represented by the uniformly shaded region in Figure 17.1a. When the piston is suddenly pushed to the right (Fig. 17.1b), the medium just in front of it is compressed (represented by the more heavily shaded region); the pressure and density in this region are higher than normal. When the piston comes to rest (Fig. 17.1c), the compressed region of the medium continues to move to the right, corresponding to a longitudinal pulse traveling down the tube with a speed v. Note that the piston speed does *not* equal v. Furthermore, the compressed region does not "stay with" the piston until the piston stops (in other words, once the pulse is created, $v_{\text{pulse}} > v_{\text{piston}}$).

EXAMPLE 17.1   Sound Waves in a Solid Bar

If a solid bar is struck at one end with a hammer, a longitudinal pulse propagates down the bar with a speed

$$v = \sqrt{\frac{Y}{\rho}}$$

where Y is the Young's modulus for the material (Section 12.4). Find the speed of sound in an aluminum bar.

Solution   From Table 12.1 we get $Y = 7.0 \times 10^{10}$ N/m² for aluminum and from Table 1.5 we get $\rho = 2.7 \times 10^3$ kg/m³. Therefore,

$$v_{\text{Al}} = \sqrt{\frac{Y}{\rho}} = \sqrt{\frac{7.0 \times 10^{10} \text{ N/m}^2}{2.7 \times 10^3 \text{ kg/m}^3}} \approx 5.1 \text{ km/s}$$

This is a typical value for the speed of sound in solids, much larger than the speed of sound in gases, as Table 17.1 shows. This difference in speeds makes sense because the molecules of a solid are bound together in a much more rigid structure than a gas.

TABLE 17.1   Speed of Sound in Various Media

Medium	v(m/s)
Gases	
Air (0°C)	331
Air (20°C)	343
Hydrogen (0°C)	1286
Oxygen (0°C)	317
Helium (0°C)	972
Liquids at 25°C	
Water	1493
Methyl alcohol	1143
Seawater	1533
Solids	
Aluminum	5100
Copper	3560
Iron	5130
Lead	1322
Vulcanized rubber	54

EXAMPLE 17.2   Speed of Sound in a Liquid

Find the speed of sound in water, which has a bulk modulus of about 2.1×10^9 N/m^2 and a density of 1.00×10^3 kg/m^3.

Solution   Using Equation 17.1, we find that

$$v_{\text{water}} = \sqrt{\frac{B}{\rho}} \approx \sqrt{\frac{2.1 \times 10^9 \text{ N/m}^2}{1.00 \times 10^3 \text{ kg/m}^3}} = \boxed{1.5 \text{ km/s}}$$

In general, sound waves travel more slowly in liquids than in solids. This is because liquids are more compressible than solids and hence have a smaller bulk modulus.

17.2   PERIODIC SOUND WAVES

As just noted in the previous section, one can produce a one-dimensional periodic sound wave in a long, narrow tube containing a gas by means of a vibrating piston at one end, as in Figure 17.2. The darker regions in this figure represent regions where the gas is compressed, and so in these regions the density and pressure are above their equilibrium values.

A compressed region is formed whenever the piston is being pushed into the tube. This compressed region, called a **condensation**, moves down the tube as a pulse, continuously compressing the layers in front of it. When the piston is withdrawn from the tube, the gas in front of it expands and the pressure and density in this region fall below their equilibrium values (represented by the lighter regions in Figure 17.2). These low-pressure regions, called **rarefactions,** also propagate along the tube, following the condensations. Both regions move with a speed equal to the speed of sound in that medium (about 343 m/s in air at 20°C).

As the piston oscillates sinusoidally, regions of condensation and rarefaction are continuously set up. The distance between two successive condensations (or two successive rarefactions) equals the wavelength, λ. As these regions travel down the tube, any small volume of the medium moves with simple harmonic motion parallel to the direction of the wave. If $s(x, t)$ is the displacement of a small volume element measured from its equilibrium position, we can express this harmonic displacement function as

$$s(x, t) = s_{\max} \cos(kx - \omega t) \tag{17.2}$$

where $s_{\max}$ is the *maximum displacement of the medium from equilibrium* (in other words, the **displacement amplitude**), k is the angular wave number, and ω is the angular frequency of the piston. Note that the displacement of the medium is along x, the direction of motion of the sound wave, which of course means we are describing a longitudinal wave.

The variation in the pressure of the gas, ΔP, measured from its equilibrium value is also periodic and given by

$$\Delta P = \Delta P_{\max} \sin(kx - \omega t) \tag{17.3}$$

The derivation of this expression follows:

The **pressure amplitude** $\Delta P_{\max}$ is the *maximum change in pressure from the equilibrium value.* As we show later, the pressure amplitude is proportional to the displacement amplitude, $s_{\max}$:

$$\Delta P_{\max} = \rho v \omega s_{\max} \tag{17.4}$$

Pressure amplitude

where $\omega s_{\max}$ is the maximum longitudinal speed of the medium in front of the piston.

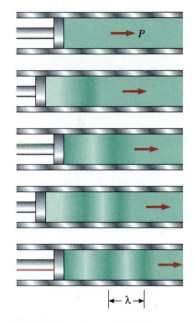

FIGURE 17.2   A sinusoidal longitudinal wave propagating down a tube filled with a compressible gas. The source of the wave is a vibrating piston at the left. The high- and low-pressure regions are dark and light, respectively.

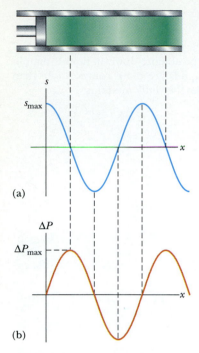

FIGURE 17.3 (a) Displacement amplitude versus position and (b) pressure amplitude versus position for a sinusoidal longitudinal wave. The displacement wave is 90° out of phase with the pressure wave.

Thus, we see that a sound wave may be considered as either a displacement wave or a pressure wave. A comparison of Equations 17.2 and 17.3 shows that *the pressure wave is 90° out of phase with the displacement wave.* Graphs of these functions are shown in Figure 17.3. Note that the pressure variation is a maximum when the displacement is zero, whereas the displacement is a maximum when the pressure variation is zero. Since pressure is proportional to density, the variation in density from the equilibrium value follows an expression similar to Equation 17.3.

We now derive Equations 17.3 and 17.4. From the definition of bulk modulus (Equation 12.8), we see that the pressure variation in a gas is

$$\Delta P = -B \frac{\Delta V}{V}$$

The volume of a medium segment that has a thickness Δx in the horizontal direction and a cross-sectional area A is $V = A \Delta x$. The change in the volume ΔV accompanying the pressure change is equal to $A \Delta s$, where Δs is the difference between the value of s at $x + \Delta x$ and the value of s at x. Hence, we can express ΔP as

$$\Delta P = -B \frac{\Delta V}{V} = -B \frac{A \; \Delta s}{A \; \Delta x} = -B \frac{\Delta s}{\Delta x}$$

As Δx approaches zero, the ratio $\Delta s/\Delta x$ becomes $\partial s/\partial x$. (The partial derivative is used here to indicate that we are interested in the variation of s with position at a *fixed* time.) Therefore,

$$\Delta P = -B \frac{\partial s}{\partial x}$$

If the displacement is the simple sinusoidal function given by Equation 17.2, we find that

$$\Delta P = -B \frac{\partial}{\partial x} [s_{max} \cos(kx - \omega t)] = B s_{max} k \sin(kx - \omega t)$$

Since the bulk modulus is given by $B = \rho v^2$ (Eq. 17.1), the pressure variation reduces to

$$\Delta P = \rho v^2 s_{max} k \; \sin(kx - \omega t)$$

Furthermore, from Equation 16.13, we can write $k = \omega/v$; hence ΔP can be expressed as

$$\Delta P = \rho \omega s_{max} v \; \sin(kx - \omega t)$$

Taking the maximum value of each side, we find

$$\Delta P_{max} = \rho v \omega s_{max}$$

which is Equation 17.4. Then, with this substitution, we arrive at Equation 17.3:

$$\Delta P = \Delta P_{max} \sin(kx - \omega t)$$

17.3   INTENSITY OF PERIODIC SOUND WAVES

In the previous chapter, we showed that a wave traveling on a taut string transports energy. The same concepts are now applied to sound waves. Consider a layer of air of mass Δm and width Δx in front of a piston oscillating with angular frequency ω,

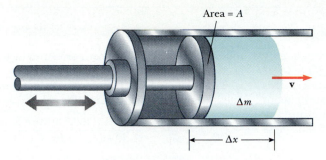

FIGURE 17.4   An oscillating piston transfers energy to the gas in the tube, causing the layer of width Δx and mass Δm to oscillate with an amplitude s_{max}.

as in Figure 17.4. The piston transmits energy to the layer of air.[1] Since the average kinetic energy equals the average potential energy in simple harmonic motion (Chapter 13), the average total energy of the mass Δm equals its maximum kinetic energy. Therefore, we can express the average energy of the moving layer of air as

$$\Delta E = \tfrac{1}{2}\Delta m(\omega s_{max})^2 = \tfrac{1}{2}(\rho A\,\Delta x)(\omega s_{max})^2$$

where $A\,\Delta x$ is the volume of the layer. The time rate at which energy is transferred to each layer—in other words the power—is

$$\text{Power} = \frac{\Delta E}{\Delta t} = \tfrac{1}{2}\rho A\left(\frac{\Delta x}{\Delta t}\right)(\omega s_{max})^2 = \tfrac{1}{2}\rho Av(\omega s_{max})^2$$

where $v = \Delta x/\Delta t$ is the speed of the disturbance to the right.

> We define the intensity I of a wave, or the power per unit area, to be the rate at which the energy being transported by the wave flows through a unit area A perpendicular to the direction of travel of the wave.

In our present case, therefore, the intensity is

$$I = \frac{\text{power}}{\text{area}} = \tfrac{1}{2}\rho(\omega s_{max})^2 v \tag{17.5}$$

Thus, we see that the intensity of a periodic sound wave is proportional to the square of the amplitude and to the square of the frequency (as in the case of a periodic string wave). This can also be written in terms of the pressure amplitude ΔP_{max}, using Equation 17.4, which gives

$$I = \frac{\Delta P_{max}^2}{2\rho v} \tag{17.6}$$

Intensity of a sound wave

---

[1] Although it is not proved here, the work done by the piston equals the energy carried away by the wave. For a detailed mathematical treatment of this concept, see Frank S. Crawford, Jr., *Waves,* New York, McGraw-Hill, 1968, Berkeley Physics Course, Volume 3, Chapter 4.

EXAMPLE 17.3   Hearing Limitations

The faintest sounds the human ear can detect at a frequency of 1000 Hz correspond to an intensity of about 1.00×10^{-12} W/m^2 (the so-called *threshold of hearing*). The loudest sounds that the ear can tolerate correspond to an intensity of about 1.00 W/m^2 *(the threshold of pain).* Determine the pressure amplitudes and maximum displacements associated with these two limits.

Solution   First, consider the faintest sounds. Using Equation 17.6 and taking $v = 343$ m/s to be the speed of sound waves in air and the density of air to be $\rho = 1.29$ kg/m^3, we get

$$\Delta P_{max} = \sqrt{2\rho v I}$$
$$= \sqrt{2(1.29 \text{ kg/m}^3)(343 \text{ m/s})(1.00 \times 10^{-12} \text{ W/m}^2)}$$
$$= 2.97 \times 10^{-5} \text{ N/m}^2$$

Since atmospheric pressure is about 10^5 N/m^2, this result tells us that the ear can discern pressure fluctuations as small as 3 parts in 10^{10}!

The corresponding maximum displacement can be calculated using Equation 17.4, recalling that $\omega = 2\pi f$:

$$s_{max} = \frac{\Delta P_{max}}{\rho \omega v} = \frac{2.97 \times 10^{-5} \text{ N/m}^2}{(1.29 \text{ kg/m}^3)(2\pi \times 10^3 \text{ s}^{-1})(343 \text{ m/s})}$$
$$= 1.07 \times 10^{-11} \text{ m}$$

This is a remarkably small number! If we compare this result for s_{max} with the diameter of an atom (about 10^{-10} m), we see that the ear is an extremely sensitive detector of sound waves.

In a similar manner, one finds that the loudest sounds the human ear can tolerate correspond to a pressure amplitude of about 30 N/m^2 and a maximum displacement of about 1.1×10^{-5} m.

The pressure amplitudes, called acoustic pressure, correspond to fluctuations taking place above and below atmospheric pressure.

Sound Level in Decibels

The previous example illustrates the wide range of intensities the human ear can detect. Because this range is so wide, it is convenient to use a logarithmic scale, where the **sound level** β is defined by the equation

$$\beta \equiv 10 \log \left(\frac{I}{I_0} \right) \qquad (17.7)$$

Sound level in decibels

The constant I_0 is the *reference intensity,* taken to be at the threshold of hearing ($I_0 = 1.00 \times 10^{-12}$ W/m^2), and I is the intensity in watts per square meter at the sound level β, where β is measured in decibels (dB).[2] On this scale, the threshold of pain ($I = 1.00$ W/m^2) corresponds to a sound level of $\beta = 10 \log(1/10^{-12}) = 10 \log(10^{12}) = 120$ dB, and the threshold of hearing corresponds to a sound level $\beta = 10 \log(1/1) = 0$ dB.

Prolonged exposure to high sound levels may produce serious damage to the ear. Ear plugs are recommended whenever sound levels exceed 90 dB. Recent evidence also suggests that "noise pollution" may be a contributing factor to high blood pressure, anxiety, and nervousness. Table 17.2 gives some typical values of the sound levels of various sources.

17.4   SPHERICAL AND PLANE WAVES

If a spherical body oscillates so that its radius varies sinusoidally with time, a spherical sound wave is produced (Fig. 17.5). The wave moves outward from the source at a constant speed if the medium is uniform.

TABLE 17.2   Sound Levels for Some Sources in Decibels

Source of Sound	β(dB)
Nearby jet airplane	150
Jackhammer; machine gun	130
Siren; rock concert	120
Subway; power mower	100
Busy traffic	80
Vacuum cleaner	70
Normal conversation	50
Mosquito buzzing	40
Whisper	30
Rustling leaves	10
Threshold of hearing	0

[2] The "bel" is named after the inventor of the telephone, Alexander Graham Bell (1847–1922). The prefix *deci-* is the metric system scale factor that stands for 10^{-1}.

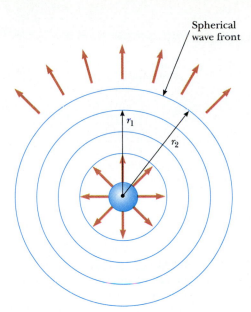

FIGURE 17.5 A spherical wave propagating radially outward from an oscillating spherical body. The intensity of the spherical wave varies as $1/r^2$.

Since all points on any given sphere behave in the same way, we conclude that the energy in a spherical wave propagates equally in all directions. That is, no one direction is preferred over any other. If P_{av} is the average power emitted by the source, then this power at any distance r from the source must be distributed over a spherical surface of area $4\pi r^2$. Hence, the wave intensity at a distance r from the source is

$$I = \frac{P_{av}}{A} = \frac{P_{av}}{4\pi r^2} \tag{17.8}$$

Since P_{av} is the same throughout any spherical surface centered at the source, we see that the intensities at distances r_1 and r_2 are

$$I_1 = \frac{P_{av}}{4\pi r_1^2} \quad \text{and} \quad I_2 = \frac{P_{av}}{4\pi r_2^2}$$

Therefore, the ratio of intensities on these two spherical surfaces is

$$\frac{I_1}{I_2} = \frac{r_2^2}{r_1^2}$$

In Equation 17.5 we found that the intensity is proportional to s_{max}^2, the square of the wave displacement amplitude. Comparing this result with Equation 17.8, we conclude that the displacement amplitude of a spherical wave must vary as $1/r$. Therefore, we can write the wave function ψ (Greek letter "psi") for an outgoing spherical wave in the form

$$\psi(r, t) = \frac{s_0}{r} \sin(kr - \omega t) \tag{17.9}$$

where s_0, the displacement amplitude at $t = 0$, is a constant.

It is useful to represent spherical waves by a series of circular arcs concentric with the source, as in Figure 17.6. Each arc represents a surface over which the phase of the wave is constant. We call such a surface of constant phase a **wave front**.

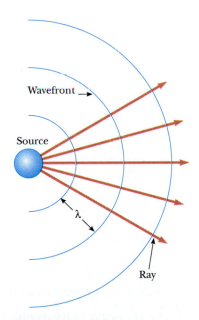

FIGURE 17.6 Spherical waves emitted by a point source. The circular arcs represent the spherical wave fronts concentric with the source. The rays are radial lines pointing outward from the source perpendicular to the wave fronts.

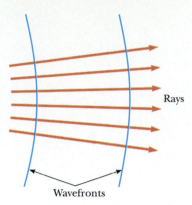

FIGURE 17.7 Far away from a point source, the wave fronts are nearly parallel planes and the rays are nearly parallel lines perpendicular to the planes. Hence, a small segment of a spherical wave front is approximately a plane wave.

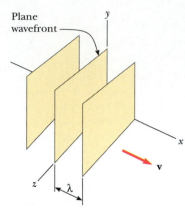

FIGURE 17.8 A representation of a plane wave moving in the positive *x* direction with a speed *v*. The wave fronts are planes parallel to the *yz* plane.

The distance between adjacent wave fronts equals the wavelength, λ. The radial lines pointing outward from the source are called **rays.**

Now consider a small portion of the wave fronts far from the source, as in Figure 17.7. In this case, the rays are nearly parallel to each other and the wave fronts are very close to being planar. Therefore, at distances from the source that are large compared with the wavelength, we can approximate the wave fronts by parallel planes. We call such a wave a **plane wave.** Any small portion of a spherical wave that is far from the source can be considered a plane wave.

Figure 17.8 illustrates a plane wave propagating along the *x* axis, which means the wave fronts are parallel to the *yz* plane. In this case, the wave function depends only on *x* and *t* and has the form

Plane wave representation

$$\psi(x,\ t) = A \sin(kx - \omega t) \tag{17.10}$$

That is, the wave function for a plane wave is identical in form to that of a one-dimensional traveling wave. The intensity is the same all over any given wave front of the plane wave.

EXAMPLE 17.4   Intensity Variations of a Point Source

A point source emits sound waves with an average power output of 80.0 W. (a) Find the intensity 3.00 m from the source.

Reasoning and Solution   A point source emits energy in the form of spherical waves (Fig. 17.5). At a distance *r* from the source, the power is distributed over the surface area of a sphere, $4\pi r^2$. Therefore, the intensity at a distance *r* from the source is given by Equation 17.8:

$$I = \frac{P_{av}}{4\pi r^2} = \frac{80.0\ \text{W}}{4\pi(3.00\ \text{m})^2} = \boxed{0.707\ \text{W/m}^2}$$

which is close to the threshold of pain.

(b) Find the distance at which the sound reduces to a level of 40 dB.

Solution We can find the intensity at the 40-dB level by using Equation 17.7 with $I_0 = 1.00 \times 10^{-12}$ W/m^2:

$$10 \log\left(\frac{I}{I_0}\right) = 40$$

$$I = 1.00 \times 10^4 \, I_0 = 1.00 \times 10^{-8} \text{ W/m}^2$$

Using this value for I in Equation 17.8 and solving for r, we get

$$r = \sqrt{\frac{P_{av}}{4\pi I}} = \sqrt{\frac{80.0 \text{ W}}{4\pi \times 1.00 \times 10^{-8} \text{ W/m}^2}}$$

$$= \boxed{2.52 \times 10^4 \text{ m}}$$

which equals about 16 miles!

*17.5 THE DOPPLER EFFECT

When a car or truck is moving while its horn is blowing, the frequency of the sound you hear is higher as the vehicle approaches you and lower as it moves away from you. This is one example of the **Doppler effect**.[3]

> In general, a Doppler effect is experienced whenever there is relative motion between source and observer. When the source and observer are moving toward each other, the frequency heard by the observer is *higher* than the frequency of the source. When the source and observer are moving away from each other, the frequency heard by the observer is *lower* than the source frequency.

"I love hearing that lonesome wail of the train whistle as the magnitude of the frequency of the wave changes due to the Doppler effect."

Although the Doppler effect is most commonly experienced with sound waves, it is a phenomenon common to all harmonic waves. For example, there is a shift in frequencies of light waves (electromagnetic waves) produced by the relative motion of source and observer. The Doppler effect is used in police radar systems to measure the speed of motor vehicles. Likewise, astronomers use the effect to determine the relative motion of stars, galaxies, and other celestial objects.

First, let us consider the case where the observer O is moving and the sound source S is stationary. For simplicity, we shall assume that the air is also stationary and that the observer moves directly toward the source. Figure 17.9 describes the situation when the observer moves with a speed v_O toward the source (considered as a point source), which is at rest ($v_S = 0$). In general, "at rest" means at rest with respect to the medium, air.

We shall take the frequency of the source to be f, the wavelength to be λ, and the speed of sound to be v. If the observer were also stationary, clearly he or she would detect f wave fronts per second. (That is, when $v_O = 0$ and $v_S = 0$, the observed frequency equals the source frequency.) When the observer moves toward the source, the speed of the waves relative to the observer is $v' = v + v_O$, but the wavelength λ is unchanged. Hence, the frequency heard by the observer is *increased* and given by

$$f' = \frac{v'}{\lambda} = \frac{v + v_O}{\lambda}$$

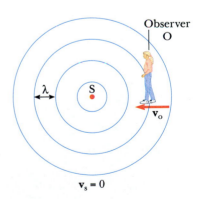

FIGURE 17.9 An observer O moving with a speed v_O toward a stationary point source S hears a frequency f' that is greater than the source frequency.

[3] Named after the Austrian physicist Christian Johann Doppler (1803–1853), who first suggested that the change in frequency observed for sound waves could also be applicable to light waves.

Since $\lambda = v/f$, we can express f' as

$$f' = f\left(1 + \frac{v_O}{v}\right) \qquad \text{(Observer moving toward the source)} \qquad \textbf{(17.11)}$$

Similarly, if the observer is moving *away* from the source, the speed of the wave relative to the observer is $v' = v - v_O$. The frequency heard by the observer in this case is *lowered* and is

$$f' = f\left(1 - \frac{v_O}{v}\right) \qquad \text{(Observer moving away from the source)} \qquad \textbf{(17.12)}$$

In general, when an observer moves with a speed v_O relative to a stationary source, the frequency heard by the observer is

$$f' = f\left(1 \pm \frac{v_O}{v}\right) \qquad \textbf{(17.13)}$$

where the *positive* sign is used when the observer moves *toward* the source and the *negative* sign holds when the observer moves *away* from the source.

Now consider the situation in which the source is in motion and the observer is at rest. If the source moves directly toward observer A in Figure 17.10a, the wave fronts seen by the observer are closer together as a result of the motion of the source in the direction of the outgoing wave. As a result, the wavelength λ' measured by observer A is shorter than the wavelength λ of the source. During each vibration, which lasts for a time T (the period), the source moves a distance $v_S T = v_S/f$ and the wave length is *shortened* by this amount. Therefore, the observed wavelength λ' is

$$\lambda' = \lambda - \Delta\lambda = \lambda - \frac{v_S}{f}$$

Since $\lambda = v/f$, the frequency heard by observer A is

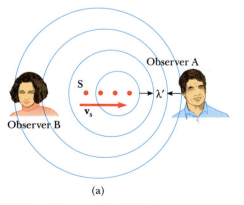

(a)

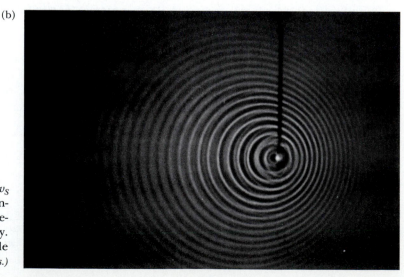

(b)

FIGURE 17.10   (a) A source S moving with a speed v_S toward a stationary observer A and away from a stationary observer B. Observer A hears an increased frequency, and observer B hears a decreased frequency. (b) The Doppler effect in water observed in a ripple tank.   *(Courtesy Educational Development Center, Newton, Mass.)*

$$f' = \frac{v}{\lambda'} = \frac{v}{\lambda - \dfrac{v_S}{f}} = \frac{v}{\dfrac{v}{f} - \dfrac{v_S}{f}}$$

$$f' = f\left(\frac{1}{1 - \dfrac{v_S}{v}}\right) \tag{17.14}$$

That is, the observed frequency is *increased* when the source moves toward the observer.

In a similar manner, when the source moves away from an observer B at rest (where observer B is to the left of the source, as in Fig. 17.10a), observer B measures a wavelength λ' that is *greater* than λ and hears a *decreased* frequency

$$f' = f\left(\frac{1}{1 + \dfrac{v_S}{v}}\right) \tag{17.15}$$

Combining Equations 17.14 and 17.15, we can express the general relationship for the observed frequency when the source is moving and the observer is at rest as

$$f' = f\left(\frac{1}{1 \mp \dfrac{v_S}{v}}\right) \tag{17.16}$$

Frequency heard with source in motion

Finally, if both the source and the observer are in motion, we find the following general relationship for the observed frequency:

$$f' = f\left(\frac{v \pm v_O}{v \mp v_S}\right) \tag{17.17}$$

Frequency heard with observer and source in motion

In this expression, the *upper* signs ($+ v_O$ and $- v_S$) refer to motion of one *toward* the other, and the lower signs ($- v_O$ and $+ v_S$) refer to motion of one *away from* the other.

A convenient rule to remember concerning signs when working with all Doppler effect problems is the following:

The word *toward* is associated with an *increase* in the observed frequency. The words *away from* are associated with a *decrease* in the observed frequency.

EXAMPLE 17.5 The Moving Train Whistle

A train moving at a speed of 40 m/s sounds its whistle, which has a frequency of 500 Hz. Determine the frequencies heard by a stationary observer as the train approaches and then recedes from the observer.

Solution We can use Equation 17.14 to get the apparent frequency as the train approaches the observer. Taking $v = 343$ m/s for the speed of sound in air gives

$$f' = f\left(\frac{1}{1 - \dfrac{v_S}{v}}\right) = (500 \text{ Hz})\left(\frac{1}{1 - \dfrac{40 \text{ m/s}}{343 \text{ m/s}}}\right) = \boxed{566 \text{ Hz}}$$

Likewise, Equation 17.15 can be used to obtain the frequency

heard as the train recedes from the observer:

$$f' = f\left(\frac{1}{1 + \dfrac{v_S}{v}}\right) = (500 \text{ Hz})\left(\frac{1}{1 + \dfrac{40 \text{ m/s}}{343 \text{ m/s}}}\right) = \boxed{448 \text{ Hz}}$$

EXAMPLE 17.6   The Noisy Siren

An ambulance travels down a highway at a speed of 33.5 m/s (75 mi/h). Its siren emits sound at a frequency of 400 Hz. What is the frequency heard by a passenger in a car traveling at 24.6 m/s (55 mi/h) in the opposite direction as the car approaches the ambulance and as the car moves away from the ambulance?

Solution   Let us take the speed of sound in air to be $v = 343$ m/s. We can use Equation 17.17 in both cases. As the ambulance and car approach each other, the observed apparent frequency is

$$f' = f\left(\frac{v + v_O}{v - v_S}\right) = (400 \text{ Hz})\left(\frac{343 \text{ m/s} + 24.6 \text{ m/s}}{343 \text{ m/s} - 33.5 \text{ m/s}}\right)$$

$$= \boxed{475 \text{ Hz}}$$

Likewise, as they recede from each other, a passenger in the

car hears a frequency

$$f' = f\left(\frac{v - v_O}{v + v_S}\right) = (400 \text{ Hz})\left(\frac{343 \text{ m/s} - 24.6 \text{ m/s}}{343 \text{ m/s} + 33.5 \text{ m/s}}\right)$$

$$= \boxed{338 \text{ Hz}}$$

The *change* in frequency as detected by the passenger in the car is $475 - 338 = 137$ Hz, which is more than 30% of the actual frequency emitted.

Exercise   Suppose that the passenger car is parked on the side of the highway as the ambulance travels down the highway at the speed of 33.5 m/s. What frequency will the passenger in the car hear as the ambulance (a) approaches the parked car and (b) recedes from the parked car?

Answer   (a) 443 Hz; (b) 364 Hz.

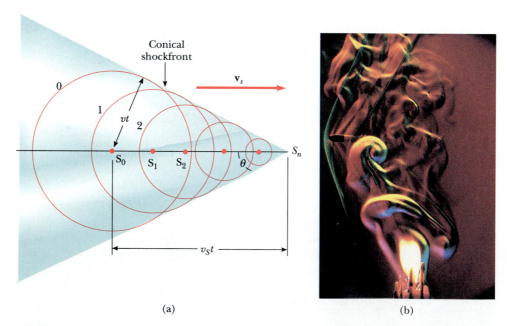

(a)   (b)

FIGURE 17.11   (a) A representation of a shock wave produced when a source moves from S_0 to S_n with a speed v_S, which is greater than the wave speed v in that medium. The envelope of the wavefronts forms a cone whose apex half-angle is given by $\sin \theta = v/v_S$. (b) A stroboscopic photograph of a bullet moving at supersonic speed through the hot air above a candle. Note the shock wave in the vicinity of the bullet.   *(©The Harold E. Edgerton 1992 Trust. Courtesy of Palm Press, Inc.)*

Shock Waves

Now let us consider what happens when the source speed v_S *exceeds* the wave speed v. This situation is described graphically in Figure 17.11. The circles represent spherical wave fronts emitted by the source at various times during its motion. At $t = 0$, the source is at S_0, and at some later time t, the source is at S_n. In the time t, the wave front centered at S_0 reaches a radius of vt. In this same interval, the source travels a distance $v_S t$ to S_n. At the instant the source is at S_n, waves are just beginning to be generated and so the wave front has zero radius at this point. The line drawn from S_n to the wave front centered on S_0 is tangent to all other wave fronts generated at intermediate times. Thus, we see that the envelope of these waves is a cone whose apex half-angle θ is

$$\sin \theta = \frac{v}{v_S}$$

The ratio v_S/v is referred to as the *Mach number*. The conical wave front produced when $v_S > v$ (supersonic speeds) is known as a *shock wave*. An interesting analogy to shock waves is the V-shaped wave fronts produced by a duck (the bow wave) when the duck's speed exceeds the speed of the surface water waves.

Jet airplanes traveling at supersonic speeds produce shock waves, which are responsible for the loud explosion, or "sonic boom," one hears. The shock wave carries a great deal of energy concentrated on the surface of the cone, with correspondingly large pressure variations. Such shock waves are unpleasant to hear and can cause damage to buildings when aircraft fly supersonically at low altitudes. In fact, an airplane flying at supersonic speeds produces a double boom because two shock fronts are formed, one from the nose of the plane and one from the tail (Fig. 17.12).

The V-shaped wavefront that follows the duck occurs because the duck travels at a speed greater than the speed of water waves. This is analogous to shock waves produced by airplanes traveling at supersonic speeds. *(© Harry Engels)*

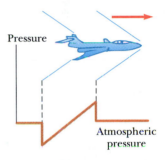

FIGURE 17.12 Two shock waves produced by the nose and tail of a jet airplane traveling at supersonic speeds.

SUMMARY

Sound waves are longitudinal and travel through a compressible medium with a speed that depends on the compressibility and inertia of that medium. The **speed of sound** in a medium having a bulk modulus B and density ρ is

$$v = \sqrt{\frac{B}{\rho}} \tag{17.1}$$

In the case of sinusoidal sound waves, the **variation in pressure** from the equilibrium value is given by

$$\Delta P = \Delta P_{\max} \sin(kx - \omega t) \tag{17.3}$$

where $\Delta P_{\max}$ is the **pressure amplitude**. The pressure wave is $90°$ out of phase with the displacement wave. If the displacement amplitude is $s_{\max}$, then $\Delta P_{\max}$ has the value

$$\Delta P_{\max} = \rho v \omega s_{\max} \tag{17.4}$$

The **intensity of a harmonic sound wave**, which is the power per unit area, is

$$I = \tfrac{1}{2}\rho(\omega s_{\max})^2 v = \frac{\Delta P_{\max}^2}{2\rho v} \tag{17.5, 17.6}$$

The **intensity of a spherical wave** produced by a point source is proportional to the average power emitted and inversely proportional to the square of the distance from the source.

The change in frequency heard by an observer whenever there is relative motion between the source and observer is called the **Doppler effect**. If the *observer moves* with a speed v_O and the source is at rest, the observed frequency f' is

$$f' = f\left(1 \pm \frac{v_O}{v}\right) \qquad (17.13)$$

where the positive sign is used when the observer moves toward the source and the negative sign refers to motion away from the source.

If the *source moves* with a speed v_S and the observer is at rest, the observed frequency is

$$f' = f\left(\frac{1}{1 \mp \dfrac{v_S}{v}}\right) \qquad (17.16)$$

where $- v_S$ refers to motion *toward* the observer and $+ v_S$ refers to motion *away* from the observer.

When the *observer and source are both moving*, the observed frequency is

$$f' = f\left(\frac{v \pm v_O}{v \mp v_S}\right) \qquad (17.17)$$

QUESTIONS

1. Why are sound waves characterized as longitudinal?
2. As a result of a distant explosion, an observer senses a ground tremor and then hears the explosion. Explain.
3. Some sound waves are harmonic, whereas others are not. Give an example of each.
4. If the distance from a point source is tripled, by what factor does the intensity decrease?
5. Explain how the Doppler effect is used with microwaves to determine the speed of an automobile.
6. If you are in a moving vehicle, explain what happens to the frequency of your echo as you move *toward* a canyon wall. What happens to the frequency as you move *away* from the wall?
7. Suppose an observer and a source of sound are both at rest and a strong wind blows toward the observer. Describe the effect of the wind (if any) on (a) the observed wavelength, (b) the observed frequency, and (c) the wave velocity.
8. Of the following sounds, which is most likely to have an intensity level of 60 dB: a rock concert, the turning of a page in this text, normal conversation, a cheering crowd at a football game, or background noise at a church?
9. Estimate the decibel level of each of the sounds in Question 10.
10. A binary star system consists of two stars revolving about each other. If we observe the light reaching us from one of these stars as it makes one complete revolution about

the other, what does the Doppler effect predict will happen to this light?
11. How could an object move with respect to an observer such that the sound from it is not shifted in frequency?
12. Why is it not possible to use sonar (sound waves) to determine the speed of an object traveling faster than the speed of sound in that medium?
13. Why is it so quiet after a snowfall?
14. Why is the intensity of an echo less than that of the original sound?
15. If the wavelength of a sound source is reduced by a factor of 2, what happens to its frequency? Its speed?
16. A sound wave travels in air at a frequency of 500 Hz. If part of the wave travels from the air into water, does its frequency change? Does its wavelength change? Justify your answers.
17. In a recent discovery, a nearby star was found to have a large planet orbiting about it, although the planet could not be seen. In terms of the concept of systems rotating about their center of mass and the Doppler shift for light (which is in many ways similar to that of sound), explain how an astronomer could determine the presence of the invisible planet.
18. Explain how the distance to a lightning bolt may be determined by counting the seconds between the flash and the sound of the thunder. Does the speed of the light signal have to be taken into account?

PROBLEMS

1. Suppose that you hear a thunder clap 16.2 s after seeing the associated lightning stroke. The speed of sound waves in air is 343 m/s and the speed of light in air is 3.0×10^8 m/s. How far are you from the lightning stroke?

2. A stone is dropped into a deep canyon and is heard to strike the bottom 10.2 s after release. The speed of sound waves in air is 343 m/s. How deep is the canyon? What would be the percentage error in the depth if the time required for the sound to reach the canyon rim were ignored?

2A. A stone is dropped into a deep canyon and is heard to strike the bottom t seconds after release. The speed of sound waves in air is v. How deep is the canyon? What would be the percentage error in the depth if the time required for the sound to reach the canyon rim were ignored?

3. Find the speed of sound in mercury, which has a bulk modulus of approximately 2.8×10^{10} N/m² and a density of 13 600 kg/m³.

4. A flower pot is knocked off a balcony 20.0 m above the sidewalk and is heading for a 1.75-m-tall man standing below. How high from the ground can the flower pot be after which it would be too late for a shouted warning to reach the man in time? Assume that the man below requires 0.300 s to respond to the warning.

5. The speed of sound in air is $v = \sqrt{\gamma P/\rho}$, where γ is a constant equal to 7/5, P is the air pressure, and ρ is the density of air. Calculate the speed of sound for $P = 1$ atm $= 1.013 \times 10^5$ Pa and $\rho = 1.29$ kg/m³.

Section 17.2   Periodic Sound Waves

(*Note:* In this section, use the following values as needed unless otherwise specified: the equilibrium density of air, $\rho = 1.29$ kg/m³; the speed of sound in air, $v = 343$ m/s. Also, pressure variations ΔP are measured relative to atmospheric pressure.)

6. The density of aluminum is 2.7×10^3 kg/m³. Use the value for the speed of sound in aluminum given in Table 17.1 to calculate Young's modulus for this material.

7. You are watching a pier being constructed on the far shore of a saltwater inlet when some blasting occurs. You hear the sound in the water 4.5 s before it reaches you through the air. How wide is the inlet? (*Hint:* See Table 17.1. Assume the air temperature is 20°C.)

8. A rescue plane flies horizontally at a constant speed searching for a disabled boat. When the plane is di-

rectly above the boat, the boat's crew blows a loud horn. By the time the plane's sound detector perceives the horn's sound, the plane has traveled a distance equal to one-half its altitude above the ocean. If it takes the sound 2.0 s to reach the plane, determine (a) the speed of the plane and (b) its altitude. Take the speed of sound to be 343 m/s.

9. The speed of sound in air (in m/s) depends on temperature according to the expression

$$v = 331.5 + 0.607T_C$$

where T_C is the Celsius temperature. In dry air the temperature decreases about 1°C for every 150-m rise in altitude. (a) Assuming this change is constant up to an altitude of 9000 m, how long will it take the sound from an airplane flying at 9000 m to reach the ground on a day when the ground temperature is 30°C? (b) Compare this to the time it would take if the air were a constant 30°C. Which time is longer?

10. Calculate the pressure amplitude of a 2.0-kHz sound wave in air if the displacement amplitude is equal to 2.0×10^{-8} m.

11. A sound wave in air has a pressure amplitude equal to 4.0×10^{-3} Pa. Calculate the displacement amplitude of the wave at a frequency of 10.0 kHz.

12. A sound wave in a cylinder is described by Equations 17.2 through 17.4. Show that $\Delta P = \pm \rho v \omega \sqrt{s_{max}^2 - s^2}$.

13. An experimenter wishes to generate in air a sound wave that has a displacement amplitude equal to 5.5×10^{-6} m. The pressure amplitude is to be limited to 8.4×10^{-1} Pa. What is the minimum wavelength the sound wave can have?

14. A sound wave in air has a pressure amplitude of 4.0 Pa and a frequency of 5.0 kHz. $\Delta P = 0$ at the point $x = 0$ when $t = 0$. (a) What is ΔP at $x = 0$ when $t = 2.0 \times 10^{-4}$ s, and (b) what is ΔP at $x = 0.020$ m when $t = 0$?

15. A sinusoidal sound wave is described by the displacement

$$s(x, t) = (2.00 \ \mu\text{m}) \cos[(15.7 \ \text{m}^{-1})x - (858 \ \text{s}^{-1})t]$$

(a) Find the amplitude, wavelength, and speed of this wave and state what material this sound wave is traveling through. (See Table 17.1.) (b) Determine the instantaneous displacement of the molecules at the position $x = 0.0500$ m at $t = 3.00$ ms. (c) Determine the maximum speed of the molecules' oscillatory motion.

16. The tensile stress in a copper rod is 99.5% of its elastic breaking point of 13×10^{10} N/m². If a 500-Hz sound wave is transmitted along the rod, (a) what displacement amplitude will cause the rod to break

□ indicates problems that have full solutions available in the Student Solutions Manual and Study Guide.

and (b) what is the maximum speed of the particles at this moment?

17. Write an expression that describes the pressure variation as a function of position and time for a sinusoidal sound wave in air if $\lambda = 0.10$ m and $\Delta P_{max} = 0.20$ Pa.

18. Write the function that describes the displacement wave corresponding to the pressure wave in Problem 17.

Section 17.3   Intensity of Periodic Sound Waves

19. Calculate the sound level in dB of a sound wave that has an intensity of 4.0 μW/m².

20. A vacuum cleaner has a measured sound level of 70 dB. What is the intensity of this sound in W/m²?

21. Show that the difference in decibel levels, β_1 and β_2, of a sound source is related to the ratio of its distances, r_1 and r_2, from the receivers by

$$\beta_2 - \beta_1 = 20 \log\left(\frac{r_1}{r_2}\right)$$

22. The intensity of a sound wave at a fixed distance from a speaker vibrating at 1.00 kHz is 0.600 W/m². (a) Determine the intensity if the frequency is increased to 2.50 kHz while a constant displacement amplitude is maintained. (b) Calculate the intensity if the frequency is reduced to 0.500 kHz and the displacement amplitude is doubled.

22A. The intensity of a sound wave at a fixed distance from a speaker vibrating at a frequency f is I. (a) Determine the intensity if the frequency is increased to f' while a constant displacement amplitude is maintained. (b) Calculate the intensity if the frequency is reduced to $f/2$ and the displacement amplitude is doubled.

23. A speaker is placed between two observers who are 110 m apart, along the line connecting them. If one observer records an intensity level of 60 dB, and the other records an intensity level of 80 dB, how far is the speaker from each observer?

24. An explosive charge is detonated at a height of several kilometers in the atmosphere. At a distance of 400 m from the explosion the acoustic pressure reaches a maximum of 10 Pa. Assuming that the atmosphere is homogeneous over the distances considered, what will be the sound level (in dB) at 4 km from the explosion? (Sound waves in air are absorbed at a rate of approximately 7 dB/km.)

25. Two small speakers emit sound waves of different frequencies. Speaker A has an output of 1.0 mW and speaker B has an output of 1.5 mW. Determine the sound intensity level (in dB) at point C (Fig. P17.25) if (a) only speaker A emits sound, (b) only speaker B emits sound, (c) both speakers emit sound.

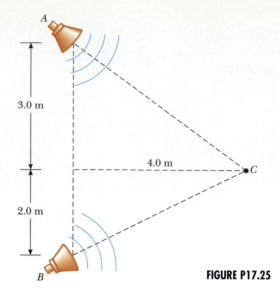

FIGURE P17.25

26. Two sources have sound levels of 75 dB and 80 dB. If they are sounding simultaneously, (a) what is the combined sound level? (b) What is their combined intensity in W/m²?

Section 17.4   Spherical and Plane Waves

27. An experiment requires a sound intensity of 1.2 W/m² at a distance of 4 m from a speaker. What power output is required?

28. A source of sound (1000 Hz) emits uniformly in all directions. An observer 3.0 m from the source measures a sound level of 40 dB. Calculate the average power output of the source.

29. The sound level at a distance of 3.0 m from a source is 120 dB. At what distance will the sound level be (a) 100 dB and (b) 10 dB?

30. A fireworks rocket explodes at a height of 100 m above the ground. An observer on the ground directly under the explosion experiences an average sound intensity of 7.0×10^{-2} W/m² for 0.20 s. (a) What is the total sound energy of the explosion? (b) What is the sound level in decibels heard by the observer?

30A. A fireworks rocket explodes at a height h above the ground. An observer on the ground directly under the explosion experiences an average sound intensity I for a time t. (a) What is the total sound energy of the explosion? (b) What is the sound level heard by the observer?

31. A rock group is playing in a studio. Sound emerging from an open door spreads uniformly in all directions. If the sound level of the music is 80.0 dB at a distance of 5.0 m from the door, at what distance is the music just barely audible to a person with a normal threshold of hearing (0 dB)? Disregard absorption.

32. A spherical wave is radiating from a point source and is described by the following:

$$y(r, t) = \left(\frac{25.0}{r}\right) \sin(1.25r - 1870t)$$

where y is in pascals, r in meters, and t in seconds. (a) What is the maximum pressure amplitude 4.00 m from the source? (b) Determine the speed of the wave and hence the material the wave is in. (c) Find the intensity of the wave in dB at a distance 4.00 m from the source. (d) Find the instantaneous pressure 5.00 m from the source at 0.0800 s.

*Section 17.5 The Doppler Effect

33. A bullet fired from a rifle travels at Mach 1.38 (that is, $v_S/v = 1.38$). What angle does the shock front make with the path of the bullet?

34. A block with a speaker bolted to it is connected to a spring having spring constant $k = 20.0$ N/m as in Figure P17.34. The total mass of the block and speaker is 5.00 kg, and the amplitude of this unit's motion is 0.500 m. (a) If the speaker emits sound waves of frequency 440 Hz, determine the range in frequencies heard by the person to the right of the speaker. (b) If the maximum intensity level heard by the person is 60 dB when he is closest to the speaker, 1.00 m away, what is the minimum intensity level heard by the observer? Assume that the speed of sound is 343 m/s.

34A. A block with a speaker bolted to it is connected to a spring having spring constant k as in Figure P17.34. The total mass of the block and speaker is m, and the amplitude of this unit's motion is A. (a) If the speaker emits sound waves of frequency f, determine the range in frequencies heard by the person to the right of the speaker. (b) If the maximum intensity level heard by the person is β when he is closest to the speaker, a distance d away, what is the minimum intensity level heard by the observer?

FIGURE P17.34

35. A jet fighter plane travels in horizontal flight at Mach 1.2 (that is, 1.2 times the speed of sound in air). At the instant an observer on the ground hears the shock wave, what is the angle her line of sight makes with the horizontal as she looks at the plane?

36. A helicopter drops a paratrooper carrying a buzzer that emits a 500-Hz signal. A sound receiver on the plane monitors the signal as the paratrooper falls. If the perceived frequency becomes constant at 450 Hz, what is the terminal speed of the paratrooper? Take the speed of sound in air to be 343 m/s and assume the paratrooper always remains below the helicopter.

37. Standing at a crosswalk, you hear a frequency of 560 Hz from the siren on an approaching police car. After the police car passes, the observed frequency of the siren is 480 Hz. Determine the car's speed from these observations.

38. A fire engine moving to the right at 40 m/s sounds its horn (frequency 500 Hz) at the two vehicles shown in Figure P17.38. The car is moving to the right at 30 m/s, while the van is at rest. (a) What frequency is heard by the passengers in the car? (b) What is the frequency as heard by the passengers in the van? (c) When the fire engine is 200 m from the car and 250 m from the van, the passengers in the car hear a sound intensity level of 90 dB. At that moment, what intensity level is heard by the passengers in the van?

Fire Engine   Car   Van

FIGURE P17.38

39. A train is moving parallel to a highway at 20 m/s. A car is traveling in the same direction as the train at 40 m/s. The car horn sounds at 510 Hz and the train whistle sounds at 320 Hz. (a) When the car is behind the train what frequency does an occupant of the car observe for the train whistle? (b) When the car is in front of the train what frequency does a train passenger observe for the car horn just after passing?

40. A tuning fork vibrating at 512 Hz falls from rest and accelerates at 9.80 m/s². How far below the point of release is the tuning fork when waves of frequency 485 Hz reach the release point? Take the speed of sound in air to be 340 m/s.

41. When high-energy, charged particles move through a transparent medium with a speed greater than the speed of light in that medium, a shock wave, or bow wave, of light is produced. This phenomenon is

called the *Cerenkov effect* and can be observed in the vicinity of the core of a swimming-pool reactor due to high-speed electrons moving through the water. In a particular case, the Cerenkov radiation produces a wavefront with an apex half-angle of 53°. Calculate the speed of the electrons in the water. (The speed of light in water is 2.25×10^8 m/s.)

42. A driver traveling northbound on a highway is driving at a speed of 25 m/s. A police car driving southbound at a speed of 40 m/s approaches with its siren sounding at a base frequency of 2500 Hz. (a) What frequency is observed by the driver as the police car approaches? (b) What frequency is detected by the driver after the police car passes him? (c) Repeat parts (a) and (b) for the case when the police car is traveling northbound.

43. A supersonic jet traveling at Mach 3 at an altitude of 20 000 m is directly overhead at time $t = 0$ as in Figure P17.43. (a) How long will it be before one encounters the shock wave? (b) Where will the plane be when it is finally heard? (Assume the speed of sound in air is uniform at 335 m/s.)

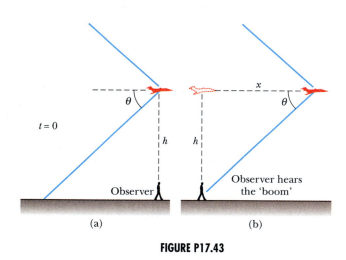

(a)   (b)

FIGURE P17.43

ADDITIONAL PROBLEMS

44. A copper rod is given a sharp compressional blow at one end. The sound of the blow, traveling through air at 0°C, reaches the opposite end of the rod 6.4 ms later than the sound transmitted through the rod. What is the length of the rod? (Refer to Table 17.1.)

45. An earthquake on the ocean floor in the Gulf of Alaska induces a *tsunami* (sometimes called a "tidal wave") that reaches Hilo, Hawaii, 4450 km distant, in a time of 9 h 30 min. Tsunamis have enormous wavelengths (100–200 km), and for such waves the propagation speed is $v \approx \sqrt{g\overline{d}}$, where $\overline{d}$ is the average depth of the water. From the information given, find the average wave speed and the average ocean depth between Alaska and Hawaii. (This method was used in 1856 to estimate the average depth of the Pacific Ocean long before soundings were made to give a direct determination.)

46. The power output of a certain stereo speaker is 6.0 W. (a) At what distance from the speaker would the sound be painful to the ear? (b) At what distance from the speaker would the sound be barely audible?

47. A jet flies toward higher altitude at a constant speed of 1963 m/s in a direction making an angle θ with the horizontal (Fig. P17.47). An observer on the ground hears the jet for the first time when it is directly overhead. Determine the value of θ if the speed of sound in air is 340.0 m/s.

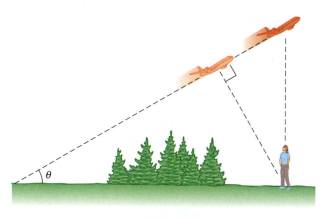

FIGURE P17.47

48. A microwave oven generates a sound level of 40.0 dB when consuming 1.00 kW of power. Estimate the fraction of this power that is converted into the energy of sound waves.

49. Two ships are moving along a line due east. The trailing vessel has a speed relative to a land-based observation point of 64 km/h, and the leading ship has a speed of 45 km/h relative to that point. The two ships are in a region of the ocean where the current is moving uniformly due west at 10 km/h. The trailing ship transmits a sonar signal at a frequency of 1200 Hz. What frequency is monitored by the leading ship? (Use 1520 m/s as the speed of sound in ocean water.)

49A. Two ships are moving along a line due east. The trailing vessel has a speed relative to a land-based observation point of v_1, and the leading ship has a speed $v_2 < v_1$ relative to that point. The two ships are in a region of the ocean where the current is moving uniformly due west at a speed v_c. The trailing ship transmits a sonar signal at a frequency f. What frequency is monitored by the leading ship? (Use v_S as the speed of sound in ocean water.)

50. Consider a longitudinal (compressional) wave of wavelength λ traveling with speed v along the x direction through a medium of density ρ. The *displacement* of the molecules of the medium from their equilibrium position is

$$s = s_{max} \sin(kx - \omega t)$$

Show that the pressure variation in the medium is

$$P = -\left(\frac{2\pi\rho v^2}{\lambda} s_{max}\right)\cos(kx - \omega t)$$

51. A meteoroid the size of a truck enters the Earth's atmosphere at a speed of 20 km/s and is not significantly slowed before entering the ocean. (a) What is the Mach angle of the shock wave from the meteoroid in the atmosphere? (Use 331 m/s as the sound speed.) (b) Assuming that the meteoroid survives the impact with the ocean surface, what is the (initial) Mach angle of the shock wave that the meteoroid produces in the water? (Use the wave speed for seawater given in Table 17.1.)

52. In the afternoon, the sound level of a busy freeway is 80 dB with 100 cars passing a given point every minute. Late at night, the traffic flow is only five cars per minute. What is the late-night sound level?

53. By proper excitation, it is possible to produce both longitudinal and transverse waves in a long metal rod. A particular metal rod is 150 cm long and has a radius of 0.20 cm and a mass of 50.9 g. Young's modulus for the material is 6.8×10^{10} N/m². What must the tension (or compression) in the rod be if the ratio of the speed of longitudinal waves to the speed of transverse waves is 8?

54. An earthquake emits both P waves and S waves that travel at different speeds through the Earth. A P wave travels at a speed of 9000 m/s and an S wave travels at 5000 m/s. If P waves are received at a seismic station 1 minute before an S wave arrives, how far away is the earthquake center?

55. A siren creates a sound level of 60.00 dB at 500.0 m from the speaker. The siren is powered by a battery that delivers a total energy of 1.00 kJ. Assuming that the efficiency of the siren is 30% (i.e., 30% of the supplied energy is transformed into sound energy), determine the total time the siren can sound.

55A. A siren creates a sound level β at a distance d from the speaker. The siren is powered by a battery that delivers a total energy E. Assuming that the efficiency of the siren is 30% (i.e., 30% of the supplied energy is transformed into sound energy), determine the total time the siren can sound.

56. The Doppler equation presented in the text is valid when the motion between the observer and the source occurs on a straight line, so that the source and observer are moving either directly toward or directly away from each other. If this restriction is relaxed, one must use the more general Doppler equation

$$f' = \left(\frac{v + v_O \cos\theta_O}{v - v_S \cos\theta_S}\right)f$$

where θ_O and θ_S are defined in Figure P17.56a. (a) If both observer and source are moving away from each other, show that the preceding equation reduces to Equation 17.17 with lower signs. (b) Use the preceding equation to solve the following problem. A train moves at a constant speed of 25.0 m/s toward the intersection shown in Figure P17.56b. A car is stopped near the intersection, 30.0 m from the tracks. If the train's horn emits a frequency of 500 Hz, what is the frequency heard by the passengers in the car when the train is 40.0 m from the intersection? Take the speed of sound to be 343 m/s.

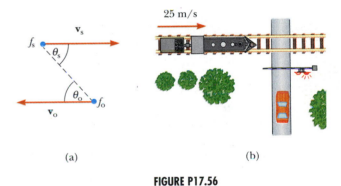

(a)   (b)

FIGURE P17.56

57. In order to be able to determine her speed, a skydiver carries a tone generator. A friend on the ground at the landing site has equipment for receiving and analyzing sound waves. While the skydiver is falling at terminal speed, her tone generator emits a steady tone of 1800 Hz. (Assume that the air is calm and that the sound speed is 343 m/s, independent of altitude.) (a) If her friend on the ground (directly beneath the skydiver) receives waves of frequency 2150 Hz, what is the skydiver's speed of descent? (b) If the skydiver were also carrying sound-receiving equipment sensitive enough to detect waves reflected from the ground, what frequency would she receive?

58. A train whistle ($f = 400$ Hz) sounds higher or lower in pitch depending on whether it approaches or re-

cedes. (a) Prove that the difference in frequency between the approaching and receding train whistle is

$$\Delta f = \frac{2f\left(\dfrac{u}{v}\right)}{1 - \dfrac{u^2}{v^2}}$$ u = speed of train
v = speed of sound

(b) Calculate this difference for a train moving at a speed of 130 km/h. Take the speed of sound in air to be 340 m/s.

59. Three metal rods are located relative to each other as shown in Figure P17.59, where $L_1 + L_2 = L_3$. Values of density and Young's modulus for the three materials are $\rho_1 = 2.7 \times 10^3$ kg/m³, $Y_1 = 7.0 \times 10^{10}$ N/m²; $\rho_2 = 11.3 \times 10^3$ kg/m³, $Y_2 = 1.6 \times 10^{10}$ N/m²; and $\rho_3 = 8.8 \times 10^3$ kg/m³, $Y_3 = 11 \times 10^{10}$ N/m². (a) If $L_3 = 1.5$ m, what must the ratio L_1/L_2 be if a sound wave is to travel the length of rods 1 and 2 in the same time for the wave to travel the length of rod 3? (b) If the frequency of the source is 4.00 kHz, determine the phase difference between the wave traveling along rods 1 and 2 and the one traveling along rod 3.

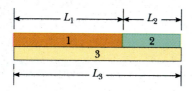

FIGURE P17.59

60. A bat, moving at 5.00 m/s, is chasing a flying insect. If the bat emits a 40.0-kHz chirp and receives back an echo at 40.4 kHz, at what speed is the insect moving toward or away from the bat? (Take the speed of sound in air to be $v = 340$ m/s.)

61. A supersonic aircraft is flying parallel to the ground. When the aircraft is directly overhead, an observer sees a rocket fired from the aircraft. Ten seconds later the observer hears the sonic boom, followed 2.8 s later by the sound of the rocket engine. What is the Mach number of the aircraft?

62. The volume knob on a radio has what is known as a "logarithmic taper." The electrical device connected to the knob (called a potentiometer) has a resistance R whose logarithm is proportional to the angular position of the knob: that is, $\log R \propto \theta$. If the intensity of the sound I (in W/m²) produced by the speaker is proportional to the resistance R, show that the sound level β (in dB) is a linear function of θ.

COMPUTER PROBLEMS

63. The program DOPPLER plots the wavefronts of a sound source moving with speed u. The only input parameter is u/v, where v is the speed of sound. The ratio u/v can be in the range -2.0 to 2.0. Select $u/v = 0$, and measure the distance between adjacent wavefronts on the computer screen. This is the wavelength λ_0 of the sound wave. Now select $u/v = 0.50$ and measure the wavelength ahead of and behind the source. Calculate the change in the wavelength $\Delta\lambda = \lambda - \lambda_0$ for both cases. Does $|\Delta\lambda/\lambda_0| = u/v$? Enter *stop* to exit the program at the prompt $u/v =$.

64. Run the program DOPPLER for a series of values of $u/v \geq 1.0$. A shock wave is produced at these speeds. Note that as u/v increases, the angle of the shock wave decreases. Verify that $\sin\theta = v/u$.

Superposition and Standing Waves

Even when silent, this organ at the Mormon Tabernacle conveys a sense of the power of sound waves.

(Courtesy of Henry Leap)

An important aspect of waves is the combined effect of two or more of them traveling in the same medium. For instance, what happens to a string when a wave traveling toward its fixed end is reflected back on itself? What is the pressure variation in the air when the instruments of an orchestra sound together?

In a linear medium, that is, one in which the restoring force of the medium is proportional to the displacement of the medium, the principle of superposition can be applied to obtain the resultant disturbance. We discussed this principle in Chapter 16 as it applies to wave pulses. The term *interference* was used to describe the effect produced by combining two wave pulses moving simultaneously through a medium.

This chapter is concerned with the superposition principle as it applies to sinusoidal waves. If the sinusoidal waves that combine in a given medium have the same frequency and wavelength, one finds that a stationary pattern, called a *standing wave,* can be produced at certain frequencies under certain circumstances. For example, a taut string fixed at both ends has a discrete set of oscillation patterns, called *modes of vibration,* that depend upon the tension and mass per unit length of the string. These modes of vibration are found in stringed muscical instruments. Other musical instruments, such as the organ and flute, make use of the natural frequencies of sound waves in hollow pipes. Such frequencies depend upon the length of the pipe and its shape and upon whether the pipe is open at both ends or open at one and closed at the other.

We also consider the superposition and interference of waves with different frequencies and wavelengths. When two sound waves with nearly the same frequency interfere, one hears variations in the loudness called *beats*. The beat frequency corresponds to the rate of alternation between constructive and destructive interference. Finally, we describe how any complex periodic wave can, in general, be described by a sum of sine and cosine functions.

18.1 SUPERPOSITION AND INTERFERENCE OF SINUSOIDAL WAVES

The superposition principle tells us that when two or more waves move in the same linear medium, the net displacement of the medium (the resultant wave) at any point equals the algebraic sum of the displacements caused by all the waves. Let us apply this principle to two sinusoidal waves traveling in the same direction in a medium. If the two waves are traveling to the right and have the same frequency, wavelength, and amplitude but differ in phase, we can express their individual wave functions as

$$y_1 = A_0 \sin(kx - \omega t) \qquad y_2 = A_0 \sin(kx - \omega t - \phi)$$

Hence, the resultant wave function y is

$$y = y_1 + y_2 = A_0 [\sin(kx - \omega t) + \sin(kx - \omega t - \phi)]$$

To simplify this expression, it is convenient to make use of the trigonometric identity

$$\sin a + \sin b = 2 \cos\left(\frac{a - b}{2}\right) \sin\left(\frac{a + b}{2}\right)$$

If we let $a = kx - \omega t$ and $b = kx - \omega t - \phi$, we find that the resultant wave function y reduces to

Resultant of two traveling sinusoidal waves

$$y = 2 A_0 \cos\left(\frac{\phi}{2}\right) \sin\left(kx - \omega t - \frac{\phi}{2}\right) \qquad (18.1)$$

There are several important features of this result. The resultant wave function y is also harmonic and has the same frequency and wavelength as the individual waves. The amplitude of the resultant wave is $2 A_0 \cos(\phi/2)$, and its phase is equal to $\phi/2$. If the phase constant ϕ equals 0, then $\cos(\phi/2) = \cos 0 = 1$ and the amplitude of the resultant wave is $2 A_0$. In other words, the amplitude of the resultant wave is twice the amplitude of either individual wave. In this case, the waves are said to be

Constructive interference

everywhere *in phase* and thus **interfere constructively.** That is, the crests and troughs of the individual waves (y_1 and y_2) occur at the same positions as the crests and troughs of the resultant wave (y), as shown by the blue curve in Figure 18.1a. In general, constructive interference occurs when $\cos(\phi/2) = \pm 1$, which means when $\phi = 0, 2\pi, 4\pi, \ldots$ rad. On the other hand, if ϕ is equal to π rad, or to any *odd* multiple of π, then $\cos(\phi/2) = \cos(\pi/2) = 0$ and the resultant wave has *zero*

Destructive interference

amplitude everywhere. In this case, the two waves **interfere destructively.** That is, the crest of one wave coincides with the trough of the second (Fig. 18.1b) and their displacements cancel at every point. Finally, when the phase constant has an arbitrary value between 0 and π rad, as in Figure 18.1c, the resultant wave has an amplitude whose value is somewhere between 0 and $2 A_0$.

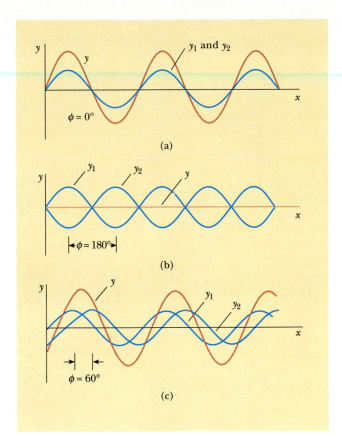

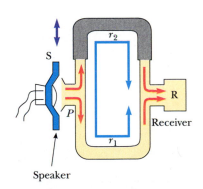

FIGURE 18.1 The superposition of two waves with amplitudes y_1 and y_2, where $y_1 = y_2$. (a) When the two waves are in phase, the result is constructive interference. (b) When the two waves are 180° out of phase, the result is destructive interference. (c) When the phase angle lies in the range $0 < \phi < 180°$, the resultant y falls somewhere between that shown in part (a) and that shown in part (b).

Interference of Sound Waves

One simple device for demonstrating interference of sound waves is illustrated in Figure 18.2. Sound from a loudspeaker S is sent into a tube at P, where there is a T-shaped junction. Half the sound power travels in one direction and half in the opposite direction. Thus, the sound waves that reach the receiver R at the other side can travel along two different paths. The distance along any path from speaker to receiver is called the *path length, r*. The lower path length r_1 is fixed but the upper path length r_2 can be varied by sliding the U-shaped tube, similar to that on a slide trombone. When the difference in the path lengths $\Delta r = |r_2 - r_1|$ is either zero or some integral multiple of the wavelength λ, the two waves reaching the receiver are in phase and interfere constructively, as in Figure 18.1a. For this case, a maximum in the sound intensity is detected at the receiver. If the path length r_2 is adjusted such that the path difference Δr is $\lambda/2, 3\lambda/2, \ldots, n\lambda/2$ (for n odd), the two waves are exactly 180° out of phase at the receiver and hence cancel each other. In this case of destructive interference, no sound is detected at the receiver. This simple experiment demonstrates that a phase difference may arise between two waves generated by the same source when they travel along paths of unequal lengths.

FIGURE 18.2 An acoustical system for demonstrating interference of sound waves. Sound from the speaker propagates into a tube and splits into two parts at P. The two waves, which superimpose at the opposite side, are detected at R. The upper path length r_2 can be varied by the sliding section.

It is often useful to express the path difference in terms of the phase difference ϕ between the two waves. Since a path difference of one wavelength corresponds to a phase difference of 2π rad, we obtain the ratio $\lambda/2\pi = \Delta r/\phi$, or

Relationship between path difference and phase angle

$$\Delta r = \frac{\lambda}{2\pi} \phi \qquad (18.2)$$

EXAMPLE 18.1   Two Speakers Driven by the Same Source

A pair of speakers placed 3.00 m apart are driven by the same oscillator (Fig. 18.3). A listener is originally at point O, which is located 8.00 m from the center of the line connecting the two speakers. The listener then walks to point P, which is a perpendicular distance 0.350 m from O before reaching the *first minimum* in sound intensity. What is the frequency of the oscillator?

Solution   The first minimum occurs when the two waves reaching the listener at P are 180° out of phase—in other words, when their pair difference equals $\lambda/2$. In order to calculate the path difference, we must first find the path lengths r_1 and r_2. Making use of the two shaded triangles in Figure 18.3, we find the path lengths to be

$$r_1 = \sqrt{(8.00 \text{ m})^2 + (1.15 \text{ m})^2} = 8.08 \text{ m}$$

$$r_2 = \sqrt{(8.00 \text{ m})^2 + (1.85 \text{ m})^2} = 8.21 \text{ m}$$

Hence, the path difference is $r_2 - r_1 = 0.13$ m. Since we require that this path difference be equal to $\lambda/2$ for the first minimum, we find that $\lambda = 0.26$ m.

To obtain the oscillator frequency, we can use $v = \lambda f$, where v is the speed of sound in air, 343 m/s:

$$f = \frac{v}{\lambda} = \frac{343 \text{ m/s}}{0.26 \text{ m}} = \boxed{1.3 \text{ kHz}}$$

Exercise   If the oscillator frequency is adjusted such that the listener hears the first minimum at a distance of 0.75 m from O, what is the new frequency?

Answer   0.63 kHz.

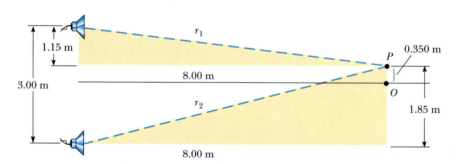

FIGURE 18.3   (Example 18.1).

18.2   STANDING WAVES

If a taut string is clamped at both ends, traveling waves are reflected from the fixed ends, creating waves traveling in both directions. The incident and reflected waves combine according to the superposition principle.

Consider two sinusoidal waves in the same medium with the same amplitude, frequency, and wavelength but traveling in opposite directions. Their wave functions can be written

$$y_1 = A_0 \sin(kx - \omega t) \qquad y_2 = A_0 \sin(kx + \omega t)$$

where y_1 represents a wave traveling to the right and y_2 represents a wave traveling to the left. Adding these two functions gives the resultant wave function y:

$$y = y_1 + y_2 = A_0 \sin(kx - \omega t) + A_0 \sin(kx + \omega t)$$

where $k = 2\pi/\lambda$ and $\omega = 2\pi f$, as usual. When we use the trigonometric identity $\sin(a \pm b) = \sin a \cos b \pm \cos a \sin b$, this expression reduces to

$$y = (2A_0 \sin kx) \cos \omega t \qquad (18.3)$$

Wave function for a standing wave

which is the wave function of a **standing wave**. A standing wave is a stationary vibration pattern formed by the superposition of two waves of the same frequency traveling in opposite directions. From Equation 18.3, we see that a standing wave has an angular frequency ω and an amplitude $2A_0 \sin kx$. That is, every particle of the string vibrates in simple harmonic motion with the same frequency. However, the amplitude of motion of a given particle depends on x. This is in contrast to the situation involving a traveling sinusoidal wave, in which all particles oscillate with both the same amplitude and the same frequency.

Because the amplitude of the standing wave at any value of x is equal to $2A_0 \sin kx$, we see that the maximum amplitude has the value $2A_0$. This maximum occurs when the coordinate x satisfies the condition $\sin kx = \pm 1$, or when

$$kx = \frac{\pi}{2}, \frac{3\pi}{2}, \frac{5\pi}{2}, \cdot \cdot \cdot$$

Since $k = 2\pi/\lambda$, the positions of maximum amplitude, called **antinodes**, are

$$x = \frac{\lambda}{4}, \frac{3\lambda}{4}, \frac{5\lambda}{4}, \cdot \cdot \cdot = \frac{n\lambda}{4} \qquad n = 1, 3, 5, \cdot \cdot \cdot \qquad (18.4)$$

Position of antinodes

Note that adjacent antinodes are separated by $\lambda/2$.

Similarly, the standing wave has a minimum amplitude of zero when x satisfies the condition $\sin kx = 0$, or when

$$kx = \pi, 2\pi, 3\pi, \cdot \cdot \cdot$$

giving

$$x = \frac{\lambda}{2}, \lambda, \frac{3\lambda}{2}, \cdot \cdot \cdot = \frac{n\lambda}{2} \qquad n = 0, 1, 2, 3, \cdot \cdot \cdot \qquad (18.5)$$

Position of nodes

These points of zero amplitude, called **nodes**, are also spaced by $\lambda/2$. The distance between a node and an adjacent antinode is $\lambda/4$.

A graphical description of the standing wave patterns produced at various times by two waves traveling in opposite directions is shown in Figure 18.4. The top and middle waves in each part of the figure represent the individual traveling waves, and the bottom waves represent the standing wave patterns. The nodes of the standing wave are labeled N, and the antinodes are labeled A. At $t = 0$ (Fig. 18.4a), the two waves are identical spatially, giving a standing wave of maximum amplitude, $2A_0$. One quarter of a period later, at $t = T/4$ (Fig. 18.4b), the individual waves have moved one quarter of a wavelength (one to the right and the other to the left). At this time, the individual displacements are equal and opposite for all values of x, and hence the resultant wave has zero displacement everywhere. At $t = T/2$ (Fig. 18.4c), the individual waves are again identical spatially, producing a standing wave pattern that is inverted relative to the $t = 0$ pattern.

It is instructive to describe the energy associated with the motion of a standing wave. To illustrate this point, consider a standing wave formed on a taut string fixed at each end, as in Figure 18.5. Except for the nodes that are stationary, all points on the string oscillate vertically with the same frequency. Furthermore, the various points have different amplitudes of motion. Figure 18.5 represents snap-

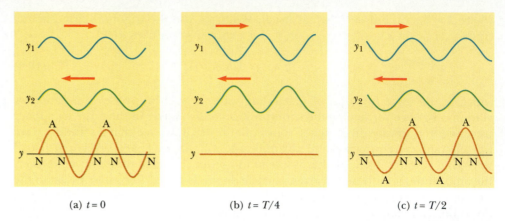

(a) $t = 0$ (b) $t = T/4$ (c) $t = T/2$

FIGURE 18.4 Standing wave patterns at various times produced by two waves of equal amplitude traveling in opposite directions. For the resultant wave y, the nodes (N) are points of zero displacement and the antinodes (A) are points of maximum displacement.

shots of the standing wave at various times over one half of a period. The nodal points do not move and may be grasped or clamped without affecting the standing wave. If dissipation of energy by the string is ignored, the oscillation is constant with time and may be called a **stationary wave;** no energy is transmitted along the wave. (In a real string, some energy is transmitted to maintain the oscillation.) Each point on the string executes simple harmonic motion in the vertical direction. That is, one can view the standing wave as a large number of oscillators vibrating parallel to each other. The energy of the vibrating string continuously alternates between elastic potential energy, at which time the string is momentarily stationary (Fig. 18.5a), and kinetic energy, at which time the string is horizontal and the particles have their maximum speed (Fig. 18.5c). At intermediate times (Figs. 18.5b and 18.5d), the string particles have both potential energy and kinetic energy.

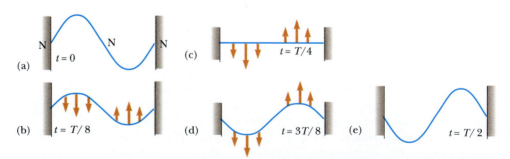

FIGURE 18.5 A standing wave pattern in a taut string showing snapshots during one-half cycle. (a) At $t = 0$, the string is momentarily at rest, and so $K = 0$ and all of the energy is potential energy U associated with the vertical displacements of the string segments. (b) At $t = T/8$, the string is in motion, and the energy is half kinetic and half potential. (c) At $t = T/4$, the string is horizontal (undeformed) and, therefore, $U = 0$; all of the energy is kinetic. The motion continues as indicated, and ultimately the initial configuration in part (a) is repeated.

EXAMPLE 18.2   Formation of a Standing Wave

Two waves traveling in opposite directions produce a standing wave. The individual wave functions are

$$y_1 = (4.0 \text{ cm}) \sin(3.0x - 2.0t)$$

$$y_2 = (4.0 \text{ cm}) \sin(3.0x + 2.0t)$$

where x and y are in centimeters. (a) Find the maximum displacement of the motion at $x = 2.3$ cm.

Solution   When the two waves are summed, the result is a standing wave whose function is given by Equation 18.3, with $A_0 = 4.0$ cm and $k = 3.0$ rad/cm:

$$y = (2A_0 \sin kx) \cos \omega t = [(8.0 \text{ cm}) \sin 3.0x] \cos \omega t$$

Thus, the maximum displacement of the motion at the position $x = 2.3$ cm is

$$y_{max} = (8.0 \text{ cm}) \sin 3.0x|_{x=2.3}$$

$$= (8.0 \text{ cm}) \sin(6.9 \text{ rad}) = \boxed{4.6 \text{ cm}}$$

(b) Find the positions of the nodes and antinodes.

Solution   Since $k = 2\pi/\lambda = 3$ rad/cm, we see that $\lambda = 2\pi/3$ cm. Therefore, from Equation 18.4 we find that the antinodes are located at

$$x = n\left(\frac{\pi}{6}\right) \text{ cm} \qquad (n = 1, 3, 5, \ldots)$$

and from Equation 18.5 we find that the nodes are located at

$$x = n\frac{\lambda}{2} = n\left(\frac{\pi}{3}\right) \text{ cm} \qquad (n = 1, 2, 3, \ldots)$$

18.3   STANDING WAVES IN A STRING FIXED AT BOTH ENDS

Consider a string of length L that is fixed at both ends, as in Figure 18.6. Standing waves are set up in the string by a continuous superposition of waves incident on and reflected from the ends. The string has a number of natural patterns of vibration, called **normal modes**. Each of these has a characteristic frequency that is easily calculated.

First, note that the ends of the string are nodes by definition because these points are *fixed*. In general, the motion of a vibrating string fixed at both ends is described by the superposition of several normal modes. The modes that are present depend on how the vibration is started. For example, when a guitar string

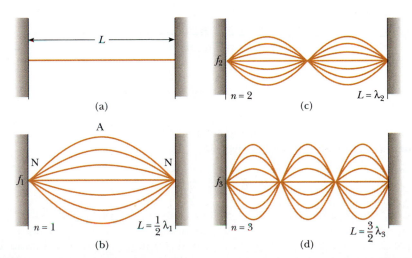

FIGURE 18.6   (a) A string of length L fixed at both ends. The normal modes of vibration form a harmonic series: (b) the fundamental frequency, or first harmonic; (c) the second harmonic; and (d) the third harmonic.

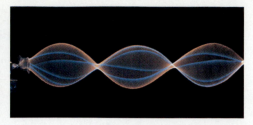

Multiflash photographs of standing wave patterns in a cord driven by a vibrator at the left end. The single loop pattern at the left represents the fundamental ($n = 1$), the two-loop pattern in the middle represents the second harmonic ($n = 2$), and the three-loop pattern at the right represents the third harmonic ($n = 3$).   *(Richard Megna 1991, Fundamental Photographs)*

is plucked near the middle, the modes shown in Figure 18.6b and 18.6d having antinodes at the center are excited, as well as other modes not shown. The first normal mode, shown in Figure 18.6b, has nodes at its ends and one antinode in the middle. This normal mode occurs when the wavelength λ_1 equals twice the length of the string, that is, when $\lambda_1 = 2L$. The next normal mode, of wavelength λ_2 (Fig. 18.6c), occurs when the wavelength equals the length of the string, that is, when $\lambda_2 = L$. The third normal mode (Fig. 18.6d) corresponds to the case where the wavelength is two-thirds the length of the string, $\lambda_3 = 2L/3$. In general, the wavelengths of the various normal modes can be conveniently expressed as

Wavelengths of normal modes	

$$\lambda_n = \frac{2L}{n} \qquad (n = 1, 2, 3, \ldots) \qquad (18.6)$$

where the index n refers to the nth normal mode of vibration. The natural frequencies associated with these modes are obtained from the relationship $f = v/\lambda$, where the *wave speed v is the same for all frequencies*. Using Equation 18.6, we find that the frequencies of the normal modes are

Frequencies of normal modes as functions of wave speed and length of string

$$f_n = \frac{v}{\lambda_n} = \frac{n}{2L} v \qquad (n = 1, 2, 3, \ldots) \qquad (18.7)$$

Because $v = \sqrt{F/\mu}$ (Eq. 16.4), where F is the tension in the string and μ is its mass per unit length, we can also express the natural frequencies of a taut string as

Frequencies of normal modes as functions of string tension and linear mass density

$$f_n = \frac{n}{2L} \sqrt{\frac{F}{\mu}} \qquad (n = 1, 2, 3, \ldots) \qquad (18.8)$$

The lowest frequency, corresponding to $n = 1$, is called the *fundamental* or the **fundamental frequency,** f_1, and is given by

Fundamental frequency of a taut string

$$f_1 = \frac{1}{2L} \sqrt{\frac{F}{\mu}} \qquad (18.9)$$

Clearly, the frequencies of the remaining normal modes (sometimes called *overtones*) are integral multiples of the fundamental frequency. These higher natural frequencies, together with the fundamental frequency, form a **harmonic series.** The fundamental, f_1, is the first harmonic; the frequency $f_2 = 2f_1$ is the second harmonic; the frequency f_n is the nth harmonic.

We can obtain the preceding results in an alternative manner. Since we require that the string be fixed at $x = 0$ and $x = L$, the wave function $y(x, t)$ given by

Equation 18.3 must be zero at these points for all times. That is, the boundary conditions require that $y(0, t) = 0$ and $y(L, t) = 0$ for all values of t. Since $y = (2A_0 \sin kx) \cos \omega t$, the first condition, $y(0, t) = 0$, is automatically satisfied because $\sin kx = 0$ at $x = 0$. To meet the second condition, $y(L, t) = 0$, we require that $\sin kL = 0$. This condition is satisfied when the angle kL equals an integral multiple of π (180°). Therefore, the allowed values of k are[1]

$$k_n L = n\pi \qquad (n = 1, 2, 3, \ldots) \qquad (18.10)$$

Since $k_n = 2\pi/\lambda_n$, we find that

$$\left(\frac{2\pi}{\lambda_n}\right) L = n\pi \qquad \text{or} \qquad \lambda_n = \frac{2L}{n}$$

which is identical to Equation 18.6.

When a taut string is distorted such that its distorted shape corresponds to any one of its harmonics, after being released it will vibrate at the frequency of that harmonic. However, if the string is struck or bowed such that its distorted shape is not just one single harmonic, the resulting vibration will include frequencies of various harmonics. In effect, the string "selects" the normal-mode frequencies when disturbed by a nonharmonic disturbance (which happens, for example, when a guitar string is plucked).

Figure 18.7 shows a taut string vibrating at its first and second harmonics simultaneously. In this figure, the combined vibration is the superposition of the vibrations shown in Figures 18.6b and 18.6c. The large loop corresponds to the fundamental frequency of vibration, f_1, and the smaller loops correspond to the second harmonic, f_2. In general, the resulting motion, or displacement, can be described by a superposition of the various harmonic wave functions, with different frequencies and amplitudes. Hence, the sound that one hears corresponds to a complex wave associated with these various modes of vibration. We shall return to this point in Section 18.8.

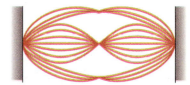

FIGURE 18.7 Multiple exposures of a string vibrating simultaneously in its first and second harmonics.

The frequency of a stringed instrument can be changed either by varying the tension F or by changing the length L. For example, the tension in guitar and violin strings is varied by a screw adjustment mechanism or by turning pegs located on the neck of the instrument. As the tension is increased, the frequency of the normal modes increases according to Equation 18.8. Once the instrument is "tuned," players vary the frequency by moving their fingers along the neck, thereby changing the length of the vibrating portion of the string. As the length is shortened, the frequency increases because the normal-mode frequencies are inversely proportional to string length.

EXAMPLE 18.3 **Give Me a C Note**

A middle C string of the C-major scale on a piano has a fundamental frequency of 264 Hz, and the A note has a fundamental frequency of 440 Hz. (a) Calculate the frequencies of the next two harmonics of the C string.

Solution Since $f_1 = 264$ Hz, we can use Equations 18.8 and 18.9 to find the frequencies f_2 and f_3:

$$f_2 = 2f_1 = \boxed{528 \text{ Hz}}$$

$$f_3 = 3f_1 = \boxed{792 \text{ Hz}}$$

(b) If the strings for the A and C notes are assumed to have the same mass per unit length and the same length, determine the ratio of tensions in the two strings.

---

[1] We exclude $n = 0$ since this corresponds to the trivial case where no wave exists ($k = 0$).

Solution Using Equation 18.8 for the two strings vibrating at their fundamental frequencies gives

$$f_{1A} = \frac{1}{2L}\sqrt{F_A/\mu} \quad \text{and} \quad f_{1C} = \frac{1}{2L}\sqrt{F_C/\mu}$$

$$f_{1A}/f_{1C} = \sqrt{F_A/F_C}$$

$$F_A/F_C = (f_{1A}/f_{1C})^2 = (440/264)^2 = \boxed{2.78}$$

(c) In a real piano, the assumption we made in part (b) is only half true. The string densities are equal, but the A string is 64% as long as the C string. What is the ratio of their tensions?

$$f_{1A}/f_{1C} = (L_C/L_A)\sqrt{F_A/F_C} = (100/64)\sqrt{F_A/F_C}$$

$$F_A/F_C = (0.64)^2(440/264)^2 = \boxed{1.14}$$

18.4 RESONANCE

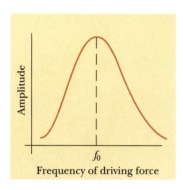

FIGURE 18.8 The amplitude (response) versus driving frequency for an oscillating system. The amplitude is a maximum at the resonance frequency, f_0.

We have seen that a system such as a taut string is capable of oscillating in one or more natural modes of vibration. *If a periodic force is applied to such a system, the resulting amplitude of motion of the system is larger when the frequency of the applied force is equal or nearly equal to one of the natural frequencies of the system* than when the driving force is applied at some other frequency. We have already discussed this phenomenon, known as *resonance*, for mechanical systems. The corresponding natural frequencies of oscillation of the system are often referred to as **resonant frequencies.**

Figure 18.8 shows the response of a vibrating system to various driving frequencies, where one of the resonant frequencies of the system is denoted by f_0. Note that the amplitude is largest when the frequency of the driving force equals the resonant frequency. The driving force continues to deliver energy to the oscillating system causing the amplitude to increase. The maximum amplitude of the motion is limited by friction in the system. Once maximum amplitude is reached, the work done by the periodic force is used only to overcome friction. A system is said to be *weakly damped* when the amount of friction to be overcome is small. Such a system has a large amplitude of motion when driven at one of its resonant frequencies and the oscillations persist for a long time after the driving force is removed. A system with considerable friction to be overcome, that is, one that is *strongly damped,* undergoes small amplitude oscillations that decrease rapidly with time once the driving force is removed.

Examples of Resonance

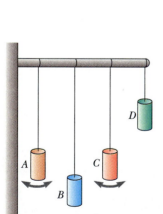

FIGURE 18.9 An example of resonance. If pendulum *A* is set into oscillation, only pendulum *C*, whose length matches that of *A*, will eventually oscillate with large amplitude, or resonate. The arrows indicate motion perpendicular to the page.

A playground swing is a pendulum with a natural frequency that depends on its length. Whenever we push a child in a swing with a series of regular impulses, the swing goes higher if the frequency of the periodic force equals the natural frequency of the swing. One can demonstrate a similar effect by suspending several pendula of different lengths from a horizontal support, as in Figure 18.9. If pendulum *A* is set into oscillation, the other pendula will soon begin to oscillate as a result of the longitudinal waves transmitted along the beam. However, you will find that a pendulum, such as *C*, whose length is close to the length of *A* oscillates with a much larger amplitude than those such as *B* and *D* whose lengths are much different from the length of A. This is because the natural frequency of C is nearly the same as the driving frequency associated with A.

Next, consider a taut string fixed at one end and connected at the opposite end to a vibrating blade as in Figure 18.10. The fixed end is a node, and the point that is near the end connected to the vibrating blade is very nearly a node, since the amplitude of the blade's motion is small compared with that of the string. As the blade oscillates, transverse waves sent down the string are reflected from the fixed

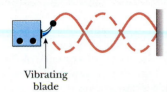

Vibrating
blade

FIGURE 18.10 Standing waves are set up in a string by connecting one end of the string to a vibrating blade. When the blade vibrates at one of the natural frequencies of the string, large-amplitude standing waves are created.

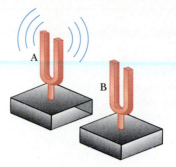

FIGURE 18.11 If tuning fork A is set into vibration, an identical tuning fork B eventually vibrates at the same frequency, or resonates.

end. As we found in Section 18.3, the string has natural frequencies of vibration that are determined by its length, tension, and mass per unit length (Eq. 18.8). When the frequency of the vibrating blade nearly equals one of the natural frequencies of the string, standing waves are produced and the string vibrates with a large amplitude. In this case, the wave being generated by the vibrating blade is in phase with the reflected wave, and so the string absorbs energy from the blade at resonance.

Once the amplitude of the standing-wave oscillations reaches a maximum, the energy delivered by the blade and absorbed by the system is lost because of the damping forces caused by friction in the system. If the applied frequency differs from one of the natural frequencies, however, energy is first transferred to the string from the blade, but later the phase of the wave becomes such that it forces the blade to receive energy from the string, thereby reducing the energy in the string.

As a final example of resonance, consider two identical tuning forks mounted on separate hollow boxes (Fig. 18.11). The hollow boxes augment the sound wave power generated by the vibrating forks. If fork A is set into vibration (say, by striking it), fork B is set into vibration as longitudinal sound waves are received from A. The frequencies of vibration of A and B are the same, assuming the forks are identical. The energy exchange, or resonance behavior, does not occur if the two have different natural frequencies of vibration.

CONCEPTUAL EXAMPLE 18.4

Some singers are able to shatter a wine glass by maintaining a certain pitch in their voice over a period of several seconds (with the help of an amplifier). What mechanism causes the glass to break?

Reasoning The wine glass has natural frequencies of vibration. If the singer's voice has a strong frequency component that corresponds to one of these natural frequencies, sound waves are coupled into the glass, setting up forced vibrations that produce large stresses in the glass causing it to shatter.

(Conceptual Example 18.4) A wine glass shattered by the amplified sound of a human voice. *(© Ben Rose 1992/The IMAGE Bank)*

CONCEPTUAL EXAMPLE 18.5

At certain speeds, an automobile driven on a washboard road will vibrate disastrously and lose traction and braking effectiveness. At other speeds, the vibration is more manageable. Explain. Why are "rumble strips" sometimes used just before stop signs?

Reasoning   At certain speeds the car crosses the ridges on the washboard road at a rate that matches one of the car's

natural frequencies of vibration. This causes the car to go into a large amplitude vibration that can be dangerous. At a speed slightly slower or faster, the natural frequency of the car and the ridges on the road are out of step, and the car moves more smoothly. Rumble strips are used to get your attention at dangerous intersections.

18.5   STANDING WAVES IN AIR COLUMNS

Standing waves can be set up in a tube of air, such as an organ pipe, as the result of interference between longitudinal waves traveling in opposite directions. The phase relationship between the incident wave and the wave reflected from one end depends on whether that end is open or closed. This is analogous to the phase relationships between incident and reflected transverse waves at the ends of a string. *The closed end of an air column is a displacement node because the wall at this end does not allow molecular motion.* As a result, at a closed end of a tube of air, the reflected wave is 180° out of phase with the incident wave. Furthermore, since the pressure wave is 90° out of phase with the displacement wave (Section 17.2), *the closed end of an air column corresponds to a pressure antinode* (that is, a point of maximum pressure variation).

The open end of an air column is approximately a displacement antinode and a pressure node. The fact that there is no pressure variation at the open end can be understood by noting that the end of an air column in a tube is open to the atmosphere, and the pressure at this end must remain constant. If the pressure at this end were to change, air would flow into or out of the column. The wave reflected from an open end is nearly in phase with the incident wave when the tube's diameter is small relative to the wavelength of the sound.

Strictly speaking, the open end of an air column is not exactly an antinode. A condensation reaching an open end does not reflect until it passes beyond the end. For a thin-walled tube of circular cross-section, this end correction is approximately $0.6R$, where R is the tube's radius. Hence, the effective length of the tube is longer than the true length L. We ignore this end correction in what follows.

The first three normal modes of vibration of a pipe open at both ends are shown in Figure 18.12a. When air is directed against an edge at the left, longitudinal standing waves are formed and the pipe resonates at its natural frequencies. All modes of vibration are excited simultaneously (although not with the same amplitude). Note that the ends are displacement antinodes (approximately). In the fundamental mode, the wavelength is twice the length of the pipe, and hence the frequency of the fundamental, $f_1 = v/2L$. The frequencies of the higher harmonics are $2f_1$, $3f_1$, ... Thus,

in a pipe open at both ends, the natural frequencies of vibration form a harmonic series, that is, the higher harmonics are integral multiples of the fundamental frequency.

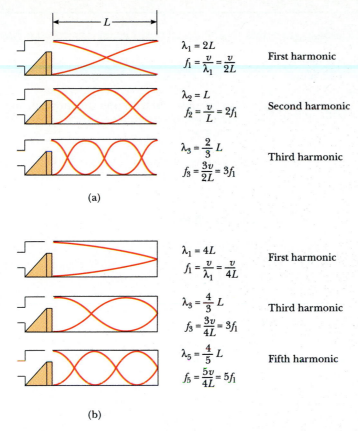

(a)

(b)

FIGURE 18.12   (a) Standing longitudinal waves in an organ pipe open at both ends. The natural frequencies that form a harmonic series are f_1, $2f_1$, $3f_1$, ... (b) Standing longitudinal waves in an organ pipe closed at one end. Only the *odd* harmonics are present, and so the natural frequencies are f_1, $3f_1$, $5f_1$, ...

Since all harmonics are present in a pipe open at both ends, we can express the natural frequencies of vibration as

$$f_n = n\frac{v}{2L} \qquad (n = 1, 2, 3, \ldots) \tag{18.11}$$

Natural frequencies of a pipe open at both ends

where v is the speed of sound in air.

If a pipe is closed at one end and open at the other, the closed end is a displacement node (Fig. 18.12b). In this case, the wavelength for the fundamental mode is four times the length of the tube. Hence, the fundamental, f_1, is equal to $v/4L$, and the frequencies of the higher harmonics are equal to $3f_1$, $5f_1$, ... That is,

in a pipe closed at one end, only odd harmonics are present, and the natural frequencies are

$$f_n = n\frac{v}{4L} \qquad (n = 1, 3, 5, \ldots) \tag{18.12}$$

Natural frequencies of a pipe closed at one end and open at the other

EXAMPLE 18.6   Resonance in a Pipe

A pipe has a length of 1.23 m. (a) Determine the frequencies of the first three harmonics if the pipe is open at each end. Take $v = 343$ m/s as the speed of sound in air.

Solution   The first harmonic of a pipe open at both ends is

$$f_1 = \frac{v}{2L} = \frac{343 \text{ m/s}}{2(1.23 \text{ m})} = \boxed{139 \text{ Hz}}$$

Since all harmonics are present, the second and third harmonics are $f_2 = 2f_1 = 278$ Hz and $f_3 = 3f_1 = 417$ Hz.

(b) What are the three frequencies determined in part (a) if the pipe is closed at one end?

Solution   The fundamental frequency of a pipe closed at one end is

$$f_1 = \frac{v}{4L} = \frac{343 \text{ m/s}}{4(1.23 \text{ m})} = \boxed{69.7 \text{ Hz}}$$

In this case, only odd harmonics are present, and so the next two resonances have frequencies $f_3 = 3f_1 = 209$ Hz and $f_5 = 5f_1 = 349$ Hz.

(c) For the pipe open at both ends, how many harmonics are present in the normal human hearing range (20 to 20 000 Hz)?

Solution   Since all harmonics are present, $f_n = nf_1$. For $f_n = 20\,000$ Hz, we have $n = 20\,000/139 = 144$, so that 144 harmonics are present in the audible range. Actually, only the first few harmonics have sufficient amplitude to be heard.

EXAMPLE 18.7   Measuring the Frequency of a Tuning Fork

A simple apparatus for demonstrating resonance in a tube is described in Figure 18.13a. A long, vertical tube open at both ends is partially submerged in a beaker of water, and a vibrating tuning fork of unknown frequency is placed near the top. The length of the air column, L, is adjusted by moving the tube vertically. The sound waves generated by the fork are reinforced when the length of the air column corresponds to one of the resonant frequencies of the tube.

For a certain tube, the smallest value of L for which a peak occurs in the sound intensity is 9.00 cm. From this measurement, determine the frequency of the tuning fork and the value of L for the next two resonant modes.

Reasoning   Although the tube is open at both ends to allow the water in, the water surface acts like a wall at one end of the length L. Therefore this setup represents a pipe closed at one end and the fundamental has a frequency of $v/4L$ (Fig. 18.13b).

Solution   Taking $v = 343$ m/s for the speed of sound in air and $L = 0.0900$ m, we get

$$f_1 = \frac{v}{4L} = \frac{343 \text{ m/s}}{4(0.0900 \text{ m})} = \boxed{953 \text{ Hz}}$$

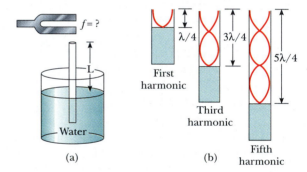

FIGURE 18.13   (Example 18.7) (a) Apparatus for demonstrating the resonance of sound waves in a tube closed at one end. The length L of the air column is varied by moving the tube vertically while it is partially submerged in water. (b) The first three normal modes of the system shown in (a).

From this information about the fundamental mode, we see that the wavelength is $\lambda = 4L = 0.360$ m. Since the frequency of the source is constant, the next two resonance modes (Fig. 18.13b) correspond to lengths of $3\lambda/4 = 0.270$ m and $5\lambda/4 = 0.450$ m.

*18.6   STANDING WAVES IN RODS AND PLATES

Standing waves can also be set up in rods and plates. If a rod is clamped in the middle and stroked at one end, it will undergo longitudinal vibrations as described in Figure 18.14a. Note that the broken lines in Figure 18.14 represent *longitudinal* displacements of various parts of the rod. The midpoint is a displacement node since it is fixed by the clamp, whereas the ends are displacement antinodes since they are free to vibrate. This setup is analogous to vibrations set up in a pipe open at each end. The broken lines in Figure 18.14a represent the fundamental mode,

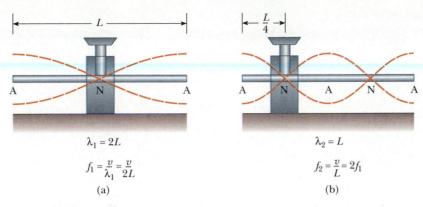

FIGURE 18.14   Normal-mode longitudinal vibrations of a rod of length L (a) clamped at the middle and (b) clamped at an approximate distance of $L/4$ from one end.

for which the wavelength is $2L$ and the frequency is $v/2L$, where v is the speed of longitudinal waves in the rod. Other modes may be excited by clamping the rod at different points. For example, the second harmonic (Fig. 18.14b) is excited by clamping the rod at a point that is a distance $L/4$ away from one end.

Two-dimensional vibrations can be set up in a flexible membrane stretched over a circular hoop, such as a drumhead. As the membrane is struck at some point, wave pulses that arrive at the fixed boundary are reflected many times. The resulting sound is not harmonic but rather explosive in nature. This is because the vibrating drumhead and the drum's hollow interior produce a disorganized set of waves that create a sound of indefinite pitch when they reach a listener's ear. This is in contrast to wind and stringed instruments, which produce sounds of definite pitch.

Some possible normal modes of oscillation of a vibrating, two-dimensional, circular membrane are shown in Figure 18.15. Note that the nodes are curves rather than points, which was the case for a vibrating string. The fixed circumference is one such nodal curve, and some other nodal curves are also indicated. The lowest mode of vibration with frequency f_1 (the fundamental) is a symmetric mode with one nodal curve, the circumference of the membrane. The other possible modes of vibration are *not* integral multiples of f_1; hence, the normal frequencies *do not* form a harmonic series. When a drum is struck, many of these modes are excited simultaneously. However, the higher-frequency modes damp out more rapidly. With this information, one can understand why the drum containing a membrane with fixed perimeters is a nonmelodious instrument. In contrast, a loudspeaker is designed to reproduce musical frequencies by unclamping the edge of the speaker diaphragm.

*18.7   BEATS: INTERFERENCE IN TIME

The interference phenomena we have been dealing with so far involve the superposition of two or more waves with the same frequency traveling in opposite directions. Since the resultant wave in this case depends on the coordinates of the disturbed medium, we can refer to the phenomenon as *spatial interference*. Standing waves in strings and pipes are common examples of spatial interference.

We now consider another type of interference, one that results from the superposition of two waves with slightly *different frequencies* traveling in the *same direction*. In this case, when the two waves are observed at a given point, they are periodically in and out of phase. That is, there is a temporal alternation between constructive

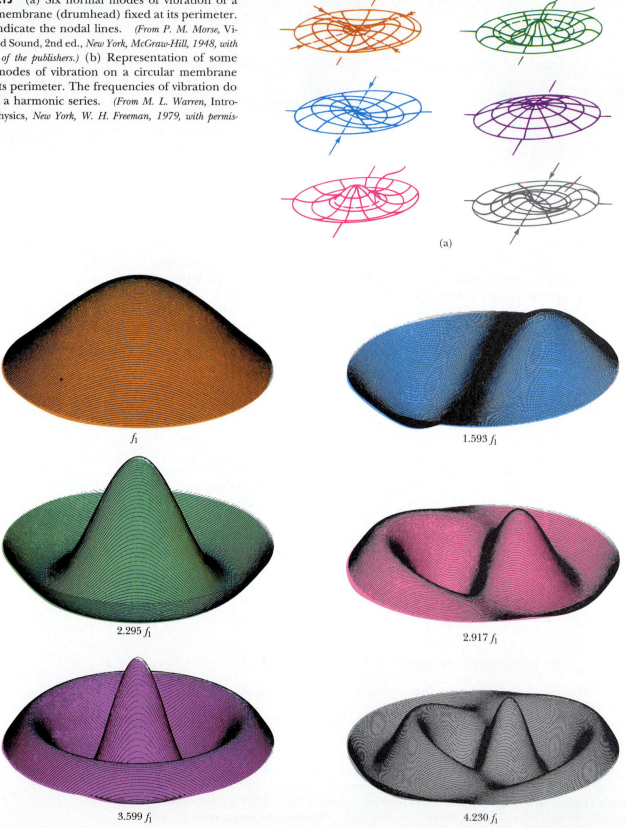

FIGURE 18.15 (a) Six normal modes of vibration of a circular membrane (drumhead) fixed at its perimeter. Arrows indicate the nodal lines. *(From P. M. Morse,* Vibration and Sound, *2nd ed., New York, McGraw-Hill, 1948, with permission of the publishers.)* (b) Representation of some natural modes of vibration on a circular membrane fixed at its perimeter. The frequencies of vibration do not form a harmonic series. *(From M. L. Warren,* Introductory Physics, *New York, W. H. Freeman, 1979, with permission.)*

(a)

f_1

$1.593 \, f_1$

$2.295 \, f_1$

$2.917 \, f_1$

$3.599 \, f_1$

$4.230 \, f_1$

(b)

and destructive interference. Thus, we refer to this phenomenon as *interference in time or temporal interference*. For example, if two tuning forks of slightly different frequencies are struck, one hears a sound of pulsating intensity, called a **beat:**

> A beat is the periodic variation in intensity at a given point due to the superposition of two waves having slightly different frequencies.

Definition of beat

The number of beats one hears per second, or *beat frequency*, equals the difference in frequency between the two sources. The maximum beat frequency that the human ear can detect is about 20 beats/s. When the beat frequency exceeds this value, it blends indistinguishably with the compound sounds producing the beats.

One can use beats to tune a stringed instrument, such as a piano, by beating a note against a reference tone of known frequency. The string can then be adjusted to equal the frequency of the reference by tightening or loosening it until the beats become too infrequent to notice.

Consider two waves of equal amplitude traveling through a medium in the same direction but with slightly different frequencies, f_1 and f_2. We can represent the displacement each wave produces at a point as

$$y_1 = A_0 \cos 2\pi f_1 t \qquad y_2 = A_0 \cos 2\pi f_2 t$$

Using the superposition principle, we find that the resultant displacement at that point is

$$y = y_1 + y_2 = A_0 (\cos 2\pi f_1 t + \cos 2\pi f_2 t)$$

It is convenient to write this equation in a form that uses the trigonometric identity

$$\cos a + \cos b = 2 \cos\left(\frac{a - b}{2}\right) \cos\left(\frac{a + b}{2}\right)$$

Letting $a = 2\pi f_1 t$ and $b = 2\pi f_2 t$, we find that

$$y = 2A_0 \cos 2\pi\left(\frac{f_1 - f_2}{2}\right)t \cos 2\pi\left(\frac{f_1 + f_2}{2}\right)t \qquad (18.13)$$

Resultant of two waves of different frequencies but equal amplitude

Graphs demonstrating the individual waves as well as the resultant wave are shown in Figure 18.16. From the factors in Equation 18.13, we see that the resultant

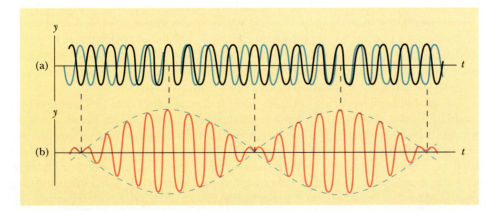

FIGURE 18.16 Beats are formed by the combination of two waves of slightly different frequencies traveling in the same direction. (a) The individual waves. (b) The combined wave has an amplitude (broken line) that oscillates in time.

vibration for a listener standing at any given point has an effective frequency equal to the average frequency, $(f_1 + f_2)/2$, and an amplitude

$$A = 2A_0 \cos 2\pi\left(\frac{f_1 - f_2}{2}\right)t \qquad (18.14)$$

That is, the *amplitude varies in time* and the frequency at which the amplitude maxima are heard is $(f_1 - f_2)/2$. When f_1 is close to f_2, this amplitude variation is slow, as illustrated by the envelope (broken line) of the resultant wave in Figure 18.16b.

Note that a beat, or a maximum in amplitude, is detected whenever

$$\cos 2\pi\left(\frac{f_1 - f_2}{2}\right)t = \pm 1$$

That is, there are *two* maxima in each cycle. Since the amplitude varies with frequency as $(f_1 - f_2)/2$, the number of beats per second, or the beat frequency f_b, is twice this value. That is,

Beat frequency

$$f_b = |f_1 - f_2| \qquad (18.15)$$

For instance, if one tuning fork vibrates at 438 Hz and a second tuning fork vibrates at 442 Hz, the resultant sound wave of the combination has a frequency of 440 Hz (the musical note A) and a beat frequency of 4 Hz. A listener would hear the 440-Hz sound wave go through an intensity maximum four times every second.

*18.8 COMPLEX WAVES

The sound wave patterns produced by most musical instruments are very complex. Characteristic patterns produced by a tuning fork, a harmonic flute, and a clarinet, each playing the same pitch, are shown in Figure 18.17. The term *pitch* refers to the human perception to the frequency of sound. Although each instrument has its own characteristic pattern, Figure 18.17 shows that each pattern is periodic. Furthermore, note that a struck tuning fork produces only one harmonic (the fundamental), whereas the flute and clarinet produce many frequencies, which include the fundamental and various harmonics. Thus, the complex wave patterns produced by a violin or clarinet, and the corresponding richness of musical tones, are the result of the superposition of various harmonics. This is in contrast to a drum, in which the overtones do not form a harmonic series.

It is interesting to investigate what happens to the frequencies of a flute and a violin during a concert as the temperature rises. A flute goes sharp (increases in frequency) as it warms up because the speed of sound increases in the warmer air inside the flute. A violin goes flat (decreases in frequency) as the strings expand thermally because the expansion causes their tension to decrease.

The problem of analyzing complex wave patterns appears at first sight to be a formidable task. However, if the wave pattern is periodic, it can be represented with arbitrary precision by the combination of a sufficiently large number of sinusoidal waves that form a harmonic series. In fact, one can represent any periodic function or any function over a finite interval as a series of sine and cosine terms by using a mathematical technique based on **Fourier's theorem**.[2] The corresponding sum of terms that represents the periodic waveform is called a **Fourier series.**

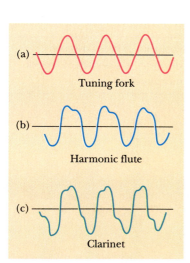

FIGURE 18.17 Waveform produced by (a) a tuning fork, (b) a harmonic flute, and (c) a clarinet, each at approximately the same frequency. *(Adapted from C. A. Culver,* Musical Acoustics, *4th ed., New York, McGraw-Hill, 1956, p. 128.)*

[2] Developed by Jean Baptiste Joseph Fourier (1786–1830).

This simulator enables you to generate complex signals using the principle of superposition or Fourier synthesis. Up to ten sinusoidal waves of varying amplitudes, frequencies, and phase angles can be added together to generate complex waveforms that may represent actual phenomena. Depending upon the computer you are using, you may be able to play back and listen to the sound that your signal generates.

Complex Waves — The Fourier Synthesizer

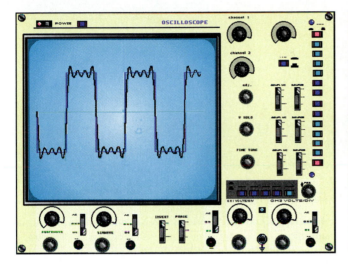
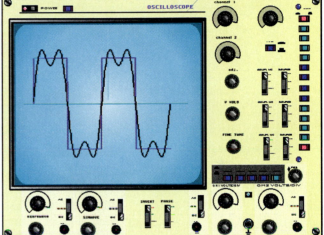

Let $y(t)$ be any function that is periodic in time with period T, such that $y(t + T) = y(t)$. Fourier's theorem states that this function can be written

$$y(t) = \sum_n (A_n \sin 2\pi f_n t + B_n \cos 2\pi f_n t) \qquad (18.16)$$

Fourier's theorem

where the lowest frequency is $f_1 = 1/T$.

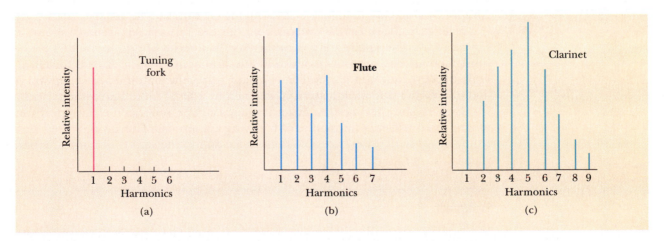

FIGURE 18.18 Harmonics of the waveforms shown in Figure 18.17. Note the variations in intensity of the various harmonics. (*Adapted from C. A. Culver,* Musical Acoustics, *4th ed., New York, McGraw-Hill, 1956.*)

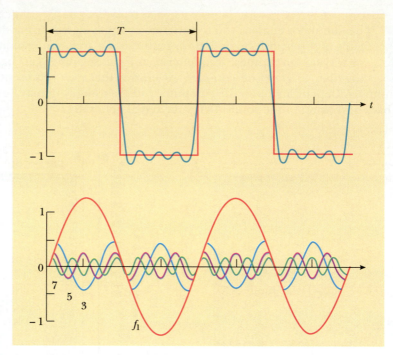

FIGURE 18.19   Harmonic synthesis of a square wave, which can be represented by the sum of odd harmonics of the fundamental.   *(From M. L. Warren,* Introductory Physics, *New York, W. H. Freeman, 1979, p. 178; by permission of the publisher.)*

The higher frequencies are integral multiples of the fundamental, so that $f_n = nf_1$. The coefficients A_n and B_n represent the amplitudes of the various waves. The amplitude of the nth harmonic is proportional to $\sqrt{A_n{}^2 + B_n{}^2}$, and its intensity is proportional to $A_n{}^2 + B_n{}^2$.

Figure 18.18 represents a harmonic analysis of the wave patterns shown in Figure 18.17. Note the variation in relative intensity of the various harmonics for the flute and clarinet. In general, any musical sound (of definite pitch) contains components that are members of a harmonic set with varying relative intensities.

As an example of *Fourier synthesis,* consider the periodic square wave shown in Figure 18.19. The square wave is synthesized by a series of odd harmonics of the fundamental. The series contains only sine functions (that is, $B_n = 0$ for all n). Only the first four odd harmonics and their respective amplitudes are shown. A better fit to the true wave pattern is obtained by adding more harmonics.

Using modern technology, musical sounds can be generated electronically by mixing any number of harmonics with varying amplitudes. These widely used electronic music synthesizers are able to produce an infinite variety of musical tones.

SUMMARY

When two waves with equal amplitudes and frequencies superimpose, the resultant wave has an amplitude that depends on the phase angle ϕ between the two waves. **Constructive interference** occurs when the two waves are in phase everywhere, corresponding to $\phi = 0, 2\pi, 4\pi, \ldots$ rad. **Destructive interference** occurs when the two waves are 180° out of phase everywhere, corresponding to $\phi = \pi$, $3\pi, 5\pi, \ldots$ rad.

Standing waves are formed from the superposition of two harmonic waves having the same frequency, amplitude, and wavelength but traveling in opposite directions. The resultant standing wave is described by the wave function

$$y = (2A_0 \sin kx) \cos \omega t \tag{18.3}$$

Hence, its amplitude varies as $\sin kx$. The maximum amplitude points (called **antinodes**) occur at $x = n\pi/2k = n\lambda/4$ (for odd n). The points of zero amplitude (called **nodes**) occur at $x = n\pi/k = n\lambda/2$ (for integral values of n).

The natural frequencies of vibration of a taut string of length L, fixed at both ends are

$$f_n = \frac{n}{2L} \sqrt{\frac{F}{\mu}} \qquad (n = 1, 2, 3, \ldots) \tag{18.8}$$

where F is the tension in the string and μ is its mass per unit length. The natural frequencies of vibration form a **harmonic series,** that is, $f_1, 2f_1, 3f_1, \ldots$.

A system capable of oscillating is said to be in **resonance** with some driving force whenever the frequency of the driving force matches one of the natural frequencies of the system. When the system is resonating, it responds by oscillating with a relatively large amplitude.

Standing waves can be produced in a tube of air. If the tube is open at both ends, all harmonics are present and the natural frequencies of vibration are

$$f_n = n\frac{v}{2L} \qquad (n = 1, 2, 3, \ldots) \tag{18.11}$$

If the tube is open at one end and closed at the other, only the odd harmonics are present, and the natural frequencies of vibration are

$$f_n = n\frac{v}{4L} \qquad (n = 1, 3, 5, \ldots) \tag{18.12}$$

QUESTIONS

1. For certain positions of the movable section in Figure 18.2, there is no sound detected at the receiver, corresponding to destructive interference. This suggests that perhaps energy is somehow lost! What happens to the energy transmitted by the speaker?

2. Does the phenomenon of wave interference apply only to sinusoidal waves?

3. When two waves interfere constructively or destructively, is there any gain or loss in energy? Explain.

4. A standing wave is set up on a string as in Figure 18.5. Explain why no energy is transmitted along the string.

5. What is common to *all* points (other than the nodes) on a string supporting a standing wave?

6. What limits the amplitude of motion of a real vibrating system that is driven at one of its resonant frequencies?

7. If the temperature of the air in an organ pipe increases, what happens to the resonance frequencies?

8. Explain why your voice seems to sound better than usual when you sing in the shower.

9. What is the purpose of the slide on a trombone or the valves on a trumpet?

10. Explain why all harmonics are present in an organ pipe open at both ends, but only the odd harmonics are present in a pipe closed at one end.

11. Explain how a musical instrument such as a piano may be tuned using the phenomenon of beats.

12. An airplane mechanic notices that the sound from a twin-engine aircraft rapidly varies in loudness when both engines are running. What could be causing this variation from loud to soft?

13. When the base of a vibrating tuning fork is placed against a chalk board, the sound becomes louder due to resonance. How does this affect the length of time for which the fork vibrates? Does this agree with conservation of energy?

14. Stereo speakers are supposed to be "phased" when set up. That is, the waves emitted from them should be in phase with each other. What would the sound be like along the center line of the speakers if one speaker were wired up backwards, that is, out of phase?

15. To keep animals away from their cars, some people mount short thin pipes on the fenders. The pipes give out a high-pitched wail when the cars are moving. How do they create the sound?

16. If you wet your fingers and lightly run them around the rim of a fine wine glass, a high-pitched sound is heard. Why? How could you produce various musical notes with a set of wine glasses?

17. When a bell is rung, standing waves are set up around the bell's circumference. What boundary conditions must be satisfied by the resonant wavelengths? How does a crack in the bell, such as in the Liberty Bell, affect the satisfying of the boundary conditions and the sound emanating from the bell?

PROBLEMS

Review Problem

For the arrangement shown in the diagram, $\theta = 30°$, the inclined plane is frictionless, the mass M rests at the bottom of the plane, and the string has a mass m that is small compared to M. When the system is in equilibrium, the vertical part of the string of length h is fixed so that the tension remains constant. If standing waves are set up in the vertical string, find (a) the tension in the string, (b) the length of the string, (c) the mass per unit length of the string, (d) the wave speed in the vertical part of the string, (e) the lowest frequency standing wave, (f) the period of the standing wave having three nodes, (g) the wavelength of the standing wave having three nodes, and (h) the frequency of the beats resulting from the interference of the sound wave of lowest frequency generated by the string with another sound wave having a frequency that is 2% greater.

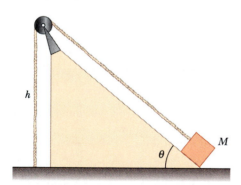

Section 18.1   Superposition and Interference of Sinusoidal Waves

1. Two harmonic waves are described by

$$y_1 = (5.0 \text{ m}) \sin[\pi(4.0x - 1200t)]$$

$$y_2 = (5.0 \text{ m}) \sin[\pi(4.0x - 1200t - 0.25)]$$

where x, y_1, and y_2 are in meters and t is in seconds.

(a) What is the amplitude of the resultant wave?
(b) What is the frequency of the resultant wave?

2. Two harmonic waves are described by

$$y_1 = (6.0 \text{ m}) \sin\left(\frac{\pi}{15} x - \frac{\pi}{0.0050} t\right)$$

$$y_2 = (6.0 \text{ m}) \sin\left(\frac{\pi}{15} x - \frac{\pi}{0.0050} t - \phi\right)$$

where x, y_1, and y_2 are in meters and t is in seconds. (a) What is the amplitude of the resultant wave when $\phi = (\pi/6)$ rad? (b) For what values of ϕ will the amplitude of the resultant wave have its maximum value?

3. Two identical harmonic waves with wavelengths of 3.0 m travel in the same direction at a speed of 2.0 m/s. The second wave originates from the same point as the first, but at a later time. Determine the minimum possible time interval between the starting moments of the two waves if the amplitude of the resultant wave is the same as that of the two initial waves.

3A. Two identical harmonic waves with wavelengths λ travel in the same direction at a speed v. The second wave originates from the same point as the first, but at a later time. Determine the minimum possible time interval between the starting moments of the two waves if the amplitude of the resultant wave is the same as that of the two initial waves.

4. Two speakers are driven by the same oscillator of frequency 200 Hz. They are located on a vertical pole a distance of 4.00 m from each other. A man walks toward one of the speakers in a direction perpendicular to the pole as shown in Figure P18.4. (a) How many times will he hear a minimum in sound intensity and (b) how far is he from the wall at these moments? Take the speed of sound to be 330 m/s and ignore any sound reflections coming off the floor.

□ indicates problems that have full solutions available in the Student Solutions Manual and Study Guide.

4A. Two speakers are driven by the same oscillator of frequency *f*. They are located on a vertical pole a distance *d* from each other. A man walks toward one of the speakers in a direction perpendicular to the pole as shown in Figure P18.4. (a) How many times will he hear a minimum in sound intensity and (b) how far is he from the wall at these moments? Take the speed of sound to be *v* and ignore any sound reflections coming off the floor.

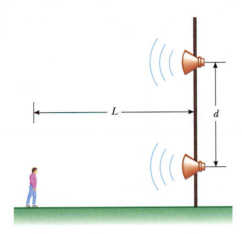

FIGURE P18.4

5. Two identical speakers separated by 10.0 m are driven by the same oscillator with a frequency of *f* = 21.5 Hz (Fig. P18.5). (a) Explain why a receiver at point *A* records a minimum in sound intensity from the two speakers. (b) If the receiver is moved in the horizontal plane of the speakers, what path should it take so that the intensity remains at a minimum? That is, determine the relationship between *x* and *y* (the coordinates of the receiver) that causes the receiver to record a minimum in sound intensity. Take the speed of sound to be 344 m/s.

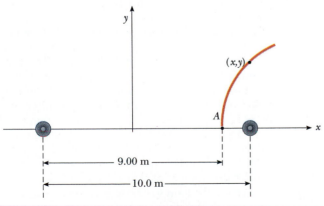

FIGURE P18.5

6. A tuning fork generates sound waves with a frequency of 246 Hz. The waves travel in opposite directions along a hallway, are reflected by walls, and return. What is the phase difference between the reflected waves when they meet? The corridor is 47 m long and the tuning fork is located 14 m from one end. The speed of sound in air is 343 m/s.

7. Two speakers are driven by a common oscillator at 800 Hz and face each other at a distance of 1.25 m. Locate the points along a line joining the two speakers where relative minima would be expected. (Use *v* = 343 m/s.)

8. For the arrangement shown in Figure 18.2, let the path length r_1 = 1.20 m and the path length r_2 = 0.80 m. (a) Calculate the three lowest speaker frequencies that will result in intensity maxima at the receiver. (b) What is the highest frequency within the audible range (20–20 000 Hz) that will result in a minimum at the receiver? (Use *v* = 340 m/s.)

Section 18.2   Standing Waves

9. Two harmonic waves are described by

$$y_1 = (3.0 \text{ cm}) \sin \pi(x + 0.60t)$$

$$y_2 = (3.0 \text{ cm}) \sin \pi(x - 0.60t)$$

where *x* is in centimeters and *t* is in seconds. Determine the *maximum* displacement of the motion at (a) *x* = 0.25 cm, (b) *x* = 0.50 cm, and (c) *x* = 1.5 cm. (d) Find the three smallest values of *x* corresponding to antinodes.

10. Use the trigonometric identity

$$\sin(a \pm b) = \sin a \cos b \pm \cos a \sin b$$

to show that the resultant of two wave functions each of amplitude A_0, angular frequency ω, and propagation number *k* and traveling in opposite directions can be written

$$y = (2A_0 \sin kx) \cos \omega t$$

11. The wave function for a standing wave in a string is

$$y = (0.30 \text{ m}) \sin(0.25x) \cos(120 \pi t)$$

where *x* is in meters and *t* is in seconds. Determine the wavelength and frequency of the interfering traveling waves.

12. Two harmonic waves traveling in opposite directions interfere to produce a standing wave described by

$$y = (1.50 \text{ m}) \sin(0.400x) \cos(200t)$$

where *x* is in meters and *t* is in seconds. Determine the wavelength, frequency, and speed of the interfering waves.

13. A standing wave is formed by the interference of two traveling waves, each of which has an amplitude *A* = π cm, angular wave number *k* = $(\pi/2)$ cm^{-1}, and

angular frequency $\omega = 10\pi$ rad/s. (a) Calculate the distance between the first two antinodes. (b) What is the amplitude of the standing wave at $x = 0.25$ cm?

14. Verify by direct substitution that the wave function for a standing wave given in Equation 18.3,

$$y = 2A_0 \sin kx \cos \omega t$$

is a solution of the general linear wave equation, Equation 16.26:

$$\frac{\partial^2 y}{\partial x^2} = \frac{1}{v^2}\frac{\partial^2 y}{\partial t^2}$$

15. Two waves given by $y_1(x, t) = A \sin(kx - \omega t)$, and $y_2(x, t) = A \sin(2kx + \omega t)$ interfere. (a) Determine all x values where there are stationary nodes. (b) Determine all x values where there are nodes that depend on the time t.

16. Two waves in a long string are given by

$$y_1 = (0.015 \text{ m}) \cos\left(\frac{x}{2} - 40t\right)$$

$$y_2 = (0.015 \text{ m}) \cos\left(\frac{x}{2} + 40t\right)$$

where the y's and x are in meters and t is in seconds. (a) Determine the positions of the nodes of the resulting standing wave. (b) What is the maximum displacement at the position $x = 0.40$ m?

Section 18.3 Standing Waves in a String Fixed at Both Ends

17. A standing wave is established in a 120-cm-long string fixed at both ends. The string vibrates in four segments when driven at 120 Hz. (a) Determine the wavelength. (b) What is the fundamental frequency?

18. A stretched string is 160 cm long and has a linear density of 0.015 g/cm. What tension in the string will result in a second harmonic of 460 Hz?

19. Consider a tuned guitar string of length L. At what point along the string (fraction of length from one end) should the string be plucked and at what point should the finger be held lightly against the string so that the second harmonic is the most prominent mode of vibration?

20. A string 50 cm long has a mass per unit length of 20×10^{-5} kg/m. To what tension should this string be stretched if its fundamental frequency is to be (a) 20 Hz and (b) 4500 Hz?

21. Find the fundamental frequency and the next three frequencies that could cause a standing wave pattern on a string that is 30 m long, has a mass per unit length 9.0×10^{-3} kg/m, and is stretched to a tension of 20 N.

22. A string of linear density 1.0×10^{-3} kg/m and length 3.0 m is stretched between two points. One end is vibrated transversely at 200 Hz. What tension in the string will establish a standing-wave pattern with three loops along the string's length?

23. A sphere having mass $M = 2.0$ kg is supported by a string that passes over a light horizontal rod of length $L = 1.0$ m (Fig. P18.23). Given the angle is $\theta = 35°$ and the fundamental frequency of standing waves in the string is $f = 50.0$ Hz, determine the mass of the string.

23A. A sphere of mass M is supported by a string that passes over a light horizontal rod of length L (Fig. P18.23). Given the angle is θ and the fundamental frequency of standing waves in the string is f, determine the mass of the string above the horizontal rod.

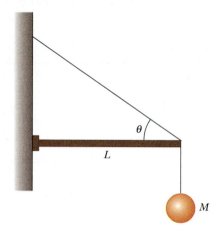

FIGURE P18.23

24. A 2.0-m-long wire having a mass of 0.10 kg is fixed at both ends. The tension in the wire is maintained at 20 N. A node is observed at a point 0.40 m from one end. What are the frequencies of the first three allowed modes of vibration?

25. A cello A-string vibrates in its fundamental mode with a frequency of 220 vibrations/s. The vibrating segment is 70 cm long and has a mass of 1.2 g. (a) Find the tension in the string. (b) Determine the frequency of the harmonic that causes the string to vibrate in three segments.

26. In the arrangement shown in Figure P18.26, a mass can be hung from a string (with linear mass density $\mu = 0.0020$ kg/m) around a light pulley. The string is connected to a vibrator (of constant frequency, f), and the length of the string between point P and the pulley is $L = 2.0$ m. When the mass m is either 16 kg or 25 kg, standing waves are observed, but no standing waves are observed for any mass between these values. (a) What is the frequency of the vibrator? (*Hint:* The greater the tension in the string, the smaller the number of nodes in the standing wave.) (b) What is the largest mass for which standing waves could be observed?

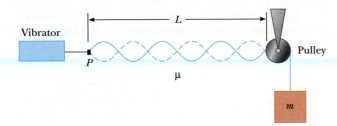

FIGURE P18.26

27. A 60-cm guitar string under a tension of 50 N has a mass per unit length of 0.10 g/cm. What is the highest resonant frequency that can be heard by a person capable of hearing frequencies up to 20 000 Hz?

28. A stretched wire vibrates in its fundamental mode at a frequency of 400 vibrations/s. What would be the fundamental frequency if the wire were half as long, with twice the diameter and with four times the tension?

29. A low C ($f = 65$ Hz) is sounded on a piano. If the length of the piano wire is 2.0 m and its linear mass density is 5.0 g/m, what is the tension in the wire?

30. A violin string has a length of 0.35 m and is tuned to concert G, $f_G = 392$ Hz. Where must the violinist place her finger to play concert A, $f_A = 440$ Hz? If this position is to remain correct to one half the width of a finger (i.e., to within 0.60 cm), by what fraction may the string tension be allowed to slip?

Section 18.5   Standing Waves in Air Columns

(In this section, unless otherwise indicated, assume that the speed of sound in air is 344 m/s.)

31. An open pipe 0.40 m in length is placed vertically in a cylindrical bucket having a bottom area of 0.10 m². Water is poured into the bucket until a sounding tuning fork of frequency 440 Hz, placed over the pipe, produces resonance. Find the mass of water in the bucket at this moment.

31A. An open pipe of length L is placed vertically in a cylindrical bucket having a bottom area A. Water is poured into the bucket until a sounding tuning fork of frequency f, placed over the pipe, produces resonance. Find the mass of water in the bucket at this moment.

32. If an organ pipe is to resonate at 20.0 Hz, what is its required length if it is (a) open at both ends and (b) closed at one end?

33. A tuning fork of frequency 512 Hz is placed near the top of the tube shown in Figure 18.13a. The water level is lowered so that the length L slowly increases from an initial value of 20.0 cm. Determine the next two values of L that correspond to resonant modes.

34. A student uses an audio oscillator of adjustable frequency to measure the depth of a water well. Two successive resonant frequencies are heard at 52.0 Hz and 60.0 Hz. What is the depth of the well?

34A. A student uses an audio oscillator of adjustable frequency to measure the depth of a water well. Two successive resonant frequencies are heard at f_1 and f_2. What is the depth of the well?

35. Calculate the length for a pipe that has a fundamental frequency of 240 Hz if the pipe is (a) closed at one end and (b) open at both ends.

36. A tunnel beneath a river is approximately 2.0 km long. At what frequencies can this tunnel resonate?

37. An organ pipe open at both ends is vibrating in its third harmonic with a frequency of 748 Hz. The length of the pipe is 0.70 m. Determine the speed of sound in air in the pipe.

38. The longest pipe on an organ that has pedal stops is often 4.88 m. What is the fundamental frequency (at 0.0°C) if the nondriven end of the pipe is (a) closed and (b) open? (c) What will be the frequencies at 20.0°C?

39. The overall length of a piccolo is 32 cm. The resonating air column vibrates as a pipe open at both ends. (a) Find the frequency of the lowest note a piccolo can play, assuming the speed of sound in air is 340 m/s. (b) Opening holes in the side effectively shortens the length of the resonant column. If the highest note a piccolo can sound is 4000 Hz, find the distance between adjacent nodes for this mode of vibration.

40. A piece of metal pipe is just the right length so that when it is cut into two pieces, their lowest resonant frequencies are 256 Hz for one and 440 Hz for the other. (a) What resonant frequency would have been produced by the original length of pipe and (b) how long was the original piece?

41. An air column 2.00 m in length is open at both ends. The frequency of a certain harmonic is 410 Hz, and the frequency of the next higher harmonic is 492 Hz. Determine the speed of sound in the air column.

42. A piece of cardboard tubing, closed at one end, is just the right length so that when it is cut into two pieces, the lowest resonant frequency is 256 Hz for the piece with the closed end and 440 Hz for the other. (a) What resonant frequency would have been produced by the original cardboard tubing and (b) how long was the original piece?

43. A glass tube is open at one end and closed at the other (by a movable piston). The tube is filled with 30.0°C air, and a 384-Hz tuning fork is held at the open end. Resonance is heard when the piston is 22.8 cm from the open end and again when it is 68.3 cm from the open end. (a) What speed of sound is implied by these data? (b) Where would the piston be for the next resonance?

44. Water is pumped into a long cylinder at a rate of 18.0 cm³/s as in Figure P18.44. The radius of the cylinder is 4.0 cm, and at the top of the cylinder there is a tuning fork vibrating with a frequency of 200 Hz. As the column rises, how much time elapses between successive resonances?

44A. Water is pumped into a long cylinder at a rate R (cm³/s) as in Figure P18.44. The radius of the cylinder is r (cm), and at the top of the cylinder there is a tuning fork vibrating with a frequency f. As the column rises, how much time elapses between successive resonances?

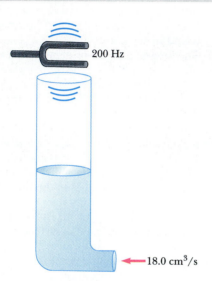

200 Hz

←18.0 cm³/s

FIGURE P18.44

45. A shower stall measures 86 cm × 86 cm × 210 cm. When you sing in the shower, which frequencies will sound the richest (resonate), assuming the shower acts as a pipe closed at both ends (nodes at both sides)? Assume also that the human voice ranges from 130 Hz to 2000 Hz (not necessarily one person's voice, however). Let the speed of sound in the hot shower stall be 355 m/s.

*Section 18.6   Standing Waves in Rods and Plates

46. An aluminum rod is clamped at the one-quarter position and set into longitudinal vibration by a variable-frequency driving source. The lowest frequency that produces resonance is 4400 Hz. The speed of sound in aluminum is 5100 m/s. Determine the length of the rod.

47. A 60.0-cm metal bar that is clamped at one end is struck with a hammer. If the speed of longitudinal (compressional) waves in the bar is 4500 m/s, what is the lowest frequency with which the struck bar will resonate?

48. Longitudinal waves move with a speed v in a bar of length L. Write an expression for the frequencies of the longitudinal vibrations of a metal bar that is (a) clamped at its center, as shown in Figure 18.14a, and (b) clamped at one-fourth the length of the bar from one end, as shown in Figure 18.14b.

49. An aluminum rod 1.6 m long is held at its center. It is stroked with a rosin-coated cloth to set up longitudinal vibrations in the fundamental mode. (a) What is the frequency of the waves established in the rod? (b) What harmonics are set up in the rod held in this manner? (c) What would be the fundamental frequency if the rod were copper?

*Section 18.7   Beats: Interference in Time

50. In certain ranges of a piano keyboard, more than one string is tuned to the same note to provide extra loudness. For example, the note at 110 Hz has two strings at this pitch. If one string slips from its normal tension of 600 N to 540 N, what beat frequency will be heard when the two strings are struck simultaneously?

51. Two waves with equal amplitude but with slightly different frequencies are traveling in the same direction through a medium. At a given point the separate displacements are described by

$$y_1 = A_0 \cos \omega_1 t \qquad \text{and} \qquad y_2 = A_0 \cos \omega_2 t$$

Use the trigonometric identity

$$\cos a + \cos b = 2 \cos\left(\frac{a-b}{2}\right) \cos\left(\frac{a+b}{2}\right)$$

to show that the resultant displacement due to the two waves is given by

$$y = 2A_0 \left[\cos\left(\frac{\omega_1 - \omega_2}{2}\right)t \right]\left[\cos\left(\frac{\omega_1 + \omega_2}{2}\right)t \right]$$

52. While attempting to tune a C note at 523 Hz, a piano tuner hears 3 beats per second between the oscillator and the string. (a) What are the possible frequencies of the string? (b) By what percentage should the tension in the string be changed to bring the string into tune?

53. A student holds a tuning fork oscillating at 256 Hz. He walks toward a wall at a constant speed of 1.33 m/s. (a) What beat frequency does he observe between the tuning fork and its echo? (b) How fast must he walk away from the wall to observe a beat frequency of 5 Hz?

ADDITIONAL PROBLEMS

54. (a) What is the fundamental frequency of a 5.00 × 10⁻³-kg steel piano wire of length 1.00 m, under a tension of 1350 N? (b) What is the fundamental frequency of an organ pipe, 1.00 m in length, closed at the bottom and open at the top?

55. Two loudspeakers are placed on a wall 2.00 m apart. A listener stands directly in front of one of the speakers, 3.00 m from the wall. The speakers are being driven by a single oscillator at a frequency of 300 Hz. (a) What is the phase difference between the two waves when they reach the observer? (b) What is the frequency closest to 300 Hz to which the oscillator may be adjusted such that the observer will hear minimal sound?

56. On a marimba, the wooden bar that sounds a tone when struck vibrates as a transverse standing wave with three antinodes and two nodes. The lowest frequency note is 87 Hz, produced by a bar 40 cm long. (a) Find the speed of transverse waves on the bar. (b) The loudness and duration of the emitted sound are enhanced by a resonant pipe suspended vertically below the center of the bar. If the pipe is open at the top end only and the speed of sound in air is 340 m/s, what is the length of the pipe required to resonate with the bar in part (a)?

57. Jane waits on a railroad platform, while two trains approach from the same direction at equal speeds of 8.0 m/s. Both trains are blowing their whistles (which have the same frequency), and one train is some distance behind the other. After the first train passes Jane, but before the second train passes her, she hears beats of frequency 4.0 Hz. What is the frequency of the train whistles?

58. A 0.010-kg and 2.0-m-long wire is fixed at both ends and vibrates in its simplest mode under tension 200 N. When a tuning fork is placed near the wire, a beat frequency of 5.0 Hz is heard. (a) What is the frequency (frequencies) of the tuning fork? (b) What should the tension in the wire be to make the beat disappear?

59. If two adjacent natural frequencies of an organ pipe are determined to be 0.55 kHz and 0.65 kHz, calculate the fundamental frequency and length of this pipe. (Use $v = 340$ m/s.)

60. Two loudspeakers are mounted on the ends of a horizontal 2.0-m stick that is set in rotation about a vertical axis through its center such that the stick makes one complete rotation every 5.0 s. A listener is situated far from the rotation axis between the speakers as shown in Figure P18.60. If the speakers each produce a middle C tone of 262 Hz, what is the beat frequency heard by the listener? Take the speed of sound to be 343 m/s.

60A. Two loudspeakers are mounted on the ends of a horizontal stick of length L that is set in rotation about a vertical axis through its center such that the stick makes one complete rotation every t seconds. A listener is situated far from the rotation axis between the speakers as shown in Figure P18.60. If the speakers each produce a frequency f, what is the beat frequency heard by the listener? Take the speed of sound to be v.

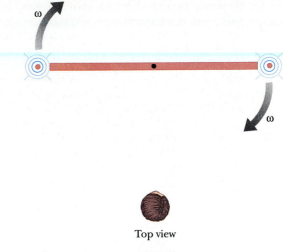

Top view

FIGURE P18.60

61. A 12-kg mass hangs in equilibrium from a string of total length $L = 5.0$ m and linear mass density $\mu = 0.0010$ kg/m. The string is wrapped around two light, frictionless pulleys that are separated by the distance $d = 2.0$ m (Fig. P18.61a). (a) Determine the tension in the string. (b) At what frequency must the string between the pulleys vibrate in order to form the standing wave pattern shown in Figure P18.61b?

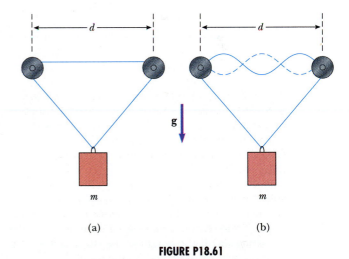

(a)   (b)

FIGURE P18.61

62. While waiting for Stan Speedy to arrive on a late passenger train, Kathy Kool notices beats occurring as a result of two trains blowing their whistles simultaneously. One train is at rest and the other is approaching her at a speed of 20 km/h. Assume that both whistles have the same frequency and that the speed of sound is 344 m/s. If Kathy hears 4 beats per second, what is the frequency of the whistles?

63. A string (mass = 4.8 g, length = 2.0 m, and tension = 48 N), fixed at both ends, vibrates in its second ($n = 2$) natural mode. What is the wavelength in air of the sound emitted by this vibrating string?

64. In an arrangement like the one shown in Figure 18.2, paths r_1 and r_2 are each 1.75 m in length. The top portion of the tube (corresponding to r_2) is filled with air at 0°C (273 K). Air in the lower portion is quickly heated to 200°C (473 K). What is the lowest speaker frequency that will produce an intensity maximum at the receiver? (You may determine the speed of sound in air in different temperatures by using the expression $v = 331(T/273)^{1/2}$ m/s, where T is in K.)

65. Two train whistles have identical frequencies of 180 Hz. When one train is at rest in the station sounding its whistle, a beat frequency of 2 Hz is heard from a moving train. What two possible speeds and directions can the moving train have?

66. In a major chord on the physical pitch musical scale, the frequencies are in the ratios $4:5:6:8$. A set of pipes, closed at one end, are to be cut so that when sounded in their fundamental mode, they will sound out a major chord. (a) What is the ratio of the lengths of the pipes? (b) What length pipes are needed if the lowest frequency of the chord is 256 Hz? (c) What are the frequencies of this chord?

67. Two wires are welded together. The wires are of the same material, but one is twice the diameter of the other one. They are subjected to a tension of 4.6 N. The thin wire has a length of 40 cm and a linear mass density of 2.0 g/m. The combination is fixed at both ends and vibrated in such a way that two antinodes are present with the central node being right at the weld. (a) What is the frequency of vibration? (b) How long is the thick wire?

68. Two identical strings, each fixed at both ends, are arranged near each other. If string A starts oscillating in its fundamental mode, it is observed that string B will begin vibrating in its third ($n = 3$) natural mode. Determine the ratio of the tension of string B to tension of string A.

69. A standing wave is set up in a string of variable length and tension by a vibrator of variable frequency. When the vibrator has a frequency f in a string of length L and tension F there are n antinodes set up in the string. (a) If the length of the string is doubled, by what factor should the frequency be changed to get the same number of antinodes? (b) If the frequency and length are held constant, what tension will produce $n + 1$ antinodes? (c) If the frequency is tripled and the length halved, by what factor should the tension be changed to get twice as many antinodes?

70. Radar detects the speed of a car, using the Doppler shift of microwaves that are reflected off the moving car, by beating the received wave with the transmitted wave and measuring the difference. The Doppler shift for light is

$$f = f_0 \sqrt{\frac{c + v}{c - v}}$$

where f_0 is the transmitted frequency, c is the speed of light (3×10^8 m/s), and v the relative speed of the two objects. (a) Show that the wave that reflects back to the source has a frequency

$$f = f_0 \frac{(c + v)}{(c - v)}$$

(b) Show that the expression for the beat frequency of the microwaves may be written as $f_b = 2 v/\lambda$. (Since the beat frequency is much smaller than the transmitted frequency, use the approximation $f + f_0 = 2f_0$.) (c) What beat frequency is measured for a speed of 30 m/s (67 mph) if the microwaves have a frequency of 10 GHz? (1 GHz = 10^9 Hz.) (d) If the beat frequency measurement is accurate to ±5 Hz, how accurate is the velocity measurement?

SPREADSHEET PROBLEMS

S1. Spreadsheet 18.1 adds two traveling waves at some fixed time t. The resultant wave function is

$$y = y_1 + y_2 = A_1 \sin(k_1 x - \omega_1 t + \phi_1) + A_2 \sin(k_2 x - \omega_2 t + \phi_2)$$

Add two waves traveling in opposite directions having the same wavelengths, the same phases, and the same speeds. (a) Use $A_1 = A_2 = 0.10$ m, $\omega_1 = -\omega_2 = 3.0$ rad/s, $\phi_1 = \phi_2 = 0$, and $k_1 = k_2 = 2.0$ rad/m. View the associated graph. Choose different values of t to see the time evolution of the resultant wave function. Do you get standing waves? (b) Repeat part (a) using $A_1 = 0.10$ m, $A_2 = 0.20$ m.

S2. Use Spreadsheet 18.1 to add two traveling waves that differ only in phase. (a) Choose $A_1 = A_2 = 0.10$ m, $\omega_1 = \omega_2 = 2.5$ rad/s, $k_1 = k_2 = 1.0$ rad/m, $\phi_1 = 0$, and $\phi_2 = 0$, $\pi/8$, $\pi/4$, $\pi/2$, and π. View the associated graphs. For which values of ϕ_2 do you get constructive interference? destructive interference? (b) Repeat part (a) using $A_2 = 0.20$ m.

S3. Use Spreadsheet 18.1 to add two traveling waves with different wavelengths. (a) Choose $A_1 = A_2 = 0.10$ m, $\omega_1 = \omega_2 = 3.0$ rad/s, $k_1 = 2.0$ rad/m, $k_2 = 1.0$ rad/m, and $\phi_1 = \phi_2 = 0$. View the associated graph. (b) Repeat part (a) with $k_2 = 3, 4, 5, 6, 7, 8, 9$, and 10. Examine the graph for each case. Explain the appearance of the graphs.

S4. Three waves with the same frequency, wavelength, and amplitude are traveling in the same direction. Each differs in phase such that $(\phi_2 - \phi_1) = (\phi_3 - \phi_2) = \Delta\phi$. Modify Spreadsheet 18.1 to add the three waves at some fixed time t. Calculate the resultant wave function at $t = 0.0$, 1.0 s, and 2.0 s. Use $A_1 = A_2 = A_3 = 0.050$ m, $\omega_1 = \omega_2 = \omega_3 = 2.5$ rad/s, and $k_1 = k_2 = k_3 = 1.0$ rad/m. Choose $\Delta\phi = \pi/6$. View the associated graph. Repeat for different values of $\Delta\phi$.

S5. Write a spreadsheet or computer program to add two traveling waves of different frequencies:

$$y_1 = y_0 \cos 2\pi f_1 t \qquad y_2 = y_0 \cos 2\pi f_2 t$$

If the two frequencies are close to each other, beats will be produced. Does your program show the beats? From your numerical results, how does the beat frequency relate to the frequencies of the two waves?

S6. The Fourier theorem states that any periodic wave of frequency f, no matter how complicated, can be expressed as a sum of even or odd harmonic functions.

That is,

$$y(t) = \sum_{n=0}^{\infty} A_n \sin(n\omega t + \phi_n)$$

where $\omega = 2\pi f$ is the fundamental angular frequency. (a) Use $A_1 = 1$, $A_2 = 1/2$, $A_3 = 1/3$, and so on. All the ϕ_n's are zero. (b) Use $A_1 = 1$, $A_3 = 1/3$, $A_5 = 1/5$, and so on. All the even A_n's and all the ϕ_n's are zero. Write a spreadsheet or computer program that will calculate this sum. You should design your spreadsheet or program so that the number of terms in the sum can be easily specified. Start with just the first term ($n = 1$). Then rerun the calculation with more terms in the sum ($n = 2, 3, 4, \ldots,$ 20). Plot the sum so that one or two periods are displayed. Note how the waveform builds up as more terms are added.

Water dripping from tree branches and mountain ledges forms ice crystals when the temperature falls below 0°C. As the crystals form, a phase transition from the liquid to the solid state occurs, and water molecules become arranged in a highly ordered structure. *(©John Beatty/Tony Stone Images)*

Thermodynamics

When dining, I had often observed that some particular dishes retained their Heat much longer than others; and that apple pies, and apples and almonds mixed (a dish in great repute in England) remained hot a surprising length of time. Much struck with this extraordinary quality of retaining Heat, which apples appeared to possess, it frequently occurred to my recollection; and I never burnt my mouth with them, or saw others meet with the same misfortune, without endeavouring, but in vain, to find out some way of accounting, in a satisfactory manner, for this surprising phenomenon.

BENJAMIN THOMPSON
(COUNT RUMFORD)

We now turn to the study of thermodynamics, which is concerned with the concepts of heat and temperature. As we shall see, thermodynamics is very successful in explaining the bulk properties of matter and the correlation between these properties and the mechanics of atoms and molecules.

Historically, the development of thermodynamics paralleled the development of the atomic theory of matter. By the 1820s, chemical experiments provided solid evidence for the existence of atoms. At that time, scientists recognized that there must be a connection between the theory of heat and temperature and the structure of matter. In 1827, the botanist Robert Brown reported that grains of pollen suspended in a liquid move erratically from one place to another, as if under constant agitation. In 1905, Albert Einstein explained this erratic motion, now called Brownian motion. Einstein explained this phenomenon by assuming that the grains of pollen are under constant bombardment by "invisible" molecules in the liquid, which themselves undergo an erratic motion. This important experiment and Einstein's insight gave scientists a means of discovering vital information concerning molecular motion. It also gave reality to the concept of the atomic constituents of matter.

Have you ever wondered how a refrigerator is able to cool its contents or what types of transformations occur in a power plant or in the engine of your automobile or what happens to the kinetic energy of an object when it falls to the ground and comes to rest? The laws of thermodynamics and the concepts of heat and temperature enable us to answer such practical questions. In general, thermodynamics is concerned with transformations of matter in all of its forms: solid, liquid, and gas.

CHAPTER 19

Temperature

After the bottle is shaken, the cork is blown off. Contrary to common belief, shaking the champagne bottle does not increase the CO_2 pressure inside. Since the temperature of the bottle and its contents remain constant, the equilibrium pressure does not change, as can be shown by replacing the cork with a pressure gauge. Shaking the bottle displaces some CO_2 from the "head space" to bubbles within the liquid that become attached to the walls. If the bubbles remain attached to the walls, when the bottle is opened, the bubbles beneath the liquid level rapidly expand, expelling liquid in the process. *(Steve Niedorf, The IMAGE Bank)*

In the study of mechanics, such concepts as mass, force, and kinetic energy were carefully defined in order to make the subject quantitative. Likewise, a quantitative description of thermal phenomena requires a careful definition of the concepts of temperature, heat, and internal energy. The laws of thermodynamics provide us with a relationship between heat flow, work, and the internal energy of a system.

The composition of a body is an important factor when dealing with thermal phenomena. For example, liquids and solids expand only slightly when heated, whereas gases expand appreciably when heated. If the gas is not free to expand, its pressure rises when heated. Certain substances may melt, boil, burn, or explode, depending on their composition and structure. Thus, the thermal behavior of a substance is closely related to its structure.

This chapter concludes with a study of ideal gases. We shall approach this study on two levels. The first will examine ideal gases on the macroscopic scale. Here we shall be concerned with the relationships among such quantities as pressure, volume, and temperature. On the second level, we shall examine gases on a microscopic scale, using a model that pictures the components of a gas as small particles.

Molten lava flowing down a mountain in Kilauea, Hawaii. In this case, the hot lava flows smoothly out of a central crater until it cools and solidifies to form the mountains. However, violent eruptions sometimes occur, as in the case of Mount St. Helens in 1980, which can cause both local and global (atmospheric) damage. *(Ken Sakomoto, Black Star)*

The latter approach, called the kinetic theory of gases, will help us to understand what happens on the atomic level to affect such macroscopic properties as pressure and temperature.

19.1 TEMPERATURE AND THE ZEROTH LAW OF THERMODYNAMICS

We often associate the concept of **temperature** with how hot or cold an object feels when we touch it. Thus, our senses provide us with a qualitative indication of temperature. However, our senses are unreliable and often misleading. For example, if we remove a metal ice tray and a cardboard box of frozen vegetables from the freezer, the ice tray feels colder to the hand than the box even though both are at the same temperature. The two objects feel different because metal is a better heat conductor than cardboard. What we need, therefore, is a reliable and reproducible method for establishing the relative hotness or coldness of bodies. Scientists have developed a variety of thermometers for making such quantitative measurements.

We are all familiar with the fact that two objects at differential initial temperatures eventually reach some intermediate temperature when placed in contact with each other. For example, a piece of meat placed on a block of ice in a well-insulated container eventually reaches a temperature near 0°C. Likewise, if an ice cube is dropped into a cup of hot coffee, the ice eventually melts and the coffee's temperature decreases.

In order to understand the concept of temperature, it is useful to first define two often-used phrases, *thermal contact* and *thermal equilibrium*. To grasp the meaning of thermal contact, imagine two objects placed in an insulated container so that they interact with each other but not with the rest of the world. If the objects are at different temperatures, energy is exchanged between them. The energy exchanged between objects because of a temperature difference is called **heat.** We shall examine the concept of heat in more detail in Chapter 20. For purposes of the current discussion, we shall assume that two objects are in **thermal contact** with each other if heat can be exchanged between them. **Thermal equilibrium** is a situation in which two objects in thermal contact with each other cease to have any exchange of heat.

Now consider two objects, A and B, which are not in thermal contact, and a third object, C, which is our thermometer. We wish to determine whether or not A and B are in thermal equilibrium with each other. The thermometer (object C) is first placed in thermal contact with A until thermal equilibrium is reached. From that moment on, the thermometer's reading remains constant, and we record it. The thermometer is then removed from A and placed in thermal contact with B, and its reading is recorded after thermal equilibrium is reached. If the two readings are the same, then A and B are in thermal equilibrium with each other.

We can summarize these results in a statement known as the **zeroth law of thermodynamics** (the law of equilibrium):

Zeroth law of thermodynamics

If objects A and B are separately in thermal equilibrium with a third object, C, then A and B are in thermal equilibrium with each other if placed in thermal contact.

This statement may easily be proven experimentally and is very important because it can be used to define temperature. We can think of temperature as the property that determines whether or not an object is in thermal equilibrium with other objects. *Two objects in thermal equilibrium with each other are at the same temperature.* Conversely, if two objects have different temperatures, they are *not* in thermal equilibrium with each other. The zeroth law can be shown to be a consequence of the second law of thermodynamics, which we meet later, in Chapter 22.

19.2 THERMOMETERS AND TEMPERATURE SCALES

Thermometers are devices used to define and measure the temperature of a system. All thermometers make use of the change in some physical property with temperature. Some of these physical properties are (1) the change in volume of a liquid, (2) the change in length of a solid, (3) the change in pressure of a gas at constant volume, (4) the change in volume of a gas at constant pressure, (5) the change in electric resistance of a conductor, and (6) the change in color of some object. For a given substance and a given temperature range, a temperature scale can be established based on any one of these physical quantities.

The most common thermometer in everyday use consists of a mass of liquid — usually mercury or alcohol — that expands into a glass capillary tube when heated (Fig. 19.1). In this case the physical property is the change in volume of a liquid. Any temperature change can be defined to be proportional to the change in length of the liquid column. The thermometer can be calibrated by placing it in thermal contact with some natural systems that remain at constant temperature. One such system is a mixture of water and ice in thermal equilibrium at atmospheric pressure, which is defined to have a temperature of zero degrees Celsius, written 0°C; this temperature is called the ice point of water. Another commonly used system is a mixture of water and steam in thermal equilibrium at atmospheric pressure; its temperature is 100°C, the steam point of water. Once the liquid levels in the thermometer have been established at these two points, the column is divided into 100 equal segments, each denoting a change in temperature of one Celsius degree.

Thermometers calibrated in this way do present problems when extremely accurate readings are needed. For instance, an alcohol thermometer calibrated at the ice and steam points of water might agree with a mercury thermometer only at the calibration points. Because mercury and alcohol have different thermal expansion properties, when one thermometer reads a temperature of 50°C, for example, the other may indicate a slightly different value. The discrepancies between thermometers are especially large when the temperatures to be measured are far from the calibration points.[1]

An additional practical problem of any thermometer is its limited temperature range. A mercury thermometer, for example, cannot be used below the freezing point of mercury, which is − 39°C, and an alcohol thermometer is not useful for temperatures above 85°C. To surmount these problems, we need a universal thermometer whose readings are independent of the substance used. The gas thermometer approaches this requirement.

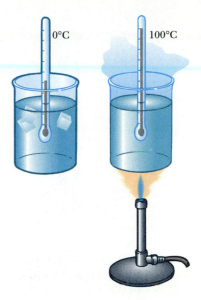

FIGURE 19.1 Schematic diagram of a mercury thermometer. As a result of thermal expansion, the level of the mercury rises as the mercury is heated from 0°C (the ice point) to 100°C (the steam point).

---

[1] Thermometers that use the same material may also give different readings. This is due in part to difficulties in constructing uniform-bore glass capillary tubes.

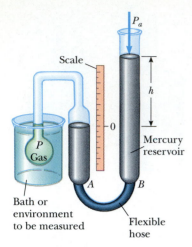

FIGURE 19.2 A constant-volume gas thermometer measures the pressure of the gas contained in the flask immersed in the bath. The volume of gas in the flask is kept constant by raising or lowering reservoir *B* such that the mercury level in column *A* remains constant.

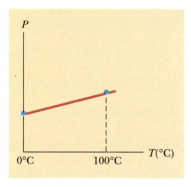

FIGURE 19.3 A typical graph of pressure versus temperature taken with a constant-volume gas thermometer. The dots represent known reference temperatures (the ice point and the steam point).

19.3 THE CONSTANT-VOLUME GAS THERMOMETER AND THE KELVIN SCALE

In a gas thermometer, the temperature readings are nearly independent of the substance used in the thermometer. One version is the constant-volume gas thermometer shown in Figure 19.2. The physical property being exploited in this device is how the pressure of a fixed volume of gas varies with temperature. When the constant-volume gas thermometer was developed, it was calibrated using the ice and steam points of water, as follows. (A different calibration procedure, to be discussed shortly, is now used.) The gas flask was inserted into an ice bath, and mercury reservoir *B* was raised or lowered until the volume of the confined gas was at some value, indicated by the zero point on the scale. The height *h*, the difference between the mercury levels in the reservoir and column *A*, indicated the pressure in the flask at 0°C. The flask was inserted into water at the steam point, and reservoir *B* was readjusted until the height in column *A* was again brought to zero on the scale, ensuring that the gas volume was the same as it had been in the ice bath (hence, the designation "constant volume"). This gave a value for pressure at 100°C. These pressure and temperature values were then plotted as in Figure 19.3. The line connecting the two points serves as a calibration curve for unknown temperatures. If we wanted to measure the temperature of a substance, we would place the gas flask in thermal contact with the substance and adjust the height of the mercury column until the gas occupied its specified volume. The height of the mercury column would tell us the pressure of the gas, and we could then find the temperature of the substance from the graph.

Now suppose that temperatures are measured with various gas thermometers containing different gases. Experiments show that the thermometer readings are nearly independent of the type of gas used, so long as the gas pressure is low and the temperature is well above the point at which the gas liquefies (Fig. 19.4). The agreement among thermometers using various gases improves as the pressure is reduced.

If you extend the curves in Figure 19.4 back toward negative temperatures, you find, in every case, that the pressure is zero when the temperature is −273.15°C. This significant temperature is used as the basis for the Kelvin temperature scale, which sets −273.15°C as its zero point (0 K). The size of a degree on the Kelvin scale is identical to the size of a degree on the Celsius scale. Thus, the relationship that enables conversion between these temperatures is

$$T_C = T - 273.15 \qquad (19.1)$$

where T_C is the **Celsius temperature** and T is the **Kelvin temperature** (sometimes called the **absolute temperature**).

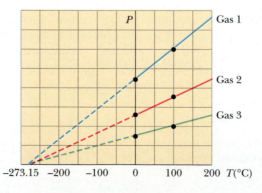

FIGURE 19.4 Pressure versus temperature for dilute gases. Note that, for all gases, the pressure extrapolates to zero at the unique temperature of −273.15°C.

TABLE 19.1 Fixed-Point Temperatures

Fixed Point	Temperature (°C)	Temperature (K)
Triple point of hydrogen	− 259.34	13.81
Boiling point of hydrogen at 33.36 kPa pressure	− 256.108	17.042
Boiling point of hydrogen	− 252.87	20.28
Triple point of neon	− 246.048	27.102
Triple point of oxygen	− 218.789	54.361
Boiling point of oxygen	− 182.962	90.188
Triple point of water	0.01	273.16
Boiling point of water	100.00	373.15
Freezing point of tin	231.9681	505.1181
Freezing point of zinc	419.58	692.73
Freezing point of silver	961.93	1235.08
Freezing point of gold	1064.43	1337.58

All values from National Bureau of Standards Special Publication 420; U.S. Department of Commerce, May 1975. All values at standard atmospheric pressure except as noted.

Early gas thermometers made use of the ice point and steam point according to the procedure just described. However, these points are experimentally difficult to duplicate. For this reason, a new temperature scale based on a single fixed point with b equal to zero was adopted in 1954 by the International Committee on Weights and Measures. The assigned temperatures of particular fixed points associated with various substances are given in Table 19.1. The *triple point of water,* which corresponds to the single temperature and pressure at which water, water vapor, and ice can coexist in equilibrium, was chosen as a convenient and reproducible reference temperature for this new scale. This triple point occurs at a temperature of approximately 0.01°C and a pressure of 4.58 mm of mercury. On the new scale, the temperature of water at the triple point was set at 273.16 kelvin, abbreviated 273.16 K.[2] This choice was made so that the old temperature scale based on the ice and steam points would agree closely with the new scale based on the triple point. This new scale is called the **thermodynamic temperature scale,** and the SI unit of thermodynamic temperature, the **kelvin,** *is defined as the fraction 1/273.16 of the temperature of the triple point of water.*

Figure 19.5 shows the Kelvin temperature for various physical processes and structures. The temperature 0 K is often referred to as **absolute zero,** and as Figure 19.5 shows, this temperature has never been achieved, although laboratory experiments have come close.

What would happen to a gas if its temperature could reach 0 K? As Figure 19.4 indicates, the pressure it exerted on the walls of its container would be zero. In Chapter 21 we show that the pressure of a gas is proportional to the kinetic energy of the molecules of that gas. Thus, according to classical physics, the kinetic energy of the gas would become zero, and molecular motion would cease; hence, the molecules would settle out on the bottom of the container. Quantum theory modifies this model and shows that there would be some residual energy, called the zero-point energy, at this low temperature.

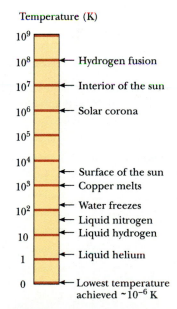

Temperature (K)

FIGURE 19.5 Absolute temperatures at which various selected physical processes occur. Note that the scale is logarithmic.

[2] We shall describe the meaning of this point in Chapter 22 when we discuss the second law of thermodynamics.

The Celsius, Fahrenheit, and Kelvin Temperature Scales[3]

Equation 19.1 shows that the Celsius temperature T_C is shifted from the absolute (Kelvin) temperature T by 273.15°. Because the size of a degree is the same on the two scales, a temperature difference of 5°C is equal to a temperature difference of 5 K. The two scales differ only in the choice of the zero point. Thus, the ice point (273.15 K) corresponds to 0.00°C, and the steam point (373.15 K) is equivalent to 100.00°C.

The most common temperature scale in everyday use in the United States is the **Fahrenheit scale.** This scale sets the temperature of the ice point at 32°F and the temperature of the steam point at 212°F. The relationship between the Celsius and Fahrenheit temperature scales is

$$T_F = \tfrac{9}{5}T_C + 32°F \tag{19.2}$$

Equation 19.2 can be used to find a relationship between changes in temperature on the Celsius and Fahrenheit scales.

$$\Delta T_C = \Delta T = \tfrac{5}{9}\Delta T_F \tag{19.3}$$

The change in temperature on the Celsius scale equals the change on the Kelvin scale. That is, $\Delta T = \Delta T_C$.

EXAMPLE 19.1 Converting Temperatures

On a day when the temperature reaches 50°F, what is the temperature in degrees Celsius and in kelvin?

Solution Substituting $T_F = 50°F$ into Equation 19.2, we get

$$T_C = \tfrac{5}{9}(T_F - 32) = \tfrac{5}{9}(50 - 32) = \boxed{10°C}$$

From Equation 19.1, we find that

$$T = T_C + 273.15 = \boxed{283\ K}$$

EXAMPLE 19.2 Heating a Pan of Water

A pan of water is heated from 25°C to 80°C. What is the change in its temperature on the Kelvin scale and on the Fahrenheit scale?

Solution From Equation 19.1, we see that the change in temperature on the Celsius scale equals the change on the

Kelvin scale. Therefore,

$$\Delta T = \Delta T_C = 80 - 25 = 55°C = \boxed{55\ K}$$

From Equation 19.3, we find

$$\Delta T_F = \tfrac{9}{5}\Delta T_C = \tfrac{9}{5}(80 - 25) = \boxed{99°F}$$

19.4 THERMAL EXPANSION OF SOLIDS AND LIQUIDS

Our discussion of the liquid thermometer made use of one of the best-known changes that occurs in a substance: As its temperature increases, its volume increases. (As we shall see shortly, in some substances the volume decreases when the temperature increases.) This phenomenon, known as **thermal expansion,** plays an important role in numerous engineering applications. For example, thermal

---

[3] Named after Anders Celsius (1701–1744), Gabriel Fahrenheit (1686–1736), and William Thomson, Lord Kelvin (1824–1907).

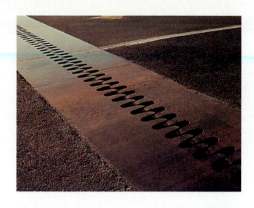

Thermal expansion joints are used to separate sections of roadways on bridges. Without these joints, the surfaces would buckle due to thermal expansion on very hot days or crack due to contraction on very cold days. *(Frank Siteman, Stock/Boston)*

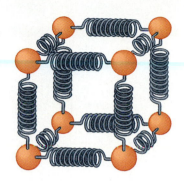

FIGURE 19.6 A mechanical model of a crystalline solid. The atoms (solid spheres) are imagined to be attached to each other by springs, which reflect the elastic nature of the interatomic forces.

expansion joints must be included in bridges and some other structures to compensate for changes in dimensions with temperature variations.

The overall thermal expansion of a body is a consequence of the change in the average separation between its constituent atoms or molecules. To understand this, consider how the atoms in a solid substance behave. Imagine that the atoms of the solid are connected by a set of stiff springs as in Figure 19.6. At ordinary temperatures, the atoms vibrate about their equilibrium positions with an amplitude of approximately 10^{-11} m and a frequency of approximately 10^{13} Hz. The average spacing between the atoms is approximately 10^{-10} m. As the temperature of the solid increases, the atoms vibrate with larger amplitudes and the average separation between them increases.[4] Consequently, the solid expands. If the thermal expansion of an object is sufficiently small compared with its initial dimensions, then the change in any dimension is, to a good approximation, dependent on the first power of the temperature change.

Suppose an object has an initial length L_0 along some direction at some temperature and that the length increases by an amount ΔL for the change in temperature ΔT. Experiments show that, when ΔT is small, ΔL is proportional to ΔT and to L_0:

$$\Delta L = \alpha L_0 \Delta T \qquad (19.4)$$

or

$$L - L_0 = \alpha L_0 (T - T_0) \qquad (19.5)$$

where L is the final length, T is the final temperature, and the proportionality constant α is called the **average coefficient of linear expansion** for a given material and has units of $(°C)^{-1}$.

It may be helpful to think of thermal expansion as an effective magnification or as a photographic enlargement of an object when it is heated. For example, as a metal washer is heated (Fig. 19.7) all dimensions, including the radius of the hole, increase according to Equation 19.4. Table 19.2 lists the average coefficient of linear expansion for various materials. Note that for these materials α is positive, indicating an increase in length with increasing temperature. This is not always the case. For example, some substances, such as calcite ($CaCO_3$), expand along one

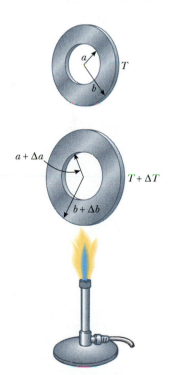

FIGURE 19.7 Thermal expansion of a homogeneous metal washer. Note that as the washer is heated, all dimensions increase. (The expansion is exaggerated in this figure.)

---

[4] Strictly speaking, thermal expansion arises from the *asymmetric* nature of the potential energy curve for the atoms in a solid. If the oscillators were truly harmonic, the average atomic separations would not change regardless of the amplitude of vibration.

TABLE 19.2   Expansion Coefficients for Some Materials Near Room Temperature

Material	Linear Expansion Coefficient $\alpha(°C)^{-1}$	Material	Volume Expansion Coefficient $\beta(°C)^{-1}$
Aluminum	24×10^{-6}	Alcohol, ethyl	1.12×10^{-4}
Brass and bronze	19×10^{-6}	Benzene	1.24×10^{-4}
Copper	17×10^{-6}	Acetone	1.5×10^{-4}
Glass (ordinary)	9×10^{-6}	Glycerine	4.85×10^{-4}
Glass (Pyrex)	3.2×10^{-6}	Mercury	1.82×10^{-4}
Lead	29×10^{-6}	Turpentine	9.0×10^{-4}
Steel	11×10^{-6}	Gasoline	9.6×10^{-4}
Invar (Ni–Fe alloy)	0.9×10^{-6}	Air at 0°C	3.67×10^{-3}
Concrete	12×10^{-6}	Helium at 0°C	3.665×10^{-3}

dimension (positive α) and contract along another (negative α) as their temperature is increased.

Because the linear dimensions of an object change with temperature, it follows that surface area and volume do so also. The change in volume at constant pressure is proportional to the original volume V and to the change in temperature according to the relationship

The change in volume of a solid at constant pressure is proportional to the change in temperature

$$\Delta V = \beta V \Delta T \qquad (19.6)$$

where β is the **average coefficient of volume expansion.** *For a solid, the coefficient of volume expansion is approximately three times the linear expansion coefficient, or $\beta = 3\alpha$.* (This assumes that the coefficient of linear expansion of the solid is the same in all directions.) Therefore, Equation 19.6 can be written

$$\Delta V = 3\alpha V \Delta T \qquad (19.7)$$

To show that $\beta = 3\alpha$ for a solid, consider an object in the shape of a box of dimensions ℓ, w, and h. Its volume at some temperature T is $V = \ell wh$. If the temperature changes to $T + \Delta T$, its volume changes to $V + \Delta V$, where each dimension changes according to Equation 19.6. Therefore,

$$\begin{aligned} V + \Delta V &= (\ell + \Delta \ell)(w + \Delta w)(h + \Delta h) \\ &= (\ell + \alpha \ell \, \Delta T)(w + \alpha w \, \Delta T)(h + \alpha h \, \Delta T) \\ &= \ell wh(1 + \alpha \, \Delta T)^3 \\ &= V[1 + 3\alpha \, \Delta T + 3(\alpha \, \Delta T)^2 + (\alpha \, \Delta T)^3] \end{aligned}$$

Hence, the fractional change in volume is

$$\frac{\Delta V}{V} = 3\alpha \, \Delta T + 3(\alpha \, \Delta T)^2 + (\alpha \, \Delta T)^3$$

Since the product $\alpha \, \Delta T$ is small compared with unity for typical values of ΔT (less than $\approx 100°C$), we can neglect the terms $3(\alpha \, \Delta T)^2$ and $(\alpha \, \Delta T)^3$. In this approximation, we see that

$$\beta = \frac{1}{V}\frac{\Delta V}{\Delta T} = 3\alpha$$

A sheet or flat plate can be described by its area. You should show (Problem 53) that the change in the area of a plate is

$$\Delta A = 2\alpha A \, \Delta T \tag{19.8}$$

As Table 19.2 indicates, each substance has its own characteristic coefficients of expansion. For example, when the temperature of a brass rod and a steel rod of equal length are raised by the same amount from some common initial value, the brass rod expands more than the steel rod because brass has a larger coefficient of expansion than steel. A simple device called a bimetallic strip that utilizes this principle is found in practical devices such as thermostats. As the temperature of the strip increases, the two metals expand by different amounts, and the strip bends as in Figure 19.8.

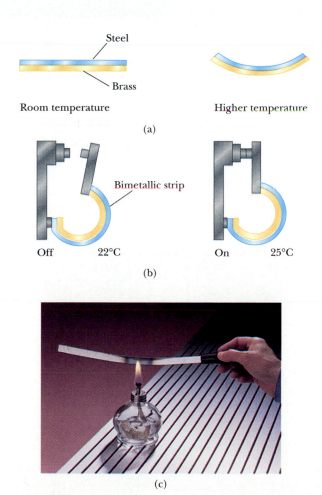

Steel

Brass

Room temperature

Higher temperature

(a)

Bimetallic strip

Off   22°C   On   25°C

(b)

(c)

FIGURE 19.8 (a) A bimetallic strip bends as the temperature changes because the two metals have different expansion coefficients. (b) A bimetallic strip used in a thermostat to break or make electrical contact. (c) The two metals that form this bimetallic strip are bonded along their longest dimension. The strip in this photograph was straight before being heated and bends when heated. Which way would it bend if it were cooled? *(Courtesy of Central Scientific Company)*

EXAMPLE 19.3   Expansion of a Railroad Track

A steel railroad track has a length of 30.0 m when the temperature is 0.0°C. (a) What is its length on a hot day when the temperature is 40.0°C?

Solution   Making use of Table 19.2 and noting that the change in temperature is 40.0°C, we find that the increase in length is

$$\Delta L = \alpha L \, \Delta T = [11 \times 10^{-6} (°C)^{-1}](30.0 \text{ m})(40.0°C)$$
$$= 0.013 \text{ m}$$

Therefore, the track's length at 40.0°C is 30.013 m.

(b) Suppose the ends of the rail are rigidly clamped at 0.0°C so as to prevent expansion. Calculate the thermal stress set up in the rail if its temperature is raised to 40.0°C.

Solution   From the definition of Young's modulus for a solid (Chapter 12), we have

$$\text{Tensile stress} = \frac{F}{A} = Y \frac{\Delta \ell}{\ell}$$

Since Y for steel is 20×10^{10} N/m² (Table 12.1), we have

$$\frac{F}{A} = \left(20 \times 10^{10} \frac{\text{N}}{\text{m}^2}\right)\left(\frac{0.013 \text{ m}}{30.0 \text{ m}}\right) = \boxed{8.7 \times 10^7 \text{ N/m}^2}$$

Thermal expansion: The extreme heat of a July day in Asbury Park, New Jersey, caused these railroad tracks to buckle. *(Wide World Photos)*

Exercise   If the rail has a cross-sectional area of 30.0 cm², calculate the force of compression in the rail.

Answer 2.6×10^5 N or 58 000 lb!

The Unusual Behavior of Water

Liquids generally increase in volume with increasing temperature and have average volume-expansion coefficients about ten times greater than those of solids. Water is an exception to this rule, as we can see from its density-versus-temperature curve in Figure 19.9. As the temperature increases from 0°C to 4°C, water con-

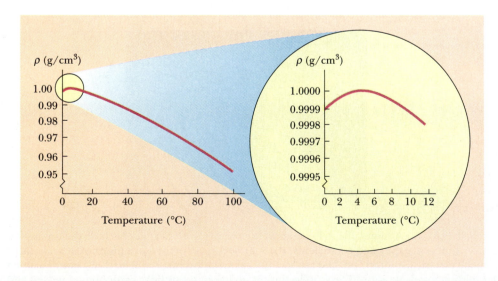

FIGURE 19.9   The variation of density with temperature for water at atmospheric pressure. The inset at the right shows that the maximum density of water occurs at 4°C.

tracts and thus its density increases. Above 4°C, water expands with increasing temperature. In other words, the density of water reaches a maximum value of 1000 kg/m³ at 4°C.

We can use this unusual thermal expansion behavior of water to explain why a pond freezes starting at the surface. When the atmospheric temperature drops from, say, 7°C to 6°C, the water at the surface of the pond also cools and consequently decreases in volume. This means that the surface water is denser than the water below it, which has not cooled and decreased in volume. As a result, the surface water sinks and warmer water from below is forced to the surface to be cooled. When the atmospheric temperature is between 4°C and 0°C, however, the surface water expands as it cools, becoming less dense than the water below it. The mixing process stops, and eventually the surface water freezes. As the water freezes, the ice remains on the surface because ice is less dense than water. The ice continues to build up at the surface, while water near the bottom remains at 4°C. If this did not happen, fish and other forms of marine life would not survive.

19.5   MACROSCOPIC DESCRIPTION OF AN IDEAL GAS

In this section we are concerned with the properties of a gas of mass m confined to a container of volume V at a pressure P and temperature T. It is useful to know how these quantities are related. In general, the equation that interrelates these quantities, called the *equation of state*, is very complicated. However, if the gas is maintained at a very low pressure (or low density), the equation of state is found experimentally to be quite simple. Such a low-density gas is commonly referred to as an **ideal gas.** Most gases at room temperature and atmospheric pressure behave approximately as ideal gases.[5]

It is convenient to express the amount of gas in a given volume in terms of the number of moles, n. As we learned in Section 1.3, **one mole** of any substance is that mass of the substance that contains Avogadro's number, $N_A = 6.022 \times 10^{23}$, of molecules. The number of moles of a substance is related to its mass m through the expression

$$n = \frac{m}{M} \tag{19.9}$$

where M is a quantity called the **molar mass** of the substance, usually expressed in grams per mole. For example, the molar mass of oxygen, O_2, is 32.0 g/mol. Therefore, the mass of one mole of oxygen is 32.0 g.

Now suppose an ideal gas is confined to a cylindrical container whose volume can be varied by means of a movable piston, as in Figure 19.10. We assume that the cylinder does not leak, and so the mass (or the number of moles) of the gas remains constant. For such a system, experiments provide the following information. First, when the gas is kept at a constant temperature, its pressure is inversely proportional to the volume (Boyle's law). Second, when the pressure of the gas is kept constant, the volume is directly proportional to the temperature (the law of Charles and Gay-Lussac). These observations can be summarized by the **equation of state for an ideal gas:**

$$PV = nRT \tag{19.10}$$

Equation of state for an ideal gas

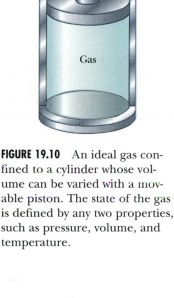

FIGURE 19.10   An ideal gas confined to a cylinder whose volume can be varied with a movable piston. The state of the gas is defined by any two properties, such as pressure, volume, and temperature.

---

[5] To be more specific, the assumption here is that the temperature of the gas must not be too low (it must not condense into a liquid) nor too high, and the pressure must be low.

In this expression, called the **ideal gas law,** R is a universal constant that is the same for all gases, and T is the absolute temperature in kelvin. Experiments on several gases show that as the pressure approaches zero, the quantity PV/nT approaches the same value R for all gases. For this reason, R is called the **universal gas constant.** In SI units, where pressure is expressed in pascals and volume in cubic meters, the product PV has units of newton·meters, or joules, and R has the value

The universal gas constant

$$R = 8.31 \text{ J/mol·K} \tag{19.11}$$

If the pressure is expressed in atmospheres and the volume in liters (1 L = 10^3 cm^3 = 10^{-3} m^3), then R has the value

$$R = 0.0821 \text{ L·atm/mol·K}$$

Using this value of R and Equation 19.10, we find that the volume occupied by 1 mole of any gas at atmospheric pressure and 0°C (273 K) is 22.4 L.

The ideal gas law is often expressed in terms of the total number of molecules, N. Since the total number of molecules equals the product of the number of moles and Avogadro's number, we can write Equation 19.10 as

$$PV = nRT = \frac{N}{N_A} RT$$

$$PV = Nk_B T \tag{19.12}$$

where k_B is called **Boltzmann's constant,** which has the value

Boltzmann's constant

$$k_B = \frac{R}{N_A} = 1.38 \times 10^{-23} \text{ J/K} \tag{19.13}$$

We have defined an ideal gas as one that obeys the equation of state, $PV = nRT$, under all conditions. In reality, an ideal gas does not exist. However, the concept of an ideal gas is very useful in view of the fact that real gases at low pressures behave as ideal gases. It is common to call quantities such as P, V, and T the **thermodynamic variables** of the system. If the equation of state is known, then one of the variables can always be expressed as some function of the other two. That is, given two of the variables, the third can be determined from the equation of state.

Other thermodynamic systems are often described with different thermodynamic variables. For example, for a wire under tension at constant pressure, the thermodynamic variables of the system are the length of the wire, the tension in it, and its temperature.

EXAMPLE 19.4 How Many Gas Molecules Are in a Container?

An ideal gas occupies a volume of 100 cm^3 at 20°C and a pressure of 100 Pa. Determine the number of moles of gas in the container.

Solution The quantities given are volume, pressure, and temperature: $V = 100$ cm^3 = 1.00×10^{-4} m^3, $P = 100$ Pa, and $T = 20$°C = 293 K. Using Equation 19.12, we get

$$n = \frac{PV}{RT} = \frac{(100 \text{ Pa})(1.00 \times 10^{-4} \text{ m}^3)}{(8.31 \text{ J/mol·K})(293 \text{ K})} = 4.11 \times 10^{-6} \text{ mol}$$

Note that you must express T as an absolute temperature (K) when using the ideal gas law.

Exercise Calculate the number of molecules in the container, using the fact that Avogadro's number is 6.02×10^{23} molecules/mol.

Answer 2.47×10^{18} molecules.

EXAMPLE 19.5 Squeezing a Tank of Gas

Pure helium gas is admitted into a tank containing a movable piston. The initial volume, pressure, and temperature of the gas are 15×10^{-3} m³, 200 kPa, and 300 K. If the volume is decreased to 12×10^{-3} m³ and the pressure increased to 350 kPa, find the final temperature of the gas. (Assume that helium behaves like an ideal gas.)

Solution If no gas escapes from the tank, the number of moles remains constant; therefore, using $PV = nRT$ at the initial and final points gives

$$\frac{P_i V_i}{T_i} = \frac{P_f V_f}{T_f}$$

where i and f refer to the initial and final values. Solving for T_f, we get

$$T_f = \left(\frac{P_f V_f}{P_i V_i}\right) T_i = \frac{(350 \text{ kPa})(12 \times 10^{-3} \text{ m}^3)}{(200 \text{ kPa})(15 \times 10^{-3} \text{ m}^3)} (300 \text{ K})$$

$$= \boxed{420 \text{ K}}$$

EXAMPLE 19.6 Heating a Bottle of Air

A sealed glass bottle containing air at atmospheric pressure (101 kPa) and having a volume of 30 cm³ is at 23°C. It is then tossed into an open fire. When the temperature of the air in the bottle reaches 200°C, what is the pressure inside the bottle? Assume any volume changes of the bottle are negligible.

Solution This example is approached in the same fashion as that used in Example 19.5. We start with the expression

$$\frac{P_i V_i}{T_i} = \frac{P_f V_f}{T_f}$$

Since the initial and final volumes of the gas are assumed equal, this expression reduces to

$$\frac{P_i}{T_i} = \frac{P_f}{T_f}$$

This gives

$$P_f = \left(\frac{T_f}{T_i}\right) P_i = \left(\frac{473 \text{ K}}{300 \text{ K}}\right)(101 \text{ kPa}) = \boxed{159 \text{ kPa}}$$

Obviously, the higher the temperature, the higher the pressure exerted by the trapped air. Of course, if the pressure rises high enough, the bottle will shatter.

Exercise In this example, we neglected the change in volume of the bottle. If the coefficient of volume expansion for glass is 27×10^{-6} (°C)$^{-1}$, find the magnitude of this volume change.

Answer 0.14 cm³.

SUMMARY

Two bodies are in **thermal equilibrium** with each other if they have the same temperature.

The **zeroth law of thermodynamics** states that if bodies A and B are separately in thermal equilibrium with a third body, C, then A and B are in thermal equilibrium with each other.

The SI unit of thermodynamic temperature is the **kelvin**, which is defined to be the fraction 1/273.16 of the temperature of the triple point of water.

When the temperature of an object is changed by an amount ΔT, its length changes by an amount ΔL that is proportional to ΔT and to its initial length L_0:

$$\Delta L = \alpha L_0 \, \Delta T \tag{19.4}$$

where the constant α is the **average coefficient of linear expansion**. The **average volume expansion coefficient**, β, for a substance is equal to 3α.

An **ideal gas** is one that obeys the *equation of state,*

$$PV = nRT \qquad (19.10)$$

where n equals the number of moles of gas, V is its volume, R is the universal gas constant (8.31 J/mol·K), and T is the absolute temperature in kelvin. A real gas behaves approximately as an ideal gas if it is far from liquefaction. An ideal gas is used as the working substance in a constant-volume gas thermometer, which defines the absolute temperature scale in kelvin. This absolute temperature T is related to temperatures on the Celsius scale by $T = T_C + 273.15$.

QUESTIONS

1. Is it possible for two objects to be in thermal equilibrium if they are not in contact with each other? Explain.
2. A piece of copper is dropped into a beaker of water. If the water's temperature rises, what happens to the temperature of the copper? Under what conditions are the water and copper in thermal equilibrium?
3. In principle, any gas can be used in a constant-volume gas thermometer. Why is it not possible to use oxygen for temperatures as low as 15 K? What gas would you use? (Look at the data in Table 19.1.)
4. Rubber has a negative average coefficient of linear expansion. What happens to the size of a piece of rubber as it is warmed?
5. Why should the amalgam used in dental fillings have the same average coefficient of expansion as a tooth? What would occur if they were mismatched?
6. Explain why a column of mercury in a thermometer first descends slightly and then rises when placed in hot water.
7. Explain why the thermal expansion of a spherical shell made of a homogeneous solid is equivalent to that of a solid sphere of the same material.
8. A steel ring bearing has an inside diameter that is 1 mm smaller than an axle. How can it be made to fit onto the axle without removing any material?
9. Markings to indicate length are placed on a steel tape in a room that has a temperature of 22°C. Are measurements made with the tape on a day when the temperature is 27°C too long, too short, or accurate? Defend your answer.
10. What would happen if the glass of a thermometer expanded more upon heating than did the liquid inside?
11. Determine the number of grams in one mole of the following gases: (a) hydrogen, (b) helium, and (c) carbon monoxide.
12. Why is it necessary to use absolute temperature when doing calculations with the ideal gas law?
13. An inflated rubber balloon filled with air is immersed in a flask of liquid nitrogen that is at 77 K. Describe what happens to the balloon, assuming that it remains flexible while being cooled.
14. Two cylinders at the same temperature each contain the same kind and quantity of gas. If the volume of cylinder A is three times greater than the volume of cylinder B, what can you say about the relative pressures in the cylinders?
15. The suspension of a certain pendulum clock is made of brass. When the temperature increases, does the period of the clock increase, decrease, or remain the same? Explain.
16. An automobile radiator is filled to the brim with water while the engine is cool. What happens to the water when the engine is running and the water is heated? What do modern automobiles have in their cooling systems to prevent the loss of coolants?
17. Metal lids on glass jars can often be loosened by running hot water over them. How is this possible?
18. When the metal ring and metal sphere in Figure 19.11 are both at room temperature, the sphere does not fit through the ring. After the ring is heated, the sphere can be passed through the ring. Explain.

FIGURE 19.11   (Question 18).   *(Courtesy Central Scientific Co.)*

PROBLEMS

Section 19.3 The Constant-Volume Gas Thermometer and the Kelvin Scale

(*Note:* A pressure of 1.00 atm = 1.01×10^5 Pa = 101 kPa.)

1. A constant-volume gas thermometer is calibrated in dry ice (which is carbon dioxide in the solid state and has a temperature of $-80.0°C$) and in boiling ethyl alcohol ($78.0°C$). The two pressures are 0.900 atm and 1.635 atm. (a) What value of absolute zero does the calibration yield? What is the pressure at (a) the freezing point of water and (b) the boiling point of water?

2. Suppose the temperature (in units of kelvin) and pressure in an ideal gas thermometer are related by a *quadratic* equation, $T = aP^2 + bP$. If the temperature and pressure at the triple point of water are T_3 and P_3, respectively, and if the temperature and pressure at the boiling point of water are T_B and P_B, respectively, determine a and b in terms of T_3, P_3, T_B, and P_B.

3. A constant-volume gas thermometer registers a pressure of 0.062 atm when it is at a temperature of 450 K. (a) What is the pressure at the triple point of water? (b) What is the temperature when the pressure reads 0.015 atm?

4. In a constant-volume gas thermometer, the pressure at $20°C$ is 0.980 atm. (a) What is the pressure at $45°C$? (b) What is the temperature if the pressure is 0.500 atm?

5. A constant-volume gas thermometer is filled with helium. When immersed in boiling liquid nitrogen (77.34 K), the absolute pressure is 25.00 kPa. (a) What is the temperature in degrees Celsius and kelvin when the pressure is 45.00 kPa? (b) What is the pressure when the thermometer is immersed in boiling liquid hydrogen?

6. The melting point of gold is $1064°C$, and the boiling point is $2660°C$. (a) Express these temperatures in kelvin. (b) Compute the difference between these temperatures in Celsius degrees and kelvin.

7. Liquid nitrogen has a boiling point of $-195.81°C$ at atmospheric pressure. Express this temperature in (a) degrees Fahrenheit and (b) kelvin.

8. The highest recorded temperature on Earth is $136°F$, at Azizia, Libya, in 1922. The lowest recorded temperature is $-127°F$, at Vostok Station, Antarctica, in 1960. Express these temperature extremes in degrees Celsius.

9. Oxygen condenses to a liquid at approximately 90 K. What temperature, in degrees Fahrenheit, does this correspond to?

10. On a Strange temperature scale, the freezing point of water is $-15°S$ and the boiling point is $+60°S$. Develop a *linear* conversion equation between this temperature scale and the Celsius scale.

11. The temperature difference between the inside and the outside of an automobile engine is $450°C$. Express this temperature difference on the (a) Fahrenheit scale and (b) Kelvin scale.

12. The normal human body temperature is $98.6°F$. A person with a fever may record $102°F$. Express these temperatures in degrees Celsius.

13. A substance is heated from $-12°F$ to $150°F$. What is its change in temperature on (a) the Celsius scale and (b) the Kelvin scale?

14. Initially, an object has a temperature, which has the same numerical value in degrees Celsius and degrees Fahrenheit. Then, its temperature is changed so that the numerical value of the new temperature in degrees Celsius is one-third as large as small as that in kelvin. Find the change in the temperature in kelvin.

15. At what temperature are the readings from a Fahrenheit thermometer and Celsius thermometer the same?

Section 19.4 Thermal Expansion of Solids and Liquids

(*Note:* Use Table 19.2.)

16. An aluminum tube is 3.0000 m long at $20.0°C$. What is its length at (a) $100.0°C$ and (b) $0.0°C$?

17. A copper telephone wire has essentially no sag between two poles 35.0 m apart on a winter's day when the temperature is $-20°C$. How much longer is the wire on a summer's day when $T = 35°C$?

18. A concrete walk is poured on a day when the temperature is $20°C$ in such a way that the ends are unable to move. (a) What is the stress in the cement on a hot day of $50°C$? (b) Does the concrete fracture? Take Young's modulus for concrete to be 7.0×10^9 N/m^2 and the tensile strength to be 2.0×10^9 N/m^2.

19. A structural steel I-beam is 15.0 m long when installed at $20.0°C$. How much does its length change over the temperature extremes $-30.0°C$ to $50.0°C$?

20. The New River Gorge Bridge in West Virginia is a steel arch bridge 518 m in length. How much does its length change between temperature extremes of $-20.0°C$ and $35.0°C$?

21. The average volume coefficient of expansion for carbon tetrachloride is 5.81×10^{-4} $(°C)^{-1}$. If a 50.0-gal steel container is filled completely with carbon tetrachloride when the temperature is $10.0°C$, how much will spill over when the temperature rises to $30.0°C$?

☐ indicates problems that have full solutions available in the Student Solutions Manual and Study Guide.

22. A steel rod undergoes a stretching force of 500 N. Its cross-sectional area is 2.00 cm². Find the change in temperature that would elongate the rod by the same amount produced by the 500-N force. (*Hint:* Refer to Tables 12.1 and 19.2.)

23. A steel rod 4.0 cm in diameter is heated so that its temperature increases by 70°C and then is fastened between two rigid supports. The rod is allowed to cool to its original temperature. Assuming that Young's modulus for the steel is 20.6×10^{10} N/m² and that its average coefficient of linear expansion is 11×10^{-6} (°C)$^{-1}$, calculate the tension in the rod.

24. A brass ring of diameter 10.00 cm at 20.0°C is heated and slipped over an aluminum rod of diameter 10.01 cm at 20.0°C. Assuming the average coefficients of linear expansion are constant, (a) to what temperature must this combination be cooled to separate them? Is this attainable? (b) What if the aluminum rod were 10.02 cm in diameter?

25. The concrete sections of a certain superhighway are designed to have a length of 25.0 m. The sections are poured and cured at 10°C. What minimum spacing should the engineer leave between the sections to eliminate buckling if the concrete is to reach a temperature of 50°C?

26. A square hole 8.0 cm along each side is cut in a sheet of copper. Calculate the change in the area of this hole if the temperature of the sheet is increased by 50 K.

27. At 20.0°C, an aluminum ring has an inner diameter of 5.000 cm and a brass rod has a diameter of 5.050 cm. (a) To what temperature must the ring be heated so that it will just slip over the rod? (b) To what temperature must both be heated so that the ring just slips over the rod? Would the latter process work?

28. A pair of eyeglass frames is made of epoxy plastic. At room temperature (assume 20.0°C), the frames have circular lens holes 2.2 cm in radius. To what temperature must the frames be heated in order to insert lenses 2.21 cm in radius? The average coefficient of linear expansion for epoxy is 1.3×10^{-4} (°C)$^{-1}$.

29. A hollow aluminum cylinder 20.0 cm deep has an internal capacity of 2.000 L at 20.0°C. It is completely filled with turpentine, and then warmed to 80.0°C. (a) How much turpentine overflows? (b) If it is then cooled back to 20.0°C, how far below the surface of the cylinder's rim is the turpentine surface?

30. A copper rod and steel rod are heated. At 0°C the copper rod has a length of L_C, the steel one has a length L_S. When the rods are being heated or cooled, a difference of 5.0 cm is maintained between their lengths. Determine the values of L_C, and L_S.

30A. A copper rod and steel rod are heated. At T(°C) the

copper rod has a length of L_C, the steel one has a length L_S. When the rods are being heated or cooled, a difference of ΔL is maintained between their lengths. Determine the values of L_C and L_S.

31. An automobile fuel tank is filled to the brim with 45 L of gasoline at 10°C. Immediately afterward, the vehicle is parked in the Sun where the temperature is 35°C. How much gasoline overflows from the tank as a result of expansion? (Neglect the expansion of the tank.)

32. A volumetric glass flask made of Pyrex is calibrated at 20.0°C. It is filled to the 100-mL mark with 35.0°C acetone. (a) What is the volume of the acetone when it cools to 20.0°C? (b) How significant is the change in volume of the flask?

33. The active element of a certain laser is made of a glass rod 30.0 cm long by 1.5 cm in diameter. If the temperature of the rod increases by 65°C, find the increase in (a) its length, (b) its diameter, and (c) its volume. (Take $\alpha = 9.0 \times 10^{-6}$ (°C)$^{-1}$.)

Section 19.5   Macroscopic Description of an Ideal Gas

34. An ideal gas is held in a container at constant volume. Initially, its temperature is 10.0°C and its pressure is 2.50 atm. What is its pressure when its temperature is 80.0°C?

35. A helium-filled balloon has a volume of 1.00 m³. As it rises in the Earth's atmosphere, its volume expands. What is its new volume (in cubic meters) if its original temperature and pressure are 20.0°C and 1.00 atm and its final temperature and pressure are -40.0°C and 0.10 atm?

36. An auditorium has dimensions 10.0 m × 20.0 m × 30.0 m. How many molecules of air are needed to fill the auditorium at 20.0°C and 101 kPa pressure?

37. A full tank of oxygen (O_2) contains 12.0 kg of oxygen under a gauge pressure of 40.0 atm. Determine the mass of oxygen that has been withdrawn from the tank when the pressure reading is 25.0 atm. Assume the temperature of the tank remains constant.

38. A car tire gauge is used to fill a tire to a gauge pressure of 32 lb/in.² on a cold morning when the temperature is -10°C. What would the tire gauge read when the tire has been heated up to 35°C?

39. The mass of a hot-air balloon and its cargo (not including the air inside) is 200 kg. The air outside is at 10.0°C and 101 kPa. The volume of the balloon is 400 m³. To what temperature must the air in the balloon be heated before the balloon will lift off? (Air density at 10.0°C is 1.25 kg/m³.)

40. A tank having a volume of 0.10 m³ contains helium gas at 150 atm. How many balloons can the tank blow up if each filled balloon is a sphere 0.30 m in diameter at an absolute pressure of 1.2 atm?

41. A room of volume 80.0 m³ contains air having an average molar mass of 29.0 g/mol. If the temperature of the room is raised from 18.0°C to 25.0°C, what mass of air (in kg) will leave the room? Assume that the air pressure in the room is maintained at 101 kPa.

41A. A room of volume V contains air having an average molar mass of M. If the temperature of the room is raised from T_1 to T_2, what mass of air will leave the room? Assume that the air pressure in the room is maintained at P_0.

42. At 25.0 m below the surface of the sea (density = 1025 kg/m³), where the temperature is 5.00°C, a diver exhales an air bubble having a volume of 1.00 cm³. If the surface temperature of the sea is 20.0°C, what is the volume of the bubble right before it breaks the surface?

42A. At a depth h below the surface of the sea (density = ρ), where the temperature is T_C, a diver exhales an air bubble having a volume V_0. If the surface temperature of the sea is T_h, what is the volume of the bubble right before it breaks the surface?

43. If 9.0 g of water is placed into a 2.0-L pressure cooker and heated to 500°C, what is the pressure inside the container?

44. In state-of-the-art vacuum systems, pressures as low as 1.00×10^{-9} Pa are being attained. Calculate the number of molecules in a 1.00-m³ vessel at this pressure if the temperature is 27°C.

45. The tire on a bicycle is filled with air to a gauge pressure of 550 kPa at 20°C. What is the gauge pressure in the tire after a ride on a hot day when the tire air temperature is 40°C? (Assume constant volume and a constant atmospheric pressure of 101 kPa.)

46. Show that 1.00 mol of any gas at atmospheric pressure (101 kPa) and standard temperature (273 K) occupies a volume of 22.4 L.

47. An automobile tire is inflated using air originally at 10°C and normal atmospheric pressure. During the process, the air is compressed to 28% of its original volume and the temperature is increased to 40°C. What is the tire pressure? After the car is driven at high speed, the tire air temperature rises to 85°C and the interior volume of the tire increases by 2%. What is the new tire pressure in pascals (absolute)?

48. A diving bell in the shape of a cylinder with a height of 2.50 m is closed at the upper end and open at the lower end. The bell is lowered from air into sea water ($\rho = 1.025$ g/cm³). The air in the bell is initially at 20.0°C. The bell is lowered to a depth (measured to the bottom of the bell) of 45.0 fathoms or 82.3 m. At this depth the water temperature is 4.0°C, and the bell is in thermal equilibrium with the water. (a) How high does sea water rise in the bell? (b) To what minimum pressure must the air in the bell be raised to expel the water that entered?

49. A bubble of marsh gas rises from the bottom of a freshwater lake at a depth of 4.2 m and a temperature of 5.0°C to the surface, where the water temperature is 12°C. What is the ratio of the bubble diameter at the two locations? (Assume that the bubble gas is in thermal equilibrium with the water at each location.)

50. An expandable cylinder has its top connected to a spring of constant 2.0×10^3 N/m (Fig. P19.50). The cylinder is filled with 5.0 L of gas with the spring relaxed at 1.0 atm and 20°C. (a) If the lid has a cross-sectional area of 0.010 m² and negligible mass, how high does the lid rise when the temperature is raised to 250°C? (b) What is the pressure of the gas at 250°C?

50A. An expandable cylinder has its top connected to a spring of constant k (Fig. P19.50). The cylinder is filled with V liters of gas with the spring relaxed at atmospheric pressure P_0 and temperature T_0. (a) If the lid has a cross-sectional area A and negligible mass, how high does the lid rise when the temperature is raised to T? (b) What is the pressure of the gas at this higher temperature?

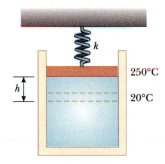

FIGURE P19.50

ADDITIONAL PROBLEMS

51. Precise temperature measurements are often made using the change in electrical resistance of a metal with temperature. The resistance varies according to the expression $R = R_0(1 + AT_C)$, where R_0 and A are constants. A certain element has a resistance of 50.0 ohms at 0°C and 71.5 ohms at the freezing point of tin (231.97°C). (a) Determine the constants A and R_0. (b) At what temperature is the resistance equal to 89.0 ohms?

52. A pendulum clock with a brass suspension system has a period of 1.000 s at 20.0°C. If the temperature increases to 30.0°C, (a) by how much does the period

change and (b) how much time does the clock gain or lose in one week?

53. The rectangular plate shown in Figure P19.53 has an area A equal to ℓw. If the temperature increases by ΔT, show that the increase in area is $\Delta A = 2\alpha A \Delta T$, where α is the average coefficient of linear expansion. What approximation does this expression assume? (*Hint:* Note that each dimension increases according to $\Delta \ell = \alpha \ell \Delta T$.)

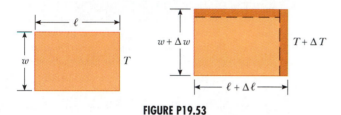

FIGURE P19.53

54. Consider an object with any one of the shapes displayed in Table 10.2. What is the percentage increase in the moment of inertia of the object when it is heated from 0°C to 100°C, if it is composed of (a) copper or (b) aluminum? (See Table 19.2. Assume that the average linear expansion coefficients do not vary between 0°C and 100°C.)

55. A mercury thermometer is constructed as in Figure P19.55. The capillary tube has a diameter of 0.0040 cm, and the bulb has a diameter of 0.25 cm. Neglecting the expansion of the glass, find the change in height of the mercury column for a temperature change of 30°C.

55A. A mercury thermometer is constructed as in Figure P19.55. The capillary tube has a diameter d_1, and the bulb has a diameter d_2. Neglecting the expansion of the glass, find the change in height of the mercury column for a temperature change ΔT.

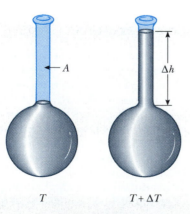

FIGURE P19.55

56. A liquid has a density ρ. (a) Show that the fractional change in density for a change in temperature ΔT is $\Delta\rho/\rho = -\beta\,\Delta T$. What does the negative sign signify? (b) Fresh water has a maximum density of 1.000 g/cm³ at 4.0°C. At 10.0°C, its density is 0.9997 g/cm³. What is β for water over this temperature interval?

57. A student measures the length of a brass rod with a steel tape at 20°C. The reading is 95.00 cm. What will the tape indicate for the length of the rod when the rod and the tape are at (a) −15°C and (b) 55°C?

58. (a) Derive an expression for the buoyant force on a spherical balloon as it is submerged in water as a function of the depth below the surface, the volume of the balloon at the surface, the pressure at the surface, and the density of the water. (Assume water temperature does not change with depth.) (b) Does the buoyant force increase or decrease as the balloon is submerged? (c) At what depth does the buoyant force decrease to one-half the surface value?

59. (a) Show that the density of an ideal gas occupying a volume V is given by $\rho = PM/RT$, where M is the molar mass. (b) Determine the density of oxygen gas at atmospheric pressure and 20.0°C.

60. Steel rails for an interurban rapid transit system form a continuous track that is held rigidly in place in concrete. (a) If the track was laid when the temperature was 0°C, what is the stress in the rails on a warm day when the temperature is 25°C? (b) What fraction of the yield strength of 52.2×10^7 N/m² does this stress represent?

61. Starting with Equation 19.12, show that the total pressure P in a container filled with a mixture of several ideal gases is $P = P_1 + P_2 + P_3 + \ldots$, where $P_1, P_2, \ldots$ are the pressures that each gas would exert if it alone filled the container (these individual pressures are called the *partial pressures* of the respective gases). This is known as *Dalton's law of partial pressures*.

62. A sample of air that has a mass of 100.00 g, collected at sea level, is analyzed and found to consist of the following gases:

nitrogen (N_2) = 75.52 g

oxygen (O_2) = 23.15 g

argon (Ar) = 1.28 g

carbon dioxide (CO_2) = 0.05 g

plus trace amounts of neon, helium, methane, and other gases. (a) Calculate the partial pressure (see Problem 61) of each gas when the pressure is 1.013×10^5 Pa. (b) Determine the volume occupied by the 100-g sample at a temperature of 15.00°C and a pressure of 1.013×10^5 Pa. What is the density of the air for these conditions? (c) What is the effective molar mass of the air sample?

63. Two concrete spans of a 250-m-long bridge are placed end to end so that there is no room for expansion (Fig. P19.63a). If a temperature increase of 20.0°C occurs, find the height, y, at which the spans buckle (Fig. P19.63b).

63A. Two concrete spans of a bridge of length L are placed end to end so that there is no room for expansion (Fig. P19.63a). If a temperature increase of ΔT occurs, find the height, y, at which the spans buckle (Fig. P19.63b).

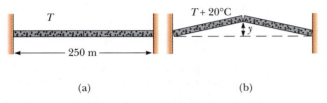

(a)   (b)

FIGURE P19.63

64. A steel ball bearing is 4.000 cm in diameter at 20.0°C. A bronze plate has a hole in it that is 3.994 cm in diameter at 20.0°C. What common temperature must they have in order that the ball just squeeze through the hole?

65. A brass pendulum on a clock is adjusted to have a period of 1.000 s at 20°C. What is the temperature of a room in which the clock loses exactly 1 minute each week?

66. A vertical cylinder of cross-sectional area A is fitted with a tight-fitting, frictionless piston of mass m (Fig. P19.66). (a) If there are n mol of an ideal gas in the cylinder at a temperature T, determine the height h at which the piston is in equilibrium under its own weight. (b) What is the value for h if $n = 0.20$ mol, $T = 400$ K, $A = 0.0080$ m², and $m = 20.0$ kg?

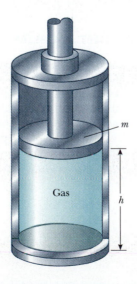

FIGURE P19.66

67. An air bubble originating from a deep sea diver has a radius of 5.0 mm at some depth h. When the bubble reaches the surface of the water, it has a radius of 7.0 mm. Assuming the temperature of the air in the bubble remains constant, determine (a) the depth h of the diver and (b) the absolute pressure at this depth.

68. Figure P19.68 shows a circular steel casting with a gap. If the casting is heated, (a) does the width of the gap increase or decrease? (b) The gap width is 1.600 cm when the temperature is 30.0°C. Determine the gap width when the temperature is 190°C.

FIGURE P19.68

69. A cylinder that has a 40.0-cm radius and is 50.0 cm deep is filled with air at 20.0°C and 1.00 atm (Fig. P19.69a). A 20.0-kg piston is now lowered into the cylinder, compressing the air trapped inside (Fig. P19.69b). Finally, a 75.0-kg man stands on the piston, further compressing the air which remains at 20°C (Fig. P19.69c). (a) How far down (Δh) does the

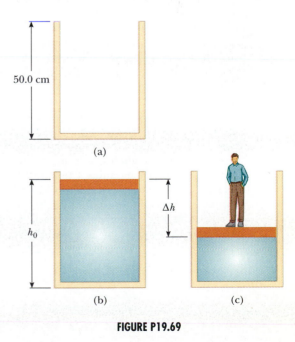

(a)

(b)   (c)

FIGURE P19.69

piston move when the man stands on it? (b) To what temperature should the gas be heated to raise the piston and man back to h_0?

70. An aluminum pot has the shape of a cylinder. The pot is initially at 4.0°C, at which temperature it has an inside diameter of 28.00 cm. The pot contains 3.000 gal of water at 4.0°C. (a) What is the depth of water in the pot? (1 gal = 3785 cm³.) (b) The pot and the water in it are heated to 90.0°C. Allowing for the expansion of the water but ignoring the expansion of the pot, what is the change in depth of the water? Express the change as a percentage of the original depth and also in millimeters. (The density of water is 1.000 g/cm³ at 4.0°C and 0.965 g/cm³ at 90.0°C.) (c) Modify your solution for part (b) to allow for the expansion of the pot. (Refer to Table 19.2.)

71. A sphere 20 cm in diameter contains an ideal gas at 1.00 atm and 20.0°C. As the sphere is heated to 100.0°C, gas is allowed to escape. The valve is closed and the sphere is placed in an ice-water bath. (a) How many moles of gas escape from the sphere as it warms? (b) What is the pressure in the sphere when it is in the ice water?

72. The relationship $L = L_0(1 + \alpha \Delta T)$ is an approximation that works when the average coefficient of expansion is small. If α is sizable, the relationship $dL/dT = \alpha L$ must be integrated to determine the final length. (a) Assuming the coefficient of linear expansion is constant, determine a general expression for the final length. (b) Given a rod of length 1.00 m and a temperature change of 100.0°C, determine the error caused by the approximation when $\alpha = 2.00 \times 10^{-5}$ (°C)$^{-1}$ (the normal value for common metals) and when $\alpha = 0.020$ (°C)$^{-1}$ (an unrealistically large value for comparison).

73. A steel guitar string with a diameter of 1.00 mm is stretched between supports 80.0 cm apart. The temperature is 0.0°C. (a) Find the mass per unit length of this string. (Use the value 7.86×10^3 kg/m³ for the density.) (b) The fundamental frequency of transverse oscillations of the string is 200 Hz. What is the tension in the string? (c) If the temperature is raised to 30.0°C, find the resulting values of the tension and the fundamental frequency. [Assume that both the Young's modulus (Table 12.1) and the average coefficient of expansion (Table 19.2) have constant values between 0.0°C and 30.0°C.]

74. A steel wire and a copper wire, each of diameter 2.000 mm, are joined end to end. At 40.0°C, each

has an unstretched length of 2.000 m; they are connected between two fixed supports 4.000 m apart on a tabletop, so that the steel wire extends from $x = -2.000$ m to $x = 0$, the copper wire extends from $x = 0$ to $x = 2.000$ m, and the tension is negligible. The temperature is then lowered to 20.0°C. At this lower temperature, find the tension in the wire and the x coordinate of the junction between the wires. (Refer to Tables 12.1 and 19.2.)

75. A bimetallic bar is made of two thin strips of dissimilar metals bonded together. As they are heated, the one with the larger average coefficient of expansion expands more than the other, forcing the bar into an arc, with the outer radius having a larger circumference (Fig. P19.75). (a) Derive an expression for the angle of bending θ as a function of the initial length of the strips, their average coefficients of linear expansion, the change in temperature, and the separation of the centers of the strips ($\Delta r = r_2 - r_1$). (b) Show that the angle of bending goes to zero when ΔT goes to zero or when the two coefficients of expansion become equal. (c) What happens if the bar is cooled?

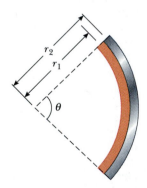

FIGURE P19.75

SPREADSHEET PROBLEM

S1. A 1.00-km steel railroad rail is fastened securely at both ends when the temperature is 20.0°C. As the temperature increases, the rail begins to buckle. If the shape of the buckle is the arc of a circle, find the height h of the center of the buckle when the temperature is 25.0°C. This problem requires you to solve a transcendental equation. Use a spreadsheet to solve the equation graphically.

CHAPTER 20

Heat and the First Law of Thermodynamics

Thermogram of a teakettle showing hot areas in white and cooler areas in purple and black. *(Gary Settles/Science Source/Photo Researchers)*

Until about 1850, the fields of heat and mechanics were considered to be two distinct branches of science and the law of conservation of energy seemed to describe only certain kinds of mechanical systems. Mid-19th century experiments performed by the Englishman James Joule (1818–1889) and others showed that energy may be added to (or removed from) a system either as thermal energy or as work done on (or by) the system. Now thermal energy is treated as a form of energy that can be transformed into mechanical energy. Once the concept of energy was broadened to include thermal energy, the law of conservation of energy emerged as a universal law of nature.

This chapter focuses on the concept of heat, the first law of thermodynamics, processes by which thermal energy is transferred, and some important applications. The first law of thermodynamics is merely the law of conservation of energy. It tells us only that an increase in one form of energy must be accompanied by a decrease in some other form of energy. The first law places no restrictions on the types of energy conversions that can occur. Furthermore, it makes no distinction between heat and work. According to the first law, a system's internal energy can be increased either by transfer of thermal energy to the system or by work done on the system. An important difference between thermal energy and mechanical energy is not evident from the first law; it is possible to convert work completely to thermal energy but impossible to convert thermal energy completely to mechanical energy in a process at constant temperature.

20.1   HEAT AND THERMAL ENERGY

A major distinction must be made between internal energy, thermal energy, and heat. **Internal energy** is all of the energy belonging to a system while it is stationary (neither translating nor rotating), including nuclear energy, chemical energy, and strain energy (as for a compressed or stretched spring), as well as thermal energy. **Thermal energy** is that portion of the internal energy that changes when the temperature of the system changes. **Thermal energy transfer** is the transfer of thermal energy caused by a temperature difference between the system and its surroundings, *which may or may not change the amount of thermal energy in the system.*

In practice, the term "**heat**" is used to mean both thermal energy and thermal energy transfer. Hence, one must always examine the context of the term *heat* to determine its intended meaning.

In the next chapter, we show that the thermal energy of a monatomic ideal gas is associated with the internal motion of its atoms. In this special case, the thermal energy is simply the kinetic energy on a microscopic scale: the higher the temperature of the gas, the greater the kinetic energy of the atoms and the greater the thermal energy of the gas. More generally, however, thermal energy includes other forms of molecular energy, such as rotational energy and vibrational kinetic and potential energy.

Thermal energy

James Prescott Joule, a British physicist, was born in Salford on December 24, 1818, into a wealthy brewing family. He received some formal education in mathematics, philosophy, and chemistry but was in large part self-educated.

Joule's most active research period, from 1837 through 1847, led to the establishment of the principle of conservation of energy and the equivalence of heat and other forms of energy. His study of the quantitative relationship between the electrical, mechanical, and chemical effects of heat culminated in his announcement in 1843 of the amount of work required to produce a unit of heat:

James Prescott Joule

| 1 8 1 8 – 1 8 8 9 |

(North Wind Picture Archives)

First: that the quantity of heat produced by the friction of bodies, whether solid or liquid, is always proportional to the quantity of energy expended. And second: that the quantity of heat capable of increasing the temperature of water . . . by 1° Fahr requires for its evolution the expenditure of a mechanical energy represented by the fall of 772 lb through the distance of one foot.

This is called the mechanical equivalent of heat (the currently accepted value is approximately 4.186 J/cal).

As an analogy, consider the distinction between work and energy that we discussed in Chapter 7. The work done on (or by) a system is a measure of energy transfer between the system and its surroundings, whereas the mechanical energy of the system (kinetic and/or potential) is a consequence of its motion and coordinates. Thus, when a person does work on a system, energy is transferred from the person to the system. It makes no sense to talk about the work of a system—one can refer only to the *work done on or by a system* when some process has occurred in which energy has been transferred to or from the system. Likewise, it makes no sense to use the term *heat* unless energy has been transferred as a result of a temperature difference.

It is also important to recognize that energy can be transferred between two systems even when no thermal energy transfer occurs. For example, when a gas is compressed by a piston, the gas is warmed and its thermal energy increases, but there is no transfer of thermal energy; if the gas then expands rapidly, it cools and its thermal energy decreases, but there is no transfer of thermal energy to the surroundings. In each case, energy is transferred to or from the system as work, but the energy appears within the system as an increase or decrease of thermal energy. The changes in internal energy in these examples are equal to changes in thermal energy and are measured by corresponding changes in temperature.

Units of Heat

Before it was understood that heat is a form of energy, scientists defined heat in terms of the temperature changes it produced in a body. Hence, the **calorie** (cal) was defined as *the amount of heat necessary to raise the temperature of 1 g of water from 14.5° C to 15.5° C.*[1] (Note that the ''Calorie,'' with a capital C, used in describing the chemical energy content of foods, is actually a kilocalorie.) Likewise, the unit of heat in the British system is the **British thermal unit** (Btu), defined as *the heat required to raise the temperature of 1 lb of water from 63°F to 64°F.*

Since heat is now recognized as a form of energy, scientists are increasingly using the SI unit of energy, the *joule,* for heat. In this textbook, heat will usually be measured in joules.

The Mechanical Equivalent of Heat

When the concept of mechanical energy was introduced in Chapters 7 and 8, we found that whenever friction is present in a mechanical system, some mechanical energy is lost, or is not conserved. Various experiments show that this lost mechanical energy does not simply disappear but is transformed into thermal energy. Although this connection between mechanical and thermal energy was first suggested by Thompson's crude cannon-boring experiment, it was Joule (pronounced ''jewel'') who first established the equivalence of the two forms of energy.

A schematic diagram of Joule's most famous experiment is shown in Figure 20.1. The system of interest is the water in a thermally insulated container. Work is done on the water by a rotating paddle wheel, which is driven by weights falling at a constant speed. The water, which is stirred by the paddles, is warmed due to the friction between it and the paddles. If the energy lost in the bearings and through the walls is neglected, then the loss in potential energy of the weights equals the

Benjamin Thompson (1753–1814). ''Being engaged, lately, in superintending the boring of cannon, in the workshops of the military arsenal at Munich, I was struck with the very considerable degree of Heat which a brass gun acquires, in a short time, in being bored; and with the still more intense Heat (much greater than that of boiling water, as I found by experiment) of the metallic chips separated from it by the borer.''   *(North Wind Picture Archives)*

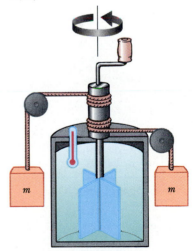

Thermal insulator

FIGURE 20.1   An illustration of Joule's experiment for determining the mechanical equivalent of heat. The falling weights rotate the paddles, causing the temperature of the water to increase.

---

[1] Originally, the calorie was defined as the heat necessary to raise the temperature of 1 g of water by 1°C. However, careful measurements showed that energy depends somewhat on temperature; hence, a more precise definition evolved.

work done by the paddle wheel on the water. If the two weights fall through a distance h, the loss in potential energy is $2mgh$, and it is this energy that is used to heat the water. By varying the conditions of the experiment, Joule found that the loss in mechanical energy, $2mgh$, is proportional to the increase in temperature of the water, ΔT. The proportionality constant was found to be equal to approximately $4.18 \, \text{J/g} \cdot {}^{\circ}\text{C}$. Hence, 4.1858 J of mechanical energy raises the temperature of 1 g of water from 14.5°C to 15.5°C. We adopt this "15 degree calorie" value:

Mechanical equivalent of heat

$$1 \, \text{cal} \equiv 4.186 \, \text{J} \qquad (20.1)$$

This is known, for purely historical reasons, as the **mechanical equivalent of heat.**

EXAMPLE 20.1   Losing Weight the Hard Way

A student eats a dinner rated at 2000 (food) Calories. He wishes to do an equivalent amount of work in the gymnasium by lifting a 50.0-kg mass. How many times must he raise the mass to expend this much energy? Assume that he raises it a distance of 2.00 m each time and that he regains no energy when it is dropped to the floor.

Solution   Since 1 Calorie = 1.00×10^3 cal, the work required is 2.00×10^6 cal. Converting this to J, we have for the total work required

$$W = (2.00 \times 10^6 \, \text{cal})(4.186 \, \text{J/cal}) = 8.37 \times 10^6 \, \text{J}$$

The work done in lifting the mass a distance h is equal to mgh, and the work done in lifting it n times is $nmgh$. We equate this to the total work required:

$$W = nmgh = 8.37 \times 10^6 \, \text{J}$$

$$n = \frac{8.37 \times 10^6 \, \text{J}}{(50.0 \, \text{kg})(9.80 \, \text{m/s}^2)(2.00 \, \text{m})} = 8.54 \times 10^3 \, \text{times}$$

If the student is in good shape and lifts the weight once every 5 s, it will take him about 12 h to perform this feat. Clearly, it is much easier to lose weight by dieting.

20.2   HEAT CAPACITY AND SPECIFIC HEAT

When heat is added to a substance (with no work done), its temperature usually rises. (An exception to this statement occurs when a substance undergoes a phase transition, say from a liquid to a gas or when a gas expands, which is discussed in the next section.) The quantity of heat energy required to raise the temperature of a given mass of a substance by some amount varies from one substance to another. For example, the heat required to raise the temperature of 1 kg of water by 1°C is 4186 J, but the heat required to raise the temperature of 1 kg of copper by 1°C is only 387 J. The **heat capacity,** C', of a particular sample of a substance is defined as the amount of heat needed to raise the temperature of that sample by one degree Celsius. From this definition, we see that if Q units of thermal energy when added to a substance produce a change in temperature of ΔT, then

Heat capacity

$$Q = C' \Delta T \qquad (20.2)$$

The **specific heat** c of a substance is the heat capacity per unit mass. Thus, if Q units of thermal energy are transferred to m kg of a substance, thereby changing its temperature by ΔT, the specific heat of the substance is

Specific heat

$$c \equiv \frac{Q}{m \, \Delta T} \qquad (20.3)$$

From this definition, we can express the thermal energy Q transferred between a substance of mass m and its surroundings for a temperature change ΔT as

$$Q = mc \, \Delta T \qquad (20.4)$$

For example, the energy required to raise the temperature of 0.5 kg of water by 3°C is equal to $(0.5 \text{ kg})(4186 \text{ J/kg} \cdot °\text{C})(3°\text{C}) = 6280 \text{ J}$. Note that when the temperature increases, Q and ΔT are taken to be positive, corresponding to thermal energy flowing into the system. When the temperature decreases, Q and ΔT are negative and thermal energy flows out of the system.

The **molar specific heat** of a substance is defined as the heat capacity per mole. Hence, if the substance contains n mol, its molar specific heat is equal to C'/n. Table 20.1 also gives the specific heats and molar specific heats of various substances.

Molar specific heat

It is important to realize that specific heat varies with temperature. If the temperature intervals are not too great, the temperature variation can be ignored and c can be treated as a constant.[2] For example, the specific heat of water varies by only about 1% from 0°C to 100°C at atmospheric pressure. Unless stated otherwise, we shall neglect such variations.

When specific heats are measured, the values obtained are also found to depend on the conditions of the experiment. In general, measurements made at constant pressure are different from those made at constant volume. For solids and liquids, the difference between the two values is usually no more than a few percent and is often neglected. The values given in Table 20.1 were measured at atmospheric pressure and room temperature. As we shall see in Chapter 21, the

TABLE 20.1 Specific Heats of Some Substances at 25°C and Atmospheric Pressure

Substance	Specific Heat, c J/kg·°C	cal/g·°C	Molar Specific Heats J/mol·°C
Elemental Solids			
Aluminum	900	0.215	24.3
Beryllium	1830	0.436	16.5
Cadmium	230	0.055	25.9
Copper	387	0.0924	24.5
Germanium	322	0.077	23.4
Gold	129	0.0308	25.4
Iron	448	0.107	25.0
Lead	128	0.0305	26.4
Silicon	703	0.168	19.8
Silver	234	0.056	25.4
Other Solids			
Brass	380	0.092	
Wood	1700	0.41	
Glass	837	0.200	
Ice (−5°C)	2090	0.50	
Marble	860	0.21	
Liquids			
Alcohol (ethyl)	2400	0.58	
Mercury	140	0.033	
Water (15°C)	4186	1.00	

[2] The definition given by Equation 20.4 assumes that the specific heat does not vary with temperature over the interval ΔT. In general, if c varies with temperature over the range T_i to T_f, the correct expression for Q is

$$Q = m \int_{T_i}^{T_f} c \, dT$$

specific heats for gases measured under constant pressure conditions are quite different from values measured under constant volume conditions.

It is interesting to note from Table 20.1 that water has the highest specific heat of common Earth materials. The high specific heat of water is responsible, in part, for the moderate temperatures found in regions near large bodies of water. As the temperature of a body of water decreases during the winter, heat is transferred from the water to the air, which in turn carries the heat landward when prevailing winds are favorable. For example, the prevailing winds of the western coast of the United States are toward the land (eastward). Hence the heat liberated by the Pacific Ocean as it cools keeps coastal areas much warmer than they would be otherwise. This explains why the western coastal states generally have more favorable winter weather than the eastern coastal states, where the prevailing winds do not tend to carry the heat toward land.

Conservation of Energy: Calorimetry

Situations in which mechanical energy is converted to thermal energy occur frequently. We shall see some in the examples following this section and in the problems at the end of the chapter, but most of our attention here will be directed toward a particular kind of conservation-of-energy situation. In problems using the procedure we shall describe, called *calorimetry* problems, only the thermal energy transfer between the system and its surroundings is considered.

One technique for measuring the specific heat of solids or liquids is simply to heat the substance to some known temperature, place it in a vessel containing water of known mass and temperature, and measure the temperature of the water after equilibrium is reached. Since a negligible amount of mechanical work is done in the process, the law of conservation of energy requires that the thermal energy that leaves the warmer substance (of unknown specific heat) equals the thermal energy that enters the water.[3] Devices in which this thermal energy transfer occurs are called **calorimeters.**

For example, suppose that m_x is the mass of a substance whose specific heat we wish to determine, c_x its specific heat, and T_x its initial temperature. Likewise, let m_w, c_w, and T_w represent corresponding values for the water. If T is the final equilibrium temperature after everything is mixed, then from Equation 20.3, we find that the thermal energy gained by the water is $m_w c_w (T - T_w)$, and the thermal energy lost by the substance of unknown c is $- m_x c_x (T - T_x)$. Assuming that the combined system (water + unknown) does not lose or gain any thermal energy, it follows that the thermal energy gained by the water must equal the thermal energy lost by the unknown (conservation of energy):

$$m_w c_w (T - T_w) = - m_x c_x (T - T_x)$$

Solving for c_x gives

$$c_x = \frac{m_w c_w (T - T_w)}{m_x (T_x - T)} \tag{20.5}$$

---

[3] For precise measurements, the container for the water should be included in our calculations, since it also exchanges heat. This would require a knowledge of its mass and composition. However, if the mass of the water is large compared with that of the container, we can neglect the heat gained by the container. Furthermore, precautions must be taken in such measurements to minimize heat transfer between the system and the surroundings.

EXAMPLE 20.2   Cooling a Hot Ingot

A 0.0500-kg ingot of metal is heated to 200.0°C and then dropped into a beaker containing 0.400 kg of water initially at 20.0°C. If the final equilibrium temperature of the mixed system is 22.4°C, find the specific heat of the metal.

Solution   Because the thermal energy lost by the ingot equals the thermal energy gained by the water, we can write

$$m_x c_x (T_i - T_f) = m_w c_w (T_f - T_i)$$

$$(0.0500 \text{ kg})(c_x)(200.0°C - 22.4°C) = (0.400 \text{ kg})$$

$$(4186 \text{ J/kg} \cdot °C)(22.4°C - 20.0°C)$$

from which we find that

$$c_x = \boxed{453 \text{ J/kg} \cdot °C}$$

The ingot is most likely iron, as can be seen by comparing this result with the data in Table 20.1.

Exercise   What is the total thermal energy transferred to the water as the ingot is cooled?

Answer   4020 J.

EXAMPLE 20.3   Fun Time for a Cowboy

A cowboy fires a silver bullet of mass 2.00 g with a muzzle speed of 200 m/s into the pine wall of a saloon. Assume that all the internal energy generated by the impact remains with the bullet. What is the temperature change of the bullet?

Solution   The kinetic energy of the bullet is

$$\tfrac{1}{2}mv^2 = \tfrac{1}{2}(2.00 \times 10^{-3} \text{ kg})(200 \text{ m/s})^2 = 40.0 \text{ J}$$

Nothing in the environment is hotter than the bullet, so the bullet gains no thermal energy. Its temperature increases because the 40.0 J of kinetic energy becomes 40.0 J of extra internal energy. The temperature change would be the same as if 40.0 J of thermal energy were transferred from a stove to

the bullet, and we imagine this process to compute ΔT from

$$Q = mc\,\Delta T$$

Since the specific heat of silver is 234 J/kg · °C (Table 20.1), we get

$$\Delta T = \frac{Q}{mc} = \frac{40.0 \text{ J}}{(2.00 \times 10^{-3} \text{ kg})(234 \text{ J/kg} \cdot °C)} = \boxed{85.5°C}$$

Exercise   Suppose the cowboy runs out of silver bullets and fires a lead bullet of the same mass and speed into the wall. What is the temperature change of the bullet?

Answer   157°C.

20.3   LATENT HEAT

A substance usually undergoes a change in temperature when thermal energy is transferred between the substance and its surroundings. There are situations, however, in which the transfer of thermal energy does not result in a change in temperature. This is the case whenever the physical characteristics of the substance change from one form to another, commonly referred to as a **phase change**. Some common phase changes are solid to liquid (melting), liquid to gas (boiling), and a change in crystalline structure of a solid. All such phase changes involve a change in internal energy.

The thermal energy required to change the phase of a given mass, m, of a pure substance is

$$Q = mL \tag{20.6}$$

where L is called the **latent heat** ("hidden" heat) of the substance[4] and depends on the nature of the phase change as well as on the properties of the substance. **Latent heat of fusion,** L_f, is the term used when the phase change is from solid to

---

[4] The word *latent* is from the Latin *latere,* meaning *hidden* or *concealed.*

TABLE 20.2	Latent Heats of Fusion and Vaporization			
Substance	Melting Point (°C)	Latent Heat of Fusion (J/kg)	Boiling Point (°C)	Latent Heat of Vaporization (J/kg)
Helium	− 269.65	5.23×10^3	− 268.93	2.09×10^4
Nitrogen	− 209.97	2.55×10^4	− 195.81	2.01×10^5
Oxygen	− 218.79	1.38×10^4	− 182.97	2.13×10^5
Ethyl alcohol	− 114	1.04×10^5	78	8.54×10^5
Water	0.00	3.33×10^5	100.00	2.26×10^6
Sulfur	119	3.81×10^4	444.60	3.26×10^5
Lead	327.3	2.45×10^4	1750	8.70×10^5
Aluminum	660	3.97×10^5	2450	1.14×10^7
Silver	960.80	8.82×10^4	2193	2.33×10^6
Gold	1063.00	6.44×10^4	2660	1.58×10^6
Copper	1083	1.34×10^5	1187	5.06×10^6

liquid ("fuse" means to melt, to liquefy), and **latent heat of vaporization,** L_v, is used when the phase change is from liquid to gas (the liquid vaporizes).[5] For example, the latent heat of fusion for ice at atmospheric pressure is 3.33×10^5 J/kg, and the latent heat of vaporization of water is 2.26×10^6 J/kg. The latent heats of various substances vary considerably, as Table 20.2 shows.

Consider, for example, the thermal energy required to convert a 1.00-g block of ice at − 30.0°C to steam (water vapor) at 120.0°C. Figure 20.2 indicates the experimental results obtained when thermal energy is gradually added to the ice. Let us examine each portion of the curve.

Part A   On this portion of the curve the temperature of the ice is changing from − 30.0°C to 0.0°C. Since the specific heat of ice is 2090 J/kg · °C, we can calculate

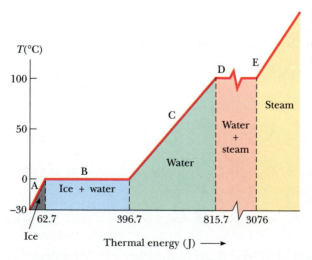

FIGURE 20.2 A plot of temperature versus thermal energy added when 1 g of ice initially at − 30°C is converted to steam.

---

[5] When a gas cools, it eventually returns to the liquid phase, or *condenses*. The heat given up per unit mass is called the *heat of condensation,* which equals the negative of the heat of vaporization. Likewise, when a liquid cools it eventually solidifies, and the *heat of solidification* equals the negative of the heat of fusion.

the amount of thermal energy added from Equation 20.4:

$$Q = m_i c_i \, \Delta T = (1.00 \times 10^{-3} \text{ kg})(2090 \text{ J/kg} \cdot {}^\circ\text{C})(30.0{}^\circ\text{C}) = 62.7 \text{ J}$$

Part B When the ice reaches 0.0°C, the ice/water mixture remains at this temperature—even though thermal energy is being added—until all the ice melts. The thermal energy required to melt 1.00 g of ice at 0.0°C is, from Equation 20.6,

$$Q = mL_f = (1.00 \times 10^{-3} \text{ kg})(3.33 \times 10^5 \text{ J/kg}) = 333 \text{ J}$$

Part C Between 0.0°C and 100.0°C, nothing surprising happens. No phase change occurs in this region. The thermal energy added to the water is being used to increase its temperature. The amount of thermal energy necessary to increase the temperature from 0.0°C to 100.0°C is

$$Q = m_w c_w \, \Delta T = (1.00 \times 10^{-3} \text{ kg})(4.19 \times 10^3 \text{ J/kg} \cdot {}^\circ\text{C})(100.0{}^\circ\text{C}) = 419 \text{ J}$$

Part D At 100.0°C, another phase change occurs as the water changes from water at 100.0°C to steam at 100.0°C. Just as in Part B, the water/steam mixture remains at 100.0°C—even though thermal energy is being added—until all of the liquid has been converted to steam. The thermal energy required to convert 1.00 g of water to steam at 100.0°C is

$$Q = mL_v = (1.00 \times 10^{-3} \text{ kg})(2.26 \times 10^6 \text{ J/kg}) = 2.26 \times 10^3 \text{ J}$$

Part E On this portion of the curve, heat is being added to the steam with no phase change occurring. The thermal energy that must be added to raise the temperature of the steam to 120.0°C is

$$Q = m_s c_s \, \Delta T = (1.00 \times 10^{-3} \text{ kg})(2.01 \times 10^3 \text{ J/kg} \cdot {}^\circ\text{C})(20.0{}^\circ\text{C}) = 40.2 \text{ J}$$

The *total amount of thermal energy* that must be added to change one gram of ice at −30.0°C to steam at 120.0°C is approximately 3.11×10^3 J. Conversely, to cool one gram of steam at 120.0°C down to the point at which we have ice at −30.0°C, we must remove 3.11×10^3 J of thermal energy.

Phase changes can be described in terms of a rearrangement of molecules when thermal energy is added or removed from a substance. Consider first the liquid-to-gas phase change. The molecules in a liquid are close together, and the forces between them are stronger than those between the more widely separated molecules of a gas. Therefore, work must be done on the liquid against these attractive molecular forces in order to separate the molecules. The latent heat of vaporization is the amount of energy that must be added to the liquid to accomplish this separation.

Similarly, at the melting point of a solid, we imagine that the amplitude of vibration of the atoms about their equilibrium position becomes large enough to allow the atoms to pass the barriers of adjacent atoms and move to their new positions. The new locations are, on the average, less symmetrical and therefore have higher energy. The latent heat of fusion is equal to the work required at the molecular level to transform the mass from the highly ordered solid phase to the less ordered liquid phase.

The average distance between atoms in the gas phase is much larger than in either the liquid or the solid phase. Each atom or molecule is removed from its neighbors, without the compensation of attractive forces to new neighbors. Therefore, it is not surprising that more work is required at the molecular level to vaporize a given mass of substance than to melt it; thus, the latent heat of vaporization is much larger than the latent heat of fusion for a given substance (Table 20.2).

Problem Solving Strategy
Calorimetry Problems

If you are having difficulty with calorimetry problems, make the following considerations.

- Be sure your units are consistent throughout. For instance, if you are using specific heats in cal/g · °C, be sure that masses are in grams and temperatures are in Celsius units throughout.
- Losses and gains in thermal energy are found by using $Q = mc\,\Delta T$ only for those intervals in which no phase changes occur. Likewise, the equations $Q = mL_f$ and $Q = mL_v$ are to be used only when phase changes *are* taking place.
- Often sign errors occur in heat loss = heat gain equations. One way to check your equation is to examine the signs of all ΔT's that appear in it.

EXAMPLE 20.4   Cooling the Steam

What mass of steam initially at 130°C is needed to warm 200 g of water in a 100-g glass container from 20.0°C to 50.0°C?

Solution   This is a heat-transfer problem in which we must equate the thermal energy lost by the steam to the thermal energy gained by the water and glass container. There are three stages as the steam loses thermal energy. In the first stage, the steam is cooled to 100°C. The thermal energy liberated in the process is

$$Q_1 = m_s c_s\,\Delta T = m_s (2.01 \times 10^3 \text{ J/kg} \cdot \text{°C})(30.0\text{°C})$$
$$= m_s (6.03 \times 10^4 \text{ J/kg})$$

In the second stage, the steam is converted to water. In this case, to find the thermal energy removed, we use the latent heat of vaporization and $Q = mL_v$:

$$Q_2 = m_s (2.26 \times 10^6 \text{ J/kg})$$

In the last stage, the temperature of the water is reduced to 50.0°C. This liberates an amount of thermal energy

$$Q_3 = m_s c_w\,\Delta T = m_s (4.19 \times 10^3 \text{ J/kg} \cdot \text{°C})(50.0\text{°C})$$
$$= m_s (2.09 \times 10^5 \text{ J/kg})$$

If we equate the thermal energy lost by the steam to the thermal energy gained by the water and glass and use the given information, we find

$$m_s (6.03 \times 10^4 \text{ J/kg}) + m_s (2.26 \times 10^6 \text{ J/kg})$$
$$+ m_s (2.09 \times 10^5 \text{ J/kg})$$
$$= (0.200 \text{ kg})(4.19 \times 10^3 \text{ J/kg} \cdot \text{°C})(30.0\text{°C})$$
$$+ (0.100 \text{ kg})(837 \text{ J/kg} \cdot \text{°C})(30.0\text{°C})$$

$$m_s = \boxed{1.09 \times 10^{-2} \text{ kg} = 10.9 \text{ g}}$$

EXAMPLE 20.5   Boiling Liquid Helium

Liquid helium has a very low boiling point, 4.2 K, and a very low heat of vaporization, 2.09×10^4 J/kg (Table 20.2). A constant power of 10.0 W is transferred to a container of liquid helium from an immersed electric heater. At this rate, how long does it take to boil away 1.00 kg of liquid helium?

Reasoning and Solution   Since $L_v = 2.09 \times 10^4$ J/kg for liquid helium, we must supply 2.09×10^4 J of energy to boil away 1.00 kg. The power supplied to the helium is 10.0 W = 10.0 J/s. That is, in 1.00 s, 10.0 J of energy is transferred to

the helium. Therefore, the time it takes to transfer an energy of 2.09×10^4 J is

$$t = \frac{2.09 \times 10^4 \text{ J}}{10.0 \text{ J/s}} = 2.09 \times 10^3 \text{ s} \approx \boxed{35 \text{ min}}$$

Exercise   If 10.0 W of power is supplied to 1.00 kg of water at 100°C, how long will it take for the water to completely boil away?

Answer   62.8 h.

20.4   WORK AND HEAT IN THERMODYNAMIC PROCESSES

In the macroscopic approach to thermodynamics we describe the *state* of a system with such variables as pressure, volume, temperature, and internal energy. The number of macroscopic variables needed to characterize a system depends on the system's nature. For a homogeneous system, such as a gas containing only one type of molecule, usually only two variables are needed. However, it is important to note that a *macroscopic state* of an isolated system can be specified only if the system is in thermal equilibrium internally. In the case of a gas in a container, internal thermal equilibrium requires that every part of the container be at the same pressure and temperature.

Consider gas contained in a cylinder fitted with a movable piston (Fig. 20.3). In equilibrium, the gas occupies a volume V and exerts a uniform pressure P on the cylinder walls and piston. If the piston has a cross-sectional area A, the force exerted by the gas on the piston is $F = PA$. Now let us assume that the gas expands **quasi-statically,** that is, slowly enough to allow the system to remain essentially in thermodynamic equilibrium at all times. As the piston moves up a distance dy, the work done by the gas on the piston is

$$dW = F\,dy = PA\,dy$$

Since $A\,dy$ is the increase in volume of the gas dV, we can express the work done as

$$dW = P\,dV \tag{20.7}$$

Since the gas expands, dV is positive and the work done by the gas is positive, whereas if the gas is compressed, dV is negative, indicating that the work done by the gas is negative.[6] (In the latter case, negative work can be interpreted as work

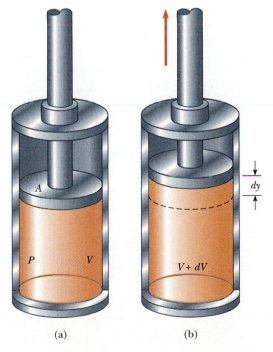

(a)   (b)

FIGURE 20.3   Gas contained in a cylinder at a pressure P does work on a moving piston as the system expands from a volume V to a volume $V + dV$.

---

[6] For historical reasons, we choose to let W represent the work done by the system here. In other parts of the text, W is the work done on the system. The change affects only the sign of W.

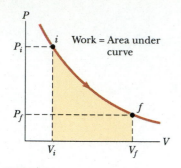

FIGURE 20.4   A gas expands reversibly (slowly) from state *i* to state *f*. The work done by the gas equals the area under the *PV* curve.

Work equals area under the curve in a *PV* diagram

being done *on* the gas.) Clearly, the work done by the gas is zero when the volume remains constant. The total work done by the gas as its volume changes from V_i to V_f is given by the integral of Equation 20.7:

$$W = \int_{V_i}^{V_f} P \, dV \tag{20.7}$$

To evaluate this integral, one must know how the pressure varies during the expansion process. (Note that a *process* is *not* specified merely by giving the initial and final states. Rather, a process is a *fully specified* change in state of a system.) In general, the pressure is not constant, but depends on the volume and temperature. If the pressure and volume are known at each step of the process, the states of the gas can then be represented as a curve on a *PV* diagram, as in Figure 20.4.

> The work done in the expansion from the initial state to the final state is the area under the curve in a *PV* diagram.

As Figure 20.4 shows, the work done in the expansion from the initial state, *i*, to the final state, *f*, depends on the path taken between these two states. To illustrate this important point, consider several different paths connecting *i* and *f* (Fig. 20.5). In the process depicted in Figure 20.5a, the pressure of the gas is first reduced from P_i to P_f by cooling at constant volume V_i, and the gas then expands from V_i to V_f at constant pressure P_f. The work done along this path is $P_f(V_f - V_i)$. In Figure 20.5b, the gas first expands from V_i to V_f at constant pressure P_i, and then its pressure is reduced to P_f at constant volume V_f. The work done along this path is $P_i(V_f - V_i)$, which is greater than that for the process described in Figure 20.5a. Finally, for the process described in Figure 20.5c, where both *P* and *V* change continuously, the work done has some value intermediate between the values obtained in the first two processes. To evaluate the work in this case, the shape of the *PV* curve must be known. Therefore, we see that the work done by a system depends on the process by which the system goes from the initial to the final state. In other words, *the work done depends on the initial, final, and intermediate states of the system.*

In a similar manner, the thermal energy transferred into or out of the system also depends on the process. This can be demonstrated by considering the situa-

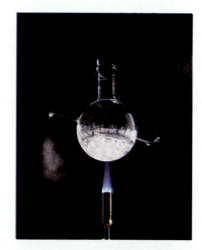

This device, called Hero's engine, was invented around 150 B.C. by Hero in Alexandria. When water is boiled in the flask, which is suspended by a cord, steam exits through two tubes at the sides of the flask (in opposite directions), creating a torque that rotates the flask. *(Courtesy of Central Scientific Co.)*

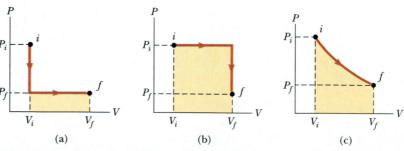

FIGURE 20.5   The work done by a gas as it is taken from an initial state to a final state depends on the path between these states.

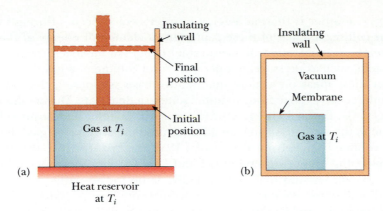

FIGURE 20.6   (a) A gas at temperature T_i expands slowly by absorbing thermal energy from a reservoir at the same temperature. (b) A gas expands rapidly into an evacuated region after a membrane is broken.

tions depicted in Figure 20.6. In each case, the gas has the same initial volume, temperature, and pressure and is assumed to be ideal. In Figure 20.6a, the gas is in thermal contact with a heat reservoir. If the pressure of the gas is infinitesimally greater than atmospheric pressure, the gas, because it absorbs thermal energy, expands and causes the piston to rise. During this expansion to some final volume V_f, just enough thermal energy to maintain a constant temperature T_i is transferred from the reservoir to the gas.

   Now consider the thermally insulated system shown in Figure 20.6b. When the membrane is broken, the gas expands rapidly into the vacuum until it occupies a volume V_f and is at a pressure P_f. In this case, the gas does no work because there is no movable piston. Furthermore, no thermal energy is transferred through the insulating wall.

   The initial and final states of the ideal gas in Figure 20.6a are identical to the initial and final states in Figure 20.6b, but the paths are different. In the first case, thermal energy is transferred slowly to the gas, and the gas does work on the piston. In the second case, no thermal energy is transferred and the work done is zero. Therefore, we conclude that *thermal energy transfer, like the work done, depends on the initial, final, and intermediate states of the system.* Furthermore, since thermal energy and work depend on the path, neither quantity is determined by the end points of a thermodynamic process.

Free expansion of a gas

20.5   THE FIRST LAW OF THERMODYNAMICS

When the law of conservation of energy was introduced in Chapter 8, it was stated that the mechanical energy of a system is constant in the absence of nonconservative forces, such as friction. That is, the changes in the internal energy of the system were not included in this mechanical model. The first law of thermodynamics, which we discuss in this section, is a generalization known as the law of conservation of energy and encompasses possible changes in internal energy. It is a universally valid law that can be applied to all kinds of processes. Furthermore, it provides us with a connection between the microscopic and macroscopic worlds.

   We have seen that energy can be transferred between a system and its surroundings in two ways. One is work done by (or on) the system, which requires that there be a macroscopic displacement of the point of application of a force (or pressure). The other is thermal energy transfer, which occurs through random molecular collisions. Each of these represents a change of energy of the system and, therefore, usually results in measurable changes in the macroscopic variables of the system, such as pressure, temperature, and volume of a gas.

To put these ideas on a more quantitative basis, suppose a thermodynamic system undergoes a change from an initial state to a final state. During this change, positive Q is the thermal energy transferred *to* the system, and positive W is the work done *by* the system. As an example, suppose the system is a gas whose pressure and volume change from P_i, V_i to P_f, V_f. If the quantity $Q - W$ is measured for various paths connecting the initial and final equilibrium states (that is, for various *processes*), we find that $Q - W$ is the same for all paths connecting the initial and final states. We conclude that the quantity $Q - W$ is determined completely by the initial and final states of the system, and we call the quantity $Q - W$ the *change in the energy of the system.* Although Q and W both depend on the path, the quantity $Q - W$ *is independent of the path.* If we represent the energy function with the letter U, then the *change* in energy, $\Delta U = U_f - U_i$, can be expressed as

| Change in energy |

$$\Delta U = U_f - U_i = Q - W \tag{20.9}$$

| First-law equation |

where all quantities must have the same energy units. Equation 20.9 is known as the **first-law equation** and is a key equation to many applications. When it is used in this form, we use the convention that Q is positive when thermal energy enters the system and negative when thermal energy leaves the system. Likewise, W is positive when the system does work on the surroundings and negative if work is done on the system.

When a system undergoes an infinitesimal change in state, where a small amount of thermal energy dQ is transferred and a small amount of work dW is done, the energy also changes by a small amount dU. Thus, for infinitesimal processes we can express the first-law equation as[7]

| First-law equation for infinitesimal changes |

$$dU = dQ - dW$$

On a microscopic level, the internal energy of a system includes the kinetic and potential energies of the molecules making up the system. Part of the energy, U, is the internal energy of the system. In thermodynamics, we do not concern ourselves with the specific form of the internal energy. An analogy can be made between the internal energy of a system and the potential energy function associated with a body moving under the influence of gravity without friction. The potential energy function is independent of the path, and it is only changes in it that is of concern. Likewise, the change in internal energy of a thermodynamic system is what matters, since only differences are defined. Because absolute values are not defined, any reference state can be chosen for the internal energy.

| The internal energy of an isolated system is constant |

Let us look at some special cases in which the only changes in energy are changes in internal energy. First consider an *isolated system,* that is, one that does not interact with its surroundings. In this case, no thermal energy transfer takes place and the work done is zero; hence, the internal energy remains constant. That is, since $Q = W = 0$, $\Delta U = 0$, and so $U_i = U_f$. We conclude that *the internal energy of an isolated system remains constant.*

---

[7] Note that dQ and dW are not true differential quantities, although dU is a true differential. In fact, dQ and dW are *inexact differentials* and are often represented by $đQ$ and $đW$. For further details on this point, see an advanced text in thermodynamics, e.g., R. P. Bauman, *Modern Thermodynamics and Statistical Mechanics,* New York, Macmillan, 1992.

Next consider a process in which a system (one not isolated from its surroundings) is taken through a **cyclic process,** that is, one that originates and ends at the same state. In this case, the change in the internal energy must again be zero and, therefore, the thermal energy added to the system must equal the work done during the cycle. That is, in a cyclic process,

$$\Delta U = 0 \quad \text{and} \quad Q = W$$

Cyclic process

Note that *the net work done per cycle equals the area enclosed by the path representing the process on a PV diagram.*

If a process occurs in which the work done is zero, then the change in internal energy equals the thermal energy entering or leaving the system. If thermal energy enters the system, Q is positive and the internal energy increases. For a gas, we can associate this increase in internal energy with an increase in the kinetic energy of the molecules. On the other hand, if a process occurs in which the thermal energy transferred is zero and work is done by the system, then the magnitude of the change in internal energy equals the negative of the work done by the system. That is, the internal energy of the system decreases. For example, if a gas is compressed with no thermal energy transferred (by a moving piston, for example), the work done by the gas is negative and the internal energy again increases. This is because kinetic energy is transferred from the moving piston to the gas molecules.

No distinction exists between thermal energy transfer and work on a microscopic scale. Both can produce a change in the internal energy of a system. Although the macroscopic quantities Q and W are *not* properties of a system, they are related to the changes of the internal energy of a stationary system through the first-law equation. Once a process, or path, is defined, Q and W can be either calculated or measured, and the change in internal energy can be found from the first-law equation. One of the important consequences of the first law is that there is a quantity called internal energy, the value of which is determined by the state of the system. The internal energy function is therefore called a *state function.*

20.6   SOME APPLICATIONS OF THE FIRST LAW OF THERMODYNAMICS

In order to apply the first law of thermodynamics to specific systems, it is useful to first define some common thermodynamic processes. An **adiabatic process** is one during which no thermal energy enters or leaves the system, that is, $Q = 0$. An adiabatic process can be achieved either by thermally insulating the system from its surroundings (as in Fig. 20.6b) or by performing the process rapidly. Applying the first law of thermodynamics to an adiabatic process, we see that

Adiabatic process

$$\Delta U = -W \tag{20.10}$$

First-law equation for an adiabatic process

From this result, we see that if a gas expands adiabatically, W is positive, so ΔU is negative and the temperature of the gas decreases. Conversely, the gas temperature rises when it is compressed adiabatically.

Adiabatic processes are very important in engineering practice. Some common examples include the expansion of hot gases in an internal combustion engine, the liquefaction of gases in a cooling system, and the compression stroke in a diesel engine.

The process described in Figure 20.6b, called an **adiabatic free expansion,** is an adiabatic process in which no work is done on or by the gas. Since $Q = 0$ and $W = 0$, we see from the first law that $\Delta U = 0$ for this process. That is, *the initial and final internal energies of a gas are equal in an adiabatic free expansion.* As we shall see in

Adiabatic free expansion

the next chapter, the internal energy of an ideal gas depends only on its temperature. Thus, we would expect no change in temperature during an adiabatic free expansion. This is in accord with experiments performed at low pressures. (Careful experiments at high pressures for real gases show a slight decrease or increases in temperature after the expansion.)

A process that occurs at constant pressure is called an **isobaric process**. When such a process occurs, the thermal energy transferred and the work done are both usually nonzero. The work done is simply $P(V_f - V_i)$.

A process that takes place at constant volume is called an **isovolumetric process**. In such an expansion process, the work done is clearly zero. Hence from the first law we see that in an isovolumetric process with $W = 0$

$$\Delta U = Q \tag{20.11}$$

This tells us that *if thermal energy is added to a system kept at constant volume, all of the thermal energy goes into increasing the internal energy of the system.* When a mixture of gasoline vapor and air explodes in the cylinder of an engine, the temperature and pressure rise suddenly because the cylinder volume doesn't change appreciably during the short duration of the explosion.

A process that occurs at constant temperature is called an **isothermal process**, and a plot of P versus V at constant temperature for an ideal gas yields a hyperbolic curve called an isotherm. The internal energy of an ideal gas is a function of temperature only. Hence, in an isothermal process of an ideal gas, $\Delta U = 0$.

Isothermal Expansion of an Ideal Gas

Suppose an ideal gas is allowed to expand quasi-statically at constant temperature as described by the PV diagram in Figure 20.7. The curve is a hyperbola, and the equation of this curve is $PV =$ constant. Let us calculate the work done by the gas in the expansion from state i to state f.

The isothermal expansion of the gas can be achieved by placing the gas in good thermal contact with a heat reservoir at the same temperature, as in Figure 20.6a.

The work done by the gas is given by Equation 20.7. Since the gas is ideal and the process is quasi-static, we can apply $PV = nRT$ for each point on the path. Therefore, we have

$$W = \int_{V_i}^{V_f} P \, dV = \int_{V_i}^{V_f} \frac{nRT}{V} \, dV$$

Since T is constant in this case, it can be removed from the integral along with n and R:

$$W = nRT \int_{V_i}^{V_f} \frac{dV}{V} = nRT \ln V \Big]_{V_i}^{V_f}$$

To evaluate the integral, we used $\int (dx/x) = \ln x$ (Table B.5 in Appendix B). Thus, we find

$$W = nRT \ln \left(\frac{V_f}{V_i} \right) \tag{20.12}$$

Numerically, this work equals the shaded area under the PV curve in Figure 20.7. Because the gas expands isothermally, $V_f > V_i$ and we see that the work done by the gas is positive, as we would expect. If the gas is compressed isothermally, then

Isobaric process

First-law equation for a constant-volume process

Isothermal process

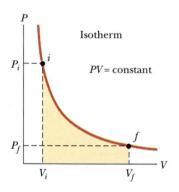

FIGURE 20.7   The PV diagram for an isothermal expansion of an ideal gas from an initial state to a final state. The curve is a hyperbola.

Work done in an isothermal process

$V_f < V_i$ and the work done by the gas is negative. Hence, for an isothermal process $\Delta U = 0$, and from the first law we conclude that the thermal energy given up by the reservoir (and transferred to the gas) equals the work done by the gas, or $Q = W$.

EXAMPLE 20.6   Work Done During an Isothermal Expansion

Calculate the work done by 1.0 mol of an ideal gas that is kept at 0.0°C during an expansion from 3.0 liters to 10.0 liters.

Solution   Substituting these values into Equation 20.12 gives

$$W = nRT \ln\left(\frac{V_f}{V_i}\right)$$

$$= (1.0 \text{ mol})(8.31 \text{ J/mol} \cdot \text{K})(273 \text{ K}) \ln\left(\frac{10.0}{3.0}\right)$$

$$= \boxed{2.7 \times 10^3 \text{ J}}$$

The thermal energy that must be supplied to the gas from the reservoir to keep T constant is also 2.7×10^3 J.

Boiling

Suppose that a liquid of mass m vaporizes at constant pressure P. Its volume in the liquid state is V_ℓ, and its volume in the vapor state is V_v. Let us find the work done in the expansion and the change in internal energy of the system.

Since the expansion takes place at constant pressure, the work done by the system is

$$W = \int_{V_\ell}^{V_v} P \, dV = P \int_{V_\ell}^{V_v} dV = P(V_v - V_\ell)$$

The thermal energy that must be transferred to the liquid to vaporize all of it is equal to $Q = mL_v$, where L_v is the latent heat of vaporization of the liquid. Using the first law and the result above, we get

$$\Delta U = Q - W = mL_v - P(V_v - V_\ell) \qquad (20.13)$$

EXAMPLE 20.7   Boiling Water

One gram of water occupies a volume of 1.00 cm³ at atmospheric pressure. When this amount of water is boiled, it becomes 1671 cm³ of steam. Calculate the change in internal energy for this process.

Solution   Since the latent heat of vaporization of water is 2.26×10^6 J/kg at atmospheric pressure, the heat required to boil 1.00 g is

$$Q = mL_v = (1.00 \times 10^{-3} \text{ kg})(2.26 \times 10^6 \text{ J/kg}) = 2260 \text{ J}$$

The work done by the system is positive and equal to

$$W = P(V_v - V_\ell)$$
$$= (1.013 \times 10^5 \text{ N/m}^2)[(1671 - 1.00) \times 10^{-6} \text{ m}^3]$$
$$= 169 \text{ J}$$

Hence, the change in internal energy is

$$\Delta U = Q - W = 2260 \text{ J} - 169 \text{ J} = \boxed{2.09 \text{ kJ}}$$

The internal energy of the system increases because ΔU is positive. We see that most (93%) of the thermal energy transferred to the liquid goes into increasing the internal energy. Only 7% goes into external work.

EXAMPLE 20.8   Heat Transferred to a Solid

A 1.0-kg bar of copper is heated at atmospheric pressure. If its temperature increases from 20°C to 50°C, (a) find the work done by the copper.

Solution   The change in volume of the copper can be calculated using Equation 19.6 and the volume expansion coefficient for copper taken from Table 19.2 (remembering that

$\beta = 3\alpha$):

$$\Delta V = \beta V \Delta T = [5.1 \times 10^{-5}(°C)^{-1}](50°C - 20°C)V$$
$$= 1.5 \times 10^{-3}\ V$$

But the volume is equal to m/ρ, and the density of copper is $8.92 \times 10^3\ \text{kg/m}^3$. Hence,

$$\Delta V = (1.5 \times 10^{-3})\left(\frac{1.0\ \text{kg}}{8.92 \times 10^3\ \text{kg/m}^3}\right) = 1.7 \times 10^{-7}\ \text{m}^3$$

Since the expansion takes place at constant pressure, the work done is

$$W = P\Delta V = (1.013 \times 10^5\ \text{N/m}^2)(1.7 \times 10^{-7}\ \text{m}^3)$$
$$= 1.9 \times 10^{-2}\ \text{J}$$

(b) What quantity of thermal energy is transferred to the copper?

Solution   Taking the specific heat of copper from Table 20.1 and using Equation 20.4, we find that the thermal energy transferred is

$$Q = mc\,\Delta T = (1.0\ \text{kg})(387\ \text{J/kg} \cdot °C)(30°C)$$
$$= 1.2 \times 10^4\ \text{J}$$

(c) What is the increase in internal energy of the copper?

Solution   From the first law of thermodynamics, the increase in internal energy is

$$\Delta U = Q - W = 1.2 \times 10^4\ \text{J}$$

Note that almost all of the thermal energy transferred goes into increasing the internal energy. The fraction of thermal energy that is used to do work against the atmosphere is only about 10^{-6}! Hence, in the thermal expansion of a solid or a liquid, the small amount of work done is usually ignored.

20.7   HEAT TRANSFER

In practice, it is important to understand the rate at which thermal energy is transferred between a system and its surroundings and the mechanisms responsible for the transfer. You may have used a Thermos bottle or some other thermally insulated vessel to store hot coffee for a length of time. The vessel reduces thermal energy transfer between the outside air and the hot coffee. Ultimately, of course, the liquid will reach air temperature because the vessel is not a perfect insulator. There is no thermal energy transfer between a system and its surroundings when they are at the same temperature.

Heat Conduction

Melted snow pattern on a parking lot indicates the presence of underground steam pipes used to aid snow removal. Heat from the steam is conducted to the pavement from the pipes, causing the snow to melt.   *(Courtesy of Dr. Albert A. Bartlett, University of Colorado, Boulder)*

The easiest thermal energy transfer process to describe quantitatively is called *conduction*. In this process, the thermal energy transfer can be viewed on an atomic scale as an exchange of kinetic energy between molecules, where the less energetic particles gain energy by colliding with the more energetic particles. For example, if you insert a metallic bar into a flame while holding one end, you will find that the temperature of the metal in your hand increases. The thermal energy reaches your hand through conduction. The manner in which thermal energy is transferred from the flame, through the bar, and to your hand can be understood by examining what is happening to the atoms and electrons of the metal. Initially, before the rod is inserted into the flame, the metal atoms and electrons are vibrating about their equilibrium positions. As the flame heats the rod, those metal atoms and electrons near the flame begin to vibrate with larger and larger amplitudes. These, in turn, collide with their neighbors and transfer some of their energy in the collisions. Slowly, metal atoms and electrons farther down the rod increase their amplitude of vibration, until the large-amplitude vibrations arrive at the end being held. The effect of this increased vibration is an increase in temperature of the metal, and possibly a burned hand.

Although the transfer of thermal energy through a metal can be partially explained by atomic vibrations and electron motion, the rate of conduction also depends on the properties of the substance being heated. For example, it is possible to hold a piece of asbestos in a flame indefinitely. This implies that very little thermal energy is being conducted through the asbestos. In general, metals are good conductors of thermal energy, and materials such as asbestos, cork, paper, and fiber glass are poor conductors. Gases also are poor conductors because of their dilute nature. Metals are good conductors of thermal energy because they contain large numbers of electrons that are relatively free to move through the metal and can transport energy from one region to another. Thus, in a good conductor, such as copper, conduction takes place via the vibration of atoms and via the motion of free electrons.

Conduction occurs only if there is a difference in temperature between two parts of the conducting medium. Consider a slab of material of thickness Δx and cross-sectional area A with its opposite faces at different temperatures T_1 and T_2, where $T_2 > T_1$ (Fig. 20.8). It is found from experiment that the thermal energy Q transferred in a time Δt flows from the hotter end to the colder end. The rate at which heat flows, $Q/\Delta t$, is found to be proportional to the cross-sectional area and the temperature difference, and inversely proportional to the thickness:

$$\frac{Q}{\Delta t} \propto A\frac{\Delta T}{\Delta x}$$

It is convenient to use the symbol H to represent the thermal energy transfer rate. That is, we take $H = Q/\Delta t$. Note that H has units of watts when Q is in joules and Δt is in seconds. For a slab of infinitesimal thickness dx and temperature difference dT, we can write the **law of heat conduction**

$$H = -kA\frac{dT}{dx} \qquad (20.14)$$

Law of heat conduction

where the proportionality constant k is called the **thermal conductivity** of the material, and dT/dx is the **temperature gradient** (the variation of temperature with position). The minus sign in Equation 20.14 denotes the fact that thermal energy flows in the direction of decreasing temperature.

Suppose a substance is in the shape of a long uniform rod of length L, as in Figure 20.9, and is insulated so that thermal energy cannot escape from its surface

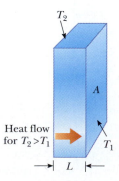

FIGURE 20.8 Heat transfer through a conducting slab of cross-sectional area A and thickness Δx. The opposite faces are at different temperatures, T_1 and T_2.

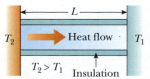

FIGURE 20.9 Conduction of heat through a uniform, insulated rod of length L. The opposite ends are in thermal contact with heat reservoirs at different temperatures.

TABLE 20.3	Thermal Conductivities of Some Substances

Substance	Thermal Conductivity W/m·°C
Metals (at 25°C)	
Aluminum	238
Copper	397
Gold	314
Iron	79.5
Lead	34.7
Silver	427
Gases (at 20°C)	
Air	0.0234
Helium	0.138
Hydrogen	0.172
Nitrogen	0.0234
Oxygen	0.0238
Nonmetals (approximate values)	
Asbestos	0.08
Concrete	0.8
Diamond	2300
Glass	0.8
Ice	2
Rubber	0.2
Water	0.6
Wood	0.08

except at the ends, which are in thermal contact with heat reservoirs having temperatures T_1 and T_2. When a steady state has been reached, the temperature at each point along the rod is constant in time. In this case, the temperature gradient is the same everywhere along the rod and is

$$\frac{dT}{dx} = \frac{T_1 - T_2}{L}$$

Thus the thermal energy transfer rate is

$$H = kA \frac{(T_2 - T_1)}{L} \tag{20.15}$$

Substances that are good thermal conductors have large thermal conductivity values, whereas good thermal insulators have low thermal conductivity values. Table 20.3 lists thermal conductivities for various substances. We see that metals are generally better thermal conductors than nonmetals.

For a compound slab containing several materials of thicknesses $L_1, L_2, \ldots$ and thermal conductivities $k_1, k_2, \ldots$, the rate of thermal energy transfer through the slab at steady state is

$$H = \frac{A(T_2 - T_1)}{\Sigma_i (L_i / k_i)} \tag{20.16}$$

where T_1 and T_2 are the temperatures of the outer extremities of the slab (which are held constant) and the summation is over all slabs. The following example is a proof of this result.

EXAMPLE 20.9   Heat Transfer Through Two Slabs

Two slabs of thickness L_1 and L_2 and thermal conductivities k_1 and k_2 are in thermal contact with each other as in Figure 20.10. The temperatures of their outer surfaces are T_1 and T_2, respectively, and $T_2 > T_1$. Determine the temperature at the interface and the rate of thermal energy transfer through the slabs in the steady-state condition.

Solution   If T is the temperature at the interface, then the rate at which thermal energy is transferred through slab 1 is

$$(1) \qquad H_1 = \frac{k_1 A(T - T_1)}{L_1}$$

Likewise, the rate at which thermal energy is transferred through slab 2 is

$$(2) \qquad H_2 = \frac{k_2 A(T_2 - T)}{L_2}$$

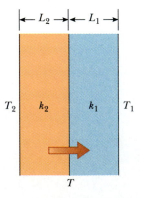

FIGURE 20.10   (Example 20.9) Heat transfer by conduction through two slabs in thermal contact with each other. At steady state, the rate of heat transfer through slab 1 equals the rate of heat transfer through slab 2.

When a steady state is reached, these two rates must be equal; hence,

$$\frac{k_1 A(T - T_1)}{L_1} = \frac{k_2 A(T_2 - T)}{L_2}$$

Solving for T gives

(3) $$T = \frac{k_1 L_2 T_1 + k_2 L_1 T_2}{k_1 L_2 + k_2 L_1}$$

Substituting (3) into either (1) or (2), we get

$$H = \frac{A(T_2 - T_1)}{(L_1/k_1) + (L_2/k_2)}$$

An extension of this model to several slabs of materials leads to Equation 20.16.

*Home Insulation

If you would like to do some calculating to determine whether or not to add insulation to a ceiling or to some other portion of a building, what you have just learned about conduction needs to be modified slightly because the insulating properties of materials used in buildings are usually expressed in engineering rather than SI units. For example, measurements stamped on a package of fiberglass insulating board will be in units such as British thermal units, feet, and degrees Fahrenheit.

In engineering practice, the term L/k for a particular substance is referred to as the R value of the material. Thus, Equation 20.16 reduces to

$$H = \frac{A(T_2 - T_1)}{\Sigma_i R_i} \qquad (20.17)$$

where $R_i = L_i/k_i$. The R values for a few common building materials are given in Table 20.4 (note the units).

Also, near any vertical surface there is a very thin, stagnant layer of air that must be considered when finding the total R value for a wall. The thickness of this stagnant layer on an outside wall depends on the speed of the wind. As a result, thermal energy loss from a house on a day when the wind is blowing hard is greater than loss on a day when the wind speed is zero. A representative R value for this stagnant layer of air is given in Table 20.4.

TABLE 20.4   *R* Values for Some Common Building Materials

Material	R Value (ft$^2 \cdot$ °F$\cdot$h/BTU)
Hardwood siding (1 in. thick)	0.91
Wood shingles (lapped)	0.87
Brick (4 in. thick)	4.00
Concrete block (filled cores)	1.93
Fiberglass batting (3.5 in. thick)	10.90
Fiberglass batting (6 in. thick)	18.80
Fiberglass board (1 in. thick)	4.35
Cellulose fiber (1 in. thick)	3.70
Flat glass (0.125 in. thick)	0.89
Insulating glass (0.25-in. space)	1.54
Vertical air space (3.5 in. thick)	1.01
Air film	0.17
Drywall (0.5 in. thick)	0.45
Sheathing (0.5 in. thick)	1.32

This thermogram of a home, made during cold weather, shows colors ranging from white and yellow (areas of greatest heat loss) to blue and purple (areas of least heat loss). *(Daedalus Enterprises, Inc./Peter Arnold, Inc.)*

EXAMPLE 20.10   The *R* Value of a Typical Wall

Calculate the total *R* value for a wall constructed as shown in Figure 20.11a. Starting outside the house (to the left in Fig. 20.11a) and moving inward, the wall consists of brick, 0.5 in. of sheathing, an air space 3.5 in. thick, and 0.5 in. of dry wall. Do not forget the air layers inside and outside the house.

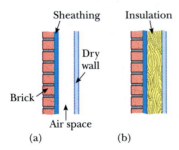

FIGURE 20.11   (Example 20.10) Cross-sectional view of an exterior wall containing (a) an air space and (b) insulation.

Solution   Referring to Table 20.4, we find the total *R* value for the wall:

R_1 (outside air film) $= 0.17 \text{ ft}^2 \cdot {}^\circ\text{F} \cdot \text{h}/\text{BTU}$

R_2 (brick) $= 4.00$

R_3 (sheathing) $= 1.32$

R_4 (air space) $= 1.01$

R_5 (dry wall) $= 0.45$

R_6 (inside air film) $= 0.17$

——————————————————

R_{total} $= 7.12 \text{ ft}^2 \cdot {}^\circ\text{F} \cdot \text{h}/\text{BTU}$

Exercise   If a layer of fiber-glass insulation 3.5 in. thick is placed inside the wall to replace the air space as in Figure 20.11b, what is the new total *R* value? By what factor is the thermal energy loss reduced?

Answer $R = 17 \text{ ft}^2 \cdot {}^\circ\text{F} \cdot \text{h}/\text{BTU}$; a factor of 2.4.

Convection

At one time or another you probably have warmed your hands by holding them over an open flame. In this situation, the air directly above the flame is heated and expands. As a result, the density of the air decreases and the air rises. This warmed mass of air heats your hands as it flows by. *Thermal energy transferred by the movement of a heated substance is said to have been transferred by* **convection.** When the movement results from differences in density, as in the example of air around a fire, it is referred to as *natural convection.* When the heated substance is forced to move by a fan or pump, as in some hot-air and hot-water heating systems, the process is called *forced convection.*

The circulating pattern of air flow at a beach is an example of convection. Likewise, the mixing that occurs as water is cooled and eventually freezes at its surface (Chapter 19) is an example of convection in nature. Recall that the mixing by convection currents ceases when the water temperature reaches 4°C. Since the water in the lake cannot be cooled by convection below 4°C, and because water is a relatively poor conductor of thermal energy, the water near the bottom remains near 4°C for a long time. As a result, fish have a comfortable temperature in which to live even in periods of prolonged cold weather.

If it were not for convection currents, it would be very difficult to boil water. As water is heated in a teakettle, the lower layers are warmed first. These heated regions expand and rise to the top because their density is lowered. At the same time, the denser cool water replaces the warm water at the bottom of the kettle so that it can be heated.

The same process occurs when a room is heated by a radiator. The hot radiator warms the air in the lower regions of the room. The warm air expands and rises to the ceiling because of its lower density. The denser regions of cooler air from above replace the warm air, setting up the continuous air current pattern shown in Figure 20.12.

FIGURE 20.12 Convection currents are set up in a room heated by a radiator.

Radiation

The third way of transferring thermal energy is through **radiation.** All objects radiate energy continuously in the form of electromagnetic waves, which we shall discuss in Chapter 34. The type of radiation associated with the transfer of thermal energy from one location to another is referred to as infrared radiation.

Through electromagnetic radiation, approximately 1340 J of thermal energy from the Sun strikes 1 m² of the top of the Earth's atmosphere every second. Some of this energy is reflected back into space and some is absorbed by the atmosphere, but enough arrives at the surface of the Earth each day to supply all of our energy needs on this planet hundreds of times over—if it could be captured and used efficiently. The growth in the number of solar houses in this country is one example of an attempt to make use of this free energy.

Radiant energy from the Sun affects our day-to-day existence in a number of ways. It influences the Earth's average temperature, ocean currents, agriculture, rain patterns, and so on. For example, consider what happens to the atmospheric temperature at night. If there is a cloud cover above the Earth, the water vapor in the clouds reflects back a part of the infrared radiation emitted by the Earth and consequently the temperature remains at moderate levels. In the absence of this cloud cover, however, there is nothing to prevent this radiation from escaping into space, and thus the temperature drops more on a clear night than when it is cloudy.

The rate at which an object emits radiant energy is proportional to the fourth power of its absolute temperature. This is known as **Stefan's law** and is expressed in equation form as

$$P = \sigma A e T^4 \qquad (20.18)$$

Stefan's law

where P is the power radiated by the body in watts, σ is a constant equal to $5.6696 \times 10^{-8}\ \text{W/m}^2 \cdot \text{K}^4$, A is the surface area of the object in square meters, e is a constant called the **emissivity,** and T is temperature in kelvin. The value of e can vary between zero and unity, depending on the properties of the surface.

An object radiates energy at a rate given by Equation 20.18. At the same time, the object also absorbs electromagnetic radiation. If the latter process did not occur, an object would eventually radiate all of its energy and its temperature would reach absolute zero. The energy that a body absorbs comes from its surroundings, which consist of other objects that radiate energy. If an object is at a temperature T and its surroundings are at a temperature T_0, the net energy gained or lost each second by the object as a result of radiation is

$$P_{net} = \sigma Ae(T^4 - T_0^4) \qquad (20.19)$$

When an object is in equilibrium with its surroundings, it radiates and absorbs energy at the same rate, and so its temperature remains constant. When an object is hotter than its surroundings, it radiates more energy than it absorbs and so it cools. An **ideal absorber** is defined as an object that absorbs all of the energy incident on it. The emissivity of an ideal absorber is equal to unity. Such an object is often referred to as a **black body**. An ideal absorber is also an ideal radiator of energy. In contrast, an object with an emissivity equal to zero absorbs none of the energy incident on it. Such an object reflects all the incident energy and so is a perfect reflector.

The Dewar Flask

The Thermos bottle, called a *Dewar flask*[8] in the scientific community, is a practical example of a container designed to minimize thermal energy losses by conduction, convection, and radiation. Such a container is used to store either cold or hot liquids for long periods of time. The standard construction (Fig. 20.13) consists of a double-walled Pyrex vessel with silvered inner walls. The space between the walls is evacuated to minimize thermal energy transfer by conduction and convection. The silvered surfaces minimize thermal energy transfer by radiation by reflecting most of the radiant heat. Very little thermal energy is lost over the neck of the flask since glass is not a very good conductor, and the cross-section of the glass in the direction of conduction is small. A further reduction in thermal energy loss is obtained by reducing the size of the neck. Dewar flasks are commonly used to store liquid nitrogen (boiling point 77 K) and liquid oxygen (boiling point 90 K).

In order to confine liquid helium, which has a very low heat of vaporization (boiling point 4.2 K), it is often necessary to use a double Dewar system in which the Dewar flask containing the liquid is surrounded by a second Dewar flask. The space between the two flasks is filled with liquid nitrogen.

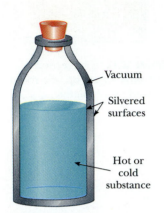

FIGURE 20.13   A cross-sectional view of a Dewar vessel, used to store hot or cold liquids or other substances.

Vacuum

Silvered surfaces

Hot or cold substance

EXAMPLE 20.11   Who Turned Down the Thermostat?

An unclothed student is in a 20°C room. If the skin temperature of the student is 37°C, how much heat is lost from his body in 10 min, assuming that the emissivity of skin is 0.90 and the surface area of the student is 1.5 m²?

Solution   Using Equation 20.19, the rate of thermal energy loss from the skin is

$$P_{net} = \sigma Ae(T^4 - T_0^4)$$

$$= (5.67 \times 10^{-8} \text{ W/m}^2 \cdot \text{K}^4)(1.5 \text{ m}^2)$$
$$\times (0.90)[(310 \text{ K})^4 - (293 \text{ K})^4]$$
$$= 140 \text{ J/s}$$

(Why is the temperature given in kelvin?) At this loss rate, the total thermal energy lost by the skin in 10 min is

$$Q = P_{net} \times \text{time} = (140 \text{ J/s})(600 \text{ s}) = \boxed{8.6 \times 10^4 \text{ J}}$$

---

[8] Invented by Sir James Dewar (1842–1923).

SUMMARY

Thermal energy transfer is a form of energy transfer that takes place as a consequence of a temperature difference. The **internal energy** of a substance is a function of its state and generally increases with increasing temperature.

The **calorie** is the amount of heat necessary to raise the temperature of 1 g of water from 14.5°C to 15.5°C. The **mechanical equivalent of heat** is 4.186 J/cal.

The **heat capacity** C of any substance is defined as the amount of thermal energy needed to raise the temperature of the substance by one degree Celsius. The thermal energy required to change the temperature of a substance by ΔT is

$$Q = mc\,\Delta T \tag{20.4}$$

where m is the mass of the substance and c is its **specific heat.**

The thermal energy required to change the phase of a pure substance of mass m is

$$Q = mL \tag{20.6}$$

The parameter L is called the **latent heat** of the substance and depends on the nature of the phase change and the properties of the substance.

The **work done** by a gas as its volume changes from some initial value V_i to some final value V_f is

$$W = \int_{V_i}^{V_f} P\,dV \tag{20.8}$$

where P is the pressure, which may vary during the process. In order to evaluate W, the nature of the process must be specified — that is, P and V must be known during each step. Since the work done depends on the initial, final, and intermediate states, it therefore depends on the path taken between the initial and final states.

The **first law of thermodynamics** states that when a system undergoes a change from one state to another, the change in its internal energy is

$$\Delta U = Q - W \tag{20.9}$$

where Q is the thermal energy transferred into (or out of) the system and W is the work done by (or on) the system. Although Q and W both depend on the path taken from the initial state to the final state, the quantity ΔU is path-independent.

In a **cyclic process** (one that originates and terminates at the same state), $\Delta U = 0$, and therefore $Q = W$. That is, the thermal energy transferred into the system equals the work done during the cycle.

An **adiabatic process** is one in which no thermal energy is transferred between the system and its surroundings ($Q = 0$). In this case, the first law gives $\Delta U = -W$. That is, the internal energy changes as a consequence of work being done by (or on) the system. In an **adiabatic free expansion** of a gas, $Q = 0$ and $W = 0$, and so $\Delta U = 0$. That is, the internal energy of the gas does not change in such a process.

An **isovolumetric process** is one that occurs at constant volume. No expansion work is done in such a process.

An **isobaric process** is one that occurs at constant pressure. The work done in such a process is $P\,\Delta V$.

An **isothermal process** is one that occurs at constant temperature. The work done by an ideal gas during a reversible isothermal process is

$$W = nRT \ln\left(\frac{V_f}{V_i}\right) \tag{20.12}$$

Heat may be transferred by conduction, convection, and radiation. **Conduction** can be viewed as an exchange of kinetic energy between colliding molecules or electrons. The rate at which heat flows by conduction through a slab of area A is

$$H = -kA\frac{dT}{dx} \tag{20.14}$$

where k is the **thermal conductivity** and dT/dx is the **temperature gradient**.

In **convection,** the heated substance moves from one place to another.

All bodies radiate and absorb energy in the form of electromagnetic waves. A body that is hotter than its surroundings radiates more energy than it absorbs, whereas a body that is cooler than its surroundings absorbs more energy than it radiates. A **black body** (or ideal absorber) is one that absorbs all energy incident on it; furthermore, a black body emits as much energy as is possible for a body of its size, shape, and temperature.

QUESTIONS

1. Ethyl alcohol has about one-half the specific heat of water. If equal masses of alcohol and water in separate beakers are supplied with the same amount of heat, compare the temperature increases of the two liquids.

2. Give one reason why coastal regions tend to have a more moderate climate than inland regions.

3. A small crucible is taken from a 200°C oven and immersed in a tub full of water at room temperature (this process is often referred to as *quenching*). What is the approximate final equilibrium temperature?

4. What is the major problem that arises in measuring specific heats if a sample with a temperature above 100°C is placed in water?

5. In a daring lecture demonstration, an instructor dips his wetted fingers into molten lead (327°C) and withdraws them quickly, without getting burned. How is this possible? (This is a dangerous experiment, which you should *not* attempt.)

6. The pioneers found that a large tub of water placed in a storage cellar would prevent their food from freezing on really cold nights. Explain why this is so.

7. What is wrong with the statement: "Given any two bodies, the one with the higher temperature contains more heat."?

8. Why is it possible to hold a lighted match, even when it is burned, to within a few millimeters of your fingertips?

9. Figure 20.14 shows a pattern formed by snow on the roof of a barn. What causes the alternating pattern of snow-covered and exposed roof?

10. Why is a person able to remove a piece of dry aluminum

FIGURE 20.14   (Question 9) Alternating pattern on a snow-covered roof.   *(Courtesy of Dr. Albert A. Bartlett, University of Colorado, Boulder)*

foil from a hot oven with bare fingers, while if there is moisture on the foil, a burn results?

11. A tile floor in a bathroom may feel uncomfortably cold to your bare feet, but a carpeted floor in an adjoining room at the same temperature will feel warm. Why?

12. Why can potatoes be baked more quickly when a metal skewer has been inserted through them?

13. Give reasons for the silvered walls and the vacuum jacket in a Thermos bottle.

14. A piece of paper is wrapped around a rod made half of wood and half of copper. When held over a flame, the paper in contact with the wood burns but the half in contact with the metal does not. Explain.

15. Why is it necessary to store liquid nitrogen or liquid oxygen in vessels equipped with either Styrofoam insulation or a double-evacuated wall?

16. Why do heavy draperies over the windows help keep a home warm in the winter and cool in the summer?

17. If you wish to cook a piece of meat thoroughly on an open fire, why should you not use a high flame? (Note that carbon is a good thermal insulator.)

18. When insulating a wood-frame house, is it better to place the insulation against the cooler outside wall or against the warmer inside wall? (In either case, there is an air barrier to consider.)

19. In an experimental house, Styrofoam beads were pumped into the air space between the double windows at night in the winter and pumped out to holding bins during the day. How would this assist in conserving heat energy in the house?

20. Pioneers stored fruits and vegetables in underground cellars. Discuss as fully as possible this choice for a storage site.

21. Why can you get a more severe burn from steam at 100°C than from water at 100°C?

22. Concrete has a higher specific heat than soil. Use this fact to explain (partially) why cities have a higher average night-time temperature than the surrounding countryside. If a city is hotter than the surrounding countryside, would you expect breezes to blow from city to country or from country to city? Explain.

23. When camping in a canyon on a still night, one notices that as soon as the Sun strikes the surrounding peaks, a breeze begins to stir. What causes the breeze?

24. Updrafts of air are familiar to all pilots. What causes these currents?

25. If water is a poor conductor of heat, why can it be heated quickly when placed over a flame?

26. The U.S. penny is now made of copper-coated zinc. Can a calorimetric experiment be devised to test for the metal content in a collection of pennies? If so, describe the procedure you would use.

27. If you hold water in a paper cup over a flame, you can bring the water to a boil without burning the cup. How is this possible?

28. When a sealed Thermos bottle full of hot coffee is shaken, what are the changes, if any, in (a) the temperature of the coffee and (b) the internal energy of the coffee?

29. Using the first law of thermodynamics, explain why the *total* energy of an isolated system is always constant.

30. Is it possible to convert internal energy to mechanical energy? Explain with examples.

31. Suppose you pour hot coffee for your guests, and one of them chooses to drink the coffee after it has been in the cup for several minutes. In order to have the warmest coffee, should the person add the cream just after the coffee is poured or just before drinking? Explain.

32. Two identical cups both at room temperature are filled with the same amount of hot coffee. One cup contains a metal spoon, while the other does not. If you wait for several minutes, which of the two will have the warmer coffee? Which heat transfer process explains your answer?

33. A warning sign often seen on highways just before a bridge is "Caution—Bridge surface freezes before road surface." Which of the three heat transfer processes is most important in causing a bridge surface to freeze before a road surface on very cold days?

PROBLEMS

Section 20.1 Heat and Thermal Energy

1. Consider Joule's apparatus described in Figure 20.1. The two masses are 1.50 kg each, and the tank is filled with 200 g of water. What is the increase in the temperature of the water after the masses fall through a distance of 3.00 m?

2. An 80-kg weight watcher wishes to climb a mountain to work off the equivalent of a large piece of chocolate cake rated at 700 (food) Calories. How high must the person climb?

3. Water at the top of Niagara Falls has a temperature of 10°C. If it falls through a distance of 50 m and all of its potential energy goes into heating the water, calculate the temperature of the water at the bottom of the falls.

Sections 20.2 and 20.3 Heat Capacity, Specific Heat, and Latent Heat

4. How many calories of heat are required to raise the temperature of 3.0 kg of aluminum from 20°C to 50°C?

5. The temperature of a silver bar rises by 10.0°C when

□ indicates problems that have full solutions available in the Student Solutions Manual and Study Guide.

it absorbs 1.23 kJ of heat. The mass of the bar is 525 g. Determine the specific heat of silver.

6. If 100 g of water at 100°C is poured into a 20-g aluminum cup containing 50 g of water at 20°C, what is the equilibrium temperature of the system?

6A. If a mass m_h of water at T_h is poured into an aluminum cup of mass m_{Al} containing m_c of cold water at T_c, where $T_h > T_c$, what is the equilibrium temperature of the system?

7. What is the final equilibrium temperature when 10 g of milk at 10°C is added to 160 g of coffee at 90°C? (Assume the heat capacities of the two liquids are the same as that of water, and neglect the heat capacity of the container.)

8. (a) A calorimeter contains 500 ml of water at 30°C and 25 g of ice at 0°C. Determine the final temperature of the system. (b) Repeat part (a) if 250 g of ice is initially present at 0°C.

9. A 1.5-kg iron horseshoe initially at 600°C is dropped into a bucket containing 20 kg of water at 25°C. What is the final temperature? (Neglect the heat capacity of the container.)

10. The air temperature above coastal areas is profoundly influenced by the large specific heat of water. One reason is that the heat released when 1 cubic meter of water cools by 1.0°C will raise the temperature of an enormously larger volume of air by 1.0°C. Estimate this volume of air. The specific heat of air is approximately 1.0 kJ/kg·°C. Take the density of air to be 1.25 kg/m³.

11. If 200 g of water is contained in a 300-g aluminum vessel at 10°C and an additional 100 g of water at 100°C is poured into the container, what is the final equilibrium temperature of the system?

12. A student inhales 22°C air and exhales 37°C air. The average volume of air in one breath is 200 cm³. Neglecting evaporation of water into the air, estimate the amount of heat absorbed in one day by the air breathed by the student. The density of air is approximately 1.25 kg/m³, and the specific heat of air is 1000 J/kg·°C.

13. How much heat must be added to 20 g of aluminum at 20°C to melt it completely?

14. An insulated vessel contains a saturated vapor being cooled as cold water flows in a pipe that passes through the vessel. The temperature of the entering water is 273 K. When the flow speed is 3.0 m/s, the temperature of the exiting water is 303 K. Determine the temperature of the exiting water when the flow speed is decreased to 2.0 m/s. Assume that the rate of condensation remains unchanged.

14A. An insulated vessel contains a saturated vapor being cooled as cold water flows in a pipe that passes through the vessel. The temperature of the entering water is T_1. When the flow speed is v_1, the temperature of the exiting water is T_{out}. Determine the temperature of the exiting water when the flow speed of the water is decreased to v_2. Assume that the rate of condensation remains unchanged.

15. A water heater is operated by solar power. If the solar collector has an area of 6.0 m² and the power delivered by sunlight is 550 W/m², how long does it take to increase the temperature of 1.0 m³ of water from 20°C to 60°C?

16. A 1.0-kg block of copper at 20°C is dropped into a large vessel of liquid nitrogen at 77 K. How many kilograms of nitrogen boil away by the time the copper reaches 77 K? (The specific heat of copper is 0.092 cal/g·°C. The latent heat of vaporization of nitrogen is 48 cal/g.)

17. How much heat is required to vaporize a 1.0-g ice cube initially at 0°C? The latent heat of fusion of ice is 80 cal/g and the latent heat of vaporization of water is 540 cal/g.

18. One liter of water at 30°C is used to make iced tea. How much ice at 0°C must be added to lower the temperature of the tea to 10°C?

19. When a driver brakes an automobile, the friction between the brake drums and brake shoes converts the car's kinetic energy to heat. If a 1500-kg automobile traveling at 30 m/s comes to a halt, how much does the temperature rise in each of the four 8.0-kg iron brake drums? (Neglect heat loss to the surroundings.)

20. If 90.0 g of molten lead at 327.3°C is poured into a 300.0-g casting made of iron initially at 20.0°C, what is the final temperature of the system? (Assume there are no heat losses.)

21. In an insulated vessel, 250 g of ice at 0°C is added to 600 g of water at 18°C. (a) What is the final temperature of the system? (b) How much ice remains when the system reaches equilibrium?

22. A 50.0-g ice cube at −20.0°C is dropped into a container of water at 0.0°C. How much water freezes onto the ice?

23. An iron nail is driven into a block of ice by a single blow of a hammer. The hammerhead has a mass of 0.50 kg and an initial speed of 2.0 m/s. Nail and hammer are at rest after the blow. How much ice melts? Assume the temperature of the nail is 0.0°C before and after.

24. Two speeding 5.0-g lead bullets, both at temperature 20°C, collide head-on when each is moving at 500 m/s. Assuming a perfectly inelastic collision and no loss of heat to the atmosphere, describe the final state of the two-bullet system.

25. A 3.0-g copper penny at 25°C drops 50 m to the ground. (a) If 60% of its initial potential energy goes into increasing the internal energy, determine its final temperature. (b) Does the result depend on the mass of the penny? Explain.

26. Lake Erie contains roughly 4.0×10^{11} m³ of water. (a) How much heat is required to raise the temperature of that volume of water from 11°C to 12°C? (b) Approximately how many years would it take to supply this amount of heat by using the full output of a 1000-MW electric power plant?

27. A 3.0-g lead bullet is traveling at 240 m/s when it embeds in a block of ice at 0°C. If all the heat generated goes into melting ice, what quantity of ice is melted? (The latent heat of fusion for ice is 80 kcal/kg, and the specific heat of lead is 0.030 kcal/kg·°C.)

Section 20.4 Work and Heat in Thermodynamic Processes

28. One mole of an ideal gas is heated slowly so that it goes from the state (P_0, V_0) to the state $(3P_0, 3V_0)$. This change occurs in such a way that the gas pressure is directly proportional to the volume. (a) How much work is done in the process? (b) How is the temperature of the gas related to its volume during this process?

29. A gas expands from *I* to *F* along three possible paths as indicated by Figure P20.29. Calculate the work in joules done by the gas along the paths *IAF*, *IF*, and *IBF*.

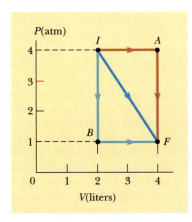

FIGURE P20.29

30. Gas in a container is at a pressure of 1.5 atm and a volume of 4.0 m³. What is the work done by the gas if (a) it expands at constant pressure to twice its initial volume and (b) it is compressed at constant pressure to one-quarter its initial volume?

31. An ideal gas is enclosed in a cylinder that has a movable piston on top. The piston has a mass of 8000 g and an area of 5.0 cm² and is free to slide up and down, keeping the pressure of the gas constant. How much work is done as the temperature of 0.20 mol of the gas is raised from 20°C to 300°C?

31A. An ideal gas is enclosed in a cylinder that has a movable piston on top. The piston has a mass m and an area A and is free to slide up and down, keeping the pressure of the gas constant. How much work is done as the temperature of n mol of the gas is raised from T_1 to T_2?

32. An ideal gas undergoes a thermodynamic process that consists of two isobaric and two isothermal steps as shown in Figure P20.32. Show that the net work done during the four steps is

$$W_{net} = P_1(V_2 - V_1) \ln\left(\frac{P_2}{P_1}\right)$$

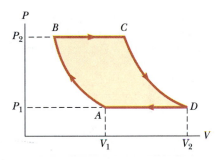

FIGURE P20.32

33. A sample of ideal gas is expanded to twice its original volume of 1.0 m³ in a quasi-static process for which $P = \alpha V^2$, with $\alpha = 5.0$ atm/m⁶, as shown in Figure P20.33. How much work was done by the expanding gas?

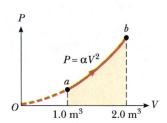

FIGURE P20.33

34. An ideal gas at STP (1 atm and 0°C) is taken through a process in which the volume is expanded from 25 L to 80 L. During this process the pressure varies inversely as the volume squared, $P = 0.5 \, aV^{-2}$. (a) Determine the constant a in standard SI units. (b) Find the final pressure and temperature. (c) Determine a general expression for the work done by the gas during this process. (d) Compute the actual work in joules done by the gas in this process.

35. One mole of an ideal gas does 3000 J of work on the surroundings as it expands isothermally to a final pressure of 1 atm and volume of 25 L. Determine

(a) the initial volume and (b) the temperature of the gas.

Section 20.5   The First Law of Thermodynamics

36. An ideal gas undergoes the cyclic process shown in Figure P20.36 from A to B to C and back to A. (a) Sketch a *PV* diagram for this cycle, and identify the steps during which heat is absorbed and those during which heat is evolved. (b) What is the overall result of the cycle in terms of *U*, *Q*, and *W*?

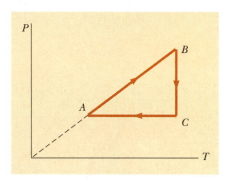

FIGURE P20.36

37. A gas is compressed at a constant pressure of 0.80 atm from 9.0 L to 2.0 L. In the process, 400 J of thermal energy leaves the gas. (a) What is the work done by the gas? (b) What is the change in its internal energy?

37A. A gas is compressed at a constant pressure *P* from V_1 to V_2. In the process, *Q* joules of thermal energy leaves the gas. (a) What is the work done by the gas? (b) What is the change in its internal energy?

38. A thermodynamic system undergoes a process in which its internal energy decreases by 500 J. If at the same time, 220 J of work is done on the system, find the thermal energy transferred to or from it.

39. A gas is taken through the cyclic process described in Figure P20.39. (a) Find the net thermal energy transferred to the system during one complete cycle.

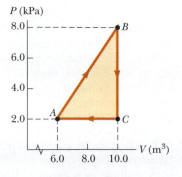

FIGURE P20.39

(b) If the cycle is reversed, that is, the process goes along *ACBA*, what is the net thermal energy transferred per cycle?

40. An ideal gas system goes through the process shown in Figure P20.40. From *A* to *B*, the process is adiabatic, and from *B* to *C*, it is isobaric with 100 kJ of heat flow into the system. From *C* to *D*, the process is isothermal, and from *D* to *A*, it is isobaric with 150 kJ of heat flow out of the system. Determine the difference in internal energy $U_B - U_A$.

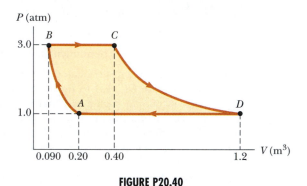

FIGURE P20.40

Section 20.6   Some Applications of the First Law of Thermodynamics

41. Five moles of an ideal gas expands isothermally at 127°C to four times its initial volume. Find (a) the work done by the gas and (b) the thermal energy transferred to the system, both in joules.

42. How much work is done by the steam when 1.0 mol of water at 100°C boils and becomes 1.0 mol of steam at 100°C at 1.0 atm pressure? Determine the change in internal energy of the steam as it vaporizes. Consider the steam to be an ideal gas.

43. Helium is heated at constant pressure from 273 K to 373 K. If the gas does 20.0 J of work during the process, what is the mass of the helium?

44. One mole of an ideal gas is heated at constant pressure so that its temperature triples. Then the gas is heated at constant temperature so that its volume triples. Find the ratio of the work done during the isothermal process to that done during the isobaric process.

45. An ideal gas initially at 300 K undergoes an isobaric expansion at 2.50 kPa. If the volume increases from 1.00 m³ to 3.00 m³ and 12.5 kJ of thermal energy is transferred to the gas, find (a) the change in its internal energy and (b) its final temperature.

46. Two moles of helium gas initially at 300 K and 0.40 atm are compressed isothermally to 1.2 atm. Find (a) the final volume of the gas, (b) the work done by the gas, and (c) the thermal energy transferred. Consider the helium to behave as an ideal gas.

47. One mole of water vapor at 373 K cools to 283 K. The heat given off by the cooling water vapor is absorbed by 10 mol of an ideal gas, and this heat absorption causes the gas to expand at a constant temperature of 273 K. If the final volume of the ideal gas is 20.0 L, determine its initial volume.

47A. One mole of water vapor at a temperature T_h cools to liquid at T_c. The heat, given off by the cooling water vapor, is absorbed by n mol of an ideal gas, and this heat absorption causes the gas to expand at a constant temperature of T_0. If the final volume of the ideal gas is V_f, determine its initial volume.

48. During a controlled expansion, the pressure of a gas is

$$P = 12e^{-bV} \text{ atm} \qquad b = \frac{1}{12 \text{ m}^3}$$

where the volume is in m³ (Fig. P20.48). Determine the work performed when the gas expands from 12 m³ to 36 m³.

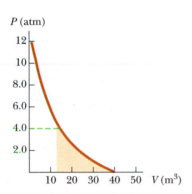

FIGURE P20.48

49. A 1.0-kg block of aluminum is heated at atmospheric pressure such that its temperature increases from 22°C to 40°C. Find (a) the work done by the aluminum, (b) the thermal energy added to it, and (c) the change in its internal energy.

50. In Figure P20.50, the change in internal energy of a gas that is taken from A to C is +800 J. The work done along path ABC is +500 J. (a) How much thermal energy has to be added to the system as it goes from A through B to C? (b) If the pressure at point A is five times that of point C, what is the work done by the system in going from C to D? (c) What is the thermal energy exchanged with the surroundings as the cycle goes from C to A? (d) If the change in internal energy in going from point D to point A is +500 J, how much thermal energy must be added to the system as it goes from point C to point D?

51. Helium with an initial volume of 1.00 liter and an initial pressure of 10.0 atm expands to a final volume of 1.00 m³. The relationship between pressure and

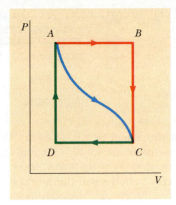

FIGURE P20.50

volume during the expansion is PV = constant. Determine (a) the value of the constant, (b) the final pressure, and (c) the work done by the helium during the expansion.

Section 20.7 Heat Transfer

52. A glass window pane has an area of 3.0 m² and a thickness of 0.60 cm. If the temperature difference between its faces is 25°C, how much heat flows through the window per hour?

53. A Thermopane window of area 6.0 m² is constructed of two layers of glass, each 4.0 mm thick separated by an air space of 5.0 mm. If the inside is at 20°C and the outside is at −30°C, what is the heat loss through the window?

54. A silver bar of length 30.0 cm and cross-sectional area 1.00 cm² is used to transfer heat from a 100.0°C reservoir to a 0.0°C reservoir. How much heat is transferred per second?

55. A bar of gold is in thermal contact with a bar of silver of the same length and area (Fig. P20.55). One end of the compound bar is maintained at 80.0°C while the opposite end is at 30.0°C. When the heat flow reaches steady state, find the temperature at the junction.

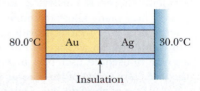

FIGURE P20.55

56. Two rods of the same length but of different materials and cross-sectional areas are placed side by side as in Figure P20.56. Determine the rate of heat flow in terms of the thermal conductivity and area of each rod. Generalize this to several rods.

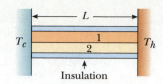

FIGURE P20.56

57. The brick wall ($k = 0.80$ W/m·°C) of a building has dimensions of 4.0 m × 10.0 m and is 15 cm thick. How much heat (in joules) flows through the wall in a 12-h period when the average inside and outside temperatures are, respectively, 20°C and 5°C?

58. A steam pipe is covered with 1.5-cm-thick insulating material of heat conductivity 0.200 cal/cm·°C·s. How much heat is lost every second when the steam is at 200°C and the surrounding air is at 20°C? The pipe has a circumference of 20 cm and a length of 50 m. Neglect losses through the ends of the pipe.

59. A box with a total surface area of 1.20 m² and a wall thickness of 4.00 cm is made of an insulating material. A 10.0-W electric heater inside the box maintains the inside temperature at 15.0°C above the outside temperature. Find the thermal conductivity k of the insulating material.

59A. A box with a total surface area A and a wall thickness t is made of an insulating material. An electric heater that delivers P watts inside the box maintains the inside temperature at a temperature T_0 above the outside temperature. Find the thermal conductivity k of the insulating material.

60. A Styrofoam box has a surface area of 0.80 m² and a wall thickness of 2.0 cm. The temperature inside is 5°C, and that outside is 25°C. If it takes 8.0 h for 5.0 kg of ice to melt in the container, determine the thermal conductivity of the Styrofoam.

61. The roof of a house built to absorb the solar radiation incident upon it has an area of 7.0 m × 10.0 m. The solar radiation at the Earth's surface is 840 W/m². On the average, the Sun's rays make an angle of 60° with the plane of the roof. (a) If 15% of the incident energy is converted to useful electrical power, how many kilowatt hours per day of useful energy does this source provide? Assume that the Sun shines for an average of 8.0 h/day. (b) If the average household user pays 6 cents/kWh, what is the monetary saving of this energy source per day?

62. The surface of the Sun has a temperature of about 5800 K. Taking the radius of the Sun to be 6.96×10^8 m, calculate the total energy radiated by the Sun each day. (Assume $e = 1$.)

63. Calculate the R value of (a) a window made of a single pane of glass 1/8 in. thick and (b) a thermal pane window made of two single panes each 1/8 in. thick

and separated by a 1/4-in. air space. (c) By what factor is the heat loss reduced by using the thermal window instead of the single pane window?

ADDITIONAL PROBLEMS

64. A cooking vessel on a slow burner contains 10.0 kg of water and an unknown mass of ice in equilibrium at 0°C at time $t = 0$. The temperature of the mixture is measured at various times, and the result is plotted in Figure P20.64. During the first 50 min, the mixture remains at 0°C. From 50 min to 60 min, the temperature increases to 2.0°C. Neglecting the heat capacity of the vessel, determine the initial mass of the ice.

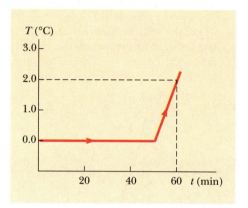

FIGURE P20.64

65. Around a crater formed by an iron meteorite, 75.0 kg of rock has melted under the impact of the meteorite. The rock has a specific heat of 0.800 kcal/kg·°C, a melting point of 500°C, and a latent heat of fusion of 48.0 kcal/kg. The original temperature of the ground was 0.0°C. If the meteorite hit the ground while moving at 600 m/s, what is the minimum mass of the meteorite? Assume no heat loss to the surrounding unmelted rock or the atmosphere during the impact. Disregard the heat capacity of the meteorite.

66. A *flow calorimeter* is an apparatus used to measure the specific heat of a liquid. The technique is to measure the temperature difference between the input and output points of a flowing stream of the liquid while adding heat at a known rate. In one particular experiment, a liquid of density 0.78 g/cm³ flows through the calorimeter at the rate of 4.0 cm³/s. At steady state, a temperature difference of 4.8°C is established between the input and output points when heat is supplied at the rate of 30 J/s. What is the specific heat of the liquid?

66A. A *flow calorimeter* is an apparatus used to measure the specific heat of a liquid. The technique is to measure the temperature difference between the input and output points of a flowing stream of the liquid while adding heat at a known rate. In one particular experiment, a liquid of density ρ flows through the calorimeter at the rate of R cm^3/s. At steady state, a temperature difference ΔT is established between the input and output points when heat is supplied at the rate of P J/s. What is the specific heat of the liquid?

67. One mole of an ideal gas initially at 300 K is cooled at constant volume so that the final pressure is one-fourth the initial pressure. Then the gas expands at constant pressure until it reaches the initial temperature. Determine the work done by the gas.

68. An electric teakettle is boiling, and the electric power consumed by the water in the kettle is 1.0 kW. Assuming that the vapor pressure in the kettle equals atmospheric pressure, determine the speed at which water vapor exits from the spout, which has a cross-sectional area of 2.0 cm^2.

68A. An electric teakettle is boiling, and the electric power consumed by the water in the kettle is P. Assuming that the vapor pressure in the kettle equals atmospheric pressure, determine the speed as which water vapor exits from the spout, which has a cross-sectional area A.

69. An aluminum rod, 0.50 m in length and of cross-sectional area 2.5 cm^2, is inserted into a thermally insulated vessel containing liquid helium at 4.2 K. The rod is initially at 300 K. (a) If one half of the rod is inserted into the helium, how many liters of helium boil off by the time the inserted half cools to 4.2 K? (Assume the upper half does not cool.) (b) If the upper portion of the rod is maintained at 300 K, what is the approximate boil-off rate of liquid helium after the lower half has reached 4.2 K? (Note that aluminum has a thermal conductivity of 31 J/s·cm·K at 4.2 K, a specific heat of 0.21 cal/g·°C, and a density of 2.7 g/cm^3. The density of liquid helium is 0.125 g/cm.^{3})

70. A heat pipe 0.025 m in diameter and 0.30 m long can transfer 3600 J of heat per second with a temperature difference across the ends of 10°C. How does this performance compare with the heat transfer of a solid silver bar of the same dimensions if the thermal conductivity of silver is k = 427 W/m·°C? (Silver is the best heat conductor of all metals.)

71. A 5.00-g lead bullet traveling at 300 m/s strikes a flat steel plate and stops. If the collision is inelastic, will the bullet melt? Lead has a melting point of 327°C, a specific heat of 0.128 J/g·°C, and a latent heat of fusion of 24.5 J/g.

72. The average thermal conductivity of the walls (including windows) and roof of the house in Figure P20.72 is 0.48 W/m·°C, and their average thickness is 21.0 cm. The house is heated with natural gas having a heat of combustion (heat given off per cubic meter of gas burned) of 9300 kcal/m^3. How many cubic meters of gas must be burned each day to maintain an inside temperature of 25.0°C if the outside temperature is 0.0°C? Disregard radiation and heat loss through the ground.

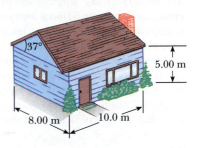

FIGURE P20.72

73. A class of 10 students taking an exam has a power output per student of about 200 W. Assume that the intitial temperature of the room is 20°C and that its dimensions are 6.0 m by 3.0 m. What is the temperature of the room at the end of 1.0 h if all the heat remains in the air in the room and none is added by an outside source? The specific heat of air is 837 J/kg·°C, and its density is about 1.3×10^{-3} g/cm^3.

74. An ideal gas initially at P_0, V_0, and T_0 is taken through a cycle as described in Figure P20.74. (a) Find the net work done by the gas per cycle. (b) What is the net heat added to the system per cycle? (c) Obtain a numerical value for the net work done per cycle for 1 mol of gas initially at 0°C.

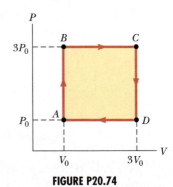

FIGURE P20.74

75. A one-person research submarine has a spherical iron hull 1.50 m in outer radius and 2.00 cm thick, lined with an equal thickness of rubber. If the submarine is used in Arctic waters (temperature 0°C) and the total rate of heat release within the sub (including the occupant's metabolic heat) is 1500 W, find the equilibrium temperature of the interior.

76. An iron plate is held against an iron wheel so that there is a sliding frictional force of 50.0 N acting between the two pieces of metal. The relative speed at which the two surfaces slide over each other is 40.0 m/s. (a) Calculate the rate at which mechanical energy is converted to thermal energy. (b) The plate and the wheel have a mass of 5.00 kg each, and each receives 50% of the thermal energy. If the system is run as described for 10.0 s and each object is then allowed to reach a uniform internal temperature, what is the resultant temperature increase?

77. A vessel in the shape of a spherical shell has an inner radius a and outer radius b. The wall has a thermal conductivity k. If the inside is maintained at a temperature T_1 and the outside is at a temperature T_2, show that the rate of heat flow between the surfaces is

$$\frac{dQ}{dt} = \left(\frac{4\pi kab}{b-a}\right)(T_1 - T_2)$$

78. The inside of a hollow cylinder is maintained at a temperature T_a while the outside is at a lower temperature, T_b (Fig. P20.78). The wall of the cylinder has a thermal conductivity k. Neglecting end effects, show that the rate of heat flow from the inner to the outer wall in the radial direction is

$$\frac{dQ}{dt} = 2\pi Lk\left[\frac{T_a - T_b}{\ln(b/a)}\right]$$

(*Hint:* The temperature gradient is dT/dr. Note that a radial heat current passes through a concentric cylinder of area $2\pi rL$.)

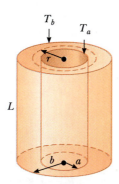

FIGURE P20.78

79. The passenger section of a jet airliner is in the shape of a cylindrical tube of length 35 m and inner radius 2.5 m. Its walls are lined with a 6.0-cm thickness of insulating material of thermal conductivity 4.0×10^{-5} cal/s·cm·°C. The inside is to be maintained at 25°C while the outside is at −35°C. What heating rate is required to maintain this temperature difference? (Use the result from Problem 78.)

80. A Thermos bottle in the shape of a cylinder has an inner radius of 4.0 cm, outer radius of 4.5 cm, and length of 30.0 cm. The insulating walls have a thermal conductivity equal to 2.0×10^{-5} cal/s·cm·°C. One liter of hot coffee at 90°C is poured into the bottle. If the outside wall remains at 20°C, how long does it take for the coffee to cool to 50°C? (Use the result from Problem 78 and assume that coffee has the same properties as water.)

81. A "solar cooker" consists of a curved reflecting mirror that focuses sunlight onto the object to be heated (Fig. P20.81). The solar power per unit area reaching the Earth at some location is 600 W/m², and the cooker has a diameter of 0.60 m. Assuming that 40% of the incident energy is converted into thermal energy, how long would it take to completely boil off 0.50 liter of water initially at 20°C? (Neglect the heat capacity of the container.)

FIGURE P20.81

82. A pond of water at 0°C is covered with a layer of ice 4.0 cm thick. If the air temperature stays constant at −10°C, how long will it be before the ice thickness is 8.0 cm? (*Hint:* To solve this problem, utilize Equation 20.14 in the form

$$\frac{dQ}{dt} = kA\frac{\Delta T}{x}$$

and note that the incremental heat dQ extracted from the water through the thickness x of ice is the amount required to freeze a thickness dx of ice. That is $dQ = L\rho A dx$, where ρ is the density of the ice, A is the area, and L is the latent heat of freezing.)

83. A student obtains the following data in a method-of-mixtures experiment designed to measure the specific heat of aluminum:

Initial temperature of water and calorimeter: 70°C
Mass of water: 0.400 kg

Mass of calorimeter: 0.040 kg
Specific heat of calorimeter: 0.63 kJ/kg·°C
Initial temperature of aluminum: 27°C
Mass of aluminum: 0.200 kg
Final temperature of mixture: 66.3°C

Use these data to determine the specific heat of aluminum. Your result should be within 15% of the value listed in Table 20.1.

SPREADSHEET PROBLEMS

S1. The rate at which an object with an initial temperature T_i cools when the surrounding temperature is T_0 is given by Newton's law of cooling:

$$\frac{dQ}{dt} = hA(T - T_0)$$

where A is the surface area of the object, and h is a constant that characterizes the rate of cooling, called the surface coefficient of heat transfer. The object's temperature T at any time t is

$$T = T_0 + (T_i - T_0)e^{-hAt/mc}$$

where m is the object's mass, and c is its specific heat.

Spreadsheet 20.1 evaluates and plots the temperature of an object as a function of time. Consider the cooling of a cup of coffee whose initial temperature is 75°C. After 3 min, it cools to a drinkable 45°C. Room temperature is 22°C, the effective area of the coffee is 125 cm², and its mass is 200 g. Take the specific heat c to be that of water. Use Spreadsheet 19.1 to find h for this cup of coffee. That is, find the value of h for which $T = 45$°C at $t = 180$ s.

S2. In the Einstein model of a crystalline solid the molar specific heat at constant volume is

$$C_V = 3R \left(\frac{T_E}{T}\right)^2 \frac{e^{T_E/T}}{(e^{T_E/T} - 1)^2}$$

where T_E is a characteristic temperature called the Einstein temperature and T is the temperature in Kelvin. (a) Verify that $C_V \approx 3R$, the Dulong-Petit law, for $T \gg T_E$, by evaluating $C_V - 3R$. (b) For diamond, T_E is approximately 1060 K. If 1 mol of diamond is heated from 300 K to 600 K, numerically integrate

$$\Delta U = \int C_V \, dT$$

to find the increase in the internal energy of the material. Use your spreadsheet to carry out a simple rectangular numeric integration. (You may also want to try to use the trapezoid and Simpson's methods to carry out the integration.)

CHAPTER 21

The Kinetic Theory of Gases

The glass vessel contains dry ice (solid carbon dioxide). The white cloud is carbon dioxide vapor, which is denser than air and hence falls from the vessel. *(R. Folwell/Science Photo Library)*

In Chapter 19 we discussed the properties of an ideal gas using such macroscopic variables as pressure, volume, and temperature. We shall now show that such large-scale properties can be described on a microscopic scale, where matter is treated as a collection of molecules. Newton's laws of motion applied in a statistical manner to a collection of particles provide a reasonable description of thermodynamic processes. In order to keep the mathematics relatively simple, we shall consider only the molecular behavior of gases, where the interactions between molecules are much weaker than in liquids or solids. In the established model of gas behavior, called the *kinetic theory* for historical reasons, gas molecules move about in a random fashion, colliding with the walls of their container and with each other. Perhaps the most important consequence of this theory is that it shows how the kinetic theory of molecular motion contributes to the internal energy of the system. Furthermore, the kinetic theory provides us with a physical basis upon which the concept of thermal energy can be understood.

In the simplest model of a gas, each molecule is considered to be a hard sphere that collides elastically with other molecules and with the container wall. The hard-sphere model assumes that the molecules do not interact with each other except during collisions and that they are not deformed by collisions. This description is adequate only for monatomic gases, where the energy is entirely translational kinetic energy. One must modify the theory for more complex molecules, such as O_2 and CO_2, to include the internal energy associated with rotations and vibrations of the molecules.

21.1 MOLECULAR MODEL OF AN IDEAL GAS

We begin this chapter by developing a microscopic model of an ideal gas. The model shows that the pressure that a gas exerts on the walls of its container is a consequence of the collisions of the gas molecules with the walls. As we shall see, the model is consistent with the macroscopic description of the preceding chapter. In developing this model, we make the following assumptions:

- *The number of molecules is large, and the average separation between them is large* compared with their dimensions. This means that the volume of the molecules is negligible when compared with the volume of the container.
- *The molecules obey Newton's laws of motion, but as a whole they move randomly.* By "randomly" we mean that any molecule can move in any direction at any speed. We also assume that the distribution of speeds does not change in time, despite the collisions between molecules. That is, at any given moment, a certain percentage of molecules moves at high speeds, a certain percentage moves at low speeds, and so on.

Assumptions of the molecular model of an ideal gas

This simulator is concerned with systems consisting of a large number of particles, rather than just the one or two objects dealt with in the previous simulators. After specifying the characteristics of a many-particle system, you will be able to observe particles colliding with each other and with the walls of a container. The results of the simulations should help you better understand the concepts of pressure, temperature, and related interesting phenomena.

Systems of Particles

- *The molecules undergo elastic collisions with each other and with the walls of the container that are elastic on the average.* Thus, in the collisions both kinetic energy and momentum are constant.
- *The forces between molecules are negligible except during a collision.* The forces between molecules are short-range, so the molecules interact with each other only during collisions.
- *The gas under consideration is a pure substance.* That is, all molecules are identical.

Although we often picture an ideal gas as consisting of single atoms, molecular gases exhibit equally good approximations to ideal gas behavior at low pressures. Effects of molecular rotations or vibrations have no effect, on the average, on the motions considered here.

Now let us derive an expression for the pressure of an ideal gas consisting of N molecules in a container of volume V. The container is a cube with edges of length d (Fig. 21.1). Consider the collision of one molecule moving with a velocity $\mathbf{v}$ toward the right-hand face of the box. The molecule has velocity components v_x, v_y, and v_z. As it collides with the wall elastically, its x component of velocity is reversed, while its y and z components of velocity remain unaltered (Fig. 21.2). Since the x component of momentum of the molecule is mv_x before the collision and $-mv_x$ afterward, the change in momentum of the molecule is

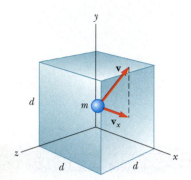

$$\Delta p_x = -mv_x - (mv_x) = -2mv_x$$

FIGURE 21.1   A cubical box with sides of length d containing an ideal gas. The molecule shown moves with velocity v.

Because the momentum of the system consisting of the wall and the molecule must be constant, we see that since the change in momentum of the molecule is $-2mv_x$, the change in momentum of the wall must be $2mv_x$. If F_1 is the magnitude of the average force exerted by a molecule on the wall in the time Δt, applying the impulse-momentum theorem to the wall gives

$$F_1 \, \Delta t = \Delta p = 2mv_x$$

In order for the molecule to make two collisions with the same wall, it must travel a distance $2d$ along the x direction in a time Δt. Therefore, the time interval between two collisions with the same wall is $\Delta t = 2d/v_x$, and the force imparted to the wall by a single collision is

$$F_1 = \frac{2mv_x}{\Delta t} = \frac{2mv_x}{2d/v_x} = \frac{mv_x^2}{d} \tag{21.1}$$

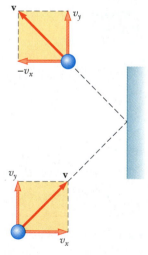

FIGURE 21.2   A molecule makes an elastic collision with the wall of the container. Its x component of momentum is reversed, thereby imparting momentum to the wall, while its y component remains unchanged. In this construction, we assume that the molecule moves in the xy plane.

The total force exerted on the wall by all the molecules is found by adding the forces exerted by the individual molecules:

$$F = \frac{m}{d} \left(v_{x1}^2 + v_{x2}^2 + \cdots \right)$$

In this equation, v_{x1} is the x component of velocity of molecule 1, v_{x2} is the x component of velocity of molecule 2, and so on. The summation terminates when we reach N molecules because there are N molecules in the container.

To proceed further, note that the average value of the square of the velocity in the x direction for N molecules is

$$\overline{v_x^2} = \frac{v_{x1}^2 + v_{x2}^2 + \cdots + v_{xN}^2}{N}$$

Thus, the total force on the wall can be written

$$F = \frac{Nm}{d} \, \overline{v_x^2}$$

Now let us focus on one molecule in the container whose velocity components are v_x, v_y, and v_z. The Pythagorean theorem relates the square of the speed to the square of these components:

$$v^2 = v_x^2 + v_y^2 + v_z^2$$

Hence, the average value of v^2 for all the molecules in the container is related to the average values of $\overline{v_x^2}$, $\overline{v_y^2}$, and $\overline{v_z^2}$ according to the expression

$$\overline{v^2} = \overline{v_x^2} + \overline{v_y^2} + \overline{v_z^2}$$

Because the motion is completely random, the average values $\overline{v_x^2}$, $\overline{v_y^2}$, and $\overline{v_z^2}$ are equal to each other. Using this fact and the above result, we find that

$$\overline{v^2} = 3\overline{v_x^2}$$

Thus, the total force on the wall is

$$F = \frac{N}{3}\left(\frac{m\overline{v^2}}{d}\right)$$

This expression allows us to find the total pressure exerted on the wall:

$$P = \frac{F}{A} = \frac{F}{d^2} = \frac{1}{3}\left(\frac{N}{d^3}\,m\overline{v^2}\right) = \frac{1}{3}\left(\frac{N}{V}\right)m\overline{v^2}$$

$$P = \frac{2}{3}\left(\frac{N}{V}\right)\left(\tfrac{1}{2}m\overline{v^2}\right) \tag{21.2}$$

Relationship between pressure and molecular kinetic energy

This result shows that *the pressure is proportional to the number of molecules per unit volume and to the average translational kinetic energy of the molecule, $\frac{1}{2}m\overline{v^2}$.* With this simplified model of an ideal gas, we have arrived at an important result that relates the large-scale quantity of pressure to an atomic quantity, the average value of the square of the molecular speed. Thus, we have a key link between the atomic world and the large-scale world.

You should note that Equation 21.2 verifies some features of pressure that are probably familiar to you. One way to increase the pressure inside a container is to increase the number of molecules per unit volume in the container. You do this when you add air to a tire. The pressure in the tire can also be increased by increasing the average translational kinetic energy of the molecules in the tire. As we shall see shortly, this can be accomplished by increasing the temperature of the gas inside the tire. That is why the pressure inside a tire increases as the tire heats up during long trips. The continuous flexing of the tires as they move along the road surface generates heat that is transferred to the air inside the tires, increasing the air's temperature, which in turn produces an increase in pressure.

Molecular Interpretation of Temperature

We can obtain some insight into the meaning of temperature by first writing Equation 21.2 in the more familiar form

$$PV = \frac{2}{3}N\left(\tfrac{1}{2}m\overline{v^2}\right)$$

Let us now compare this with the empirical equation of state for an ideal gas (Eq. 19.12):

$$PV = Nk_B T$$

These colorful balloons rise as a large gas burner heats the air inside them. Because warm air is less dense than cool air, the bouyant force upward can exceed the total force downward, causing the balloons to rise. The vertical motion can also be controlled by venting hot air at the top of the balloon. *(Richard Megna/Fundamental Photographs)*

Recall that the equation of state is based on experimental facts concerning the macroscopic behavior of gases. Equating the right sides of these expressions, we find that

$$T = \frac{2}{3k_B} \left(\tfrac{1}{2}m\overline{v^2} \right) \tag{21.3}$$

That is, *temperature is a direct measure of average molecular kinetic energy.*

By rearranging Equation 21.3, we can relate the translational molecular kinetic energy to the temperature:

$$\tfrac{1}{2}m\overline{v^2} = \tfrac{3}{2}k_B T \tag{21.4}$$

That is, the average translational kinetic energy per molecule is $\tfrac{3}{2}k_B T$. Since $\overline{v_x^2} = \tfrac{1}{3}\overline{v^2}$, it follows that

$$\tfrac{1}{2}m\overline{v_x^2} = \tfrac{1}{2}k_B T \tag{21.5}$$

In a similar manner, for the y and z motions it follows that

$$\tfrac{1}{2}m\overline{v_y^2} = \tfrac{1}{2}k_B T \quad \text{and} \quad \tfrac{1}{2}m\overline{v_z^2} = \tfrac{1}{2}k_B T$$

Thus, each translational degree of freedom contributes an equal amount of energy to the gas, namely, $\tfrac{1}{2}k_B T$. (In general, "degrees of freedom" refers to the number of independent means by which a molecule can possess energy.) A generalization of this result, known as the **theorem of equipartition of energy,** says that, with certain restrictions,

the energy of a system in thermal equilibrium is equally divided among all degrees of freedom.

The total translational kinetic energy of N molecules of gas is simply N times the average energy per molecule, which is given by Equation 21.4:

$$E = N(\tfrac{1}{2}m\overline{v^2}) = \tfrac{3}{2}Nk_B T = \tfrac{3}{2}nRT \tag{21.6}$$

where we have used $k_B = R/N_A$ for Boltzmann's constant and $n = N/N_A$ for the number of moles of gas. This result, together with Equation 21.2, implies that the pressure exerted by an ideal gas depends only on the number of molecules per unit volume and the temperature.

The square root of $\overline{v^2}$ is called the *root mean square* (rms) *speed* of the molecules. From Equation 21.4 we get, for the rms speed,

$$v_{\text{rms}} = \sqrt{\overline{v^2}} = \sqrt{\frac{3k_B T}{m}} = \sqrt{\frac{3RT}{M}} \tag{21.7}$$

where M is the molar mass in kg/mol. This expression shows that, at a given temperature, lighter molecules move faster, on the average, than heavier molecules. For example, at a given temperature, hydrogen, with a molar mass of 2×10^{-3} kg/mol, has an average speed of four times that of oxygen, whose molar mass is 32×10^{-3} kg/mol.

Table 21.1 lists the rms speeds for various molecules at 20°C.

TABLE 21.1   Some rms Speeds

Gas	Molecular Mass (g/mol)	v_{rms} at 20°C (m/s)
H_2	2.02	1902
He	4.0	1352
H_2O	18	637
Ne	20.1	603
N_2 or CO	28	511
NO	30	494
CO_2	44	408
SO_2	64	338

EXAMPLE 21.1   A Tank of Helium

A tank of volume 0.300 m³ contains 2.00 mol of helium gas at 20.0°C. Assuming the helium behaves like an ideal gas, (a) find the total thermal energy of the system.

Solution   Using Equation 21.6 with $n = 2.00$ and $T = 293$ K, we get

$$E = \tfrac{3}{2}nRT = \tfrac{3}{2}(2.00 \text{ mol})(8.31 \text{ J/mol·K})(293 \text{ K})$$

$$= \boxed{7.30 \times 10^3 \text{ J}}$$

(b) What is the average kinetic energy per molecule?

Solution   From Equation 21.4, we see that the average kinetic energy per molecule is

$$\tfrac{1}{2}m\overline{v^2} = \tfrac{3}{2}k_B T = \tfrac{3}{2}(1.38 \times 10^{-23} \text{ J/K})(293 \text{ K})$$

$$= \boxed{6.07 \times 10^{-21} \text{ J}}$$

Exercise   Using the fact that the molar mass of helium is 4.00×10^{-3} kg/mol, determine the rms speed of the atoms at 20.0°C.

Answer 1.35×10^3 m/s.

21.2   SPECIFIC HEAT OF AN IDEAL GAS

The heat required to raise the temperature of n moles of gas from T_i to T_f depends on the path taken between the initial and final states. To understand this, consider an ideal gas undergoing several reversible processes such that the change in temperature is $\Delta T = T_f - T_i$ for all processes. The temperature change can be achieved by traveling along a variety of paths from one isotherm to another, as in Figure 21.3. Because ΔT is the same for each path, the change in internal energy ΔU is the same for all paths. However, the first law, $Q = \Delta U + W$, tells us that the heat Q required for each path will be different because W (the area under the curves) is different for each path. Thus the heat required to produce a given change in temperature does not have a unique value.

This difficulty is resolved by defining specific heats for only two processes that frequently occur: changes at constant volume and changes at constant pressure. Because the number of moles is a convenient measure of the amount of gas, we define the **molar specific heats** associated with these processes with the following equations:

$$Q = nC_V \, \Delta T \qquad \text{(constant volume)} \qquad (21.8)$$

$$Q = nC_P \, \Delta T \qquad \text{(constant pressure)} \qquad (21.9)$$

where C_v is the **molar specific heat at constant volume**, while C_p is the **molar specific heat at constant pressure**.

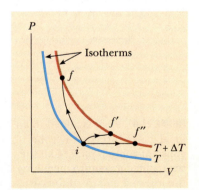

FIGURE 21.3   An ideal gas is taken from one isotherm at temperature T to another at temperature $T + \Delta T$ along three different paths.

In the previous section, we found that the temperature of a gas is a measure of the average translational kinetic energy of the gas molecules. This kinetic energy is associated with the motion of the center of mass of each molecule. It does not include the energy associated with the internal motion of the molecule, namely, vibrations and rotations about the center of mass.

In view of this, let us first consider the simplest case of an ideal monatomic gas, that is, a gas containing one atom per molecule, such as helium, neon, or argon. All of the kinetic energy of such molecules is associated with the motion of their centers of mass. When energy is added to a monatomic gas in a container of fixed volume (by heating, for example), all of the added energy goes into increasing the translational kinetic energy of the atoms. There is no other way to store the energy in a monatomic gas. Therefore, from Equation 21.6 we see that the total thermal energy U of N molecules (or n mol) of an ideal monatomic gas is

Internal energy of an ideal monatomic gas is proportional to its temperature

$$U = \tfrac{3}{2}Nk_B T = \tfrac{3}{2}nRT \tag{21.10}$$

It is important to note that for an ideal gas (only), U is a function of T only. If heat is transferred to the system at *constant volume*, the work done by the system is zero. That is, $W = \int P\, dV = 0$ for a constant volume process. Hence, from the first-law equation we see that

$$Q = \Delta U = \tfrac{3}{2}nR\,\Delta T \tag{21.11}$$

In other words, all of the heat transferred goes into increasing the internal energy (and temperature) of the system. The constant-volume process from i to f is described in Figure 21.4, where ΔT is the temperature difference between the two isotherms. Substituting the value for Q given by Equation 21.8 into Equation 21.11, we get

$$nC_V\,\Delta T = \tfrac{3}{2}nR\,\Delta T$$

$$C_V = \tfrac{3}{2}R \tag{21.12}$$

Note that this expression predicts a value of $\tfrac{3}{2}R = 12.5\,\text{J/mol}\cdot\text{K}$ for all monatomic gases. This is in excellent agreement with measured values of molar specific heats for such gases as helium, neon, argon, xenon, etc. over a wide range of temperatures (Table 21.2).

The change in internal energy for an ideal gas can be expressed as

$$\Delta U = nC_V\,\Delta T \tag{21.13}$$

In the limit of infinitesimal changes, we find the molar specific heat at constant volume to be

$$C_V = \frac{1}{n}\frac{dU}{dT} \tag{21.14}$$

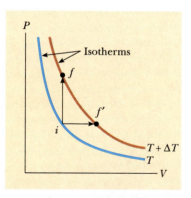

FIGURE 21.4 Heat is added to an ideal gas in two ways. For the constant-volume path *if*, all the heat added goes into increasing the internal energy of the gas since no work is done. Along the constant-pressure path *if'*, part of the heat added goes into work done by the gas. Note that the internal energy is constant along any isotherm.

Now suppose that the gas is taken along the constant-pressure path $i \rightarrow f'$ in Figure 21.4. Along this path, the temperature again increases by ΔT. The heat that must be transferred to the gas in this process is $Q = nC_P\,\Delta T$. Since the volume increases in this process, the work done by the gas is $W = P\,\Delta V$. Applying the first-law equation to this process gives

$$\Delta U = Q - W = nC_P\,\Delta T - P\,\Delta V \tag{21.15}$$

In this case, the energy added to or removed from the gas is transferred in two forms. Part of it does work on the surroundings by moving a piston, and the

TABLE 21.2   Molar Specific Heats of Various Gases

	Molar Specific Heat (J/mol·K)			
	C_p	C_v	$C_p - C_v$	$\gamma = C_p/C_v$
Monatomic Gases				
He	20.8	12.5	8.33	1.67
Ar	20.8	12.5	8.33	1.67
Ne	20.8	12.7	8.12	1.64
Kr	20.8	12.3	8.49	1.69
Diatomic Gases				
H_2	28.8	20.4	8.33	1.41
N_2	29.1	20.8	8.33	1.40
O_2	29.4	21.1	8.33	1.40
CO	29.3	21.0	8.33	1.40
Cl_2	34.7	25.7	8.96	1.35
Polyatomic Gases				
CO_2	37.0	28.5	8.50	1.30
SO_2	40.4	31.4	9.00	1.29
H_2O	35.4	27.0	8.37	1.30
CH_4	35.5	27.1	8.41	1.31

Note: All values obtained at 300 K.

remainder is transferred as thermal energy to the gas. But the change in internal energy for the process $i \rightarrow f'$ is equal to the change for the process $i \rightarrow f$, because U depends only on temperature for an ideal gas and ΔT is the same for each process. In addition, since $PV = nRT$, we note that for a constant-pressure process, $P\,\Delta V = nR\,\Delta T$. Substituting this value for $P\,\Delta V$ into Equation 21.15 with $\Delta U = nC_V\,\Delta T$ (Eq. 21.13) gives

$$nC_V\,\Delta T = nC_P\,\Delta T - nR\,\Delta T$$

$$C_P - C_V = R \tag{21.16}$$

This expression applies to *any* ideal gas. It shows that the molar specific heat of an ideal gas at constant pressure is greater than the molar specific heat at constant volume by an amount R, the universal gas constant (which has the value 8.31 J/mol·K). This result is in good agreement with real gases under standard conditions, that is, 0°C and atmospheric pressure (Table 21.2).

Since $C_V = \frac{3}{2}R$ for a monatomic ideal gas, Equation 21.16 predicts a value $C_P = \frac{5}{2}R = 20.8$ J/mol·K for the molar specific heat of a monatomic gas at constant pressure. The ratio of these heat capacities is a dimensionless quantity γ:

$$\gamma = \frac{C_P}{C_V} = \frac{\frac{5}{2}R}{\frac{3}{2}R} = \frac{5}{3} = 1.67 \tag{21.17}$$

Ratio of molar specific heats for a monatomic ideal gas

The values of C_P and γ are in excellent agreement with experimental values for monatomic gases, but in serious disagreement with the values for the more complex gases (Table 21.2). This is not surprising because the value $C_V = \frac{3}{2}R$ was derived for a monatomic ideal gas, and we expect some additional contribution to the molar specific heat from the internal structure of the more complex molecules. In Section 21.4, we describe the effect of molecular structure on the specific heat of a gas. We shall find that the internal energy and hence the specific heat of a complex gas must include contributions from the rotational and vibrational motions of the molecule.

We have seen that the specific heats of gases at constant pressure are greater than the specific heats at constant volume. This difference is a consequence of the fact that in a constant-volume process, no work is done and all of the heat goes into increasing the internal energy (and temperature) of the gas, whereas in a constant-pressure process some of the thermal energy is transformed into work done by the gas. In the case of solids and liquids heated at constant pressure, very little work is done since the thermal expansion is small. Consequently C_P and C_V are approximately equal for solids and liquids.

EXAMPLE 21.2   Heating a Cylinder of Helium

A cylinder contains 3.00 mol of helium gas at a temperature of 300 K. (a) How much heat must be transferred to the gas to increase its temperature to 500 K if it is heated at constant volume?

Solution   For the constant-volume process, the work done is zero. Therefore from Equation 21.11, we get

$$Q_1 = \tfrac{3}{2}nR\,\Delta T = nC_V\,\Delta T$$

But $C_V = 12.5\ \text{J/mol·K}$ for He and $\Delta T = 200$ K; therefore,

$$Q_1 = (3.00\ \text{mol})(12.5\ \text{J/mol·K})(200\ \text{K}) = \boxed{7.50 \times 10^3\ \text{J}}$$

(b) How much thermal energy must be transferred to the gas at constant pressure to raise the temperature to 500 K?

Solution   Making use of Table 21.2, we get

$$Q_2 = nC_P\,\Delta T = (3.00\ \text{mol})(20.8\ \text{J/mol·K})(200\ \text{K})$$

$$= \boxed{12.5 \times 10^3\ \text{J}}$$

Exercise   What is the work done by the gas in this process?

Answer $W = Q_2 - Q_1 = 5.00 \times 10^3\ \text{J}$.

21.3   ADIABATIC PROCESSES FOR AN IDEAL GAS

As you will recall, an adiabatic process is one in which there is no thermal energy transfer between a system and its surroundings. In reality, true adiabatic processes cannot occur because there is no such thing as a perfect thermal insulator. However, there are processes that are nearly adiabatic. For example, if a gas is compressed (or expanded) very rapidly, very little thermal energy flows into (or out of) the system, and so the process is nearly adiabatic. Such processes occur in the cycle of a gasoline engine, which we discuss in detail in the next chapter.

Another example of an adiabatic process is the very slow expansion of a gas that is thermally insulated from its surroundings. In general,

Definition of a reversible adiabatic process

> a **reversible adiabatic process** is one that is slow enough to allow the system to always be near equilibrium but fast compared with the time it takes the system to exchange thermal energy with its surroundings.

Suppose that an ideal gas undergoes a reversible adiabatic expansion. At any time during the process, we assume that the gas is in an equilibrium state, so that the equation of state, $PV = nRT$, is valid. The pressure and volume at any time during the process are related by the expression

Relationship between P and V for a reversible adiabatic process involving an ideal gas

$$PV^{\gamma} = \text{constant} \tag{21.18}$$

where $\gamma = C_P/C_V$ is assumed to be constant during the process. Thus, we see that all the thermodynamic variables—P, V, and T—change during a reversible adiabatic process.

Proof That PV^γ = Constant for a Reversible Adiabatic Process

When a gas expands adiabatically in a thermally insulated cylinder, there is no thermal energy transferred between the gas and its surroundings, and so $Q = 0$. Let us take the infinitesimal change in volume to be dV and the infinitesimal change in temperature to be dT. The work done by the gas is $P\,dV$. Since the internal energy of an ideal gas depends only on temperature, the change in internal energy is $dU = nC_V\,dT$. Hence, the first-law equation, $\Delta U = Q - W$, becomes

$$dU = nC_V\,dT = -P\,dV$$

Taking the total differential of the equation of state of an ideal gas, $PV = nRT$, we see that

$$P\,dV + V\,dP = nR\,dT$$

Eliminating dT from these two equations, we find that

$$P\,dV + V\,dP = -\frac{R}{C_V}P\,dV$$

Substituting $R = C_P - C_V$ and dividing by PV, we get

$$\frac{dV}{V} + \frac{dP}{P} = -\left(\frac{C_P - C_V}{C_V}\right)\frac{dV}{V} = (1 - \gamma)\frac{dV}{V}$$

$$\frac{dP}{P} + \gamma\frac{dV}{V} = 0$$

Integrating this expression gives

$$\ln P + \gamma \ln V = \text{constant}$$

which is equivalent to Equation 21.18:

$$PV^\gamma = \text{constant}$$

The PV diagram for a reversible adiabatic expansion is shown in Figure 21.5. Because $\gamma > 1$, the PV curve is steeper than that for an isothermal expansion. As the gas expands adiabatically, no thermal energy is transferred in or out of the system. Hence, from the first law, we see that ΔU is negative and so ΔT is also negative. Thus, we see that the gas cools ($T_f < T_i$) during an adiabatic expansion. Conversely, the temperature increases if the gas is compressed adiabatically. Applying Equation 21.18 to the initial and final states, we see that

$$P_i V_i^\gamma = P_f V_f^\gamma \tag{21.19}$$

Using the ideal gas law, Equation 21.19 can also be expressed as

$$T_i V_i^{\gamma - 1} = T_f V_f^{\gamma - 1} \tag{21.20}$$

Note that the above analysis is valid only in a reversible adiabatic process.

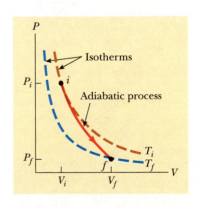

FIGURE 21.5 The PV diagram for a reversible adiabatic expansion. Note that $T_f < T_i$ in this process.

Reversible adiabatic process

EXAMPLE 21.3   A Diesel Engine Cylinder

Air in the cylinder of a diesel engine at 20.0°C is compressed from an initial pressure of 1.00 atm and volume of 800.0 cm³ to a volume of 60.0 cm³. Assuming that air behaves as an ideal gas ($\gamma = 1.40$) and that the compression is adiabatic and reversible, find the final pressure and temperature.

Solution   Using Equation 21.19, we find that

$$P_f = P_i \left(\frac{V_i}{V_f} \right)^{\gamma} = (1.00 \text{ atm}) \left(\frac{800.0 \text{ cm}^3}{60.0 \text{ cm}^3} \right)^{1.4} = \boxed{37.6 \text{ atm}}$$

Since $PV = nRT$ is always valid during the process and since no gas escapes from the cylinder,

$$\frac{P_i V_i}{T_i} = \frac{P_f V_f}{T_f}$$

$$T_f = \frac{P_f V_f}{P_i V_i} T_i = \frac{(37.6 \text{ atm})(60.0 \text{ cm}^3)}{(1.00 \text{ atm})(800.0 \text{ cm}^3)} (293 \text{ K})$$

$$= \boxed{826 \text{ K} = 553°C}$$

21.4   THE EQUIPARTITION OF ENERGY

We have found that model predictions based on specific heat agree quite well with the behavior of monatomic gases but not with the behavior of complex gases (Table 21.2). Furthermore, the value predicted by the model for the quantity $C_P - C_V = R$ is the same for all gases. This is not surprising, because this difference is the result of the work done by the gas, which is independent of its molecular structure.

In order to explain the variations in C_V and C_P in going from monatomic gases to the more complex gases, let us explain the origin of the specific heat. So far, we have assumed that the sole contribution to the thermal energy of a gas is the translational kinetic energy of the molecules. However, the thermal energy of a gas actually includes contributions from the translational, vibrational, and rotational motion of the molecules. The rotational and vibrational motions of molecules can be activated by collisions and therefore are "coupled" to the translational motion of the molecules. The branch of physics known as *statistical mechanics* has shown that, for a large number of particles obeying the laws of Newtonian mechanics, the available energy is, on the average, shared equally by each independent degree of freedom. Recall from Section 21.1 that the equipartition theorem states that, at equilibrium, each degree of freedom contributes, on the average, $\frac{1}{2}k_B T$ of energy per molecule.

Let us consider a diatomic gas, in which the molecules have the shape of a dumbbell (Fig. 21.6). In this model, the center of mass of the molecule can translate in the *x*, *y*, and *z* directions (Fig. 21.6a). In addition, the molecule can rotate about three mutually perpendicular axes (Fig. 21.6b). We can neglect the rotation about the *y* axis for reasons that are discussed later. If the two atoms are taken to be point masses, then I_y is identically zero. Thus, there are five degrees of freedom: three associated with the translational motion and two associated with the rotational motion. Since *each degree of freedom contributes, on the average, $\frac{1}{2}k_B T$ of energy per molecule,* the total energy for *N* molecules is

$$U = 3N(\tfrac{1}{2}k_B T) + 2N(\tfrac{1}{2}k_B T) = \tfrac{5}{2}Nk_B T = \tfrac{5}{2}nRT$$

We can use this result and Equation 21.14 to get the molar specific heat at constant volume:

$$C_V = \frac{1}{n}\frac{dU}{dT} = \frac{1}{n}\frac{d}{dT}(\tfrac{5}{2}nRT) = \tfrac{5}{2}R$$

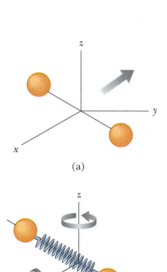

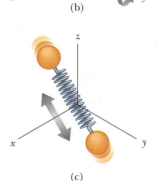

FIGURE 21.6   Possible motions of a diatomic molecule: (a) translational motion of the center of mass, (b) rotational motion about the various axes, and (c) vibrational motion along the molecular axis.

From Equations 21.16 and 21.17 we find that

$$C_P = C_V + R = \tfrac{7}{2}R$$

$$\gamma = \frac{C_P}{C_V} = \frac{\tfrac{7}{2}R}{\tfrac{5}{2}R} = \frac{7}{5} = 1.40$$

These results agree quite well with most of the data given in Table 21.2 for diatomic molecules. This is rather surprising because we have not yet accounted for the possible vibrations of the molecule. In the vibratory model, the two atoms are joined by an imaginary spring. The vibrational motion adds two more degrees of freedom, corresponding to the kinetic and potential energies associated with vibrations along the length of the molecule. Hence, the equipartition theorem predicts a thermal energy of $\tfrac{7}{2}nRT$ and a higher specific heat than is observed. Examination of the experimental data (Table 21.2) suggests that some diatomic molecules, such as H_2 and N_2, do not vibrate at room temperature, and others, such as Cl_2, do. For molecules with more than two atoms, the number of degrees of freedom is even larger and the vibrations are more complex. This results in an even higher predicted specific heat, which is in qualitative agreement with experiment.

We have seen that the equipartition theorem is successful in explaining some features of the specific heat of gas molecules with structure. However, the equipartition theorem does not explain the observed temperature variation in specific heats. As an example of such a temperature variation, C_V for the hydrogen molecule is $\tfrac{5}{2}R$ from about 250 K to 750 K and then increases steadily to about $\tfrac{7}{2}R$ well above 750 K (Fig. 21.7). This suggests that vibrations occur at very high temperatures. At temperatures well below 250 K, C_V has a value of about $\tfrac{3}{2}R$, suggesting that the molecule has only translational energy at low temperatures.

A Hint of Energy Quantization

The failure of the equipartition theorem to explain such phenomena is due to the inadequacy of classical mechanics when applied to molecular systems. For a more satisfactory description, it is necessary to use a quantum-mechanical model in

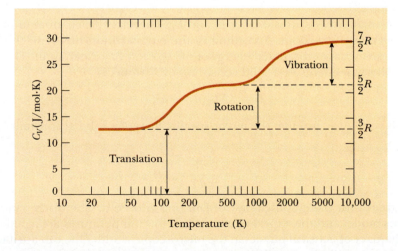

FIGURE 21.7   The molar specific heat of hydrogen as a function of temperature. The horizontal scale is logarithmic. Note that hydrogen liquefies at 20 K.

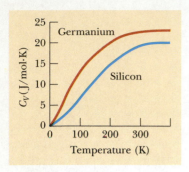

FIGURE 21.8 Molar specific heat of silicon and germanium. As *T* approaches zero, the specific heat also approaches zero. *(From C. Kittel,* Introduction to Solid State Physics, *New York, John Wiley, 1971.)*

which the energy of an individual molecule is quantized. The energy separation between adjacent vibrational energy levels for a molecule such as H_2 is about ten times as great as the average kinetic energy of the molecule at room temperature. Consequently, collisions between molecules at low temperatures do not provide enough energy to change the vibrational state of the molecule. It is often stated that such degrees of freedom are "frozen out." This explains why the vibrational energy does not contribute to the specific heats of molecules at low temperatures.

The rotational energy levels are also quantized, but the spacing of most levels at ordinary temperatures is small compared with $k_B T$. The major exception is rotation of atoms or of linear molecules about their linear axes, for which the moments of inertia are very small and therefore the rotational spacings are very large. If the spacing between rotational levels is so small compared with $k_B T$, the system behaves classically. However, at sufficiently low temperatures (typically less than 50 K), where $k_B T$ is small compared with the spacing between rotational levels, intermolecular collisions may not be energetic enough to alter the rotational states. This explains why C_V reduces to $\frac{3}{2}R$ for H_2 in the range from 20 K to approximately 100 K.

The Specific Heat of Solids

Measurements of the specific heats of solids also show a marked temperature dependence. Solids have specific heats that generally decrease in a nonlinear manner with decreasing temperature and approach zero as the absolute temperature approaches zero. At high temperatures (usually above 300 K), the specific heats approach the value of about $3R \approx 25$ J/mol·K, a result known as the *DuLong-Petit law.* The typical data shown in Figure 21.8 demonstrate the temperature dependence of the molar specific heats for two semiconducting solids, silicon and germanium.

The specific heat of a solid at high temperatures can be explained using the equipartition theorem. For small displacements of an atom from its equilibrium position, each atom executes simple harmonic motion in the *x*, *y*, and *z* directions. The energy associated with vibrational motion in the *x* direction is

$$E_x = \tfrac{1}{2}mv_x^2 + \tfrac{1}{2}kx^2$$

There are analogous expressions for E_y and E_z. Therefore, each atom of the solid has six degrees of freedom. According to the equipartition theorem, this corresponds to an average vibrational energy of $6(\tfrac{1}{2}k_B T) = 3k_B T$ per atom. Therefore, the total thermal energy of a solid consisting of *N* atoms is

Total thermal energy of a solid

$$U = 3Nk_B T = 3nRT \tag{21.21}$$

From this result, we find that the molar specific heat of a solid at constant volume is

Molar specific heat of a solid at constant volume

$$C_V = \frac{1}{n}\frac{dU}{dT} = 3R \tag{21.22}$$

which agrees with the empirical law of DuLong and Petit. The discrepancies between this model and the experimental data at low temperatures are again due to the inadequacy of classical physics in the microscopic world. One can attribute the decrease in specific heat with decreasing temperature to a "freezing out" of various vibrational excitations.

*21.5 THE BOLTZMANN DISTRIBUTION LAW

Thus far we have neglected the fact that not all molecules in a gas have the same speed and energy. Their motion is extremely chaotic. Any individual molecule is colliding with others at the enormous rate of typically a billion times per second. Each collision results in a change in the speed and direction of motion of each of the participant molecules. From Equation 21.10, we see that average molecular speeds increase with increasing temperature. What we would like to know now is the distribution of molecular speeds. For example, how many molecules of a gas have a speed in the range from, say, 400 to 410 m/s? Intuitively, we expect that the speed distribution depends on temperature. Furthermore, we expect that the distribution peaks in the vicinity of v_{rms}. That is, few molecules are expected to have speeds much less than or much greater than v_{rms}, since these extreme speeds will result only from an unlikely chain of collisions.

As we examine the distribution of particles in space, we shall find that the particles distribute themselves among states of different energy in a specific way that depends exponentially on the energy, as first noted by Maxwell and extended by Boltzmann.

The Exponential Atmosphere

We begin by considering the distribution of molecules in our atmosphere. Specifically, we determine how the number of molecules per unit volume varies with altitude. Our model assumes that the atmosphere is at a constant temperature T. (This assumption is not correct because the temperature of our atmosphere decreases by about 2°C per 300 m of altitude, but the model illustrates the basic features of the distribution.)

According to the ideal gas law that we studied in Chapter 19, a gas of N particles in thermal equilibrium obeys the relationship $PV = Nk_B T$. It is convenient to rewrite this equation in terms of the number of particles per unit volume of gas, $n_V = N/V$. This quantity is important because it can vary from one point to another. In fact, our goal is to determine how n_V changes in our atmosphere. We can express the ideal gas law in terms of n_V as $P = n_V k_B T$. Thus, if the number density n_V is known, we can find the pressure and vice versa. The pressure in the atmosphere decreases as the altitude increases because a given layer of air has to support the weight of all the atmosphere above it—the greater the altitude, the less the weight of the air above that layer, and the lower the pressure.

To determine the variation in pressure with altitude, consider an atmospheric layer of thickness dy and cross-sectional area A as in Figure 21.9. Because the air is in static equilibrium, the upward force on the bottom of this layer, PA, must exceed the downward force on the top of the layer, $(P + dP)A$, by an amount equal to the weight of gas in this thin layer. If the mass of a gas molecule in the layer is m, and there are a total of N molecules in the layer, then the weight of the layer is $w = mgN = mgn_V V = mgn_V A\, dy$. Thus, we see that

$$PA - (P + dP)A = mgn_V A\, dy$$

which reduces to

$$dP = -mgn_V\, dy$$

Because $P = n_V k_B T$, and T is assumed to remain constant, we see that $dP = k_B T\, dn_V$. Substituting this into the above expression for dP and rearranging gives

$$\frac{dn_V}{n_V} = -\frac{mg}{k_B T}\, dy$$

Ludwig Boltzmann (1844–1906), an Austrian theoretical physicist, made many important contributions to the development of the kinetic theory of gases, electromagnetism, and thermodynamics. His pioneering work in the field of kinetic theory led to the branch of physics known as statistical mechanics. *(Courtesy of AIP Niels Bohr Library, Lande Collection)*

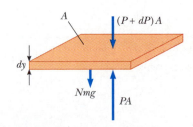

FIGURE 21.9 An atmospheric layer of gas in equilibrium.

Integrating this expression, we find

$$n(y) = n_0 e^{-mgy/k_B T} \qquad (21.23)$$

where the constant n_0 is the density at $y = 0$. This result is known as the **law of atmospheres.**

According to Equation 21.23, the density in thermal equilibrium decreases exponentially with increasing altitude. The density of our atmosphere at sea level is about $n_0 = 2.69 \times 10^{25}$ molecules/m^3. Because the pressure is $P = nk_B T$, we see from Equation 21.23 that the pressure varies with altitude as

$$P = P_0 e^{-mgy/k_B T} \qquad (21.24)$$

where $P_0 = n_0 k_B T$. A comparison of this model with the actual atmospheric pressure as a function of altitude shows that the exponential form is a reasonable approximation to the Earth's atmosphere.

Because our atmosphere contains different gases, each with a different molecular mass, one finds a higher concentration of heavier molecules at the lower altitudes, while the lighter molecules are more concentrated at the higher altitudes.

EXAMPLE 21.4   Distribution of Molecules in the Atmosphere

What is the density of air at an altitude of 12.0 km compared with the density at sea level?

Solution   The density of our atmosphere decreases exponentially with altitude according to the law of atmospheres, Equation 21.23. We assume a temperature of 0°C ($T = 273$ K) and an average molecular mass of 28.8 u $= 4.78 \times 10^{-26}$ kg. Taking $y = 12.0$ km, the power of the exponential in Equation 21.23 is calculated to be

$$\frac{mgy}{k_B T} = \frac{(4.78 \times 10^{-26} \text{ kg})(9.80 \text{ m/s}^2)(12\ 000 \text{ m})}{(1.38 \times 10^{-23} \text{ J/K})(273 \text{ K})} = 1.49$$

Thus, Equation 21.23 gives

$$n = n_0 e^{-mgy/k_B T} = n_0 e^{-1.49} = \boxed{0.225\, n_0}$$

That is, the density of air at an altitude of 12.0 km is only 22.5% of the density at sea level.

Computing Average Values

The exponential function $e^{-mgy/k_B T}$ that appears in Equation 21.23 can be interpreted as a probability distribution that gives the relative probability of finding a gas molecule at some height y. Thus, the probability distribution $p(y)$ is proportional to the density distribution $n(y)$. This concept allows us to determine many properties of the gas, such as the fraction of molecules below a certain height or the average potential energy of a molecule.

As an example, let us determine the average height $\bar{y}$ of a molecule in the atmosphere at temperature T. The expression for this average height is

$$\bar{y} = \frac{\int_0^\infty y\, n(y)\ dy}{\int_0^\infty n(y)\ dy} = \frac{\int_0^\infty y e^{-mgy/k_B T}\ dy}{\int_0^\infty e^{-mgy/k_B T}\ dy}$$

where the height of a molecule can range from 0 to ∞. The numerator in the preceding expression represents the sum of the heights of the particles times their number, while the denominator is the sum of the numbers of particles. The denominator is needed to give the correct average value. After performing the indicated integrations, we find

$$\bar{y} = \frac{(k_B T/mg)^2}{k_B T/mg} = \frac{k_B T}{mg}$$

This tells us that the average height of a molecule increases as T increases, as expected.

We can use a similar procedure to determine the average potential energy of a gas molecule. Because the gravitational potential energy of a molecule at height y is $U = mgy$, the average potential energy is equal to $mg\bar{y}$. Since $\bar{y} = k_B T/mg$, we see that $\overline{U} = mg(k_B T/mg) = k_B T$. This important result shows that the average gravitational potential energy of a molecule depends only on temperature, and not on m or g. Thus, the energy source that distributes the molecules in the atmosphere is thermal energy.

The Boltzmann Distribution

Because the gravitational potential energy of a molecule at height y is $U = mgy$, we can express the distribution law (Eq. 21.23) as

$$n = n_0 e^{-U/k_B T}$$

This means that gas molecules in thermal equilibrium are distributed in space with a probability that depends on gravitational potential energy according to the exponential factor $e^{-U/k_B T}$.

This can be extended to three dimensions, noting that the gravitational potential energy of a particle depends in general on three coordinates. That is, $U = U(x, y, z)$, hence the distribution of particles in space is

$$n(x, y, z) = n_0 e^{-U(x, y, z)/k_B T}$$

where n_0 is the number of particles where $U = 0$.

This kind of distribution applies to any energy the particles have, such as kinetic energy. In general, the relative number of particles having energy E is

$$n(E) = n_0 e^{-E/k_B T} \qquad (21.25)$$

Boltzmann distribution law

This is called the **Boltzmann distribution law** and is important in describing the statistical mechanics of a large number of particles. It states that *the probability of finding the particles in a particular energy state varies exponentially as the negative of the energy divided by $k_B T$*. All the particles would fall into the lowest energy level, except that the thermal energy $k_B T$ tends to excite the particles to higher energy levels.

EXAMPLE 21.5 Thermal Excitation of Atomic Energy Levels

As we discussed briefly in Chapter 8, the electrons of an atom can occupy only certain discrete energy levels. Consider a gas at a temperature of 2500 K whose atoms can occupy only two energy levels separated by 1.5 eV (Fig. 21.10). Determine the ratio of the number of atoms in the higher energy level to the number in the lower energy level.

Solution Equation 21.25 gives the relative number of atoms in a given energy level. In our case, the atom has two possible energies, E_1 and E_2, where E_1 is the lower energy level. Hence, the ratio of the number of atoms in the higher

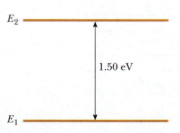

FIGURE 21.10 A gas whose atoms can occupy two energy levels.

energy level to the number in the lower level is

$$\frac{n(E_2)}{n(E_1)} = \frac{n_0 e^{-E_2/k_B T}}{n_0 e^{-E_1/k_B T}} = e^{-(E_2 - E_1)/k_B T}$$

In this problem, $E_2 - E_1 = 1.5$ eV and since 1 eV $= 1.60 \times 10^{-19}$ J,

$$k_B T = (1.38 \times 10^{-23} \text{ J/K})(2500 \text{ K})/1.60 \times 10^{-19} \text{ J/eV}$$

$$= 0.216 \text{ eV}$$

Therefore, the required ratio is

$$\frac{n(E_2)}{n(E_1)} = e^{-1.50\,\text{eV}/0.216\,\text{eV}} = e^{-6.94} = \boxed{9.64 \times 10^{-4}}$$

This result shows that at $T = 2500$ K, only a small fraction of the atoms are in the higher energy level. In fact, for every atom in the higher energy level, there are about 1000 atoms in the lower level. The number of atoms in the higher level increases at even higher temperatures, but the distribution law tells us that at equilibrium, there are always more atoms in the lower level than in the higher level.

*21.6   DISTRIBUTION OF MOLECULAR SPEEDS

In 1860 James Clerk Maxwell (1831–1879) derived an expression that describes the distribution of molecular speeds in a very definite manner. His work and developments by other scientists shortly thereafter were highly controversial, since experiments at that time were not capable of directly detecting molecules. However, about 60 years later, experiments were devised that confirmed Maxwell's predictions.

The observed speed distribution of gas molecules in thermal equilibrium is shown in Figure 21.11. The quantity N_v, called the **Maxwell-Boltzmann distribution function**, is defined as follows: If N is the total number of molecules, then the number of molecules with speeds between v and $v + dv$ is $dN = N_v\,dv$. This number is also equal to the area of the shaded rectangle in Figure 21.11. Furthermore, the fraction of molecules with speeds between v and $v + dv$ is $N_v\,dv/N$. This fraction is also equal to the probability that a molecule has a speed in the range from v to $v + dv$.

The fundamental expression that describes the most probable distribution of speeds of N gas molecules is

$$N_v = 4\pi N \left(\frac{m}{2\pi k_B T}\right)^{3/2} v^2 e^{-mv^2/2k_B T} \tag{21.26}$$

where m is the mass of a gas molecule, k_B is Boltzmann's constant, and T is the absolute temperature.[1]

As indicated in Figure 21.11, the average speed, $\bar{v}$, is somewhat lower than the rms speed. The most probable speed, v_{mp}, is the speed at which the distribution curve reaches a peak. Using Equation 21.26, one finds that

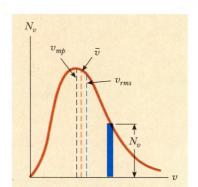

FIGURE 21.11   The speed distribution of gas molecules at some temperature. The number of molecules in the range Δv is equal to the area of the shaded rectangle, $N_v \Delta v$. The function N_v approaches zero as v approaches infinity.

rms speed

$$v_{\text{rms}} = \sqrt{\overline{v^2}} = \sqrt{3k_B T/m} = 1.73\sqrt{k_B T/m} \tag{21.27}$$

Average speed

$$\bar{v} = \sqrt{8k_B T/\pi m} = 1.60\sqrt{k_B T/m} \tag{21.28}$$

Most probable speed

$$v_{mp} = \sqrt{2k_B T/m} = 1.41\sqrt{k_B T/m} \tag{21.29}$$

---

[1] For the derivation of this expression, see any text on thermodynamics, such as that by R. P. Bauman, *Modern Thermodynamics and Statistical Mechanics*, New York, Macmillan, 1992.

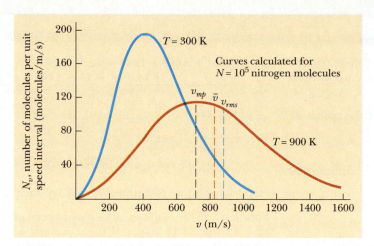

FIGURE 21.12   The speed distribution function for 10^5 nitrogen molecules at 300 K and 900 K. The total area under either curve is equal to the total number of molecules, which in this case equals 10^5. Note that $v_{rms} > \bar{v} > v_{mp}$.

The details of these calculations are left for the student (Problems 44 and 64), but from these equations we see that $v_{rms} > \bar{v} > v_{mp}$.

Figure 21.12 represents speed distribution curves for nitrogen molecules. The curves were obtained by using Equation 21.26 to evaluate the distribution function at various speeds and at two temperatures. Note that the curve shifts to the right as T increases, indicating that the average speed increases with increasing temperature, as expected. The asymmetric shape of the curves is due to the fact that the lowest speed possible is zero while the upper classical limit of the speed is infinity. Furthermore, as temperature increases the distribution curve broadens and the range of speeds also increases.

Equation 21.26 shows that the distribution of molecular speeds in a gas depends on mass as well as temperature. At a given temperature, the fraction of particles with speeds exceeding a fixed value increases as the mass decreases. This explains why lighter molecules, such as hydrogen and helium, escape more readily from the Earth's atmosphere than heavier molecules, such as nitrogen and oxygen. (See the discussion of escape speed in Chapter 14. Gas molecules escape even more readily from the Moon's surface because the escape speed on the Moon is lower.)

The speed distribution of molecules in a liquid is similar to that shown in Figure 21.12. The phenomenon of evaporation of a liquid can be understood from this distribution in speeds using the fact that some molecules in the liquid are more energetic than others. Some of the faster-moving molecules in the liquid penetrate the surface and leave the liquid even at temperatures well below the boiling point. The molecules that escape the liquid by evaporation are those that have sufficient energy to overcome the attractive forces of the molecules in the liquid phase. Consequently, the molecules left behind in the liquid phase have a lower average kinetic energy, causing the temperature of the liquid to decrease. Hence evaporation is a cooling process. For example, an alcohol-soaked cloth is often placed on a feverish head to cool and comfort the patient.

The evaporation process

EXAMPLE 21.6   A System of Nine Particles

Nine particles have speeds of 5.00, 8.00, 12.0, 12.0, 12.0, 14.0, 14.0, 17.0, and 20.0 m/s. (a) Find the average speed.

Solution   The average speed is the sum of the speeds divided by the total number of particles:

$$\bar{v} = \frac{\begin{array}{c}(5.00 + 8.00 + 12.0 + 12.0 + 12.0 \\ + 14.0 + 14.0 + 17.0 + 20.0) \text{ m/s}\end{array}}{9}$$

$$= \boxed{12.7 \text{ m/s}}$$

(b) What is the rms speed?

Solution   The average value of the square of the speed is

$$\overline{v^2} = \frac{\begin{array}{c}(5.00^2 + 8.00^2 + 12.0^2 + 12.0^2 + 12.0^2 \\ + 14.0^2 + 14.0^2 + 17.0^2 + 20.0^2)\text{ m}^2/\text{s}^2\end{array}}{9}$$

$$= 178 \text{ m}^2/\text{s}^2$$

Hence, the rms speed is

$$v_{\text{rms}} = \sqrt{\overline{v^2}} = \sqrt{178 \text{ m}^2/\text{s}^2} = \boxed{13.3 \text{ m/s}}$$

(c) What is the most probable speed of the particles?

Solution   Three of the particles have a speed of 12 m/s, two have a speed of 14 m/s, and the remaining have different speeds. Hence, we see that the most probable speed, v_{mp}, is 12 m/s.

---

*21.7   MEAN FREE PATH

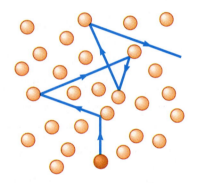

FIGURE 21.13   A molecule moving through a gas collides with other molecules in a random fashion. This behavior is sometimes referred to as a *random-walk process*. The mean free path increases as the number of molecules per unit volume decreases. Note that the motion is not limited to the plane of the paper.

Most of us are familiar with the fact that the strong odor associated with a gas such as ammonia may take a fraction of a minute to diffuse through a room. However, since average molecular speeds are typically several hundred meters per second at room temperature, we might expect a time much less than one second. To understand this apparent contradiction, we note that molecules collide with each other, because they are not geometrical points. Therefore, they do not travel from one side of a room to the other in a straight line. Between collisions, the molecules move with constant speed along straight lines. The average distance between collisions is called the **mean free path.** The path of individual molecules is random and resembles that shown in Figure 21.13. As we would expect from this description, the mean free path is related to the diameter of the molecules and the density of the gas.

We now describe how to estimate the mean free path for a gas molecule. For this calculation we assume that the molecules are spheres of diameter *d*. We see from Figure 21.14a that no two molecules collide unless their centers are less than a distance *d* apart as they approach each other. An equivalent description of the collisions is to imagine that one of the molecules has a diameter 2*d* and the rest are geometrical points (Fig. 21.14b). In a time *t*, the molecule having the speed that

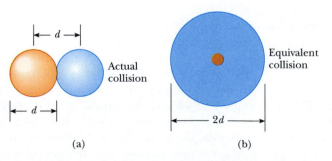

(a)   (b)

FIGURE 21.14   (a) Two spherical molecules, each of diameter *d*, collide if their centers are within a distance *d* of each other. (b) The collision between the two molecules is equivalent to a point mass colliding with a molecule having an effective diameter of 2*d*.

we take to be the average speed, $\bar{v}$, travels a distance $\bar{v}t$. In this same time interval, our molecule with equivalent diameter $2d$ sweeps out a cylinder having a cross-sectional area πd^2 and a length $\bar{v}t$ (Fig. 21.15). Hence, the volume of the cylinder is $\pi d^2 \bar{v}t$. If n_V is the number of molecules per unit volume, then the number of molecules in the cylinder is $(\pi d^2 \bar{v}t)\, n_V$. The molecule of equivalent diameter $2d$ collides with every molecule in this cylinder in the time t. Hence, the number of collisions in the time t is equal to the number of molecules in the cylinder, $(\pi d^2 \bar{v}t)\, n_V$.

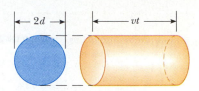

FIGURE 21.15 In a time t, a molecule of effective diameter $2d$ sweeps out a cylinder of length $\bar{v}t$, where $\bar{v}$ is its average speed. In this time, it collides with every molecule within this cylinder.

The **mean free path**, ℓ, which is the mean distance between collisions, equals the average distance $\bar{v}t$ traveled in a time t divided by the number of collisions that occurs in the time:

$$\ell = \frac{\bar{v}t}{(\pi d^2 \bar{v}t)\, n_V} = \frac{1}{\pi d^2 n_V}$$

Since the number of collisions in a time t is $(\pi d^2 \bar{v}t)\, n_V$, the number of collisions per unit time, or **collision frequency** f, is

$$f = \pi d^2 \bar{v} n_V$$

The inverse of the collision frequency is the average time between collisions, called the **mean free time**.

Our analysis has assumed that molecules in the cylinder are stationary. When the motion of these molecules is included in the calculation, the correct results are

$$\ell = \frac{1}{\sqrt{2}\,\pi d^2 n_V} \qquad\qquad (21.30)$$

Mean free path

$$f = \sqrt{2}\,\pi d^2 \bar{v} n_V = \frac{\bar{v}}{\ell} \qquad\qquad (21.31)$$

Collision frequency

EXAMPLE 21.7 A Collection of Nitrogen Molecules

Calculate the mean free path and collision frequency for nitrogen molecules at 20.0°C and 1.00 atm. Assume a molecular diameter of 2.00×10^{-10} m.

Solution Assuming the gas is ideal, we can use the equation $PV = NkT$ to obtain the number of molecules per unit volume under these conditions:

$$n_V = \frac{N}{V} = \frac{P}{kT} = \frac{1.01 \times 10^5 \text{ N/m}^2}{(1.38 \times 10^{-23} \text{ J/K})(293 \text{ K})}$$

$$= 2.50 \times 10^{25} \,\frac{\text{molecules}}{\text{m}^3}$$

Hence, the mean free path is

$$\ell = \frac{1}{\sqrt{2}\,\pi d^2 n_V}$$

$$= \frac{1}{\sqrt{2}\,\pi (2.00 \times 10^{-10} \text{ m})^2 \left(2.50 \times 10^{25}\, \dfrac{\text{molecules}}{\text{m}^3}\right)}$$

$$= 2.25 \times 10^{-7} \text{ m}$$

This is approximately 10^3 times greater than the molecular diameter. Since the average speed of a nitrogen molecule at 20.0°C is about 511 m/s (Table 21.1), the collision frequency is

$$f = \frac{\bar{v}}{\ell} = \frac{511 \text{ m/s}}{2.25 \times 10^{-7} \text{ m}} = 2.27 \times 10^9 /\text{s}$$

The molecule collides with other molecules at the average rate of about two billion times each second!

The mean free path, ℓ, is *not* the same as the average separation between particles. In fact, the average separation, d, between particles is given approximately by $n_V^{-1/3}$. In this example, the average molecular separation is

$$d = \frac{1}{n_V^{1/3}} = \frac{1}{(2.5 \times 10^{25})^{1/3}} = 3.4 \times 10^{-9} \text{ m}$$

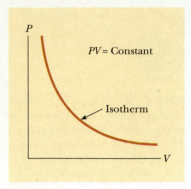

FIGURE 21.16 The *PV* diagram of an isothermal process for an ideal gas. In this case, the pressure and volume are related by *PV* = constant.

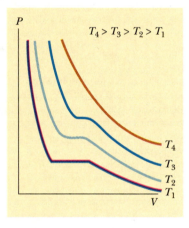

FIGURE 21.17 Isotherms for a real gas at various temperatures. At higher temperatures, such as T_4, the behavior is nearly ideal. The behavior is not ideal at the lower temperatures.

van der Waals' equation of state

*21.8 VAN DER WAALS' EQUATION OF STATE

Thus far we have assumed all gases to be ideal, that is, to obey the equation of state, $PV = nRT$. To a good approximation, real gases behave as ideal gases at ordinary temperatures and pressures. In the kinetic theory derivation of the ideal gas law, we neglected the volume occupied by the molecules and assumed that intermolecular forces were negligible. Now let us investigate the qualitative behavior of real gases and the conditions under which deviations from ideal gas behavior are expected.

Consider a gas contained in a cylinder fitted with a movable piston. As noted in Chapter 20, if the temperature is kept constant while the pressure is measured at various volumes, a plot of P versus V yields a hyperbolic curve (an *isotherm*) as predicted by the ideal gas law (Fig. 21.16).

Now let us describe what happens to a real gas. Figure 21.17 gives some typical experimental curves taken on a gas at various temperatures. At the higher temperatures, the curves are approximately hyperbolic and the gas behavior is close to ideal. However, as the temperature is lowered, the deviations from the hyperbolic shape are very pronounced.

There are two major reasons for this behavior. First, we must account for the volume occupied by the gas molecules. If V is the volume of the container and b is the volume occupied by the molecules, then $V - nb$ is the empty volume available to the gas, where b is a constant. As V decreases for a given quantity of gas, the fraction of the volume occupied by the molecules increases.

The second important effect concerns the intermolecular forces when the molecules are close together. At close separations, the molecules attract each other, as we might expect, since gases can condense to form liquids. When weak attractions between molecules are considered, the short-range forces alter the trajectories of molecules, introducing curvature and thereby increasing traversal time. This results in a decrease in the frequency of wall collisions and hence a decrease in pressure exerted on the walls. The magnitude of the effect depends on the number of collisions, and thus on the square of the number density. The average kinetic energy of the molecules is unchanged, so temperature is not affected. The net pressure is reduced by a factor proportional to the square of the density, which varies as $1/V^2$. Hence, the actual pressure, P, must be supplemented to give a corrected pressure, $P + na/V^2$, that fits the form of the ideal gas law; a is (approximately) a constant.

The two effects just described can be incorporated into a modified equation of state proposed by J. D. van der Waals (1837–1923) in 1873. **Van der Waals' equation of state** is

$$\left(P + \frac{na}{V^2}\right)(V - b) = RT \tag{21.32}$$

The constants a and b are empirical and are chosen to provide the best fit to the experimental data for a particular gas.

The experimental curves in Figure 21.18 for CO_2 are described quite accurately by van der Waals' equation at the higher temperatures (T_3, T_4, and T_5) and outside the shaded regions. Within the yellow region there are major discrepancies. If the van der Waals' equation of state is used to predict the PV relationship at a temperature such as T_1, then a nonlinear curve is obtained that is unlike the observed flat portion of the curve in the figure.

The departure from the predictions of van der Waals' equation at the lower temperatures and higher densities is due to the onset of liquefaction. That is, the gas begins to liquefy at the pressure P_c, called the **critical pressure**. In the region within the dotted line below P_c, the gas is partially liquefied and the gas vapor and liquid coexist. However, liquefaction cannot occur (even at very high pressures) unless the temperature is below a critical value called the **critical temperature**. In the flat portions of the low-temperature isotherms, as the volume is decreased more gas liquefies and the pressure remains constant. At even lower volumes, the gas is completely liquefied. Any further decrease in volume leads to large increases in pressure because liquids are not easily compressed.

It is now realized that, because of the complex nature of the intermolecular forces, a real gas cannot be rigorously described by any simple equation of state, such as Equation 21.32. Nevertheless, the basic concepts involved in Equation 21.32 are correct. At very low temperatures, the low-energy molecules attract each other and the gas tends to liquefy. A further increase in pressure accelerates the rate of liquefaction. At the higher temperatures, the average kinetic energy is large enough to overcome the attractive intermolecular forces; hence, the molecules do not bind together at the higher temperatures and the gas phase is maintained.

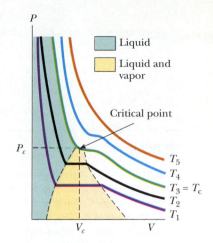

FIGURE 21.18 Isotherms for CO_2 at various temperatures. Below the critical temperature, T_c, the substance could be in the liquid state, the liquid-vapor equilibrium state, or the gaseous state, depending on the pressure and volume. *(Adapted from K. Mendelssohn, The Quest for Absolute Zero, New York, McGraw-Hill, World University Library, 1966.)*

SUMMARY

The pressure of N molecules of an ideal gas contained in a volume V is

$$P = \tfrac{2}{3} \frac{N}{V} \left(\tfrac{1}{2} m \overline{v^2} \right) \tag{21.2}$$

where $\tfrac{1}{2} m \overline{v^2}$ is the average kinetic energy per molecule.

The temperature of an ideal gas is related to the average kinetic energy per molecule through the expression

$$T = \frac{2}{3 k_B} \left(\tfrac{1}{2} m \overline{v^2} \right) \tag{21.3}$$

where k_B is Boltzmann's constant.

The average translational kinetic energy per molecule of a gas is

$$\tfrac{1}{2} m \overline{v^2} = \tfrac{3}{2} k_B T \tag{21.4}$$

Each translational degree of freedom (x, y, or z) has $\tfrac{1}{2} k_B T$ of energy associated with it.

The **equipartition of energy theorem** states that, with certain restrictions, the energy of a system in thermal equilibrium is equally divided among all degrees of freedom.

The total energy of N molecules (or n mol) of an ideal monatomic gas is

$$U = \tfrac{3}{2} N k_B T = \tfrac{3}{2} n R T \tag{21.10}$$

The change in internal energy for n mol of any ideal gas that undergoes a change in temperature ΔT is

$$\Delta U = n C_V \Delta T \tag{21.13}$$

where C_V is the molar specific heat at constant volume.

The molar specific heat of an ideal monatomic gas at constant volume is $C_V = \frac{3}{2}R$; the molar specific heat at constant pressure is $C_P = \frac{5}{2}R$. The ratio of specific heats is $\gamma = C_P/C_V = 5/3$.

If an ideal gas undergoes an adiabatic expansion or compression, the first law of thermodynamics together with the equation of state, shows that

$$PV^\gamma = \text{constant} \tag{21.18}$$

QUESTIONS

1. Dalton's law of partial pressures states: *The total pressure of a mixture of gases is equal to the sum of the partial pressures of gases making up the mixture.* Give a convincing argument of this law based on the kinetic theory of gases.
2. One container is filled with helium gas and another with argon gas. If both containers are at the same temperature, which molecules have the higher rms speed?
3. If you wished to manufacture an after-shave lotion with a scent that is less "likely to get there before you do," would you use a high- or low-molecular-mass lotion?
4. A gas consists of a mixture of He and N_2 molecules. Do the lighter He molecules travel faster than the N_2 molecules? Explain.
5. Although the average speed of gas molecules in thermal equilibrium at some temperature is greater than zero, the average velocity is zero. Explain.
6. Why does a fan make you feel cooler on a hot day?
7. Alcohol taken internally makes you feel warmer. Yet when it is rubbed on your body, it lowers body temperature. Explain the latter effect.
8. A liquid partially fills a container. Explain why the temperature of the liquid decreases when the container is partially evacuated. (It is possible to freeze water using this technique.)
9. A vessel containing a fixed volume of gas is cooled. Does the mean free path of the gas molecules increase, decrease, or remain constant in the cooling process? What about the collision frequency?
10. A gas is compressed at a constant temperature. What hap-

pens to the mean free path of the molecules in this process?
11. If a helium-filled balloon initially at room temperature is placed in a freezer, will its volume increase, decrease, or remain the same?
12. What happens to a helium-filled balloon released into the air? Will it expand or contract? Will it stop rising at some height?
13. Why does a diatomic gas have a greater thermal energy content per mole than a monatomic gas at the same temperature?
14. What happens to the van der Waals' equation (Eq. 21.32) as the volume per mole, V, increases?
15. Ideal gas is contained in a vessel at 300 K. If the temperature is increased to 900 K, (a) by what factor does the rms speed of each molecule change? (b) By what factor does the pressure in the vessel change?
16. At room temperature, the average speed of an air molecule is several hundred meters per second. A molecule traveling at this speed should travel across a room in a small fraction of a second. In view of this, why does it take the odor of perfume (or other smells) several minutes to travel across the room?
17. A vessel is filled with gas at some equilibrium pressure and temperature. Can all gas molecules in the vessel have the same speed?
18. In our model of the kinetic theory of gases, molecules were viewed as hard spheres colliding elastically with the walls of the container. Is this model realistic?

PROBLEMS

Section 21.1   Molecular Model of an Ideal Gas

1. Find the rms speed of nitrogen molecules under standard conditions, 0.0°C and 1.00 atm pressure. Recall that 1 mol of any gas occupies a volume of 22.4 liters under standard conditions.
2. Two moles of oxygen gas is confined to a 5.00-liter vessel at a pressure of 8.00 atm. Find the average translational kinetic energy of an oxygen molecule under these conditions. (The mass of an O_2 molecule is 5.31×10^{-26} kg.)

3. A high-altitude research balloon contains helium gas. At its maximum altitude of 20.0 km, the outside temperature is -50.0°C and the pressure has dropped to $\frac{1}{19}$ atm. The volume of the balloon at this location is 800 m³. Assuming the helium has the same temperature and pressure as the surrounding atmosphere, find the number of moles of helium in the balloon.
4. For the previous problem, find (a) the mass of the helium and (b) the volume of the balloon when it

---

□ indicates problems that have full solutions available in the Student Solutions Manual and Study Guide.

was launched from the ground at standard pressure and temperature (1.00 atm, 0.0°C). (c) What volume tank at 27.0°C and 170 atm will supply this much helium?

5. A spherical balloon of volume 4000 cm³ contains helium at an (inside) pressure of 120 kPa. How many moles of helium are in the balloon, if each helium atom has an average kinetic energy of 3.6×10^{-22} J?

6. In a 30-s interval, 500 hailstones strike a glass window of area 0.60 m² at an angle of 45° to the window surface. Each hailstone has a mass of 5.0 g and a speed of 8.0 m/s. If the collisions are elastic, find the average force and pressure on the window.

6A. In a time t, N hailstones strike a glass window of area A at an angle θ to the window surface. Each hailstone has a mass m and a speed v. If the collisions are elastic, find the average force and pressure on the window.

7. A cylinder contains a mixture of helium and argon gas in equilibrium at 150°C. What is the average kinetic energy of each gas molecule?

8. Calculate the rms speed of an H_2 molecule at 250°C.

9. (a) Determine the temperature at which the rms speed of an He atom equals 500 m/s. (b) What is the rms speed of He on the surface of the Sun, where the temperature is 5800 K?

10. Gaseous helium is in thermal equilibrium with liquid helium at 4.20 K. Determine the most probable speed of a helium atom (mass $= 6.65 \times 10^{-27}$ kg).

11. If the rms speed of a helium atom at room temperature is 1350 m/s, what is the rms speed of an oxygen (O_2) molecule at this temperature? (The molar mass of O_2 is 32, and the molar mass of He is 4.)

12. A 5.00-liter vessel contains nitrogen gas at 27.0°C and 3.00 atm. Find (a) the total translational kinetic energy of the gas molecules and (b) the average kinetic energy per molecule.

13. One mole of xenon gas at 20.0°C occupies 0.0224 m³. What is the pressure exerted by the Xe atoms on the walls of a container?

14. (a) How many atoms of helium gas are required to fill a balloon to diameter 30.0 cm at 20.0°C and 1.00 atm? (b) What is the average kinetic energy of each helium atom? (c) What is the rms speed of each helium atom?

Section 21.2   Specific Heat of an Ideal Gas

(*Note:* Use data in Table 21.2.)

15. Calculate the change in internal energy of 3.0 mol of helium gas when its temperature is increased by 2.0 K.

16. One mole of a diatomic gas has pressure P and volume V. By heating the gas, its pressure is tripled and

its volume is doubled. If this heating process includes two steps, one at constant pressure and the other at constant volume, determine the amount of heat transferred to the gas.

17. One mole of an ideal monatomic gas is at an initial temperature of 300 K. The gas undergoes an isovolumetric process acquiring 500 J of heat. It then undergoes an isobaric process losing this same amount of heat. Determine (a) the new temperature of the gas and (b) the work done on the gas.

17A. One mole of an ideal monatomic gas is at an initial temperature T_0. The gas undergoes an isovolumetric process acquiring heat Q. It then undergoes an isobaric process losing this same amount of heat. Determine (a) the new temperature of the gas and (b) the work done on the gas.

18. One mol of air ($C_v = 5R/2$) at 300 K confined in a cylinder under a heavy piston occupies a volume of 5.0 liters. Determine the new volume of the gas if 4.4 kJ of heat is transferred to the air.

19. One mole of hydrogen gas is heated at constant pressure from 300 K to 420 K. Calculate (a) the heat transferred to the gas, (b) the increase in its internal energy, and (c) the work done by the gas.

20. In a constant-volume process, 209 J of heat is transferred to 1 mol of an ideal monatomic gas initially at 300 K. Find (a) the increase in internal energy of the gas, (b) the work it does, and (c) its final temperature.

21. What is the thermal energy of 100 g of He gas at 77 K? How much more energy must be supplied to heat this gas to 24°C?

22. A container has a mixture of two gases: n_1 mol of gas 1 having molar specific heat C_1 and n_2 mol of gas 2 of molar specific heat C_2. (a) Find the molar specific heat of the mixture. (b) What is the molar specific heat if the mixture has m gases having moles, n_1, n_2, n_3, . . . , n_m, and molar specific heats C_1, C_2, C_3, . . . , C_m, respectively?

23. A room of a well-insulated house has a volume of 100 m³ and is filled with air at 300 K. (a) Estimate the energy required to increase the temperature of this volume of air by 1.0°C. (b) If this energy could be used to lift an object of mass m to a height of 2.0 m, calculate the value of m.

24. How much thermal energy is in the air in a 20.0 m³ room at (a) 0.0°C and (b) 20.0°C? Assume that the pressure remains at 1.00 atm.

Section 21.3   Adiabatic Processes for an Ideal Gas

25. Two moles of an ideal gas ($\gamma = 1.40$) expands slowly and adiabatically from a pressure of 5.00 atm and a volume of 12.0 liters to a final volume of 30.0 liters. (a) What is the final pressure of the gas? (b) What are the initial and final temperatures?

26. Four liters of a diatomic ideal gas ($\gamma = 1.40$) confined to a cylinder are put through a closed cycle. The gas is initially at 1.0 atm and at 300 K. First, its pressure is tripled under constant volume. It then expands adiabatically to its original pressure and finally is compressed isobarically to its original volume. (a) Draw a *PV* diagram of this cycle. (b) Determine the volume at the end of the adiabatic expansion. Find (c) the temperature of the gas at the start of the adiabatic expansion and (d) the temperature at the end of this process. (e) What was the net work done for this cycle?

26A. A diatomic ideal gas ($\gamma = 1.40$) confined to a cylinder is put through a closed cycle. Initially the gas is at P_0, V_0, and T_0. First, its pressure is tripled under constant volume. It then expands adiabatically to its original pressure and finally is compressed isobarically to its original volume. (a) Draw a *PV* diagram of this cycle. (b) Determine the volume at the end of the adiabatic expansion. Find (c) the temperature of the gas at the start of the adiabatic expansion and (d) the temperature at the end of this process. (e) What was the net work done for this cycle?

27. Air ($\gamma = 1.4$) at 27°C and atmospheric pressure is drawn into a bicycle pump that has a cylinder with an inner diameter of 2.5 cm and length 50.0 cm. The down stroke adiabatically compresses the air, which reaches a gauge pressure of 800 kPa before entering the tire. Determine (a) the volume of the compressed air and (b) the temperature of the compressed air. (c) The pump is made of steel and has an inner wall which is 2.0 mm thick. Assume that 4.0 cm of the cylinder's length is allowed to come to thermal equilibrium with the air. What will be the increase in wall temperature?

28. During the power stroke in a four-stroke automobile engine, the piston is forced down as the mixture of gas and air undergoes a reversible adiabatic expansion. Find the average power generated during the expansion by assuming (a) the engine is running at 2500 rpm, (b) the gauge pressure right before the expansion is 20 atm, (c) the volumes of the mixture right before and after the expansion are 50 and 400 cm³, respectively (Fig. P21.28), (d) the time involved in the expansion is one-fourth that of the total cycle, and (e) the mixture behaves like an ideal diatomic gas.

29. During the compression stroke of a certain gasoline engine, the pressure increases from 1.00 atm to 20.0 atm. Assuming that the process is adiabatic and reversible and the gas is ideal with $\gamma = 1.40$, (a) by what factor does the volume change and (b) by what factor does the temperature change?

30. Helium gas at 20.0°C is compressed reversibly without heat loss to one-fifth its initial volume. (a) What

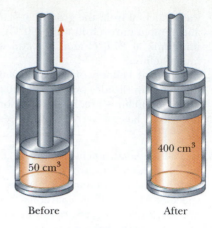

Before   After

FIGURE P21.28

is its temperature after compression? (b) What if the gas is dry air (77% N_2, 23% O_2)?

31. Air in a thundercloud expands as it rises. If its initial temperature was 300 K, and no heat is lost on expansion, what is its temperature when the initial volume is doubled?

32. How much work is required to compress 5.00 mol of air at 20.0°C and 1.00 atm to 1/10 of the original volume by (a) an isothermal process and (b) a reversible adiabatic process? (c) What are the final pressures for the two cases?

33. One mole of an ideal diatomic gas occupies a volume of one liter at a pressure of 0.10 atm. The gas undergoes a process in which the pressure is proportional to the volume, and at the end of the process, it is found that the speed of sound in the gas has doubled from its initial value. Determine the amount of heat transferred to the gas.

33A. One mole of an ideal diatomic gas occupies a volume V_0 at a pressure P_0. The gas undergoes a process in which the pressure is proportional to the volume, and at the end of the process, it is found that the speed of sound in the gas has doubled from its initial value. Determine the amount of heat transferred to the gas.

Section 21.4   The Equipartition of Energy

34. If a molecule has f degrees of freedom, show that a gas consisting of such molecules has the following properties: (1) its total thermal energy is $fnRT/2$; (2) its molar specific heat at constant volume is $fR/2$; (3) its molar specific heat at constant pressure is $(f+2)R/2$; and (4) the ratio $\gamma = C_P/C_V = (f+2)/f$.

35. A 5.00-liter vessel contains 0.125 mol of an ideal gas at 1.50 atm. What is the average translational kinetic energy of a single molecule?

36. Inspecting the magnitudes of C_V and C_P for the diatomic and polyatomic gases in Table 21.2, we find that the values increase with increasing molecular mass. Give a qualitative explanation of this observation.

37. In a crude model (Fig. P21.37) of a rotating diatomic molecule of chlorine (Cl_2), the two Cl atoms are 2.0×10^{-10} m apart and rotate about their center-of-mass with angular speed $\omega = 2.0 \times 10^{12}$ rad/s. What is the rotational kinetic energy of one molecule of Cl_2, which has a molar mass of 70?

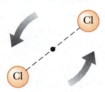

FIGURE P21.37

38. Consider 2 mol of an ideal diatomic gas. Find the total heat capacity at constant volume and at constant pressure if (a) the molecules rotate but do not vibrate and (b) the molecules rotate and vibrate.

*Section 21.5   The Boltzmann Distribution Law
*Section 21.6   Distribution of Molecular Speeds

39. The latent heat of vaporization for water at room temperature is 2430 J/g. (a) How much kinetic energy does each water molecule that evaporates possess before evaporating? (b) Find the average speed before evaporating of a water molecule that is evaporating. (c) What is the effective temperature of these molecules? Why don't these molecules burn you?

40. Fifteen identical particles have the following speeds: one has speed 2.0 m/s; two have speed 3.0 m/s; three have speed 5.0 m/s; four have speed 7.0 m/s; three have speed 9.0 m/s; two have speed 12.0 m/s. Find (a) the average speed, (b) the rms speed, and (c) the most probable speed of these particles.

41. It is reported that there is only one particle per cubic meter in deep space. Using the average temperature of 3.0 K and assuming the particle is H_2 with a diameter of 0.20 nm, (a) determine the mean free path of the particle and the average time between collisions. (b) Repeat part (a) assuming only one particle per cubic centimeter.

42. The chemical composition of the atmosphere changes slightly with altitude because the various molecules have different masses. Use the law of the atmospheres to determine how the ratio of oxygen to nitrogen molecules changes between sea level and 10 km. Assume a temperature of 300 K and take the masses to be 32 u for oxygen (O_2) and 28 u for nitrogen (N_2).

43. (a) Find the ratio of speeds for the two isotopes of chlorine, ^{35}Cl and ^{37}Cl, as they diffuse through air. (b) Which isotope moves faster?

44. Show that the most probable speed of a gas molecule is given by Equation 21.29. Note that the most probable speed corresponds to the point where the slope of the speed distribution curve, dN_v/dv, is zero.

45. At what temperature would the average speed of helium atoms equal (a) the escape speed from Earth, 1.12×10^4 m/s and (b) the escape speed from the Moon, 2.37×10^3 m/s? (See Chapter 14 for a discussion of escape speed, and note that the mass of a helium atom is 6.65×10^{-27} kg.)

46. A gas is at 0°C. To what temperature must it be heated to double the rms speed of its molecule?

*Section 21.7   Mean Free Path

47. In an ultrahigh vacuum system, the pressure is measured to be 1.00×10^{-10} torr (where 1 torr = 133 Pa). If the gas molecules have a molecular diameter of 3.00×10^{-10} m and the temperature is 300 K, find (a) the number of molecules in a volume of 1.00 m³, (b) the mean free path of the molecules, and (c) the collision frequency, assuming an average speed of 500 m/s.

48. Show that the mean free path for the molecules of an ideal gas is

$$\ell = \frac{k_B T}{\sqrt{2}\,\pi d^2 P}$$

where d is the molecular diameter.

49. In a tank full of oxygen, how many molecular diameters d (on average) will an oxygen molecule travel (at 1.00 atm and 20.0°C) before colliding with another O_2 molecule? (The diameter of the O_2 molecule is approximately 3.60×10^{-10} m.)

50. Argon gas at atmospheric pressure and 20.0°C is confined in a 1.00-m³ vessel. The effective hard-sphere diameter of the argon atom is 3.10×10^{-10} m. (a) Determine the mean free path ℓ. (b) Find the pressure when $\ell = 1.00$ m. (c) Find the pressure when $\ell = 3.10 \times 10^{-10}$ m.

*Section 21.8   Van der Waals' Equation of State

51. The constant b that appears in van der Waals' equation of state for oxygen is measured to be 31.8 cm³/mol. Assuming a spherical shape, estimate the diameter of the molecule.

52. Show that the work done in expanding 1 mol of a van

der Waals' gas from an initial volume V_i to final volume V_f at constant temperature is

$$W = RT \ln\left(\frac{V_f - b}{V_i - b}\right) + a(V_f^{-1} - V_i^{-1})$$

ADDITIONAL PROBLEMS

53. A mixture of two gases will diffuse through a filter at rates proportional to their rms speeds. If the molecules of the two gases have masses m_1 and m_2, show that the ratio of their rms speeds (or the ratio of diffusion rates) is

$$\frac{(v_1)_{rms}}{(v_2)_{rms}} = \sqrt{\frac{m_2}{m_1}}$$

54. A cylinder containing n mol of an ideal gas undergoes a reversible adiabatic process. (a) Starting with the expression $W = \int P\,dV$ and using $PV^\gamma =$ constant, show that the work done is

$$W = \left(\frac{1}{\gamma - 1}\right)(P_i V_i - P_f V_f)$$

(b) Starting with the first-law equation in differential form, prove that the work done is also equal to $nC_V(T_i - T_f)$. Show that this result is consistent with the equation in part (a).

55. Twenty particles, each of mass m and confined to a volume v, have the following speeds: two have speed v; three have speed $2v$; five have speed $3v$; four have speed $4v$; three have speed $5v$; two have speed $6v$; one has speed $7v$. Find (a) the average speed, (b) the rms speed, (c) the most probable speed, (d) the pressure they exert on the walls of the vessel, and (e) the average kinetic energy per particle.

56. A vessel contains 1.00×10^4 oxygen molecules at 500 K. (a) Make an accurate graph of the Maxwell speed distribution function versus speed with points at speed intervals of 100 m/s. (b) Determine the most probable speed from this graph. (c) Calculate the average and rms speeds for the molecules and label these points on your graph. (d) From the graph, estimate the fraction of molecules with speeds in the range 300 m/s to 600 m/s.

57. One cubic meter of atomic hydrogen at 0°C at atmospheric pressure contains approximately 2.7×10^{25} atoms. The first excited state of the hydrogen atom has an energy of 10.2 eV above the lowest energy level called the ground state. Use the Boltzmann factor to find the number of atoms in the first excited state at 0°C and at 10 000°C.

58. On a day when the atmospheric pressure is 1.00 atm and the temperature is 20.0°C, a diving bell in the shape of a cylinder 4.0 m tall, closed at the upper end, is lowered into water to aid in the construction of an underground foundation for a bridge tower.

The water inside the diving bell rises to within 1.5 m of the top, and the temperature drops to 8.0°C. (a) Find the air pressure inside the bell. (b) How far below the surface of the water is the bell located? (In actual use, additional air is pumped in, forcing the water out to provide working space for the construction workers.)

58A. On a day when the atmospheric pressure is P_0 and the temperature is T_0, a diving bell in the shape of a cylinder of height h, closed at the upper end, is lowered into water to aid in the construction of an underground foundation for a bridge tower. The water inside the diving bell rises to within a distance $d < h$ of the top, and the temperature drops to T. (a) Find the air pressure inside the bell. (b) How far below the surface of the water is the bell located?

59. Oxygen at pressures much above 1 atm becomes toxic to lung cells. What ratio, by weight, of helium gas (He) to oxygen (O_2) must be used by a scuba diver who is to descend to an ocean depth of 50.0 m?

60. The compressibility, κ, of a substance is defined as the fractional change in volume of that substance for a given change in pressure:

$$\kappa = -\frac{1}{V}\frac{dV}{dP}$$

(a) Explain why the negative sign in this expression ensures that κ is always positive. (b) Show that if an ideal gas is compressed isothermally, its compressibility is given by $\kappa_1 = 1/P$. (c) Show that if an ideal gas is compressed adiabatically, its compressibility is given by $\kappa_2 = 1/\gamma P$. (d) Determine values for κ_1 and κ_2 for a monatomic ideal gas at a pressure of 2.00 atm.

61. One mole of a gas obeying van der Waals' equation of state is compressed isothermally. At some critical temperature, T_c, the isotherm has a point of zero slope, as in Figure 21.17. That is, at $T = T_c$,

$$\frac{\partial P}{\partial V} = 0 \quad \text{and} \quad \frac{\partial^2 P}{\partial V^2} = 0$$

Using Equation 21.32 and these conditions, show that at the critical point, $P_c = a/27b^2$, $V_c = 3b$, and $T_c = 8a/27Rb$.

62. Consider the particles in a gas centrifuge, a device used to separate particles of different mass by whirling them in a circular path of radius r at angular speed ω. The central force acting on a particle is $mr\omega^2$. (a) Discuss how a gas centrifuge can be used to separate particles of different mass. (b) Show that the density of the particles as a function of r is

$$n(r) = n_0 e^{-mr^2\omega^2/2k_BT}$$

63. Consider a system of 1.00×10^4 oxygen molecules at a temperature T. Write a program that will enable

you to calculate the Maxwell distribution function N_v as a function of the speed of the molecules and the temperature. Use your program to evaluate N_v for speeds ranging from $v = 0$ to $v = 2000$ m/s (in intervals of 100 m/s) at temperatures of (a) 300 K and (b) 1000 K. (c) Make graphs of your results (N_v versus v) and use the graph at $T = 1000$ K to calculate the number of molecules having speeds between 800 m/s and 1000 m/s at $T = 1000$ K.

64. Verify Equations 21.27 and 21.28 for the rms and average speeds of the molecules of a gas at a temperature T. Note that the average value of v^n is

$$\overline{v^n} = \frac{1}{N} \int_0^\infty v^n N_v \, dv$$

and make use of the integrals

$$\int_0^\infty x^3 e^{-ax^2} \, dx = \frac{1}{2a^2} \qquad \int_0^\infty x^4 e^{-ax^2} \, dx = \frac{3}{8a^2} \sqrt{\frac{\pi}{a}}$$

65. (a) Show that the fraction of particles below an altitude h in the atmosphere is

$$f = 1 - e^{-mgh/k_B T}$$

(b) Use this result to show that half the particles are below the altitude $h' = k_B T \ln(2) / mg$. What is the value of h' for Earth? (Assume a uniform temperature of 270 K and note that the average molecular mass for air is 28.8 u.)

66. By volume, air is composed of approximately 78% nitrogen (N_2), 21% oxygen (O_2), and 1% other gases. Ignoring the 1% other gases, (a) use these facts to find the mass of a cubic meter of air at standard conditions (1.00 atm, 0.0°C). (b) Given this result, calculate the lifting force on a helium-filled balloon with a volume of 1.00 m^3 at a pressure of 1.00 atm. (c) Show that a helium-filled balloon has 92.6% the lifting force of a similar hydrogen-filled balloon.

67. There are roughly 10^{59} neutrons and protons in an average star and about 10^{11} stars in a typical galaxy. Galaxies tend to form in clusters of (on the average) about 10^3 galaxies, and there are about 10^9 clusters in the known part of the Universe. (a) Approximately how many neutrons and protons are there in the known Universe? (b) Suppose all this matter were compressed into a sphere of nuclear matter such that each nuclear particle occupied a volume of 10^{-45} m^3 (about the "volume" of a neutron or proton). What would be the radius of this sphere of nuclear matter? (c) How many moles of nuclear particles are there in the observable Universe?

68. (a) If it has enough kinetic energy, a molecule at the surface of the Earth can escape the Earth's gravitation. Using energy conservation, show that the minimum kinetic energy needed to escape is mgR, where

m is the mass of the molecule, g is the free-fall acceleration at the surface, and R is the radius of the Earth. (b) Calculate the temperature for which the minimum escape kinetic energy equals ten times the average kinetic energy of an oxygen molecule.

69. Using multiple laser beams, physicists have been able to cool and trap sodium atoms in a small region. In one experiment the temperature of the atoms was reduced to 2.4×10^{-4} K. (a) Determine the rms speed of the sodium atoms at this temperature. The atoms can be trapped for about 1.0 s. The trap has a linear dimension of roughly 1.0 cm. (b) Approximately how long would it take an atom to wander out of the trap region if there were no trapping action?

70. For a Maxwellian gas, use a computer or programmable calculator to find the numerical value of the ratio $\{N_v(v) / N_v(v_{mp})\}$ for the following values of v: $v = (v_{mp}/50)$, $(v_{mp}/10)$, $(v_{mp}/2)$, $2v_{mp}$, $10v_{mp}$, $50v_{mp}$. Give your results to three significant figures.

SPREADSHEET PROBLEMS

S1. For a gas consisting of N molecules, the number of molecules that have speeds in the range between v and $v + dv$ is $dN = N_v \, dv$, where N_v is the Maxwell-Boltzmann distribution function given by Equation 21.26. If the interval dv is small enough, N_v/N can be considered to be constant over the interval. Spreadsheet 21.1 calculates the fraction of the total number of molecules $dN/N = N_v \, dv/N$ having a particular speed in a narrow range of speeds dv and plots dN/N versus v. The spreadsheet also calculates the most probable speed and the rms speed of the molecules. (a) Use Spreadsheet 21.1 to find the fraction of molecules of hydrogen gas with speeds between 1000 m/s and 1100 m/s and between 3000 m/s and 3100 m/s at 100 K. (b) Repeat part (a) for $T = 273$ K. (c) Repeat part (a) for $T = 1000$ K.

S2. Repeat Problem S1. for oxygen molecules.

S3. Spreadsheet 21.2 calculates and plots three isotherms for the van der Waals' equation of state:

$$\left(P + \frac{an^2}{v^2} \right)(V - bn) = nRT$$

It also calculates the critical temperature, critical pressure, and critical volume for the gas. The constants a and b and the experimental values for the critical temperature and pressure of several real gases shown in the table below. For a van der Waals gas, it can be shown that the critical temperature is $8a/27Rb$, the critical pressure is $a/27b^2$, and the critical volume is $3bn$. Use Spreadsheet 21.2 to plot the isotherms for each of the gases in the table. Compare the calculated and experimental values for the critical temperatures and pressures for each gas.

Gas	$a(L^2 \cdot atm/mol^2)$	$b(L/mol)$	$T_c(K)$	$P_c(atm)$
N_2	1.390	0.03913	126	33.5
O_2	1.360	0.03183	155	50.1
NO_2	5.284	0.04424	431	100
He	0.03412	0.02370	5.25	2.26
C_6H_6	18.00	0.11540	562	48.6

S4. Modify Spreadsheet 21.1 so that it plots dN/N versus v for three temperatures on the same graph. Plot the distributions for hydrogen molecules at $T = 273$ K, 293 K, and 393 K. Compare the fraction of molecules having speeds between 3000 m/s and 3050 m/s for the three temperatures.

B.C.   **By John Hart**

By permission of John Hart and Field Enterprises, Inc.

CHAPTER 22

Heat Engines, Entropy, and the Second Law of Thermodynamics

This steam-driven locomotive runs from Durango to Silverton, Colorado. Early steam-driven locomotives obtained their energy by burning wood or coal. The generated heat produces the steam, which powers the locomotive. Modern trains use electricity or diesel fuel to power their locomotives. All heat engines extract heat from a burning fuel and convert only a fraction of this energy to mechanical energy. (© Lois Moulton, Tony Stone Worldwide, Ltd.)

The first law of thermodynamics, studied in Chapter 20, is a statement of conservation of energy, generalized to include heat as a form of energy transfer. This law tells us only that an increase in one form of energy must be accompanied by a decrease in some other form of energy. It places no restrictions on the types of energy conversions that can occur. Furthermore, it makes no distinction between heat and work. According to the first law, the internal (thermal) energy of a body may be increased either by adding heat to it or by doing work on it. An important distinction exists between heat and work, however, that is not evident from the first law. One manifestation of this difference is the fact that it is impossible to convert thermal energy to mechanical energy in an isothermal process.

Contrary to what the first law implies, only certain types of energy conversions can take place. The second law of thermodynamics establishes which processes in nature can and which cannot occur. The following are examples of processes that are consistent with the first law of thermodynamics but proceed in an order governed by the second law.

- When two objects at different temperatures are placed in thermal contact with each other, thermal energy always flows from the warmer to the cooler object, never from the cooler to the warmer.
- A rubber ball that is dropped to the ground bounces several times and eventually comes to rest, but a ball lying on the ground never begins bouncing on its own.
- An oscillating pendulum eventually comes to rest because of collisions with air molecules and friction at the point of suspension. The initial mechanical energy is converted to thermal energy; the reverse conversion of energy never occurs.

These are all *irreversible* processes, that is, processes that occur naturally in only one direction. No irreversible process ever runs backward, because if it did, it would violate the second law of thermodynamics.[1]

From an engineering viewpoint, perhaps the most important application of the second law of thermodynamics is the limited efficiency of heat engines. The second law says that a machine capable of continuously converting thermal energy completely to other forms of energy in a cyclic process cannot be constructed.

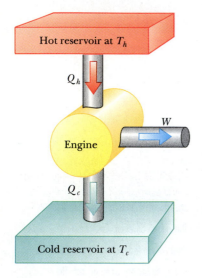

FIGURE 22.1   Schematic representation of a heat engine. The engine absorbs thermal energy Q_h from the hot reservoir, expels thermal energy Q_c to the cold reservoir, and does work W.

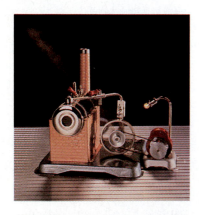

A model steam engine equipped with a built-in horizontal boiler. The water is heated electrically, which generates steam that is used to power the electric generator at the right.   *(Courtesy of Central Scientific Co.)*

22.1   HEAT ENGINES AND THE SECOND LAW OF THERMODYNAMICS

As we learned in Chapter 20, a heat engine is a device that converts thermal energy to other useful forms, such as mechanical and electrical energy. In a typical process for producing electricity in a power plant, for instance, coal or some other fuel is burned and the thermal energy produced is used to convert water to steam. This steam is directed at the blades of a turbine, setting it into rotation. Finally, the mechanical energy associated with this rotation is used to drive an electric generator. Another heat engine, the internal combustion engine in your automobile, extracts thermal energy from a burning fuel and converts a fraction of this energy to mechanical energy.

A heat engine carries some working substance through a cyclic process during which (1) thermal energy is absorbed from a source at a high temperature, (2) work is done by the engine, and (3) thermal energy is expelled by the engine to a source at a lower temperature. As an example, consider the operation of a steam engine in which the working substance is water. The water is carried through a cycle in which it first evaporates to steam in a boiler and then expands against a piston. After the steam is condensed with cooling water, the liquid water produced is returned to the boiler and the process is repeated. It is useful to represent a heat engine schematically as in Figure 22.1. The engine absorbs a quantity of heat Q_h from the hot reservoir, does work W, and then gives up heat Q_c to the cold reservoir. Because the working substance goes through a cycle, its initial and final internal energies are equal, so $\Delta U = 0$. Hence, from the first law of thermodynamics we see that *the net work W done by a heat engine equals the net heat flowing into it.*

---

[1] To be more precise, we should say that the set of events in the time-reversed sense is highly improbable. From this viewpoint, events occur with a vastly higher probability in one direction than in the opposite direction.

As we can see from Figure 22.1, $Q_{net} = Q_h - Q_c$; therefore,

$$W = Q_h - Q_c \qquad (22.1)$$

where Q_h and Q_c are taken to be positive quantities.

The net work done for a cyclic process is the area enclosed by the curve representing the process on a PV diagram. This is shown for an arbitrary cyclic process in Figure 22.2.

> The **thermal efficiency**, *e*, of a heat engine is defined as the ratio of the net work done to the thermal energy absorbed at the higher temperature during one cycle:
>
> $$e = \frac{W}{Q_h} = \frac{Q_h - Q_c}{Q_h} = 1 - \frac{Q_c}{Q_h} \qquad (22.2)$$

We can think of the efficiency as the ratio of what you get (mechanical work) to what you give (thermal energy at the higher temperature). Equation 22.2 shows that a heat engine has 100% efficiency ($e = 1$) only if $Q_c = 0$—that is, if no thermal energy is expelled to the cold reservoir. In other words, a heat engine with perfect efficiency would have to convert all of the absorbed thermal energy to mechanical work. One of the consequences of the second law of thermodynamics is that this is impossible.

In practice, it is found that all heat engines convert only a fraction of the absorbed thermal energy to mechanical work. For example, a good automobile engine has an efficiency of about 20% and diesel engines have efficiencies ranging from 35% to 40%. On the basis of this fact, the **Kelvin-Planck** form of the **second law of thermodynamics** states the following:

> It is impossible to construct a heat engine that, operating in a cycle, produces no other effect than the absorption of thermal energy from a reservoir and the performance of an equal amount of work.

This form of the second law is useful in understanding the operation of heat engines. With reference to Equation 22.2, the second law says that, during the operation of a heat engine, *W* can never be equal to Q_h or, alternatively, that some thermal energy Q_c must be rejected to the environment. As a result, it is theoretically impossible to construct an engine that works with 100% efficiency. Figure 22.3 is a schematic diagram of the impossible "perfect" heat engine.

Our assessment of the first two laws of thermodynamics can be summed up as follows: The first law says that *we cannot get more energy out of a cyclic process than the amount of thermal energy we put in*, and the second law says that *we cannot break even because we must put more thermal energy in, at the higher temperature, than the net amount of work output.*

Refrigerators and Heat Pumps

Refrigerators and heat pumps are heat engines running in reverse (Fig. 22.4). The engine absorbs thermal energy Q_c from the cold reservoir and expels thermal energy Q_h to the hot reservoir. This can be accomplished only if work is done *on*

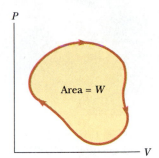

FIGURE 22.2 The *PV* diagram for an arbitrary cyclic process. The net work done equals the area enclosed by the curve.

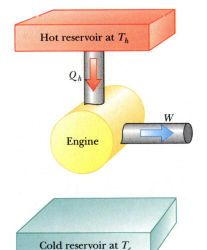

The impossible engine

FIGURE 22.3 Schematic diagram of a heat engine that absorbs thermal energy Q_h from a hot reservoir and does an equivalent amount of work. It is impossible to construct such a perfect engine.

Lord Kelvin, British physicist and mathematician (1824–1907). Born William Thomson in Belfast, Kelvin was the first to propose the use of an absolute scale of temperature. The kelvin scale, named in his honor, is discussed in Section 19.3. Kelvin's study of Carnot's theory led to the idea that heat cannot pass spontaneously from a colder body to a hotter body.

(*J. L. Charmet/SPL/Photo Researchers*)

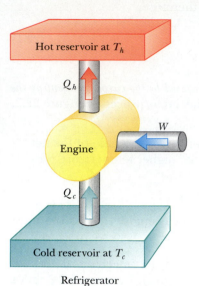

FIGURE 22.4 Schematic diagram for a refrigerator, which absorbs thermal energy Q_c from the cold reservoir and expels thermal energy Q_h to the hot reservoir. Work W is done *on* the refrigerator.

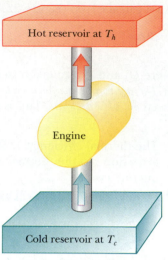

Impossible refrigerator

FIGURE 22.5 Schematic diagram of the impossible refrigerator, that is, one that absorbs thermal energy Q_c from a cold reservoir and expels an equivalent amount of thermal energy to the hot reservoir with $W = 0$.

the refrigerator. From the first law, we see that the thermal energy given up to the hot reservoir must equal the sum of the work done and the thermal energy absorbed from the cold reservoir. Therefore, we see that the refrigerator transfers thermal energy from a colder body (the contents of the refrigerator) to a hotter body (the room). In practice, it is desirable to carry out this process with a minimum of work. If it could be accomplished without doing any work, we would have a "perfect" refrigerator (Fig. 22.5). Again, this is in violation of the second law of thermodynamics, which in the form of the **Clausius statement**[2] says the following:

> It is impossible to construct a machine operating in a cycle that produces no other effect than to transfer thermal energy continuously from one object to another object at a higher temperature.

In simpler terms, *thermal energy does not flow spontaneously from a cold object to a hot object.* For example, homes are cooled in summer by pumping thermal energy out; the work done on the air conditioner is supplied by the power company.

The Clausius and Kelvin-Planck statements of the second law appear, at first sight, to be unrelated. They are, in fact, equivalent in all respects. Although we do not prove it here, it can be shown that if either statement is false, so is the other.[3]

---

[2] First expressed by Rudolf Clausius (1822–1888).

[3] See, for example, R. P. Bauman, *Modern Thermodynamics and Statistical Mechanics*, New York, Macmillan, 1992.

EXAMPLE 22.1   The Efficiency of an Engine

Find the efficiency of an engine that introduces 2000 J of heat during the combustion phase and loses 1500 J at exhaust.

Solution   The efficiency of the engine is given by Equation 22.2:

$$e = 1 - \frac{Q_c}{Q_h} = 1 - \frac{1500\ \text{J}}{2000\ \text{J}} = 0.25,\ \text{or}\ \boxed{25\%}$$

22.2 REVERSIBLE AND IRREVERSIBLE PROCESSES

In the next section we shall discuss a theoretical heat engine that is the most efficient engine possible. In order to understand its nature, we must first examine the meaning of reversible and irreversible processes. A **reversible process** is one that can be performed so that, at its conclusion, both the system and its surroundings have been returned to their exact initial conditions. A process that does not satisfy these requirements is **irreversible.**

All natural processes are known to be irreversible. From the endless number of examples that can be selected, let us examine the free expansion of a gas, already discussed in Section 20.6, and show that it cannot be reversible. The gas is in an insulated container, as in Figure 22.6, with a membrane separating the gas from a vacuum. If the membrane is punctured, the gas expands freely into the vacuum. Because the gas does not exert a force through a distance on the surroundings, it does no work as it expands. In addition, no thermal energy is transferred to or from the gas because the container is insulated from its surroundings. Thus, in this adiabatic process, the system has changed but the surroundings have not. Now imagine that we try to reverse the process by first compressing the gas to its original volume. Let's say an engine is being used to force the piston inward. This action is changing both the system and its surroundings. The surroundings are changing because work is being done by an outside agent on the system, and the system is changing because the compression is increasing the temperature of the gas. We can lower the temperature of the gas by allowing it to come into contact with an external heat reservoir. Although this second procedure returns the gas to its original state, the surroundings are again affected because thermal energy is being added to the surroundings. If this thermal energy could somehow be used to drive the engine and compress the gas, the system and its surroundings could be returned to their initial states. However, our statement of the second law says that this extracted thermal energy cannot be completely converted to mechanical energy isothermally. We must conclude that a reversible process has not occurred.

Although real processes are always irreversible, some are *almost* reversible. If a real process occurs very slowly so that the system is virtually always in equilibrium, the process can be considered reversible. For example, imagine compressing a gas very slowly by dropping some grains of sand onto a frictionless piston as in Figure 22.7. The process is made isothermal by placing the gas in thermal contact with a heat reservoir, and just enough thermal energy is transferred from the gas to the reservoir during the process to keep the temperature constant. The pressure, volume, and temperature of the gas are well defined during the isothermal compression. Each time a grain of sand is added to the piston, the volume decreases slightly while the pressure increases slightly. Each added grain represents a change to a

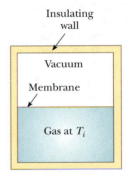

FIGURE 22.6   Free expansion of a gas.

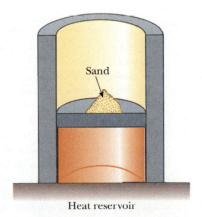

FIGURE 22.7   A gas in thermal contact with a heat reservoir is compressed slowly by dropping grains of sand onto the piston. The compression is isothermal and reversible.

new equilibrium state. The process can be reversed by slowly removing grains from the piston. Although adding and removing grains of sand is an irreversible process for the surroundings, changes in the system have occurred as if the entire process were reversible.

A general characteristic of a reversible process is that there can be no dissipative effects present, such as turbulence or friction, that convert mechanical energy to thermal energy. In reality, such effects are impossible to eliminate completely, and hence it is not surprising that real processes in nature are irreversible.

22.3 THE CARNOT ENGINE

In 1824 a French engineer named Sadi Carnot (1796–1832) described a theoretical engine, now called a **Carnot engine,** that is of great importance from both practical and theoretical viewpoints. He showed that a heat engine operating in an ideal, reversible cycle—called a **Carnot cycle**—between two heat reservoirs is the most efficient engine possible. Such an ideal engine establishes an upper limit on the efficiencies of all engines. That is, the net work done by a working substance taken through the Carnot cycle is the greatest amount of work possible for a given amount of thermal energy supplied to the substance at the upper temperature. **Carnot's theorem** can be stated as follows:

> No real heat engine operating between two heat reservoirs can be more efficient than a Carnot engine operating between the same two reservoirs.

Let us briefly describe some aspects of this theorem. First, we assume that the second law is valid. Next, imagine two heat engines operating between the same heat reservoirs, one which is a Carnot engine with efficiency e_c, and the other whose efficiency e is greater than e_c. If the more efficient engine is used to drive the Carnot engine as a refrigerator, the net result is a transfer of heat from the cold to the hot reservoir. According to the second law, this is impossible. Hence, the assumption that $e > e_c$ must be false.

Sadi Carnot, a French physicist, was the first to show the quantitative relationship between work and heat. Carnot was born in Paris on June 1, 1796, and was educated at the École Polytechnique in Paris and at the École Genie in Metz. His interests included mathematics, tax reform, industrial development, and the fine arts.

In 1824 he published his only work —*Reflections on the Motive Power of Heat*—which reviewed the industrial, political, and economic importance of the steam engine. In it he defined work as "weight lifted through a height." Carnot began to study the physical properties of gases in 1831, particularly the relationship between temperature and pressure.

On August 24, 1832, he died suddenly of cholera. In accordance with the custom of his time, all of his personal effects were burned, but some of his notes fortunately escaped destruction. Carnot's notes led Lord Kelvin to confirm and extend the science of thermodynamics in 1850.

Sadi Carnot

| 1 7 9 6 – 1 8 3 2 | (FPG)

To describe the Carnot cycle, we assume that the substance working between temperatures T_c and T_h is an ideal gas contained in a cylinder with a movable piston at one end. The cylinder walls and the piston are thermally nonconducting. Four stages of the Carnot cycle are shown in Figure 22.8, and the PV diagram for the cycle is shown in Figure 22.9. The Carnot cycle consists of two adiabatic and two isothermal processes, all reversible.

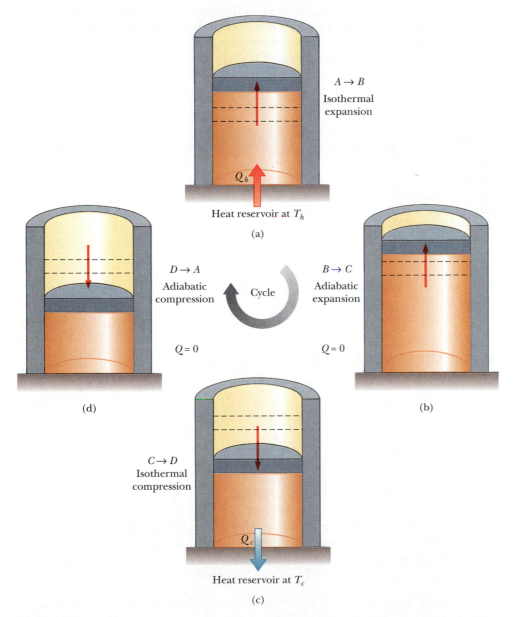

$A \rightarrow B$
Isothermal expansion

Q_h

Heat reservoir at T_h

(a)

$D \rightarrow A$
Adiabatic compression

Cycle

$B \rightarrow C$
Adiabatic expansion

$Q = 0$

$Q = 0$

(d)

(b)

$C \rightarrow D$
Isothermal compression

Q_c

Heat reservoir at T_c

(c)

FIGURE 22.8 The Carnot cycle. In process $A \rightarrow B$, the gas expands isothermally while in contact with a reservoir at T_h. In process $B \rightarrow C$, the gas expands adiabatically ($Q = 0$). In process $C \rightarrow D$, the gas is compressed isothermally while in contact with a reservoir at $T_c < T_h$. In process $D \rightarrow A$, the gas is compressed adiabatically. The upward arrows on the piston indicate weights being removed during the expansions, and the downward arrows indicate the addition of weights during the compressions.

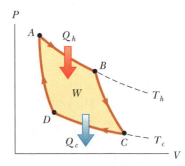

P

A Q_h

B

W

T_h

D

Q_c C T_c

V

FIGURE 22.9 The PV diagram for the Carnot cycle. The net work done, W, equals the net heat received in one cycle, $Q_b - Q_c$. Note that $\Delta U = 0$ for the cycle.

- The process $A \rightarrow B$ is an isothermal expansion at temperature T_h, in which the gas is placed in thermal contact with a heat reservoir at temperature T_h (Fig. 22.8a). During the expansion, the gas absorbs thermal energy Q_h from the reservoir through the base of the cylinder and does work W_{AB} in raising the piston.
- In the process $B \rightarrow C$, the base of the cylinder is replaced by a thermally nonconducting wall and the gas expands adiabatically; that is, no thermal energy enters or leaves the system (Fig. 22.8b). During the expansion, the temperature falls from T_h to T_c and the gas does work W_{BC} in raising the piston.
- In the process $C \rightarrow D$, the gas is placed in thermal contact with a heat reservoir at temperature T_c (Fig. 22.8c) and is compressed isothermally at temperature T_c. During this time, the gas expels thermal energy Q_c to the reservoir and the work done on the gas by an external agent is W_{CD}.
- In the final stage, $D \rightarrow A$, the base of the cylinder is replaced by a nonconducting wall (Fig. 22.8d) and the gas is compressed adiabatically. The temperature of the gas increases to T_h and the work done on the gas by an external agent is W_{DA}.

The net work done in this reversible, cyclic process is equal to the area enclosed by the path *ABCDA* of the *PV* diagram (Fig. 22.9). As we showed in Section 22.1, the net work done in one cycle equals the net heat transferred into the system, $Q_h - Q_c$, since the change in internal energy is zero. Hence, the thermal efficiency of the engine is given by Equation 22.2:

$$e = \frac{W}{Q_h} = 1 - \frac{Q_c}{Q_h}$$

In Example 22.2, we will show that for a Carnot cycle,

Ratio of heats for a Carnot cycle

$$\frac{Q_c}{Q_h} = \frac{T_c}{T_h} \tag{22.3}$$

Hence, the thermal efficiency of a Carnot engine is

Efficiency of a Carnot engine

$$e_c = 1 - \frac{T_c}{T_h} \tag{22.4}$$

According to this result,

all Carnot engines operating between the same two temperatures in a reversible manner have the same efficiency. From Carnot's theorem, the efficiency of any reversible engine operating in a cycle between two temperatures is greater than the efficiency of any irreversible (real) engine operating between the same two temperatures.

Equation 22.4 can be applied to any working substance operating in a Carnot cycle between two heat reservoirs. According to this result, the efficiency is zero if $T_c = T_h$, as one would expect. The efficiency increases as T_c is lowered and as T_h increases. However, the efficiency can be unity (100%) only if $T_c = 0$ K. Such reservoirs are not available, and so the maximum efficiency is always less than 100%. In most practical cases, the cold reservoir is near room temperature, about

300 K. Therefore, one usually strives to increase the efficiency by raising the temperature of the hot reservoir. *All real engines are less efficient than the Carnot engine since they are subject to such practical difficulties as friction and heat losses by conduction.*

EXAMPLE 22.2   Efficiency of the Carnot Engine

Show that the efficiency of a heat engine operating in a Carnot cycle using an ideal gas is given by Equation 22.4.

Solution   During the isothermal expansion, $A \rightarrow B$ (Fig. 22.8a), the temperature does not change and so the internal energy remains constant. The work done by the gas is given by Equation 20.12. According to the first law, the thermal energy absorbed, Q_h, equals the work done, so that

$$Q_h = W_{AB} = nRT_h \ln \frac{V_B}{V_A}$$

In a similar manner, the thermal energy rejected to the cold reservoir during the isothermal compression $C \rightarrow D$ is

$$Q_c = |W_{CD}| = nRT_c \ln \frac{V_C}{V_D}$$

Dividing these expressions, we find that

$$(1) \qquad \frac{Q_c}{Q_h} = \frac{T_c \ln(V_C/V_D)}{T_h \ln(V_B/V_A)}$$

We now show that the ratio of the logarithmic quantities is unity by obtaining a relation between the ratio of volumes.

For any quasi-static, adiabatic process, the pressure and volume are related by Equation 21.18:

$$PV^\gamma = \text{constant}$$

During any reversible, quasi-static process, the ideal gas must also obey the equation of state, $PV = nRT$. Substituting this into the above expression to eliminate the pressure, we find that

$$TV^{\gamma-1} = \text{constant}$$

Applying this result to the adiabatic processes $B \rightarrow C$ and $D \rightarrow A$, we find that

$$T_h V_B{}^{\gamma-1} = T_c V_C{}^{\gamma-1}$$
$$T_h V_A{}^{\gamma-1} = T_c V_D{}^{\gamma-1}$$

Dividing these equations, we obtain

$$(V_B/V_A)^{\gamma-1} = (V_C/V_D)^{\gamma-1}$$

$$(2) \qquad \frac{V_B}{V_A} = \frac{V_C}{V_D}$$

Substituting (2) into (1), we see that the logarithmic terms cancel and we obtain the relationship

$$\frac{Q_c}{Q_h} = \frac{T_c}{T_h}$$

Using this result and Equation 22.2, we see that the thermal efficiency of the Carnot engine is

$$e_c = 1 - \frac{Q_c}{Q_h} = 1 - \frac{T_c}{T_h} = \frac{T_h - T_c}{T_h}$$

EXAMPLE 22.3   The Steam Engine

A steam engine has a boiler that operates at 500 K. The heat changes water to steam, and this steam then drives the piston. The exhaust temperature is that of the outside air, approximately 300 K. What is the maximum thermal efficiency of this steam engine?

Solution   From the expression for the efficiency of a Carnot engine, we find the maximum thermal efficiency for any engine operating between these temperatures:

$$e_c = 1 - \frac{T_c}{T_h} = 1 - \frac{300 \text{ K}}{500 \text{ K}} = 0.4, \text{ or } \boxed{40\%}$$

You should note that this is the highest theoretical efficiency of the engine. In practice, the efficiency is considerably lower.

Exercise   Determine the maximum work the engine can perform in each cycle of operation if it absorbs 200 J of heat from the hot reservoir during each cycle.

Answer   80 J.

EXAMPLE 22.4   The Carnot Efficiency

The highest theoretical efficiency of a gasoline engine, based on the Carnot cycle, is 30%. If this engine expels its gases into the atmosphere, which has a temperature of 300 K, what is the temperature in the cylinder immediately after combustion?

Solution   The Carnot efficiency is used to find T_h:

$$e_c = 1 - \frac{T_c}{T_h}$$

$$T_h = \frac{T_c}{1 - e_c} = \frac{300 \text{ K}}{1 - 0.3} = \boxed{429 \text{ K}}$$

Exercise   If the heat engine absorbs 837 J of heat from the hot reservoir during each cycle, how much work can it perform in each cycle?

Answer   251 J.

22.4   THE ABSOLUTE TEMPERATURE SCALE

In Chapter 19 we defined temperature scales in terms of how certain physical properties of materials change as the temperature changes. It is desirable to define a temperature scale that is independent of material properties. The Carnot cycle provides us with the basis for such a temperature scale. Equation 22.3 tells us that the ratio Q_c/Q_h depends *only* on the temperatures of the two heat reservoirs. The ratio T_c/T_h can be obtained by operating a reversible heat engine in a Carnot cycle between these two temperatures and carefully measuring Q_c and Q_h. A temperature scale can be determined with reference to some fixed-point temperatures. The **absolute,** or **kelvin, temperature scale** is defined by choosing 273.16 K as the temperature of the triple point of water.

The temperature of any substance can be obtained in the following manner: (1) take the substance through a Carnot cycle; (2) measure the thermal energy Q absorbed or expelled by the system at some temperature T; and (3) measure the thermal energy Q_3 absorbed or expelled by the system when it is at the temperature of the triple point of water. From Equation 22.3 and this procedure, we find that the unknown temperature is

$$T = (273.16 \text{ K}) \frac{Q}{Q_3}$$

The absolute temperature scale is identical to the ideal gas temperature scale and is independent of the properties of the working substance. Therefore it can be applied even at very low temperatures.

In the previous section, we found that the thermal efficiency of any Carnot engine is given by $e_c = 1 - (T_c/T_h)$. This result shows that a 100% efficient engine is possible only if a temperature of absolute zero is maintained for T_c. If this were possible, any Carnot engine operating between T_h and $T_c = 0$ K would convert all of the absorbed thermal energy to work.[4] Using this idea, Lord Kelvin defined absolute zero as follows: *Absolute zero is the temperature of a reservoir at which a Carnot engine will expel no heat.*

---

[4] Experimentally, it is not possible to reach absolute zero. Temperatures as low as about 10^{-5} K have been achieved with enormous difficulties using a technique called *nuclear demagnetization*. Even lower temperatures have recently been obtained using lasers to stop the motion of atoms of a gas. The fact that absolute zero may be approached but never reached is a consequence of a law of nature known as the *third law of thermodynamics*.

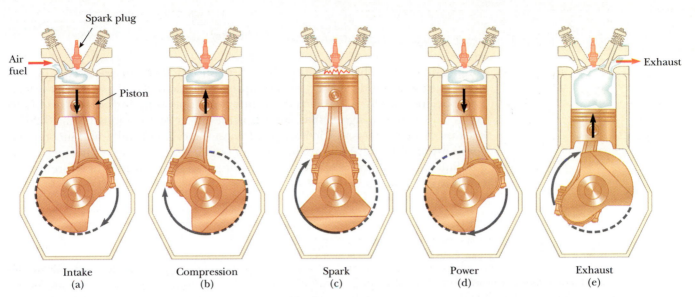

Intake	Compression	Spark	Power	Exhaust
(a)	(b)	(c)	(d)	(e)

FIGURE 22.10 The four-stroke cycle of a conventional gasoline engine. (a) In the intake stroke, air is mixed with fuel. (b) The intake valve is then closed, and the air-fuel mixture is compressed by the piston. (c) The mixture is ignited by the spark plug raising it to a higher temperature. (d) In the power stroke, the gas expands against the piston. (e) Finally, the residual gases are expelled and the cycle repeats.

22.5 THE GASOLINE ENGINE

In this section we discuss the efficiency of the common gasoline engine. Five successive processes occur in each cycle, as illustrated in Figure 22.10. These processes can be approximated by the **Otto cycle**, a *PV* diagram of which is illustrated in Figure 22.11:

- During the intake stroke $O \rightarrow A$, air is drawn into the cylinder at atmospheric pressure and the volume increases from V_2 to V_1.
- In the process $A \rightarrow B$ (compression stroke), the air-fuel mixture is compressed adiabatically from volume V_1 to volume V_2, and the temperature increases from T_A to T_B. The work done on the gas is the area under the curve *AB*.
- In the process $B \rightarrow C$, combustion occurs and thermal energy Q_h is added to the gas. This is not an inflow of thermal energy but rather a release of thermal energy from the combustion process. During this time the pressure and temperature rise rapidly, but the volume remains approximately constant. No work is done on the gas.
- In the process $C \rightarrow D$ (power stroke), the gas expands adiabatically from V_2 to V_1, causing the temperature to drop from T_C to T_D. The work done by the gas equals the area under the curve *CD*.
- In the process $D \rightarrow A$, thermal energy Q_c is extracted from the gas as its pressure decreases at constant volume when an exhaust valve is opened. No work is done during this process.
- In the final process of the exhaust stroke $A \rightarrow O$, the residual gases are exhausted at atmospheric pressure, and the volume decreases from V_1 to V_2. The cycle then repeats itself.

If the air-fuel mixture is assumed to be an ideal gas, the efficiency of the Otto cycle that is shown in Example 22.7 will be

Model of a Stirling engine. This device uses a "regenerator," which recycles some of the heat that would normally be lost back into the system; for this reason, the Stirling cycle has an efficiency equal to that of the Carnot cycle. *(Courtesy of Central Scientific Co.)*

$$e = 1 - \frac{1}{(V_1/V_2)^{\gamma-1}} \tag{22.5}$$

where γ is the ratio of the molar specific heats C_p/C_v and V_1/V_2 is called the **compression ratio.** This expression shows that the efficiency increases with increasing compression ratios. For a typical compression ratio of 8 and $\gamma = 1.4$, a theoretical efficiency of 56% is predicted for an engine operating in the idealized Otto cycle. This is much higher than what is achieved in real engines (15% to 20%) because of such effects as friction, heat loss to the cylinder walls, and incomplete combustion of the air-fuel mixture. Diesel engines have higher efficiencies than gasoline engines because of their higher compression ratios and therefore higher combustion temperatures.

CONCEPTUAL EXAMPLE 22.5

A diesel engine operates with a compression ratio of about 16. A gasoline engine has a compression ratio of about 8. Which engine runs hotter? Which engine (at least theoretically) can operate more efficiently?

Reasoning An internal-combustion engine takes in fuel and air at the environmental temperature and then compresses it. During the rapid (adiabatic) compression, work is delivered to the fuel-air mixture (the system). Heat has no time to leave the system, so the internal energy and temperature rise. The increase in temperature depends only on the nature of the molecules and the decrease in volume, or the so-called compression ratio. Because a diesel engine has a larger compression ratio, its fuel and air mixture will be hotter before and after combustion, compared with a gasoline engine with spark plugs. Its Carnot efficiency limit is set by the higher temperature at which it can take in heat from the chemical energy in the fuel, so the diesel engine can have higher efficiency.

CONCEPTUAL EXAMPLE 22.6

Electrical energy can be converted to thermal energy with an efficiency of 100%. Why is this number misleading with regard to heating a home? That is, what other factors must be considered in comparing the cost of electric heating with the cost of hot-air or hot-water heating?

Reasoning Most electric power plants produce their electricity by burning fossil fuels, with an efficiency of approximately 30%. When you pay your electric bill, you are indirectly paying for the oil or coal used to generate the electricity, in addition to the other services provided. In conventional hot-air or hot-water heating systems, oil or gas is used to directly heat the air or water. You must compare the lower cost of this direct heating with the cost of maintenance, initial installation, and so forth.

EXAMPLE 22.7   Efficiency of the Otto Cycle

Show that the thermal efficiency of an engine operating in an idealized Otto cycle (Fig. 22.11) is given by Equation 22.5. Treat the working substance as an ideal gas.

Solution First, let us calculate the work done by the gas during each cycle. No work is done during the processes $B \rightarrow C$ and $D \rightarrow A$. Work is done on the gas during the adiabatic compression $A \rightarrow B$, and work is done by the gas during the adiabatic expansion $C \rightarrow D$. The net work done equals the area bounded by the closed curve in Figure 22.11. Because the change in internal energy is zero for one cycle, we see from the first law that the net work done for each cycle equals the net thermal energy into the system:

$$W = Q_h - Q_c$$

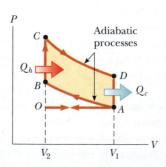

FIGURE 22.11 The *PV* diagram for the Otto cycle, which approximately represents the processes in the internal combustion engine.

Because the processes $B \rightarrow C$ and $D \rightarrow A$ take place at constant volume and since the gas is ideal, we find from the definition of specific heat that

$$Q_h = nC_v(T_C - T_B) \quad \text{and} \quad Q_c = nC_v(T_D - T_A)$$

Using these expressions together with Equation 22.2, we obtain for the thermal efficiency

$$(1) \qquad e = \frac{W}{Q_h} = 1 - \frac{Q_c}{Q_h} = 1 - \frac{T_D - T_A}{T_C - T_B}$$

We can simplify this expression by noting that the processes $A \rightarrow B$ and $C \rightarrow D$ are adiabatic and hence obey the relationship $TV^{\gamma-1} = $ constant. Using this condition and the facts that $V_A = V_D = V_1$ and $V_B = V_C = V_2$, we find

$$(2) \qquad \frac{T_D - T_A}{T_C - T_B} = \left[\frac{V_2}{V_1}\right]^{\gamma-1}$$

Substituting (2) into (1) gives for the thermal efficiency

$$(3) \qquad e = 1 - \frac{1}{(V_1/V_2)^{\gamma-1}}$$

This can also be expressed in terms of a ratio of temperatures by noting that since $T_A V_1^{\gamma-1} = T_B V_2^{\gamma-1}$, it follows that

$$\left[\frac{V_2}{V_1}\right]^{\gamma-1} = \frac{T_A}{T_B} = \frac{T_D}{T_C}$$

Therefore, (3) becomes

$$(4) \qquad e = 1 - \frac{T_A}{T_B} = 1 - \frac{T_D}{T_C}$$

During this cycle, the lowest temperature is T_A and the highest temperature is T_C. Therefore the efficiency of a Carnot engine operating between reservoirs at these two temperatures, given by $e_c = 1 - (T_A/T_C)$, would be *greater* than the efficiency of the Otto cycle, given by (4).

22.6   HEAT PUMPS AND REFRIGERATORS

A **heat pump** is a mechanical device that moves thermal energy from a region at lower temperature to a region at higher temperature. Heat pumps have long been popular for cooling and are now becoming increasingly popular for heating purposes as well. In the heating mode, a circulating fluid absorbs thermal energy from the outside and releases it to the interior of the building. The fluid is usually in the form of a low-pressure vapor when in the coils of a unit outside the building, where it absorbs heat from either the air or the ground. This gas is then compressed and enters the building as a hot, high-pressure vapor and enters the interior part of the unit, where it condenses to a liquid and releases its stored thermal energy. An air conditioner is simply a heat pump installed backward, with "exterior" and "interior" interchanged.

Figure 22.12 is a schematic representation of a heat pump. The outside temperature is T_c, the inside temperature is T_h, and the thermal energy absorbed by the circulating fluid is Q_c. The heat pump does work W on the fluid, and the thermal energy transferred from the pump into the building is Q_h.

The effectiveness of a heat pump, in its heating mode, is described in terms of a number called the **coefficient of performance, COP**. This is defined as the ratio of the heat transferred into the hot reservoir and the work required to transfer that heat:

$$\text{COP (heat pump)} \equiv \frac{\text{heat transferred}}{\text{work done by pump}} = \frac{Q_h}{W} \qquad (22.6)$$

If the outside temperature is 25°F or higher, the COP for a heat pump is about 4. That is, the heat transferred into the house is about four times greater than the work done by the motor in the heat pump. However, as the outside temperature decreases, it becomes more difficult for the heat pump to extract sufficient heat from the air and the COP drops. In fact, the COP can fall below unity for temperatures below the midteens.

A Carnot-cycle heat engine run in reverse constitutes a heat pump—in fact, the heat pump with the highest possible coefficient of performance for

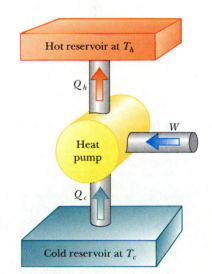

FIGURE 22.12   Schematic diagram of a heat pump, which absorbs thermal energy Q_c from the cold reservoir and expels thermal energy Q_h to the hot reservoir.

Rudolph Clausius (1822–1888). "I propose . . . to call *S* the entropy of a body, after the Greek word 'transformation.' I have designedly coined the word 'entropy' to be similar to energy, for these two quantities are analogous in their physical significance, that an analogy of denominations seems to be helpful." *(SPL/Photo Researchers)*

the temperatures between which it operates. The maximum coefficient of performance is

$$\text{COP}_c \text{ (heat pump)} = \frac{T_h}{T_h - T_c}$$

The refrigerator works much like a heat pump; it cools its interior by pumping thermal energy from the food storage compartments into the warmer air outside. During its operation, a refrigerator removes a quantity of thermal energy Q_c from the interior of the refrigerator, and in the process (like the heat pump) its motor does work W. The coefficient of performance of a refrigerator or of a heat pump is defined in terms of Q_c:

$$\text{COP (refrigerator)} = \frac{Q_c}{W} \tag{22.7}$$

An efficient refrigerator is one that removes the greatest amount of thermal energy from the cold reservoir for the least amount of work. Thus, a good refrigerator should have a high coefficient of performance, typically 5 or 6.

The highest possible coefficient of performance is again that of a refrigerator whose working substance is carried through a Carnot heat-engine cycle in reverse:

$$\text{COP}_c \text{ (refrigerator)} = \frac{T_c}{T_h - T_c}$$

As the difference between temperatures of the two reservoirs approaches zero, the theoretical coefficient of performance of a Carnot heat pump approaches infinity. In practice, the low temperature of the cooling coils and the high temperature at the compressor limit the COP to values below 10.

22.7   ENTROPY

The zeroth law of thermodynamics involves the concept of temperature, and the first law involves the concept of internal energy. Temperature and internal energy are both state functions; that is, they can be used to describe the thermodynamic state of a system. Another state function, this one related to the second law of thermodynamics, is the **entropy function,** *S*. In this section we define entropy on a macroscopic scale as it was first expressed by Rudolph Clausius in 1865.

Consider any infinitesimal process for a system between two equilibrium states. If dQ_r is the amount of thermal energy that would be transferred *if the system had followed a reversible path,* then the change in entropy dS, *regardless of the actual path followed,* is equal to this amount of thermal energy transferred along the reversible path divided by the absolute temperature of the system:

Clausius definition of change in entropy

$$dS = \frac{dQ_r}{T} \tag{22.8}$$

The subscript r on the term dQ_r, is a reminder that the thermal energy transfer is to be measured along a reversible path, even though the system may actually have followed some irreversible path. When thermal energy is absorbed by the system, dQ_r is positive and hence the entropy increases. When thermal energy is expelled by the system, dQ_r is negative and the entropy decreases. Note that Equation 22.8

defines not entropy, but rather the *change* in entropy. This is consistent with the fact that a change in state always accompanies heat transfer. Hence, the meaningful quantity in describing a process is the *change* in entropy.

Entropy originally found its place in thermodynamics, but its importance grew tremendously as the field of statistical mechanics developed because this method of analysis provided an alternative way of interpreting entropy. In statistical mechanics, the behavior of a substance is described in terms of the statistical behavior of the atoms and molecules contained in the substance. One of the main results of this treatment is that

isolated systems tend toward disorder, and entropy is a measure of this disorder.

For example, consider the molecules of a gas in the air in your room. If all the gas molecules moved together like soldiers marching in step, this would be a very ordered state. It is also an unlikely state. If you could see the molecules, you would see that they moved haphazardly in all directions, bumping into one another, changing speed upon collision, some going fast, some slowly. This is a highly disordered state. It is also highly possible that the actual state of the system is such a highly disordered state.

The cause of this perpetual drive toward disorder is easily seen. For any given energy of the system, only certain states are possible, or accessible. Among those states, it is assumed that all are equally probable. However, when such possible states are examined, it is found that far more of them are disordered states than ordered states. Because each of the states is equally probable, it is highly probable that the actual state will be one of the highly disordered states, or more precisely, that the actual state will be one in which the system moves between states of equivalent amounts of disorder. In the following discussion, therefore, the term *state* indicates a collection of states of equivalent amounts of disorder.

All physical processes tend toward more probable states for the system and its surroundings. The more probable state is always one of higher disorder. Because entropy is a measure of disorder, an alternative way of saying this is that

the entropy of the Universe increases in all processes.

This statement is yet another way of stating the second law of thermodynamics.

In judging the relative order or disorder of two states, it is important to examine the spatial order. For example, a perfect crystal has the highest possible spatial order. But one must also consider the order of velocities. At 0 K, all particles have the same speed (in classical physics), but as the temperature increases, the variation in speeds increases, corresponding to an increase in disorder and therefore in entropy. An isolated supercooled liquid may convert to a crystal (corresponding to a decrease in spatial entropy), but only if the temperature rises (corresponding to an increase in velocity entropy), for a net increase in entropy of the system.

To calculate the change in entropy for a finite process, we must recognize that T is generally not constant. If dQ_r is the thermal energy transferred when the system is at a temperature T, then the change in entropy in an arbitrary reversible process between an initial state and a final state is

Changes in entropy for a finite process

$$\Delta S = \int_i^f dS = \int_i^f \frac{dQ_r}{T} \qquad \text{(reversible path)} \qquad (22.9)$$

Because entropy is a state function, the change in entropy of a system in going from one state to another has the same value for *all* paths connecting the two states. That is, *the change in entropy of a system depends only on the properties of the initial and final equilibrium state.*

In the case of a *reversible, adiabatic* process, no thermal energy is transferred between the system and its surroundings, and therefore $\Delta S = 0$ in this case. Since there is no change in entropy, such a process is often referred to as an **isentropic process.**

Consider the changes in entropy that occur in a Carnot heat engine operating between the temperatures T_c and T_h. In one cycle, the engine absorbs thermal energy Q_h from the hot reservoir and rejects thermal energy Q_c to the cold reservoir. Thus, the total change in entropy for one cycle is

$$\Delta S = \frac{Q_h}{T_h} - \frac{Q_c}{T_c}$$

where the negative sign represents the fact that thermal energy Q_c is expelled by the system. In Example 22.2 we showed that for a Carnot cycle,

$$\frac{Q_c}{Q_h} = \frac{T_c}{T_h}$$

Using this result in the previous expression for ΔS, we find that the total change in entropy for a Carnot engine operating in a cycle is *zero:*

$$\Delta S = 0$$

Change in entropy for a Carnot cycle is zero

Now consider a system taken through an arbitrary cycle. Since the entropy function is a state function and hence depends only on the properties of a given equilibrium state, we conclude that $\Delta S = 0$ for *any* cycle. In general, we can write this condition in the mathematical form

$\Delta S = 0$ for any reversible cycle

$$\oint \frac{dQ_r}{T} = 0 \qquad (22.10)$$

where the symbol $\oint$ indicates that the integration is over a closed path.

Another important property of entropy is the fact that the entropy of the Universe is not changed by a reversible process. This can be understood by noting that two bodies A and B that interact with each other reversibly must always be in thermal equilibrium with each other. That is, their temperatures must always be equal. Therefore, when a small amount of thermal energy dQ is transferred from A to B, the increase in entropy of B is dQ/T, while the corresponding change in entropy of A is $-dQ/T$. Thus the total change in entropy of the system (A + B) is zero, and the entropy of the Universe is unaffected by the reversible process.

Quasi-Static, Reversible Process for an Ideal Gas

An ideal gas undergoes a quasi-static, reversible process from an initial state T_i, V_i to a final state T_f, V_f. Let us calculate the change in entropy for this process.

According to the first law, $dQ_r = dU + dW$, where $dW = P \, dV$. For an ideal gas, recall that $dU = nC_V \, dT$ and from the ideal gas law, we have $P = nRT/V$. Therefore, we can express the thermal energy transferred as

$$dQ_r = dU + P\,dV = nC_V\,dT + nRT\,\frac{dV}{V}$$

We cannot integrate this expression as it stands because the last term contains two variables, T and V. However, if we divide each term by T, we can integrate both terms on the right-hand side:

$$\frac{dQ_r}{T} = nC_V\,\frac{dT}{T} + nR\,\frac{dV}{V} \tag{22.11}$$

Assuming that C_V is constant over the interval in question, and integrating Equation 22.11 from T_i, V_i to T_f, V_f, we get

$$\Delta S = \int_i^f \frac{dQ_r}{T} = nC_V \ln\frac{T_f}{T_i} + nR \ln\frac{V_f}{V_i} \tag{22.12}$$

This expression shows that ΔS *depends only on the initial and final states and is independent of the reversible path.* Furthermore, ΔS can be positive or negative depending on whether the gas absorbs or expels thermal energy during the process. Finally, for a cyclic process ($T_i = T_f$ and $V_i = V_f$), we see that $\Delta S = 0$.

EXAMPLE 22.8   **Change in Entropy—Melting Process**

A solid substance that has a latent heat of fusion l_f melts at a temperature T_m. Calculate the change in entropy that occurs when m grams of this substance is melted.

$$\Delta S = \int \frac{dQ_r}{T} = \frac{1}{T_m}\int dQ = \frac{Q}{T_m} = \boxed{\frac{mL_f}{T_m}}$$

Note that we were able to remove T_m from the integral because the process is isothermal.

Solution   Let us assume that the melting occurs so slowly that it can be considered a reversible process. In that case the temperature can be regarded as constant and equal to T_m. Making use of Equations 22.9 and the fact that the latent heat of fusion $Q = mL_f$, (Eq. 20.6), we find that

Exercise   Calculate the change in entropy when 0.30 kg of lead melts at 327°C. Lead has a latent heat of fusion equal to 24.5 kJ/kg.

Answer $\Delta S = 12.3 \text{ J/K}$.

22.8 ENTROPY CHANGES IN IRREVERSIBLE PROCESSES

By definition, calculation of the change in entropy requires information about a reversible path connecting the initial and final equilibrium states. In order to calculate changes in entropy for real (irreversible) processes, we must first recognize that the entropy function (like internal energy) depends only on the *state* of the system. That is, entropy is a state function. Hence, the change in entropy when a system moves between any two equilibrium states depends only on the initial and final states. Experimentally it is found that the entropy change is the same for all processes that can occur between a given set of initial and final states. It is possible to show that if this were not the case, the second law of thermodynamics would be violated.

We now calculate entropy changes for irreversible processes between two equilibrium states by devising a reversible process (or series of reversible processes) between the same two states and computing $\int dQ_r/T$ for the reversible process. The entropy change for the irreversible process is the same as that of the reversible process between the same two equilibrium states. In irreversible processes, it is critically important to distinguish between Q, the actual thermal energy transfer in the process, and Q_r, the thermal energy that would have been transferred along a reversible path. It is only the second process that gives the correct value for the entropy change.

An illustration from Flammarion's novel *La Fin du Monde*, depicting the "heat death" of the Universe.

As we shall see in the following examples, the change in entropy for the system plus its surroundings is always positive for an irreversible process. In general, the total entropy (and disorder) always increases in an irreversible process. From these considerations, the second law of thermodynamics can be stated as follows: *The total entropy of an isolated system that undergoes a change cannot decrease.* Furthermore, if the process is *irreversible,* the total entropy of an isolated system always *increases.* On the other hand, in a reversible process, the total entropy of an isolated system remains constant.

When dealing with interacting objects that are not isolated from the environment, remember that the increase of entropy applies to the system *and* its surroundings. When two objects interact in an irreversible process, the increase in entropy of one part of the system is greater than the decrease in entropy of the other part. Hence, we conclude that the change in entropy of the Universe must be greater than zero for an irreversible process and equal to zero for a reversible process. Ultimately, the entropy of the Universe should reach a maximum value. At this point the Universe will be in a state of uniform temperature and density. All physical, chemical, and biological processes will cease, since a state of perfect disorder implies that no energy is available for doing work. This gloomy state of affairs is sometimes referred to as the heat death of the Universe.

CONCEPTUAL EXAMPLE 22.9

A living system, such as a tree, combines unorganized molecules (CO_2, H_2O), using sunlight, to produce leaves and branches. Is this reduction of entropy in the tree a violation of the second law of thermodynamics?

Reasoning   Living things survive on the flow of solar radiation from Sun to Earth. A snake basking in the Sun diverts thermal energy temporarily, before the energy is finally radiated into outer space. A tree builds organized cellulose molecules and humans construct buildings, all out of a stream of energy from the Sun. The second law is not violated, because the local reductions in entropy are created at the expense of increase in the total entropy of the Universe.

Heat Conduction

Consider the reversible transfer of thermal energy Q from a hot reservoir at temperature T_h to a cold reservoir at temperature T_c. Since the cold reservoir absorbs thermal energy Q, its entropy increases by Q/T_c. At the same time, the hot reservoir loses thermal energy Q and its entropy decreases by Q/T_h. The increase in entropy of the cold reservoir is greater than the decrease in entropy of the hot reservoir since T_c is less than T_h. Therefore, the total change in entropy of the system (and of the Universe) is greater than zero:

$$\Delta S_u = \frac{Q}{T_c} - \frac{Q}{T_h} > 0$$

EXAMPLE 22.10   Which Way Does the Heat Flow?

A large cold object is at 273 K, and a large hot object is at 373 K. Show that it is impossible for a small amount of thermal energy, say 8.00 J, to be transferred from the cold object to the hot one without decreasing the entropy of the Universe.

Reasoning   We assume that, during the heat transfer, the two objects do not undergo a temperature change. This is not a necessary assumption; it is used to avoid reliance

on the techniques of integral calculus. The process as described is irreversible, so we must find an equivalent reversible process. It is sufficient to assume that the hot and cold objects are connected by a poor thermal conductor whose temperature spans the range from 273 K to 373 K, which transfers thermal energy but whose state does not change during the process. Then, the thermal energy transfer to or from each object is reversible, and we may set $Q = Q_r$.

Solution   The entropy change of the hot object is

$$\Delta S_h = \frac{Q}{T_h} = \frac{8.00 \text{ J}}{373 \text{ K}} = 0.0214 \text{ J/K}$$

The cold object loses thermal energy, and its entropy change is

$$\Delta S_c = \frac{Q}{T_c} = \frac{-8.00 \text{ J}}{273 \text{ K}} = -0.0293 \text{ J/K}$$

The net entropy change of the Universe is

$$\Delta S_u = \Delta S_c + \Delta S_h = -0.0079 \text{ J/K}$$

This is in violation of the concept that the entropy of the Universe always increases in natural processes. That is, *the spontaneous transfer of thermal energy from a cold to a hot object cannot occur.*

Exercise   In the preceding example, suppose that 8.00 J of thermal energy is transferred from the hot to the cold object. What is the net entropy change of the Universe?

Answer $+0.0079$ J/K.

Free Expansion

An ideal gas in an insulated container initially occupies a volume V_i (Fig. 22.13). A partition separating the gas from an evacuated region is suddenly broken so that the gas expands (irreversibly) to a volume V_f. Let us find the change in entropy of the gas and the Universe.

The process is clearly neither reversible nor quasi-static. The work done by the gas against the vacuum is zero, and since the walls are insulating, no thermal energy is transferred during the expansion. That is, $W = 0$ and $Q = 0$. Using the first law, we see that the change in internal energy is zero, therefore $U_i = U_f$. Since the gas is ideal, U depends on temperature only, so we conclude that $T_i = T_f$.

To apply Equation 22.9, we must find Q_r; that is, we must find an equivalent reversible path that shares the same initial and final states. A simple choice is an isothermal, reversible expansion in which the gas pushes slowly against a piston. Since T is constant in this process, Equation 22.9 gives

$$\Delta S = \int \frac{dQ_r}{T} = \frac{1}{T}\int_i^f dQ_r$$

But $\int dQ_r$ is simply the work done by the gas during the isothermal expansion from V_i to V_f, which is given by Equation 20.12. Using this result, we find that

$$\Delta S = nR \ln \frac{V_f}{V_i} \tag{22.13}$$

Since $V_f > V_i$, we conclude that ΔS is positive, and this positive result tells us that both the entropy and the disorder of the gas (and the Universe) increase as a result of the irreversible, adiabatic expansion.

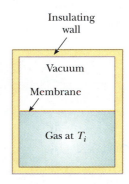

FIGURE 22.13   Free expansion of a gas. When the partition separating the gas from the evacuated region is ruptured, the gas expands freely and irreversibly so that it occupies a greater final volume. The container is thermally insulated from its surroundings, so $Q = 0$.

EXAMPLE 22.11   Free Expansion of a Gas

Calculate the change in entropy of 2.00 mol of an ideal gas that undergoes a free expansion to three times its initial volume.

Solution   Using Equation 22.13 with $n = 2.00$ and $V_f = 3V_i$, we find that

$$\Delta S = nR \ln \frac{V_f}{V_i} = (2.00 \text{ mol})(8.31 \text{ J/mol}\cdot\text{K}) \ln 3$$

$$= 18.3 \text{ J/K}$$

Irreversible Heat Transfer

A substance of mass m_1, specific heat c_1, and initial temperature T_1 is placed in thermal contact with a second substance of mass m_2, specific heat c_2, and initial temperature T_2, where $T_2 > T_1$. The two substances are contained in an insulated box so no heat is lost to the surroundings. The system is allowed to reach thermal equilibrium. What is the total entropy change for the system?

First, let us calculate the final equilibrium temperature, T_f. Energy conservation requires that the heat lost by one substance equals the heat gained by the other. Since by definition, $Q = mc\,\Delta T$ for each substance, we get $Q_1 = -Q_2$, or

$$m_1 c_1 \,\Delta T_1 = -m_2 c_2 \,\Delta T_2$$

$$m_1 c_1 (T_f - T_1) = -m_2 c_2 (T_f - T_2)$$

Solving for T_f gives

$$T_f = \frac{m_1 c_1 T_1 + m_2 c_2 T_2}{m_1 c_1 + m_2 c_2} \tag{22.14}$$

Note that $T_1 < T_f < T_2$, as expected.

The process is irreversible because the system goes through a series of non-equilibrium states. During such a transformation, the temperature at any time is not well defined. However, we can imagine that the hot substance at the initial temperature T_i is slowly cooled to the temperature T_f by placing it in contact with a series of reservoirs differing infinitesimally in temperature, the first reservoir being at T_i and the last at T_f. Such a series of very small changes in temperature would approximate a reversible process. Applying Equation 22.9 and noting that $dQ = mc\,dT$ for an infinitesimal change, we get

$$\Delta S = \int_1 \frac{dQ_1}{T} + \int_2 \frac{dQ_2}{T} = m_1 c_1 \int_{T_1}^{T_f} \frac{dT}{T} + m_2 c_2 \int_{T_2}^{T_f} \frac{dT}{T}$$

where we have assumed that the specific heats remain constant. Integrating, we find

Change in entropy for a heat transfer process

$$\Delta S = m_1 c_1 \ln \frac{T_f}{T_1} + m_2 c_2 \ln \frac{T_f}{T_2} \tag{22.15}$$

where T_f is given by Equation 22.14. If Equation 22.14 is substituted into Equation 22.15, we can show that one of the terms in Equation 22.15 is always positive and the other always negative. (You may want to verify this for yourself.) However, the positive term is always larger than the negative term, resulting in a positive value for ΔS. Thus, we conclude that the entropy of the Universe increases in this irreversible process.

Finally, you should note that Equation 22.15 is valid only when no mixing occurs. If the substances are liquids or gases and mixing occurs, the result applies only if the two fluids are identical, as in the following example.

EXAMPLE 22.12   Calculating ΔS for a Mixing Process

Suppose 1.00 kg of water at 0°C is mixed with an equal mass of water at 100°C. After equilibrium is reached, the mixture has a uniform temperature of 50°C. What is the change in entropy of the system?

Solution   The change in entropy can be calculated from Equation 22.15 using the values $m_1 = m_2 = 1.00$ kg, $c_1 =$

$c_2 = 4186$ J/kg·K, $T_1 = 0$°C $(= 273$ K$)$, $T_2 = 100$°C $(= 373$ K$)$, and $T_f = 50$°C $(= 323$ K$)$.

$$\Delta S = m_1 c_1 \ln \frac{T_f}{T_1} + m_2 c_2 \ln \frac{T_f}{T_2}$$

$$= (1.00 \text{ kg})(4186 \text{ J/kg·K}) \ln\left(\frac{323 \text{ K}}{273 \text{ K}}\right)$$

$$= + (1.00 \text{ kg})(4186 \text{ J/kg·K}) \ln\left(\frac{323 \text{ K}}{373 \text{ K}}\right)$$

$$= 704 \text{ J/K} - 602 \text{ J/K} = \boxed{102 \text{ J/K}}$$

That is, as a result of this irreversible process, the increase in entropy of the cold water is greater than the decrease in entropy of the warm water. Consequently, the increase in entropy of the system is 102 J/K.

*22.9 ENTROPY ON A MICROSCOPIC SCALE[5]

As we have seen, entropy can be approached via macroscopic concepts, using parameters such as pressure and temperature. Entropy can also be treated from a microscopic viewpoint through statistical analysis of molecular motions. We shall use a microscopic model to investigate the free expansion of an ideal gas discussed in the preceding section.

In the kinetic theory of gases, gas molecules are taken to be particles moving in a random fashion. Suppose the gas is confined to the volume V_i (Fig. 22.14a). When the partition separating V_i from the larger container is removed, the molecules eventually are distributed in some fashion throughout the greater volume, V_f (Fig. 22.14b). The exact nature of the distribution is a matter of probability.

Such a probability can be determined by first finding the probabilities for the variety of molecular locations involved in the free expansion process. The instant after the partition is removed (and before the molecules have had a chance to rush into the other half of the container), all the molecules are in the initial volume. Let us estimate the probability of the molecules arriving at a particular configuration through natural random motions into some larger volume V. Assume each molecule occupies some microscopic volume V_m. Then the total number of possible locations of that molecule, in a macroscopic initial volume V_i, is the ratio $W_i = (V_i/V_m)$, which is a huge number. We assume each of these locations is equally probable. (W_i is not to be confused with work.)

As more molecules are added to the system, the number of possible states multiply together. Neglecting the very small probability of two molecules trying to occupy the same location, each molecule may go into any of the (V_i/V_m) locations, so the number of possibilities is $W_i^N = (V_i/V_m)^N$. Although a large number of possible states would be considered unusual, the total number of states is so enormous that these unusual states have a negligible probability of occurring. Hence, the number of equivalent states is proportional to $(V_i/V_m)^N$. Similarly, if the volume is increased to V_f, the number of equivalent states increases to $W_f^N = (V_f/V_m)^N$. Hence, the probabilities are $P_i = cW_i^N$ and $P_f = cW_f^N$, where the constant, c, has been left undetermined. The ratio of these probabilities is

$$\frac{W_f^N}{W_i^N} = \frac{(V_f/V_m)^N}{(V_i/V_m)^N} = \left(\frac{V_f}{V_i}\right)^N$$

If we now take the natural logarithm of this equation and multiply by Boltzmann's constant, we find

$$N k_B \ln\left(\frac{W_f}{W_i}\right) = n N_A k_B \ln\left(\frac{V_f}{V_i}\right)$$

where we write the number of molecules, N, as $n N_A$, the number of moles times Avogadro's number (Section 19.5). Furthermore, we know that $N_A k_B$ is the uni-

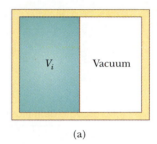

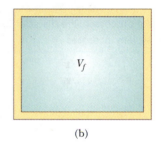

(b)

FIGURE 22.14 In a free expansion, the gas is allowed to expand into a larger volume that was previously a vacuum.

A full house is a very good hand in the game of poker. Can you calculate the probability of being dealt this hand in a standard deck of 52 cards? *(Tom Mareschel, The IMAGE Bank)*

[5] This section was adapted from A. Hudson and R. Nelson, *University Physics*, Philadelphia, Saunders College Publishing, 1990, used with permission of the publisher.

versal gas constant, R, so this equation may be written as

$$Nk_B \ln W_f - Nk_B \ln W_i = nR \ln\left(\frac{V_f}{V_i}\right) \tag{22.16}$$

From thermodynamic considerations (specifically, Eq. 22.13) we know that when n mol of a gas undergoes a free expansion from V_i to V_f, the change in entropy is

$$S_f - S_i = nR \ln\left(\frac{V_f}{V_i}\right) \tag{22.17}$$

Note that the right-hand sides of Equations 22.16 and 22.17 are identical. We thus make the following important connection between *entropy* and *probability*.

Entropy (microscopic definition)

$$S \equiv Nk_B \ln W \tag{22.18}$$

Although our discussion used the specific example of the free expansion of an ideal gas, a more rigorous development of the statistical interpretation of entropy leads to the same conclusion. *Entropy is a measure of microscopic disorder.*

In real processes, the disorder in the system increases

Imagine that the container of gas in Figure 22.15a has molecules with speeds above the mean value on the left side, and molecules with lower than the mean value on the right side (an ordered arrangement). Compare this with an even mixture of fast- and slow-moving molecules as in Figure 22.15b (a disordered arrangement). You might expect the ordered arrangement to be very unlikely because random motions tend to mix the slow- and fast-moving molecules uniformly. Yet each of these arrangements, *individually,* is equally probable. However, there are far more disordered arrangements than ordered arrangements, so *collectively,* the ordered arrangements are much less probable.

Let us estimate this probability for 100 molecules. The chance of any one molecule being in the left part of the container as a result of random motion is $1/2$. If the molecules move independently, the probability of 50 faster molecules being found in the left part at any instant is $(1/2)^{50}$. Likewise, the probability of the remaining 50 slower molecules being found in the right part at any instant is $(1/2)^{50}$. Therefore, the probability of finding this fast-slow separation through random motion is the product $(1/2)^{50}(1/2)^{50} = (1/2)^{100}$, which corresponds to about 1 chance in 10^{30}. When this calculation is extrapolated from 100 molecules to 1 mol of gas (6.02×10^{23} molecules), the ordered arrangement is found to be *extremely* improbable!

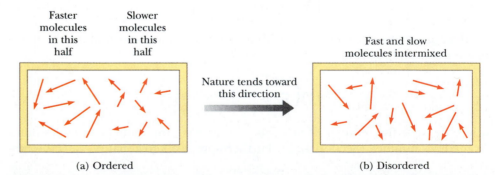

Faster molecules in this half   Slower molecules in this half

Nature tends toward this direction

Fast and slow molecules intermixed

(a) Ordered   (b) Disordered

FIGURE 22.15   A box of gas in two equally probable states of molecular motion that may exist. (a) An ordered arrangement; one of a few, and therefore a collectively unlikely set. (b) A disordered arrangement; one of many, and therefore a collectively likely set.

EXAMPLE 22.13 Free Expansion of an Ideal Gas—Revisited

Again consider the free expansion of an ideal gas. Let us verify that the macroscopic and microscopic approaches lead to the same conclusion. Suppose that 1 mol of gas undergoes a free expansion to four times its initial volume. The initial and final temperatures are, of course, the same. (a) Using a macroscopic approach, calculate the entropy change. (b) Find the probability, P_i, that all molecules will, through random motions, be found simultaneously in the original volume. (c) Using the probability considerations of part (b), calculate the change in entropy for the free expansion and show that it agrees with part (a).

Solution (a) From Equation 22.13, we obtain

$$\Delta S = nR \ln\left(\frac{V_f}{V_i}\right) = (1) R \ln\left(\frac{4V_i}{V_i}\right) = \boxed{R \ln 4}$$

(b) The number of states available to a single molecule in the initial volume V_i is $W_i = (V_i/V_m)$. For one mole (N_A molecules), the number of available states is

$$P_i = cW_i{}^{N_A} = \boxed{c\left(\frac{V_i}{V_m}\right)^{N_A}}$$

(c) The number of states for all N_A molecules in the volume $V_f = 4V_i$ is

$$P_f = cW_f{}^{N_A} = c\left(\frac{V_f}{V_m}\right)^{N_A} = c\left(\frac{4V_i}{V_m}\right)^{N_A}$$

From Equation 22.18 we obtain

$$\Delta S = k_B \ln W_f{}^{N_A} - k_B \ln W_i{}^{N_A} = k_B \ln\left(\frac{W_f{}^{N_A}}{W_i{}^{N_A}}\right) = k_B \ln(4^{N_A})$$

$$= N_A k_B \ln 4 = \boxed{R \ln 4}$$

The answer is the same as that in part (a), which dealt with microscopic parameters.

CONCEPTUAL EXAMPLE 22.14

Suppose you have a bag of 100 marbles, of which 50 are red and 50 are green. You are allowed to draw four marbles from the bag according to the following rules: Draw one marble, record its color, return it to the bag, and draw another. Continue this process until four marbles have been drawn. What are the possible outcomes for this set of events?

Reasoning Because each marble is returned to the bag before the next one is drawn, the probability of drawing a red marble is always the same as that of drawing a green one. All the possible outcomes are shown in Table 22.1. As this table indicates, there is only one way to draw four red marbles. However, there are four possible sequences that could produce one green and three red marbles, six sequences that could produce two green and two red, four sequences that could produce three green and one red, and one sequence that could produce all green. The most likely outcome — two red and two green — corresponds to the most disordered

TABLE 22.1 Possible Results of Drawing Four Marbles from a Bag

End Result	Possible Draws	Total Number of Same Results
All R	RRRR	1
1G, 3R	RRRG, RRGR, RGRR, GRRR	4
2G, 2R	RRGG, RGRG, GRRG, RGGR, GRGR, GGRR	6
3G, 1R	GGGR, GGRG, GRGG, RGGG	4
All G	GGGG	1

state. The probability that you would draw four red or four green marbles, these being the most ordered states, is much lower.

SUMMARY

A **heat engine** is a device that converts thermal energy to other useful forms of energy. The net work done by a heat engine in carrying a substance through a cyclic process ($\Delta U = 0$) is

$$W = Q_h - Q_c \tag{22.1}$$

where Q_h is the thermal energy absorbed from a hot reservoir and Q_c is the thermal energy rejected to a cold reservoir.

The **thermal efficiency**, e, of a heat engine is

$$e = \frac{W}{Q_h} = 1 - \frac{Q_c}{Q_h} \qquad (22.2)$$

The **second law of thermodynamics** can be stated in many ways:

- It is impossible to construct a heat engine that, operating in a cycle, produces no effect other than the absorption of thermal energy from a reservoir and the performance of an equal amount of work (Kelvin-Planck statement).
- It is impossible to construct a cyclical machine whose sole effect is to transfer heat continuously from one body to another body at a higher temperature (Clausius statement).

A **reversible cyclic process** is one that can be performed so that, at its conclusion, both the system and its surroundings have been returned to their exact initial conditions. A process that does not satisfy these requirements is **irreversible.**

Carnot's theorem states that no real heat engine operating (irreversibly) between the temperatures T_c and T_h can be more efficient than an engine operating reversibly in a Carnot cycle between the same two temperatures.

The *efficiency of a heat engine* operating in the **Carnot cycle** is

$$e_c = 1 - \frac{T_c}{T_h} \qquad (22.4)$$

The second law of thermodynamics states that when real (irreversible) processes occur, the degree of disorder in the system plus the surroundings increases. When a process occurs in an isolated system, an ordered state is converted into a more disordered state. The measure of disorder in a system is called **entropy,** S.

The **change in entropy,** dS, of a system moving between two equilibrium states is

$$dS = \frac{dQ_r}{T} \qquad (22.8)$$

The change in entropy of a system in an arbitrary reversible process moving between an initial state and a final state is

$$\Delta S = \int_i^f \frac{dQ_r}{T} \qquad (22.9)$$

The value of ΔS is the same for all paths connecting the initial and final states. The change in entropy for any reversible, cyclic process is zero, and when such a process occurs, the entropy of the Universe remains constant.

Entropy is a state function; that is, it depends on the state of the system. The change in entropy for a system undergoing a real (irreversible) process between two equilibrium states is the same as that of a reversible process between the same states.

In an irreversible process, the total entropy of an isolated system always increases. In general, the total entropy (and disorder) always increases in any irreversible process. Furthermore, the change in entropy of the Universe is greater than zero for an irreversible process.

From a microscopic viewpoint, **entropy,** S, is defined as

$$S \equiv Nk_B \ln W \qquad (22.18)$$

where k_B is Boltzmann's constant and W is the number of (microscopic) states available to the system in its (macroscopic) state. Because of the statistical tendency of systems to proceed toward states of greater probability and greater disorder, all natural processes are irreversible and entropy increases. Thus, *entropy is a measure of microscopic disorder.*

QUESTIONS

1. Distinguish clearly among temperature, heat (Q) and internal energy.

2. When a sealed Thermos bottle full of hot coffee is shaken, what are the changes, if any, in (a) the temperature of the coffee and (b) its internal energy?

3. Use the first law of thermodynamics to explain why the total energy of an isolated system is always constant.

4. Is it possible to convert internal energy to mechanical energy?

5. What are some factors that affect the efficiency of automobile engines?

6. In practical heat engines, which do we have more control of, the temperature of the hot reservoir or the temperature of the cold reservoir? Explain.

7. A steam-driven turbine is one major component of an electric power plant. Why is it advantageous to have the temperature of the steam as high as possible?

8. Is it possible to construct a heat engine that creates no thermal pollution?

9. Discuss three common examples of natural processes that involve an increase in entropy. Be sure to account for all parts of each system under consideration.

10. Discuss the change in entropy of a gas that expands (a) at constant temperature and (b) adiabatically.

11. In solar ponds constructed in Israel, the Sun's energy is concentrated near the bottom of a salty pond. With the proper layering of salt in the water, convection is prevented, and temperatures of 100°C may be reached. Can you guess the maximum efficiency with which useful energy can be extracted from the pond?

12. The vortex tube (Fig. 22.16) is a T-shaped device that takes in compressed air at 20 atm and 20°C and produces cold air at −20°C out one flared end and 60°C hot air out the other flared end. Does the operation of this device violate the second law of thermodynamics?

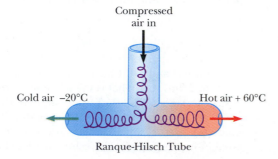

Compressed air in

Cold air −20°C   Hot air + 60°C

Ranque-Hilsch Tube

FIGURE 22.16   (Question 12).

13. Why does your automobile burn more gas in winter than in summer?

14. Can a heat pump have a coefficient of performance less than unity? Explain.

15. Give some examples of irreversible processes that occur in nature.

16. Give an example of a process in nature that is nearly reversible.

17. A thermodynamic process occurs in which the entropy of a system changes by −8.0 J/K. According to the second law of thermodynamics, what can you conclude about the entropy change of the environment?

18. If a supersaturated sugar solution is allowed to slowly evaporate, sugar crystals form in the container. Hence, sugar molecules go from a disordered form (in solution) to a highly ordered crystalline form. Does this process violate the second law of thermodynamics? Explain.

19. How could you increase the entropy of 1 mol of a metal that is at room temperature? How could you decrease its entropy?

20. A heat pump is to be installed in a region where the average outdoor temperature in the winter months is −20°C. In view of this, why would it be advisable to place the outdoor compressor unit deep in the ground? Why are heat pumps not commonly used for heating in cold climates?

21. Suppose your roommate is ''Mr. Clean'' and tidies up your messy room after a big party. Since more order is being created by your roommate, does this represent a violation of the second law of thermodynamics?

22. Discuss the entropy changes that occur when you (a) bake a loaf of bread and (b) consume the bread.

23. The device in Figure 22.17, called a thermoelectric converter, uses a series of semiconductor cells to convert thermal energy to electrical energy. In the photograph at the left, both legs of the device are at the same temperature, and no electrical energy is produced. However, when one leg is at a higher temperature than the other, as in the photograph on the right, electrical energy is produced as the device extracts energy from the hot reservoir and drives a small electric motor. (a) Why does the temperature differential produce electrical energy in this demonstration? (b) In what sense does this intriguing experiment demonstrate the second law of thermodynamics?

FIGURE 22.17   (Question 23). (*Courtesy of PASCO Scientific Co.*)

PROBLEMS

Review Problem

One mole of a monatomic ideal gas is at an initial pressure P and an initial volume V. The gas is taken through the cycle described in the PV diagram. Find (a) the heat absorbed during the cycle, (b) the heat rejected during the cycle, (c) the work done by the gas during the cycle, (d) the efficiency of the cycle, (e) the minimum temperature of the gas during the cycle, (f) the maximum temperature of the gas during the cycle, (g) the efficiency of a Carnot engine operating between the minimum and maximum temperatures, (h) the coefficient of performance of a Carnot refrigerator operating between the minimum and maximum temperatures, (i) the maximum work that can be done by a Carnot engine that absorbs the heat found in (a), and (j) the change in entropy during the process 1–2.

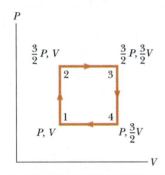

Section 22.1   Heat Engines and the Second Law of Thermodynamics

1. A heat engine absorbs 360 J of thermal energy and performs 25 J of work in each cycle. Find (a) the efficiency of the engine and (b) the thermal energy expelled in each cycle.

2. A heat engine performs 200 J of work in each cycle and has an efficiency of 30%. For each cycle, how much thermal energy is (a) absorbed and (b) expelled?

3. The heat absorbed by an engine is three times greater than the work it performs. (a) What is its thermal efficiency? (b) What fraction of the heat absorbed is expelled to the cold reservoir?

4. Determine the change in the internal energy of a system that (a) absorbs 500 cal of thermal energy while doing 800 J of external work, (b) absorbs 500 cal of thermal energy while 500 J of external work is done on the system, and (c) is maintained at a constant volume while 1000 cal is removed from the system.

5. An ideal gas is compressed to half its original volume while its temperature is held constant. (a) If 1000 J of energy is removed from the gas during the compression, how much work is done on the gas? (b) What is the change in the internal energy of the gas during the compression?

6. A particular engine has a power output of 5.0 kW and an efficiency of 25%. If the engine expels 8000 J of thermal energy in each cycle, find (a) the heat absorbed in each cycle and (b) the time for each cycle.

7. An engine absorbs 1600 J from a hot reservoir and expels 1000 J to a cold reservoir in each cycle. (a) What is the efficiency of the engine? (b) How much work is done in each cycle? (c) What is the power output of the engine if each cycle lasts for 0.30 s?

Section 22.3   The Carnot Engine

8. A heat engine operates between two reservoirs at 20°C and 300°C. What is the maximum efficiency possible for this engine?

9. A power plant operates at a 32% efficiency during the summer when the sea water for cooling is at 20°C. The plant uses 350°C steam to drive turbines. Assuming that the plant's efficiency changes in the same proportion as the ideal efficiency, what is the plant's efficiency in the winter when the sea water is at 10°C?

10. A Carnot engine has a power output of 150 kW. The engine operates between two reservoirs at 20°C and 500°C. (a) How much thermal energy is absorbed per hour? (b) How much thermal energy is lost per hour?

11. A power plant has been proposed that would make use of the temperature gradient in the ocean. The system is to operate between 20°C (surface water temperature) and 5°C (water temperature at a depth of about 1 km). (a) What is the maximum efficiency of such a system? (b) If the power output of the plant is 75 MW, how much thermal energy is absorbed per hour? (c) What compensating factor made this proposal of interest despite the value calculated in part (a)?

12. A heat engine operates in a Carnot cycle between 80°C and 350°C. It absorbs 2.0×10^4 J of thermal energy per cycle from the hot reservoir. The duration of each cycle is 1.0 s. (a) What is the maximum power

☐ indicates problems that have full solutions available in the Student Solutions Manual and Study Guide.

output of this engine? (b) How much thermal energy does it expel in each cycle?

13. One of the most efficient engines ever built (42%) operates between 430°C and 1870°C. (a) What is its maximum theoretical efficiency? (b) How much power does the engine deliver if it absorbs 1.4×10^5 J of thermal energy each second?

14. Steam enters a turbine at 800°C and is exhausted at 120°C. What is the maximum efficiency of this turbine?

15. The efficiency of a 1000-MW nuclear power plant is 33%; that is, 2000 MW of heat is rejected to the environment for every 1000 MW of electrical energy produced. If a river of flow rate 10^6 kg/s were used to transport the excess thermal energy away, what would be the average temperature increase of the river?

16. At point A in a Carnot cycle, 2.34 mol of a monatomic gas has a pressure of 1400 kPa, a volume of 10.0 liters, and a temperature of 720 K. It expands isothermally to point B, and then expands adiabatically to point C, where its volume is 24.0 liters. An isothermal compression brings it to point D, where its new volume is 15.0 liters. An adiabatic process returns the gas to point A. (a) Determine all the unknown pressures, volumes, and temperatures by filling in the following table.

	P	V	T
A	1400 kPa	10.0 liters	720 K
B			
C		24.0 liters	
D		15.0 liters	

(b) Find the heat added, work done, and the change in internal energy for each of the steps, AB, BC, CD, and DA. (c) Show that $W_{net}/Q_{in} = 1 - T_C/T_A$, the Carnot efficiency.

17. A steam engine is operated in a cold climate where the exhaust temperature is 0°C. (a) Calculate the theoretical maximum efficiency of the engine using an intake steam temperature of 100°C. (b) If, instead, superheated steam at 200°C is used, find the maximum possible efficiency.

18. A Carnot engine has an efficiency of 25% when the hot reservoir temperature is 500°C. If we want to improve the efficiency to 30%, what should be the temperature of the hot reservoir, assuming everything else remains unchanged?

19. A 20%-efficient engine is used to speed up a train from rest to 5.0 m/s. It is known that an ideal (Carnot) engine using the same cold and hot reservoirs would accelerate the same train from rest to a speed of 6.5 m/s using the same amount of fuel. If the engines use air at 300 K as a cold reservoir, find the temperature of the steam serving as the hot reservoir.

Section 22.5 The Gasoline Engine

20. A gasoline engine has a compression ratio of 6 and uses a gas for which $\gamma = 1.4$. (a) What is the efficiency of the engine if it operates in an idealized Otto cycle? (b) If the actual efficiency is 15%, what fraction of the fuel is wasted as a result of friction and unavoidable heat losses? (Assume complete combustion of the air-fuel mixture.)

21. A 1.6-liter gasoline engine with a compression ratio of 6.2 has a power output of 102 hp. If the engine operates in an idealized Otto cycle, find the heat absorbed and exhausted by the engine each second. Assume the fuel-air mixture behaves like an ideal diatomic gas.

22. In a cylinder of an automobile engine, just after combustion, the gas is confined to a volume of 50 cm³ and has an initial pressure of 3.0×10^6 Pa. The piston moves outward to a final volume of 300 cm³ and the gas expands without heat loss. If $\gamma = 1.40$ for the gas, what is the final pressure?

23. How much work is done by the gas in Problem 22 in expanding from $V_1 = 50$ cm³ to $V_2 = 300$ cm³?

Section 22.6 Heat Pumps and Refrigerators

24. An ideal refrigerator or ideal heat pump is equivalent to a Carnot engine running in reverse. That is, heat Q_c is absorbed from a cold reservoir and heat Q_h is rejected to a hot reservoir. (a) Show that the work that must be supplied to run the refrigerator or pump is

$$W = \frac{T_h - T_c}{T_c} Q_c$$

(b) Show that the coefficient of performance of the ideal refrigerator is

$$COP = \frac{T_c}{T_h - T_c}$$

25. A refrigerator has a coefficient of performance equal to 5. If the refrigerator absorbs 120 J of thermal energy from a cold reservoir in each cycle, find (a) the work done in each cycle and (b) the thermal energy expelled to the hot reservoir.

26. What is the coefficient of performance of a heat pump that brings heat from outdoors at -3°C into a 22°C house? (*Hint:* The heat pump does work W, which is also available to heat the house.)

27. How much work is required, using an ideal Carnot refrigerator, to remove 1.0 J of thermal energy from helium gas at 4.0 K and reject this thermal energy to a room-temperature (293 K) environment?

27A. How much work is required, using an ideal Carnot refrigerator, to remove Q joules of thermal energy from helium gas at T_c and reject this thermal energy to a room-temperature environment at T_h?

Section 22.7   Entropy

28. What is the change in entropy when 1 mol of silver (108 g) is melted at 961°C?

29. What is the entropy decrease in 1 mol of helium gas which is cooled at 1 atm from room temperature 293 K to a final temperature 4 K? (C_p of helium = 21 J/mol·K.)

30. An airtight freezer contains air at 25°C and 1.0 atm. The air is then cooled to -18°C. (a) What is the change in entropy of the air if the volume is held constant? (b) What would the change be if the pressure were maintained at 1 atm during the cooling?

31. Calculate the change in entropy of 250 g of water heated slowly from 20°C to 80°C. (*Hint:* Note that $dQ = mc\,dT$.)

32. An ice tray contains 500 g of water at 0°C. Calculate the change in entropy of the water as it freezes completely and slowly at 0°C.

33. A 1.0-kg iron horseshoe is taken from a furnace at 900°C and dropped into 4.0 kg of water at 10°C. If no heat is lost to the surroundings, determine the total entropy change.

34. At a pressure of 1 atm, liquid helium boils at 4.2 K. The latent heat of vaporization is 20.5 kJ/kg. Determine the entropy change (per kilogram) resulting from vaporization.

Section 22.8   Entropy Changes in Irreversible Processes

35. The surface of the Sun is approximately 5700 K, and the temperature of the Earth's surface is approximately 290 K. What entropy change occurs when 1000 J of thermal energy is transferred from the Sun to the Earth?

36. A 100 000-kg iceberg at -5°C breaks away from the polar ice shelf and floats away into the ocean at 5°C. What is the final change in the entropy of the system when the iceberg has completely melted? (The specific heat of ice is 2010 J/kg·°C.)

37. One mole of H_2 gas is contained in the left-hand side of the container shown in Figure P22.37, which has equal volumes left and right. The right-hand side is evacuated. When the valve is opened, the gas streams into the right side. What is the final entropy change? Does the temperature of the gas change?

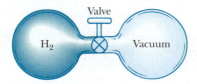

Valve

H_2 Vacuum

FIGURE P22.37

38. A 2.0-liter container has a center partition that divides it into two equal parts, as shown in Figure

P22.38. The left side contains H_2 gas and the right side contains O_2 gas. Both gases are at room temperature and at atmospheric pressure. The partition is removed and the gases are allowed to mix. What is the entropy increase?

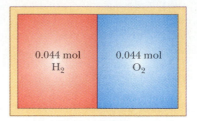

0.044 mol
H_2

0.044 mol
O_2

FIGURE P22.38

39. One mole of an ideal monatomic gas, initially at a pressure of 1.000 atm and a volume of 0.025 m³, is heated to a final state where the pressure is 2.000 atm and the volume is 0.040 m³. Determine the change in entropy for this process.

40. One mole of a diatomic ideal gas, initially having pressure P and volume V, expands to having pressure $2P$ and volume $2V$. If, during the expansion, the pressure remains directly proportional to the volume, determine the entropy change in the process.

ADDITIONAL PROBLEMS

41. An 18-g ice cube at 0.0°C is heated until it vaporizes as steam. (a) How much does the entropy increase? (See Table 20.2.) (b) How much energy was required to vaporize the ice cube?

42. If a 35%-efficient Carnot heat engine (Fig. 22.1) is run in reverse so as to form a refrigerator (Fig. 22.4), what would be this refrigerator's coefficient of performance?

43. A house loses thermal energy through the exterior walls and roof at a rate of 5000 J/s = 5 kW when the interior temperature is 22°C and the outside temperature is -5°C. Calculate the electric power required to maintain the interior temperature at 22°C for the following two cases: (a) The electric power is used in electric resistance heaters (which convert all of the electricity supplied to thermal energy). (b) The electric power is used to operate the compressor of a heat pump (which has a coefficient of performance equal to 60% of the Carnot cycle value).

44. Calculate the increase in entropy of the Universe when you add 20 g of 5°C cream to 200 g of 60°C coffee. The specific heat of cream and coffee is 4.2 J/g·°C.

45. One mole of an ideal monatomic gas is taken through the cycle shown in Figure P22.45. The process *AB* is a reversible isothermal expansion. Calcu-

late (a) the net work done by the gas, (b) the thermal energy added to the gas, (c) the thermal energy expelled by the gas, and (d) the efficiency of the cycle.

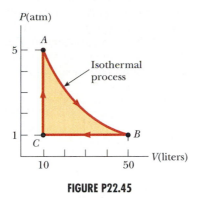

FIGURE P22.45

46. Using an ideal Carnot refrigerator, how much work is required to change 0.50 kg of tap water at 10°C into ice at −20°C? Assume the freezer compartment is held at −20°C and the refrigerator exhausts heat into a room at 20°C.

47. Figure P22.47 represents n mol of an ideal monatomic gas being taken through a reversible cycle consisting of two isothermal processes at temperatures $3T_0$ and T_0 and two constant-volume processes. For each cycle, determine in terms of n, R, and T_0 (a) the net thermal energy transferred to the gas and (b) the efficiency of an engine operating in this cycle.

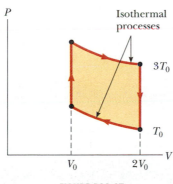

FIGURE P22.47

48. An ideal (Carnot) freezer has a constant temperature of 260 K, while the external air has a constant temperature of 300 K. Suppose the insulation for the freezer is not perfect so that some heat flows into the freezer at a rate of 0.15 W. Determine the average power of the freezer's motor that is needed to maintain the constant temperature in the freezer.

49. One mole of a monatomic ideal gas is taken through the reversible cycle shown in Figure P22.49. At point A, the pressure, volume, and temperature are P_0, V_0, and T_0, respectively. In terms of R and T_0, find

(a) the total heat entering the system per cycle, (b) the total heat leaving the system per cycle, (c) the efficiency of an engine operating in this reversible cycle, and (d) the efficiency of an engine operating in a Carnot cycle between the same temperature extremes.

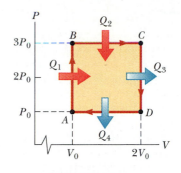

FIGURE P22.49

50. One mole of an ideal gas expands isothermally. (a) If the gas doubles its volume, show that the work of expansion is $W = RT \ln 2$. (b) Since the internal energy U of an ideal gas depends solely on its temperature, there is no change in U during the expansion. It follows from the first law that the thermal energy absorbed by the gas during the expansion is converted completely to work. Why does this *not* violate the second law?

51. A system consisting of n mol of an ideal gas undergoes a reversible, *isobaric* process from a volume V_0 to a volume $3V_0$. Calculate the change in entropy of the gas. (*Hint:* Imagine that the system goes from the initial state to the final state first along an isotherm and then along an adiabatic path—there is no change in entropy along the adiabatic path.)

52. An electrical power plant has an overall efficiency of 15%. The plant is to deliver 150 MW of power to a city, and its turbines use coal as the fuel. The burning coal produces steam which drives the turbines. This steam is then condensed to water at 25°C by passing it through cooling coils in contact with river water. (a) How many metric tons of coal does the plant consume each day (1 metric ton = 10^3 kg)? (b) What is the total cost of the fuel per year if the delivered price is $8/metric ton? (c) If the river water is delivered at 20°C, at what minimum rate must it flow over the cooling coils in order that its temperature not exceed 25°C? (*Note:* The heat of combustion of coal is 33 kJ/g.)

53. A power plant, having a Carnot efficiency, produces 1000 MW of electrical power from turbines that take in steam at 500 K and reject water at 300 K into a flowing river. If the water downstream is 6 K warmer

due to the output of the power plant, determine the flow rate of the river.

53A. A power plant, having a Carnot efficiency, produces electrical power P (in MW) from turbines that take in steam at T_h and reject water at T_c into a flowing river. If the water downstream is ΔT warmer due to the output of the power plant, determine the flow rate of the river.

54. Suppose you are working in a patent office, and an inventor comes to you with the claim that her heat engine, which employs water as a working substance, has a thermodynamic efficiency of 0.61. She explains that it operates between heat reservoirs at 4°C and 0°C. It is a very complicated device, with many pistons, gears, and pulleys, and the cycle involves freezing and melting. Does her claim that $e = 0.61$ warrant serious consideration? Explain.

55. An idealized Diesel engine operates in a cycle known as the *air-standard Diesel cycle*, shown in Figure P22.55. Fuel is sprayed into the cylinder at the point of maximum compression, B. Combustion occurs during the expansion $B \rightarrow C$, which is approximated as an isobaric process. The rest of the cycle is the same as in the gasoline engine, described in Figure 22.11. Show that the efficiency of an engine operating in this idealized Diesel cycle is

$$e = 1 - \frac{1}{\gamma}\left(\frac{T_D - T_A}{T_C - T_B}\right)$$

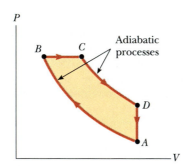

FIGURE P22.55

56. One mole of an ideal gas ($\gamma = 1.4$) is carried through the Carnot cycle described in Figure 22.9. At point A, the pressure is 25 atm and the temperature is 600 K. At point C, the pressure is 1 atm and the temperature is 400 K. (a) Determine the pressures and volumes at points A, B, C, and D. (b) Calculate the net work done per cycle. (c) Determine the efficiency of an engine operating in this cycle.

57. A typical human has a mass of 70 kg and produces about 2000 kcal (2.0×10^6 cal) of metabolic heat per day. (a) Find the rate of heat production in watts and in calories per hour. (b) If none of the metabolic heat were lost, and assuming that the specific heat of the human body is 1.0 cal/g·°C, find the rate at which body temperature would rise. Give your answer in °C per hour and in °F per hour.

58. Suppose 1.00 kg of water at 10.0°C is mixed with 1.00 kg of water at 30.0°C at constant pressure. When the mixture has reached equilibrium, (a) what is the final temperature? (b) Take $C_p = 4.19$ kJ/kg·K for water and show that the entropy of the system increases by

$$\Delta S = 4.19 \ln\left[\left(\frac{293}{283}\right)\left(\frac{293}{303}\right)\right] \text{kJ/K}$$

(c) Verify numerically that $\Delta S > 0$. (d) Is the mixing an irreversible process?

59. The Stirling engine described in Figure P22.59 operates between the isotherms T_1 and T_2, where $T_2 > T_1$. Assuming that the operating gas is an ideal monatomic gas, calculate the efficiency of an engine whose constant-volume processes occur at the volumes V_1 and V_2.

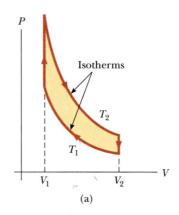

(a)

FIGURE P22.59

60. Suppose a heat engine is connected to two heat reservoirs, one a pool of molten aluminum (660°C) and the other a block of solid mercury (-38.9°C). The engine runs by freezing 1.00 g of aluminum and melting 15.0 g of mercury during each cycle. The latent heat of fusion of aluminum is 3.97×10^5 J/kg, and that of mercury is 1.18×10^4 J/kg. (a) What is the efficiency of this engine? (b) How does the efficiency compare with that of a Carnot engine?

61. A gas is taken through the cyclic process described by Figure P22.61. (a) If Q is negative for the process BC and ΔU is negative for the process CA, determine the signs of Q, W, and ΔU associated with each process. (b) Find the net heat tranferred to the system during one complete cycle. (c) If the cycle is reversed—that is, the process follows the path $ACBA$—what is the net heat transferred per cycle?

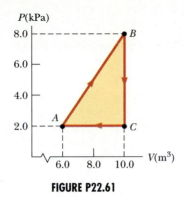

FIGURE P22.61

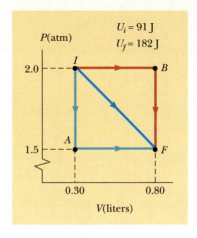

FIGURE P22.62

62. One mole of gas is initially at a pressure of 2.0 atm and a volume of 0.30 L and has an internal energy equal to 91 J. In its final state the gas is at a pressure of 1.5 atm and a volume of 0.80 L, and its internal energy equals 180 J. For the paths *IAF*, *IBF*, and *IF* in Figure P22.62, calculate (a) the work done by the gas and (b) the net heat-transferred to the gas in the process.

63. Powdered steel is placed in a container filled with oxygen and fitted with a piston that moves to keep the pressure in the container constant at one atmo-sphere. A chemical reaction occurs that produces heat. To keep the contents at a constant temperature of 22°C, it is necessary to remove 8.3×10^5 J of heat from the container as it contracts. During the chemical reaction, it is found that 1.5 mol of oxygen is consumed. Find the internal energy change for the system of iron and oxygen.

TABLE A.1   Conversion Factors

Length

	m	cm	km	in.	ft	mi
1 meter	1	10^2	10^{-3}	39.37	3.281	6.214×10^{-4}
1 centimeter	10^{-2}	1	10^{-5}	0.3937	3.281×10^{-2}	6.214×10^{-6}
1 kilometer	10^3	10^5	1	3.937×10^4	3.281×10^3	0.6214
1 inch	2.540×10^{-2}	2.540	2.540×10^{-5}	1	8.333×10^{-2}	1.578×10^{-5}
1 foot	0.3048	30.48	3.048×10^{-4}	12	1	1.894×10^{-4}
1 mile	1609	1.609×10^5	1.609	6.336×10^4	5280	1

Mass

	kg	g	slug	u
1 kilogram	1	10^3	6.852×10^{-2}	6.024×10^{26}
1 gram	10^{-3}	1	6.852×10^{-5}	6.024×10^{23}
1 slug	14.59	1.459×10^4	1	8.789×10^{27}
1 atomic mass unit	1.660×10^{-27}	1.660×10^{-24}	1.137×10^{-28}	1

Time

	s	min	h	day	year
1 second	1	1.667×10^{-2}	2.778×10^{-4}	1.157×10^{-5}	3.169×10^{-8}
1 minute	60	1	1.667×10^{-2}	6.994×10^{-4}	1.901×10^{-6}
1 hour	3600	60	1	4.167×10^{-2}	1.141×10^{-4}
1 day	8.640×10^4	1440	24	1	2.738×10^{-3}
1 year	3.156×10^7	5.259×10^5	8.766×10^3	365.2	1

Speed

	m/s	cm/s	ft/s	mi/h
1 meter/second	1	10^2	3.281	2.237
1 centimeter/second	10^{-2}	1	3.281×10^{-2}	2.237×10^{-2}
1 foot/second	0.3048	30.48	1	0.6818
1 mile/hour	0.4470	44.70	1.467	1

Note: 1 mi/min = 60 mi/h = 88 ft/s.

Force

	N	dyn	lb
1 newton	1	10^5	0.2248
1 dyne	10^{-5}	1	2.248×10^{-6}
1 pound	4.448	4.448×10^5	1

TABLE A.1 *Continued*

Work, Energy, Heat

	J	erg	ft·lb
1 joule	1	10^7	0.7376
1 erg	10^{-7}	1	7.376×10^{-8}
1 ft·lb	1.356	1.356×10^7	1
1 eV	1.602×10^{-19}	1.602×10^{-12}	1.182×10^{-19}
1 cal	4.186	4.186×10^7	3.087
1 Btu	1.055×10^3	1.055×10^{10}	7.779×10^2
1 kWh	3.600×10^6	3.600×10^{13}	2.655×10^6

	eV	cal	Btu	kWh
1 joule	6.242×10^{18}	0.2389	9.481×10^{-4}	2.778×10^{-7}
1 erg	6.242×10^{11}	2.389×10^{-8}	9.481×10^{-11}	2.778×10^{-14}
1 ft·lb	8.464×10^{18}	0.3239	1.285×10^{-3}	3.766×10^{-7}
1 eV	1	3.827×10^{-20}	1.519×10^{-22}	4.450×10^{-26}
1 cal	2.613×10^{19}	1	3.968×10^{-3}	1.163×10^{-6}
1 Btu	6.585×10^{21}	2.520×10^2	1	2.930×10^{-4}
1 kWh	2.247×10^{25}	8.601×10^5	3.413×10^2	1

Pressure

	Pa	dyn/cm²	atm
1 pascal	1	10	9.869×10^{-6}
1 dyne/centimeter²	10^{-1}	1	9.869×10^{-7}
1 atmosphere	1.013×10^5	1.013×10^6	1
1 centimeter mercury*	1.333×10^3	1.333×10^4	1.316×10^{-2}
1 pound/inch²	6.895×10^3	6.895×10^4	6.805×10^{-2}
1 pound/foot²	47.88	4.788×10^2	4.725×10^{-4}

	cm Hg	lb/in.²	lb/ft²
1 newton/meter²	7.501×10^{-4}	1.450×10^{-4}	2.089×10^{-2}
1 dyne/centimeter²	7.501×10^{-5}	1.450×10^{-5}	2.089×10^{-3}
1 atmosphere	76	14.70	2.116×10^3
1 centimeter mercury*	1	0.1943	27.85
1 pound/inch²	5.171	1	144
1 pound/foot²	3.591×10^{-2}	6.944×10^{-3}	1

* At 0°C and at a location where the acceleration due to gravity has its "standard" value, 9.80665 m/s².

TABLE A.2 Symbols, Dimensions, and Units of Physical Quantities

Quantity	Common Symbol	Unit*	Dimensions†	Unit in Terms of Base SI Units
Acceleration	**a**	m/s^2	L/T^2	m/s^2
Amount of substance	n	mole		mol
Angle	θ, ϕ	radian (rad)	1	
Angular acceleration	$\boldsymbol{\alpha}$	rad/s^2	T^{-2}	s^{-2}
Angular frequency	ω	rad/s	T^{-1}	s^{-1}
Angular momentum	L	$kg \cdot m^2/s$	ML^2/T	$kg \cdot m^2/s$
Angular velocity	$\boldsymbol{\omega}$	rad/s	T^{-1}	s^{-1}
Area	A	m^2	L^2	m^2
Atomic number	Z			
Capacitance	C	farad (F) $(= Q/V)$	Q^2T^2/ML^2	$A^2 \cdot s^4/kg \cdot m^2$
Charge	q, Q, e	coulomb (C)	Q	$A \cdot s$
Charge density				
Line	λ	C/m	Q/L	$A \cdot s/m$
Surface	σ	C/m^2	Q/L^2	$A \cdot s/m^2$
Volume	ρ	C/m^3	Q/L^3	$A \cdot s/m^3$
Conductivity	σ	$1/\Omega \cdot m$	Q^2T/ML^3	$A^2 \cdot s^3/kg \cdot m^3$
Current	I	AMPERE	Q/T	A
Current density	J	A/m^2	Q/T^2	A/m^2
Density	ρ	kg/m^3	M/L^3	kg/m^3
Dielectric constant	κ			
Displacement	s	METER	L	m
Distance	d, h			
Length	ℓ, L			
Electric dipole moment	p	$C \cdot m$	QL	$A \cdot s \cdot m$
Electric field	E	V/m	ML/QT^2	$kg \cdot m/A \cdot s^3$
Electric flux	Φ	$V \cdot m$	ML^3/QT^2	$kg \cdot m^3/A \cdot s^3$
Electromotive force	$\mathcal{E}$	volt (V)	ML^2/QT^2	$kg \cdot m^2/A \cdot s^3$
Energy	E, U, K	joule (J)	ML^2/T^2	$kg \cdot m^2/s^2$
Entropy	S	J/K	$ML^2/T^2 \cdot K$	$kg \cdot m^2/s^2 \cdot K$
Force	F	newton (N)	ML/T^2	$kg \cdot m/s^2$
Frequency	f, ν	hertz (Hz)	T^{-1}	s^{-1}
Heat	Q	joule (J)	ML^2/T^2	$kg \cdot m^2/s^2$
Inductance	L	henry (H)	ML^2/Q^2	$kg \cdot m^2/A^2 \cdot s^2$
Magnetic dipole moment	$\boldsymbol{\mu}$	$N \cdot m/T$	QL^2/T	$A \cdot m^2$
Magnetic field	B	tesla (T) $(= Wb/m^2)$	M/QT	$kg/A \cdot s^2$
Magnetic flux	Φ_m	weber (Wb)	ML^2/QT	$kg \cdot m^2/A \cdot s^2$
Mass	m, M	KILOGRAM	M	kg
Molar specific heat	C	$J/mol \cdot K$		$kg \cdot m^2/s^2 \cdot mol \cdot K$
Moment of inertia	I	$kg \cdot m^2$	ML^2	$kg \cdot m^2$
Momentum	p	$kg \cdot m/s$	ML/T	$kg \cdot m/s$
Period	T	s	T	s
Permeability of space	μ_0	N/A^2 $(= H/m)$	ML/Q^2T	$kg \cdot m/A^2 \cdot s^2$
Permittivity of space	ϵ_0	$C^2/N \cdot m^2$ $(= F/m)$	Q^2T^2/ML^3	$A^2 \cdot s^4/kg \cdot m^3$

continued

TABLE A.2 *Continued*

Quantity	Common Symbol	Unit*	Dimensions†	Unit in Terms of Base SI Units
Potential (voltage)	V	volt (V) $(=J/C)$	ML^2/QT^2	$kg \cdot m^2/A \cdot s^3$
Power	P	watt (W) $(=J/s)$	ML^2/T^3	$kg \cdot m^2/s^3$
Pressure	P, p	pascal (Pa) $= (N/m^2)$	M/LT^2	$kg/m \cdot s^2$
Resistance	R	ohm $(\Omega)(=V/A)$	ML^2/Q^2T	$kg \cdot m^2/A^2 \cdot s^3$
Specific heat	c	$J/kg \cdot K$	$L^2/T^2 \cdot K$	$m^2/s^2 \cdot K$
Temperature	T	KELVIN	K	K
Time	t	SECOND	T	s
Torque	τ	$N \cdot m$	ML^2/T^2	$kg \cdot m^2/s^2$
Speed	v	m/s	L/T	m/s
Volume	V	m^3	L^3	m^3
Wavelength	λ	m	L	m
Work	W	joule (J) $(=N \cdot m)$	ML^2/T^2	$kg \cdot m^2/s^2$

* The base SI units are given in upper case letters.

† The symbols M, L, T, and Q denote mass, length, time, and charge, respectively.

TABLE A.3 **Table of Selected Atomic Masses***

Atomic Number Z	Element	Symbol	Mass Number A	Atomic Mass†	Percent Abundance, or Decay Mode (if radioactive)‡	Half-Life (if radioactive)
0	(Neutron)	n	1	1.008665	β^-	10.6 min
1	Hydrogen	H	1	1.007825	99.985	
	Deuterium	D	2	2.014102	0.015	
	Tritium	T	3	3.016049	β^-	12.33 y
2	Helium	He	3	3.016029	0.00014	
			4	4.002603	≈ 100	
3	Lithium	Li	6	6.015123	7.5	
			7	7.016005	92.5	
4	Beryllium	Be	7	7.016930	EC, γ	53.3 days
			8	8.005305	2α	6.7×10^{-17} s
			9	9.012183	100	
5	Boron	B	10	10.012938	19.8	
			11	11.009305	80.2	
6	Carbon	C	11	11.011433	β^+, EC	20.4 min
			12	12.000000	98.89	
			13	13.003355	1.11	
			14	14.003242	β^-	5730 y
7	Nitrogen	N	13	13.005739	β^+	9.96 min
			14	14.003074	99.63	
			15	15.000109	0.37	
8	Oxygen	O	15	15.003065	$\beta+$, EC	122 s
			16	15.994915	99.759	
			18	17.999159	0.204	
9	Fluorine	F	19	18.998403	100	

continued

TABLE A.3 *Continued*

Atomic Number Z	Element	Symbol	Mass Number A	Atomic Mass†	Percent Abundance, or Decay Mode (if radioactive)‡	Half-Life (if radioactive)
10	Neon	Ne	20	19.992439	90.51	
			22	21.991384	9.22	
11	Sodium	Na	22	21.994435	β^+, EC, γ	2.602 y
			23	22.989770	100	
			24	23.990964	β^-, γ	15.0 h
12	Magnesium	Mg	24	23.985045	78.99	
13	Aluminum	Al	27	26.981541	100	
14	Silicon	Si	28	27.976928	92.23	
			31	30.975364	β^-, γ	2.62 h
15	Phosphorus	P	31	30.973763	100	
			32	31.973908	β^-	14.28 days
16	Sulfur	S	32	31.972072	95.0	
			35	34.969033	β^-	87.4 days
17	Chlorine	Cl	35	34.968853	75.77	
			37	36.965903	24.23	
18	Argon	Ar	40	39.962383	99.60	
19	Potassium	K	39	38.963708	93.26	
			40	39.964000	β^-, EC, γ, β^+	1.28×10^9 y
20	Calcium	Ca	40	39.962591	96.94	
21	Scandium	Sc	45	44.955914	100	
22	Titanium	Ti	48	47.947947	73.7	
23	Vanadium	V	51	50.943963	99.75	
24	Chromium	Cr	52	51.940510	83.79	
25	Manganese	Mn	55	54.938046	100	
26	Iron	Fe	56	55.934939	91.8	
27	Cobalt	Co	59	58.933198	100	
			60	59.933820	β^-, γ	5.271 y
28	Nickel	Ni	58	57.935347	68.3	
			60	59.930789	26.1	
			64	63.927968	0.91	
29	Copper	Cu	63	62.929599	69.2	
			64	63.929766	β^-, β^+	12.7 h
			65	64.927792	30.8	
30	Zinc	Zn	64	63.929145	48.6	
			66	65.926035	27.9	
31	Gallium	Ga	69	68.925581	60.1	
32	Germanium	Ge	72	71.922080	27.4	
			74	73.921179	36.5	
33	Arsenic	As	75	74.921596	100	
34	Selenium	Se	80	79.916521	49.8	
35	Bromine	Br	79	78.918336	50.69	
36	Krypton	Kr	84	83.911506	57.0	
			89	88.917563	β^-	3.2 min
37	Rubidium	Rb	85	84.911800	72.17	
38	Strontium	Sr	86	85.909273	9.8	
			88	87.905625	82.6	
			90	89.907746	β^-	28.8 y
39	Yttrium	Y	89	88.905856	100	

continued

TABLE A.3   *Continued*

Atomic Number Z	Element	Symbol	Mass Number A	Atomic Mass†	Percent Abundance, or Decay Mode (if radioactive)‡	Half-Life (if radioactive)
40	Zirconium	Zr	90	89.904708	51.5	
41	Niobium	Nb	93	92.906378	100	
42	Molybdenum	Mo	98	97.905405	24.1	
43	Technetium	Tc	98	97.907210	β^-, γ	4.2×10^6 y
44	Ruthenium	Ru	102	101.904348	31.6	
45	Rhodium	Rh	103	102.90550	100	
46	Palladium	Pd	106	105.90348	27.3	
47	Silver	Ag	107	106.905095	51.83	
			109	108.904754	48.17	
48	Cadmium	Cd	114	113.903361	28.7	
49	Indium	In	115	114.90388	95.7; β^-	5.1×10^{14} y
50	Tin	Sn	120	119.902199	32.4	
51	Antimony	Sb	121	120.903824	57.3	
52	Tellurium	Te	130	129.90623	34.5; β^-	2×10^{21} y
53	Iodine	I	127	126.904477	100	
			131	130.906118	β^-, γ	8.04 days
54	Xenon	Xe	132	131.90415	26.9	
			136	135.90722	8.9	
55	Cesium	Cs	133	132.90543	100	
56	Barium	Ba	137	136.90582	11.2	
			138	137.90524	71.7	
			144	143.922673	β^-	11.9 s
57	Lanthanum	La	139	138.90636	99.911	
58	Cerium	Ce	140	139.90544	88.5	
59	Praseodymium	Pr	141	140.90766	100	
60	Neodymium	Nd	142	141.90773	27.2	
61	Promethium	Pm	145	144.91275	EC, α, γ	17.7 y
62	Samarium	Sm	152	151.91974	26.6	
63	Europium	Eu	153	152.92124	52.1	
64	Gadolinium	Gd	158	157.92411	24.8	
65	Terbium	Tb	159	158.92535	100	
66	Dysprosium	Dy	164	163.92918	28.1	
67	Holmium	Ho	165	164.93033	100	
68	Erbium	Er	166	165.93031	33.4	
69	Thulium	Tm	169	168.93423	100	
70	Ytterbium	Yb	174	173.93887	31.6	
71	Lutecium	Lu	175	174.94079	97.39	
72	Hafnium	Hf	180	179.94656	35.2	
73	Tantalum	Ta	181	180.94801	99.988	
74	Tungsten (wolfram)	W	184	183.95095	30.7	
75	Rhenium	Re	187	186.95577	62.60, β^-	4×10^{10} y
76	Osmium	Os	191	190.96094	β^-, γ	15.4 days
			192	191.96149	41.0	
77	Iridium	Ir	191	190.96060	37.3	
			193	192.96294	62.7	
78	Platinum	Pt	195	194.96479	33.8	
79	Gold	Au	197	196.96656	100	
80	Mercury	Hg	202	201.97063	29.8	

continued

TABLE A.3 *Continued*

Atomic Number Z	Element	Symbol	Mass Number A	Atomic Mass†	Percent Abundance, or Decay Mode (if radioactive)‡	Half-Life (if radioactive)
81	Thallium	Tl	205	204.97441	70.5	
			208	207.981988	β^-, γ	3.053 min
82	Lead	Pb	204	203.973044	β^-, 1.48	1.4×10^{17} y
			206	205.97446	24.1	
			207	206.97589	22.1	
			208	207.97664	52.3	
			210	209.98418	α, β^-, γ	22.3 y
			211	210.98874	β^-, γ	36.1 min
			212	211.99188	β^-, γ	10.64 h
			214	213.99980	β^-, γ	26.8 min
83	Bismuth	Bi	209	208.98039	100	
			211	210.98726	α, β^-, γ	2.15 min
84	Polonium	Po	210	209.98286	α, γ	138.38 days
			214	213.99519	α, γ	164 μs
85	Astatine	At	218	218.00870	α, β^-	≈ 2 s
86	Radon	Rn	222	222.017574	α, γ	3.8235 days
87	Francium	Fr	223	223.019734	α, β^-, γ	21.8 min
88	Radium	Ra	226	226.025406	α, γ	1.60×10^3 y
			228	228.031069	β^-	5.76 y
89	Actinium	Ac	227	227.027751	α, β^-, γ	21.773 y
90	Thorium	Th	228	228.02873	α, γ	1.9131 y
			232	232.038054	100, α, γ	1.41×10^{10} y
91	Protactinium	Pa	231	231.035881	α, γ	3.28×10^4 y
92	Uranium	U	232	232.03714	α, γ	72 y
			233	233.039629	α, γ	1.592×10^5 y
			235	235.043925	0.72; α, γ	7.038×10^8 y
			236	236.045563	α, γ	2.342×10^7 y
			238	238.050786	99.275; α, γ	4.468×10^9 y
			239	239.054291	β^-, γ	23.5 min
93	Neptunium	Np	239	239.052932	β^-, γ	2.35 days
94	Plutonium	Pu	239	239.052158	α, γ	2.41×10^4 y
95	Americium	Am	243	243.061374	α, γ	7.37×10^3 y
96	Curium	Cm	245	245.065487	α, γ	8.5×10^3 y
97	Berkelium	Bk	247	247.07003	α, γ	1.4×10^3 y
98	Californium	Cf	249	249.074849	α, γ	351 y
99	Einsteinium	Es	254	254.08802	α, γ, β^-	276 days
100	Fermium	Fm	253	253.08518	EC, α, γ	3.0 days
101	Mendelevium	Md	255	255.0911	EC, α	27 min
102	Nobelium	No	255	255.0933	EC, α	3.1 min
103	Lawrencium	Lr	257	257.0998	α	≈ 35 s
104	Unnilquadium	Rf	261	261.1087	α	1.1 min
105	Unnilpentium	Ha	262	262.1138	α	0.7 min
106	Unnilhexium		263	263.1184	α	0.9 s
107	Unnilseptium		261	261	α	1–2 ms

* Data are taken from *Chart of the Nuclides,* 12th ed., General Electric, 1977, and from C. M. Lederer and V. S. Shirley, eds., *Table of Isotopes,* 7th ed., New York, John Wiley & Sons, Inc., 1978.

† The masses given in column (5) are those for the neutral atom, including the Z electrons.

‡ The process EC stands for "electron capture."

Mathematics Review

These appendices in mathematics are intended as a brief review of operations and methods. Early in this course, you should be totally familiar with basic algebraic techniques, analytic geometry, and trigonometry. The appendices on differential and integral calculus are more detailed and are intended for those students who have difficulty applying calculus concepts to physical situations.

B.1   SCIENTIFIC NOTATION

Many quantities that scientists deal with often have very large or very small values. For example, the speed of light is about 300 000 000 m/s and the ink required to make the dot over an i in this textbook has a mass of about 0.000 000 001 kg. Obviously, it is very cumbersome to read, write, and keep track of numbers such as these. We avoid this problem by using a method dealing with powers of the number 10:

$$10^0 = 1$$

$$10^1 = 10$$

$$10^2 = 10 \times 10 = 100$$

$$10^3 = 10 \times 10 \times 10 = 1000$$

$$10^4 = 10 \times 10 \times 10 \times 10 = 10\ 000$$

$$10^5 = 10 \times 10 \times 10 \times 10 \times 10 = 100\ 000$$

and so on. The number of zeros corresponds to the power to which 10 is raised, called the **exponent** of 10. For example, the speed of light, 300 000 000 m/s, can be expressed as 3×10^8 m/s.

In this method, some representative numbers smaller than unity are

$$10^{-1} = \frac{1}{10} = 0.1$$

$$10^{-2} = \frac{1}{10 \times 10} = 0.01$$

$$10^{-3} = \frac{1}{10 \times 10 \times 10} = 0.001$$

$$10^{-4} = \frac{1}{10 \times 10 \times 10 \times 10} = 0.0001$$

$$10^{-5} = \frac{1}{10 \times 10 \times 10 \times 10 \times 10} = 0.00001$$

In these cases, the number of places the decimal point is to the left of the digit 1 equals the value of the (negative) exponent. Numbers expressed as some power of 10 multiplied by another number between 1 and 10 are said to be in **scientific notation.** For example, the scientific notation for 5 943 000 000 is 5.943×10^9 and that for 0.0000832 is 8.32×10^{-5}.

When numbers expressed in scientific notation are being multiplied, the following general rule is very useful:

$$10^n \times 10^m = 10^{n+m} \tag{B.1}$$

where n and m can be *any* numbers (not necessarily integers). For example, $10^2 \times 10^5 = 10^7$. The rule also applies if one of the exponents is negative: $10^3 \times 10^{-8} = 10^{-5}$.

When dividing numbers expressed in scientific notation, note that

$$\frac{10^n}{10^m} = 10^n \times 10^{-m} = 10^{n-m} \tag{B.2}$$

EXERCISES

With help from the above rules, verify the answers to the following:

1. $86\,400 = 8.64 \times 10^4$
2. $9\,816{,}762.5 = 9.8167625 \times 10^6$
3. $0.0000000398 = 3.98 \times 10^{-8}$
4. $(4 \times 10^8)(9 \times 10^9) = 3.6 \times 10^{18}$
5. $(3 \times 10^7)(6 \times 10^{-12}) = 1.8 \times 10^{-4}$
6. $\dfrac{75 \times 10^{-11}}{5 \times 10^{-3}} = 1.5 \times 10^{-7}$
7. $\dfrac{(3 \times 10^6)(8 \times 10^{-2})}{(2 \times 10^{17})(6 \times 10^5)} = 2 \times 10^{-18}$

B.2   ALGEBRA

Some Basic Rules

When algebraic operations are performed, the laws of arithmetic apply. Symbols such as x, y, and z are usually used to represent quantities that are not specified, what are called the **unknowns.**

First, consider the equation

$$8x = 32$$

If we wish to solve for x, we can divide (or multiply) each side of the equation by the same factor without destroying the equality. In this case, if we divide both sides by 8, we have

$$\frac{8x}{8} = \frac{32}{8}$$

$$x = 4$$

Next consider the equation

$$x + 2 = 8$$

In this type of expression, we can add or subtract the same quantity from each side. If we subtract 2 from each side, we get

$$x + 2 - 2 = 8 - 2$$

$$x = 6$$

In general, if $x + a = b$, then $x = b - a$.

Now consider the equation

$$\frac{x}{5} = 9$$

If we multiply each side by 5, we are left with x on the left by itself and 45 on the right:

$$\left(\frac{x}{5}\right)(5) = 9 \times 5$$

$$x = 45$$

In all cases, *whatever operation is performed on the left side of the equality must also be performed on the right side.*

The following rules for multiplying, dividing, adding, and subtracting fractions should be recalled, where a, b, and c are three numbers:

	Rule	**Example**
Multiplying	$\left(\dfrac{a}{b}\right)\left(\dfrac{c}{d}\right) = \dfrac{ac}{bd}$	$\left(\dfrac{2}{3}\right)\left(\dfrac{4}{5}\right) = \dfrac{8}{15}$
Dividing	$\dfrac{(a/b)}{(c/d)} = \dfrac{ad}{bc}$	$\dfrac{2/3}{4/5} = \dfrac{(2)(5)}{(4)(3)} = \dfrac{10}{12}$
Adding	$\dfrac{a}{b} \pm \dfrac{c}{d} = \dfrac{ad \pm bc}{bd}$	$\dfrac{2}{3} - \dfrac{4}{5} = \dfrac{(2)(5) - (4)(3)}{(3)(5)} = -\dfrac{2}{15}$

EXERCISES

In the following exercises, solve for x:

Answers

1. $a = \dfrac{1}{1 + x}$ $\qquad x = \dfrac{1 - a}{a}$

2. $3x - 5 = 13$ $\qquad x = 6$

3. $ax - 5 = bx + 2$ $\qquad x = \dfrac{7}{a - b}$

4. $\dfrac{5}{2x + 6} = \dfrac{3}{4x + 8}$ $\qquad x = -\dfrac{11}{7}$

Powers

When powers of a given quantity x are multiplied, the following rule applies:

$$x^n x^m = x^{n+m} \tag{B.3}$$

For example, $x^2 x^4 = x^{2+4} = x^6$.

When dividing the powers of a given quantity, the rule is

$$\frac{x^n}{x^m} = x^{n-m} \qquad \text{(B.4)}$$

For example, $x^8/x^2 = x^{8-2} = x^6$.

A power that is a fraction, such as $\frac{1}{3}$, corresponds to a root as follows:

$$x^{1/n} = \sqrt[n]{x} \qquad \text{(B.5)}$$

For example, $4^{1/3} = \sqrt[3]{4} = 1.5874$. (A scientific calculator is useful for such calculations.)

Finally, any quantity x^n raised to the mth power is

$$(x^n)^m = x^{nm} \qquad \text{(B.6)}$$

Table B.1 summarizes the rules of exponents.

TABLE B.1 Rules of Exponents

$$x^0 = 1$$
$$x^1 = x$$
$$x^n x^m = x^{n+m}$$
$$x^n/x^m = x^{n-m}$$
$$x^{1/n} = \sqrt[n]{x}$$
$$(x^n)^m = x^{nm}$$

EXERCISES

Verify the following:

1. $3^2 \times 3^3 = 243$
2. $x^5 x^{-8} = x^{-3}$
3. $x^{10}/x^{-5} = x^{15}$
4. $5^{1/3} = 1.709975$ (Use your calculator.)
5. $60^{1/4} = 2.783158$ (Use your calculator.)
6. $(x^4)^3 = x^{12}$

Factoring

Some useful formulas for factoring an equation are

$$ax + ay + az = a(x + y + x) \qquad \text{common factor}$$
$$a^2 + 2ab + b^2 = (a + b)^2 \qquad \text{perfect square}$$
$$a^2 - b^2 = (a + b)(a - b) \qquad \text{differences of squares}$$

Quadratic Equations

The general form of a quadratic equation is

$$ax^2 + bx + c = 0 \qquad \text{(B.7)}$$

where x is the unknown quantity and a, b, and c are numerical factors referred to as **coefficients** of the equation. This equation has two roots, given by

$$x = \frac{-b \pm \sqrt{b^2 - 4ac}}{2a} \qquad \text{(B.8)}$$

If $b^2 \geq 4ac$, the roots are real.

EXAMPLE 1

The equation $x^2 + 5x + 4 = 0$ has the following roots corresponding to the two signs of the square-root term:

$$x = \frac{-5 \pm \sqrt{5^2 - (4)(1)(4)}}{2(1)} = \frac{-5 \pm \sqrt{9}}{2} = \frac{-5 \pm 3}{2}$$

$$x_+ = \frac{-5 + 3}{2} = \boxed{-1} \qquad x_- = \frac{-5 - 3}{2} = \boxed{-4}$$

where x_+ refers to the root corresponding to the positive sign and x_- refers to the root corresponding to the negative sign.

EXERCISES

Solve the following quadratic equations:

Answers

1. $x^2 + 2x - 3 = 0$	$x_+ = 1$	$x_- = -3$
2. $2x^2 - 5x + 2 = 0$	$x_+ = 2$	$x_- = \frac{1}{2}$
3. $2x^2 - 4x - 9 = 0$	$x_+ = 1 + \sqrt{22}/2$	$x_- = 1 - \sqrt{22}/2$

Linear Equations

A linear equation has the general form

$$y = mx + b \tag{B.9}$$

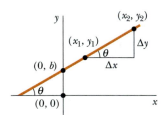

FIGURE B.1

where m and b are constants. This equation is referred to as being linear because the graph of y versus x is a straight line, as shown in Figure B.1. The constant b, called the *y*-**intercept,** represents the value of y at which the straight line intersects the y axis. The constant m is equal to the **slope** of the straight line and is also equal to the tangent of the angle that the line makes with the x axis. If any two points on the straight line are specified by the coordinates (x_1, y_1) and (x_2, y_2), as in Figure B.1, then the **slope** of the straight line can be expressed as

$$\text{Slope} = \frac{y_2 - y_1}{x_2 - x_1} = \frac{\Delta y}{\Delta x} = \tan \theta \tag{B.10}$$

Note that m and b can have either positive or negative values. If $m > 0$, the straight line has a *positive* slope, as in Figure B.1. If $m < 0$, the straight line has a *negative* slope. In Figure B.1, both m and b are positive. Three other possible situations are shown in Figure B.2.

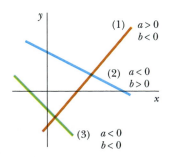

FIGURE B.2

EXERCISES

1. Draw graphs of the following straight lines:
(a) $y = 5x + 3$ (b) $y = -2x + 4$ (c) $y = -3x - 6$
2. Find the slopes of the straight lines described in Exercise 1.

Answers (a) 5   (b) -2 (c) -3

3. Find the slopes of the straight lines that pass through the following sets of points:

(a) $(0, -4)$ and $(4, 2)$, (b) $(0, 0)$ and $(2, -5)$, and (c) $(-5, 2)$ and $(4, -2)$

Answers (a) $3/2$ (b) $-5/2$ (c) $-4/9$

Solving Simultaneous Linear Equations

Consider the equation $3x + 5y = 15$, which has two unknowns, x and y. Such an equation does not have a unique solution. Instead, $(x = 0, y = 3)$, $(x = 5, y = 0)$, and $(x = 2, y = 9/5)$ are all solutions to this equation.

If a problem has two unknowns, a unique solution is possible only if we have *two* equations. In general, if a problem has n unknowns, its solution requires n equations. In order to solve two simultaneous equations involving two unknowns, x and y, we solve one of the equations for x in terms of y and substitute this expression into the other equation.

EXAMPLE 2

Solve the following two simultaneous equations:

$$(1) \quad 5x + y = -8$$

$$(2) \quad 2x - 2y = 4$$

Solution From (2), $x = y + 2$. Substitution of this into (1) gives

$$5(y + 2) + y = -8$$

$$6y = -18$$

$$y = -3$$

$$x = y + 2 = \boxed{-1}$$

Alternate Solution Multiply each term in (1) by the factor 2 and add the result to (2):

$$10x + 2y = -16$$

$$\underline{2x - 2y = 4}$$

$$12x = -12$$

$$x = -1$$

$$y = x - 2 = \boxed{-3}$$

Two linear equations containing two unknowns can also be solved by a graphical method. If the straight lines corresponding to the two equations are plotted in a conventional coordinate system, the intersection of the two lines represents the solution. For example, consider the two equations

$$x - y = 2$$

$$x - 2y = -1$$

These are plotted in Figure B.3. The intersection of the two lines has the coordinates $x = 5$, $y = 3$. This represents the solution to the equations. You should check this solution by the analytical technique discussed above.

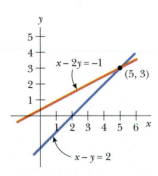

FIGURE B.3

EXERCISES

Solve the following pairs of simultaneous equations involving two unknowns:

		Answers
1.	$x + y = 8$	$x = 5, y = 3$
	$x - y = 2$	
2.	$98 - T = 10a$	$T = 65, a = 3.27$
	$T - 49 = 5a$	
3.	$6x + 2y = 6$	$x = 2, y = -3$
	$8x - 4y = 28$	

Logarithms

Suppose that a quantity x is expressed as a power of some quantity a:

$$x = a^y \tag{B.11}$$

The number a is called the **base** number. The **logarithm** of x with respect to the base a is equal to the exponent to which the base must be raised in order to satisfy the expression $x = a^y$:

$$y = \log_a x \tag{B.12}$$

Conversely, the **antilogarithm** of y is the number x:

$$x = \text{antilog}_a y \tag{B.13}$$

In practice, the two bases most often used are base 10, called the *common* logarithm base, and base $e = 2.718$ . . . , called the *natural* logarithm base. When common logarithms are used,

$$y = \log_{10} x \qquad (\text{or } x = 10^y) \tag{B.14}$$

When natural logarithms are used,

$$y = \ln_e x \qquad (\text{or } x = e^y) \tag{B.15}$$

For example, $\log_{10} 52 = 1.716$, so that $\text{antilog}_{10} 1.716 = 10^{1.716} = 52$. Likewise, $\ln_e 52 = 3.951$, so that $\text{antiln}_e 3.951 = e^{3.951} = 52$.

In general, note that you can convert between base 10 and base e with the equality

$$\ln_e x = (2.302585) \log_{10} x \tag{B.16}$$

Finally, some useful properties of logarithms are

$$\log(ab) = \log a + \log b$$
$$\log(a/b) = \log a - \log b$$
$$\log(a^n) = n \log a$$
$$\ln e = 1$$
$$\ln e^a = a$$
$$\ln\left(\frac{1}{a}\right) = -\ln a$$

B.3   GEOMETRY

The **distance** d between two points having coordinates (x_1, y_1) and (x_2, y_2) is

$$d = \sqrt{(x_2 - x_1)^2 + (y_2 - y_1)^2} \tag{B.17}$$

Radian measure: The arc length s of a circular arc (Fig. B.4) is proportional to the radius r for a fixed value of θ (in radians):

$$s = r\theta$$

$$\theta = \frac{s}{r} \qquad \text{(B.18)}$$

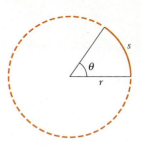

FIGURE B.4

Table B.2 gives the areas and volumes for several geometric shapes used throughout this text:

TABLE B.2	Useful Information for Geometry		
Shape	**Area or Volume**	**Shape**	**Area or Volume**

Rectangle: $\text{Area} = \ell w$

Sphere: $\text{Surface area} = 4\pi r^2$, $\text{Volume} = \dfrac{4\pi r^3}{3}$

Circle: $\text{Area} = \pi r^2$ $(\text{Circumference} = 2\pi r)$

Cylinder: $\text{Volume} = \pi r^2 \ell$

Triangle: $\text{Area} = \frac{1}{2}bh$

Rectangular box: $\text{Area} = 2(\ell h + \ell w + hw)$, $\text{Volume} = \ell w h$

The equation of a **straight line** (Fig. B.5) is

$$y = mx + b \qquad \text{(B.19)}$$

where b is the y-intercept and m is the slope of the line.

The equation of a **circle** or radius R centered at the origin is

$$x^2 + y^2 = R^2 \qquad \text{(B.20)}$$

The equation of an **ellipse** having the origin at its center (Fig. B.6) is

$$\frac{x^2}{a^2} + \frac{y^2}{b^2} = 1 \qquad \text{(B.21)}$$

where a is the length of the semi-major axis (the longer one) and b is the length of the semi-minor axis (the shorter one).

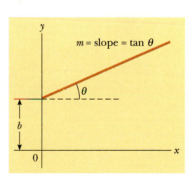

FIGURE B.5

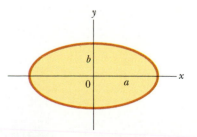

FIGURE B.6

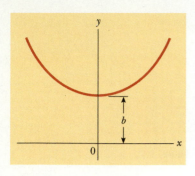

FIGURE B.7

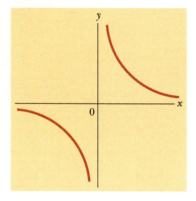

FIGURE B.8

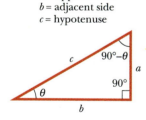

a = opposite side
b = adjacent side
c = hypotenuse

FIGURE B.9

The equation of a **parabola** the vertex of which is at $y = b$ (Fig. B.7) is

$$y = ax^2 + b \tag{B.22}$$

The equation of a **rectangular hyperbola** (Fig. B.8) is

$$xy = \text{constant} \tag{B.23}$$

B.4 TRIGONOMETRY

That portion of mathematics based on the special properties of the right triangle is called trigonometry. By definition, a right triangle is one containing at 90° angle. Consider the right triangle shown in Figure B.9, where side a is opposite the angle θ, side b is adjacent to the angle θ, and side c is the hypotenuse of the triangle. The three basic trigonometric functions defined by such a triangle are the sine (sin), cosine (cos), and tangent (tan) functions. In terms of the angle θ, these functions are defined by

$$\sin \theta \equiv \frac{\text{side opposite } \theta}{\text{hypotenuse}} = \frac{a}{c} \tag{B.24}$$

$$\cos \theta \equiv \frac{\text{side adjacent to } \theta}{\text{hypotenuse}} = \frac{b}{c} \tag{B.25}$$

$$\tan \theta \equiv \frac{\text{side opposite } \theta}{\text{side adjacent to } \theta} = \frac{a}{b} \tag{B.26}$$

The Pythagorean theorem provides the following relationship between the sides of a right triangle:

$$c^2 = a^2 + b^2 \tag{B.27}$$

From the above definitions and the Pythagorean theorem, it follows that

$$\sin^2 \theta + \cos^2 \theta = 1$$

$$\tan \theta = \frac{\sin \theta}{\cos \theta}$$

The cosecant, secant, and cotangent functions are defined by

$$\csc \theta \equiv \frac{1}{\sin \theta} \qquad \sec \theta \equiv \frac{1}{\cos \theta} \qquad \cot \theta \equiv \frac{1}{\tan \theta}$$

The relationship below follow directly from the right triangle shown in Figure B.9:

$$\sin \theta = \cos(90° - \theta)$$

$$\cos \theta = \sin(90° - \theta)$$

$$\cot \theta = \tan(90° - \theta)$$

Some properties of trigonometric functions are

$$\sin(-\theta) = -\sin \theta$$

$$\cos(-\theta) = \cos \theta$$

$$\tan(-\theta) = -\tan \theta$$

The following relationships apply to *any* triangle, as shown in Figure B.10:

$$\alpha + \beta + \gamma = 180°$$

Law of cosines
$$a^2 = b^2 + c^2 - 2bc \cos \alpha$$
$$b^2 = a^2 + c^2 - 2ac \cos \beta$$
$$c^2 = a^2 + b^2 - 2ab \cos \gamma$$

Law of sines
$$\frac{a}{\sin \alpha} = \frac{b}{\sin \beta} = \frac{c}{\sin \gamma}$$

Table B.3 lists a number of useful trigonometric identities.

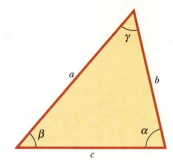

FIGURE B.10

TABLE B.3 **Some Trigonometric Identities**

$$\sin^2 \theta + \cos^2 \theta = 1 \qquad\qquad \csc^2 \theta = 1 + \cot^2 \theta$$

$$\sec^2 \theta = 1 + \tan^2 \theta \qquad\qquad \sin^2 \frac{\theta}{2} = \tfrac{1}{2}(1 - \cos \theta)$$

$$\sin 2\theta = 2 \sin \theta \cos \theta \qquad\qquad \cos^2 \frac{\theta}{2} = \tfrac{1}{2}(1 + \cos \theta)$$

$$\cos 2\theta = \cos^2 \theta - \sin^2 \theta \qquad\qquad 1 - \cos \theta = 2 \sin^2 \frac{\theta}{2}$$

$$\tan 2\theta = \frac{2 \tan \theta}{1 - \tan^2 \theta} \qquad\qquad \tan \frac{\theta}{2} = \sqrt{\frac{1 - \cos \theta}{1 + \cos \theta}}$$

$$\sin(A \pm B) = \sin A \cos B \pm \cos A \sin B$$
$$\cos(A \pm B) = \cos A \cos B \mp \sin A \sin B$$
$$\sin A \pm \sin B = 2 \sin[\tfrac{1}{2}(A \pm B)]\cos[\tfrac{1}{2}(A \mp B)]$$
$$\cos A + \cos B = 2 \cos[\tfrac{1}{2}(A + B)]\cos[\tfrac{1}{2}(A - B)]$$
$$\cos A - \cos B = 2 \sin[\tfrac{1}{2}(A + B)]\sin[\tfrac{1}{2}(B - A)]$$

EXAMPLE 3

Consider the right triangle in Figure B.11, in which $a = 2$, $b = 5$, and c is unknown. From the Pythagorean theorem, we have

$$c^2 = a^2 + b^2 = 2^2 + 5^2 = 4 + 25 = 29$$

$$c = \sqrt{29} = \boxed{5.39}$$

To find the angle θ, note that

$$\tan \theta = \frac{a}{b} = \frac{2}{5} = 0.400$$

From a table of functions or from a calculator, we have

$$\theta = \tan^{-1}(0.400) = \boxed{21.8°}$$

where $\tan^{-1}(0.400)$ is the notation for "angle whose tangent is 0.400," sometimes written as $\arctan(0.400)$.

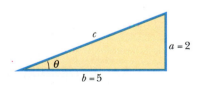

FIGURE B.11

EXERCISES

1. In Figure B.12, identify (a) the side opposite θ and (b) the side adjacent to ϕ and then find (c) $\cos \theta$, (d) $\sin \phi$, and (e) $\tan \phi$.

Answers (a) 3, (b) 3, (c) $\frac{4}{5}$, (d) $\frac{4}{5}$, and (e) $\frac{4}{3}$

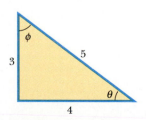

FIGURE B.12

2. In a certain right triangle, the two sides that are perpendicular to each other are 5 m and 7 m long. What is the length of the third side?

Answer 8.60 m

3. A right triangle has a hypotenuse of length 3 m, and one of its angles is 30°. What is the length of (a) the side opposite the 30° angle and (b) the side adjacent to the 30° angle?

Answers (a) 1.5 m, (b) 2.60 m

B.5   SERIES EXPANSIONS

$$(a + b)^n = a^n + \frac{n}{1!} a^{n-1}b + \frac{n(n-1)}{2!} a^{n-2}b^2 + \cdots$$

$$(1 + x)^n = 1 + nx + \frac{n(n-1)}{2!} x^2 + \cdots$$

$$e^x = 1 + x + \frac{x^2}{2!} + \frac{x^3}{3!} + \cdots$$

$$\ln(1 \pm x) = \pm x - \tfrac{1}{2}x^2 \pm \tfrac{1}{3}x^3 - \cdots$$

$$\left.\begin{array}{l} \sin x = x - \dfrac{x^3}{3!} + \dfrac{x^5}{5!} - \cdots \\[2ex] \cos x = 1 - \dfrac{x^2}{2!} + \dfrac{x^4}{4!} - \cdots \\[2ex] \tan x = x + \dfrac{x^3}{3} + \dfrac{2x^5}{15} + \cdots \qquad |x| < \pi/2 \end{array}\right\} x \text{ in radians}$$

For $x \ll 1$, the following approximations can be used:

$$(1 + x)^n \approx 1 + nx \qquad \sin x \approx x$$

$$e^x \approx 1 + x \qquad \cos x \approx 1$$

$$\ln(1 \pm x) \approx \pm x \qquad \tan x \approx x$$

B.6   DIFFERENTIAL CALCULUS

In various branches of science, it is sometimes necessary to use the basic tools of calculus, invented by Newton, to describe physical phenomena. The use of calculus is fundamental in the treatment of various problems in newtonian mechanics, electricity, and magnetism. In this section, we simply state some basic properties and "rules of thumb" that should be a useful review to the student.

First, a **function** must be specified that relates one variable to another (such as a coordinate as a function of time). Suppose one of the variables is called y (the dependent variable), the other x (the independent variable). We might have a function relationship such as

$$y(x) = ax^3 + bx^2 + cx + d$$

If a, b, c, and d are specified constants, then y can be calculated for any value of x. We usually deal with continuous functions, that is, those for which y varies "smoothly" with x.

The **derivative** of y with respect to x is defined as the limit, as Δx approaches zero, of the slopes of chords drawn between two points on the y versus x curve. Mathematically, we write this definition as

$$\frac{dy}{dx} = \lim_{\Delta x \to 0} \frac{\Delta y}{\Delta x} = \lim_{\Delta x \to 0} \frac{y(x + \Delta x) - y(x)}{\Delta x} \qquad \text{(B.28)}$$

where Δy and Δx are defined as $\Delta x = x_2 - x_1$ and $\Delta y = y_2 - y_1$ (Fig. B.13). It is important to note that dy/dx *does not* mean dy divided by dx, but is simply a notation of the limiting process of the derivative as defined by Equation B.28.

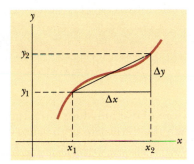

FIGURE B.13

A useful expression to remember when $y(x) = ax^n$, where a is a *constant* and n is *any* positive or negative number (integer or fraction), is

$$\frac{dy}{dx} = nax^{n-1} \qquad \text{(B.29)}$$

If $y(x)$ is a polynomial or algebraic function of x, we apply Equation B.29 to *each* term in the polynomial and take $da/dx = 0$. In Examples 4 through 7, we evaluate the derivatives of several functions.

EXAMPLE 4

Suppose $y(x)$ (that is, y as a function of x) is given by

$$y(x) = ax^3 + bx + c$$

where a and b are constants. Then it follows that

$$y(x + \Delta x) = a(x + \Delta x)^3$$
$$+ b(x + \Delta x) + c$$

$$y(x + \Delta x) = a(x^3 + 3x^2\,\Delta x + 3x\,\Delta x^2 + \Delta x^3)$$
$$+ b(x + \Delta x) + c$$

so

$$\Delta y = y(x + \Delta x) - y(x) = a(3x^2\,\Delta x + 3x\,\Delta x^2 + \Delta x^3)$$
$$+ b\,\Delta x$$

Substituting this into Equation B.28 gives

$$\frac{dy}{dx} = \lim_{\Delta x \to 0} \frac{\Delta y}{\Delta x} = \lim_{\Delta x \to 0} [3ax^2 + 3x\,\Delta x + \Delta x^2] + b$$

$$\boxed{\frac{dy}{dx} = 3ax^2 + b}$$

EXAMPLE 5

$$y(x) = 8x^5 + 4x^3 + 2x + 7$$

Solution Applying Equation B.29 to each term independently, and remembering that d/dx (constant) $= 0$, we have

$$\frac{dy}{dx} = 8(5)x^4 + 4(3)x^2 + 2(1)x^0 + 0$$

$$\frac{dy}{dx} = 40x^4 + 12x^2 + 2$$

Special Properties of the Derivative

A. **Derivative of the product of two functions** If a function $f(x)$ is given by the product of two functions, say, $g(x)$ and $h(x)$, then the derivative of $f(x)$ is defined as

$$\frac{d}{dx}f(x) = \frac{d}{dx}[g(x)h(x)] = g\frac{dh}{dx} + h\frac{dg}{dx} \qquad \text{(B.30)}$$

B. **Derivative of the sum of two functions** If a function $f(x)$ is equal to the sum of two functions, then the derivative of the sum is equal to the sum of the derivatives:

$$\frac{d}{dx}f(x) = \frac{d}{dx}[g(x) + h(x)] = \frac{dg}{dx} + \frac{dh}{dx} \qquad \text{(B.31)}$$

C. **Chain rule of differential calculus** If $y = f(x)$ and $x = f(z)$, then dy/dx can be written as the product of two derivatives:

$$\frac{dy}{dx} = \frac{dy}{dz}\frac{dz}{dx} \qquad \text{(B.32)}$$

D. **The second derivative** The second derivative of y with respect to x is defined as the derivative of the function dy/dx (the derivative of the derivative). It is usually written

$$\frac{d^2y}{dx^2} = \frac{d}{dx}\left(\frac{dy}{dx}\right) \qquad \text{(B.33)}$$

EXAMPLE 6

Find the derivative of $y(x) = x^3/(x+1)^2$ with respect to x.

Solution We can rewrite this function as $y(x) = x^3(x+1)^{-2}$ and apply Equation B.30:

$$\frac{dy}{dx} = (x+1)^{-2}\frac{d}{dx}(x^3) + x^3\frac{d}{dx}(x+1)^{-2}$$

$$= (x+1)^{-2}3x^2 + x^3(-2)(x+1)^{-3}$$

$$\frac{dy}{dx} = \frac{3x^2}{(x+1)^2} - \frac{2x^3}{(x+1)^3}$$

EXAMPLE 7

A useful formula that follows from Equation B.30 is the derivative of the quotient of two functions. Show that

$$\frac{d}{dx}\left[\frac{g(x)}{h(x)}\right] = \frac{h\dfrac{dg}{dx} - g\dfrac{dh}{dx}}{h^2}$$

Solution   We can write the quotient as gh^{-1} and then apply Equations B.29 and B.30:

$$\frac{d}{dx}\left(\frac{g}{h}\right) = \frac{d}{dx}(gh^{-1}) = g\frac{d}{dx}(h^{-1}) + h^{-1}\frac{d}{dx}(g)$$

$$= -gh^{-2}\frac{dh}{dx} + h^{-1}\frac{dg}{dx}$$

$$= \frac{h\dfrac{dg}{dx} - g\dfrac{dh}{dx}}{h^2}$$

Some of the more commonly used derivatives of functions are listed in Table B.4.

B.7   INTEGRAL CALCULUS

We think of integration as the inverse of differentiation. As an example, consider the expression

$$f(x) = \frac{dy}{dx} = 3ax^2 + b \qquad \text{(B.34)}$$

which was the result of differentiating the function

$$y(x) = ax^3 + bx + c$$

in Example 4. We can write Equation B.34 as $dy = f(x)\,dx = (3ax^2 + b)\,dx$ and obtain $y(x)$ by "summing" over all values of x. Mathematically, we write this inverse operation

$$y(x) = \int f(x)\,dx$$

For the function $f(x)$ given by Equation B.34, we have

$$y(x) = \int (3ax^2 + b)\,dx = ax^3 + bx + c$$

where c is a constant of the integration. This type of integral is called an *indefinite integral* because its value depends on the choice of c.

A general **indefinite integral** $I(x)$ is defined as

$$I(x) = \int f(x)\,dx \qquad \text{(B.35)}$$

where $f(x)$ is called the *integrand* and $f(x) = \dfrac{dI(x)}{dx}$.

For a *general continuous* function $f(x)$, the integral can be described as the area under the curve bounded by $f(x)$ and the x axis, between two specified values of x, say, x_1 and x_2, as in Figure B.14.

The area of the blue element is approximately $f_i\,\Delta x_i$. If we sum all these area elements from x_1 to x_2 and take the limit of this sum as $\Delta x_i \to 0$, we obtain the *true* area under the curve bounded by $f(x)$ and x, between the limits x_1 and x_2:

$$\text{Area} = \lim_{\Delta x_i \to 0} \sum_i f(x_i)\,\Delta x_i = \int_{x_1}^{x_2} f(x)\,dx \qquad \text{(B.36)}$$

Integrals of the type defined by Equation B.36 are called **definite integrals**.

TABLE B.4	Derivatives for Several Functions

$$\frac{d}{dx}(a) = 0$$

$$\frac{d}{dx}(ax^n) = nax^{n-1}$$

$$\frac{d}{dx}(e^{ax}) = ae^{ax}$$

$$\frac{d}{dx}(\sin ax) = a\cos ax$$

$$\frac{d}{dx}(\cos ax) = -a\sin ax$$

$$\frac{d}{dx}(\tan ax) = a\sec^2 ax$$

$$\frac{d}{dx}(\cot ax) = -a\csc^2 ax$$

$$\frac{d}{dx}(\sec x) = \tan x \sec x$$

$$\frac{d}{dx}(\csc x) = -\cot x \csc x$$

$$\frac{d}{dx}(\ln ax) = \frac{1}{x}$$

Note: The letters a and n are constants.

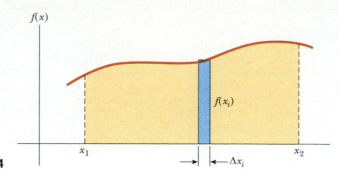

FIGURE B.14

One common integral that arises in practical situations has the form

$$\int x^n \, dx = \frac{x^{n+1}}{n+1} + c \qquad (n \ne -1) \tag{B.37}$$

This result is obvious, being that differentiation of the right-hand side with respect to x gives $f(x) = x^n$ directly. If the limits of the integration are known, this integral becomes a *definite integral* and is written

$$\int_{x_1}^{x_2} x^n \, dx = \frac{x_2^{n+1} - x_1^{n+1}}{n+1} \qquad (n \ne -1) \tag{B.38}$$

EXAMPLES

1. $\displaystyle\int_0^a x^2 \, dx = \frac{x^3}{3}\bigg]_0^a = \frac{a^3}{3}$

2. $\displaystyle\int_0^b x^{3/2} \, dx = \frac{x^{5/2}}{5/2}\bigg]_0^b = \frac{2}{5} b^{5/2}$

3. $\displaystyle\int_3^5 x \, dx = \frac{x^2}{2}\bigg]_3^5 = \frac{5^2 - 3^2}{2} = 8$

Partial Integration

Sometimes it is useful to apply the method of *partial integration* (also called "integrating by parts") to evaluate certain integrals. The method uses the property that

$$\int u \, dv = uv - \int v \, du \tag{B.39}$$

where u and v are *carefully* chosen so as to reduce a complex integral to a simpler one. In many cases, several reductions have to be made. Consider the function

$$I(x) = \int x^2 e^x \, dx$$

This can be evaluated by integrating by parts twice. First, if we choose $u = x^2$, $v = e^x$, we get

$$\int x^2 e^x \, dx = \int x^2 \, d(e^x) = x^2 e^x - 2 \int e^x x \, dx + c_1$$

Now, in the second term, choose $u = x$, $v = e^x$, which gives

$$\int x^2 e^x \, dx = x^2 e^x - 2xe^x + 2 \int e^x \, dx + c_1$$

or

$$\int x^2 e^x \, dx = x^2 e^x - 2xe^x + 2e^x + c_2$$

The Perfect Differential

Another useful method to remember is the use of the *perfect differential*, in which we look for a change of variable such that the differential of the function is the differential of the independent variable appearing in the integrand. For example, consider the integral

$$I(x) = \int \cos^2 x \sin x \, dx$$

This becomes easy to evaluate if we rewrite the differential as $d(\cos x) = -\sin x \, dx$. The integral then becomes

$$\int \cos^2 x \sin x \, dx = -\int \cos^2 x \, d(\cos x)$$

If we now change variables, letting $y = \cos x$, we get

$$\int \cos^2 x \sin x \, dx = -\int y^2 dy = -\frac{y^3}{3} + c = -\frac{\cos^3 x}{3} + c$$

Table B.5 lists some useful indefinite integrals. Table B.6 gives Gauss's probability integral and other definite integrals. A more complete list can be found in various handbooks, such as *The Handbook of Chemistry and Physics,* CRC Press.

TABLE B.5   Some Indefinite Integrals (an arbitrary constant should be added to each of these integrals)

$\displaystyle\int x^n \, dx = \frac{x^{n+1}}{n+1}$ (provided $n \neq -1$)	$\displaystyle\int xe^{ax} \, dx = \frac{e^{ax}}{a^2}(ax - 1)$
$\displaystyle\int \frac{dx}{x} = \int x^{-1} \, dx = \ln x$	$\displaystyle\int \frac{dx}{a + be^{cx}} = \frac{x}{a} - \frac{1}{ac}\ln(a + be^{cx})$
$\displaystyle\int \frac{dx}{a + bx} = \frac{1}{b}\ln(a + bx)$	$\displaystyle\int \sin ax \, dx = -\frac{1}{a}\cos ax$
$\displaystyle\int \frac{dx}{(a + bx)^2} = -\frac{1}{b(a + bx)}$	$\displaystyle\int \cos ax \, dx = \frac{1}{a}\sin ax$
$\displaystyle\int \frac{dx}{a^2 + x^2} = \frac{1}{a}\tan^{-1}\frac{x}{a}$	$\displaystyle\int \tan ax \, dx = -\frac{1}{a}\ln(\cos ax) = \frac{1}{a}\ln(\sec ax)$
$\displaystyle\int \frac{dx}{a^2 - x^2} = \frac{1}{2a}\ln\frac{a + x}{a - x}$ $(a^2 - x^2 > 0)$	$\displaystyle\int \cot ax \, dx = \frac{1}{a}\ln(\sin ax)$
$\displaystyle\int \frac{dx}{x^2 - a^2} = \frac{1}{2a}\ln\frac{x - a}{x + a}$ $(x^2 - a^2 > 0)$	$\displaystyle\int \sec ax \, dx = \frac{1}{a}\ln(\sec ax + \tan ax) = \frac{1}{a}\ln\left[\tan\left(\frac{ax}{2} + \frac{\pi}{4}\right)\right]$
$\displaystyle\int \frac{x \, dx}{a^2 \pm x^2} = \pm\tfrac{1}{2}\ln(a^2 \pm x^2)$	$\displaystyle\int \csc ax \, dx = \frac{1}{a}\ln(\csc ax - \cot ax) = \frac{1}{a}\ln\left(\tan\frac{ax}{2}\right)$
$\displaystyle\int \frac{dx}{\sqrt{a^2 - x^2}} = \sin^{-1}\frac{x}{a} = -\cos^{-1}\frac{x}{a}$ $(a^2 - x^2 > 0)$	$\displaystyle\int \sin^2 ax \, dx = \frac{x}{2} - \frac{\sin 2ax}{4a}$
$\displaystyle\int \frac{dx}{\sqrt{x^2 \pm a^2}} = \ln(x + \sqrt{x^2 \pm a^2})$	$\displaystyle\int \cos^2 ax \, dx = \frac{x}{2} + \frac{\sin 2ax}{4a}$

continued

TABLE B.5 *Continued*

$$\int \frac{x\,dx}{\sqrt{a^2 - x^2}} = -\sqrt{a^2 - x^2}$$

$$\int \frac{dx}{\sin^2 ax} = -\frac{1}{a}\cot ax$$

$$\int \frac{x\,dx}{\sqrt{x^2 \pm a^2}} = \sqrt{x^2 \pm a^2}$$

$$\int \frac{dx}{\cos^2 ax} = \frac{1}{a}\tan ax$$

$$\int \sqrt{a^2 - x^2}\,dx = \tfrac{1}{2}\left(x\sqrt{a^2 - x^2} + a^2 \sin^{-1}\frac{x}{a} \right)$$

$$\int \tan^2 ax\,dx = \frac{1}{a}(\tan ax) - x$$

$$\int x\sqrt{a^2 - x^2}\,dx = -\tfrac{1}{3}(a^2 - x^2)^{3/2}$$

$$\int \cot^2 ax\,dx = -\frac{1}{a}(\cot ax) - x$$

$$\int \sqrt{x^2 \pm a^2}\,dx = \tfrac{1}{2}[x\sqrt{x^2 \pm a^2} \pm a^2 \ln(x + \sqrt{x^2 \pm a^2})]$$

$$\int \sin^{-1} ax\,dx = x(\sin^{-1} ax) + \frac{\sqrt{1 - a^2 x^2}}{a}$$

$$\int x(\sqrt{x^2 \pm a^2})\,dx = \tfrac{1}{3}(x^2 \pm a^2)^{3/2}$$

$$\int \cos^{-1} ax\,dx = x(\cos^{-1} ax) - \frac{\sqrt{1 - a^2 x^2}}{a}$$

$$\int e^{ax}\,dx = \frac{1}{a}e^{ax}$$

$$\int \frac{dx}{(x^2 + a^2)^{3/2}} = \frac{x}{a^2\sqrt{x^2 + a^2}}$$

$$\int \ln ax\,dx = (x \ln ax) - x$$

$$\int \frac{x\,dx}{(x^2 + a^2)^{3/2}} = -\frac{1}{\sqrt{x^2 + a^2}}$$

TABLE B.6 **Gauss's Probability Integral and Related Integrals**

$$I_0 = \int_0^\infty e^{-\alpha x^2}\,dx = \tfrac{1}{2}\sqrt{\frac{\pi}{\alpha}} \qquad \text{(Gauss's probability integral)}$$

$$I_1 = \int_0^\infty x e^{-\alpha x^2}\,dx = \frac{1}{2\alpha}$$

$$I_2 = \int_0^\infty x^2 e^{-\alpha x^2}\,dx = -\frac{dI_0}{d\alpha} = \tfrac{1}{4}\sqrt{\frac{\pi}{\alpha^3}}$$

$$I_3 = \int_0^\infty x^3 e^{-\alpha x^2}\,dx = -\frac{dI_1}{d\alpha} = \frac{1}{2\alpha^2}$$

$$I_4 = \int_0^\infty x^4 e^{-\alpha x^2}\,dx = \frac{d^2 I_0}{d\alpha^2} = \tfrac{3}{8}\sqrt{\frac{\pi}{\alpha^5}}$$

$$I_5 = \int_0^\infty x^5 e^{-\alpha x^2}\,dx = \frac{d^2 I_1}{d\alpha^2} = \frac{1}{\alpha^3}$$

$$\vdots$$

$$I_{2n} = (-1)^n \frac{d^n}{d\alpha^n} I_0$$

$$I_{2n+1} = (-1)^n \frac{d^n}{d\alpha^n} I_1$$

Periodic Table of the Elements*

Legend:

Symbol — **Ca** 20 — Atomic number
Atomic mass † — 40.08
$4s^2$ — Electron configuration

Group I	Group II	Transition elements						
H 1 $\quad$ 1.0080 $\quad$ $1s^1$								
Li 3 $\quad$ 6.94 $\quad$ $2s^1$	**Be** 4 $\quad$ 9.012 $\quad$ $2s^2$							
Na 11 $\quad$ 22.99 $\quad$ $3s^1$	**Mg** 12 $\quad$ 24.31 $\quad$ $3s^2$							
K 19 $\quad$ 39.102 $\quad$ $4s^1$	**Ca** 20 $\quad$ 40.08 $\quad$ $4s^2$	**Sc** 21 $\quad$ 44.96 $\quad$ $3d^14s^2$	**Ti** 22 $\quad$ 47.90 $\quad$ $3d^24s^2$	**V** 23 $\quad$ 50.94 $\quad$ $3d^34s^2$	**Cr** 24 $\quad$ 51.996 $\quad$ $3d^54s^1$	**Mn** 25 $\quad$ 54.94 $\quad$ $3d^54s^2$	**Fe** 26 $\quad$ 55.85 $\quad$ $3d^64s^2$	**Co** 27 $\quad$ 58.93 $\quad$ $3d^74s^2$
Rb 37 $\quad$ 85.47 $\quad$ $5s^1$	**Sr** 38 $\quad$ 87.62 $\quad$ $5s^2$	**Y** 39 $\quad$ 88.906 $\quad$ $4d^15s^2$	**Zr** 40 $\quad$ 91.22 $\quad$ $4d^25s^2$	**Nb** 41 $\quad$ 92.91 $\quad$ $4d^45s^1$	**Mo** 42 $\quad$ 95.94 $\quad$ $4d^55s^1$	**Tc** 43 $\quad$ (99) $\quad$ $4d^55s^2$	**Ru** 44 $\quad$ 101.1 $\quad$ $4d^75s^1$	**Rh** 45 $\quad$ 102.91 $\quad$ $4d^85s^1$
Cs 55 $\quad$ 132.91 $\quad$ $6s^1$	**Ba** 56 $\quad$ 137.34 $\quad$ $6s^2$	57-71*	**Hf** 72 $\quad$ 178.49 $\quad$ $5d^26s^2$	**Ta** 73 $\quad$ 180.95 $\quad$ $5d^36s^2$	**W** 74 $\quad$ 183.85 $\quad$ $5d^46s^2$	**Re** 75 $\quad$ 186.2 $\quad$ $5d^56s^2$	**Os** 76 $\quad$ 190.2 $\quad$ $5d^66s^2$	**Ir** 77 $\quad$ 192.2 $\quad$ $5d^76s^2$
Fr 87 $\quad$ (223) $\quad$ $7s^1$	**Ra** 88 $\quad$ (226) $\quad$ $7s^2$	89-103**	**Unq** 104 $\quad$ (261) $\quad$ $6d^27s^2$	**Unp** 105 $\quad$ (262) $\quad$ $6d^37s^2$	**Unh** 106 $\quad$ (263)	**Uns** 107 $\quad$ (262)	**Uno** 108 $\quad$ (265)	**Une** 109 $\quad$ (266)

*Lanthanide series

La 57 $\quad$ 138.91 $\quad$ $5d^16s^2$	**Ce** 58 $\quad$ 140.12 $\quad$ $5d^14f^16s^2$	**Pr** 59 $\quad$ 140.91 $\quad$ $4f^36s^2$	**Nd** 60 $\quad$ 144.24 $\quad$ $4f^46s^2$	**Pm** 61 $\quad$ (147) $\quad$ $4f^56s^2$	**Sm** 62 $\quad$ 150.4 $\quad$ $4f^66s^2$

**Actinide series

Ac 89 $\quad$ (227) $\quad$ $6d^17s^2$	**Th** 90 $\quad$ (232) $\quad$ $6d^27s^2$	**Pa** 91 $\quad$ (231) $\quad$ $5f^26d^17s^2$	**U** 92 $\quad$ (238) $\quad$ $5f^36d^17s^2$	**Np** 93 $\quad$ (239) $\quad$ $5f^46d^17s^2$	**Pu** 94 $\quad$ (239) $\quad$ $5f^66d^07s^2$

□ Atomic mass values given are averaged over isotopes in the percentages in which they exist in nature.
† For an unstable element, mass number of the most stable known isotope is given in parentheses.

Group III	Group IV	Group V	Group VI	Group VII	Group 0
				H 1 1.0080 $1s^1$	**He** 2 4.0026 $1s^2$
B 5 10.81 $2p^1$	**C** 6 12.011 $2p^2$	**N** 7 14.007 $2p^3$	**O** 8 15.999 $2p^4$	**F** 9 18.998 $2p^5$	**Ne** 10 20.18 $2p^6$
Al 13 26.98 $3p^1$	**Si** 14 28.09 $3p^2$	**P** 15 30.97 $3p^3$	**S** 16 32.06 $3p^4$	**Cl** 17 35.453 $3p^5$	**Ar** 18 39.948 $3p^6$

Ni 28 58.71 $3d^8 4s^2$	**Cu** 29 63.54 $3d^{10}4s^2$	**Zn** 30 65.37 $3d^{10}4s^2$	**Ga** 31 69.72 $4p^1$	**Ge** 32 72.59 $4p^2$	**As** 33 74.92 $4p^3$	**Se** 34 78.96 $4p^4$	**Br** 35 79.91 $4p^5$	**Kr** 36 83.80 $4p^6$
Pd 46 106.4 $4d^{10}$	**Ag** 47 107.87 $4d^{10}5s^1$	**Cd** 48 112.40 $4d^{10}5s^2$	**In** 49 114.82 $5p^1$	**Sn** 50 118.69 $5p^2$	**Sb** 51 121.75 $5p^3$	**Te** 52 127.60 $5p^4$	**I** 53 126.90 $5p^5$	**Xe** 54 131.30 $5p^6$
Pt 78 195.09 $5d^9 6s^1$	**Au** 79 196.97 $5d^{10}6s^1$	**Hg** 80 200.59 $5d^{10}6s^2$	**Tl** 81 204.37 $6p^1$	**Pb** 82 207.2 $6p^2$	**Bi** 83 208.98 $6p^3$	**Po** 84 (210) $6p^4$	**At** 85 (218) $6p^5$	**Rn** 86 (222) $6p^6$

Eu 63 152.0 $4f^7 6s^2$	**Gd** 64 157.25 $5d^1 4f^7 6s^2$	**Tb** 65 158.92 $5d^1 4f^8 6s^2$	**Dy** 66 162.50 $4f^{10}6s^2$	**Ho** 67 164.93 $4f^{11}6s^2$	**Er** 68 167.26 $4f^{12}6s^2$	**Tm** 69 168.93 $4f^{13}6s^2$	**Yb** 70 173.04 $4f^{14}6s^2$	**Lu** 71 174.97 $5d^1 4f^{14}6s^2$
Am 95 (243) $5f^7 6d^0 7s^2$	**Cm** 96 (245) $5f^7 6d^1 7s^2$	**Bk** 97 (247) $5f^8 6d^1 7s^2$	**Cf** 98 (249) $5f^{10}6d^0 7s^2$	**Es** 99 (254) $5f^{11}6d^0 7s^2$	**Fm** 100 (253) $5f^{12}6d^0 7s^2$	**Md** 101 (255) $5f^{13}6d^0 7s^2$	**No** 102 (255) $6d^0 7s^2$	**Lr** 103 (257) $6d^1 7s^2$

SI Units

TABLE D.1 SI Base Units

Base Quantity	SI Base Unit	
	Name	Symbol
Length	Meter	m
Mass	Kilogram	kg
Time	Second	s
Electric current	Ampere	A
Temperature	Kelvin	K
Amount of substance	Mole	mol
Luminous intensity	Candela	cd

TABLE D.2 Some Derived SI Units

Quantity	Name	Symbol	Expression in Terms of Base Units	Expression in Terms of Other SI Units
Plane angle	Radian	rad	m/m	
Frequency	Hertz	Hz	s^{-1}	
Force	Newton	N	$kg \cdot m/s^2$	J/m
Pressure	Pascal	Pa	$kg/m \cdot s^2$	N/m^2
Energy: work	Joule	J	$kg \cdot m^2/s^2$	$N \cdot m$
Power	Watt	W	$kg \cdot m^2/s^3$	J/s
Electric charge	Coulomb	C	$A \cdot s$	
Electric potential (emf)	Volt	V	$kg \cdot m^2/A \cdot s^3$	W/A
Capacitance	Farad	F	$A^2 \cdot s^4/kg \cdot m^2$	C/V
Electric resistance	Ohm	Ω	$kg \cdot m^2/A^2 \cdot s^3$	V/A
Magnetic flux	Weber	Wb	$kg \cdot m^2/A \cdot s^2$	$V \cdot s$
Magnetic field intensity	Tesla	T	$kg/A \cdot s^2$	Wb/m^2
Inductance	Henry	H	$kg \cdot m^2/A^2 \cdot s^2$	Wb/A

Nobel Prizes

All Nobel Prizes in physics are listed (and marked with a P), as well as relevant Nobel Prizes in Chemistry (C). The key dates for some of the scientific work are supplied; they often antedate the prize considerably.

1901 (P) *Wilhelm Roentgen* for discovering x-rays (1895).

1902 (P) *Hendrik A. Lorentz* for predicting the Zeeman effect and *Pieter Zeeman* for discovering the Zeeman effect, the splitting of spectral lines in magnetic fields.

1903 (P) *Antoine-Henri Becquerel* for discovering radioactivity (1896) and *Pierre* and *Marie Curie* for studying radioactivity.

1904 (P) *Lord Rayleigh* for studying the density of gases and discovering argon.
      (C) *William Ramsay* for discovering the inert gas elements helium, neon, xenon, and krypton, and placing them in the periodic table.

1905 (P) *Philipp Lenard* for studying cathode rays, electrons (1898–1899).

1906 (P) *J. J. Thomson* for studying electrical discharge through gases and discovering the electron (1897).

1907 (P) *Albert A. Michelson* for inventing optical instruments and measuring the speed of light (1880s).

1908 (P) *Gabriel Lippmann* for making the first color photographic plate, using interference methods (1891).
      (C) *Ernest Rutherford* for discovering that atoms can be broken apart by alpha rays and for studying radioactivity.

1909 (P) *Guglielmo Marconi* and *Carl Ferdinand Braun* for developing wireless telegraphy.

1910 (P) *Johannes D. van der Waals* for studying the equation of state for gases and liquids (1881).

1911 (P) *Wilhelm Wien* for discovering Wien's law giving the peak of a blackbody spectrum (1893).
      (C) *Marie Curie* for discovering radium and polonium (1898) and isolating radium.

1912 (P) *Nils Dalén* for inventing automatic gas regulators for lighthouses.

1913 (P) *Heike Kamerlingh Onnes* for the discovery of superconductivity and liquefying helium (1908).

1914 (P) *Max T. F. von Laue* for studying x-rays from their diffraction by crystals, showing that x-rays are electromagnetic waves (1912).
      (C) *Theodore W. Richards* for determining the atomic weights of sixty elements, indicating the existence of isotopes.

1915 (P) *William Henry Bragg* and *William Lawrence Bragg*, his son, for studying the diffraction of x-rays in crystals.

1917 (P) *Charles Barkla* for studying atoms by x-ray scattering (1906).

1918 (P) *Max Planck* for discovering energy quanta (1900).

1919 (P) *Johannes Stark,* for discovering the Stark effect, the splitting of spectral lines in electric fields (1913).

1920   (P) *Charles-Édouard Guillaume* for discovering invar, a nickel-steel alloy with low coefficient of expansion.
(C) *Walther Nernst* for studying heat changes in chemical reactions and formulating the third law of thermodynamics (1918).

1921   (P) *Albert Einstein* for explaining the photoelectric effect and for his services to theoretical physics (1905).
(C) *Frederick Soddy* for studying the chemistry of radioactive substances and discovering isotopes (1912).

1922   (P) *Niels Bohr* for his model of the atom and its radiation (1913).
(C) *Francis W. Aston* for using the mass spectrograph to study atomic weights, thus discovering 212 of the 287 naturally occurring isotopes.

1923   (P) *Robert A. Millikan* for measuring the charge on an electron (1911) and for studying the photoelectric effect experimentally (1914).

1924   (P) *Karl M. G. Siegbahn* for his work in x-ray spectroscopy.

1925   (P) *James Franck* and *Gustav Hertz* for discovering the Franck-Hertz effect in electron-atom collisions.

1926   (P) *Jean-Baptiste Perrin* for studying Brownian motion to validate the discontinuous structure of matter and measure the size of atoms.

1927   (P) *Arthur Holly Compton* for discovering the Compton effect on x-rays, their change in wavelength when they collide with matter (1922), and *Charles T. R. Wilson* for inventing the cloud chamber, used to study charged particles (1906).

1928   (P) *Owen W. Richardson* for studying the thermionic effect and electrons emitted by hot metals (1911).

1929   (P) *Louis Victor de Broglie* for discovering the wave nature of electrons (1923).

1930   (P) *Chandrasekhara Venkata Raman* for studying Raman scattering, the scattering of light by atoms and molecules with a change in wavelength (1928).

1932   (P) *Werner Heisenberg* for creating quantum mechanics (1925).

1933   (P) *Erwin Schrödinger* and *Paul A. M. Dirac* for developing wave mechanics (1925) and relativistic quantum mechanics (1927).
(C) *Harold Urey* for discovering heavy hydrogen, deuterium (1931).

1935   (P) *James Chadwick* for discovering the neutron (1932).
(C) *Irène* and *Frédéric Joliot-Curie* for synthesizing new radioactive elements.

1936   (P) *Carl D. Anderson* for discovering the positron in particular and antimatter in general (1932) and *Victor F. Hess* for discovering cosmic rays.
(C) *Peter J. W. Debye* for studying dipole moments and diffraction of x-rays and electrons in gases.

1937   (P) *Clinton Davisson* and *George Thomson* for discovering the diffraction of electrons by crystals, confirming de Broglie's hypothesis (1927).

1938   (P) *Enrico Fermi* for producing the transuranic radioactive elements by neutron irradiation (1934–1937).

1939   (P) Ernest O. Lawrence for inventing the cyclotron.

1943   (P) *Otto Stern* for developing molecular-beam studies (1923), and using them to discover the magnetic moment of the proton (1933).

1944   (P) *Isidor I. Rabi* for discovering nuclear magnetic resonance in atomic and molecular beams.
(C) *Otto Hahn* for discovering nuclear fission (1938).

1945   (P) *Wolfgang Pauli* for discovering the exclusion principle (1924).

1946   (P) *Percy W. Bridgman* for studying physics at high pressures.

1947   (P) *Edward V. Appleton* for studying the ionosphere.

1948   (P) *Patrick M. S. Blackett* for studying nuclear physics with cloud-chamber photographs of cosmic-ray interactions.

1949   (P) *Hideki Yukawa* for predicting the existence of mesons (1935).

1950   (P) *Cecil F. Powell* for developing the method of studying cosmic rays with photographic emulsions and discovering new mesons.

1951   (P) *John D. Cockcroft* and *Ernest T. S. Walton* for transmuting nuclei in an accelerator (1932).
        (C) *Edwin M. McMillan* for producing neptunium (1940) and *Glenn T. Seaborg* for producing plutonium (1941) and further transuranic elements.

1952   (P) *Felix Bloch* and *Edward Mills Purcell* for discovering nuclear magnetic resonance in liquids and gases (1946).

1953   (P) *Frits Zernike* for inventing the phase-contrast microscope, which uses interference to provide high contrast.

1954   (P) *Max Born* for interpreting the wave function as a probability (1926) and other quantum-mechanical discoveries and *Walther Bothe* for developing the coincidence method to study subatomic particles (1930–1931), producing, in particular, the particle interpreted by Chadwick as the neutron.

1955   (P) *Willis E. Lamb, Jr.,* for discovering the Lamb shift in the hydrogen spectrum (1947) and *Polykarp Kusch* for determining the magnetic moment of the electron (1947).

1956   (P) *John Bardeen, Walter H. Brattain,* and *William Shockley* for inventing the transistor (1956).

1957   (P) *T.-D. Lee* and *C.-N. Yang* for predicting that parity is not conserved in beta decay (1956).

1958   (P) *Pavel A. Čerenkov* for discovering Čerenkov radiation (1935) and *Ilya M. Frank* and *Igor Tamm* for interpreting it (1937).

1959   (P) *Emilio G. Segrè* and *Owen Chamberlain* for discovering the antiproton (1955).

1960   (P) *Donald A. Glaser* for inventing the bubble chamber to study elementary particles (1952).
        (C) *Willard Libby* for developing radiocarbon dating (1947).

1961   (P) *Robert Hofstadter* for discovering internal structure in protons and neutrons and *Rudolf L. Mössbauer* for discovering the Mössbauer effect of recoilless gamma-ray emission (1957).

1962   (P) *Lev Davidovich Landau* for studying liquid helium and other condensed matter theoretically.

1963   (P) *Eugene P. Wigner* for applying symmetry principles to elementary-particle theory and *Maria Goeppert Mayer* and *J. Hans D. Jensen* for studying the shell model of nuclei (1947).

1964   (P) *Charles H. Townes, Nikolai G. Basov,* and *Alexandr M. Prokhorov* for developing masers (1951–1952) and lasers.

1965   (P) *Sin-itiro Tomonaga, Julian S. Schwinger,* and *Richard P. Feynman* for developing quantum electrodynamics (1948).

1966   (P) *Alfred Kastler* for his optical methods of studying atomic energy levels

1967   (P) *Hans Albrecht Bethe* for discovering the routes of energy production in stars (1939).

1968   (P) *Luis W. Alvarez* for discovering resonance states of elementary particles.

1969   (P) *Murray Gell-Mann* for classifying elementary particles (1963).

1970   (P) *Hannes Alfvén* for developing magnetohydrodynamic theory and *Louis Eugène Félix Néel* for discovering antiferromagnetism and ferrimagnetism (1930s).

1971   (P) *Dennis Gabor* for developing holography (1947).

(C) *Gerhard Herzberg* for studying the structure of molecules spectroscopically.

1972   (P) *John Bardeen, Leon N. Cooper,* and *John Robert Schrieffer* for explaining superconductivity (1957).

1973   (P) *Leo Esaki* for discovering tunneling in semiconductors, *Ivar Giaever* for discovering tunneling in superconductors, and *Brian D. Josephson* for predicting the Josephson effect, which involves tunneling of paired electrons (1958–1962).

1974   (P) *Anthony Hewish* for discovering pulsars and *Martin Ryle* for developing radio interferometry.

1975   (P) *Aage N. Bohr, Ben R. Mottelson,* and *James Rainwater* for discovering why some nuclei take asymmetric shapes.

1976   (P) *Burton Richter* and *Samuel C. C. Ting* for discovering the J/psi particle, the first charmed particle (1974).

1977   (P) *John H. Van Vleck, Nevill F. Mott,* and *Philip W. Anderson* for studying solids quantum-mechanically.

(C) *Ilya Prigogine* for extending thermodynamics to show how life could arise in the face of the second law.

1978   (P) *Arno A. Penzias* and *Robert W. Wilson* for discovering the cosmic background radiation (1965) and *Pyotr Kapitsa* for his studies of liquid helium.

1979   (P) *Sheldon L. Glashow, Abdus Salam,* and *Steven Weinberg* for developing the theory that unified the weak and electromagnetic forces (1958–1971).

1980   (P) *Val Fitch* and *James W. Cronin* for discovering CP (charge-parity) violation (1964), which possibly explains the cosmological dominance of matter over antimatter.

1981   (P) *Nicolaas Bloembergen* and *Arthur L. Schawlow* for developing laser spectroscopy and *Kai M. Siegbahn* for developing high-resolution electron spectroscopy (1958).

1982   (P) *Kenneth G. Wilson* for developing a method of constructing theories of phase transitions to analyze critical phenomena.

1983   (P) *William A. Fowler* for theoretical studies of astrophysical nucleosynthesis and *Subramanyan Chandrasekhar* for studying physical processes of importance to stellar structure and evolution, including the prediction of white dwarf stars (1930).

1984   (P) *Carlo Rubbia* for discovering the W and Z particles, verifying the electroweak unification, and *Simon van der Meer,* for developing the method of stochastic cooling of the CERN beam that allowed the discovery (1982–1983).

1985   (P) *Klaus von Klitzing* for the quantized Hall effect, relating to conductivity in the presence of a magnetic field (1980).

1986   (P) *Ernst Ruska* for inventing the electron microscope (1931), and *Gerd Binnig* and *Heinrich Rohrer* for inventing the scanning-tunneling electron microscope (1981).

1987   (P) *J. Georg Bednorz* and *Karl Alex Müller* for the discovery of high temperature superconductivity (1986).

1988   (P) *Leon M. Lederman, Melvin Schwartz,* and *Jack Steinberger* for a collaborative experiment that led to the development of a new tool for studying the weak nuclear force, which affects the radioactive decay of atoms.

1989   (P) *Norman Ramsay* (U.S.) for various techniques in atomic physics; and

Hans Dehmelt (U.S.) and *Wolfgang Paul* (Germany) for the development of techniques for trapping single charge particles.

1990   (P) *Jerome Friedman, Henry Kendall* (both U.S.), and *Richard Taylor* (Canada) for experiments important to the development of the quark model.

1991   (P) *Pierre-Gilles de Gennes* for discovering that methods developed for studying order phenomena in simple systems can be generalized to more complex forms of matter, in particular to liquid crystals and polymers.

1992   (P) *George Charpak* for developing detectors that trace the paths of evanescent subatomic particles produced in particle accelerators.

1993   (P) *Russell Hulse* and *Joseph Taylor* for discovering evidence of gravitational waves.

Spreadsheet Problems

OVERVIEW

Students come to introductory physics courses with a wide variety of computing experience. Many are already accomplished programmers in one or more programming languages (BASIC, Pascal, FORTRAN, and so forth). Others have never even turned on a computer. To further complicate matters, a wide variety of hardware environments exists, although most can be classified as IBM/compatible (MS-DOS) or Macintosh environments. We have designed the end-of-chapter spreadsheet problems and the text ancillary, *Spreadsheet Investigations in Physics,* to be usable by and useful to students in all these diverse situations. Our goal is to enable students to investigate a range of physical phenomena and obtain a feel for the physics. Merely "getting the right answer" by plugging numbers into a formula and comparing the result to the answer in the back of the book is discouraged.

Spreadsheets are particularly valuable in exploratory investigations. Once you have constructed a spreadsheet, you can simply vary the parameters and see instantly how things change. Even more important is the ease with which you can construct accurate graphs of relations between physical variables. When you change a parameter, you can view the effects of the change upon the graphs simply by pressing a key. "What if" questions can be easily addressed and depicted graphically.

HOW TO USE THE TEMPLATES

The computer spreadsheet problems are arranged by level of difficulty. The least difficult problems are coded in black. For most of these problems, spreadsheets are provided on disk, and only the input parameters need to be changed. Problems of moderate difficulty, coded in blue, require additional analysis, and the provided spreadsheets must be modified to solve them. The most challenging problems are coded in magenta. For most of these, you must develop your own spreadsheets. The emphasis should be on understanding what the results mean rather than just getting an answer. For example, one spreadsheet problem explores how the distance of the horizon varies with height above the ground. You can explore why you can see farther distances when you're on top of a tall building than when you are on the ground. Why were lookouts on sailing ships placed in the crow's nest at the top of the mast?

SOFTWARE REQUIREMENTS

The spreadsheet templates are provided on a high-density (1.44 Megabyte) MS-DOS diskette using the Lotus 1-2-3 WK1 format. This format was introduced with versions 2.x of the Lotus 1-2-3 program and can be read by all subsequent versions. It can also be read directly by all the other major spreadsheet programs, including the latest Windows versions of Lotus 1-2-3, Microsoft Excel, Microsoft Works, and Novell/Wordperfect Quattro Pro as well as Microsoft Excel for the Macintosh. The program f(g) Scholar can import WK1 spreadsheets; however, some minor format changes of the templates are needed.

The Lotus WK1 format was chosen so that the templates will be usable in the widest possible variety of computing environments. Even though most spreadsheet programs operate in basically the same way, many of the latest spreadsheet programs have very powerful formatting and graphing capabilities along with many other useful features. The user of these powerful programs can exploit these capabilities to improve the appearance of their spreadsheets.

HARDWARE REQUIREMENTS

You will need a microcomputer that can run one of the spreadsheet programs in one of their many versions. Your computer should be connected to a printer that can print text and graphics. Older versions of the software will run on a 8086/8088 MS-DOS system with just a single floppy disk drive or on a 512K Macintosh with two floppy drives. Newer versions require a more powerful computer to run effectively. For example, to run Excel for Windows, Version 5.0, you must have a hard disk with about 15 megabytes of disk memory available and four to eight megabytes of RAM. Your software manuals will tell you exactly what you need to run the particular version of your spreadsheet program. However, all problems require only a minimal computer system.

There are many different software versions and many different computer configurations. You might have one floppy drive, two floppy drives, a hard disk, a local area network, and so on. The combinations are almost endless. Our best suggestion is to read your software manual and ask your instructor or computer laboratory personnel how to start your spreadsheet program.

SPREADSHEET TUTORIAL

Some students will have the required computer and spreadsheet skills to start working with the templates immediately. Other students, who have not had experience with a spreadsheet program, will need some instruction. We have written an ancillary entitled *Spreadsheet Investigations in Physics* for both these groups of students. The first part of this ancillary contains a spreadsheet tutorial that the novice student can use *independently* to gain the required spreadsheet skills. A two- or three-hour initial session with the tutorial and the computer is all that most students need to get started. Once the students have mastered the basic operations of the spreadsheet program, they should try one or two of the easier problems.

Because very few introductory physics students have studied numerical methods, we have also included a brief introduction to numerical methods in this ancillary. This section covers numerical interpolation, differentiation, integration, and the solution of simple differential equations. The student should not try to master all of this material at one time; only study the sections that are needed to solve the currently assigned problems.

The templates supplied on the distribution diskettes with *Spreadsheet Investigations in Physics* constitute an outline. You must enter the appropriate data and parameters. The parameters must be adjusted to fit the needs of your problem. Feel free to change any parameters, to expand or decrease the number of rows of output, and to change the size of increments (for example, in time or distance).

Most templates have graphs associated with them. You may have to adjust the ranges of variables plotted and the scales for the axes. See the tutorials for how to adjust the appearance of your graphs.

COPY YOUR DISTRIBUTION DISKETTES

Since you will be modifying the spreadsheets on the distribution diskettes, copy the distribution diskettes and place the originals in a safe place.

Answers to Odd-Numbered Problems

Chapter 1

1. 2.80 g/cm^3
3. 184 g
3A. $m = \frac{4}{3}\pi\rho(r_2^3 - r_1^3)$
5. (a) $4 \text{ u} = 6.64 \times 10^{-27} \text{ kg}$
   (b) $56 \text{ u} = 9.30 \times 10^{-26} \text{ kg}$
   (c) $207 \text{ u} = 3.44 \times 10^{-25} \text{ kg}$
7. (a) 72.58 kg (b) 7.82×10^{26} atoms
9. It is.
11. It is.
13. (b) only
15. L^3/T^3, L^3T
17. The units of G are $\text{m}^3/(\text{kg}\cdot\text{s}^2)$.
19. $1.39 \times 10^3 \text{ m}^2$
21. $8.32 \times 10^{-4} \text{ m/s}$
23. $11.4 \times 10^3 \text{ kg/m}^3$
25. (a) $6.31 \times 10^4 \text{ AU}$ (b) $1.33 \times 10^{11} \text{ AU}$
27. (a) 127 y (b) $15\,500$ times
29. (a) $1 \text{ mi/h} = 1.609 \text{ km/h}$ (b) 88.5 km/h
    (c) 16.1 km/h
31. $1.51 \times 10^{-4} \text{ m}$
33. $1.00 \times 10^{10} \text{ lb}$
35. 5 m
37. $5.95 \times 10^{24} \text{ kg}$
39. 2.86 cm
41. $\approx 10^6$
43. $1.78 \times 10^{-9} \text{ m}$
45. $3.84 \times 10^8 \text{ m}$
47. 34.1 m
49. $\approx 10^2$
51. (a) $(346 \pm 13) \text{ m}^2$ (b) $(66.0 \pm 1.3) \text{ m}$
53. $195.8 \text{ cm}^2 \pm 0.7\%$
55. $3, 4, 3, 2$
57. $5.2 \text{ m}^3 \pm 3\%$
59. It is not.
61. 0.449%
63. (a) 1000 kg (b) $5 \times 10^{-16} \text{ kg}$, 300 g, 0.01 g
65. (a) 10^6 (b) 10^7 (c) 10^3
S1. (a) Jud's horizon is 5.060 m and Spud's horizon is 4.382 m. It makes no difference whether you use d, s, or l. (b) On the Moon, Jud's horizon is 2.638 m and Spud's horizon is 2.285 m.
S3. (a) A plot of log T versus log L for the given data shows an approximate linear relationship. Applying the least-squares method to the data yields $1.017\,647$ for the slope and $0.890\,686$ for the intercept of the straight line that best fits the data. The data points and the best-fit line, log $T = 1.017647$ log $m + 0.890686$ are shown in the figure. Paying attention to significant figures, a reasonable interpretation of this result is that $n = 1.1$. In terms of T and m, our best-fit equation is $T = Cm^n$, where $C = 10^{1.1} = 12.6$.

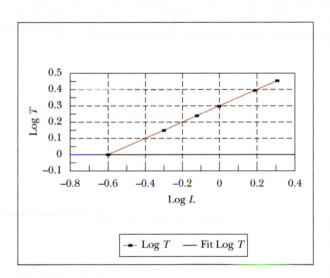

Chapter 2

1. (a) 2.3 m/s (b) 16.1 m/s (c) 11.5 m/s
3. (a) 5.0 m/s (b) 1.2 m/s (c) -2.5 m/s
   (d) -3.3 m/s (e) 0
5. (a) 3.75 m/s (b) 0
7. 50.0 km/h
9. (a) -2.5 m/s (b) -3.4 m/s (c) 4.2 s
11. (b) 1.60 m/s
13. (a) 5.0 m/s (b) -2.5 m/s (c) 0
    (d) 5.0 m/s
15. -4.00 m/s^2
17. 20.0 m/s; 6.00 m/s^2
19. (a) 5.00 m/s (b) 262 m
21. (a) 2.00 m (b) -3.00 m/s (c) -2.00 m/s^2
23. (a) 1.3 m/s^2 (b) 2.2 m/s^2 at 3.0 s
    (c) At $t = 6.2 \text{ s}$ and for the interval $10 \text{ s} < t < 12 \text{ s}$
    (d) -2.0 m/s^2 at 8.2 s

25. -16.0 cm/s^2
27. 160 ft
29. (a) 12.7 m/s   (b) -2.30 m/s
31. (a) 20.0 s   (b) no
33. (a) 8.94 s   (b) 89.4 m/s
35. (a) 0.444 m/s^2 (b) 1.33 m/s   (c) 2.12 s
    (d) 0.943 m/s
37. (a) 45.7   (b) 574 m   (c) 12.6 m/s   (d) 765 s
39. (a) $x = 30t - t^2, v = 30 - 2t$ (b) 225 m
41. 11.4 s, 212 m
43. (a) -662 ft/s^2 (b) 649 ft
45. (a) -96 ft/s   (b) 3.08×10^3 ft/s^2
    (c) 3.12×10^{-2} s
47. (a) 10.0 m/s   (b) -4.68 m/s
49. (a) 1.53 s   (b) 11.5 m
    (c) -4.60 m/s, -9.80 m/s^2
51. (a) 29.4 m/s   (b) 44.1 m
53. 7.96 s
55. (a) 7.82 m   (b) 0.782 s
57. (a) 7.00 m/s   (b) -5.35 m/s   (c) -9.8 m/s^2
59. $a(t) = a_0 + Jt, v(t) = v_0 + a_0t + \frac{1}{2}Jt^2,$
    $x(t) = x_0 + v_0t + \frac{1}{2}a_0t^2 + \frac{1}{6}Jt^3$
61. 0.509 s
63. (a) 3.00 s   (b) -15.3 m/s
    (c) -31.4 m/s, -34.8 m/s
65. (a) -6.26 m/s   (b) 6.02 m/s   (c) 1.25 s
67. 4.63 m
69. (a) 6.46 s   (b) 73.0 m
    (c) $v_{\text{STAN}} = 22.6$ m/s, $v_{\text{KATHY}} = 26.7$ m/s
71. (a) 12.5 s   (b) -2.29 m/s^2 (c) 13.1 s
73. (a) 5.28 m/s   (b) 5.86×10^{-4} m/s^2
75. $t_{AB} = t_{CD} = 2.00$ min, $t_{BC} = 1.00$ min
77. (a) 5.43 m/s^2, 3.83 m/s^2 (b) 10.9 m/s, 11.5 m/s
    (c) Maggie, 2.5 m
79. 155 s; 129 s
81. $0.577v$
S4. (a) At approximately $t = 37.0$ s after the police car
    starts, or 42.0 s after the first policeman sees the
    speeder.   (b) 74.0 m/s   (c) 1370 m

Chapter 3

1. (a) 8.60 m   (b) 4.47 m at 297°; 4.24 m at 135°
3. $(-2.75$ m, -4.76 m)
5. 2.24 m
7. (2.05 m, 1.43 m)
9. 70.0 m
11. 310 km at 57° south of west
13. (a) 10.0 m   (b) 15.7 m   (c) 0
15. (a) 5.20 m at 60.0°   (b) 3.00 m at 330°
    (c) 3.00 m at 150°   (d) 5.20 m at 300°
17. 421 ft at $-2.63°$
19. 86.6 m, -50.0 m
21. 5.83 m at 59.0° to the right of first direction
23. 47.2 units at 122°

25. (b) $5\mathbf{i} + 4\mathbf{j} = 6.40$ at 38.7°;
    $-\mathbf{i} + 8\mathbf{j} = 8.06$ at 97.2°
27. 7.21 m at 56.3°
29. (a) $2\mathbf{i} - 6\mathbf{j}$ (b) $4\mathbf{i} + 2\mathbf{j}$ (c) 6.32   (d) 4.47
    (e) 2.88°; 26.6°
31. (a) $(-11.1$ m$)\mathbf{i} + (6.40$ m$)\mathbf{j}$
    (b) $(1.65$ cm$)\mathbf{i} + (2.86$ cm$)\mathbf{j}$
    (c) $(-18.0$ in.$)\mathbf{i} - (12.6$ in.$)\mathbf{j}$
33. 9.48 m at 166°
35. 390 mph at 7.37° north of east
37. $(2.60$ m$)\mathbf{i} + (4.50$ m$)\mathbf{j}$
39. (a) $8\mathbf{i} + 12\mathbf{j} - 4\mathbf{k}$ (b) $2\mathbf{i} + 3\mathbf{j} - \mathbf{k}$
    (c) $-24\mathbf{i} - 36\mathbf{j} + 12\mathbf{k}$
41. (a) 5.92   (b) 19.0
43. (a) $-3\mathbf{i} + 2\mathbf{j}$ (b) 3.61 at 146°   (c) $3\mathbf{i} - 6\mathbf{j}$
47. (a) 49.5, 27.1   (b) 56.4 at 28.7°
49. 240 m at 237°
51. (10.0 m, 20.0 m)

Chapter 4

1. $(8a + 2b)\mathbf{i} + 2c\mathbf{j}$
3. (a) 4.87 km at 209° from east   (b) 23.3 m/s
   (c) 13.5 m/s at 209°
5. (a) $(2\mathbf{i} + 3\mathbf{j})$ m/s^2
   (b) $(3t + t^2)\mathbf{i} + (-2t + 1.5t^2)\mathbf{j}$
7. (a) $(0.8\mathbf{i} - 0.3\mathbf{j})$ m/s^2 (b) at 339°
   (c) $(360\mathbf{i} - 72.8\mathbf{j})$ m; at 345°
9. (a) $\mathbf{v} = (-5\mathbf{i} + 0\mathbf{j})$ m/s, $\mathbf{a} = (0\mathbf{i} - 5\mathbf{j})$ m/s^2
   (b) $\mathbf{r} = -5\mathbf{i} \sin t + 4\mathbf{j} - 5\mathbf{j} \cos t$
       $\mathbf{v} = -5\mathbf{i} \cos t + 5\mathbf{j} \sin t$
       $\mathbf{a} = 5\mathbf{i} \sin t + 5\mathbf{j} \cos t$
   (c) a circle of radius 5 m centered at (0, 4 m)
11. (a) 3.34 m/s at 0°   (b) 309°
13. 7.00 s at 143 m/s
15. 9.91 m/s
17. (a) clears by 0.89 m   (b) while falling 13.3 m/s
19. (a) 1.69 km/s   (b) 6490 s
21. (a) 277 km   (b) 284 s
23. 53.1°
25. 22.4° or 89.4°
27. (a) 7.90 km/s   (b) 5070 s
29. 377 m/s^2
31. (a) 1.02 km/s   (b) 2.72 mm/s^2
33. 54.4 m/s^2
35. (a) 13.0 m/s^2 (b) 5.70 m/s   (c) 7.50 m/s^2
37. 1.48 m/s^2
39. (a) $-30.8\mathbf{j}$ m/s^2 (b) $70.4\mathbf{j}$ m/s^2
41. (a) 26.9 m/s   (b) 67.3 m   (c) $(2\mathbf{i} - 5\mathbf{j})$ m/s^2
43. 2.02×10^3 s; 21.0% longer
45. 2.50 m/s
47. 153 km/h at 11.3° north of west
49. (a) 57.7 km/h at 210°   (b) 28.9 km/h down
51. (a) 10.1 m/s^2 south at 75.7° below horizontal
    (b) 9.80 m/s^2 down
53. (a) $4.00\mathbf{i}$ m/s   (b) (4.00 m, 6.00 m)

55. (a) 20.0 m/s, 5.00 s   (b) 31.4 m/s at 301°
    (c) 6.53 s   (d) 24.5 m away
57. 3.14 m
59. 20.0 m
61. (a) 0.60 m   (b) 0.40 m
    (c) 1.87 m/s^2 toward center   (d) 9.80 m/s^2 down
63. 4.12 m
65. (a) $\sqrt{gR}$ (b) $(\sqrt{2} - 1)R$
67. 10.8 m
69. (a) 6.80 km   (b) directly above explosion
    (c) 66.2°
71. (a) $(-7.05$ cm$)\mathbf{j}$ (b) $(7.61$ cm$)\mathbf{i} - (6.48$ cm$)\mathbf{j}$
    (c) $(10.0$ cm$)\mathbf{i} - (7.05$ cm$)\mathbf{j}$
73. (a) 22.2° or 67.8°   (b) 235 cm   (c) 235 m
75. (a) 407 km/h at 10.6° N of E   (b) 10.8° S of E
77. (a) 5.15 s   (b) 4.85 m/s at 74.5° N of W
    (c) 19.4 m
79. (a) 43.2 m   (b) $(9.66$ m/s$)\mathbf{i} - (25.6$ m/s$)\mathbf{j}$
81. $(18.8$ m, -17.4 m$)$
83. (a) 46.5 m/s   (b) $-77.6°$ (c) 6.34 s
85. 2.98 km/s forward and 1.96° inward
S1. There are an infinite number of solutions to the prob-
    lem; however, there is one practical limitation. We can
    estimate that the maximum speed that the punter can
    give the ball is about 20 to 30 m/s. Use a speed in this
    range.
S3. There are an infinite number of solutions. For example
    if $v_0 = 22.32$ m/s at 45°, then the ball clears the cross-
    bar in 3.01 s. Or if $v_0 = 24.60$ m/s at 30°, then the ball
    clears the crossbar in 2.23 s. The time it takes the ball to
    clear the crossbar is immaterial, since in all likelihood
    time will run out before the clock is stopped.

Chapter 5

1. (a) 1/3   (b) 0.750 m/s^2
3. (a) 3.00 s   (b) 20.1 m   (c) $(18.0\mathbf{i} - 9.0\mathbf{j})$ m
5. 312 N
7. $(6\mathbf{i} + 15\mathbf{j})$ N, 16.2 N
9. (a) 556 N   (b) 56.7 kg
11. (a) 4.47×10^{15} m/s^2 outward
    (b) 2.09×10^{-10} N inward
13. 1.72 kg, 6.97 m/s^2
15. (c) Forces of brick on spring and spring on brick; of
    spring on table and table on spring; of table on Earth
    and Earth on table; of brick on Earth and of Earth on
    brick; of spring on Earth and of Earth on spring.
17. (a) 5.10 kN   (b) 2.65×10^3 kg
19. (a) 2.0 m/s^2 (b) 170 N   (c) 2.93 m/s^2
21. (a) $(2.50$ N$)\mathbf{i} + (5.00$ N$)\mathbf{j}$ (b) 5.59 N
23. 1.24 ft/s^2
25. 640 N for $0 \leq t \leq 1.0$ s, 627 N at $t = 1.3$ s, 589 N at $t = 2.0$ s
27. $(14.7$ N$)\mathbf{i} - (2.5$ N$)\mathbf{j}$
29. 613 N
31. (b) $T_1 = 513.5$ N, $T_2 = 557.4$ N, $T_3 = 325.0$ N
33. (a) 33.9 N   (b) 39.0 N

35. (a) $g \tan \theta$ (b) 4.16 m/s^2
37. (a) $F_x > 19.6$ N   (b) $F_x \leq -78.4$ N
39. survival chance better with force on larger mass
41. (a) 4.90 m/s^2 (b) 3.13 m/s   (c) 1.35 m
    (d) 1.14 s   (e) no
43. (a) 706 N   (b) 814 N   (c) 706 N   (d) 648 N
45. 21.8 m/s
47. 36.9 N
49. 81.0 m/s
51. $\mu = 0.077$
53. (a) 0.161   (b) 1.01 m/s^2
55. (b) 27.2 N, 1.286 m/s^2
57. (a) 1.78 m/s^2 (b) 0.368   (c) 9.37 N
    (d) 2.67 m/s
59. (a) $a_1 = 2.31$ m/s^2 down, $a_2 = 2.31$ m/s^2 left,
    $a_3 = 2.31$ m/s^2 up,
    (b) $T_{\text{Left}} = 30.0$ N, $T_{\text{Right}} = 24.2$ N
61. 0.293
63. 182.5 m
65. (a) $Mg/2$, $Mg/2$, $Mg/2$, $3Mg/2$, Mg (b) $Mg/2$
67. (a) $\mu_s = \dfrac{h}{\sqrt{L^2 - h^2}}$ (b) $a = \dfrac{2L}{t^2}$

    (c) $\sin \theta = h/L$ (d) $\mu_k = \dfrac{h - \dfrac{2L^2}{gt^2}}{\sqrt{L^2 - h^2}}$

69. (a) 0.232 m/s^2 (b) 9.68 N
71. (a) F forward   (b) $3F/2Mg$
    (c) $F/(M + m_1 + m_2 + m_3)$
    (d) $m_1 F/(M + m_1 + m_2 + m_3)$,
    $(m_1 + m_2)F/(M + m_1 + m_2 + m_3)$,
    $(m_1 + m_2 + m_3)F/(M + m_1 + m_2 + m_3)$
    (e) $m_2 F/(M + m_1 + m_2 + m_3)$
73. (a) friction between the two blocks   (b) 34.7 N
    (c) 0.306
75. (a) 0.408 m/s^2 (b) 83.3 N
77. (a) 4.00 m   (b) 3.72 m/s
81. (a) $(-45\mathbf{i} + 15\mathbf{j})$ m/s   (b) at 162°
    (c) $(-225\mathbf{i} + 75\mathbf{j})$ m   (d) $(-227$ m, 79 m$)$
83. $(M + m_1 + m_2)m_2 g/m_1$
85. $T_1 = 74.5$ N, $T_2 = 34.7$ N, $\mu_k = 0.572$
87. (a) 2.20 m/s^2 (b) 27.37 N
89. (a) 30.7°   (b) 0.843 N
91. 6.00 cm
S1. (a) A plot of F as the independent variable and L as the
    dependent variable shows that the last four points tend
    to vary the most from a straight line. However, because
    the first point has the largest percentage deviation from
    a straight line, we probably are not justified in throwing
    any of the data points out. The slope of the best straight
    line obtained from the least-squares fit is 8.654 545 5
    mm/N or $k = 0.116$ N/mm $= 116$ N/m. (b) The least-
    squares fit we used was of the form $L = aF + b$; solving
    for F gives $F = L/a - b/a = 0.116 L - 0.561$. Hence, for
    $L = 105$ mm, $F = 12.7$ mm.

Chapter 6

1. (a) 8.0 m/s   (b) 3.02 N
3. (a) 5.40 kN down   (b) 1.60 kN down
     (c) seatbelt tension plus gravity
5. $0 < v < 8.08$ m/s
7. (a) 1.52 m/s^2 (b) 1.66 km/s   (c) 6820 s
9. $v \leqslant 14.3$ m/s
11. (a) 9.80 N   (b) 9.80 N   (c) 6.26 m/s
13. (a) static friction   (b) 0.085
15. 3.13 m/s
17. (a) 4.81 m/s   (b) 700 N up
19. (a) $(-0.163$ m/s$^2)\mathbf{i} + (0.233$ m/s$^2)\mathbf{j}$
     (b) 6.53 m/s   (c) $(-0.181$ m/s$^2)\mathbf{i} + (0.181$ m/s$^2)\mathbf{j}$
21. no
23. (a) 0.822 m/s^2 (b) 37.0 N   (e) 0.0839
25. (a) 17.0°   (b) 5.12 N
27. (a) 491 N   (b) 50.1 kg   (c) 2.00 m/s
31. (a) 3.47×10^{-2} s^{-1} (b) 2.50 m/s   (c) $a = -cv$
33. (a) 1.47 N·s/m   (b) 2.04×10^{-3} s
     (c) 2.94×10^{-2} N
35. (a) 8.32×10^{-8} N   (b) 9.13×10^{22} m/s^2
     (c) 6.61×10^{15} rev/s
37. (a) $\mathbf{T} = (68.6$ N$)\mathbf{i} + (784$ N$)\mathbf{j}$ (b) 0.857 m/s^2
39. (a) The true weight is greater than the apparent weight.
     (b) $w = w' = 735$ N at the poles; $w' = 732.4$ N at the
     equator
41. 780 N
43. 12.8 N
45. (a) 967 lb   (b) 647 lb up
47. (a) 6.67 kN   (b) 20.3 m/s

47A. (a) $mg - \dfrac{mv^2}{R}$ (b) $\sqrt{gR}$

49. (b) 2.54 s, 23.6 rev/min
51. (a) 1.58 m/s^2 (b) 455 N   (c) 329 N
     (d) 397 N upward and 9.15° inward
53. (a) 0.0132 m/s   (b) 1.03 m/s   (c) 687 m/s
S1. Spreadsheet 6.1 typifies the solution of second-order differential equations. In this spreadsheet, we want to solve $a = dv/dt$, where $v = dx/dt$. To find v as a function of t, we assume that a is constant over the time interval dt. Hence, we can use Euler's method to integrate $dv/dt = a$. Therefore, $v_{i+1} = v_i + a_i \Delta t$. In the Lotus 1-2-3 Spreadsheet 6.1, cells F35 and down implement this equation. To find x as a function of t, we know that v varies over the time interval, so we use Euler's modified method to integrate $dx/dt = v$. Or, $x_{i+1} = x_i + 1/2(v_{i+1} + v_i) \Delta t$. Cells F35 and down implement this equation.
S2. All objects fall faster when there is no air resistance than when there is air resistance. As the mass of the objects increases, the difference between their positions at the same time with and without air resistance becomes smaller.
S4. $F_{\min} \cong 252$ N at 31°.
S5. Try large positive speeds, that is $v_0 > 100$ m/s. You may need to increase Δt.

S7. If the terminal speeds are to be equal, then the relationship between b_1 (for $n = 1$) and b_2 (for $n = 2$) is $b_1 = b_2 \, mg$.

Chapter 7

1. 30.6 m
3. (a) 31.9 J   (b) 0   (c) 0   (d) 31.9 J
5. 5.88 kJ
7. (a) 900 J   (b) -900 kJ   (c) 0.383
9. (a) 137 W   (b) -137 W
11. (a) 79.4 N   (b) 1.49 kJ   (c) -1.49 kJ
11A. (a) $\mu_k \, mg/(\cos\theta + \mu_k \sin\theta)$
     (b) $\mu_k \, mg \, d \cos\theta/(\cos\theta + \mu_k \sin\theta)$
     (c) $-\mu_k \, mg \, d \cos\theta/(\cos\theta + \mu_k \sin\theta)$
13. 14.0
13A. $r_1 r_2 \cos(\theta_1 - \theta_2)$
17. (a) 16.0 J   (b) 36.9°
19. 32 J
21. (a) 11.3°   (b) 156°   (c) 82.3°
23. (a) 7.50 J   (b) 15.0 J   (c) 7.50 J   (d) 30.0 J
25. (a) 575 N/m   (b) 46.0 J
25A. (a) F/d (b) $\frac{1}{2}Fd$
27. 0.299 m/s
29. (b) mgR
31. 12.0 J
31A. $3W$
33. (a) 4.10×10^{-18} J
     (b) 1.14×10^{-17} N   (c) 1.25×10^{13} m/s^2
     (d) 240 ns
35. (a) 2.00 m/s   (b) 200 N
35A. (a) $v = \sqrt{2W/m}$ (b) $\overline{F} = W/d$
37. (a) 650 J   (b) -588 J   (c) 62.0 J
     (d) 1.76 m/s
39. 6.34 kN

41. (a) $\sqrt{\dfrac{m}{M+m}} \, \sqrt{2gh}$ (b) $\sqrt{\dfrac{2}{M+m}} \, \sqrt{mgh - \mu_k M gh}$

43. 1.25 m/s
45. 2.04 m
47. (a) 168 J   (b) 184 J   (c) 500 J   (d) 148 J
     (e) 5.65 m/s
49. (a) 4.51 m   (b) no, since $f > mg \sin\theta$
51. (a) 63.9 J   (b) -35.4 J   (c) -9.51 J
     (d) 19.0 J
53. 875 W
55. (a) 7.92 hp   (b) 14.9 hp
57. (a) 7.5×10^4 J   (b) 2.50×10^4 W (33.5 hp)
     (c) 3.33×10^4 W (44.7 hp)

57A. (a) $\dfrac{1}{2} mv^2$ (b) $\dfrac{mv^2}{2t}$ (c) $\dfrac{mv^2 t_1}{t^2}$

59. 220 ft·lb/s
61. 685
63. 80.0 hp
65. (a) 1.35×10^{-2} gal   (b) 73.8   (c) 8.08 kW
67. 5.90 km/liter

69. (a) 5.37×10^{-11} J   (b) 1.33×10^{-9} J
71. 3.70 m/s
75. (a) $(2 + 24t^2 + 72t^4)$ J   (b) $a = 12t$ m/s^2; $F = 48t$ N
    (c) $(48t + 288t^3)$ W   (d) 1.25×10^3 J
77. 878 kN
79. (a) 4.12 m   (b) 3.35 m
81. (a) -5.60 J   (b) 0.152   (c) 2.29 rev
83. (a) $W = mgh$ (b) $\Delta K = mgh$
    (c) $K_f = mgh + mv_0^2/2$
85. 1.94 kJ
87. (b) 8.49×10^5 kg/s   (c) 7.34×10^7 m^3
    (d) 1.53 km
89. 1.68 m/s

Chapter 8

1. (a) -147 J   (b) -147 J   (c) -147 J
   The force is conservative.
3. (b) conservative 62.7 J, nonconservative 20.7 J
   (c) $\mu = 0.330$
5. (a) 125 J   (b) 50.0 J   (c) 66.7 J
   (d) nonconservative, since W is path-dependent
7. (a) 40.0 J   (b) -40.0 J   (c) 62.5 J
9. (a) -9.00 J; No. A constant force is conservative.
   (b) 3.39 m/d   (c) 9.00 J
11. $v_A = \sqrt{3gR}$; 0.098 N downward
13. (a) $v = (gh + v_0^2)^{1/2}$ (b) $v_x = 0.6v_0$;
    $v_y = -(0.64v_0^2 + gh)^{1/2}$
15. (a) 18.5 km, 51.0 km   (b) 10.0 MJ
17. (a) 4.43 m/s   (b) 5.00 m
17A. (a) $\sqrt{2(m_1 - m_2)gh/(m_1 + m_2)}$
    (b) $2m_1h/(m_1 + m_2)$
19. (a) -160 J   (b) 73.5 J   (c) 28.8 N
    (d) 0.679
21. 489 kJ
23. (a) -4.1 MJ   (b) 9.97 m/s   (c) 50.8 m
    (d) It is better to keep the engine with the train.
25. 3.74 m/s
27. 0.721 m/s
29. (a) 0.400 J   (b) 0.225 J   (c) 0 J
31. (a) -28.0 J   (b) 0.446 m
33. 10.2 m
33A. $(kd^2/2mg) - d$
35. 0.327
37. (a) $F_r = A/r^2$
39. $F = (7 - 9x^2y)i - 3x^3j$
41. (c) $v = 0.894$ m/s
43. (a) $v_B = 5.94$ m/s; $v_C = 7.67$ m/s   (b) 147 J
45. (a) 1.50×10^{-10} J   (b) 1.07×10^{-9} J
    (c) 9.15×10^{-10} J
47. (a) 0.225 J   (b) 0.363 J
    (c) No. The normal force varies with position, and so
    the frictional force also varies.
49. $\dfrac{h}{5}(4 \sin^2 \theta + 1)$

51. (a) 349 J, 676 J, 741 J   (b) 174 N, 338 N, 370 N
    (c) yes
53. (a) $\Delta U = -\dfrac{ax^2}{2} - \dfrac{bx^3}{3}$ (b) $\Delta U = \dfrac{A}{\alpha}(1 - e^{\alpha x})$
55. 0.115
59. 1.24 m/s
61. (b) 7.42 m/s
63. (a) 3.19 m   (b) 2.93 m/s
65. (a) 0.400 mm   (b) 4.10 m/s
    (c) It reaches the top.
67. $m_1 gd(m_2 - \mu_k m_1 \cos \theta - m_1 \sin \theta)/(m_1 + m_2)$
69. (a) 0.378 m   (b) 2.30 m/s   (c) 1.08 m
S1. If the particle has an initial energy less than 2471 J, it
    will be trapped in the potential well; for example, if
    $E_T = 1000$ J, it will be confined approximately to
    -3.15 m $\leq x \leq 4.58$ m.
S2. $x = 0$ is a point of stable equilibrium; $x = 8.33$ m is a
    point of unstable equilibrium.

Chapter 9

1. $(9.00i - 12.0j)$ kg·m/s, 15.0 kg·m/s
3. 1.60 kN
5. (a) 12.0 kg·m/s   (b) 6.00 m/s   (c) 4.00 m/s
7. (a) 13.5 kg·m/s   (b) 9.00×10^3 N
   (c) 18.0×10^3 N
9. 87.5 N
11. (a) 7.50 kg·m/s   (b) 375 N
13. (a) 13.5 kg·m/s toward the pitcher
    (b) 6.75×10^3 N toward the pitcher
15. 260 N toward the left in the diagram
15A. $\dfrac{-2mv \sin \theta}{t}i$
17. (a) 0.125 m/s   (b) 8 times
19. 120 m
21. (a) 1.15 m/s   (b) -0.346 m/s
23. 4.01×10^{-20} m/s
25. 301 m/s
27. (a) 20.9 m/s east   (b) 8.74 kJ into thermal energy
29. (a) 0.284, or 28.4%   (b) $K_n = 1.15 \times 10^{-13}$ J,
    $K_c = 4.54 \times 10^{-14}$ J
31. 3.75 kN; no
33. (a) 0.571 m/s   (b) 28.6 J   (c) 0.003 97
35. 91.2 m/s
37. 0.556 m
39. 497 m/s
41. $v = (3.00i - 1.20j)$ m/s
43. (a) $v_x = -9.33 \times 10^6$ m/s, $v_y = -8.33 \times 10^6$ m/s
    (b) 4.39×10^{-13} J
45. 3.01 m/s, 3.99 m/s
47. 2.50 m/s at $-60.0°$
51. -0.429 m
53. CM = 454 km, well within the Sun
55. (4.50 m, 2.79 m)
57. 70/6 cm, 80/6 cm

59. (a) $(1.40i + 2.40j)$ m/s   (b) $(7.00i + 12.0j)$ kg·m/s
61. (a) 2.10 m/s, 0.900 m/s   (b) 6.30×10^{-3} kg·m/s, -6.30×10^{-3} kg·m/s
61A. (a) $m_2 v_1/(m_1 + m_2)$ and $m_1 v_1/(m_1 + m_2)$
   (b) $m_1 m_2 v_1/(m_1 + m_2)$ toward the CM
63. 200 kN
65. 2150 kg
67. 0.595 m³/s
67A. $F/\rho v$
69. 291 N
71. (a) 1.80 m/s to the left   (b) 257 N to the left
   (c) larger than part b
73. (a) 4160 N   (b) 4.17 m/s
75. 32.0 kN; 7.13 MW
77. (a) 6.81 m/s   (b) 1.00 m
79. 240 s
81. $(3Mgx/L)j$
83. (a) As the child walks to the right, the boat moves to the left, but the center of mass remains fixed.
   (b) 5.55 m from the pier   (c) Since the turtle is 7 m from the pier, the boy will not be able to reach the turtle, even with a 1 m reach.
85. (a) 100 m/s   (b) 374 J
85A. (a) $v_0 - d\sqrt{\dfrac{km}{m}}$ (b) $v_0 d\sqrt{km} - \dfrac{1}{2}kd^2\left(1 + \dfrac{m}{m}\right)$
87. $2v_0$ and 0
89. (a) 3.8 kg·m/s²   (b) 3.8 N   (c) 3.8 N
   (d) 2.8 J   (e) 1.4 J
   (f) Friction between sand and belt converts half of the input work into thermal energy.
S1. (a) The maximum acceleration is 100 m/s². It occurs at the end of the burn time of 80 s, when the rocket has its smallest mass. The maximum speed that the rocket reaches is 3.22 km/s. (b) The speed reaches half its maximum after 55.5 s; if the acceleration were constant, it would reach half its maximum speed at 40 s (half the burn time), but the acceleration is always increasing during the burn time.
S3. (b) The disadvantages are that the ship has to withstand twice the acceleration and that it has only traveled half as far when the fuel burns out. The advantage is that it takes half as long to reach its final speed; hence, it travels farther in 100 s.

Chapter 10

1. (a) 4.00 rad/s²   (b) 18.0 rad
3. (a) 1.99×10^{-7} rad/s   (b) 2.66×10^{-6} rad/s
5. (a) 5.24 s   (b) 27.4 rad
7. 13.7 rad/s²
9. (a) 0.18 rad/s   (b) 8.10 m/s² toward the center of the track
9A. (a) v/R (b) v^2/R toward center
11. (a) 8.00 rad/s   (b) 8.00 m/s, $a_r = -64.0$ m/s², $a_t = 4.00$ m/s²   (c) 9.00 rad

13. (a) 126 rad/s   (b) 3.77 m/s   (c) 1.26 km/s²
   (d) 20.1 m
15. 29.4 m/s², 9.80 m/s²
15A. $-2g\dfrac{(h-R)}{R}i - gj$
17. (a) 143 kg·m²   (b) 2.57×10^3 J
19. (a) 92.0 kg·m², 184 J   (b) 6.00 m/s, 4.00 m/s, 8.00 m/s, 184 J
23. (a) $(3/2)MR^2$ (b) $(7/5)MR^2$
25. -3.55 N·m
27. 2.79%
29. (a) 0.309 m/s²   (b) $T_1 = 7.67$ N, $T_2 = 9.22$ N
29A. (a) $\dfrac{m_2 \sin\theta - \mu_k(m_1 + m_2 \cos\theta)}{m_1 + M/2 + m_2} g$
   (b) $T_1 = \mu_k mg + m_1 a$, $T_2 = T_1 + \dfrac{1}{2}Ma$
31. (a) 56.3 J   (b) 8.38 rad/s   (c) 2.35 m/s
   (d) 1.4% greater
33. (a) $2(Rg/3)^{1/2}$ (b) $4(Rg/3)^{1/2}$ (c) $(Rg)^{1/2}$
35. (a) 11.4 N, 7.57 m/s², 9.53 m/s down   (b) 9.53 m/s
37. (a) 1.03 s   (b) 10.3 rev
39. 168 N·m (clockwise)
41. (a) 4.00 J   (b) 1.60 s   (c) yes
43. (a) $\omega = \sqrt{3g/L}$ (b) $\alpha = 3g/2L$
   (c) $-\frac{3}{2}gi - \frac{3}{4}gj$ (d) $-\frac{3}{4}Mgi + \frac{1}{4}Mgj$
45. (a) $0.707R$ (b) $0.289L$ (c) $0.632R$
49. (a) 2.60×10^{29} J   (b) -1.65×10^{17} J/day
51. (a) 118 N, 156 N   (b) 1.17 kg·m²
51A. (a) $T_1 = m_1(a + g\sin\theta)$, $T_2 = m_2(g - a)$
   (b) $m_2 R^2 g/a - m_1 R^2 - m_2 R^2 - m_1 R^2(g/a)\sin\theta$
53. (a) -0.176 rad/s²   (b) 1.29 rev   (c) 9.26 rev
S1. The answer is not unique because the torque $\tau = FR$.
S3. Replace X^2 with $(X - H)^2$ in line 110.

Chapter 11

1. (a) 500 J   (b) 250 J   (c) 750 J
3. (a) $a_{CM} = \frac{2}{3}g\sin\theta$ (disk), $a_{CM} = \frac{1}{2}g\sin\theta$ (hoop)
   (b) $\frac{1}{3}\tan\theta$
5. 44.8 J
5A. $0.7Mv^2$
7. (a) $-17k$ (b) 70.5°
9. (a) negative z direction   (b) positive z direction
11. 45.0°
13. $|F_3| = |F_1| + |F_2|$, no
15. $(17.5$ kg·m²/s$)k$
15A. $\frac{1}{2}(m_1 + m_2)vd$
17. $(60$ kg·m²/s$)k$
19. $mvR\left[\cos\left(\dfrac{vt}{R}\right) + 1\right]k$
21. $-mg\ell t\cos\theta\, k$
23. (a) zero   (b) $[-mv_0^3 \sin^2\theta\cos\theta/2g]k$
   (c) $[-2mv_0^3 \sin^2\theta\cos\theta/g]k$ (d) The downward force of gravity exerts a torque in the $-z$ direction.

25. (a) $0.433 \text{ kg} \cdot \text{m}^2/\text{s}$ (b) $1.73 \text{ kg} \cdot \text{m}^2/\text{s}$

27. (a) $\omega = \omega_0 I_1/(I_1 + I_2)$ (b) $I_1/(I_1 + I_2)$

29. (a) 0.360 rad/s in the counterclockwise direction
    (b) 99.9 J

31. (a) 6.05 rad/s (b) 114 J

33. (a) $mv\ell$ down   (b) $M/(M + m)$

35. (a) $2.19 \times 10^6 \text{ m/s}$ (b) $2.18 \times 10^{-18} \text{ J}$
    (c) $4.13 \times 10^{16} \text{ rad/s}$

37. 0.91 km/s

41. (a) The net torque around this axis is zero.
    (b) Since $\tau = 0$, $\mathbf{L} = \text{const.}$ But initially, $\mathbf{L} = 0$, hence it remains zero throughout the motion. Consequently, the monkey and bananas move upward with the same speed at any instant. The distance between the monkey and bananas stays constant. Hence, the monkey will not reach the bananas.

47. 30.3 rev/s

49. (a) $v_0 r_0/r$ (b) $T = (mv_0^2 r_0^2) r^{-3}$
    (c) $\frac{1}{2}mv_0^2\left(\dfrac{r_0^2}{r^2} - 1\right)$ (d) 4.50 m/s, 10.1 N, 0.450 J

51. (a) $F_y = \dfrac{W}{L}\left(d - \dfrac{ah}{g}\right)$ (b) 0.306 m
    (c) $(-306\mathbf{i} + 553\mathbf{j}) \text{ N}$

53. (a) $3.75 \times 10^3 \text{ kg} \cdot \text{m}^2/\text{s}$ (b) 1.875 kJ
    (c) $3.75 \times 10^3 \text{ kg} \cdot \text{m}^2/\text{s}$ (d) 10.0 m/s
    (e) 7.50 kJ (f) 5.625 kJ

53A. (a) Mvd (b) Mv^2 (c) Mvd (d) $2v$
    (e) $4Mv^2$ (f) $3Mv^2$

55. $\frac{1}{3}L$

57. (c) $(8Fd/3M)^{1/2}$

61. $v_0 = [ag(16/3)(\sqrt{2} - 1)]^{1/2}$

63. F_1 clockwise torque, F_2 zero torque, F_3 and F_4 counterclockwise torque

65. (a) 0.80 m/s^2, 0.40 m/s^2
    (b) 0.60 N (top), 0.20 N (bottom)

Chapter 12

1. 10.0 N up; $6.00 \text{ N} \cdot \text{m}$ counterclockwise

3. $[(w_1 + w)d + w_1\ell/2]/W_2$

5. $F_W = 480 \text{ N}$, $F_v = 1200 \text{ N}$

5A. $\left(\dfrac{w_1}{2} + \dfrac{w_2 X}{L}\right)\left(\dfrac{d}{\sqrt{L^2 - d^2}}\right)$; $w_1 + w_2$

9. -1.50 m, -1.50 m

11. (a) 859 N (b) 1040 N, left and upward at $36.9°$

13. 0.789

15. $F_f = 4410 \text{ N}$, $F_r = 2940 \text{ N}$

17. $2R/5$

19. $\frac{1}{3}$ by the left string, $\frac{2}{3}$ by the right string

21. $x = \frac{3}{4}L$

23. 4.90 mm

25. (a) 73.6 kN (b) 2.50 mm

27. $29.2 \ \mu\text{m}$

27A. $\dfrac{8m_1 m_2 gL}{\pi d^2 Y(m_1 + m_2)}$

29. (a) $3.14 \times 10^4 \text{ N}$ (b) 62.8 kN

31. 1800 atm

33. $N_A = 5.98 \times 10^5 \text{ N}$, $N_B = 4.80 \times 10^5 \text{ N}$

35. (b) 69.8 N (c) 0.877ℓ

37. (a) 160 N right   (b) 13.2 N right   (c) 292 N up
    (d) 192 N

39. (a) $T = w(\ell + d)/\sin\theta(2\ell + d)$ and
    (b) $R_x = w(\ell + d)\cot\theta/(2\ell + d)$; $R_y = w\ell/(2\ell + d)$

41. (a) $F_x = 268 \text{ N}$, $F_y = 1300 \text{ N}$ (b) 0.324

41A. (a) $\dfrac{m_1 g}{2 \tan\theta} + \dfrac{m_2 gx}{L \tan\theta}$; $(m_1 + m_2)g$
    (b) $\dfrac{m_1/2 + m_2 d/L}{(m_1 + m_2)\tan\theta}$

43. 5.08 kN, $R_x = 4.77 \text{ kN}$, $R_y = 8.26 \text{ kN}$

45. $T = 2.71 \text{ kN}$, $R_x = 2.65 \text{ kN}$, $R_y = -12.01 \text{ N}$

47. (a) 20.1 cm to the left of the front edge, $\mu = 0.571$
    (b) 0.501 m

47A. (a) $\dfrac{F\cos\theta}{mg - F\sin\theta}$; $\dfrac{mgw/2 - Fh\cos\theta}{(mg - F\sin\theta)}$ (b) $\dfrac{mgw}{2F\cos\theta}$

49. (a) $W = \dfrac{w}{2}\left(\dfrac{2\mu_s \sin\theta - \cos\theta}{\cos\theta - \mu_s \sin\theta}\right)$
    (b) $R = (w + W)\sqrt{1 + \mu_s^2}$, $F = \sqrt{W^2 + \mu_s^2(w + W)^2}$

51. (a) 133 N (b) $N_A = 429 \text{ N}$, $N_B = 257 \text{ N}$
    (c) $R_x = 133 \text{ N}$, $R_y = -257 \text{ N}$

53. 66.7 N

55. $F = \frac{3}{8}w$

57. (a) 1.67 N, 3.33 N, 1.67 N (b) 2.36 N

59. (a) 4500 N (b) $4.50 \times 10^6 \text{ N/m}^2$
    (c) This is more than sufficient to break the board.

61. $y_{cg} = 16.7 \text{ cm}$

Chapter 13

1. (a) 1.50 Hz, 0.667 s (b) 4.00 m (c) $\pi \text{ rad}$
    (d) 2.83 m

3. (b) 1.81 s (c) no

5. (a) 13.9 cm/s, 16.0 cm/s^2 (b) 16.0 cm/s, 1.83 s
    (c) 32.0 cm/s^2, 1.05 s

7. (b) $6\pi \text{ cm/s}$, 0.333 s (c) $18\pi^2 \text{ cm/s}^2$, 0.500 s
    (d) 12.0 cm

9. (a) 0.542 kg (b) 1.81 s (c) 1.20 m/s^2

11. 40.9 N/m

13. (a) 2.40 s (b) 0.417 Hz (c) 2.62 rad/s

15. (a) 0.400 m/s, 1.60 m/s^2
    (b) $\pm 0.320 \text{ m/s}$, -0.960 m/s^2 (c) 0.232 s

17. (a) 0.750 m (b) $x = -(0.75 \text{ m})\sin(2.0t)$

17A. (a) v/ω (b) $(v/\omega)\cos(\omega t + \pi/2)$

19. 2.23 m/s

21. (a) quadrupled   (b) doubled
    (c) doubled   (d) no change

23. $\pm 2.60 \text{ cm}$

25. (a) 1.55 m   (b) 6.06 s
27. (a) 3.65 s   (b) 6.41 s   (c) 4.24 s
27A. (a) $2\pi\sqrt{L/(g+a)}$ (b) $2\pi\sqrt{L/(g-a)}$
     (c) $2\pi L^{1/2}(g^2 + a^2)^{-1/4}$
29. (a) 0.817 m/s   (b) 2.57 rad/s²   (c) 0.641 N
31. increases by 1.78×10^{-3} s
33. 0.944 kg·m²
35. (a) 5.00×10^{-7} kg·m²
     (b) 3.16×10^{-4} N·m/rad
39. 1.00×10^{-3} s⁻¹
41. (a) 1.42 Hz   (b) 0.407 Hz
43. 318 N

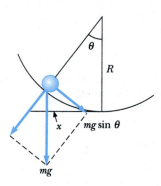

45. 1.57 s
47. (a) $E = \frac{1}{2}mv^2 + mgL(1 - \cos\theta)$ (b) $U = \frac{1}{2}m\omega^2 s^2$
49. Referring to the sketch, we have $F = -mg\sin\theta$ and
tan $\theta = x/R$. For small displacements, tan $\theta \approx \sin\theta$ and
$F = -(mg/R)x = -kx$ and $\omega = (k/m)^{1/2} = (g/R)^{1/2}$.

51. (a) $2Mg$, $T_P = Mg\left(1 + \dfrac{y}{L}\right)$

     (b) $\dfrac{4\pi}{3}\sqrt{\dfrac{2L}{g}} = 2.68$ s

55. $f = \dfrac{1}{2\pi}\sqrt{\dfrac{MgL + kh^2}{ML^2}}$

57. 0.0662 m

57A. $\dfrac{\mu_s g}{4\pi^2 f^2}$

59. (a) 3.00 s   (b) 14.3 J   (c) 0.441 rad
61. 9.19×10^{13} Hz
63. (a) 15.8 rad/s   (b) 5.23 cm   (c) 1.31 cm, π
67. (a) $k = 1.74$ N/m ± 6%   (b) $k = 1.82$ N/m ± 3%;
the values of k agree   (c) $m_s = 8$ g ± 12%, in
agreement with 7.4 g
S2. Since $E_{tot} = K(t) + U(t) = \frac{1}{2}kA^2[\cos^2(\omega t + \delta) + \sin^2(\omega t + \delta)] = \frac{1}{2}kA^2$, the total energy is independent of
ω and δ. It depends only on k and A.
S3. The periods increase as ϕ_0 increases. These periods are
always greater than those calculated using $T_0 = 2\pi\sqrt{L/g}$.
For example, if $\phi_0 = 45° = \pi/4$ rad, and $L = 1$ m, then
$T_0 = 2.00\ 709$ s, but the actual period is 2.08 s. (*Note:*
Due to numerical errors in the integration of the differential equations, the amplitudes may tend to increase.
If this occurs, try smaller time steps.)

Chapter 14

1. (a) 3.46×10^8 m   (b) 3.34×10^{-3} m/s
3. (1.00 m − 61.3 nm)
5. $\dfrac{GM}{\ell^2}\left(\dfrac{2\sqrt{2}+1}{2}\right)$ toward the opposite corner

7. 35.0 N toward the Moon
9. 3.73 m/s²
11. 12.6×10^{31} kg
11A. $\dfrac{2v^3 T}{\pi G}$
13. (a) 4.39×10^{20} N   (b) 1.99×10^{20} N
     (c) 3.55×10^{22} N
15. 1.90×10^{27} kg
17. Y has completed 1.30 revolutions
19. 8.98×10^7 m
21. $2GMr/(r^2 + a^2)^{3/2}$, to the left
23. 3.84×10^4 km from the Moon's center
25. 2.82×10^9 J
27. (a) 1.84×10^9 kg/m³   (b) 3.27×10^6 m/s²
     (c) -2.08×10^{13} J
29. 1.66×10^4 m/s
31. 11.8 km/s
35. 1.58×10^{10} J
35A. $\dfrac{mGM_E(R_E + 2h)}{2R_E(R_E + h)} - \frac{1}{2}mv^2$
37. (a) 42.1 km/s relative to the Sun   (b) 2.20×10^{11} m
39. (a) 1.31×10^{14} N/kg   (b) 2.62×10^{12} N/kg
39A. (a) $\dfrac{GM}{(d + \ell/2)^2}$ (b) $\dfrac{GM\ell(2\ell + d)}{d^2(d + \ell)^2}$
41. (a) 7.41×10^{-10} N   (b) 1.04×10^{-8} N
     (c) 5.21×10^{-9} N
43. 2.26×10^{-7}
45. 0.0572 rad/s = 32.7 rev/h
45A. $\omega = \sqrt{2g/d}$
47. 7.41×10^{-10} N
49. (a) $k = \dfrac{GmM_E}{R_E^3}$, $A = \dfrac{L}{2}$

     (b) $\dfrac{L}{2}\left(\dfrac{GM_E}{R_E}\right)^{1/2}$, at the middle of the tunnel

     (c) 1.55×10^3 m/s
51. $\dfrac{2\sqrt{2}\,Gm}{a^2}(-\mathbf{i})$

53. 2.99×10^3 rev/min
55. (a) $v_1 = m_2\left[\dfrac{2G}{d(m_1 + m_2)}\right]^{1/2}$

     $v_2 = m_1\left[\dfrac{2G}{d(m_1 + m_2)}\right]^{1/2}$

     $v_{rel} = \left[\dfrac{2G(m_1 + m_2)}{d}\right]^{1/2}$

     (b) $K_1 = 1.07 \times 10^{32}$ J, $K_2 = 2.67 \times 10^{31}$ J
57. (a) 7.34×10^{22} kg   (b) 1.63×10^3 m/s
     (c) 1.32×10^{10} J

59. 119 km

61. (a) 5300 s   (b) 7.79 km/s   (c) 6.45×10^9 J

61A. (a) $2\pi(GM_E)^{1/2}\,(R_E + h)^{3/2}$

(b) $\sqrt{GM_E/(R_E + h)}$ (c) $\dfrac{mGM_E(R_E + 2h)}{2R_E(R_E + h)}$

63. (a) $M/\pi R^4$ (b) $-GmM/r^2$

(c) $-(GmM/R^4)\,r^2$

65. 1.48×10^{22} kg

67. (b) 981 kg/m^3

69. (b) $GMm/2R$

S5. The kinetic energy is

$$K = \tfrac{1}{2}mv^2 = \tfrac{1}{2}m(v_x{}^2 + v_y{}^2)$$

The potential energy is

$$U = \frac{GM_E m}{R_E} - \frac{GM_E m}{r}$$

where G is the universal gravitation constant, M_E is the mass of the Earth, R_E is the radius of the Earth, m is the mass of the satellite, and r is the distance from the center of the Earth to the satellite. The total energy, $K + U$, is constant. There may be a small change in the numerical value of the total energy, because of numerical errors during the integration of the differential equations.

S6. According to Kepler's laws of planetary motion the angular momentum, L, is a constant. There may be a small change in the numerical value of L, because of numerical errors during the integration of the differential equations.

Chapter 15

1. 0.111 kg

3. 3.99×10^{17} kg/m^3. Matter is mostly free space.

5. 6.24×10^6 Pa

7. 4.77×10^{17} kg/m^3

9. $P_{\text{ATM}} + \rho\sqrt{g^2 + a^2}\,(L/\sqrt{2})\cos\left(45° - \arctan\dfrac{a}{g}\right)$

11. 1.62 m

13. 77.4 cm^2

15. 0.722 mm

17. 9.12 MPa

19. 12.6 cm

21. 10.5 m; no, a little alcohol and water evaporate

23. 1.08 cm

25. 1470 N

27. (a) 7.00 cm   (b) 2.80 kg

29. $\rho_{\text{oil}} = 1250$ kg/m^3; $\rho_{\text{sphere}} = 500$ kg/m^3

31. 0.611 kg

33. 1.07 m^2

33A. $\dfrac{m}{h(\rho_w - \rho_s)}$

35. 1430 m^3

37. (a) 17.7 m/s   (b) 1.73 mm

37A. (a) $\sqrt{2gh}$ (b) $(R/\pi)^{1/2}(8/gh)^{1/4}$

39. 0.0128 m^3/s

41. 31.6 m/s

43. (a) 28.0 m/s   (b) 392 kPa

45. Av/a

47. 103 m/s

49. $2[h(h_0 - h)]^{1/2}$

51. 4.83 kW at 20°C

53. (b) $\tfrac{1}{2}\rho Av^3$ if the mill could make the air stop; the same

55. 0.258 N

57. 1.91 m

59. 455 kPa

63. 8 cm/s

65. 2.01×10^6 N

69. 90.04% Zn

71. 5.02 GW

73. 4.43 m/s

75. (a) 1.25 cm   (b) 13.8 m/s

77. (a) 18.3 mm   (b) 14.3 mm   (c) 8.56 mm

Chapter 16

1. $y = \dfrac{6}{(x - 4.5t)^2 + 3}$

3. (a) longitudinal   (b) 666 s

5.

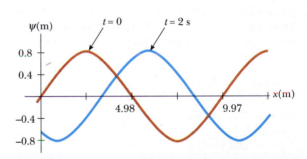

7. (a) 5.00 rad   (b) 0.858 cm

9. (a) Wave 1 travels in the $+x$ direction; wave 2 travels in the $-x$ direction.   (b) 0.75 s   (c) $x = 1.00$ cm

11. 520 m/s

13. 13.5 N

15. 586 m/s

17. 0.329 s

19. (b) 0.125 s

21. 0.319 m

23. (b) $k = 18.0$ rad/m, $T = 0.0833$ s, $\omega = 75.4$ rad/s, $v = 4.20$ m/s

(c) $y(x, t) = (0.20\text{ m})\sin(18.0x + 75.4t - 0.151)$

25. $y_1 + y_2 = 11.2\sin(2.0x - 10t + 63.4°)$

27. (a) $y = (0.0800\text{ m})\sin(7.85x + 6\pi t)$

(b) $y = (0.0800\text{ m})\sin(7.85x + 6\pi t - 0.785)$

29. (a) 2.15 cm   (b) 0.379 rad   (c) 541 cm/s

(d) $y(x, t) = (2.15\text{ cm})\cos(80\pi t + 8\pi x/3 + 0.379)$

31. $A = 2.0$ cm; $k = 2.11$ rad/m; $\lambda = 2.98$ m; $\omega = 3.62$ rad/s; $v = 1.72$ m/s; $f = 0.756$ Hz

33. (a) $y = (0.200 \text{ mm}) \sin[16.0x - 3140t]$
   (b) $T = 158$ N
35. 1.07 kW
35A. $2\pi^2 M v^3 A^2 / L\lambda^2$
37. (a) remains constant   (b) remains constant
   (c) remains constant   (d) p is quadrupled
39. (a) 62.5 m/s   (b) 7.85 m   (c) 7.96 Hz
   (d) 21.1 W
45. (a) 3.33 m/s in the positive x direction
   (b) -5.48 cm   (c) 0.667 m, 5.00 Hz
   (d) 11.0 m/s
47. (a) 179 m/s   (b) 17.7 kW
49. (a) 39.2 N   (b) 89.2 cm   (c) 83.6 m/s
49A. (a) $2Mg$ (b) $L_0 + 2Mg/k$
   (c) $(2MgL_0/m + 4M^2g^2/km)^{1/2}$
51. (a) 5.00 m/s $+ x$ (b) 5.00 m/s $- x$
   (c) 7.5 m/s $- x$ (d) 24.0 m/s $+ x$
55. 3.86×10^{-4} (at 5.93°C)
57. $(2L/g)^{1/2}$, $L/4$

59. (a) $\dfrac{\mu\omega^3}{2k} A_0^2 e^{-2bx}$ (b) $\dfrac{\mu\omega^3}{2k} A_0^2$ (c) e^{-2bx}

Chapter 17

1. 5.56 km
3. 1430 m/s
5. 332 m/s
7. 1.988 km
9. (a) 27.2 s   (b) 25.7 s   It is shorter by 5.30%.
11. 1.55×10^{-10} m
13. 5.81 m
15. (a) 2.00 μm, 0.400 m, 54.6 m/s   (b) -0.433 μm
   (c) 1.72 mm/s
17. $(0.200 \text{ Pa}) \sin(62.8x - 2.16 \times 10^4 t)$
19. 66.0 dB
23. 100.0 m and 10.0 m
25. (a) 65.0 dB   (b) 67.8 dB   (c) 69.6 dB
27. 241 W
29. (a) 30.0 m   (b) 9.49×10^5 m
31. 50.0 km
33. 46.4°
35. 56.4°
37. 26.4 m/s
39. (a) 338 Hz   (b) 483 Hz
41. 2.82×10^8 m/s
43. (a) 56.3 s
   (b) $(56.6 \text{ km})\mathbf{i} + (20.0 \text{ km})\mathbf{j}$ from the observer
45. 130 m/s, 1.73 km
47. 80.0°
49. 1204 Hz
51. (a) 0.948°   (b) 4.40°
53. 1.34×10^4 N
55. 95.5 s

55A. $\dfrac{0.3E}{4\pi d^2 I_0} 10^{-\beta/10}$

57. (a) 55.8 m/s   (b) 2500 Hz

59. (a) 6.45   (b) 0
61. 1.60
63. The measured size of the wavelengths will depend on your monitor. However, the wavelengths you measure should be proportional to those given here. For $u/v = 0.0$, $\lambda_0 = 0.75$ cm. For $u/v = 0.5$, $\lambda_{\text{front}} = 0.37$ cm and $\lambda_{\text{back}} = 1.13$ cm. Therefore,

$$\left| \frac{\Delta\lambda_{\text{front}}}{\lambda_0} \right| = \frac{0.37 \text{ cm} - 0.75 \text{ cm}}{0.75 \text{ cm}} = 0.5$$

and

$$\left| \frac{\Delta\lambda_{\text{back}}}{\lambda_0} \right| = \frac{1.13 \text{ cm} - 0.75 \text{ cm}}{0.75 \text{ cm}} = 0.5$$

We see that $\dfrac{\Delta\lambda}{\lambda_0} = \dfrac{u}{v}$.

Chapter 18

1. (a) 9.24 m   (b) 600 Hz
3. 0.500 s
5. (a) The path difference to A is $\lambda/2$.
   (b) $9x^2 - 16y^2 = 144$
7. at 0.0891 m, 0.303 m, 0.518 m, 0.732 m, 0.947 m, and 1.16 m
9. (a) 4.24 cm   (b) 6.00 cm   (c) 6.00 cm
   (d) $x = 0.5$ cm, 1.5 cm, 2.5 cm
11. 25.1 m, 60.0 Hz
13. (a) 2.00 cm   (b) 2.40 cm
15. (a) $0, \pm 2\pi/3k, \pm 4\pi/3k, \ldots$
   (b) $(\pm \pi - 2\omega t)/k, (\pm 3\pi - 2\omega t)/k, (\pm 5\pi - 2\omega t)/k$
17. (a) 60.0 cm   (b) 30.0 Hz
19. $L/4$, $L/2$
21. 0.786 Hz, 1.57 Hz, 2.36 Hz, 3.14 Hz
23. 2.80 g

23A. $m = \dfrac{Mg}{4Lf_1^2 \tan \theta}$

25. (a) $T = 163$ N   (b) 660 Hz
27. 19.976 kHz
29. 338 N
31. 20.5 kg

31A. $m_w = \rho_w A \left(L - \dfrac{v_s}{4f} \right)$

33. 50.4 cm, 84.0 cm
35. 35.8 cm, 71.7 cm
37. 349 m/s
39. (a) 531 Hz   (b) 4.25 cm
41. 328 m/s
43. (a) 350 m/s   (b) 114 cm
45. $n(206 \text{ Hz})$ and $n(84.5 \text{ Hz})$, where $n = 1, 2, 3, \ldots$
47. 1.88 kHz
49. (a) 1.59 kHz   (b) odd   (c) 1.11 kHz
51. It is.
53. (a) 1.99 Hz   (b) 3.38 m/s
55. (a) 3.33 rad   (b) 283 Hz

57. 85.7 Hz

59. $f = 50.0$ Hz; $L = 1.70$ m

61. (a) 70.7 N   (b) 199 Hz

63. $\lambda = 4.86$ m

65. 3.87 m/s *away* from the station *or* 3.78 m/s *toward* the station

67. (a) 59.9 Hz   (b) 20 cm

69. (a) 0.5   (b) $\dfrac{n^2 F}{(n+1)^2}$ (c) $\dfrac{F'}{F} = \dfrac{9}{16}$

S4. The following steps can be used to modify the spreadsheet.
   1. MOVE the entire Y_T column—one column to the right.
   2. COPY the Y_2 column—one column to the right. EDIT the heading label to Y_3.
   3. COPY the second wave input data block to the right and EDIT the labels to reflect the third wave.
   4. EDIT the Y_3 column to reflect the third wave data block addresses.
   5. EDIT the Y_T column to include Y_1, Y_2, and Y_3 in the sum.

S6. Plot $Y(t)$ versus ωt. You may want to start with three terms in the series and then add additional terms and watch how $Y(t)$ changes.

Chapter 19

1. (a) $-273.5°C$ (b) 1.27 atm, 1.74 atm

3. (a) 30.4 mm Hg   (b) 18.0 K

5. 139 K, $-134°C$ (b) 6.56 kPa

7. (a) $-321°F$ (b) $139°R$ (c) 77.3 K

9. $-297°F$

11. (a) $810°F$ (b) 450 K

13. (a) $90.0°C$ (b) 90.0 K

15. $-40.0°C$

17. 3.27 cm

19. 1.32 cm

21. 0.548 gal

23. 217 kN

25. 1.20 cm

27. (a) $437°C$ (b) $2100°C$. No; they melt first.

29. (a) 99.4 cm³   (b) 0.943 cm

31. 1.08 L

33. (a) 0.176 mm   (b) 8.78 μm   (c) 93.0 mm³

35. 7.95 m³

37. 4.39 kg

39. 472 K

41. 2.28 kg

41A. $M n_1 \left(1 - \dfrac{T_1}{T_2} \right)$, where $n_1 = \dfrac{V}{0.0224} \left(\dfrac{273}{T_1} \right)$

43. 1.61 MPa = 16.1 atm

45. 594 kPa

47. 400 kPa, 448 kPa

49. 1.13

51. (a) $A = 1.85 \times 10^{-3}$ (1/°C), $R_0 = 50.0 \ \Omega$
    (b) 421 °C

53. $\alpha \Delta T \ll 1$

55. 3.55 cm

55A. $\Delta h = \dfrac{V}{A} \beta \Delta T$

57. (a) 94.97 cm   (b) 95.03 cm

59. (b) 1.33 kg/m³

63. 2.74 m

63A. $y = \sqrt{(L + \Delta L)^2 - L^2}$, where $\Delta L = \alpha L \Delta T$

65. $30.4°C$

67. (a) 18.0 m   (b) 277 kPa

69. (a) 7.06 mm   (b) 297 K

71. (a) 0.0374 mol   (b) 0.732 atm

73. (a) 6.17×10^{-3} kg/m   (b) 632 N
    (c) 580 N; 192 Hz

75. (a) $(\alpha_2 - \alpha_1) L \Delta T / (r_2 - r_1)$
    (c) It bends the other way.

S1. From the figure below, note that

$$h = \frac{L}{20} (1 - \cos \theta)$$

and

$$\frac{L}{L_0} = (1 + \alpha \Delta T) = \frac{\theta}{\sin \theta} = 1.000 \ 055$$

Solving this transcendental equation, $\theta = 0.018 \ 165$ rad and $h = 4.54$ m.

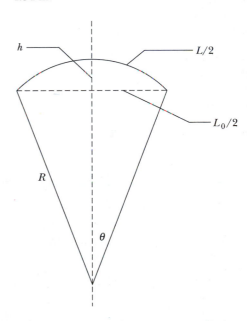

Chapter 20

1. $0.105°C$

3. $10.117°C$

5. 0.234 kJ/kg · °C

7. $85.3°C$

9. $29.6°C$

11. $34.7°C$

13. 19.5 kJ

15. 50.7 kg
17. 720 cal
19. 47.1°C
21. (a) 0°C   (b) 115 g
23. 2.99 mg
25. (a) 25.760°C   (b) no
27. 0.258 g if the bullet is at 0°C
29. 810 J, 506 J, 203 J
31. 466 J
31A. $nR \Delta T$
33. 1.18 MJ
35. (a) 7.65 L   (b) 305 K
37. (a) -567 J   (b) 167 J
37A. (a) $W = P \Delta V$ (b) $\Delta U = Q - W$
39. (a) 12.0 kJ   (b) -12.0 kJ
41. (a) 23.1 kJ   (b) 23.1 kJ
43. 0.096 g
45. (a) 7.50 kJ   (b) 900 K
47. 2.47 L
47A. $V_i = V_f e^{-W/nRT_0}$, where $W = m_w C_w (T_h - T_c)$
49. (a) 48.6 mJ   (b) 16.2 kJ   (c) 16.2 kJ
51. (a) 10.0 L·atm   (b) 0.0100 atm   (c) 7.00 kJ
53. 1.34 kW
55. 51.0°C
57. 138 MJ
59. 0.0222 W/m·°C

59A. $k = \dfrac{Pt}{A \Delta T}$

61. (a) 61.1 kWh   (b) $3.67
63. (a) 0.89 ft²·°F·h/BTU
    (b) 1.85 ft²·°F·h/BTU   (c) 2.08
65. 781 kg
67. 1.87 kJ
69. (a) 16.8 L   (b) 0.351 L/s
71. The bullet will partially melt.
73. 45°C
75. 5.46°C
79. 9.32 kW
81. 5.31 h
83. 800 J/kg·°C
S1. By varying h until $T = 45$ °C at $t = 180$ s, one finds $h = 0.010\ 855$ cal/s·cm²·°C. Examine the associated graph for each choice of h.
S2. (b) $\Delta U = 4669$ J

17. (a) 316 K   (b) 200 J
17A. (a) $dT = \dfrac{Q}{C_V} - \dfrac{Q}{C_p}$ (b) $\dfrac{Q}{C_p}$
19. (a) 3.46 kJ   (b) 2.45 kJ   (c) 1.01 kJ
21. 24.0 kJ; 68.7 kJ
23. (a) 118 kJ   (b) 6.03 km
25. (a) 1.39 atm   (b) 366 K, 254 K
27. (a) 2.06×10^{-4} m³   (b) 560 K   (c) 12.9°C
29. (a) 0.118, so the compression ratio $V_i / V_f = 8.50$
    (b) 2.35
31. 227 K
33. 91.2 J
33A. $c = \sqrt{\gamma R T / M}$
35. 1.51×10^{-20} J
37. 2.33×10^{-21} J
41. (a) 5.63×10^{18} m, 1.00×10^9 y
    (b) 5.63×10^{12} m, 1.00×10^3 y
43. (a) 1.028   (b) ^{35}Cl
45. (a) 2.02×10^4 K   (b) 904 K
47. (a) 3.21×10^{12} molecules
    (b) 778 km   (c) 6.42×10^{-4} s^{-1}
49. 193
51. 4.65×10^{-8} cm
55. (a) $3.65v$ (b) $3.99v$ (c) $3.00\,v$
    (d) $106\,mv^2/V$ (e) $7.98\,mv^2$
57. zero, 2.70×10^{20}
59. 0.625
63. (c) 2.0×10^3
65. (b) 5.47 km
67. (a) 10^{82} (b) 10^{12} m   (c) 10^{58} moles
69. (a) 0.510 m/s   (b) 20 ms
S1. The time it takes for the temperature to drop to 1/2 of its original value is approximately 0.7 ms. This time is independent of the number of particles. In all cases the curve resembles a decreasing exponential. As the number of particles increases the curve becomes smoother.
S2. (a) At $T = 100$ K, $f(1000)\ dv = 0.0896$ and $f(3000)\ dv = 0.000\ 05$.
    (b) At $T = 273$ K, $f(1000)\ dv = 0.0428$ and $f(3000)\ dv = 0.011\ 09$.
    (c) At $T = 1000$ K, $f(1000)\ dv = 0.0084$ and $f(3000)\ dv = 0.028\ 77$.

Chapter 21

1. 2.43×10^5 m²/s²
3. 2.30 kmol
5. 3.32 mol
7. 8.76×10^{-21} J
9. (a) 40.1 K   (b) 6.01 km/s
11. 477 m/s
13. 109 kPa
15. 75.0 J

Chapter 22

1. (a) 6.94%   (b) 335 J
3. (a) 0.333   (b) 0.667
5. (a) 1.00 kJ   (b) 0
7. (a) 0.375   (b) 600 J   (c) 2.00 kW
9. 0.330
11. (a) 5.12%   (b) 5.27 TJ
13. (a) 0.672   (b) 58.8 kW
15. 0.478°C

17. (a) 0.268   (b) 0.423
19. 453 K
21. 146 kW, 70.8 kW
23. 192 J
25. (a) 24.0 J   (b) 144 J
27. 72.2 J
27A. $W = \dfrac{\Delta T}{T_c}$
29. $\Delta S = -90.2$ J/K
31. 195 J/K
33. 3.59 J/K
35. 3.27 J/K
37. 5.76 J/K, no temperature change
39. 18.4 J/K
41. (a) 154.5 J/K   (b) 54.2 kJ

43. (a) 5.00 kW   (b) 763 W
45. (a) 4.10 kJ   (b) 14.2 kJ
    (c) 10.1 kJ   (d) 28.8%
47. (a) $2nRT_0 \ln 2$ (b) 0.273
49. (a) $10.5nRT_0$ (b) $8.5nRT_0$ (c) 0.190
    (d) 0.833
51. $nC_p \ln 3$
53. 5.97×10^4 kg/s
53A. $\dfrac{dm}{dt} = \dfrac{P}{C_w \Delta T}\left(\dfrac{T_h}{T_h - T_c}\right)$
57. (a) 96.9 W   (b) 1.19°C/h
59. $e = \dfrac{2(T_2 - T_1)\ln (V_2/V_1)}{3(T_2 - T_1) + 2T_2 \ln (V_2/V_1)}$

Index

Page numbers in *italics* indicate illustrations; page numbers followed by an n indicate footnotes; page numbers followed by t indicate tables.